FARADAY DISCUSSIONS
NO. 119   2001

Combustion Chemistry: Elementary Reactions to Macroscopic Processes

The Faraday Division
The Royal Society of Chemistry
London

Organising Committee
Professor M. J. Pilling (*Chairman*)
Professor D. A. Greenhalgh
Professor A. N. Hayhurst
Professor R. P. Lindstedt
Dr D. B. Smith
Professor I. W. M. Smith
Professor J. Wolfrum

ISBN: 0-85404-981-9

ISSN: 1359-6640

Typeset by Santype International Ltd., Netherhampton Road, Salisbury, Wiltshire and printed and bound in Great Britain by Black Bear Press, Cambridge, UK.

A General Discussion

on

Combustion Chemistry: Elementary Reactions to Macroscopic Processes

9th, 10th and 11th July, 2001

A General Discussion on Combustion Chemistry was held at the University of Leeds, UK on 9th, 10th and 11th July, 2001.

Contents

1 Introductory Lecture: Advanced laser spectroscopy in combustion chemistry: From elementary steps to practical devices
Jürgen Wolfrum

27 Crossed beam studies of elementary reactions of N and C atoms and CN radicals of importance in combustion
Piergiorgio Casavecchia, Nadia Balucani, Laura Cartechini, Giovanni Capozza, Astrid Bergeat and **Gian Gualberto Volpi**

51 A combined crossed molecular beam and *ab initio* investigation of C_2 and C_3 elementary reactions with unsaturated hydrocarbons—pathways to hydrogen deficient hydrocarbon radicals in combustion flames
Ralf I. Kaiser, Trung N. Le, Thanh L. Nguyen, Alexander M. Mebel, Nadia Balucani, Yuan T. Lee, Frank Stahl, Paul v. R. Schleyer and **Henry F. Schaefer III**

67 Determination of the $CH + O_2$ product channels
Astrid Bergeat, Teresa Calvo, Françoise Caralp, Jean-Hugues Fillion, Gérard Dorthe and **Jean-Christophe Loison**

79 A theoretical analysis of the reaction between propargyl and molecular oxygen
David K. Hahn, Stephen J. Klippenstein and **James A. Miller**

101 Infrared frequency-modulation probing of product formation in alkyl + O_2 reactions. Part IV. Reactions of propyl and butyl radicals with O_2
John D. DeSain, Craig A. Taatjes, James A. Miller, Stephen J. Klippenstein and **David K. Hahn**

121 General Discussion

145 Rotational effects in broadening factors of fall-off curves of unimolecular dissociation reactions
Jürgen Troe and **Vladimir G. Ushakov**

159 Time-dependent master equation simulation of complex elementary reactions in combustion: Application to the reaction of $^1CH_2$ with C_2H_2 from 300–2000 K
Terry J. Frankcombe and **Sean C. Smith**

173 Use of quantum methods for a consistent approach to combustion modelling: Hydrocarbon bond dissociation energies
Juan P. Senosiain, Joseph H. Han, Charles B. Musgrave and **David M. Golden**

191 Low-energy paths for the unimolecular decomposition of CH_3OH: A G2M/statistical theory study
W. S. Xia, R. S. Zhu, M. C. Lin and **A. M. Mebel**

This journal is © The Royal Society of Chemistry 2001

207 A direct transition state theory based analysis of the branching in $NH_2 + NO$
De-Cai Fang, Lawrence B. Harding, Stephen J. Klippenstein and **James A. Miller**

223 An experimental and theoretical study of the product distribution of the reaction CH_2 ($\tilde{X}\,^3B_1$) + NO
Mustapha Fikri, Stefan Meyer, Jan Roggenbuck and **Friedrich Temps**

243 Reactions of methyl radicals with propene at temperatures between 750 and 1000 K
Elke Goos, Horst Hippler, Karlheinz Hoyermann and **Bettina Jürges**

255 General Discussion

275 Energy transfer in combustion diagnostics: Experiment and modeling
Andreas Brockhinke and **Katharina Kohse-Höinghaus**

287 Temperature fields during the development of autoignition in a rapid compression machine
John F. Griffiths, John P. MacNamara, Caroline Mohamed, Benjamin J. Whitaker, Jinfeng Pan and **Christopher G. W. Sheppard**

305 NO reburning study based on species quantification obtained by coupling LIF and cavity ring-down spectroscopy
Xavier Mercier, Laure Pillier, Abderrahman El Bakali, Michel Carlier, Jean-François Pauwels and **Pascale Desgroux**

321 Laser absorption spectroscopy diagnostics of nitrogen-containing radicals in low-pressure hydrocarbon flames doped with nitrogen oxides
V. A. Lozovsky, I. Rahinov, N. Ditzian and **S. Cheskis**

337 Experimental and modelling study of sulfur and nitrogen doped premixed methane flames at low pressure
Kevin J. Hughes, Alison S. Tomlin, Valerie A. Dupont and **Mohammed Pourkashanian**

353 General Discussion

371 Detailed surface reaction mechanism in a three-way catalyst
Daniel Chatterjee, Olaf Deutschmann and **Jürgen Warnatz**

385 A study of the reaction of oxygen with graphite: Model chemistry
Raymond Backreedy, Jenny M. Jones, Mohammad Pourkashanian and **Alan Williams**

395 Small-angle X-ray studies of soot inception and growth
Jan P. Hessler, Soenke Seifert, Randall E. Winans and **Thomas H. Fletcher**

409 Thermodynamic and kinetic issues in the formation and oxidation of aromatic species
Peter Lindstedt, Lourdes Maurice and **Michael Meyer**

433 The influence of fuel additives on the behaviour of gaseous alkali-metal compounds during pulverised coal combustion
H. Schürmann, S. Unterberger, K. R. G. Hein, P. B. Monkhouse and **U. Gottwald**

445 General Discussion

461 Concluding Remarks

477 List of Posters

479 List of Participants

481 Index of Contributors

■ Electronic supplementary information is available on http://www.rsc.org/esi See article for further information.

Introductory Lecture
Advanced laser spectroscopy in combustion chemistry: From elementary steps to practical devices

Jürgen Wolfrum

Universität Heidelberg, Physikalisch-Chemisches Institut, Im Neuenheimer Feld 253, Heidelberg, Germany. E-mail: wolfrum@urz.uni-heidelberg.de

Received 31st August 2001
First published as an Advance Article on the web 6th November 2001

In recent years a large number of linear and nonlinear laser-based diagnostic techniques for nonintrusive measurements of species concentrations, temperatures, and gas velocities in a wide pressure and temperature range with high temporal and spatial resolution have been developed and have become extremely valuable tools to study many aspects of combustion. Beside the nonintrusive diagnostics of technical combustion devices the kinetics and microscopic dynamics of elementary chemical combustion reactions can be investigated in great detail by laser spectroscopy. These investigations show, that a small number of relatively simple elementary steps like $H + O_2 \rightarrow OH + O$, $H_2O_2 \rightarrow 2OH$, $O + N_2 \rightarrow NO + N$, $NH_2 + NO \rightarrow H_2O + N_2$, $OH + N_2H$ control a large variety of combustion phenomena and pollutant formation processes. Laminar flames are ideal objects to develop the application of laser spectroscopic methods for practical combustion systems and to test and improve the gas-phase reaction mechanism in combustion models. Nonintrusive laser point and field measurements are of basic importance in the validation and further development of turbulent combustion models. Nonlinear laser spectroscopic techniques using infrared–visible sum-frequency generation can now bridge the pressure and materials gap to provide kinetic data for catalytic combustion. Finally, the potential of laser techniques for active combustion control in municipal waste incinerators is illustrated.

Introduction

Although it is well known that combustion of fossil fuels leads to the release of unwanted pollutants like *e.g.* carbon monoxide, unburned and oxygenated hydrocarbons, soot, sulfur and nitric oxides, which affect our environment, it is still the most important technology providing the energy supply for our modern society. Therefore considerable efforts are undertaken to optimize combustion processes, to reduce pollutant formation while, at the same time, maintaining a high efficiency for the conversion of chemical to thermal, mechanical and electrical energy. However, even today the construction of technical combustion devices in general is still an essentially empirical process, calling to a large degree upon the experience of engineers and technicians employing methods based on cut and trial processes and global performance measurements. With the growing number of performance and environmental protection requirements that must be met, this kind of approach is reaching its limits. Combustion processes consist of a complex interaction of homogeneous and heterogeneous chemistry with various physical transport processes.[1] As a consequence, the development of numerical simulation tools that can accurately predict combustion and its characteristics for practical applications has to be based on a detailed understanding of the underlying chemical reaction systems and physical processes. New laser spectroscopic methods with high sensitivity for nonintrusive measurements of multidimensional temperature, velocities and species distributions in technical systems[2] can play an important role in the validation of the combustion models for laminar and turbulent reactive flows as well as in the development of active combustion control applications. As shown in Fig. 1, the principle of stimulated emission has now been realized in all states of matter. This allows generation of coherent radiation from the far-infrared to the X-ray region. Particularly, the introduction of tunable lasers and the development of nonlinear optical techniques[3] greatly expanded the possibilities of combustion spectroscopy using conventional light sources. Even though higher order polarizations responsible for nonlinear optical phenomena are usually small, signal intensities in higher order wave mixing processes can be large, due to their strong dependence on laser intensities, concentration and interaction length. Nonlinear susceptibilities are enhanced whenever one or more of the frequencies of the interacting laser beams are close to quantum mechanically allowed one- or multiphoton transitions. In microscopic terms the induced oscillating nonlinear polarization arises from the nonlinear mixing of the input waves to produce the same or new frequencies at various sum and difference frequency combinations of the input frequencies. The spatial resolution of these techniques is determined by the intersection volume of the interacting beams. Another common feature is that all techniques need line-of-sight access to the combustion event, although large window openings, as are often necessary for incoherent methods, are not required. The direction of the emitted coherent signal beam is determined by the phase matching condition for efficient signal generation. The various linear and nonlinear coherent methods which have most successfully been used so far in combustion analysis are depicted in Fig. 2. On the left, the relevant energy level diagrams are given. Upward-facing arrows indicate photon absorption (or conversion); downward-facing arrows depict photon emission; $\omega_i = 2\pi\nu$ are the frequencies of the fields involved. The middle column depicts a simplified experimental set-up for each method with corresponding beam and detection arrangements. The right column shows approximate analytic expressions for the signal intensity. The detection sensitivities of the various techniques listed in

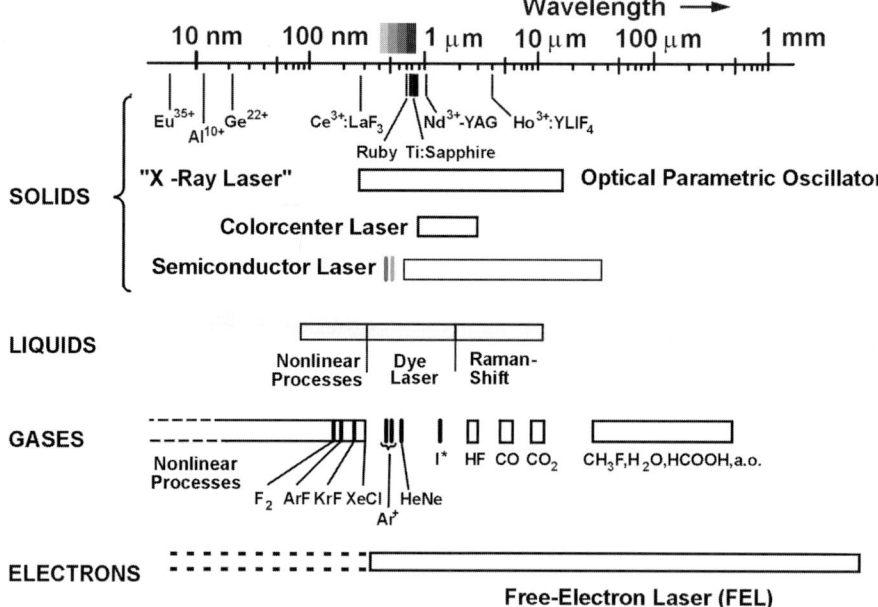

Fig. 1 Wavelength range of various available laser light sources.

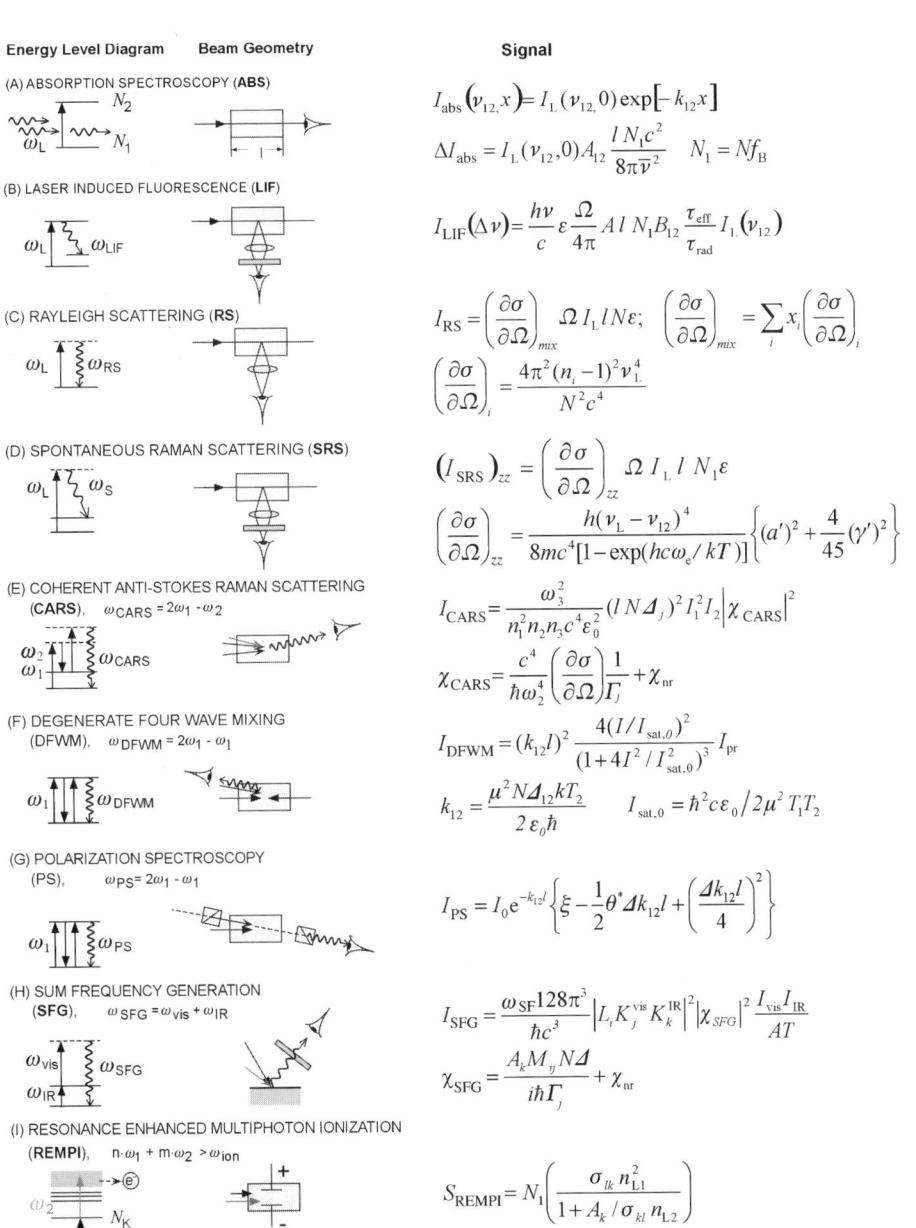

Fig. 2 Survey of linear and nonlinear laser spectroscopic techniques applied in combustion diagnostics (symbols see ref. 4).

Fig. 2 depend on the details of the experimental set-ups, laser parameter (temporal and spectral width, beam energy, crossing angle for multiple beam techniques, state of polarization) and detection schemes (optical throughput, photodetectors, averaging time *etc.*). Typically, laser-induced fluorescence (LIF) is a very sensitive technique with a detection limit in flames at atmospheric pressure in the ppm (10^{-6}) range for small radicals, whereas the spontaneous or coherent Raman methods are limited to the 10^{-3} range. Extremely high sensitivities in the ppt (10^{-12}) level can be achieved with the REMPI method using molecular beam sampling.

Laser spectroscopic techniques can now be applied in two directions. Beside the nonintrusive diagnostics of technical or model combustion devices the kinetics and microscopic dynamics of elementary chemical combustion reactions can be investigated in great detail. In the discussion on the following pages both aspects will be considered.

Laminar and turbulent flames

Laminar premixed flames at low pressure on a flat flame burner constitute an ideal experimental arrangement for studying the interaction of the large number of elementary chemical reactions, as first described by the Warnatz mechanism for hydrocarbon flames,[5] with various physical transport processes for mass, momentum, and energy. Several examples presented at this meeting show how new experimental methods like cavity ring-down laser absorption spectroscopy (CRLAS)[6] can be used to test and improve gas-phase reaction mechanism in combustion models. However, such flat flame studies are not completely free from deviations from a pure one-dimensional behaviour. Therefore, multidimensional diagnostic concepts are more appropriate. An elegant example in this direction is shown in Fig. 3. Michael Faraday visualized in his famous "Lectures on the Chemical History of a Candle" the temperature distribution inside a candle flame by the burning pattern on a thin paper sheet, showing that the temperature maximum is not in the center of the flame as was assumed previously. Instead the temperature maximum is localized at the region of the laminar diffusion flame. With the development of modern laser spectroscopic methods, the paper sheet can be replaced by a laser light sheet. Fig. 4 shows schematically the experimental set-up of a laser experiment for simultaneous measurement of temperature and hydroxyl radical concentrations in three adjacent planes (200 μm thickness, 30 mm height, 1 mm distance). The three temporally separated laser beams are generated with two tunable KrF-excimer lasers,[7] one of which is beam-split and delayed *via* an optical delay line. The temperature distribution in a swirling, unconfined 150 kW natural gas flame is obtained from imaging the signal from Rayleigh scattering in three image intensified CCD cameras. Simultaneously the flame front surface is visualized by laser-induced fluorescence (LIF) of OH radicals. By measuring the Rayleigh intensity distributions in the nonreacting swirl flow, it could be shown that the rapid fuel–air mixing process is complete shortly (10–20 mm) above the nozzle. Under these conditions for a stoichiometric flame the effective Rayleigh scattering cross section remains constant within 2% in the unburned and burned part. Thus, if the very small pressure changes in the atmospheric flame are neglected, the Rayleigh signal is inversely proportional to temperature. The 3D-visualization of the turbulent flame front is shown in Fig. 5. The flame front position is determined from the steepest gradient of the OH concentration. Between the three measurement planes the flame front positions and temperatures are interpolated to obtain a 3D-representation.

Similar to the laminar case, turbulent reactive flows, as shown in Fig. 5, can be described by solving the system of conservation equations for mass, momentum, energy and a species conserva-

Fig. 3 Michael Faraday "Lectures on the Chemical History of a Candle". [*reprinta historica didactica*, Band 3, Bad Salzdetfurth, 1980.] Reproduced with the permission of the publisher.

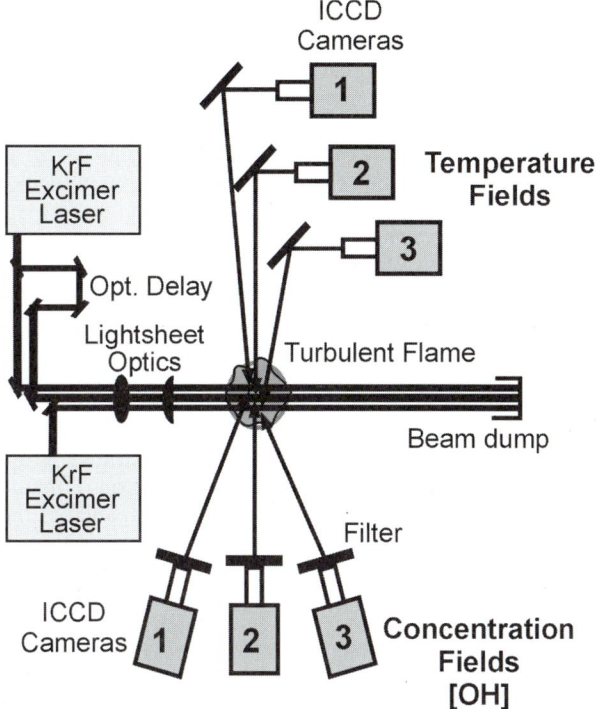

Fig. 4 Experimental set-up of a laser experiment for the simultaneous measurement of temperature and OH concentration in a turbulent flame in three adjacent planes (distance 1 mm). Reproduced with permission from ref. 7.

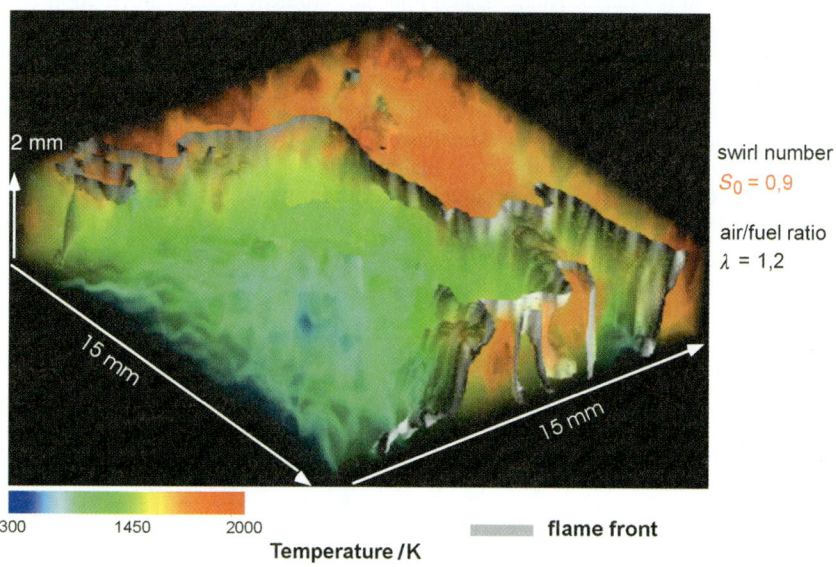

Fig. 5 3-D visualization of temperature and flamefront position in a turbulent swirlflame based on simultaneous measurements of Rayleigh temperature and OH-LIF in three adjacent planes. Flamefront surface and the temperature volume-visualization are superimposed. Reproduced with permission from ref. 7.

tion equation for each chemical species. The resulting Navier–Stokes equations describe the changes in temperature and species mass fraction in the form of a differential equation with three terms (convection, transport and chemical reaction). However, to solve the Navier–Stokes equations a large variety of different length-, time- and velocity-scales have to be considered. The theory of homogeneous isotropic turbulence assumes that a turbulent flow field can be described as a system of turbulent eddies in which the turbulent energy is being consumed at the small scales (Kolmogorov scale η) by viscous dissipation. For the local description of the flow field between the large eddies in the range from the geometric dimensions of the combustion system to the Kolmogorov scale one needs, in each direction at least 3 orders of magnitude resolution, *i.e.* 10^9 grid points. Due to the nonlinear nature of the chemical reaction source terms the time steps for integration of the chemical kinetic equations range from 10^{-7} s at atmospheric pressure for fast radical–radical reactions to 10^{-2} s for slow reactions with high activation barriers like O + N$_2$ → NO + N. Since for each species one conservation equation must be solved, depending on the complexity of the combusted fuel, 10^2 to 10^4 equations are necessary. In addition one is interested, for practical applications, not only in one particular situation produced by a single direct numerical simulation (DNS), but in global structures (averaged temperatures and compositions) with varied input parameters. Therefore, as illustrated in Fig. 6, a model-free description of the turbulent reactive flow in practical devices like spark ignition engines and gas turbines requires many orders of magnitude higher computing capabilities than the best current platforms offer. Although substantial improvements (hardware progress according to Moore's law of one order of magnitude within seven years, massively parallel computing, methods for simplification of chemical kinetics, better algorithmic techniques *etc.*) can be expected in the near future, it will take 10–20 years before we can close this gap. In addition this consideration does not take into account the severe problems of describing two-phase flows in spray combustion, the inclusion of radiation heat transfer as well as very complex chemical processes like soot formation. Thus, for the moment and the near future, practical combustion situations can only be described by the development of sophisticated models. However, DNS provides insight into the detailed coupling of molecular transport and chemical kinetics with turbulent flow fields and is therefore of great value in developing and validating such models.

Damköhler[8] discussed theoretically the interaction of a flame front from a premixed flame with a turbulent flow field. He considered the ratio of two characteristic times, which was later called the Damköhler number Da. $Da = \tau_t/\tau_c$ is obtained from the rotation time (τ_t) of a turbulent eddy (calculated from the eddy integral length scale l and the turbulent velocity fluctuation v') divided by a characteristic time τ_c of the chemistry (calculated from the inverse of a global rate coefficient). For values of $Da < 1$ we have fast turbulent mixing and slower chemistry which results in the well known situation of the 'stirred reactor' (see Fig. 7). For $Da \gg 1$ combustion occurs in thin 'flamelets', which are stretched and wrinkled by the turbulent flow field. The turbulent flame front can then be constructed from a library of flamelets calculated from a laminar flame. With increasing turbulent Reynolds number $Re_t = lv'/v$ (v is the temperature dependent kinematic viscosity) the eddies with the Kolmogorov size η can become smaller than the reactive zone thickness $\delta_l = v/S_L$ (S_L is the laminar flame speed) and might penetrate into the reaction zone and enhance the trans-

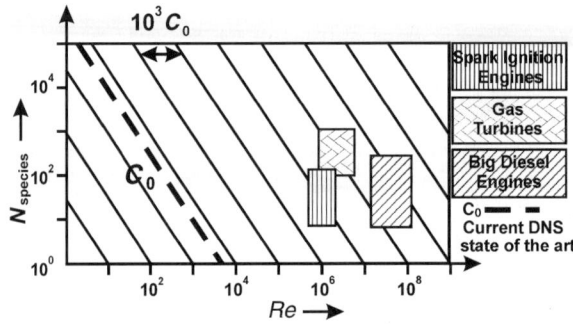

Fig. 6 Direct numerical simulation (DNS) of reactive flows.

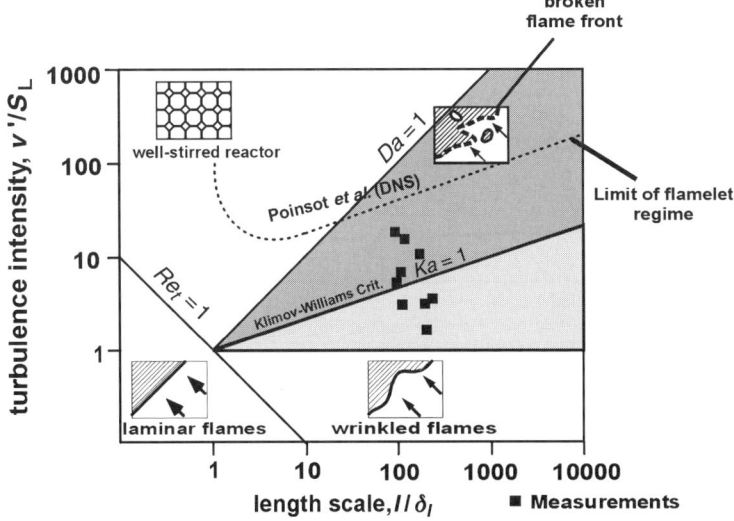

Fig. 7 Phase diagram showing different regimes in premixed turbulent combustion.[9,10]

port of heat and species. The transition between the two regimes is described in the phase diagram of premixed turbulent combustion[9,10] by the Klimov–Williams line at the Karlovitz number $Ka = (\delta_l/\eta)^2 = 1$ (see Fig. 7). Laser experiments with simultaneous 2D-UV-Rayleigh thermometry and 2D-LIF of OH radicals[11] show for flames with $Ka < 1$ local flame-front structures similar to those calculated for laminar premixed flames. For $Ka > 1$ the flame thickness based on the OH concentration increase remains independent of Ka while the mean thermal flame front thickness is significantly thinned compared with calculations in laminar, unstrained flames with the same stoichiometry, in contrast to the Klimov–Williams criterion but confirmed by DNS calculations.[12] Therefore Peters[13] has suggested using another definition of the flame thickness in form of the inner layer thickness l_δ, which is about one tenth of the preheat zone thickness. This corresponds to a $Ka_\delta \sim Ka/10^2$ and agrees with DNS calculations by Poinsot et al.[14] for the limit of the flamelet regime, where flame quenching is observed.

As shown in Fig. 7 the laminar flame speed S_L is a very important parameter also in the characterization of turbulent combustion, which is the basis of most technical combustion devices. If one now relates S_L to the elementary chemical reactions taking place during the oxidation of hydrocarbon fuels, Warnatz[15] has shown, that the S_L of a stoichiometric propane–air flame is mainly determined by the two simple elementary reactions

$$H + O_2 \rightarrow O + OH \tag{1}$$

$$OH + CO \rightarrow CO_2 + H \tag{2}$$

Such elementary reactions should therefore be characterized in great detail and with high accuracy. The final goal is to have for these elementary steps a consistent picture of the results of quasiclassical trajectory (QCT) and quantum mechanical scattering (QMS) calculations on accurate *ab initio* potential energy surfaces (PES) with experimental data from reaction dynamics and kinetic investigations. The dynamics of elementary gas-phase reactions can be studied in microscopic detail by combining short-pulse laser photolysis (LP) to generate translationally excited reactants with a time-, state- and orientation-resolved product detection by LIF spectroscopy. As illustrated in Fig. 2, LIF uses a tunable laser which is scanned through an absorption band of a molecule or over the absorption line of an atom. Once a transition between specific internal energy levels is excited, the resulting total fluorescence (excitation spectrum) or the spectrally resolved fluorescence (fluorescence spectrum) is detected. From relative line intensity measurements, species concentrations and internal quantum state distributions can be determined with high sensitivity (*e.g.* 10^3 particles per OH quantum state or single atoms). Fig. 8 shows a schematic

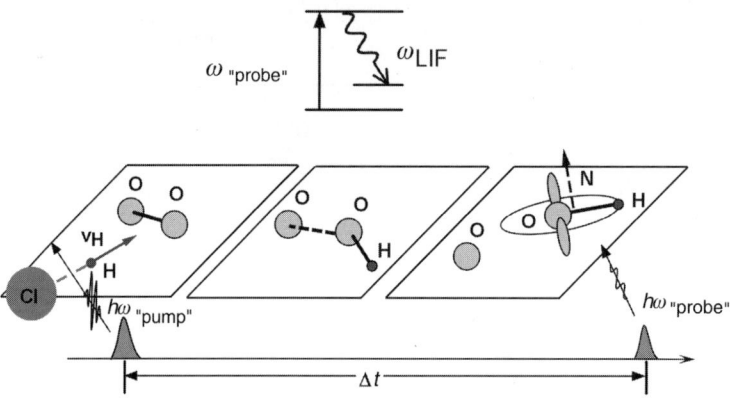

Fig. 8 Schematic description of the LP/LIF "pump-and-probe" measurement technique to study the microscopic dynamics of reaction (1).

description of the application of the LP/LIF "pump-and-probe" method[16] to reaction (1). In such dynamics experiments it was observed that with increasing collision energy, the nascent OH product rotational state distribution continuously changes from a statistical to a non-statistical one. This behavior could be well reproduced by QCT calculations and is attributed to a change in the underlying reaction mechanism from a HO_2 complex-forming pathway at lower energies to a direct one at higher collision energies. From the measured OH λ-doublet distributions a preference for the Π^+ component with increasing relative collision energy is observed, which suggests a less important out of plane rotation of the HO_2 complex at high energy. The alignment of the OH formed relative to the flight direction of the OH atoms can be measured by polarizing the analysis and photolysis laser beams.[17] Beside providing insights into the microscopic reaction dynamics, absolute reaction cross sections (σ) can be measured directly with this laser technique. If the complete dependence of T on the relative kinetic energy (E) is determined (excitation function) the thermal rate constant $k(T)$ can be obtained.

Fig. 9 shows a comparison of absolute reactive cross sections for reaction (1) from laser-based reaction dynamics studies and QCT and QMS.[18] Calculations carried out by Varandas[19] are also shown. Varandas extrapolated QMS results for total angular momentum equal to zero ($J = 0$) to

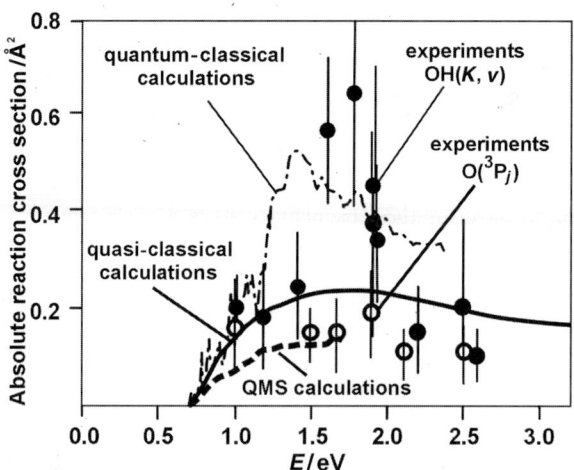

Fig. 9 Comparison between experimental and theoretical absolute reactive cross sections as a function of translational energy E for reaction (1). Reproduced with permission from ref. 18.

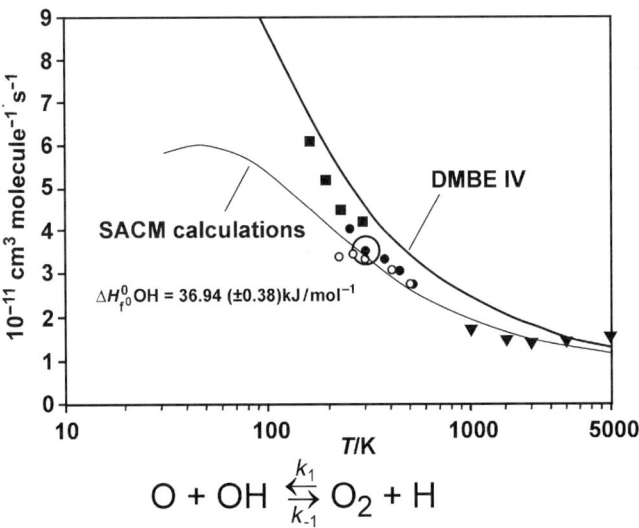

Fig. 10 Measured and calculated thermal rate constants for reaction (1). Reproduced with permission from ref. 26.

higher J-values. Total angular momentum values up to $J = 57$ had to be included. This, as a consequence, makes rigorous quantum mechanical calculations a formidable task. Recently QMS studies became available and showed a decrease in the reactive cross section with increasing J values.[20] As illustrated in Fig. 9 this gives now a better agreement of QCT and QMS results with more recent pump–probe experiments detecting the $O(^3P)$ atoms formed in reaction (1). However, a final comparison of theory and experiment on $\sigma(E)$ can only be made if exact quantum dynamics calculations for all involved states of the reactants on an accurate *ab initio* calculated potential energy surface at full configuration interaction level with a complete basis set are carried out.[21,22] In addition, up to now all the reactive scattering calculations were carried out on a single-valued $^2A''$ ground-state surface which in C_s geometry correlates directly *via* the ground state of the HO_2 radical to the products $O(^3P)$ and $OH(^2P)$. However, there are several energetically low-lying electronically excited states which correlate with the same products and could therefore be involved in the real reaction process. The true ground-state surface should have two conical intersections, one for C_{2v} and one for $C_{\infty h}$ geometry. In dynamics experiments by Hall and co-workers[23] Doppler-resolved OH LIF was used to determine differential cross sections for reaction (1) at different collision energies. The observed energy dependence of the differential cross section was proposed to be a consequence of the dynamics in the vicinity of the C_{2v} conical intersection.[24]

Fig. 10 depicts the present situation for the measured and calculated thermal rate constants, shown for the reverse direction k_{-1}. Using a revised enthalpy of formation for the hydroxyl radical[25] (down by about 2 kJ mol^{-1}) and a new *ab initio* potential quantum statistical adiabatic channel, model calculations gave agreement with experimental data better than 20% between 300 and 5000 K.[26] Nevertheless, as already mentioned above a better PES is still needed to resolve the question of non-RRKM dynamics of the HO_2 complex[27] and the policy to solve the zero point energy problem for the most important elementary step in combustion kinetics.[28] For the hydrogen exchange reaction the surprising close agreement of classical, quantum and experimental data[29] apparently is the result of a compensation of opposing errors (tunneling, zero-point energy constraints).

Ignition processes

Optimal control of ignition processes is one of the key factors in improving the performance of many combustion devices. To this end, a better detailed understanding of unsteady combustion phenomena is required. In order to develop quantitative mathematical models for complete simulation of ignition processes, which include detailed chemistry, experimental studies of simple systems are particularly useful. The experimental techniques should allow visualisation of the ignition process in time and space. Among different fuels, methanol has an interesting potential for such studies. It can be produced in large quantities from biomass or natural gas, leads to a decreased pollutant formation and the kinetics can be described with a limited set of elementary steps. An experimental set-up to investigate the laser induced ignition of methanol is shown in Fig. 11. The coincidence of the 9P12 CO_2-laser line in the (001)–(020) band with the R12-CO stretch fundamental band of the methanol molecule at 9.6 μm allows controlled heating and ignition of the mixture. For temperature measurements, hydroxyl radicals formed during flame propagation can be excited in two rotational states of the (3–0) vibrational band in the A $^2\Sigma^+$–X $^2\Pi$ transition at 248 nm using either two separate KrF excimer lasers or a single two-wavelength KrF excimer laser.[30] Each excimer laser beam is formed into a light sheet 30 mm in height and 400 μm thick using quartz lenses. The sheets are spatially overlapped with a time delay of 100 ns to separate the signals excited by the different pulses. Reflexion filters consisting of four narrowband dielectric mirrors (297 ± 6 nm, transmission >90%, blocking 5×10^4) were used to isolate the (3–2) fluorescence bands of OH. The fluorescence was detected by gated image-intensified CCD cameras. The excitation of two different rotational transitions of OH radicals, starting from $N'' = 8$ and 11, allowed the measurement of spatially corresponding LIF image pairs. Using the Boltzmann distribution, the ratio of these images can be converted into a temperature field. Calculating temperatures from a two-line LIF measurement requires a careful consideration of a number of effects. The influence of fluorescence quenching, usually the dominating deactivation process for excited OH radicals under atmospheric pressure conditions, is reduced in the experiments to less than 3% by the predissociating nature of the excited state.[31,32] The tunable excimer laser emits polarized light, which in turn induces LIF signals with a spatial preference, depending on the transition that has been excited.[33] Collisions can redistribute the spatial alignment; this means that the ratio of the LIF images can depend on the nature of the collider gas composition. This effect, as well as the variation of the RET rates in the ground state,[34] can be accounted for with a calibration procedure, where temperatures measured by the two-line LIF method are compared to pointwise CARS temperature measurements.[35] For the direct numerical simulation, the two-dimensional system of coupled ordinary differential and algebraic conservation equations was solved numerically by spatial discretization using finite differences.[36] Two-dimensional imaging of the temperature as well as the numerical simulation[37] show a conical flame front (see Fig. 12).

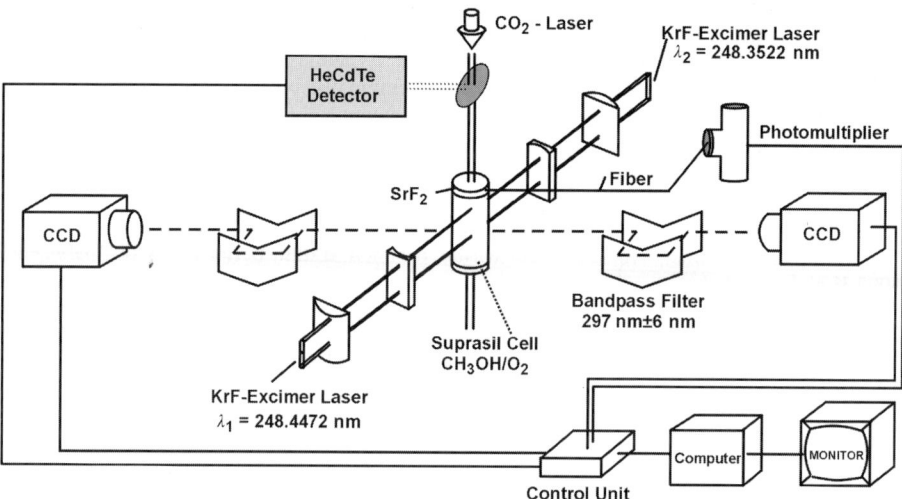

Fig. 11 Experimental set-up for studying thermal ignition processes in methanol–oxygen mixtures. The mixtures were ignited with a CO_2 laser. The flame propagation is followed by 2D laser-induced fluorescence of hydroxyl radicals.

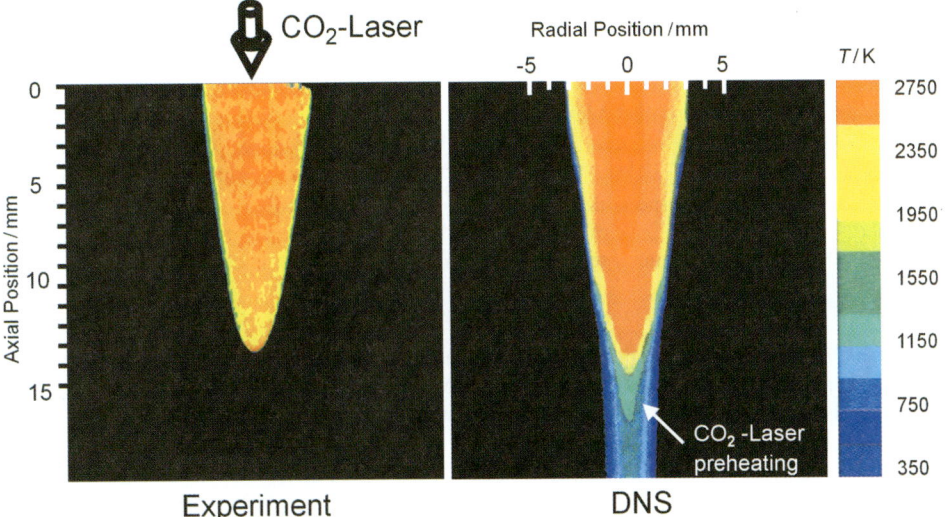

Fig. 12 The development of the two-dimensional temperature distribution during the ignition of a methanol–oxygen mixture ($\Phi = 0.9$, $p = 300$ mbar).

This can be explained by the fact that one channel is preheated by the CO_2-laser beam. Therefore the flame propagates faster in the axial than in the radial direction. The fast axial flame propagation is caused mainly by successive ignition along the cell axis, due to different induction times which follow the axial temperature gradient. Typical propagation speeds of the ignited CH_3OH–O_2 mixtures investigated are 30 m s^{-1} in the radial direction and 130 m s^{-1} in the axial direction. This observation gives an important clue for the understanding of knock phenomena in spark ignition (SI) engines. Unwanted self-ignition occurs here due to local temperature fluctuations as small as 20 to 30 K ('hot spots') during the adiabatic compression phase. This local ignition, which can be monitored by PLIF of formaldehyde,[38] forms pressure waves which produce temperature jumps, that can accelerate flame propagation in the same way as shown in Fig. 12. Later on, the combustion wave reaches the pressure wave, and the system is subject to the transition to detonation.

For the investigation of ignition processes in more complex fuels rapid compression machines (RCM) have been used for many years[39] and are described also at this meeting.[40] Fig. 13 shows computed temperature and concentration profiles of a typical RCM experiment on the ignition of neopentane.[41] The two-stage ignition process observed for such hydrocarbons is clearly visible. At the first stage, occurring 5 ms after the temperature rise due to rapid compression, ignition is the

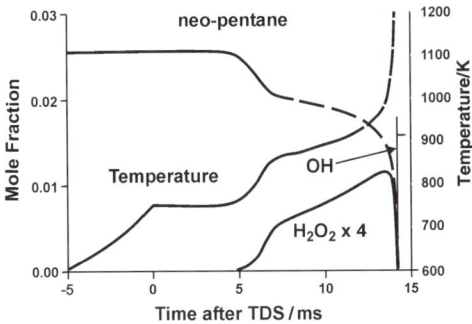

Fig. 13 Computed temperature and concentration of neopentane, H_2O_2, and OH in a RCM experiment. Reproduced with permission from ref. 41.

result of a complex kinetic process.[42]

$$RH + O_2 \longrightarrow R + HO_2 QOOH \longrightarrow Q + HO_2$$

$$R + O_2 \xrightarrow{M} RO_2 \longrightarrow QO + OH$$

$$RO_2 \longrightarrow QOOH \xrightarrow{O_2} O_2QOOH$$

A detailed scheme for n-heptane is depicted in Fig. 14. A large number of intermediate products are formed, even oxygen heterocycles (see Fig. 15). During the time before the second stage ignition, the H_2O_2 concentration increases steadily until around 1000 K the decomposition

$$H_2O_2 \xrightarrow{M} 2OH \qquad (3)$$

becomes fast and the hydroxyl radical concentration increases rapidly, which allows fast consumption of the remaining fuel and rapid temperature increase (see Fig. 13). Thus, similar to the kinetic analysis of the laminar flame velocity, we can identify the simple elementary step (3) as the key

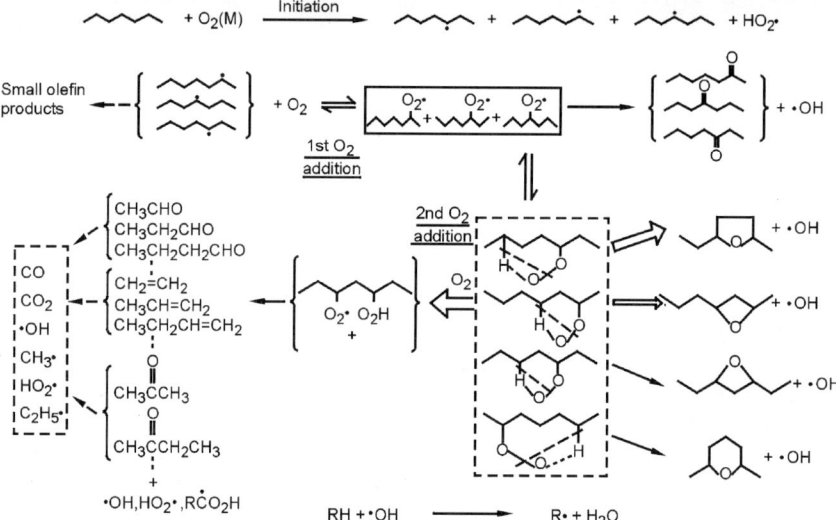

Fig. 14 Reaction mechanism for the thermal oxidation of n-heptane in the cool flame region.

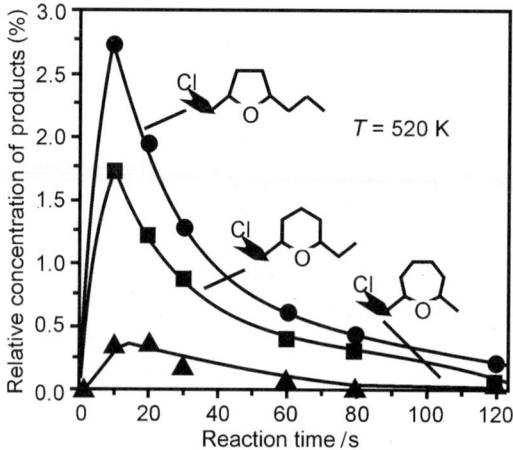

Fig. 15 Formation of O-heterocyclic products by laser induced oxidation of 1-chlorooctane. Reproduced with permission from ref. 43.

reaction controlling the second stage ignition process at temperatures of 900–1000 K. Examining the rate of reaction (3) one must keep in mind the second channel of reaction (3)[44]

$$H_2O_2 \xrightarrow{M} O + H_2O \tag{3'}$$

Engine combustion

Laser techniques are widespread in modern investigations of practical combustion systems. Engine related studies illuminate different aspects of mixture formation, droplet and spray characteristics, flow dynamics, temperature and species concentration as well as pollutant formation (nitric oxide, soot, emissions of unburned hydrocarbons). Many of these data, acquired experimentally in technical systems, are used to validate and improve combustion models that are frequently applied in the development of engine design.

Fluorescent diagnostic systems, based on the photophysics of organic molecules, are used for LIF measurements of fuel distribution, mixing, evaporation and flow visualization. Fluorescing molecules interact with their surroundings, either by concentration-dependent effects like quenching or physical effects like temperature and polarity. Commercial standard fuels contain a number of strongly fluorescing aromatic compounds which can easily be used for qualitative imaging of mixing and fuel distribution. This method has been applied to detailed studies of crevice release and mixture formation in a two-stroke direct-injection engine.[45] When using non-fluorescing fuels, well-defined fluorescing dopants can be added to quantify the fuel concentration as long as the vaporization properties of fuel and tracer are reasonably matched. Carbonyl components exhibit a much lower sensitivity to oxygen quenching than aromatic compounds. Therefore the use of carbonyl components like acetone and 3-pentanone is becoming increasingly widespread as a diagnostic tool for mixing in both reacting and non-reacting flows.[46–48] The temperature dependence of ketone fluorescence offers also the opportunity to measure gas-phase temperatures[49–51] in the range up to 1000 K, which are the relevant temperatures prior to ignition in internal combustion (IC). For simultaneous measurements of the fuel distribution between liquid and gas phase and for quantitative measurements of fuel vapor distribution in the presence of droplets an exciplex-based system can be applied.[52]

Design of gas turbine combustors and direct-injection IC engines requires exact knowledge of the evaporation and heating behavior of fuel sprays. Attempts to model heat transfer and evaporation as well as local concentrations within droplets have led to numerous laser studies of isolated droplets under well-defined high-temperature and ambient conditions. The properties of fuel sprays directly affect performance and emission characteristics of the engines. The spherical or nearly spherical nature of the droplets provides several unique optical features. Thresholds for stimulated processes like lasing or stimulated Raman scattering within droplets may be reduced by some orders of magnitude if compared to bulk liquids. From the mode structure of the emitted fluorescence light, droplet diameters can be obtained.[53] The synopsis of the results of laser techniques yielding information on liquid fuel distribution, vapor–fuel/air mixing, polycyclic aromatic hydrocarbon (PAH) distribution, soot concentrations and size distributions, flame structure measurements and localization of autoignition sites yielded a model for diesel combustion which differs significantly from what was thought to occur before laser techniques were applied.[54] Autoignition and the premixed burn were considered to be localized in regions with nearly stoichiometric mixing around the jet periphery. Furthermore, ignition was thought to occur at only a few points followed by a rapid spread of the flame. Laser measurements revealed that ignition starts progressively at multiple points across the downstream region of the jet. The liquid-fuel penetration depth was found to be significantly shorter than expected whereas the fuel vaporization was attributed primarily to the entrainment of hot air and not to combustion heating.

Despite the development of improved exhaust-gas after-treatment systems, it is necessary to investigate possibilities to reduce the formation of nitric oxide during the combustion process itself, especially in view of very lean burning engines, where catalytic converters are difficult to operate. Various strategies using different laser excitation wavelengths, laser sources and fluorescence detection bands were investigated as tools for in-cylinder engine NO diagnostics. NO concentration measurements in spark ignition (SI) and Diesel engines, exciting NO with tunable ArF excimer lasers at 193 nm (ε-bands)[55,56] and at 225 nm (γ-bands) using the Raman-shifted output of tunable KrF excimer lasers,[57] are reported. However, to circumvent the strong attenuation of these short excitation wavelengths in many technical combustion systems, excitation of NO with tunable KrF excimer lasers at 248 nm was suggested.[58] Since then, this technique has frequently been applied for quantitative measurements of NO in IC and Diesel engines.[59,60] This detection scheme was characterized by measurements in stabilized high-pressure flames.[61] The excitation laser is tuned to the O_{12} bandhead of the NO A–X(0,2)-transition where the fluorescence excitation spectrum of molecular oxygen has a local minimum. NO fluorescence emitted in the A–X(0,1) and A–X(0,0) bands at shorter wavelengths is detected, which further minimizes the influence of interfering species. Attenuation of signal light is weak as long as the illuminated volume is close to the detection window and, thus, the optical path length through the burned gases at high pressure is short. Quantification of LIF signal intensities requires detailed knowledge of several temperature- and pressure-dependent factors (see Fig. 16). Variation of absorption spectra due to line broadening and shifting[62] changes the spectral overlap of laser profile and absorption line. To account for the effect of collisional quenching, data obtained for many combustion-relevant colliders are available.[63] Using modeling results for the change in mixture composition around the flame front, local quenching rates were calculated. The collisional quenching rate scales linearly with pressure, and is a function of temperature and gas composition. It turns out, however, that behind the flame front, i.e., the region where NO is predominantly formed in engines, the quenching rates are nearly constant due to compensation effects of composition changes and temperature.

Fig. 17 shows an experimental set-up for the simultaneous single shot measurement of NO concentration and temperature fields in a transparent SI engine.[59] Optical access in the single-cylinder engine was possible for the entire combustion chamber. This was achieved by building the combustion chamber walls from polished quartz plates, mounted on a small Hatz single-cylinder engine. The piston is not lubricated, thus allowing extended measurements without obscuring layers on the windows. An electrical motor and brake permits the engine to run at different speeds and loads. The engine could be operated with gaseous or liquid fuels. The beam from a tunable, narrow band ($\Delta \nu = 1$ cm^{-1}) KrF excimer laser at 248 nm (Lambda Physik, LPX 150) was formed into a vertical light sheet (65 mJ (pulse)$^{-1}$ in a 12 × 0.5 mm cross section) using a cylindrical lens, and aligned through the cylinder in a plane directly below the spark plug between the intake and outlet valve. The pulse energies were measured on a single shot basis behind the engine. Perpendicular to the light sheet plane, the slow scan cameras were mounted on opposite sides of the engine.

Quantitative results are obtained by calibrating the overall detection efficiency. In the case of NO, calibration is possible by doping known amounts of NO into the burning gases. Results from the simultaneous application of Rayleigh scattering and LIF for measurements of temperatures and NO concentrations in an internal combustion engine are shown in Fig. 18. From the image pairs with corresponding NO concentrations and temperature fields it can be seen that NO is formed in high-temperature areas. While the overall spatial distribution of NO and temperature is

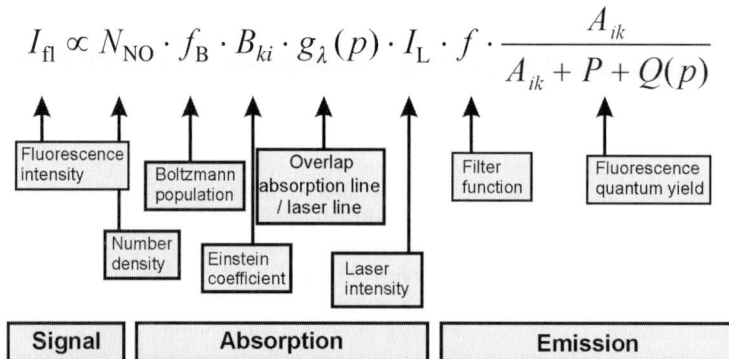

$$I_{fl} \propto N_{NO} \cdot f_B \cdot B_{ki} \cdot g_\lambda(p) \cdot I_L \cdot f \cdot \frac{A_{ik}}{A_{ik} + P + Q(p)}$$

Fig. 16 Requirements for quantifying the NO-LIF signal intensity.

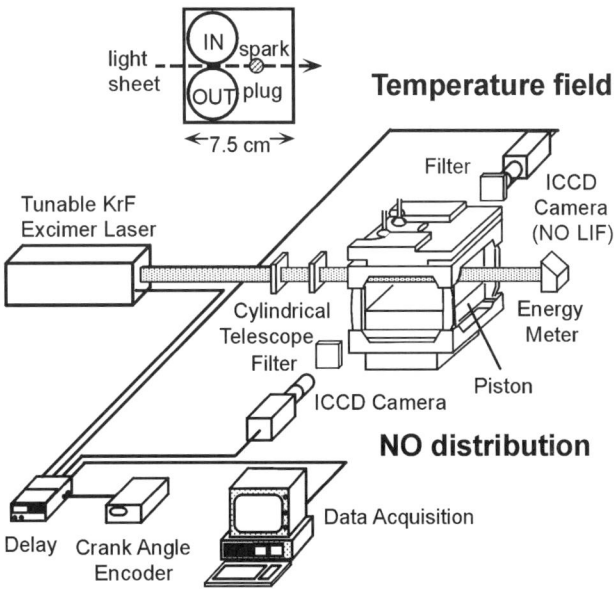

Fig. 17 Experimental set-up for the simultaneous single-shot measurement of NO concentration and temperature fields in a transparent SI engine. Reproduced with permission from ref. 59.

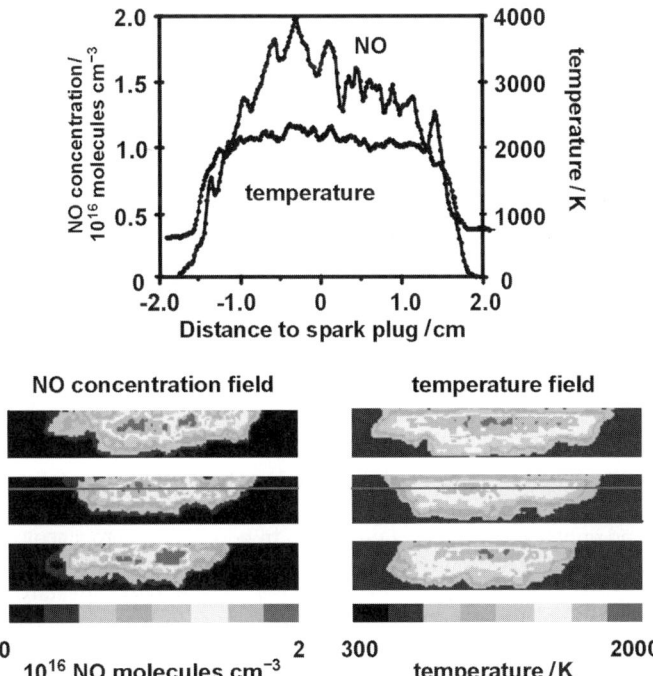

Fig. 18 Simultaneous single-shot absolute NO concentration and temperature fields in a transparent SI engine fuelled with propane/air ($\Phi = 1$) using 2-D LIF and Rayleigh scattering. Reproduced with permission from ref. 59.

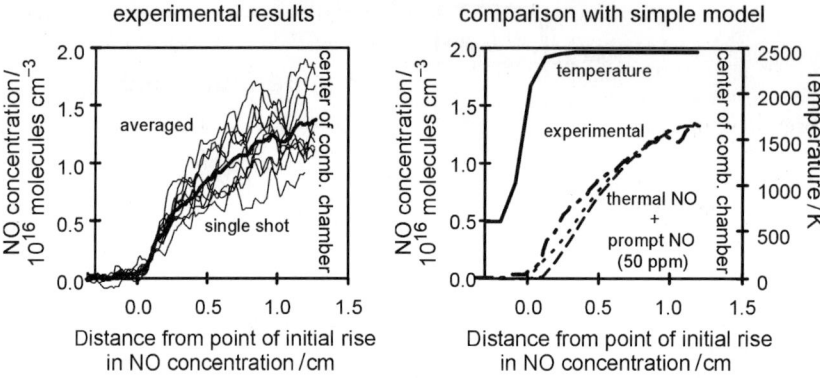

Fig. 19 Comparison of experimentally acquired NO profiles and results of 1-D simulation calculations. Reproduced with permission from ref. 64.

strongly correlated, the profiles in Fig. 18 indicate that the temperature distribution is more uniform than the NO concentration distribution. Due to the strong nonlinear temperature dependence of the NO production rate, careful control and homogenization of the combustion conditions should allow reduction of primary NO formation. A comparison of the experimental results with a detailed chemical kinetics modeling using a 1-D model[64] shows (see Fig. 19) that the pathway of NO formation relevant for IC engine combustion processes can predominantly be described by the Zeldovich mechanism.[65] The rate determining step is the reaction:[66]

$$O + N_2 \rightarrow NO + N \qquad (4)$$

The high activation energy of 319 kJ mol^{-1} of this reaction results in an extreme temperature dependence for the Zeldovich NO formation rate. At temperatures below 1500 K, this process is usually negligible so that in lean burning gas turbines a major contribution comes from the sequence of reactions.[67]

$$O + N_2 \xrightarrow{M} N_2O \qquad (5)$$

$$O + N_2O \longrightarrow 2NO \qquad (6)$$

Due to the high activation energy of reaction (4) the overall amount of NO is strongly influenced by the residence time of the hot burned gases. The nonlinear temperature dependence causes a strong contribution of NO formed in hot spots, that is, inhomogeneous temperature distributions. High temperature and pressure and relatively long residence times are characteristic of IC engine combustion, causing much higher overall NO concentrations[68] compared to laminar or turbulent nonenclosed flames. The prompt-NO contribution[69] occurs directly in the flame front. Hydrocarbon radicals, mainly CH, CH$_2$ and C,[70] react with N$_2$ forming HCN, from which further oxidizing steps lead to NO. As shown in Fig. 19 this is only a minor pathway for NO formation in IC-engine combustion.

Heterogenous combustion

As demonstrated by Pfefferle and Pfefferle,[71] heterogeneously catalyzed combustion is a promising method for burning fuel in lean fuel–air mixtures which can significantly reduce pollutants, improve ignition, and enhance the stability of flames. In addition, heterogeneous catalysts play a key role in various pollutant emission control processes such as the three-way catalyst (TWC) used for the after-treatment of automobile exhaust gases. In the TWC, a catalyst containing Pt/Pd/Rh converts the two reducing pollutants, CO and unburned hydrocarbons (HC), as well as the oxidizing pollutant, NO, to the stable products H$_2$O, CO$_2$ and N$_2$. In this process the global surface reaction 2NO + 2CO → N$_2$ + 2CO$_2$ occurs.

However, compared with gas-phase combustion, the situation is much more complex, because besides the dynamics of the transport processes in the boundary layer above the catalytic surface, the effects of the topology of the active surface on reactivity, including adsorption and desorption processes, also have to be considered. As a consequence, the development of appropriate mathematical models for the simulation of surface reactions and their coupling to the surrounding gas phase is essential to an understanding of heterogeneous catalysis under technically relevant conditions. Computational tools for the description of different catalytic combustion systems have recently been developed and will be described at this Discussion.[72] These tools include detailed surface chemistry as well as detailed models for molecular species transport. However, surface reaction mechanisms derived so far are based mainly on studies of elementary surface reaction steps carried out under ultra-high vacuum (UHV) conditions and on well-defined single-crystal surfaces. Use of this kind of surface kinetics data in the modeling of technical processes that usually take place at high pressure ("pressure gap") and, for example, on polycrystalline catalyst material ("materials gap") emphasizes the importance of developing *in situ* diagnostics techniques that can be applied under practical pressure and temperature conditions and on realistic catalysts (see Fig. 20). Optical diagnostic methods that probe interface electronic and vibrational resonances offer significant advantages over conventional surface spectroscopic methods in which, for example, beams of charged particles are used as a probe. In particular, three-wave mixing techniques[73] such as the optical second-harmonic generation (SHG) and the sum-frequency generation (SFG) (see Fig. 2) have already become important tools for the *in situ* study of catalytic combustion, although they are still in their infancy. Both methods are potentially surface-sensitive at nondestructive power densities, and their application is not restricted to UHV conditions.[74] Hence both diagnostic methods can be used to bridge the "pressure gap". SHG, however, suffers from a serious drawback namely, its lack of molecular selectivity. As a consequence, SHG cannot be used for identifying unknown surface species. On the other hand, the application of infrared–visible (IR–VIS) SFG allows surface vibrational spectroscopic measurements with submonolayer sensitivity with the help of a tunable IR laser.[75,76] Such investigations represent important steps toward developing validated adsorption/desorption and surface reaction mechanisms for catalysts with higher structural complexity. These studies can be regarded as intermediate between the UHV surface science studies on well-defined single crystals as pioneered by Ertl and co-workers[77] and investigations of supported nanoparticles[78] as used in practical catalysts.

Fig. 21 reproduces an experimental set-up that allows catalytic combustion studies over a wide pressure range from UHV conditions (base pressure about 3×10^{-10} mbar) up to atmospheric pressure conditions. In the high-pressure regime, this arrangement permits studies of adsorption/desorption[74] and heterogeneous reaction processes of well-defined stagnation point flows of reactant mixtures on the catalyst surface.[79] The reaction cell is equipped with a quadrupole mass spectrometer (QMS) for thermal programmed desorption (TPD) measurements, an Ar^+ ion sputter source, a retarding field analyzer (RFA) for Auger electron spectroscopy (AES) and low-energy electron diffraction (LEED) measurements. Another differentially pumped QMS is connected to the vacuum line behind the reaction chamber for on-line monitoring of stable reaction products in the exhaust gas. For the optical IR–VIS SFG measurements a 30 ps mode-locked Nd : YAG laser system was used. Part of its output was frequency doubled to 532 nm and used as

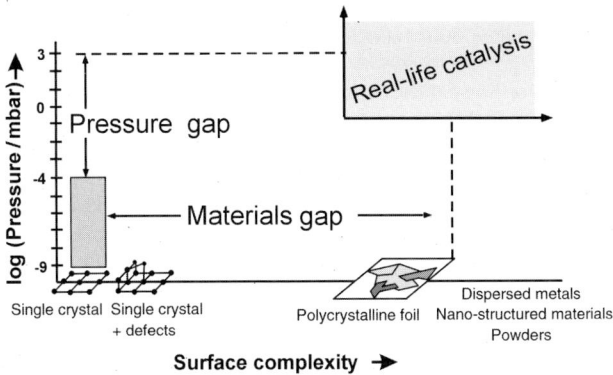

Fig. 20 Schematic illustration of the so-called "pressure" and "materials" gaps that separate UHV single-crystal model studies from technical catalytic combustion investigations.

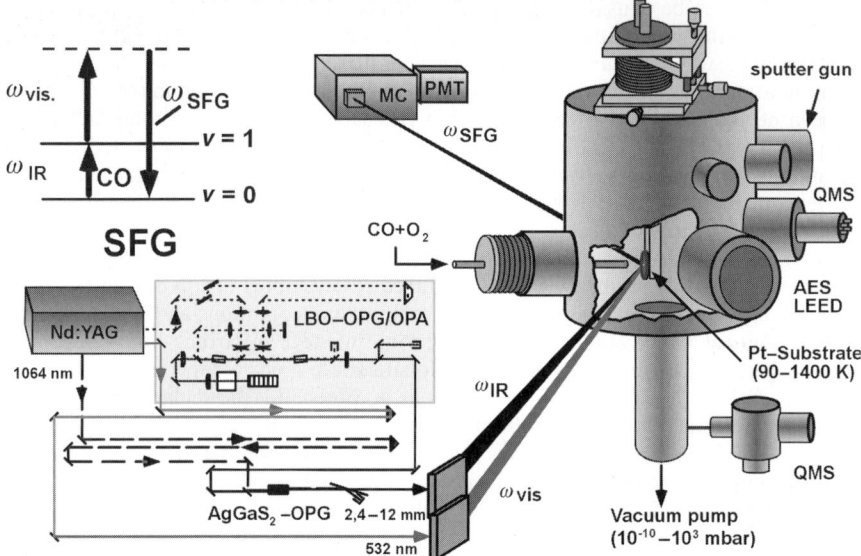

Fig. 21 Experimental set-up for the *in situ* detection of chemisorbed CO during heterogeneous combustion of CO on a Pt catalyst using optical infrared–visible sum-frequency generation. LBO: lithium–barium–borate crystal for OPG/OPA: optical parameter generator/amplifier, MC: monochromator, PMT: photomultiplier, AES: Auger electron spectrometer, LEED: low-energy electron diffraction spectrometer, QMS: quadrupole mass spectrometers for CO thermal desorption (TD) and CO_2 production rate measurements. Reproduced with the permission of the publisher.

the visible input frequency (ω_{VIS}) for the SFG process. The other part was used to pump an LBO–OPG/OPA system to generate in a first step IR radiation tunable in the wavelength region 1.25–1.39 µm. In a second step, this IR radiation is mixed in a $AgGaS_2$ crystal with the fundamental radiation of the Nd : YAG laser (1064 nm) to obtain *via* difference-frequency generation the tunable IR radiation (ω_{IR} = 1400–2200 cm^{-1} with a bandwidth of about 7 cm^{-1}). The IR and VIS laser pulses were temporally aligned, and the beams were spatially overlapped at the surface. The generated SFG signal is reflected from the Pt surface, according to the phase-matching condition at an angle θ_{SFG} with respect to the surface normal. The SFG signal was detected after filtering off the scattered light by a monochromator with a photomultiplier (PM) and a gated integrator and transferred to a laboratory computer. It was corrected for fluctuations in the VIS and IR beam by dividing the SFG signal by the VIS and IR laserbeam energies. Every point in the measured SFG spectra shown in Fig. 22 was obtained by averaging over 120 laser shots. The catalyst (in the present case either a Pt(111) single-crystal disk or a polycrystalline Pt foil) could be resistively heated in the temperature range 300–1600 K and translated, tilted, and rotated by means of a manipulator.

Adsorption of molecules on the catalyst can be regarded as the principal first step in any heterogeneously catalyzed reaction. As a consequence, knowledge of sticking coefficients S, and in particular information about their variation with surface coverage θ and catalyst temperature T_c is of great importance in the development of any elementary surface reaction mechanism. Although initial (for bare surfaces) and coverage-dependent sticking coefficients have been measured for various gas/single-crystal surface systems using different UHV methods such as the King–Wells technique,[81] there is still considerable lack of information regarding the temperature dependence as far as practical catalytic combustion temperatures are concerned. As a consequence, models such as the temperature-dependent Kisliuk model[82] are usually employed for $S(\theta, T_c)$ in order to calculate adsorption rates at practical catalytic combustion temperatures. At this point, it should be noted that also the desorption rate coefficients k_d, which are usually expressed as $k_d = v_d \exp\{-E_d/kT_c\}$, can, even for quite simple adsorbate/surface systems, exhibit a rather complicated dependence on the surface coverage. For example, for the NO/Pt(111) system both the pre-

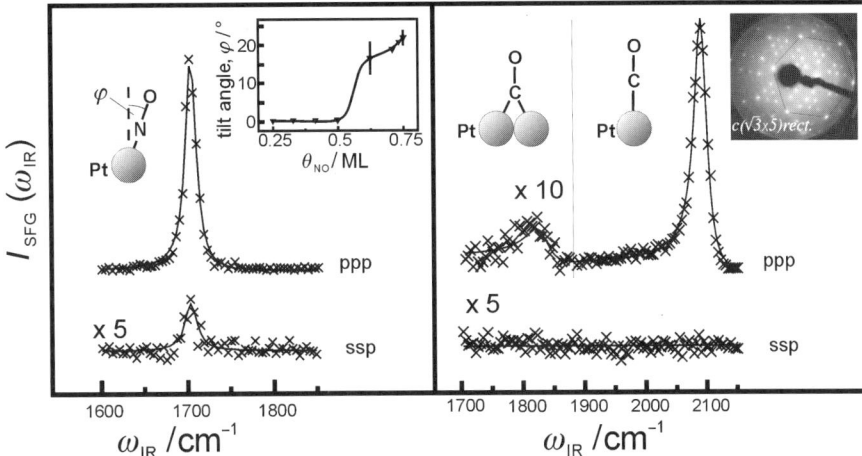

Fig. 22 (a) SFG spectra of NO terminally adsorbed on Pt(111) in a tilted adsorption geometry at a surface coverage of $\theta_{NO} = 0.75$ ML. The measured spectra (×) were recorded for two different polarisation combinations: ppp (p-polarised sum frequency signal, p-polarised visible and p-polarised IR laser beam) and ssp (s-polarised sum frequency signal, s-polarised visible and p-polarised IR laser beam) (b) SFG spectra of bridge-bound and terminally-bound CO on Pt(111). SFG spectra were recorded at room temperature for the ppp and ssp polarization combinations. LEED diffraction pattern[80] from which an absolute CO coverage of $\theta_{CO} = 0.6$ ML can be derived is shown as an insert. Reproduced with the permission of the publisher.

exponential factor v_d and the activation energy of desorption E_d were found to vary as a function of the NO coverage.[83]

In the following, results from SFG *in situ* CO coverage measurements performed under adsorption/desorption equilibrium conditions are presented, which together with the frequency information derived from the CO SFG spectra can be used to develop a validated adsorption/desorption mechanism for the CO/polycrystalline Pt foil system. In Fig. 23, CO coverages (solid squares) derived from the depicted SFG spectra are plotted against the temperature of the polycrystalline Pt catalyst.[84] In addition, the corresponding CO vibrational frequencies derived from the numerical least-squares fit analysis of the SFG spectra are shown as solid circles. The CO coverage *vs.* substrate temperature plot clearly reveals the presence of two different adsorption/desorption regimes. The first is in the temperature range $300 \leq T_c/K \leq 600$ in which the CO equilibrium coverage decreases almost linearly with substrate temperature; the second is at $T_c > 600$ K in which the CO coverage decreases more rapidly with increasing substrate temperature. This behavior indicates the presence of at least two distinct CO adsorption sides on the Pt catalyst with markedly different activation energies for desorption. Further insight into the nature of the CO adsorption sites can be gained from a comparison of the coverage dependence of the measured CO vibrational frequencies with the results of studies carried out under UHV conditions on different Pt single-crystal surfaces. As outlined in detail in ref. 76 and 79, the coverage dependence of the CO vibrational frequency observed on the polycrystalline Pt catalyst closely resembles the coverage dependence observed on a Pt[4(111) × (100)] single-crystal surface.[85] This resemblance indicates similarity in the topology between the latter single-crystal surface and the polycrystalline Pt foil. The Pt[4(111) × (100)] single-crystal surface consists of (111) terraces that are four atoms wide, separated by monatomic steps with 25% of its exposed Pt atoms coordinated to 7 nearest neighbors (compared to 9 nearest neighbors for the remaining (111) terrace site Pt atoms as shown in Fig. 23). Based on the similarity between the Pt[4(111) × (100)] single-crystal and the polycrystalline Pt foil surface topology, a surface adsorption/desorption model could be developed.[79] This model which accounts for two distinct adsorption sites (denoted as A- and B-sites) is based on surface adsorption kinetics data obtained on Pt(111) (for the A-sites) surfaces and on single-crystal surfaces exhibiting seven-fold coordinated adsorption sites (for the B-sites). The solid line shown in Fig. 23 represents the results of a numerical simulation, assuming that the polycrystalline Pt catalyst surface consists of 80% A-sites and 20% B-sites. The latter adsorption site

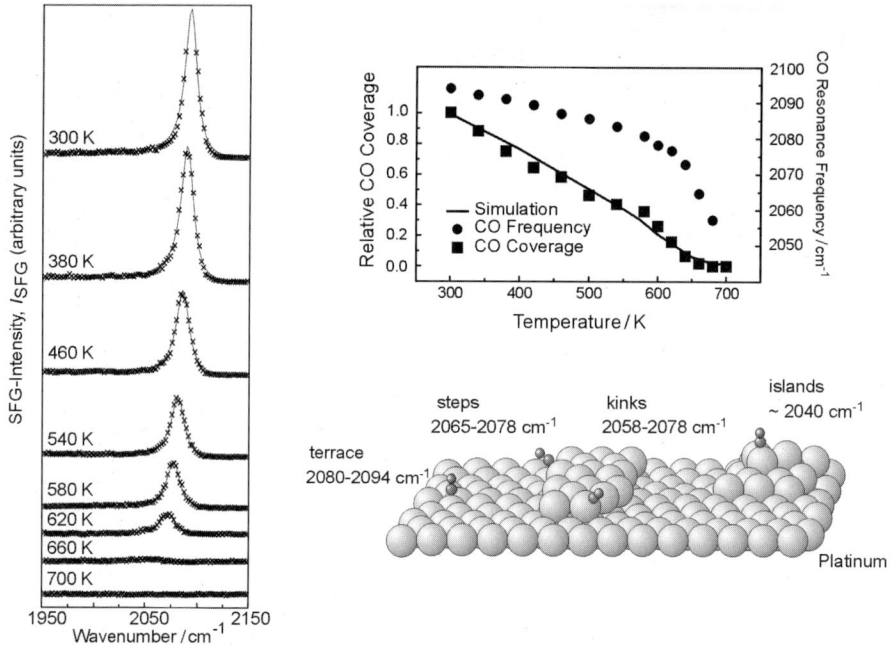

Fig. 23 CO coverages and CO vibrational resonance frequencies obtained from *in situ* SFG measurements on a polycrystalline Pt catalyst under adsorption/desorption equilibrium conditions at a CO gas-phase pressure of 1 mbar. The solid line represents CO coverages obtained in a numerical simulation.[79] Schematic description of the surface topology of the polycrystalline Pt catalyst together with three possible adsorption sites of terminally bound CO (the coordination number of the Pt atom on which CO is adsorbed is given together with the corresponding typical CO vibrational resonance frequencies).

ratio is supported by the TPD spectra measured for the used Pt foil.[84] In the numerical simulation for the CO sticking coefficient $S(\theta_{CO}, T_c)$, a temperature-dependent Kisliuk model with an initial sticking probability of 0.7 at $T_c = 300$ K was applied. For the A-site desorption rate coefficient k_d, a coverage-independent preexponential factor of $v_d = 3 \times 10^{19}$ s^{-1} was employed in combination with a linearly coverage-dependent activation energy for desorption, $E_d(\theta_{CO}) = E_d(\theta_{CO} \to 0) - (112$ kJ mol$^{-1} \times \theta_{CO})$. For $E_d(\theta_{CO} \to 0)$, which represents the activation energy for desorption in the limit of zero CO coverage, a value of 183 kJ mol^{-1} was used. This value is in good agreement with the heat of adsorption value of 187 ± 11 kJ mol^{-1} measured for a Pt(111) single-crystal surface in the limit of zero CO coverage.[86] For the desorption rate coefficient of the more strongly bound B-sites of the polycrystalline Pt-foil, a coverage independent pre-exponential factor of $v_d = 5 \times 10^{21}$ s^{-1} was used along with a linearly coverage dependent activation energy for desorption. A value of $E_d(\theta_{CO} \to 0) = 220$ kJ mol^{-1} closely matches the initial heat of adsorption of 210 ± 7 kJ mol^{-1}, which was recently measured for the Pt(311) surface on which CO adsorbs predominantly at seven-fold coordinated Pt atoms.[87] A moderate linear coverage dependence was used for the B-sites, $E_d(\theta_{CO}) = E_d(\theta_{CO} \to 0) - (24$ kJ mol$^{-1} \times \theta_{CO})$, which also closely resembles the coverage dependence measured for the Pt(311) surface.

In Fig. 24, CO coverages (solid squares) derived from the depicted SFG spectra are plotted against the temperature T_c of the polycrystalline Pt catalyst. The corresponding CO vibrational frequencies derived from the numerical least-squares fit analysis of the SFG spectra are also shown as solid circles. Comparison with the CO coverages measured during the CO adsorption/ desorption studies (see Fig. 23) reveals, that if O_2 is present in the stagnation flow, the total CO coverage at $T_c = 300$ K is considerably reduced. However, although the CO coverage is reduced by about 30%, the CO vibrational frequency remains at a value of 2095 cm^{-1}, which is close to the saturation coverage vibrational frequency of CO terminally adsorbed at Pt(111) terrace sites (A-sites). In addition, in contrast to the observation in the CO adsorption/desorption measure-

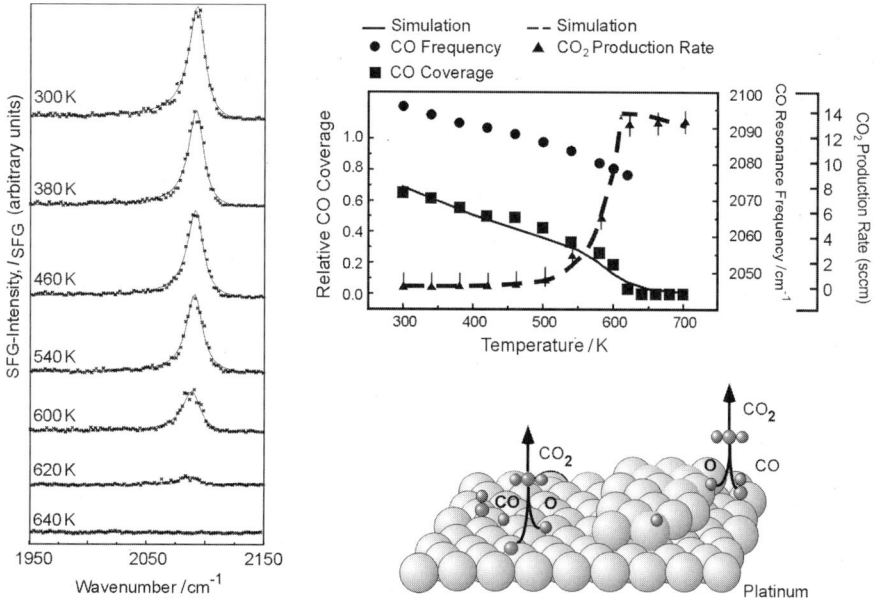

Fig. 24 CO coverages and vibrational resonance frequencies obtained from *in situ* SFG measurements during CO combustion on a polycrystalline Pt catalyst. The CO_2 production rate is simultaneously measured by mass-spectrometry. Solid lines represent results of a numerical simulation.[79]

ments, during CO oxidation over the whole temperature range where adsorbed CO could be detected, the frequency remains in the range 2095–2078 cm^{-1} which is typical for CO adsorbed on Pt(111) terrace sites (see Fig. 23). Therefore, the major difference between the CO adsorption/desorption and the CO combustion data is the absence of the low-frequency contribution from the step sites (B-sites) in the latter case. This observation can be attributed to a higher sticking probability of O_2 at the step sites, which blocks CO step sites adsorption when both CO and O_2 are present in the gas phase. The preferential adsorption of O_2 on step sites has already been observed during studies[88] of CO oxidation on Pt[4(111) × (100)]. The solid line depicted in Fig. 24 represents the result of a numerical reactive flow simulation of the CO combustion experiment. This simulation is based on the Langmuir–Hinshelwood (LH) scheme for CO oxidation, including molecular adsorption and desorption of CO (using the adsorption/desorption kinetics data described above), dissociative adsorption of O_2, as well as the formation of CO_2 through the reaction of the adsorbed CO and O species. Further details about the mean-field surface reaction model and the rate parameters employed for the A- and B-sites (in the CO combustion simulation, the same CO adsorption site ratio as in the previous section was used) are given in ref. 79. Although the results of the CO combustion simulation shown in Fig. 24, with the two distinct CO coverage *vs.* temperature regimes, look quite similar to that shown in Fig. 23, the actual origins of the CO coverage curve shapes are quite different. The sudden decrease in the CO coverage in the temperature range T_c = 550–600 K observed under combustion conditions does not result from the difference in the desorption kinetics of the A- and B-sites as in the reaction model, B-sites are actually blocked by adsorbed oxygen. The rapid decrease in CO coverage on the A-sites in this temperature interval is due to a transition from a CO-covered to an O-covered state, resulting from an increase in O_2 adsorption with decreasing CO coverage and an increase in the LH-reaction rate. A comparison of the simultaneously measured CO_2 production rate (solid triangles) with the result of the simulation (solid line) is also shown in Fig. 24. The sharp increase in the CO_2 production rate in the temperature interval $600 > T_c/K > 550$, followed by an almost temperature-independent CO_2 production rate for $T_c > 600$ K, is typical for a reaction system that undergoes a transition from a state which is predominantly determined by the surface chemical kinetics (300 ⩽ T_c/K < 550) to a state which is predominantly controlled by transport limitations in the gas phase (T_c ⩾ 600 K).

Combustion control using NIR-diode lasers

Combustion of municipal waste can be a valuable source of energy which makes use of the relatively high energy content of the waste. Since fossil fuels are replaced, waste combustion has no negative effects on the CO_2 budget, whereas waste disposal in landfills, the most common disposal path in many countries, is responsible for long term emission of CH_4 into the atmosphere. CH_4 has a nearly 30-fold higher global warming potential than CO_2. To take advantage of the thermal conversion process, however, it is compelling to fulfil strict legal emission regulations while maximizing heat release and conversion efficiency of the overall combustion process. This can be achieved by ensuring homogeneous combustion throughout the entire furnace. Since waste shows very strong variations in its fuel parameters, *e.g.* calorific value and waste humidity, fast, *in situ* species and temperature sensors and efficient combustion control strategies, are needed, to allow optimization of the primary process parameters. Over the last years several *in situ* sensors were developed in our institute and tested under realistic conditions in a full-scale industrial waste incinerator[89–92] as depicted schematically in Fig. 25. Primary air injection, grate movement, and fuel input are under closed-loop control through an IR-scanner-camera (TACCOS).[90] A second sensor allows closed-loop control of the selective noncatalytic reduction (SNCR) process, commonly used on stationary combustion systems to control NO_x emissions. Our laser *in situ* ammonia monitor (LISA)[89] which is based on a $^{13}CO_2$ laser allows precise control of the necessary additive injection (NH_3 or urea), minimized additive consumption and ensured up to 90% NO_x removal. The SNCR process is based on the fast reaction:

$$NO + NH_2 \rightarrow N_2 + H_2O \tag{7a}$$

$$\rightarrow N_2H + OH \tag{7b}$$

which gives at room temperature nitrogen and highly vibrationally excited water molecules as products,[93,94] while the contribution of channel (7b) rises with increasing temperature as shown in many experimental and theoretical investigations.[95] As shown in Fig. 26 the addition of hydrocar-

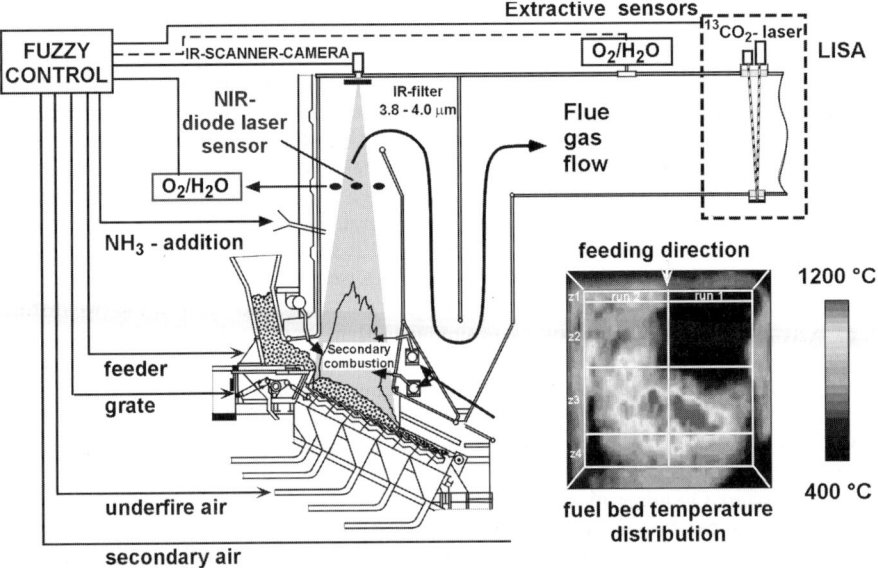

Fig. 25 Intersection through a waste incinerator with a selection of important sensors. Diode laser *in situ* sensor for O_2 and H_2O detection, laser *in situ* ammonia sensor (LISA), IR-scanner-camera (TACCOS) as well as the possible actors that are connected to the combustion control system (CCS).

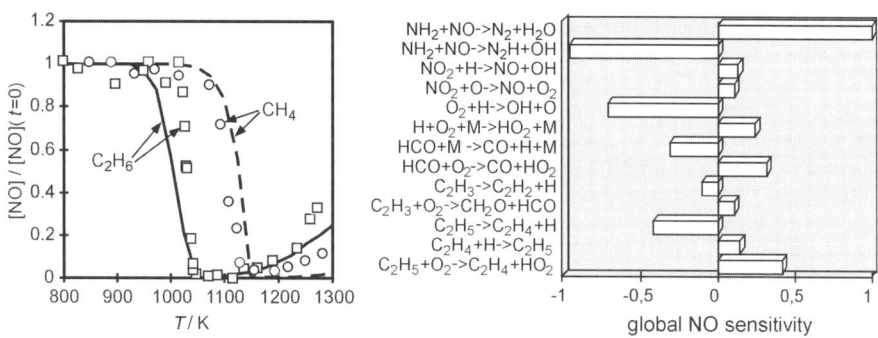

Fig. 26 (a) Influence of hydrocarbons on the temperature windows of the NO–NH_3–O_2 SNCR-process. (b) Global sensitivity coefficients for NO concentrations in the temperature window of the NO–NH_3–O_2 SNCR-process with hydrocarbon addition.

bons together with NH_3 or urea accelerates the chain reaction triggered by channel (7b) and adjust the "temperature window" of the SNCR process according to the actual flue gas temperature of the incinerator.[96]

For fast *in situ* detection of fuel quality fluctuations, and an active control of the homogeneous combustion process above the solid fuel bed near-infrared diode lasers (NIRDL) and high resolution absorption spectroscopy are employed for simultaneous *in situ* detection of O_2 and H_2O along a single absorption path at the end of the secondary combustion zone of the full-scale municipal waste-to-energy plant (see Fig. 25). NIRDL are used because they offer a unique combination of advantages (low cost, size and power requirement; room temperature operation; high spectral resolution; long lifetime), which make them most suitable for industrial applications. Their versatile tuning properties—slow and coarse (range: several nm; speed: some GHz s^{-1}) by a change of laser temperature, and extremely fast fine tuning (range: tens of GHz; speed: 10^5 GHz s^{-1}) by a modulation of the laser current—enable the detection of the complete profile of an absorption line and application of very efficient noise reduction strategies, so that excellent sensitivities down to the ppm-level could be demonstrated under well-controlled laboratory conditions. For the application in the incinerator it was most important to correct for any changes in the effective transmission of the complete light path between laser and detector. These changes can be caused by the quickly fluctuating dust load of the flue gases, window degradation by dust, heat, or chemical attack, vibrations or mechanical deformation of the boiler walls, and refractive index gradients or thermal lensing effects. For this purpose an all-analog electronic on-line transmission correction circuit was developed.[92] The circuit, which uses a double divider configuration, corrects each individual scan in real time and has to be adapted to the results of a careful characterization of the absolute magnitude, frequency and temporal behaviour of the transmission fluctuations of the specific incinerator. The idea to correct these fluctuations makes use of the diode lasers high-speed tuning capabilities, and requires that the scan frequency is chosen well above the fastest transmission variations. Since in this way all interferences are frozen and only the molecular signal component changes within the scan time, an electronic correction by division through the quasi-static background transmission signal is possible. An additional noise reduction is achieved by averaging consecutive absorption profiles and extracting the absorption signal *i.e.* the area underneath the absorption line by a fast fitting algorithm.

As a last example the combustion of natural gas in the combined-cycle processes will be considered. Due to the high conversion efficiency of this process, its very low emission of pollutants like NO_x or SO_x and especially its low overall costs it can even compete with electricity production from nuclear reactors. A common requirement for most energy production processes is the ability for rapid load variations to adapt the total power production of the plant to the public demand. As a consequence there is a need for power plants which can be turned on and off very frequently and rapidly to cover the peak power variations of the distributing net. In principle gas-fired power plants are very well suited for this purpose, but there is a proven risk of severe ignition delays. Then a large amount of premixed gas is injected in the combustion chamber, so that the next successful ignition in the multi-burner system will cause a so-called "puff". Severe damage to the combustion chamber often in the multi-10-million-dollar range can be the consequence of such an event. The cause for these ignition delays is still under investigation, but may be partly associated with flame quenching by excess humidity and hence high heat capacity of the primary air. Despite the use of suction-probe-type extractive methane sensors these events could not be avoided because of the slow response time of these sensors and their lack of ability to gain a methane signal which is representative for the large combustion chamber. In order to maintain the positive properties of large-scale natural gas combustion systems under all process conditions, *i.e.* under frequent and strong variations in total power, and to enable secure ignition and precise active control of the huge multi-burner systems, it would be most helpful to have a versatile *in situ* sensor with fast response time for gaseous species and the temperature. We have developed such a multi-purpose *in situ* absorption spectrometer based upon room temperature near infrared diode lasers, which is able to determine simultaneously multiple species concentrations, namely all major combustion gas constituents (CH_4, H_2O, CO_2 and O_2) and the temperature on a single absorption path.[97]

Conclusions

After three decades of research and development in the area of laser-based concepts for combustion diagnostics many of the spectroscopic methods have matured now from qualitative to quantitative techniques. They give valuable insights into the details of the interaction of chemical kinetics with various transport processes, allow the efficient development of combustion models and show that relatively simple elementary steps control a large variety of combustion phenomena like flame propagation, ignition, pollutant formation and destruction.

Acknowledgements

Our work was supported by the State of Baden-Württemberg and the Bundesministerium für Bildung und Forschung (TECFLAM, COSI), the Deutsche Forschungsgemeinschaft (SFB 359 "Reaktive Strömung, Diffusion und Transport"), the European Comission (JOULE, JOULE-THERMIE, TOPDEC-project), and the Heidelberger Akademie der Wissenschaften. Thanks are due to F. Behrendt, R. Kissel-Osterrieder, and J. Warnatz (Interdisziplinäres Zentrum für Wissenschaftliches Rechnen) and H. R. Volpp (University of Heidelberg) for stimulating discussions on catalytic combustion.

References

1 J. Warnatz, U. Maas and R. W. Dibble, *Combustion*, Springer, Heidelberg, 3rd edn., 2001.
2 A. Eckbreth, *Laser Diagnostics for Combustion Temperature and Species*, Gordon & Breach, Amsterdam, 2nd edn., 1996.
3 W. Demtröder, *Laser Spectroscopy-Basic Concepts and Instrumentation*, Springer, New York, 2nd edn., 1997.
4 J. Wolfrum, *Proc. Combust. Inst.*, 1998, **27**, 1.
5 J. Warnatz, *Proc. Combust. Inst.*, 1981, **18**, 369.
6 W. S. Xia, R. S. Zhu, M. C. Lin and A. M. Mebel, *Faraday Discuss.*, 2001, **119**, 191; V. A. Lozovsky, I. Rahinov, N. Ditzian and S. Cheskis, *Faraday Discuss.*, 2001, **119**, 321.
7 T. Landenfeld, A. Kremer, E. P. Hassel, J. Janicka, T. Schäfer, J. Kazenwadel, C. Schulz and J. Wolfrum, *Proc. Combust. Inst.*, 1998, **27**, 1023.
8 G. Damköhler, *Z. Elektrochem.*, 1940, **46**, 601.
9 R. Borghi, in *Recent Advances in Aerospace Science*, ed. C. Casci, Plenum, New York, 1985, p. 117.
10 N. Peters, *Proc. Combust. Inst.*, 1986, **21**, 1231.
11 A. Buschmann, F. Dinkelacker, T. Schäfer, M. Schäfer and J. Wolfrum, *Proc. Combust. Inst.*, 1996, **26**, 437.
12 M. Baum, T. S. Poinsot, D. C. Haworth and N. Darabiha, *J. Fluid Mech.*, 1994, 281.
13 N. Peters, *Turbulent Combustion*, Cambridge University Press, Cambridge, 2000, p. 78.
14 T. J. Poinsot, D. Veynante and S. Candel, *J. Fluid Mech.*, 1991, **228**, 561.
15 J. Warnatz, *Proc. Combust. Inst.*, 1992, **24**, 553.
16 K. Kleinermanns and J. Wolfrum, *J. Chem. Phys.*, 1984, **80**, 1446.
17 K. Kleinermanns and J. Wolfrum, in *Photophysics and Photochemistry above 6eV*, ed. F. Lahmani, Elsevier, Amsterdam, 1985, p. 605.

18 M. A. Bajeh, E. M. Goldfield, A. Hanf, Ch. Kappel, H. R. Volpp and J. Wolfrum, *J. Phys. Chem. A*, 2001, **105**, 3359.
19 A. J. Varandas, *Mol. Phys.*, 1995, **186**, 1159.
20 A. J. H. M. Meijer and E. M. Goldfield, *J. Chem. Phys.*, 1999, **110**, 870.
21 M. R. Pastrana, L. A. M. Quintãles, J. Brandao and A. J. C. Varandas, *J. Phys. Chem.*, 1990, **94**, 8073.
22 L. B. Harding, J. Troe and V. G. Ushakov, *Phys. Chem. Chem. Phys.*, 2000, **2**, 631.
23 R. Fei, S. X. Zheng and G. E. Hall, *J. Phys. Chem.*, 1997, **106**, 2541.
24 B. Kendrick and R. T. Pack, *J. Chem. Phys.*, 1997, **106**, 3519.
25 B. Ruscic, D. Feller, D. A. Dixon, K. A. Peterson, L. B. Harding, R. L. Asher and A. F. Wagner, *J. Phys. Chem. A*, 2001, **105**, 1.
26 J. Troe and V. G. Ushakov, *J. Chem. Phys.*, 2001, **115**, 3621.
27 J. A. Miller and S. J. Klippenstein, *Int. J. Chem. Kinet.*, 1999, **13**, 753.
28 J. A. Miller, *Proc. Combust. Inst.*, 1996, **26**, 461.
29 R. A. Brownsword, M. Hillenkamp, T. Laurent, H.-R. Volpp, J. Wolfrum, R. K. Vatsa and H. S. Yoo, *J. Phys. Chem. A*, 1997, **101**, 6448.
30 W. Ketterle, A. Arnold and M. Schäfer, *Appl. Phys. B*, 1990, **51**, 91.
31 J. A. Gray and R. L. Farrow, *J. Chem. Phys.*, 1991, **95**, 7054.
32 D. E. Heard, D. R. Crosley, J. B. Jeffries, G. P. Smith and A. Hirano, *J. Chem. Phys.*, 1992, **96**, 4366.
33 P. M. Doherty and D. R. Crosley, *Appl. Opt.*, 1984, **23**, 713.
34 E. W. Rothe, Y. Gu, A. Chryssostomou, P. Andresen and F. Bormann, *Appl. Phys. B*, 1998, **66**, 251.
35 A. Arnold, B. Lange, T. Couché, T. Heitzmann, G. Schiff, W. Ketterle, P. Monkhouse and J. Wolfrum, *Ber. Bunsen-Ges. Phys. Chem.*, 1992, **96**, 1388.
36 T. Heitzmann, J. Wolfrum, U. Maas and J. Warnatz, *Z. Phys. Chem. NF*, 1995, **188**, 177.
37 A. Dreizler, V. Sick and J. Wolfrum, *Ber. Bunsen-Ges. Phys. Chem.*, 1997, **101**, 771.
38 B. Bäuerle, F. Hoffmann, F. Behrendt and J. Warnatz, *Proc. Combust. Inst.*, 1994, **25**, 135.
39 J. F. Griffiths, in *Comprehensive Chemical Kinetics*, ed. M. J. Pilling, Elsevier, Amsterdam, 1997, vol. 35.
40 J. F. Griffiths, J. P. MacNamara, C. Mohamed, B. J. Whitaker, J. Pan and C. G. W. Sheppard, *Faraday Discuss.*, 2001, **119**, 202.
41 M. Ribaucour, R. Minetti, L. R. Sochet, H. J. Curran, W. J. Pitz and C. K. Westbrook, *Proc. Combust. Inst.*, 2000, **28**, 1671.
42 C. K. Westbrook, *Proc. Combust. Inst.*, 2000, **28**, 1563.
43 C. Hao, Dissertation, Universität Heidelberg, 1992.
44 D. Fulle, H. F. Hamann, H. Hippler and J. Troe, *J. Chem. Phys.*, 1996, **105**, 1001.
45 M. C. Drake, D. T. French and T. D. Fansler, *Proc. Combust. Inst.*, 1996, **26**, 2581.
46 A. Lozano, B. Yip and R. K. Hanson, *Exp. Fluids*, 1992, **13**, 369.
47 A. Arnold, A. Buschmann, B. Cousyn, M. Decker, F. Vannobel, V. Sick and J. Wolfrum, *SAE Technical Paper*, 93-2696.
48 M. Berckmüller, N. P. Tait and D. A. Greenhalgh, *SAE Technical Paper*, 97-0826.
49 F. Grossmann, P. B. Monkhouse, M. Ridder, V. Sick and J. Wolfrum, *J. Appl. Phys. B*, 1996, **62**, 249.
50 M. C. Thurber, F. Grisch and R. K. Hanson, *Opt. Lett.*, 1997, **22**, 251.
51 S. Einecke, C. Schulz and V. Sick, *Appl. Phys. B*, 2000, **71**, 717.
52 L. A. Melton and J. F. Verdieck, *Proc. Combust. Inst.*, 1984, **20**, 1283.
53 S.-X. Quian, J. B. Snow, H.-M. Tzeng and R. K. Chang, *Science*, 1986, **231**, 486.
54 J. E. Dec, *SAE Technical Paper*, 97-0873.
55 A. Arnold, F. Dinkelacker, T. Heitzmann, P. Monkhouse, M Schäfer, V. Sick, J. Wolfrum, W. Hentschel and K.-P. Schindler, *Proc. Combust. Inst.*, 1992, **24**, 1605.
56 T. M. Brugman, R. Klein-Douwel, G. Huigen, E. van Walwijk and J. ter Meulen, *J. Appl. Phys. B*, 1993, **57**, 405.
57 A. Bräumer, V. Sick, J. Wolfrum, V. Drewes, R. R. Maly and M. Zahn, *SAE Technical Paper*, 95-2462.
58 C. Schulz, B. Yip, V. Sick and J. Wolfrum, *Chem. Phys. Lett.*, 1995, **242**, 259.
59 C. Schulz, V. Sick, J. Wolfrum, V. Drewes, M. Zahn and R. R. Maly, *Proc. Combust. Inst.*, 1996, **26**, 2597.
60 F. Hildenbrand, C. Schulz, J. Wolfrum, F. Keller and E. Wagner, *Proc. Combust. Inst.*, 2000, **28**, 1137.
61 C. Schulz, V. Sick, J. Heinze and W. Stricker, *Appl. Opt.*, 1997, **36**, 3227.
62 M. D. Rosa and R. K. Hanson, *J. Quant. Spectrosc. Radiat. Transfer*, 1994, **52**, 515.
63 M. R. Furlanetto, J. W. Thoman, Jr., J. A. Gray, P. H. Paul and J. L. Durant, Jr., *J. Chem. Phys.*, 1994, **101**, 10452.
64 C. Schulz, J. Wolfrum and V. Sick, *Proc. Combust. Inst.*, 1998, **27**, 2077.
65 Y. B. Zeldovich, *Acta Physicochim. U.R.S.S.*, 1947, **21**, 577.
66 W. C. Gardiner, Jr., *Gas-Phase Combustion Chemistry*, Springer, New York, 1999.
67 J. Wolfrum, *Chem.-Ing-Tech.*, 1972, **44**, 656.
68 J. B. Heywood, *Internal Combustion Engine Fundamentals*, McGraw-Hill, New York, 1988.
69 C. P. Fenimore, *Proc. Combust. Inst.*, 1971, **13**, 373.
70 J. Blauwens, B. Smets and J. Peeters, *Proc. Combust. Inst.*, 1977, **16**, 1055.
71 W. C. Pfefferle and L. D. Pfefferle, *Prog. Energy Combust. Sci.*, 1986, **12**, 25.
72 D. Chatterjee, O. Deutschmann and J. Warnatz, *Faraday Discuss.*, 2001, **119**, 371.

73　Y. R. Shen, *The Principles of Nonlinear Optics*, Wiley, New York, 1984.
74　H. Härle, U. Metka, H.-R. Volpp and J. Wolfrum, *Phys. Chem. Chem. Phys.*, 1999, **1**, 5059.
75　G. A. Somorjai and G. Rupprechter, *J. Phys. Chem. B*, 1999, **103**, 1623.
76　H. Härle, A. Lehnert, U. Metka, H.-R. Volpp, L. Willms and J. Wolfrum, *Chem. Phys. Lett.*, 1998, **293**, 26.
77　G. Ertl, in *Gas Phase Chemical Reaction Systems–Experiments and Models 100 Years After Max Bodenstein*, ed. J. Wolfrum, H.-R. Volpp, R. Rannacher and J. Warnatz, Springer Verlag, Heidelberg, 1996, p. 245.
78　A. Sandell, J. Libuda, M. Bäumer and H.-J. Freund, *Surf. Sci.*, 1996, **346**, 108.
79　R. Kissel-Osterrieder, F. Behrendt, J. Warnatz, U. Metka, H.-R. Volpp and J. Wolfrum, *Proc. Combust. Inst.*, 2000, **28**, 1341.
80　G. Ertl, M. Neumann and K. M. Streit, *Surf. Sci.*, 1977, **64**, 393.
81　X.-C. Guo and D. A. King, *Surf. Sci. Lett.*, 1994, **302**, L251.
82　P. Kisliuk, *J. Phys. Chem. Solids*, 1957, **3**, 95.
83　U. Metka, M. G. Schweitzer, H.-R. Volpp, J. Wolfrum and J. Warnatz, *Z. Phys. Chem. NF*, 2000, **214**, 865.
84　H. Härle, A. Lehnert, U. Metka, H.-R. Volpp, L. Willms and J. Wolfrum, *Appl. Phys. B*, 1999, **68**, 567.
85　J. Xu and J. T. Yates, Jr., *Surf. Sci.*, 1995, **327**, 193.
86　W. A. Brown, R. Kose and D. A. King, *Chem. Rev.*, 1998, **98**, 797.
87　R. Kose and D. A. King, *Chem. Phys. Lett.*, 1999, **313**, 1.
88　A. Szabó, M. A. Henderson and J. T. Yates, Jr., *J. Chem. Phys.*, 1992, **96**, 6191.
89　W. Meienburg and J. Wolfrum, *Proc. Combust. Inst.*, 1990, **23**, 231.
90　F. Schuler, F. Rampp, J. Martin and J. Wolfrum, *Combust. Flame*, 1994, **99**, 431.
91　V. Ebert, J. Fitzer, I. Gerstenberg, K. U. Pleban, H. Pitz, J. Wolfrum, M. Jochem and J. Martin, *Proc. Combust. Inst.*, 1998, **27**, 1301.
92　V. Ebert, J. Fitzer, I. Gerstenberg, K.-U. Pleban, H. Pitz, J. Wolfrum, M. Jochem and J. Martin, in *5th International Symposium on Gas Analysis by Tunable Diode Lasers*, VDI-Berichte, Düsseldorf, 1998, No. 1366.
93　M. Gehring, K. Hoyermann, H. Schacke and J. Wolfrum, *Proc. Combust. Inst.*, 1973, **14**, 99.
94　T. Dreier and J. Wolfrum, *Proc. Combust. Inst.*, 1985, **20**, 695.
95　D.-C. Fang, L. B. Harding, S. J. Klippenstein and J. A. Miller, *Faraday Discuss.*, 2001, **119**, 207.
96　R. Hemberger, S. Muris, K.-U. Pleban and J. Wolfrum, *Combust. Flame*, 1994, **99**, 660.
97　V. Ebert, T. Fernholz, C. Giesemann, H. Pitz, H. Teichert, J. Wolfrum and H. Jaritz, *Proc. Combust. Inst.*, 2000, **28**, 423.

Crossed beam studies of elementary reactions of N and C atoms and CN radicals of importance in combustion

Piergiorgio Casavecchia,* Nadia Balucani, Laura Cartechini, Giovanni Capozza, Astrid Bergeat† and Gian Gualberto Volpi

Dipartimento di Chimica, Università di Perugia, 06123 Perugia, Italy. E-mail: piero@dyn.unipg.it

Received 21st March 2001
First published as an Advance Article on the web 31st October 2001

The dynamics of some elementary reactions of $N(^2D)$, $C(^3P,^1D)$ and $CN(X\,^2\Sigma^+)$ of importance in combustion have been investigated by using the crossed molecular beam scattering method with mass spectrometric detection. The novel capability of producing intense, *continuous* beams of the radical reagents by a radio-frequency discharge beam source was exploited. From angular and velocity distribution measurements obtained in the laboratory frame, primary reaction products have been identified and their angular and translational energy distributions in the center-of-mass system, as well as branching ratios, have been derived. The dominant N/H exchange channel has been examined in the reaction $N(^2D) + CH_4$, which is found to lead to $H + CH_2NH$ (methylenimine) and $H + CH_3N$ (methylnitrene); no H_2 elimination is observed. In the reaction $N(^2D) + H_2O$ the N/H exchange channel has been found to occur *via* two competing pathways leading to $HNO + H$ and $HON + H$, while formation of $NO + H_2$ is negligible. Formation of $H + H_2CCCH$ (propargyl) is the dominant pathway, at low collision energy (E_c), of the $C(^3P) + C_2H_4$ reaction, while at high E_c formation of the less stable C_3H_3 isomers (cyclopropenyl and/or propyn-1-yl) also occurs; the H_2 elimination channel is negligible. The H elimination channel has also been found to be the dominant pathway in the $C(^3P,^1D) + CH_3CCH$ reaction leading to C_4H_3 isomers and, again, no H_2 elimination has been observed to occur. In contrast, both H and H_2 elimination, leading in comparable ratio to $C_3H + H$ and $C_3(X\,^1\Sigma_g^+) + H_2(X\,^1\Sigma_g^+)$, respectively, have been observed in the reaction $C(^3P) + C_2H_2(X\,^1\Sigma_g^+)$. The occurrence of the spin-forbidden molecular pathway in this reaction, never detected before, has been rationalized by invoking the occurrence of intersystem crossing between triplet and singlet manifolds of the C_3H_2 potential energy surfaces. The reaction $CN(X\,^2\Sigma^+) + C_2H_2$ has been found to lead to internally excited HCCCN (cyanoacetylene) + H. For all the reactions the dynamics have been discussed in the light of recent theoretical calculations on the relevant potential energy surfaces. Previous, lower resolution studies on C and CN reactions carried out using pulsed beams are noted. Finally, throughout the paper the relevance of these results to combustion chemistry is considered.

† Permanent address: Département de Mesures Physiques, IUT A, Université Bordeaux I, 33405 Talence Cedex, France.

DOI: 10.1039/b102634h

I. Introduction

Over the last decades there has been continuing investigation of the kinetics of elementary radical reactions which has increased the reliability of models of complex combustion reaction networks.[1,2] More recently, there has also been significant progress in the investigation of the dynamics of combustion related reactions (see, for instance, ref. 3 and 4); these studies, by identifying the primary reaction products and their dynamics of formation, have been complementing the kinetic work and contributing significantly to our endeavour of understanding complex reaction systems. Such investigations at the microscopic level have benefitted from remarkable developments in molecular beam and laser techniques as well as in theoretical and computational methods.[3] In particular, the classic crossed molecular beam (CMB) scattering technique with "universal" mass spectrometric detection has continued to be, for both simple and complex polyatomic multichannel reactions,[3,5,6] one of the most powerful techniques for providing information on: (a) the nature of primary reaction products, (b) the branching ratios of competing reaction pathways, (c) the microscopic reaction mechanisms, (d) the product energy partitioning and hence (e) the underlying potential energy surfaces (PESs) governing the transformation from reactants to products.

Among the reactions of importance in combustion, those of oxygen, nitrogen and carbon atoms and those of hydroxyl and cyano radicals are of particular relevance.[1,2] By using the CMB method with mass spectrometric detection, in our laboratory we have investigated the dynamics of some combustion relevant reactions, such as those of atomic oxygen in the ground and first excited states, $O(^3P, ^1D)$, with halogenated compounds[7-9] and H_2S,[10,11] and of $OH(X\ ^2\Pi)$ with H_2[11,12] and CO.[11,13] More recently, we have started investigating elementary reactions of atomic nitrogen in its first excited state, $N(^2D)$,[14-17] of atomic carbon in its ground and first excited state, $C(^3P, ^1D)$[18] and of $CN(X\ ^2\Sigma^+)$ radicals. In all these studies we have exploited our novel capability of generating *continuous* supersonic beams of these transient species by a high-pressure radio-frequency (rf) discharge source,[11,19] which optimally couples with the *continuous* "universal" detector of our CMB apparatus.

In this paper, after a brief description of the experimental technique, we will present and discuss results on the dynamics of some selected reactions of nitrogen and carbon atoms, and of cyano radicals with molecules of relevance in combustion, such as water, methane, acetylene, methylacetylene and ethylene.

II. Experimental method

The experiments reported here were performed by using the crossed molecular beam apparatus sketched in Fig. 1 and described in detail elsewhere.[11] Briefly, two supersonic beams of the reactants—doubly differentially pumped and well collimated in angle and velocity—are crossed at 90° under single collision conditions in a large scattering chamber kept below 5×10^{-7} mbar. The angular and velocity distributions of the reaction products are recorded by a triply differentially pumped, ultra-high-vacuum (10^{-11} mbar) electron impact ionizer followed by a quadrupole mass filter. The whole detector unit can be rotated in the plane of the two beams around their intersection axis ($\Theta = 0°$ represents the direction of the atomic beam and $\Theta = 90°$ represents that of the secondary beam). The velocity of reactants and products is derived by using time-of-flight (TOF) analysis.

The main advantage of CMB experiments is that the reactive events are investigated under single collision conditions. In fact, differently from bulk experiments, the reactants are confined into distinct supersonic beams which cross each other at a specific angle and collision energy; the species of each beam are made to collide only with the molecules of the other beam, allowing us to observe the consequences of well defined molecular collisions and preventing the effects of secondary collisions. The measurable quantities by this technique contain basic information.[5] Any species can be ionized at the electron energy typically used in the ionizer (100 eV) which precedes the mass filter and, in principle, it is possible to determine the mass (and the gross formula) of all the possible products of a bimolecular reaction by varying the mass-to-charge ratio, m/z, in the mass filter. Therefore, despite some problems (such as dissociative ionization and background noise) which restrict the sensitivity of the method, the use of mass spectrometric detection is

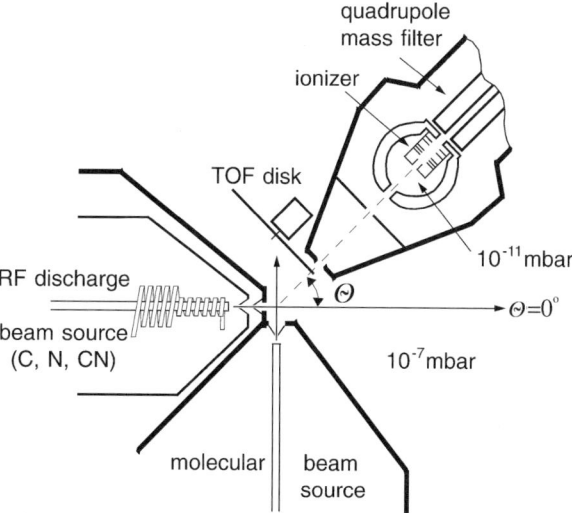

Fig. 1 Schematic top-view of the crossed molecular beam apparatus with rotating mass-spectrometer detector and time-of-flight chopper.

advantageous if compared to spectroscopic techniques; their applicability, in fact, requires the knowledge of the optical properties of the products, while, in many cases, their nature itself is unknown. Furthermore, the measurement of the product velocity distributions allows one to directly derive the amount of the total energy available to the products and, therefore, the energetics of the reaction. This is crucial when more isomers with the same gross formula, but different enthalpy of formation, can be produced as exemplified in Fig. 2 for an ideal experiment relative to a reaction involving an atom A and a generic hydrocarbon RH. If two product isomers with the

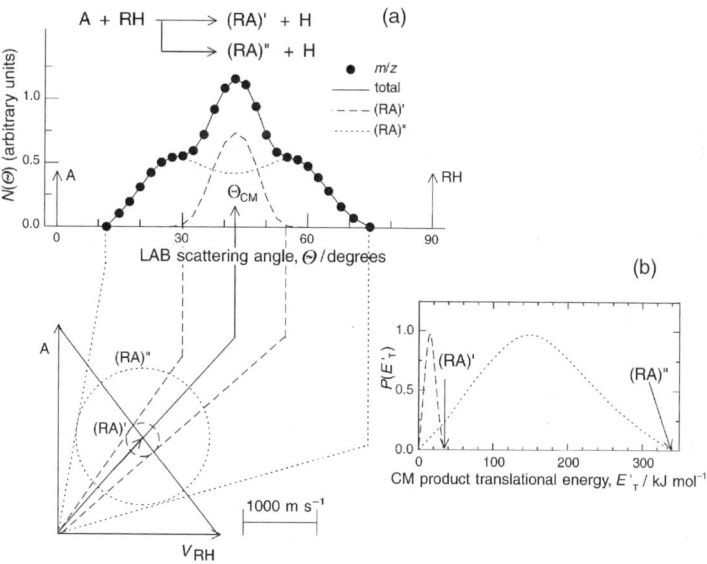

Fig. 2 (a) Angular distributions (schematic) for the (RA)′ and (RA)″ isomeric products from a generic A + RH reaction together with velocity vector ("Newton") diagram. The circles delimit the maximum speeds of the two isomeric products and the corresponding ranges of scattering angles. The solid circles indicate the total angular distribution corresponding to the sum of the two isomer angular distributions. (b) RA′ and RA″ translational energy distributions reflecting the large difference in exoergicity between the two isomer channels.

same gross formula RA are formed and if the energetics of the two channels are significantly different, the two product isomers will be scattered within different angular ranges. In the example of Fig. 2 the enthalpy of reaction for the channels leading to the two isomers (RA)′ and (RA)″ is very different and so is the total energy available to product translational motion (because of the energy conservation rule, the total energy, E_{tot}, is given by the sum of the initial collision energy, E_c and the heat of reaction, $E_{tot} = E_c - \Delta H°$). As a consequence, the maximum speed in the centre-of-mass frame that the two isomers can reach is different and the limiting circles in the Newton diagram will define laboratory angular ranges of different extent. When the difference is pronounced, the distinct contributions to the observed signal will be easily separated during the data analysis; when the difference is not sharp, accurate measurements of product velocity distributions as a function of scattering angle usually allow us to discriminate between different contributions (see below).

The recent technological improvements in the production of beams of unstable species[3] have allowed the extension of the CMB technique to the study of elementary reactions of practical interest, such as those reported here. In the present work, intense continuous supersonic beams of atomic (C or N) and radical (CN) species are produced by means of a rf discharge in a water-cooled quartz nozzle ($\phi = 0.3$ mm)[11,19] which operates at high rf power (300 W) and relatively high pressure (~ 300 mbar) on dilute (a few percent) mixtures of suitable molecular precursors (CO_2, N_2, CO_2/N_2) in He or Ne. The atomic and radical beams had typical peak velocities of 1500–2800 m s^{-1} and speed ratios of 6–8 as measured from single-shot TOF analysis. In all cases the beam angular divergence was 2.3°. When using N_2/He(Ne) mixtures to produce atomic nitrogen, we have verified that more than 60% of molecular N_2 is dissociated and that atomic nitrogen is formed with a distribution of electronic states: 72% of the N atoms were found in the ground ^4S state, and 21% and 7% in the first excited ^2D and ^2P states (lying 2.39 and 3.56 eV, respectively, above the ground state).[19] Atomic carbon is produced mainly in the ground ^3P state, with also a significant concentration (10–20%) in the excited ^1D state (lying 1.26 eV above the ground state).[18] The C beams contain only 2–3% C_2 and no detectable C_3. CN radicals are, in contrast, in their ground $^2\Sigma^+$ electronic state (since electronically excited states possibly formed in the discharge are very short lived and decay to the ground state before reaching the collision region) and in the lowest rovibrational levels because of the supersonic expansion.

The secondary beams were produced by expanding neat or seeded gases through a resistively heated stainless steel nozzle ($\phi = 0.1$ mm); the peak velocities and speed ratios were typically in the range 760–2200 m s^{-1} and 6–15, respectively, depending on the nature of the molecules, the temperature of the nozzle and the presence of carrier gases. To vary the collision energy we either change the carrier gas of the primary beam and the discharge conditions and/or the carrier and the nozzle temperature of the secondary beam.

Product angular distributions were recorded by modulating the secondary beam at 160 Hz for background subtraction and taking at least 4–5 scans at each mass (typical counting times were 50 s at each angle). Product TOF distributions were recorded at selected angles using the pseudo-random TOF technique[11] at 5 μs per channel (typical counting times varied from 15 to 120 min depending on signal intensity).

For the physical interpretation of the experimental distributions it is necessary to transform the laboratory (LAB) distributions into the centre-of-mass (CM) system;[5,11] because of the finite experimental resolution (*i.e.*, finite angular and velocity spread of the reactant beams and angular resolution of the detector) the LAB–CM transformation is not single-valued and, therefore, analysis of the laboratory data has been performed by forward convolution over CM trial angular, $T(\theta)$ and translational energy, $P(E'_T)$, distributions. The LAB distributions are then calculated taking into account the transformation Jacobian and the averaging over the experimental parameters. The procedure is repeated until a satisfying fit of the LAB distributions is achieved and the CM functions so determined are the best-fit functions.

III. Reactions of nitrogen atoms

In general, chemical reactions involving nitrogen-containing species are of relevance in the production of pollutants during combustion processes;[20] for instance, formation of NO_x is constantly observed whenever combustion takes place with air and both prompt and thermal NO formation mechanisms involve atomic nitrogen reactions.[20] Although kinetic data exist[2,20] and reaction mechanisms have been suggested to model high-temperature processes involving atomic nitrogen, a better knowledge of the reactive behaviour requires a dynamical investigation of nitrogen atom reactions. The capability of generating intense continuous beams of atomic nitrogen[19] has opened up the possibility of studying the reactive scattering of this species under single collision conditions.

So far, we have mainly focused on the reactions of N atoms in their first excited state, because, when the reactive partner is a closed shell molecule, $N(^2D)$ is much more reactive than the ground state $N(^4S)$ and than the higher electronic state $N(^2P)$, which is mainly removed by physical quenching.[21,22] In spite of the $N(^2D)$ high energy content, the study of its reactions with molecules of relevance in combustion is of great interest. The presence of $N(^2D)$ in the flame front has been invoked to explain chemiionization in ammonia–oxygen flames;[23] furthermore, $N(^2D)$ was found to be the main product of the reaction $O(^3P) + CN$,[24] and, since few reactive systems which form atomic nitrogen adiabatically correlate with the quartet state, chemical production of $N(^2D)$ can probably occur to a significant extent. For instance, the initiation chain reaction of the prompt NO mechanism, $CH + N_2 \rightarrow HCN + N$, can form $N(^4S)$ only through a doublet–quartet intersystem crossing (ISC),[20] which was found to have low probability,[25] while the same reactants adiabatically correlate with $N(^2D)$. In other processes, such as plasma, laser and upper atmospheric chemistry, the role of $N(^2D)$ reactions is well established:[21,22] its metastable character and the much larger rate constants of its reactions with respect to $N(^4S)$ reactions make its chemical behaviour similar to that of $O(^1D)$ as opposed to $O(^3P)$.

We have already reported on the investigation of the reaction dynamics of $N(^2D)$ with the unsaturated hydrocarbons acetylene[16] and ethylene.[17] For the two systems, we have been able to establish the main reaction primary products, that is cyanomethylene and its cyclic isomer in the case of the $N(^2D) + C_2H_2$ reaction[16] and 2H-azirine and ketenimine in the case of the $N(^2D) + C_2H_4$ reaction.[17] Interestingly, from the observed energy release, we have learnt that in the case of the $N(^2D) + C_2H_4$ reaction, the two primary products are formed with enough internal energy to significantly tautomerize to the most stable acetonitrile.[17] For both reactions, the micromechanism involves the formation of bound intermediates, initially formed by the addition of the electrophilic $N(^2D)$ to the π cloud of the unsaturated hydrocarbon. We also studied the reaction $N(^2D) + H_2 \rightarrow NH + H$ because of its prototypical nature for understanding the chemistry of atomic nitrogen in the ^2D state.[15]

Here we present an extension of the experimental investigation to two other $N(^2D)$ reactions, those involving another hydrocarbon, the simplest alkane CH_4, and an inorganic molecule, H_2O.

$N(^2D) + CH_4$

$N(^2D) + CH_4$ is a reaction of interest in all processes occurring at high temperature and in the presence of methane and air, such as chemical vapour deposition. Recently, the rate constants as a function of temperature[26] and the NH product rovibrational distributions[27,28] have been determined for the $N(^2D) + CH_4/CD_4$ reactions. According to recent *ab initio* calculations[29] on the reactive potential energy surface, several reactive channels are open:

$$N(^2D) + CH_4(^1A_1) \rightarrow (CH_3NH)^\dagger \rightarrow NH_2 + CH_2, \quad \Delta H_0^0 = -40 \text{ kJ mol}^{-1} \quad (1a)$$

$$\rightarrow CH_3N(^3A_2) + H, \quad \Delta H_0^0 = -83 \text{ kJ mol}^{-1} \quad (1b)$$

$$\rightarrow NH + CH_3, \quad \Delta H_0^0 = -118 \text{ kJ mol}^{-1} \quad (1c)$$

$$\rightarrow CHNH_2 + H, \quad \Delta H_0^0 = -151 \text{ kJ mol}^{-1} \quad (1d)$$

$$\rightarrow CH_2NH + H, \quad \Delta H_0^0 = -308 \text{ kJ mol}^{-1} \quad (1e)$$

All were found to be accessible after $N(^2D)$ has inserted into one of the C–H bonds forming a first intermediate, CH_3NH, which is bound by about 427 kJ mol^{-1} with respect to the reactants; this first intermediate can directly dissociate into the products $CH_2NH + H$, $NH + CH_3$ and $CH_3N + H$ or can rearrange to its isomer CH_2NH_2 which is the global minimum of the PES (-452 kJ mol^{-1} with respect to reactants), and which in turn can fragment to products; in this second case, the possible products are $NH_2 + CH_2$, $CHNH_2 + H$ and again $CH_2NH + H$ (see Fig. 3). A

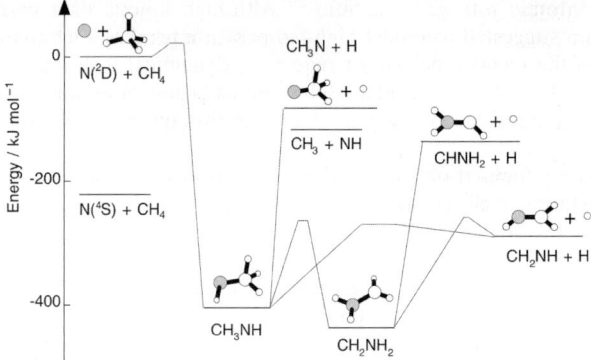

Fig. 3 Schematic potential energy surface for the CH$_4$N system (adapted from ref. 29).

recent, accurate *ab initio* study has concluded that CH$_3$NH should fragment directly to the most energetically favoured channel (1e), rather than isomerize,[30] since the energy gain is not significant and the transition state for the isomerization is located at a higher energy value than that leading to CH$_2$NH + H. A recent experimental study[28] identified NH and H as the main reaction products with an absolute yield of 0.3 ± 0.1 and 0.8 ± 0.2, respectively. Amongst the three possible H-displacement channels the authors suggested the channel (1e) to be the dominant one on the basis of the energy release derived from H/D Doppler profile measurements. According to their measurements, indeed, the average translational energy is about 80 kJ mol^{-1}, that is too high a value to be entirely produced from the channels (1b) or (1d). However, in the authors' assumption that only channel (1e) contributes to the H yield, the fraction of energy released as product translational energy is relatively small (about 25%) for a channel where the C–H bond breaking is thought to occur promptly (for instance, a fraction of ∼40% was derived from crossed beam experiments[6] on the related reaction O(^{1}D) + CH$_4$ → CH$_2$OH + H).

With the aim of understanding whether the channel leading to CH$_2$NH + H is the only active H-displacement pathway we have performed a systematic study of the N(^{2}D) + CH$_4$ reaction as a function of collision energy. Because of the best signal-to-noise (S/N) ratio, all the product distributions were measured at $m/z = 27$, corresponding to the daughter ion CHN$^+$; we have verified, though, that the parent ion ($m/z = 29$) distributions were identical. Interestingly, we have found some evidence that the channel leading to one (or both) of the two other possible isomers (methylnitrene, CH$_3$N and CHNH$_2$) must also be open and that its contribution changes with the total energy available to the reactive system. The analysis of the complete set of data is still under way; here we have reported the results obtained at $E_c = 37.2$ kJ mol^{-1} (see Fig. 4 and 5) with preliminary best-fit CM functions (see Fig. 6). In order to reach a reasonable fit of the data for all the collision energies investigated (from about 20 to 60 kJ mol^{-1}) it is necessary to consider two different mechanisms which contribute differently to the observed signal; the relative importance of the two mechanisms was found to change considerably with E_c. The energetics of the two mechanisms are quite different as is well visible from the shapes of the best-fit $P(E'_T)$s (see Fig. 6): in particular, the energy cut-off and the slope in its proximity are very different for the two mechanisms. Also, the CM angular distribution (Fig. 6) associated with the $P(E'_T)$ with the higher energy cut-off is anisotropic, being confined in the forward direction, while that associated with the $P(E'_T)$ with the lower energy cut-off is isotropic, that is the intensity is the same over the whole angular range. The most obvious interpretation of these experimental findings is that two channels are contributing: one can be associated with the formation of methylenimine and the other with methylnitrene. The former is characterised by a forward distribution (see Fig. 6), as expected for a strongly exothermic channel that implies a short-living complex; also the fraction of energy released as product translational energy is relatively high (50%), because the complex does not have enough time to completely randomise its internal energy and there is an exit barrier. The mechanism leading to CH$_3$N is instead characterised by an isotropic distribution because, due to the reduced exothermicity, the system will experience longer the potential energy well associated with

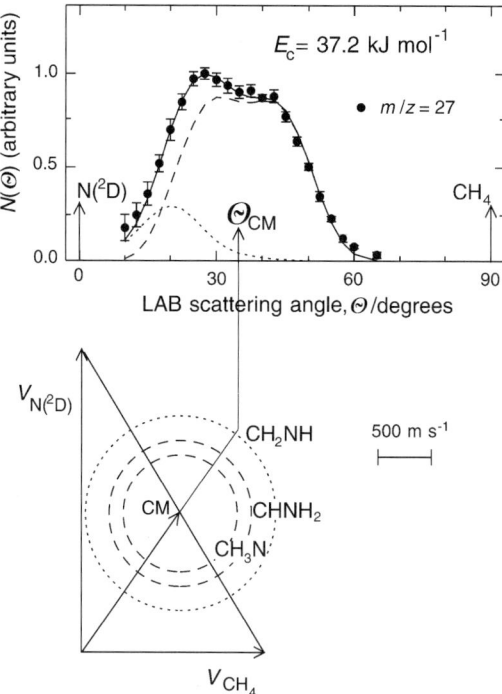

Fig. 4 Laboratory angular distribution (●) of the CH_3N/CH_2NH products (detected at $m/z = 27$) from the reaction $N(^2D) + CH_4$ at $E_c = 37.2$ kJ mol^{-1}. The circles in the Newton diagram delimit the maximum velocity that the three isomeric products, CH_3N, $CHNH_2$ and CH_2NH can attain on the basis of energy conservation if all the available energy for the corresponding channels goes into product translational energy. The separate contributions to the total LAB angular distribution (solid line) from the CH_3N channel (dashed line) and the CH_2NH channel (dotted line) are shown.

the CH_3NH intermediate. At $E_c = 37.2$ kJ mol^{-1} the ratio of cross sections $\sigma_{1e}/(\sigma_{1b} + \sigma_{1e})$ is 0.1. The large yield of 0.9 for CH_3N is not surprising, even if we consider the large difference in the reaction enthalpy between channels (1e) and (1b); in fact, the excess energy released during the formation of the intermediate CH_3NH will be concentrated mainly on the two new bonds formed by N insertion and the fission of one C–H bond—not directly involved in the insertion process—requires an energy rearrangement which is not necessary in the case of N–H/C–N bond breaking.

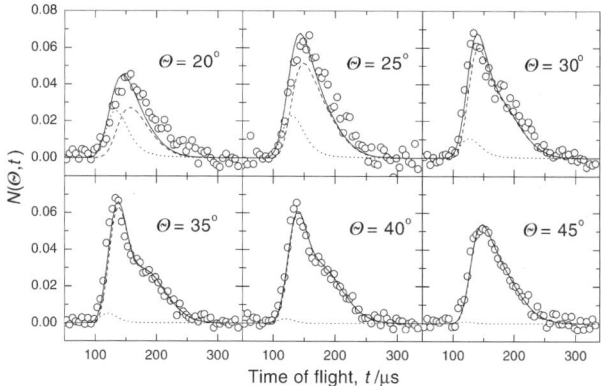

Fig. 5 Time-of-flight distributions of the $m/z = 27$ product from the reaction $N(^2D) + CH_4$ at the indicated LAB angles for $E_c = 37.2$ kJ mol^{-1}. Symbols for best-fit global and partial distributions are as in Fig. 4.

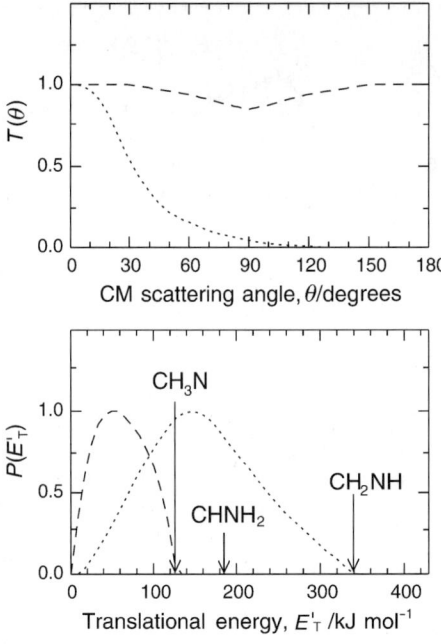

Fig. 6 Best-fit CM product angular (top) and translational energy (bottom) distributions for the methylnitrene (dashed line) and the methylenimine (dotted line) forming channels at $E_c = 37.2$ kJ mol^{-1}. The arrows indicate the total available energy for the various isomer channels.

Indeed, the channel leading to NH + CH$_3$ was observed to occur at thermal collision energies[27,28] and we expect about the same yield for the N–H bond breaking, since the energetics are quite similar. At the lowest E_c investigated (21.3 kJ mol^{-1}) $\sigma_{1e}/(\sigma_{1b} + \sigma_{1e})$ is 0.6 and reduces to 0.04 at the highest E_c (63.2 kJ mol^{-1}). This trend can be well explained by assuming that the yield of the most exothermic channel reduces when E_{tot} increases. At this stage, we cannot rule out that, in addition to methylnitrene and methylenimine, CHNH$_2$ is also partly formed. However, no argument, either energetic or dynamic, appears to be in favor of this reaction channel. A final consideration can be made on whether the methylnitrene isomer has enough internal energy to tautomerize to the most stable methylenimine. Extensive studies on this subject have been reported,[31] since the conversion of nitrenes to the corresponding imines is of relevance in different chemical processes, such as the decomposition of organic azides. The ground state of methylnitrene is a triplet state, $^3A_2$, which is also the state formed from the present reaction. Tautomerization of singlet methylnitrene to methylenimine occurs promptly and therefore it is the occurrence of the triplet–singlet ISC that determines whether the process is open. The seam of crossing has been identified[31] at about 156 kJ mol^{-1} above CH$_3$N($^3A_2$) and therefore is not accessible in the conditions of our experiment.

N(^{2}D) + H$_2$O

N(^{2}D) + H$_2$O is one of the asymptotes of the H$_2$NO isomeric system—the simplest member of the family of nitroxide radicals—of great relevance in combustion chemistry: the other asymptotes are the O + NH$_2$, NO + H$_2$ and HNO + H reactions, which play an important role in processes such as the inhibition of combustion by NO, the catalysed recombination of H and OH radicals in flames and the combustion of nitrogen-containing fuels.[32] While kinetic studies have been performed on NO + H$_2$,[33] HNO + H[34] and both kinetic[35] and dynamic studies[36] on O + NH$_2$, no experimental studies on N(^{2}D) + H$_2$O were available until the recent spectroscopic studies reported by Umemoto, which provided rate constants[26b] and NH product rovibrational distributions.[37]

The H$_2$NO potential energy surface has been the subject of a large number of *ab initio* electronic structure calculations.[38] When N(^{2}D) and H$_2$O are the reagents, the possible reaction channels are:[39]

$$N(^2D) + H_2O(^1A_1) \rightarrow (H_2NO)^\dagger \rightarrow HON(^3A'') + H(^2S), \quad \Delta H_0^0 = -26 \text{ kJ mol}^{-1}$$
$$\rightarrow O(^3P) + NH_2(^2B_1), \quad \Delta H_0^0 = -26 \text{ kJ mol}^{-1}$$
$$\rightarrow OH(^2\Pi) + NH(X\,^3\Sigma^-), \quad \Delta H_0^0 = -61 \text{ kJ mol}^{-1}$$
$$\rightarrow HNO(^1A') + H(^2S), \quad \Delta H_0^0 = -144 \text{ kJ mol}^{-1}$$
$$\rightarrow NO(^2\Pi) + H_2(X\,^1\Sigma_g^+), \quad \Delta H_0^0 = -369 \text{ kJ mol}^{-1}$$

Spectroscopic studies[37] have shown that in the OH + NH formation channel, the old O–H bond acts as a spectator, being formed with no vibrational excitation. Also, by recording the Doppler profile of H atoms, the same authors[37] found that about 45% of the available energy is released as translational energy in the H-displacement channel, the absolute yield of which was 0.11 ± 0.03 and 0.17 ± 0.04 in the reaction of $N(^2D)$ with H_2O and $D_2O)$ respectively.[37]

According to a recent *ab initio* study of the $N(^2D) + H_2O$ reaction[39] (for a schematic representation of the PES see Fig. 7) the long range attractive forces favour the addition of the N atom to the out-of-plane lone pair of O, producing an addition intermediate, NOH_2, which is not a bound intermediate and is better described as a van der Waals adduct; NOH_2 easily rearranges to *cis*- or *trans*-HONH, that is the same intermediates which would be formed from direct insertion of $N(^2D)$ into one of the σ bonds of water. *Trans*- and *cis*-HONH can directly fragment to products, that is HON + H, OH + NH and HNO + H; also, *trans*-HONH can rearrange to the most stable isomer H_2NO, which in turn can fragment to products, which in this case are $O + NH_2$, HNO + H and also $NO + H_2$. Even though its value is affected by some uncertainty, the exit barrier to $NO + H_2$ was found to be very high.[38,39]

The channels which can be conveniently investigated with our technique are those of H-displacement, leading to the two possible isomers HNO and HON and that of H_2-elimination, leading to the formation of NO. The $NO + H_2$ channel was not found to be open under our experimental conditions, as predictable from the presence of the very high exit barrier.

We have performed reactive scattering experiments at two different collision energies ($E_c = 30.5$ and 49.0 kJ mol^{-1}) for $N(^2D) + H_2O$ and at $E_c = 41.8$ kJ mol^{-1} for the isotopic variant $N(^2D) + D_2O$. The laboratory angular distribution, together with the Newton diagram showing the kinematics of the high E_c experiment, is reported in Fig. 8. The experimental data were fitted by using two different sets of CM functions, shown in Fig. 9, which reflect two reaction mechanisms. Interestingly, for the lowest energy experiment, we could reproduce the experimental data by simply using one set of $P(E_T')$ and $T(\theta)$, which are very similar to those reported with dashed lines in Fig. 9, and this suggests that at low E_c only one reaction mechanism is involved. Since the angular distribution associated with this first mechanism is forward peaked, we can suggest that this reaction channel proceeds through the formation of an osculating complex, that is a complex whose lifetime is comparable to its rotational period. The associated $P(E_T')$ is consistent with the energetics of the HNO + H channel. At higher collision energy, a second mechanism characterised

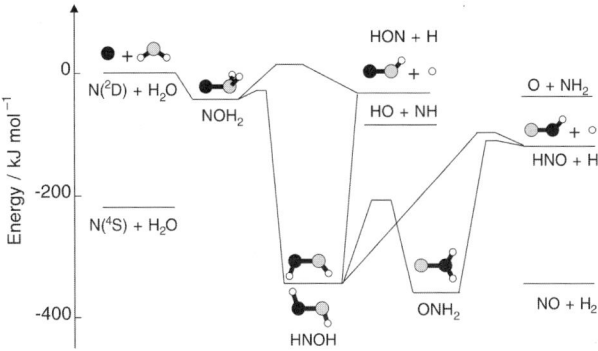

Fig. 7 Schematic potential energy surface for the H_2NO system (adapted from ref. 38 and 39).

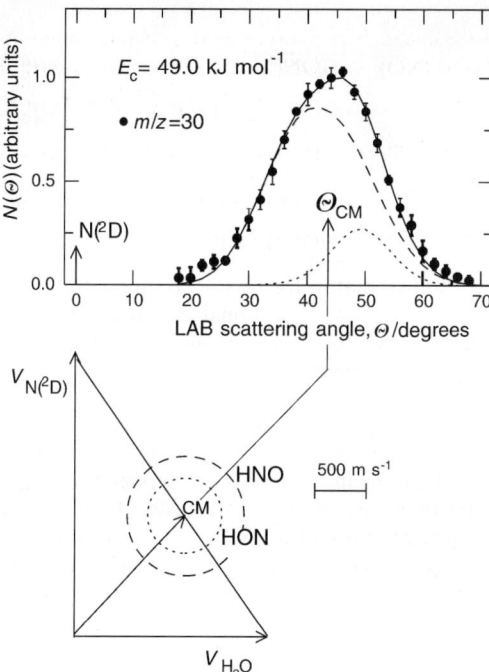

Fig. 8 Laboratory angular distribution (●) of the HNO/HON products (detected at $m/z = 30$) at $E_c = 49.0$ kJ mol^{-1} from the reaction N(^{2}D) + H$_2$O. The circles in the Newton diagram delimit the maximum velocity that the two isomeric products, HNO and HON can attain. The separate contributions to the total LAB angular distribution (solid line) from the HNO (dashed line) and HON (dotted line) channels are shown.

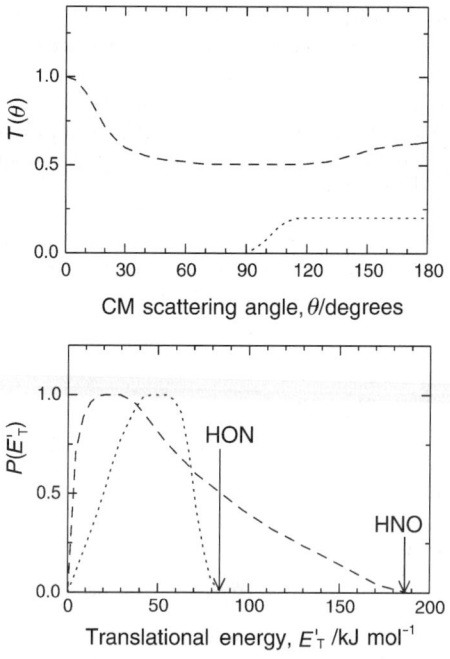

Fig. 9 Best-fit CM product angular (top) and translational energy (bottom) distributions for the HNO (dashed line) and HON (dotted line) forming channels from the N(^{2}D) + H$_2$O reaction at $E_c = 49.0$ kJ mol^{-1}. The arrows indicate the total ($E_c - \Delta H$) available energy for the two different isomer channels.

by a backward peaked CM angular distribution (see dotted line in Fig. 9) becomes active and its importance increases with increasing available energy (the second contribution is visible also in the experiment with perdeuterated water). In this second case the cut-off of the associated $P(E'_T)$ is quite different, being consistent with the energetics of the channel leading to the isomer HON and H. The whole picture gains full support from the *ab initio* calculations of the H_2NO potential energy surface, since a barrier of about 20–40 kJ mol^{-1} (with respect to the reactant asymptote) has been predicted[37] for the direct dissociation of the addition van der Waals adduct, NOH_2. Our experimental findings can therefore be explained by invoking the onset of this second reaction pathway when the collision energy overcomes the critical energy to activate it. The ratio $\sigma_{HON}/(\sigma_{HON} + \sigma_{HNO})$ is zero at $E_c = 30.5$ kJ mol^{-1}, 0.05 at $E_c = 41.8$ kJ mol^{-1} and 0.13 at $E_c = 49.0$ kJ mol^{-1}.

IV. Reactions of carbon atoms

Gas-phase reactions of carbon atoms are of basic chemical interest and of great relevance from combustion to astrochemistry. Kinetic studies at room temperature of a variety of $C(^3P)$ reactions with unsaturated hydrocarbons found them to be very fast ($k \approx 2$–3×10^{-10} cm^3 molecule^{-1} s^{-1}) and suggested that these are barrierless reactions dominated by long range attractive forces.[40] Recent, novel kinetic measurements[41] down to 15 K have concluded that the reactions of $C(^3P)$ with acetylene, ethylene, methylacetylene and allene are very fast ($k \approx 4 \times 10^{-10}$ cm^3 molecule^{-1} s^{-1}) down to this very low temperature and hence have confirmed the above suggestion. Those kinetic studies followed the decay of $C(^3P)$ and were not able to provide information on the primary reaction products. The development of pulsed beams of $C(^3P)$ by laser ablation of graphite[42] made it possible to apply CMB techniques to the investigation of C atom reactions using LIF detection.[42,43] Very recently the above kinetic work has been confirmed and complemented by pulsed CMB experiments with LIF detection yielding integral cross sections for H-atom production in the range of relative translational energies of 0.5–30 kJ mol^{-1}.[44,45] In particular, the energy dependence of the integral cross sections for H production was found to be in excellent agreement with the temperature dependence of the rate coefficients for $C(^3P)$ decay in the case of the reactions $C(^3P)$ + ethylene, methylacetylene and allene,[45] but not, interestingly, for the reaction $C(^3P)$ + acetylene.[44,46]

Starting from the mid 1990s, pulsed C beams generated by laser ablation have also been used in classic CMB experiments with mass spectrometric detection to investigate the reactive scattering of $C(^3P)$ with a variety of unsaturated hydrocarbons.[47–51] In all these studies, often supported by synergistic theoretical calculations on the relevant potential energy surfaces, only the C/H exchange channel was observed. Because of the relatively low S/N in the determination of the laboratory angular distribution from measured TOF spectra, the derived CM angular and translational energy distributions were often characterized by a considerable uncertainty.[49] Moreover, these experiments were complicated by the presence of a large concentration of C_2 and C_3 clusters in the pulsed C beams; in fact, C_2 and C_3 are known to react with unsaturated hydrocarbons and, for instance, in the case of the $C(^3P) + C_2H_2$ reaction[47] the presence of C_3 in the beam apparently affected the capability for detection of the important $C_3 + H_2$ formation channel (see below).

We have undertaken the investigation of the dynamics of $C(^3P)$ and $C(^1D)$ reactions with unsaturated hydrocarbons using continuous beams, in a synergistic fashion with the low temperature (energy) kinetic (dynamic) work[41,44–46] within an EU Network on Astrophysical Chemistry. By studying reactions of $C(^3P)$ and $C(^1D)$ under the same experimental conditions, as previously done in related studies of $O(^3P)$ and $O(^1D)$ reactions,[9,11] we have been able to explore the effect of electronic excitation on the dynamics of C atom reactions. It should be noted that in general, as in the case of O atom reactions,[9,11] in C atom reactions with hydrocarbons the singlet complex intermediates are usually lower in energy than the triplet intermediates, so that there are always crossings between triplet and singlet surfaces with the possibility of intersystem crossing.

We have looked in detail at some of the most fundamental reactions, such as those with acetylene, methylacetylene, and ethylene. An interesting aspect of these reactions is that the intermediate collision complexes are fascinating isomers that challenge our understanding of structure and bonding (as for instance C_3H_2 and C_3H_4)[52] and often are key intermediates in the decomposition of stable molecules (singlet C_3H_2 is the key intermediate in the production of $C_3(X\,^1\Sigma_g^+)$

during the photodissociation of propyne and allene[53]). The thermal and photochemical decomposition of ground state C_3H_2 and C_3H_4 have been extensively studied, both experimentally and theoretically.[53–59] From a fundamental point of view, many of the mechanisms for the thermal interconversion of isomers on the singlet C_3H_4 surface play a central role in organic chemistry because, besides serving as models for much larger systems, they also provide a fundamental basis for reactions of hydrocarbons.[54,59]

In addition to reactions which are exoergic for both $C(^3P)$ and $C(^1D)$ we have also studied two reactions which are endoergic for ground state carbon: $C(^1D) + H_2$ and $C(^1D) + CH_4$. The first reaction ($\Delta H_0^0 = (-26.4 \text{ kJ mol}^{-1}$) is found to proceed by insertion through a singlet methylene (CH_2) intermediate which dissociates into methylidene, $CH(v')$ and H.[18] The reaction $C(^1D) + CH_4$ is the simplest carbon chain lenghtening reaction. At $E_c = 30.1 \text{ kJ mol}^{-1}$ it is found to proceed by insertion forming an excited ethylene/ethylidene molecule which decomposes following two energetically and dynamically quite different reaction pathways: simple bond fission leads to radical products, $H + CH_2CH$ (vinyl), and concerted elimination to molecular products, H_2 + acetylene.[60] Information on reactions involving the vinyl radicals is very important in combustion; the competition between the formation of $C_2H_2 + H_2$ and C_2H_4 in the reaction $C_2H_3 + H$ is an example.

$C(^3P,^1D) + C_2H_2$

The $C(^3P,^1D) + C_2H_2$ reaction occurs over the triplet/singlet C_3H_2 PESs which contain several isomers: propynylidene is the simplest acetylenic carbene, propadienylidene is the simplest vinylidene carbene, cyclopropenylidene is the smallest cyclic alkyne. These highly reactive molecules are fundamentally important not only within the context of organic chemistry, but also within the context of the chemistry of the interstellar medium and hydrocarbon combustion.

Energetically allowed reaction channels in the reaction $C(^3P) + C_2H_2$ are those leading to H and H_2 elimination:

$$C(^3P) + C_2H_2(X\,^1\Sigma^+) \rightarrow [C_3H_2]^\dagger \rightarrow l\text{-}C_3H(X\,^2\Pi) + H(^2S), \quad \Delta H_0^0 = -1.5 \text{ kJ mol}^{-1}$$
$$\rightarrow c\text{-}C_3H(X\,^2B_2) + H(^2S), \quad \Delta H_0^0 = -8.6 \text{ kJ mol}^{-1}$$
$$\rightarrow C_3(X\,^1\Sigma_g^+) + H_2(X\,^1\Sigma_g^+), \quad \Delta H_0^0 = -105.1 \text{ kJ mol}^{-1}$$

The analogous reactions of $C(^1D)$ are more exoergic by 121.9 kJ mol^{-1} (corresponding to the ^1D–^3P splitting). Both C_3 and C_3H have been observed in space (C_3 also in comets)[61] and may well be produced by the above reactions. C_3H_2 intermediates have been observed in C_2H_2/O_2 flames.[62] Many of the PESs, both singlet and triplet, and the isomerization and dissociation processes of C_3H_2 molecules have been explored extensively using *ab initio* methods.[52,57,58,63,64] The essential results of these efforts are shown in Fig. 10(a), which will serve as basis for the discussion of our experimental results.

We have investigated the reaction $C(^3P,^1D) + C_2H_2(X\,^1\Sigma^+)$ at the collision energy of 29.3 kJ mol^{-1} by measuring product angular and TOF distributions at $m/z = 37$ (C_3H^+) and 36 (C_3^+). Fig. 11(a) (l.h.s.) depicts the $m/z = 37$ and 36 angular distributions, while Fig. 11(a) (r.h.s.) shows the TOF distribution for $m/z = 36$ at $\Theta = 25°$. As can be seen, the $m/z = 36$ angular distribution is different from that at $m/z = 37$, being significantly wider, and also the TOF at $m/z = 36$ is different from that at $m/z = 37$ (not shown) being somewhat faster. This indicates that the $m/z = 36$ signal comes partly from dissociative ionization of the C_3H product in the ionizer and partly from a different channel. The $m/z = 37$ product corresponds to the H-displacement channel leading to $C_3H + H$; moreover, the angular distribution and TOF data indicate that the C_3H product originates from both the $C(^3P)$ and the $C(^1D)$ reactions. From the best-fit of the $m/z = 37$ distributions, the CM product angular and translational energy distributions for the triplet and singlet H-displacement channels were derived. The $P(E'_T)$ distributions are depicted in Fig. 12(a); as can be seen, the distributions extend to the maximum total available energy for the triplet and singlet reactions and it is not possible to distinguish whether the linear or cyclic C_3H isomer is formed because of the very small energy difference.[63,64] About 43% and 53% of the total available energy is channelled in product translation for the ^3P and ^1D reaction, respectively, which indicates that the C_3H radical is formed with a significant degree of internal excitation. The CM angular dis-

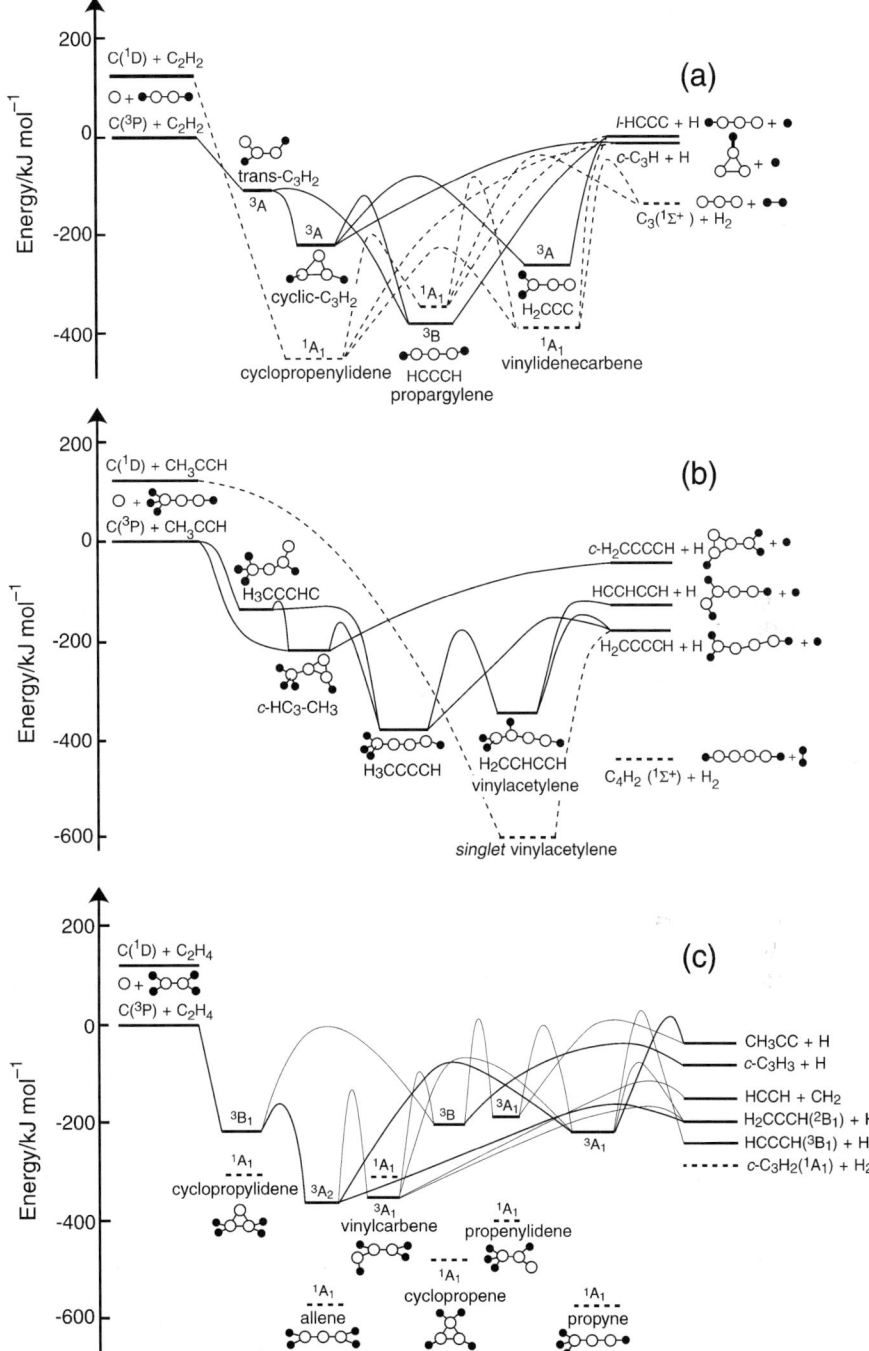

Fig. 10 Schematic representation of the C_3H_2 (a), C_4H_4 (b) and C_3H_4 (c) potential energy surfaces (adapted from ref. 52, 57–59, 64, 66, 67).

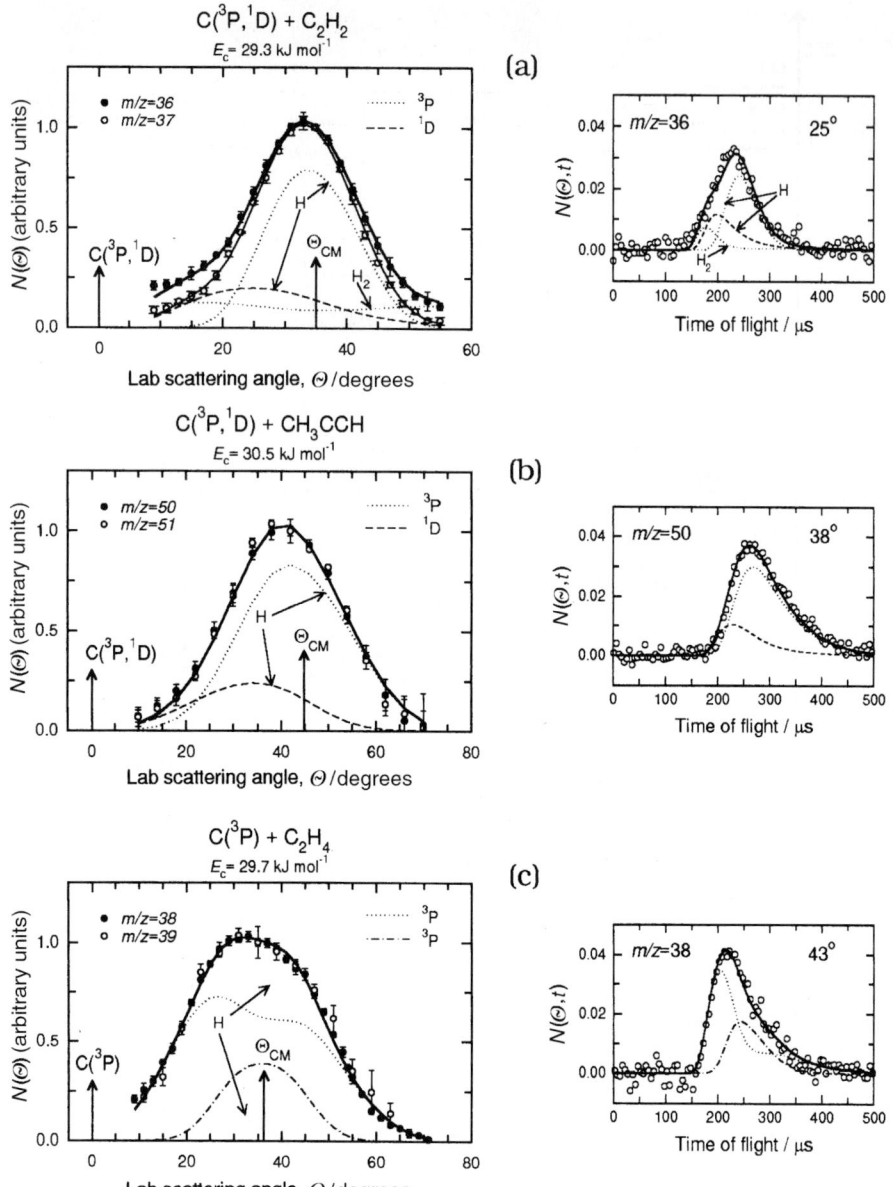

Fig. 11 (a, l.h.s.) Lab angular distributions of products at $m/z = 36$ (solid circles) and $m/z = 37$ (open circles) from the reaction $C(^3P,^1D) + C_2H_2$ at $E_c = 29.3$ kJ mol^{-1}. Dotted lines are the calculated contributions from the ^3P reaction to the H and H$_2$ elimination channels; the dashed line is the ^1D contribution to the H forming channel; the light continuous line is the total best-fit to the $m/z = 37$ data (H channel from both $C(^3P,^1D)$); the heavy continuous line is the total best fit to the $m/z = 36$ data (H channel from both $C(^3P,^1D)$ and H$_2$ channel from $C(^3P)$). (a, r.h.s.) $m/z = 36$ TOF distribution at $\Theta = 25°$ (line symbols are as for angular distribution at $m/z = 36$). (b) Angular distributions at $m/z = 50$ and 51 and TOF distribution at $m/z = 50$ ($\Theta = 38°$) for the $C(^3P,^1D) + CH_3CCH$ reactions at $E_c = 30.5$ kJ mol^{-1}. Solid lines are the total best-fit, while dotted and dashed lines are the separate contributions from the $C(^3P)$ and $C(^1D)$ reactions to the H forming channel. (c) Angular distributions at $m/z = 39$ and 38 and TOF distribution at $m/z = 38$ ($\Theta = 43°$) for the $C(^3P) + C_2H_4$ reaction at $E_c = 29.7$ kJ mol^{-1}. Solid lines are the total best-fit, while dotted and dotted-dashed lines are the separate contributions of the two isomer product channels (dotted: propargyl formation; dotted-dashed: cyclopropenyl/propyn-1-yl formation).

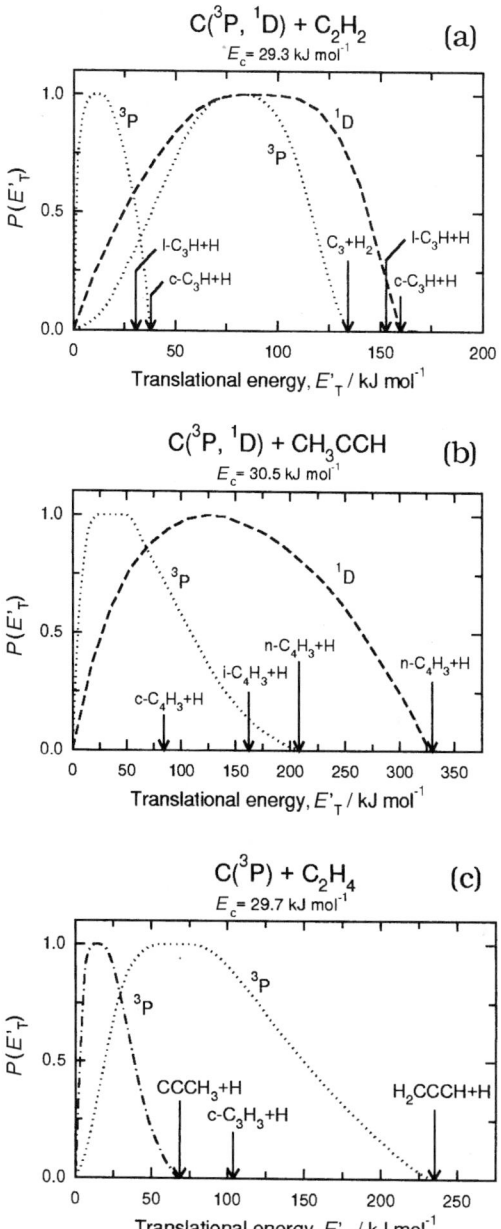

Fig. 12 Best-fit CM product translational energy distributions for the $C(^3P,^1D) + C_2H_2$ (a), $C(^3P,^1D) + CH_3CCH$ (b) and $C(^3P) + C_2H_4$ (c) reactions at the indicated collision energies. The $P(E'_T)$ for the different reaction channels observed from the $C(^3P)$ and $C(^1D)$ reactions are marked with different lines; the total available energy for the H and H_2 forming channels in the $C(^3P) + C_2H_2$ reaction are specified.

tribution (not shown) for the ^3P reaction is nearly backward–forward symmetric with more intensity in the forward direction, indicating that the reaction is proceeding through an osculating complex. The angular distribution of the ^1D reaction shows a much larger forward-to-backward scattering ratio than the ^3P reaction consistently with an expected singlet complex lifetime significantly shorter than the triplet complex lifetime because of the much larger exoergicity of the ^1D reaction.

The wider angular distribution and the faster TOF distributions at $m/z = 36$ indicate that a more exoergic reaction pathway also contributes to the overall reaction. This can only be the H_2 elimination channel leading to $C_3(X\ ^1\Sigma_g^+) + H_2(X\ ^1\Sigma_g^+)$ formation, which is the only other energetically allowed pathway for the $C(^3P, ^1D)$ reactions with acetylene (see Fig. 10(a)). Hence, three different reactive pathways contribute to the $m/z = 36$ signal: the $C_3H + H$ channel coming from both the triplet and singlet reactions, whose C_3H product partly cracks in the ionizer and the $C_3 + H_2$ channel. The CM angular distribution for the $C_3 + H_2$ channel is nearly backward–forward symmetric with only a slight preference for forward scattering, indicating the intermediacy of a long-lived complex which just starts to osculate. The corresponding $P(E'_T)$ is shown in Fig. 12(a): interestingly, the extent of the distribution is consistent with the energetics of the $C(^3P) + C_2H_2 \rightarrow C_3(X\ ^1\Sigma_g^+) + H_2$ reaction (although a small contribution from $C(^1D)$ also cannot be ruled out). The ratio of cross sections $\sigma(C_3 + H_2)/[\sigma(C_3 + H_2) + \sigma(C_3H + H)]$ is 0.3. This branching ratio is expected to be collision energy dependent. The only way to rationalize the formation of $C_3(X\ ^1\Sigma_g^+) + H_2(X\ ^1\Sigma_g^+)$ from the $C(^3P) + C_2H_2$ reaction, a spin-forbidden pathway, is to invoke the occurrence of ISC from the triplet to the singlet PES (see below). Interestingly, very recent kinetic work at room temperature[65] on the reaction $C(^3P) + C_2H_2$ found a branching ratio for H production of 0.53 ± 0.04, suggesting that $C_3 + H_2$ are indeed also formed. ISC was also invoked to rationalize the formation of $C(^3P) + C_2H_2$ in the photodissociation of singlet C_3H_2.[53] Finally, a recent, detailed theoretical study[57] on the C_3H_2 singlet and triplet PESs found evidence of ISC between triplet propargylene and singlet cyclo-propylidene which we think offers a rationale of the present experimental results. In conclusion, the mechanism of H and H_2 formation in the $C(^3P)$ reaction can be summarized as follows: $C(^3P)$ adds to the acetylenic π-bond leading to *trans*-HCCHC/cyclic-C_3H_2 which isomerizes to the more stable triplet propargylene; the latter can undergo CH bond cleavage to l-$C_3H + H$ (c-C_3H can also be formed from c-C_3H_2 by CH bond rupture before isomerization) or ISC to singlet cyclopropenylidene, which can isomerize to singlet vinylidene and lead, by (1,1) H_2 elimination, to $C_3(X\ ^1\Sigma_g^+) + H_2$ formation. ISC is expected to be strongly facilitated by the long-lived nature[58] of the strongly bound triplet propargylene intermediate which is deeply embedded in the singlet PES manifold (see Fig. 10(a)).

In the previous CMB study of this reaction at comparable E_c, the $m/z = 36$ and $m/z = 37$ angular and TOF distributions were found to be identical and this led to the conclusion that the reaction $C(^3P) + C_2H_2$ forms only $C_3H + H$.[47] The observation of the C_3 channel in our study was possible because of the absence of C_3 in the C beam and of the high S/N of the experiment (see Fig. 11(a)).

$C(^3P, ^1D) + CH_3CCH$

This reaction occurs over the C_4H_4 PES which has been recently investigated by *ab initio* means;[66] a simplified representation of it is given in Fig. 10(b). The main energetically allowed reaction channels are:

$C(^3P) + CH_3CCH(X\ ^1A_1) \rightarrow [C_4H_4]^\dagger$

$\rightarrow H_2CCCCH + H,$ $\quad \Delta H_0^0 = -177$ kJ mol^{-1}

$\rightarrow HCCHCCH + H,$ $\quad \Delta H_0^0 = -132$ kJ mol^{-1}

$\rightarrow c\text{-}HC_3CH_2 + H,$ $\quad \Delta H_0^0 = -54$ kJ mol^{-1}

$\rightarrow HCCCCH(X\ ^1\Sigma_g^+) + H_2(X\ ^1\Sigma_g^+),$ $\quad \Delta H_0^0 = -444$ kJ mol^{-1}

Again, the corresponding reactions of $C(^1D)$ are more exoergic by 121.9 kJ mol^{-1} and in this case more channels become energetically accessible. The n-C_4H_3 (H_2CCCCH) and i-C_4H_3 (HCCHCCH) isomers are believed to play an important role in formation of the first aromatic ring (the phenyl radical) *via* the reaction with acetylene in oxygen-deficient combustion.[62]

In the light of the results obtained for the reaction $C + C_2H_2$, we wanted to explore whether ISC would also occur in the $C + CH_3CCH$ reaction leading to the H_2 elimination channel with formation of singlet diacetylene (see Fig. 10(b)). We have therefore measured high-resolution angular and velocity distributions at $m/z = 50$ and 51 for $E_c = 30.5$ kJ mol^{-1}: both types of distributions were found to be perfectly superimposable, as can be seen from the angular distribution of Fig. 11(b). This indicates that the HCCCCH(X $^1\Sigma_g^+$) + H$_2$(X $^1\Sigma_g^+$) pathway does not occur and that the $m/z = 50$ (C$_4$H$_2^+$) signal originates completely from dissociative ionization of the primary C$_4$H$_3$ product(s) in the ionizer. Similarly to what has been observed in the case of the C(^{3}P) + HCCH reaction, contributions from both C(^{3}P) and C(^{1}D) to the C$_4$H$_3$ + H channel(s) are observed, as can be seen from the best-fit of the angular and TOF distributions in Fig. 11(b). The CM angular distribution for the C(^{3}P) reaction is backward–forward symmetric with some more intensity in the forward direction, indicating the intermediacy of an osculating C$_4$H$_4$ complex. Instead, the CM angular distribution for the C(^{1}D) reaction exhibits a considerably more pronounced forward/backward asymmetry, consistent with a shorter singlet C$_4$H$_4$ complex lifetime. The corresponding product translational energy distributions are consistent with the energetics of the H formation channel (see Fig. 12(b)). Although the dominant product isomer should be n-C$_4$H$_3$, some contribution also to the $m/z = 51$ signal from the less stable i-C$_4$H$_3$ and c-C$_4$H$_3$ isomers cannot be ruled out. From the extent of the $P(E'_T)$ distributions we derive that about 32% and 45% of the total available energy goes into product translational energy (assuming n-C$_4$H$_3$ to be the product) for the C(^{3}P) and C(^{1}D) reactions, respectively. The fact that the spin-forbidden H$_2$ elimination channel does not occur to any appreciable extent suggests that ISC is negligible in this system and/or that the H$_2$ elimination pathway is characterized by a very high potential barrier.

The present results on the C(^{3}P) + CH$_3$CCH reaction corroborate those obtained in a recent lower resolution CMB study with pulsed C beams at comparable E_c.[50]

C(^{3}P) + C$_2$H$_4$

The C(^{3}P) + ethylene reaction occurs over the C$_3$H$_4$ PES. This PES is quite complex, with many minima at thermally accessible energies, which permits extensive isomerization and opens up a large number of possible dissociation routes:

C(^{3}P) + C$_2$H$_4$(X ^{1}A$_g$) + → [C$_3$H$_4$]†

→ H$_2$CCCH(X ^{2}B$_1$) + H(^{2}S), $\quad \Delta H_0^0 = -205.9 \pm 8.4$ kJ mol^{-1}

→ c-C$_3$H$_3$(X ^{2}E″) + H(^{2}S), $\quad \Delta H_0^0 = -63.7 \pm 25$ kJ mol^{-1}

→ H$_3$CCC(X ^{2}E) + H(^{2}S), $\quad \Delta H_0^0 = -39 \pm 10$ kJ mol^{-1}

→ c-C$_3$H$_2$(X ^{1}A$_1$) + H$_2$(X $^1\Sigma_g^+$), $\quad \Delta H_0^0 = -289$ kJ mol^{-1}

→ C$_2$H$_2$(X $^1\Sigma_g^+$) + CH$_2$(X ^{3}B$_2$), $\quad \Delta H_0^0 = -170$ kJ mol^{-1}

→ C$_2$H(X ^{2}Π) + CH$_3$(X ^{2}A″$_2$), $\quad \Delta H_0^0 = -58$ kJ mol^{-1}

Both triplet and singlet C$_3$H$_4$ PESs have been explored using *ab initio* methods.[59,67] The essential results are shown in Fig. 10(c), where at least five minima are involved, which may be identified as allene, propyne, propenylidene (methylvinylidene), cyclopropene, and vinylcarbene. The previous CMB study[49] concluded that formation of H + H$_2$CCCH (propargyl) is the main reaction pathway and this was speculated to occur *via* two competing microchannels. We recall that the 2-propynyl radical (propargyl), because of its high stability, can be present in relatively large concentrations in flames. The significant role of propargyl radicals in the formation, *via* recombination, of the first aromatic rings in combustion of aliphatic fuels has been widely discussed;[68,69] this process is also thought to represent the first step in the formation of soot and polyaromatic hydrocarbons (PAHs).[70]

We have carried out accurate measurements of product angular and velocity distributions at $m/z = 39$ and $m/z = 38$ at three collision energies ($E_c = 8.8$, 15.9 and 29.7 kJ mol^{-1}) which clearly show that the C/H exchange channel is the dominant pathway and no H$_2$ formation was found to occur within our sensitivity. In addition, the contribution from C(^{1}D) to this reaction channel is negligible. In fact, as Fig. 11(c) shows, the $m/z = 38$ distribution is perfectly superimposable on that at $m/z = 39$ indicating that the $m/z = 38$ signal (C$_3$H$_2^+$) originates from fragmentation of C$_3$H$_3$ in the ionizer. Interestingly, while at $E_c = 8.8$ kJ mol^{-1} formation of the propargyl radical is the dominant pathway, at the two higher E_c formation of the cyclopropenyl (and/or propyn-1-yl)

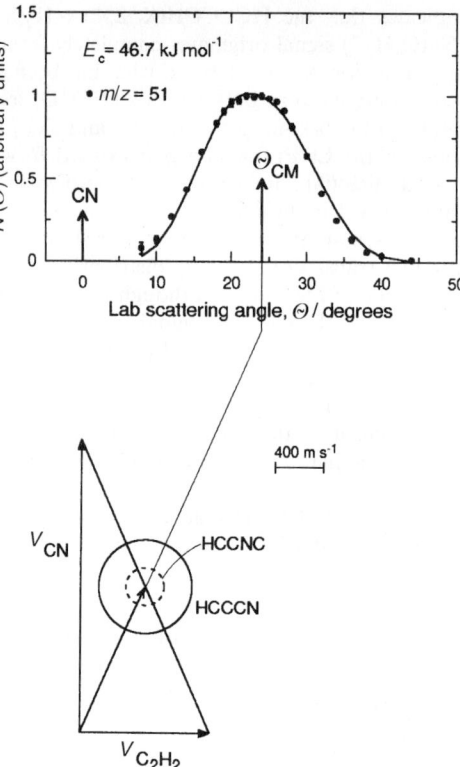

Fig. 13 Laboratory angular distribution (●) of the HCCCN product (detected at $m/z = 51$) from the reaction $CN(^2\Sigma^+) + C_2H_2$ at $E_c = 46.7$ kJ mol^{-1}. The circles in the Newton diagram delimit the maximum velocity that the two isomer products, HCCCN and HCCNC, can attain. Solid line: calculated distribution from best-fit CM functions.

radical is also inferred. Fig. 11(c) shows the two separate contributions to the best-fit curve at the highest E_c of 29.7 kJ mol^{-1}. The translational energy distributions are consistent with the formation of $H_2CCCH + H$, and $CCCH_3 + H$ and/or c-$C_3H_3 + H$ (see Fig. 12(c)). The ratio of cross sections $\sigma_{(c-C_3H_3/H_3CCC)}/(\sigma_{c-C_3H_3/H_3CCC}) + \sigma_{H_2CCCH})$ has been found to be 0.02 at the lowest E_c and to rise to ~0.05 at $E_c = 15.9$ kJ mol^{-1} and ~0.14 at $E_c = 29.7$ kJ mol^{-1}. While the CM angular distributions for the dominant propargyl channel is backward–forward symmetric with some more intensity in the forward direction, witnessing the formation of an oscillating complex, that for formation of the less stable isomer(s) is isotropic, reflecting the formation of a long-lived complex. The fraction of energy in product translation is 39% for propargyl formation and somewhat lower (34%) for the less stable isomer.

According to recent theoretical work[67] (see Fig. 10(c)), $C(^3P)$ attacks the π-bond of ethylene leading to triplet cyclopropylidene which can rapidly isomerize to the more stable allene, which easily undergoes C–H bond cleavage to $H_2CCCH + H$ because of its high energy content. The contribution also from vinylcarbene dissociation to propargyl + H formation cannot be ruled out (this would imply isomerization of allene to vinylcarbene). As E_c rises our experimental results suggest that the initial cyclopropylidene adduct may have enough internal energy to also isomerize to cyclopropene which can undergo C–H bond rupture to cyclopropenyl + H. However, isomerization of allene to propyne leading to formation of propyn-1-yl + H by cleavage of the acetylenic C–H bond cannot be ruled out. Notably, in the photodissociation of ground state propyne, formation of propyn-1-yl + H, following rupture of the stronger acetylenic CH bond with respect to the weaker aliphatic CH bond, was found to be dominant, indicating a strong bond selectivity.[55]

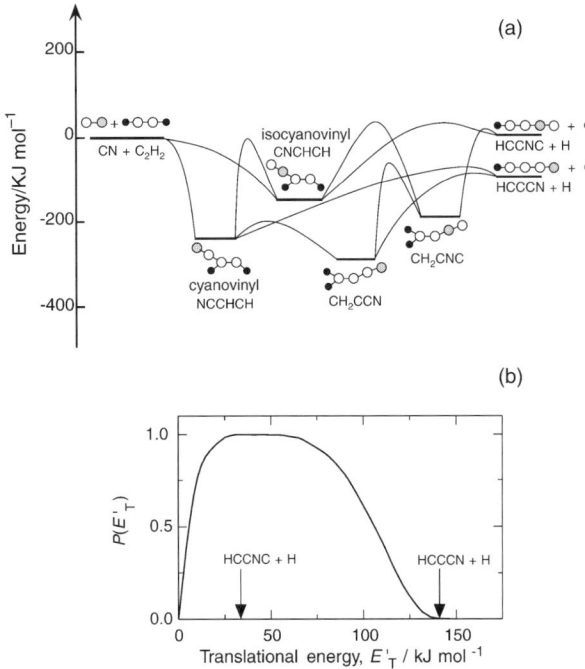

Fig. 14 (a) Schematic potential energy surface for the HCCHN system (adapted from ref. 74 and 75). (b) Best-fit CM product translational energy distribution; the total available energy for the cyanoacetylene and isocyanoacetylene isomers are indicated with arrows.

In contrast to the reaction $C(^3P) + C_2H_2$, no H_2 elimination is observed in $C(^3P) + C_2H_4$. Since theoretical calculations[67] indicate that H_2 elimination from triplet propyne to $HCCCH(^3B_1)$ + H_2 has a very high energy barrier (see Fig. 10(c)) and the only other pathway for forming H_2 would require ISC from triplet to singlet PESs with formation of c-$C_3H_2(^1A_1) + H_2$, our results indicate that ISC in the C_3H_4 system is not an efficient process, in contrast to the C_3H_2 system. The reason for that can likely be traced back to the fact that the H elimination channel is strongly exoergic in the $C(^3P) + C_2H_4$ reaction, while in $C(^3P) + C_2H_2$ it is nearly thermoneutral (see Fig. 10(a) and (c)). Because of this, the C_3H_4 system will spend a significantly shorter time in the triplet allene potential than does C_3H_2 in the triplet propargylene potential well, and this should determine a lower probability of ISC from triplet to singlet surfaces in C_3H_4. A shorter C_3H_4 lifetime with respect to C_3H_2 is indicated by the fact that at the comparable E_c of about 30 kJ mol^{-1} the CM angular distribution for $C_3H_3 + H$ formation from $C(^3P) + C_2H_4$ is significantly more forward peaked than that for $C_3H + H$ formation from $C(^3P) + C_2H_2$.

V. Reactions of CN radicals

In recent years, there has been much interest in the reactions of the cyanogen radical with various compounds because of its important role in many processes, from combustion chemistry to astrochemistry.[71] In particular, the reactions of CN with unsaturated hydrocarbons are very fast down to very low temperarures ($k \approx 3$–5×10^{-10} cm^3 molecule^{-1} s^{-1}).[72] Recently, a large variety of CN reactions with unsaturated hydrocarbons have been studied in CMB experiments using pulsed CN beams with velocity ranging from 900 to 1900 m s^{-1}.[73] Exploiting our continuous CN beams with variable velocity from about 1500 up to 2500 m s^{-1} we have also been able to tackle, with high S/N, CN reactions with unsaturated hydrocarbons at higher collision energies, pertinent to combustion conditions. The first reaction we have looked at is that with acetylene at $E_c = 46.7$ kJ mol^{-1}. This study extends the characterization of the reaction dynamics to much higher collision energies with respect to previous CMB work obtained at $E_c = 21.1$ and 27.0 kJ mol^{-1}.[74]

According to recent theoretical work,[74,75] two isomer products are energetically accessible at $E_c = 46.7$ kJ mol^{-1}:

$$CN(X\,^2\Sigma^+) + C_2H_2(X\,^1\Sigma_g^+) \rightarrow [C_3H_2N]^\dagger \rightarrow HCCCN + H, \quad \Delta H_0^0 = -94 \text{ kJ mol}^{-1}$$
$$\rightarrow HCCNC + H, \quad \Delta H_0^0 = +13 \text{ kJ mol}^{-1}$$

Fig. 13 shows the product angular distribution measured at $m/z = 51$, together with the relative Newton diagram. The outer and inner circles in the diagram delimit the maximum speed that the two energetically allowed isomer products (cyanoacetylene and isocyanoacetylene) can attain on the basis of energy and linear momentum conservation. Analysis of the LAB angular and TOF distributions has led to a CM product translational energy distribution which is fully consistent with the total energy available to the HCCCN + H products (see Fig. 14(b)); however, a minor contribution from the HCCNC + H channel also cannot be ruled out (experiments at more E_c are needed to clarify this point). It is interesting to note that statistical calculations[74] at $E_c = 27$ kJ mol^{-1} suggest a negligible yield for the HCCNC channel; this reflects both a slight endoergicity and the presence of exit barriers. The fraction of energy in translation is about 40%, indicating that the cyanoacetylene molecule is formed highly internally excited. The angular distribution is backward–forward symmetric with a slight forward bias indicating that the reaction proceeds *via* an osculating complex. This interpretation is supported by the results of electronic structure calculations[74,75] which show that CN adds to the π-bond of acetylene forming a strongly bound cyanovinyl intermediate which dissociates to HCCCN + H directly or *via* prior isomerization to the more stable H$_2$CCCN intermediate (see Fig. 14(a)). The latter can also lead to HCCNC + H; however, isocyanoacetylene can also be formed directly from dissociation of the isocyanovinyl intermediate following CN attack of acetylene on the N side. This may become relevant at higher collision energies and, in general, in high energy processes.

VI. Conclusion and outlook

In this paper we have reported on the results of CMB experiments on elementary reactions of N(^{2}D), C(^{3}P,^{1}D) and CN(X$\,^2\Sigma^+$) radicals of relevance in combustion, as made possible by the recent development in our laboratory of *continuous* supersonic beams of these transient species. The results on product identification, branching ratios and reaction dynamics have been discussed in the light of the most recent theoretical developments on the potential energy surfaces for the systems.

The work on the dynamics of N atom reactions with hydrocarbons and water helps to elucidate the formation of CN-containing molecules/radicals and HNO/NO in nitrogen-containing combustion systems. The CMB investigations of the reactions of C(^{3}P,^{1}D) with unsaturated hydrocarbons have confirmed that the C/H exchange channel is the main reaction pathway and represents a barrierless route to the synthesis of highly reactive hydrocarbon radicals in combustion processes. These are C$_3$H (tricarbon hydride) in the C + C$_2$H$_2$ reaction, H$_2$CCCH (propargyl) and, to a minor extent, c-C$_3$H$_3$ (cyclopropen-1-yl) and/or H$_3$CCC (propyn-1-yl) in the C + C$_2$H$_4$ reaction and C$_4$H$_3$ in the case of the C + CH$_3$CCH reaction. These hydrocarbon radicals can start important secondary reactions leading ultimately to soot formation.

Very interestingly, molecular products, C$_3$(X$\,^1\Sigma_g^+$) + H$_2$, can also be formed from the C(^{3}P) + C$_2$H$_2$ reaction, indicating that ISC is facile in the C$_3$H$_2$ system. This had not been observed before. The observation that the C$_3$ + H$_2$ channel is comparable in importance to the C$_3$H + H channel may have important consequences in establishing the role of this reaction in the modelling of combustion as well as interstellar chemistry networks. In fact, C$_3$(X$\,^1\Sigma_g^+$) can undergo subsequent reactions with other hydrocarbon molecules that, although known to be characterized by significant activation barriers, may occur readily at the high temperatures of combustion environments, leading to the formation of hydrogen-deficient hydrocarbon molecules/radicals.[76]

Finally, as an outlook, we note that another class of very important reactions in combustion are radical–radical reactions. On these, kinetic measurements are difficult and hence kinetic property data are rather scarce; dynamics experiments are even more difficult and hence rare to date. Yet, thanks to the recent development of supersonic beams of a variety of radicals,[3] the investigation of radical–radical reactions (such as N + OH, C + OH, O + CN, O + hydrocarbon radicals, *etc.*) in crossed beams may be expected to become feasible in the near future and this should contribute to increasing our understanding of complex combustion systems.

Acknowledgements

Support by the Italian "Consiglio Nazionale delle Ricerche (CNR)", "Ministero Università e Ricerca Scientifica (MURST)", "Ente per le Nuove Tecnologie, l'Energia e l'Ambiente (ENEA)" and the European Commission through contracts FMRX-CT97-0132 and HPRN-CT-1999-00007 is gratefully acknowledged.

References

1. *Gas Phase Combustion Chemistry*, ed. W. C. Gardiner, Jr., Springer, New York, 2000, and references therein; V. Ebert, C. Schultz, H.-R. Volpp, J. Wolfrum and P. Monkhouse, *Israel J. Chem.*, 1999, **39**, 1; D. R. Crosley, J. B. Jeffries and G. P. Smith, *Israel J. Chem.*, 1999, **39**, 41. See also the other papers in *Israel J. Chem.*, 1999, **39**.
2. D. L. Baulch, C. J. Cobos, R. A. Cox, P. Frank, G. Hayman, Th. Just, J. A. Kerr, T. Murrelis, M. J. Pilling, J. Troe, R. W. Walker and J. Warnatz, *J. Phys. Chem. Ref. Data*, 1994, **23**, 847, and references therein.
3. P. Casavecchia, *Rep. Prog. Phys.*, 2000, **63**, 355, and references therein.
4. P. Casavecchia, N. Balucani and G. G. Volpi, *Annu. Rev. Phys. Chem.*, 1999, **50**, 347; F. Ausfelder and K. G. McKendrick, *Prog. React. Kinet. Mec.*, 2000, **25**, 299; and references therein.
5. Y. T. Lee, in *Atomic and Molecular Beam Methods*, ed. G. Scoles, Oxford University Press, New York, 1987, vol. 1.
6. J. J. Lin, J. Shu, Y. T. Lee and X. Yang, *J. Chem. Phys.*, 2000, **113**, 5287; J. Shu, J. J. Lin, Y. T. Lee and X. Yang, *J. Chem. Phys.*, 2001, **114**, 4.
7. N. Balucani, L. Beneventi, P. Casavecchia and G. G. Volpi, *Chem. Phys. Lett.*, 1991, **180**, 34; N. Balucani, L. Beneventi, P. Casavecchia, G. G. Volpi, E. J. Kruus, P. Aker and J. J. Sloan, *Can. J. Chem.*, 1994, **72**, 888; M. Alagia, N. Balucani, P. Casavecchia and G. G. Volpi, *J. Phys. Chem. A*, 1997, **101**, 6455.
8. M. Alagia, N. Balucani, P. Casavecchia, A. Laganà, G. Ochoa de Aspuru, E. H. van Kleef and G. G. Volpi, *Chem. Phys. Lett.*, 1996, **258**, 323.
9. M. Alagia, N. Balucani, L. Cartechini, P. Casavecchia, M. van Beek, G. G. Volpi, L. Bonnet and J.-C. Rayez, *Faraday Discuss.*, 1999, **113**, 133.
10. N. Balucani, L. Beneventi, P. Casavecchia, D. Stranges and G. G. Volpi, *J. Chem. Phys.*, 1991, **94**, 8611.
11. M. Alagia, N. Balucani, P. Casavecchia, D. Stranges and G. G. Volpi, *J. Chem. Soc., Faraday Trans.*, 1995, **91**, 575.
12. M. Alagia, N. Balucani, P. Casavecchia, D. Stranges, G. G. Volpi, D. C. Clary, A. Kliesch and H.-J. Werner, *Chem. Phys.*, 1996, **207**, 389.
13. M. Alagia, N. Balucani, P. Casavecchia, D. Stranges and G. G. Volpi, *J. Chem. Phys.*, 1993, **98**, 8341.
14. P. Casavecchia, N. Balucani, M. Alagia, L. Cartechini and G. G. Volpi, *Acc. Chem. Res.*, 1999, **32**, 503.
15. N. Balucani, M. Alagia, L. Cartechini, P. Casavecchia, G. G. Volpi, L. A. Pederson and G. C. Schatz, *J. Phys. Chem. A*, 2001, **105**, 2414.
16. N. Balucani, M. Alagia, L. Cartechini, P. Casavecchia, G. G. Volpi, K. Sato, T. Takayanagi and Y. Kurosaki, *J. Am. Chem. Soc.*, 2000, **122**, 4443.
17. N. Balucani, L. Cartechini, M. Alagia, P. Casavecchia and G. G. Volpi, *J. Phys. Chem. A*, 2000, **104**, 5655.
18. A. Bergeat, L. Cartechini, N. Balucani, G. Capozza, L. F. Phillips, P. Casavecchia, G. G. Volpi, L. Bonnet and J.-C. Rayez, *Chem. Phys. Lett.*, 2000, **327**, 197.
19. M. Alagia, V. Aquilanti, D. Ascenzi, N. Balucani, D. Cappelletti, L. Cartechini, P. Casavecchia, F. Pirani, G. Sanchini and G. G. Volpi, *Israel J. Chem.*, 1997, **37**, 329.
20. J. A. Miller and C. T. Bowman, *Prog. Energy Combust. Sci.*, 1989, **15**, 287.
21. J. T. Herron, *J. Phys. Chem. Ref. Data*, 1999, **28**, 1453.
22. K. Schoefield, *J. Phys. Chem. Ref. Data*, 1979, **8**, 723.
23. C. Bertrand and P. J. van Tiggelen, *J. Phys. Chem.*, 1974, **78**, 2320; P. Guillaume, C. François and P. J. van Tiggelen, *Bull. Soc. Chim. Belg.*, 1983, **92**, 633.
24. K. J. Schmatjko and J. Wolfrum, *Ber. Bunsen-Ges. Phys. Chem.*, 1978, **82**, 419.
25. M. R. Manaa and D. R. Yarkony, *Chem. Phys. Lett.*, 1992, **188**, 352.
26. (*a*) T. Takayanagi, Y. Kurosaki, K. Sato, K. Misawa, Y. Kobayashi and S. Tsunashima, *J. Phys. Chem. A*, 1999, **103**, 250; (*b*) H. Umemoto, N. Hachiya, E. Matsunaga, A. Suda and M. Kawasaki, *Chem. Phys. Lett.*, 1998, **296**, 203.
27. H. Umemoto, Y. Kimura and T. Asai, *Chem. Phys. Lett.*, 1997, **264**, 215; H. Umemoto, Y. Kimura and T. Asai, *Bull. Chem. Soc. Jpn.*, 1997, **70**, 2951.
28. H. Umemoto, T. Nakae, H. Hashimoto, K. Kongo and M. Kawasaki, *J. Chem. Phys.*, 1998, **109**, 5844.

29 Y. Kurosaki, T. Takayanagi, K. Sato and S. Tsunashima, *J. Phys. Chem. A*, 1998, **102**, 254.
30 B. S. Jursic, *Int. J. Quantum Chem.*, 1999, **71**, 481.
31 J. F. Arenas, J. I. Marcos, J. C. Otero, A. Sanchez-Galvez and J. Soto, *J. Chem. Phys.*, 1999, **111**, 551; C. R. Kemnitz, G. B. Ellison, W. L. Karney and W. T. Borden, *J. Am. Chem. Soc.*, 2000, **122**, 1098.
32 J. W. Bozzelli and A. M. Dean, in *Gas-phase Combustion Chemistry*, ed. W. C. Gardiner Jr., Springer-Verlag, New York, 2nd edn, 2000, ch. 2.
33 E. W. Diau, M. J. Halbgewachs, A. R. Smith and M. C. Lin, *Int. J. Chem. Kinet.*, 1995, **27**, 867, and references therein.
34 M. R. Soto and M. J. Page, *J. Chem. Phys.*, 1992, **97**, 7287.
35 J. D. Adamson, S. K. Farhat, C. L. Morter, G. P. Glass, R. F. Curl and L. F. Phillips, *J. Phys. Chem.*, 1994, **98**, 5665.
36 D. P. Misra, D. G. Sanders and P. J. Dagdigian, *J. Chem. Phys.*, 1991, **95**, 955; D. P. Misra and P. J. Dagdigian, *Chem. Phys. Lett.*, 1991, **185**, 387; S. Duan and M. Page, *J. Chem. Phys.*, 1995, **102**, 6121.
37 H. Umemoto, T. Asai, H. Hashimoto and T. Nakae, *J. Phys. Chem. A*, 1999, **103**, 700.
38 R. Sumathi, D. Sengupta and M. T. Nguyen, *J. Phys. Chem. A*, 1998, **102**, 3175, and references therein.
39 Y. Kurosaki and T. Takayanagi, *J. Phys. Chem. A*, 1999, **103**, 436.
40 N. Haider and D. Husain, *J. Chem. Soc., Faraday Trans.*, 1993, **89**, 7; D. C. Clary, N. Haider, D. Husain and M. Kabir, *Astrophys. J.*, 1994, **422**, 416.
41 D. Chastaing, S. D. Le Picard, I. R. Sims and I. W. M. Smith, *Astron. Astrophys.*, 2001, **365**, 241; D. Chastaing, P. L. James, I. R. Sims and I. W. M. Smith, *Phys. Chem. Chem. Phys.*, 1999, **1**, 2447.
42 M. Costes, C. Naulin, G. Dorthe, G. Daleau, J. Joussot-Dubien, C. Lalaude, M. Vinckert, A. Destor, C. Vaucamps and G. Nouchi, *J. Phys. E*, 1989, **22**, 1017.
43 M. R. Scholefield, J. H. Choi, S. Goyal and H. Reisler, *Chem. Phys. Lett.*, 1998, **288**, 487, and references therein.
44 W. D. Geppert, C. Naulin and M. Costes, *Chem. Phys. Lett.*, 2001, **333**, 51.
45 D. Chastaing, S. D. Le Picard, I. R. Sims, I. W. M. Smith, W. D. Geppert, C. Naulin and M. Costes, *Chem. Phys. Lett.*, 2000, **331**, 170.
46 W. D. Geppert, C. Naulin, M. Costes, L. Cartechini, A. Bergeat, G. Capozza, P. Casavecchia and G. G. Volpi, in preparation.
47 R. I. Kaiser, C. Ochsenfeld, D. Stranges, M. Head-Gordon, Y. T. Lee and A. G. Suits, *Science*, 1996, **274**, 1508; R. I. Kaiser, C. Ochsenfeld, M. Head-Gordon, Y. T. Lee and A. G. Suits, *J. Chem. Phys.*, 1997, **106**, 1729.
48 R. I. Kaiser, A. M. Mebel and Y. T. Lee, *J. Chem. Phys.*, 2001, **114**, 231.
49 R. I. Kaiser, Y. T. Lee and A. G. Suits, *J. Chem. Phys.*, 1996, **105**, 8705.
50 R. I. Kaiser, D. Stranges, Y. T. Lee and A. G. Suits, *J. Chem. Phys.*, 1996, **105**, 8721.
51 R. I. Kaiser, C. Ochsenfeld, D. Stranges, M. Head-Gordon and Y. T. Lee, *Faraday Discuss.*, 1998, **109**, 183.
52 R. A. Seburg, E. V. Patterson, J. F. Stants and R. J. McMahon, *J. Am. Chem. Soc.*, 1997, **119**, 5847.
53 W. M. Jackson, D. S. Anex, R. E. Continetti and Y. T. Lee, *J. Chem. Phys.*, 1991, **95**, 7327.
54 J. H. Kiefer, P. S. Mudipalli, S. S. Sidhu, R. D. Kern, B. S. Jursic, K. Xie and H. Chen, *J. Phys. Chem. A*, 1997, **101**, 4057.
55 W. Sun, K. Yokoyama, J. C. Robinson, A. G. Suits and D. M. Neumark, *J. Chem. Phys.*, 1999, **110**, 4363.
56 S. Harich, J. J. Lin, Y. T. Lee and X. Yang, *J. Chem. Phys.*, 2000, **112**, 6656.
57 A. M. Mebel, W. M. Jackson, A. H. H. Chang and S. H. Lin, *J. Am. Chem. Soc.*, 1998, **120**, 5751.
58 R. Guadagnini, G. C. Schatz and S. P. Walch, *J. Phys. Chem. A*, 1998, **102**, 5857.
59 N. Honjou, J. Pacansky and M. Yoshimine, *J. Am. Chem. Soc.*, 1985, **107**, 5332; M. Yoshimine, J. Pacansky and N. Honjou, *J. Am. Chem. Soc.*, 1989, **111**, 4198.
60 P. Casavecchia, L. Cartechini, A. Bergeat, G. Capozza and G. G. Volpi, in preparation.
61 B. E. Turner, E. Herbst and R. Terziera, *Astrophys. J. Suppl. Ser.*, 2000, **126**, 427.
62 M. Hausmann and K. H. Homann, *Ber. Bunsen-Ges. Phys. Chem.*, 1990, **94**, 1308.
63 Ochsenfeld, R. I. Kaiser, A. G. Suits, Y. T. Lee and M. Head-Gordon, *J. Chem. Phys.*, 1997, **106**, 4141.
64 J. Takahashi and K. Yashimata, *J. Chem. Phys.*, 1996, **104**, 6613.
65 A. Bergeat and J.-C. Loison, *Phys. Chem. Chem. Phys.*, 2001, **3**, 2038.
66 A. M. Mebel, R. I. Kaiser and Y. T. Lee, *J. Am. Chem. Soc.*, 2000, **122**, 1776.
67 T. N. Le, H.-Y. Lee, A. M. Mebel and R. I. Kaiser, *J. Phys. Chem. A*, 2001, **105**, 1847; for an evaluation of reaction exoergicities, see: T. L. Nguyen, A. M. Mebel and R. I. Kaiser, *J. Phys. Chem. A*, 2001, **105**, 3284, and references therein.
68 C. F. Melius, J. A. Miller and E. M. Evleth, *24th Symposium (International) on Combustion*, The Combustion Institute, Pittsburgh, 1992, p. 621; R. D. Kern, H. Chen, J. H. Kiefer and P. S. Mudipali, *Combust. Flame*, 1995, **100**, 177; J. D. Adamson, L. Morter, J. D. De Sain, G. P. Glas and R. F. Curl, *J. Phys. Chem.*, 1996, **100**, 2125.
69 L. Vereecken, K. Pierloot and J. Peeters, *J. Chem. Phys.*, 1998, **108**, 1068.
70 H. Richter and J. B. Howard, *Prog. Energy Combust. Sci.*, 2000, **26**, 565.
71 D. L. Yang and M. C. Lin, in *The Chemical Dynamics and Kinetics of Small Radicals*, ed. K. Liu and A. F. Wagner, World Scientific, Singapore, 1995, p. 164.

72 I. W. M. Smith, I. R. Sims and B. R. Rowe, *Chem. Eur. J.*, 1997, **3**, 1925.
73 N. Balucani, O. Asvany, Y. Osamura, L. C. L. Huang, Y. T. Lee and R. I. Kaiser, *Planet. Space Sci.*, 2000, **48**, 447.
74 L. C. L. Huang, O. Asvany, A. H. H. Chang, N. Balucani, S. H. Lin, Y. T. Lee, R. I. Kaiser and Y. Osamura, *J. Chem. Phys.*, 2000, **113**, 8656.
75 K. Fukuzawa and Y. Osamura, *Astrophys. J.*, 1997, **489**, 113.
76 R. I. Kaiser, T. N. Le, T. L. Nguyen, A. M. Mebel, N. Balucani, Y. T. Lee, F. Stahl, P. v. R. Schleyer and H. F. Schaefer III, *Faraday Discuss.*, 2001, **119**, (DOI: 10.1039/b101967h).

A combined crossed molecular beam and *ab initio* investigation of C_2 and C_3 elementary reactions with unsaturated hydrocarbons—pathways to hydrogen deficient hydrocarbon radicals in combustion flames

Ralf I. Kaiser,[abc] Trung N. Le,[c] Thanh L. Nguyen,[c] Alexander M. Mebel,[c] Nadia Balucani,[cd] Yuan T. Lee,[c] Frank Stahl,[ef] Paul v. R. Schleyer[ef] and Henry F. Schaefer III[f]

[a] *Department of Chemistry, University of York, York, UK YO10 5DD*
[b] *Department of Physics, Technical University Chemnitz-Zwickau, 09107 Chemnitz, Germany*
[c] *Institute of Atomic and Molecular Sciences, Academia Sinica, 107 Taipei, Taiwan, ROC*
[d] *Dipartimento di Chimica, Università di Perugia, 06123 Perugia, Italy*
[e] *Institut für Organische Chemie der Universität Erlangen-Nürnberg, 91054 Erlangen, Germany*
[f] *Center for Computational Quantum Chemistry, The University of Georgia, Athens, GA, 30602, USA*

Received 1st March 2001
First published as an Advance Article on the web 26th September 2001

Crossed molecular beam experiments on dicarbon and tricarbon reactions with unsaturated hydrocarbons acetylene, methylacetylene, and ethylene were performed to investigate the dynamics of channels leading to hydrogen-deficient hydrocarbon radicals. In the light of the results of new *ab initio* calculations, the experimental data suggest that these reactions are governed by an initial addition of C_2/C_3 to the π molecular orbitals forming highly unsaturated cyclic structures. These intermediates are connected *via* various transition states and are suggested to ring open to chain isomers which decompose predominantly by displacement of atomic hydrogen, forming C_4H, C_5H, $HCCCCH_2$, $HCCCCCH_3$, H_2CCCCH and $H_2CCCCCH$. The $C_2(^1\Sigma_g^+) + C_2H_4$ reaction has no entrance barrier and the channel leading to the H_2CCCCH product is strongly exothermic. This is in strong contrast with the $C_3(^1\Sigma_g^+) + C_2H_4$ reaction as this is characterized by a 26.4 kJ mol^{-1} threshold to form a $HCCCCCH_2$ isomer. Analogous to the behavior with ethylene, preliminary results on the reactions of C_2 and C_3 with C_2H_2 and CH_3CCH showed the H-displacement channels of these systems to share many similarities such as the absence/presence of an entrance barrier and the reaction mechanism. The explicit identification of the C_2/C_3 *vs.* hydrogen displacement demonstrates that hydrogen-deficient hydrocarbon radicals can be formed easily in environments like those of combustion processes. Our work is a first step towards a systematic database of the intermediates and the reaction products which are involved in this important class of reactions. These findings should be included in future models of PAH and soot formation in combustion flames.

DOI: 10.1039/b101967h

1. Introduction

Laboratory investigations on the reactions of small carbon molecules, dicarbon, C_2 and tricarbon, C_3, with hydrocarbons are expected to elucidate possible formation pathways of hydrogen-deficient carbon chains and soot in a variety of terrestrial and astrophysical environments: high-temperature combustion flames, chemical vapor deposition (CVD) of diamond, outflow of asymptotic giant branch (AGB) carbon stars, comets, and interstellar clouds.[1] The spectral lines of dicarbon in its $^1\Sigma_g^+$ electronic ground state were first detected in comets more then a century ago[2] and in terrestrial hydrocarbon flames fifty years later.[3] In the decades following it became clear that $C_2(X\,^1\Sigma_g^+)$ is ubiquitous in the interstellar medium.[4] Transitions were observed towards warm carbons stars like IRC+10126,[5] post AGB stars such as HD 56126,[6] and towards the HII region W40 IRS. The tricarbon molecule was first assigned at the turn of the 19th century in comets[7] prior to an identification in fuel flames,[8] explosions,[9] and during arc vaporization of carbon. Likewise, Hinkle et al. observed vibration–rotation transitions in the circumstellar shell of the carbon star IRC+10216, whereas Haffner and Meyer[10] detected the infrared spectrum of $C_3(X\,^1\Sigma_g^+)$ in the translucent cloud HD147889.

Since the formation of small carbon clusters is thought to be strongly linked to the synthesis of polycyclic aromatic hydrocarbons (PAHs) and ultimately soot production in combustion flames,[11] laboratory studies are needed to assess the role of these species and to include them into the pertinent chemical models. Three kinds of laboratory results are crucial to describe the reactivity of small carbon clusters, namely the measurement of temperature-dependent rate constants, the identification of the reaction products, and an assignment of reaction intermediates. The kinetics of reactions involving the singlet ground state and the electronically excited triplet state of dicarbon have been extensively investigated[12] (recall that the electronically excited triplet state, $a\,^3\Pi_u$, lies only 718.32 cm^{-1} above the ground state $X\,^1\Sigma_g^+$). In these studies, the disappearance of C_2 in the two electronic states ($X\,^1\Sigma_g^+$ and $a\,^3\Pi_u$) was followed; the reactions of $C_2(X\,^1\Sigma_g^+)$ were found to be quite fast (of the gas kinetic order when the molecular partner is an unsaturated hydrocarbon) whereas the $C_2(a\,^3\Pi_u)$ reactions were suggested to be systematically slower. Rate constants of C_3 reactions with several alkenes, alkynes, and allenes were found to be much smaller and seldom reached 10^{-12} cm^3 s^{-1} range at room temperature.[12]

Despite these extensive kinetic investigation, information on the reaction products and intermediates involved is still lacking. In some cases, primary products and reaction mechanisms were postulated on the basis of the observed temperature dependence of the reactions. For instance, from the measured removal rate constants of both $C_2(X\,^1\Sigma_g^+)$ and $C_2(a\,^3\Pi_u)$ by ethylene and by halogen substituted ethylene, the favored approach was suggested to be an addition of the electrophilic C_2 (both singlet and triplet states) to the olefinic π bond. Nevertheless, $C_2(X\,^1\Sigma_g^+)$ reacts faster than $C_2(a\,^3\Pi_u)$ and is quite insensitive to halogen substitution. The reduced reactivity of $C_2(a\,^3\Pi_u)$ with the halogen-substituted ethylene was attributed to the hindrance of the bulky halogen atoms which hinders the approach to the π system. This implies that $C_2(X\,^1\Sigma_g^+)$ also reacts through alternative pathways. A few reactions of C_2 also were investigated at 77 and 10 K in the condensed phase and ab initio calculations were performed in order to help understand the reaction mechanism.[13] Interestingly, the reaction products of the reaction of dicarbon with ethylene have been always assumed to be C_2H and C_2H_3 ($\Delta_rH = -75.5$ kJ mol^{-1}) or C_2H_2 and $C_2H_2(\Delta_rH = -435.8$ kJ mol^{-1}). The formation of complex carbon chains such as HCCCH$_2$ and H, which actually is a strongly exothermic channel for both C_2 singlet and triplet states ($\Delta_rH = -156.3$ and -165.5 kJ mol^{-1}, respectively), has never been considered before. In contrast with the reaction of C_2 with C_2H_4, the reaction C_3 with C_2H_4 was found to be slower, since the disappearance rate of C_3 resulted to be $<10^{-15}$ cm^3 s^{-1} at $T = 294$ K and an estimate of 27 kJ mol^{-1} for the activation energy was given. The reaction products were not identified in this reaction either.

A systematic study of the related reactions of atomic carbon, $C(^3P_j)$, with some unsaturated hydrocarbons (acetylene,[14] methylacetylene,[15] dimethylacetylene,[16] propargyl,[17] vinyl,[29] ethylene,[18] propylene,[19] buta-1,2-diene,[20] buta-1,3-diene,[21] allene,[22] and benzene[23]) established that a C/H exchange channel is available in all cases. The reactions occur predominantly through the formation of a bound intermediate according to the general scheme:

$$C(^3P_j) + C_nH_m \rightarrow [C_{n+1}H_m]^* \rightarrow C_{n+1}H_{m-1} + H, \qquad (1)$$

Analogously, we expect that C_2 and C_3 can react with unsaturated hydrocarbons in a similar way through bound intermediates:

$$C_2 + C_nH_m \rightarrow [C_{n+2}H_m]^* \rightarrow C_{n+2}H_{m-1} + H, \qquad (2)$$

$$C_3 + C_nH_m \rightarrow [C_{n+3}H_m]^* \rightarrow C_{n+3}H_{m-1} + H. \qquad (3)$$

The occurrence of such reaction channels has been considered recently and investigated in crossed beam experiments with mass spectrometric detection.[24] The aim was to establish whether the formation of $C_{n+2}H_{m-1}$ + H and $C_{n+3}H_{m-1}$ + H is an open reactive channel or not. In order to elucidate the nature of the primary products, it is important to perform such experiments under single collision conditions. Thus in a binary reaction C_k ($k = 2, 3$) + $C_nH_m \rightarrow [C_{n+k}H_m]^* \rightarrow C_{n+k}H_{m-1}$ + H, one carbon cluster C_k reacts with only one hydrocarbon molecule C_nH_m without collisional stabilization of the $[C_{n+k}H_m]^*$ complex(es) or successive reaction of the nascent reaction products. As implied from the general schemes (2) and (3), the primary products are highly hydrogen-deficient carbon clusters. These are extremely reactive and their spectroscopic properties are often unknown. Therefore, most of the combustion-relevant radicals are difficult to probe by optical detection methods. Resorting to a 'universal' detector is crucial in experiments when the nature itself of the product is obscure.

The crossed molecular beam technique with mass spectrometric detection has been established as a powerful technique to achieve these requirements and to observe radical product formation under well-characterized experimental conditions in the gas phase. Since investigations are performed at the molecular, microscopic level in a collision free environment—where it is possible to observe the consequences of a single reactive event—this approach provides a complete insight into the reaction mechanism as the nature of the primary reaction products can be inferred. When the products are polyatomic molecules, the crossed beam technique with mass spectrometric detection has proved to be essential in identifying the relevant reaction pathways.[25] In fact, when distinct structural isomers—molecules with the same chemical formula but different arrangements of atoms—might be formed, knowledge of chemical reaction dynamics is crucial in order to assign the isomers produced.

The first reactions we have investigated, involved C_2/C_3 and acetylene, C_2H_2, ethylene, C_2H_4, and methylacetylene, CH_3CCH. In order to assign the reaction mechanisms and products explicitly, the crossed molecular beam experiments were combined with high level *ab initio* calculations. The later yield structures and energies of possible intermediate collision complexes as well as reaction energies.[26] In this paper, we present experimental and theoretical results on the reaction of $C_2(X\,^1\Sigma_g^+, a\,^3\Pi_u)$ with C_2H_4 and report preliminary results on the reaction of $C_3(X\,^1\Sigma_g^+)$ with C_2H_4. We also discuss other possible formation routes to the tricarbon molecule, *i.e.* the reaction of $C(^3P_j)$ with acetylene, and report on astrophysical and combustion implications of the reactions investigated.

2. Experimental and data analysis

All reactive scattering experiments were performed by using the 35″ universal crossed molecular beam apparatus with mass spectrometric detection.[27] This machine consists of two source chambers fixed at 90° crossing angle, a stainless steel scattering chamber, and a ultra-high-vacuum (8×10^{-13} mbar) triply differentially pumped quadrupole mass spectrometric detector, *cf.* Fig. 1. The scattering chamber is evacuated by two oil free, magnetically suspended turbo pumps to a pressure in the range of 10^{-7} mbar. In the primary source, a pulsed supersonic beam of $C_2(X\,^1\Sigma_g^+/a\,^3\Pi_u)$ and $C_3(X\,^1\Sigma_g^+)$ is generated by laser ablation of a graphite rod at 266 nm and subsequent seeding of the liberated species in helium carrier gas.[28] Since dicarbon and tricarbon are simultaneously produced in the primary beam, in principle both species might react. However, this does not complicate the present experiments since at lower collision energies the reactions of $C_3(X\,^1\Sigma_g^+)$ cannot occur (they are either endothermic or have high entrance barriers, *cf.* Section 4). Consequently, the observed signals are attributable to C_2 reactions. On the other hand, when studying C_3 reactions, the H-displacement channels lead to products larger by 12 u than those

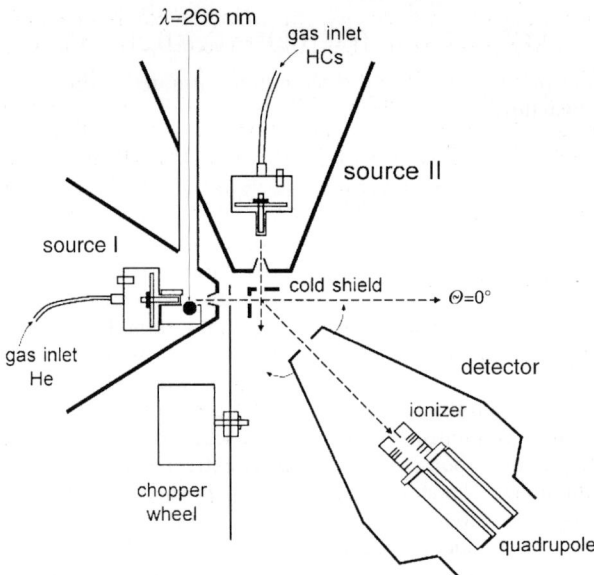

Fig. 1 Schematic top view of the 35″ crossed molecular beams machine with source chambers, chopper wheel, and interaction region.

produced by C_2 reactions. Also, the relative contributions of C_2 vs. C_3 can be controlled by varying the laser output carefully between 20–95 mJ pulse^{-1}, adjusting the focus diameter, and changing the delay time between the pulsed valve and the laser trigger. The primary beam passes through a skimmer into the main chamber. A four-slot chopper wheel is located after the skimmer and rotates at 240 Hz. It selects a 9 μs slice of well-defined velocity and speed ratio from the primary beam which reaches the interaction region. At this point, C_2 and C_3 collide at a right angle with the second pulsed beam of the unsaturated hydrocarbon at distinct collision energies selected between 18 and 120 kJ mol^{-1}.

The reactively scattered species are analyzed at different laboratory angles, Θ, and at distinct mass-to-charge ratios, m/z, in the time-of-flight (TOF) mode. The intensity of an ion is recorded at a certain m/z vs. the flight time to yield a TOF spectrum at the laboratory angle Θ; the pulse of a photodiode attached to the chopper wheel is taken as a well-defined time zero for the experiments. We can integrate the TOF spectra taken at different laboratory angles and thus obtain the angular distribution of the product in the laboratory frame (LAB). The detector consists of a liquid-nitrogen-cooled electron impact ionizer followed by a quadrupole mass selector, and a Daly type ion detector; the ionizer is located in innermost part of a triply differentially pumped ultra-high-vacuum chamber. The whole detector unit can be rotated in the collision plane defined by the two intersecting beams. Despite the triply differential pumping set-up, molecules desorbing from wall surfaces lying on a straight line to the electron impact ionizer cannot be avoided. Their mean free path is in the order of 10^3 m compared to maximum detector dimensions of about 1 m. To reduce this background, a copper plate attached to a two stage closed cycle helium refrigerator is placed right before the collision center and cooled to 4.5 K. Thus the ionizer views a cooled surface which traps all the species with the exception of molecular hydrogen and helium.

For the physical interpretation of the reactive scattering data, it is necessary to transform the laboratory data into the center-of-mass (CM) system.[9] Here, we employ a forward-convolution routine to fit the TOF spectra $N(\Theta,t)$ at different laboratory angles and the LAB distribution. This procedure initially guesses the angular flux distribution, $T(\theta)$, and the translational energy flux distribution, $P(E_T)$, in the CM frame assuming mutual independence (θ is the scattering angle in the CM system measured with respect to the primary beam direction taken as $\theta = 0°$ and E_T is the CM product translational energy). Then, TOF spectra and LAB distribution are calculated from these $T(\theta)$ and $P(E_T)$ by a routine which takes into account the velocity and angular spread of both beams, the detector acceptance angle, and the ionizer length. Both $T(\theta)$ and $P(E_T)$ are refined until a satisfying fit of the experimental data is achieved.[29] The ultimate output data of our experiments is the generation of a product flux contour map which reports the differential cross section (the intensity of the reactively scattered products), $I(\theta,u) = P(u) \times T(\theta)$, as the intensity as a function of angle θ and product center-of-mass velocity u. The limiting circle is the maximum translational energy releases of the fragment if we assume that all the available energy channels only into the translational degrees of freedom of the products. The contour map serves as an image of the reaction and contains all the information of the scattering process.

3. The computational approach

All *ab initio* calculations on the singlet C_4H_4, triplet C_4H_4 and singlet C_5H_4 PESs were carried out using the G2M(RCC,MP2) method,[30] which approximates the RCCSD(T)/6-311+G(3df,2p) energy.[31] The geometries of the reactants, reaction intermediates, transition states, and products were optimized at the density functional B3LYP/6-311G** level.[32] Vibrational frequencies calculated at this level were used for characterization of stationary points as minima and transition states, for zero-point energy (ZPE) corrections, and for RRKM calculations of reaction rate constants. The GAUSSIAN 94,[33] MOLPRO 98,[34] ACES-II,[35] and DALTON[36] *ab initio* program packages were employed.

Regarding the C_2/C_2H_2 system, geometries were optimized at UB3LYP/6-311+G** using the GAUSSIAN 98 program.[37] Each structure was confirmed as a minimum by frequency calculations at the same level and the unscaled UB3LYP/6-311+G** zero-point energies (ZPE) were used to correct the energies. Single point CCSD(T)/cc-pVTZ energies for the reactants and products were obtained using the ACES II program.[38]

4. Results and discussion

4.1 The $C_2 + C_2H_4$ reaction

We have performed reactive scattering experiments at two different collision energies, $E_c = 14.7$ and 28.9 kJ mol^{-1}, by selecting different slices of the pulsed beams characterized by a peak velocity of 1200 ± 8 m s^{-1} and 1940 ± 15 m s^{-1} and a speed ratio of 9.0 ± 0.3 and 6.2 ± 0.2, respectively. At both collision energies, signals were observed both at $m/z = 51$ ($C_4H_3^+$) and $m/z = 50$ ($C_4H_2^+$). However, since the TOF spectra at the two mass-to-charge ratios are superimposable, we conclude that under our experimental conditions the only product in this range of masses is C_4H_3 and that the ion $C_4H_2^+$ is actually formed in the ionizer by dissociative ionization. Fig. 2 displays selected TOF spectra recorded at the two collision energies, while Fig. 3 reports the laboratory angular distribution measured at the highest collision energy. The solid lines superimposed on the experimental distributions are the calculated curves when using the best-fit CM functions shown in Fig. 4 and 5. At these collision energies, tricarbon does not react with ethylene (*cf.* Section 4.2), and therefore cannot contribute to the observed signal. At all the collision energies investigated, the product CM angular distributions are symmetric with respect to $\theta = 90°$; the best-fit angular distributions are reported for the lowest and the highest energy experiments in Fig. 5. The observed symmetry suggests that the reaction involves bound C_4H_4 intermediate(s) which has(have) a lifetime longer than its(their) rotational period. Interestingly, if we examine the product translational energy distributions reported in Fig. 4, we can see how at the lowest E_c the $P(E_T)$ peaks very close to zero translational energy, suggesting that the C_4H_4 complex dissociates to $C_4H_3 + H$ without exit barrier; at the higher collision energy a broad plateau from 0 to 45.3 kJ mol^{-1} is clearly visible. The dramatic change in shape might well imply the onset of a second reaction channel with the increase of the available energy. The tails of both distributions extend up to 159–201 and 151–210 kJ mol^{-1}, respectively.

The *ab initio* electronic structure calculations of the singlet potential energy surface (PES) show that $C_2(^1\Sigma_g^+)$ attacks the π-bond of C_2H_4 without an entrance barrier (Fig. 6), forming a three-member ring intermediate **i1**. In contrast, the initial addition of $C_2(^3\Pi_u)$ to the π-system of ethylene on the triplet surface involves a barrier of less then 15.5 kJ mol^{-1} calculated at the G2M(RCC,MP2) level with MP2/6-311G** optimized geometry (this theoretical value probably is

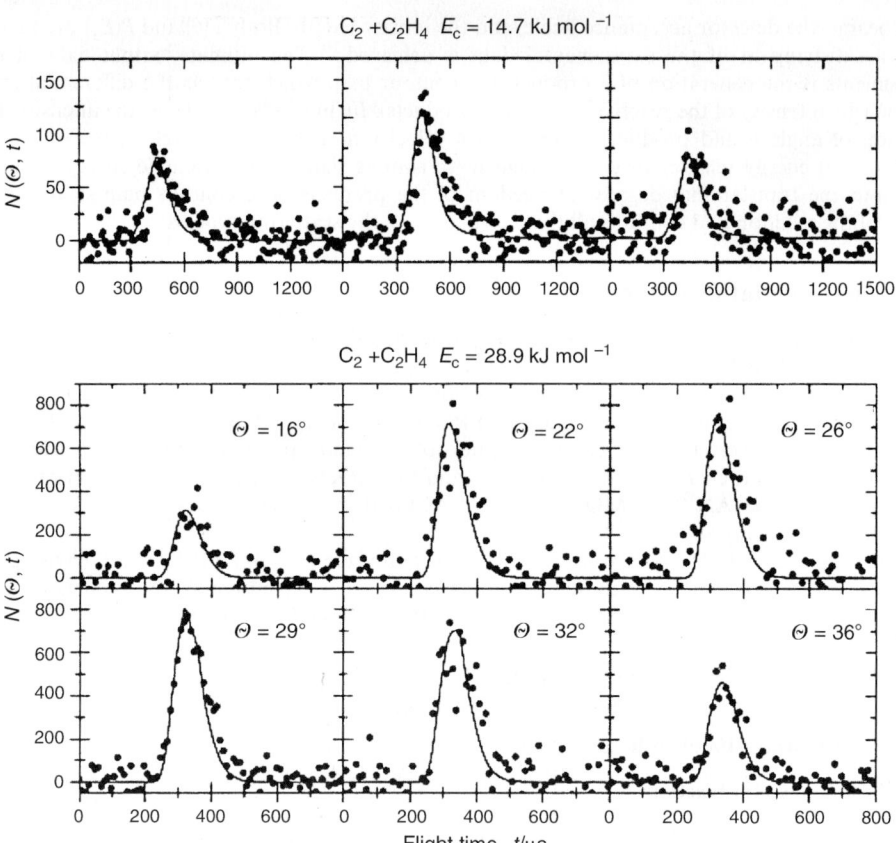

Fig. 2 Time-of-flight data of distinct laboratory angles of the reaction of dicarbon with ethylene to HCCCCH$_2$ + H at a collision energy of 28.9 kJ mol^{-1}. The dots indicate the experimental data, the solid lines the calculated fit.

overestimated since an activation barrier <4 kJ mol^{-1} has been estimated from kinetic measurements). This result is in contrast with earlier lower level HF/3-21G calculations which indicated that the addition of C$_2$($^3\Pi_u$) to ethylene occurs without a barrier. The singlet C$_4$H$_4$ PES reveals how the initially formed **i1** complex (bound by 359 kJ mol^{-1} with respect to the reactants)

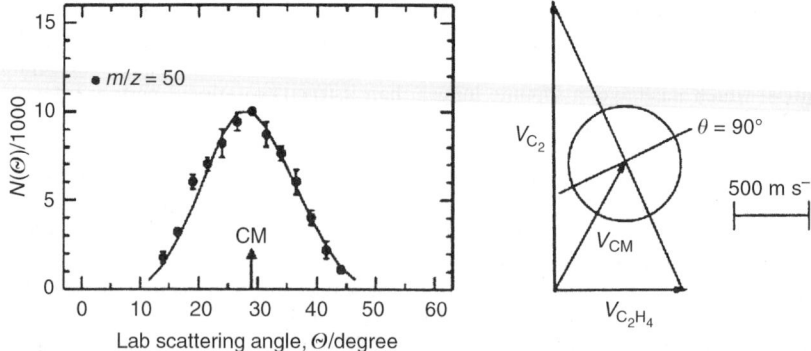

Fig. 3 Laboratory angular distribution of product channel at m/z = 50. Circles and 1σ error bars indicate experimental data, the solid lines the calculated distributions.

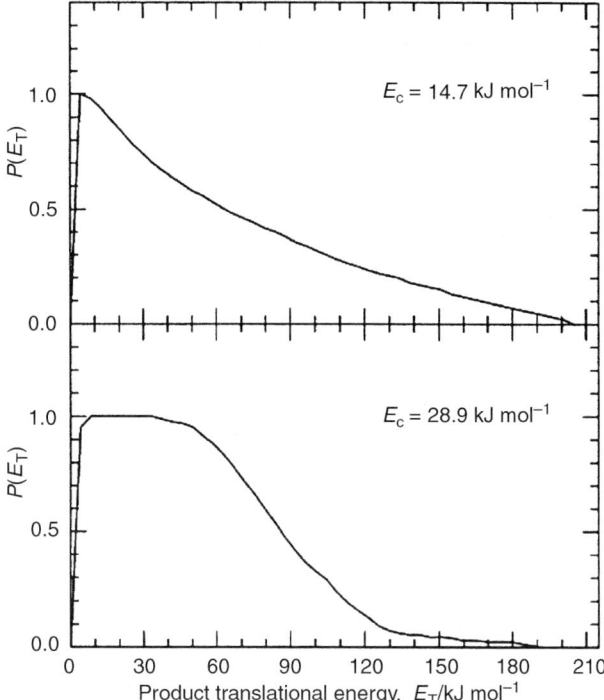

Fig. 4 Center-of-mass translational energy flux distributions at 14.7 kJ mol^{-1} (top) and 28.9 kJ mol^{-1} (bottom) of the reaction of C$_2$ with ethylene to C$_4$H$_3$ and atomic hydrogen.

might lose a hydrogen atom to **p4** or isomerizes to butatriene **i2**. The **i1** → **i2** isomerization has a relatively low 59.5 kJ mol^{-1} barrier and is exothermic by 197 kJ mol^{-1}. Intrinsic reaction coordinate (IRC) calculations at the B3LYP/6-311G** level confirmed that the transition state actually connects **i1** and **i2**. Butatriene can eliminate a H atom to form the major reaction products, the HCCCCH$_2$ radical (**p1**) + H, without an exit barrier. The total reaction exothermicity to form **p1** + H from the reactants is calculated to be 156.3 kJ mol^{-1}.

The reaction of the C$_2$ triplet with ethylene is quite complex (Fig. 7). The addition of C$_2$($^3\Pi_u$) to the ethylene π-system forms intermediate **i1** bound by 127 kJ mol^{-1} with respect to the reactants. The initial intermediate can undergo ring closure to form a three-member ring species **i2**, which in turn would rearrange to a branched isomer **i4**, and then to another three-member cyclic intermediate, **i5**. Ring-opening of **i5** gives a linear triplet butatriene **i6** (391 kJ mol^{-1} below the reactants). **i6** loses a hydrogen atom to produce n-C$_4$H$_3$ **p1** with an overall exothermicity of 165 kJ mol^{-1}. The highest barrier on this pathway is located between **i2** and **i4** at 94 kJ mol^{-1} above **i2** and 33 kJ mol^{-1} below the reactants. A second route from **i1** to **i6** involves the four-member ring isomer **i3**, triplet cyclobutyne. The ring closure from **i1** to **i3** is characterized by a barrier of 63 kJ mol^{-1} and the corresponding transition state is 63.7 kJ mol^{-1} lower in energy than the reactants. Subsequently, **i3** undergoes ring opening along the CH$_2$–CH$_2$ bond to yield **i6** with a barrier of 86 kJ mol^{-1}. Both pathways leading from **i1** to **i6** are expected to compete. Therefore, we conclude that like C$_2$(X $^1\Sigma_g^+$) + C$_2$H$_4$, n-C$_4$H$_3$ (**p1**) is the major C$_4$H$_3$ product also for the C$_2$($^3\Pi_u$) + C$_2$H$_4$ reaction.

The trend of our experimental data with increasing collision energy provides clear evidence of the involvement of a second reaction mechanism. Even though a spectroscopic characterization of the beam is needed in order to establish its exact ratio of triplet *vs.* singlet dicarbon, we expect that both the ground and the first electronically excited states of C$_2$ are present in the fast part of the beam because of the very small energy of the excited state. Therefore we can rationalize our observations in the light of the *ab initio* calculations of the singlet and triplet PES assuming

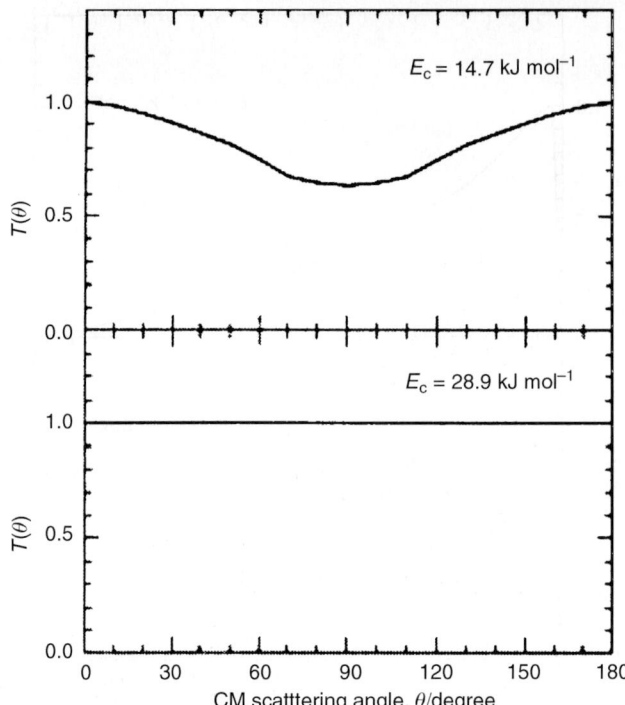

Fig. 5 Center-of-mass angular distributions at 14.7 kJ mol^{-1} (top) and 28.9 kJ mol^{-1} (bottom) of the reaction of C$_2$ with ethylene to C$_4$H$_3$ and atomic hydrogen.

that (i) at the lower collision energy we have a contribution to the recorded signal only from the C$_2$($^1\Sigma_g^+$) reaction and that (ii) an appreciable contribution from the triplet state reaction becomes evident only at the higher collision energy. In fact, although at this stage we do not know the exact barrier height for the triplet reaction, we expect an appreciable contribution from the triplet state reaction only when the collision energy is significantly larger than the reaction barrier. Hence, we can claim that the results of the low energy experiment are attributable for the most

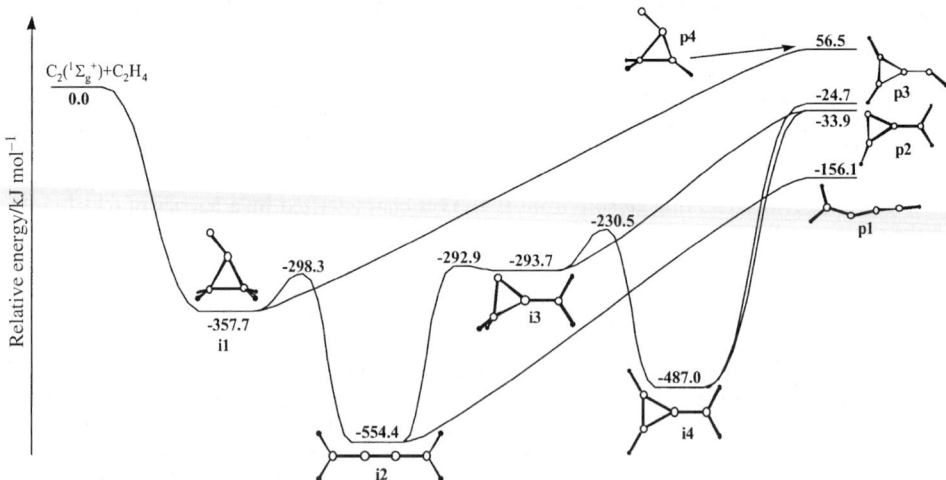

Fig. 6 Potential energy surface involved in the reaction of C$_2$(X $^1\Sigma_g^+$) with ethylene.

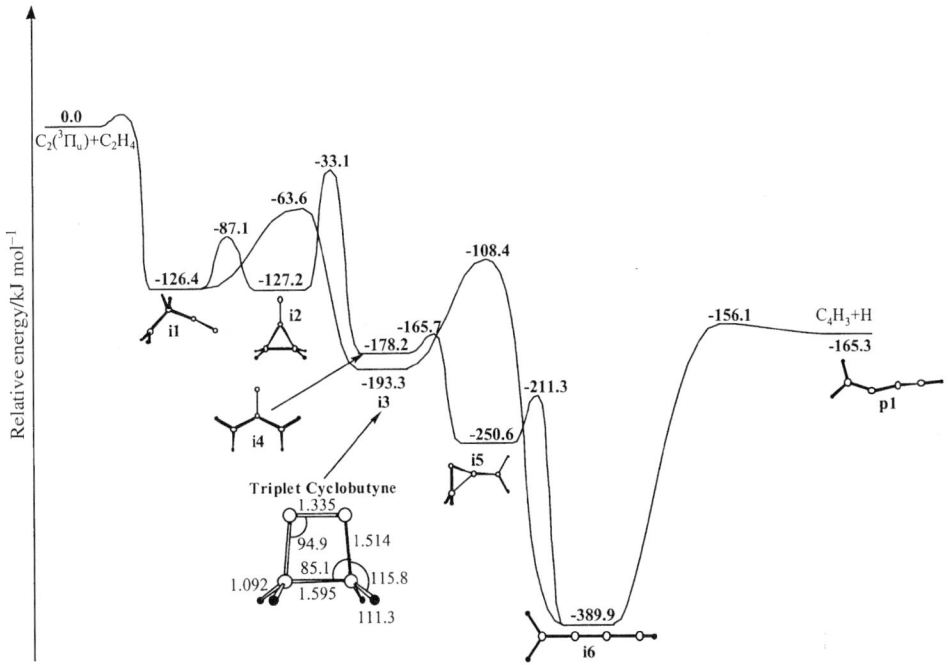

Fig. 7 Potential energy surface involved in the reaction of $C_2(a\ ^3\Pi_u)$ with ethylene.

part to the $C_2(^1\Sigma_g^+)$ reaction. This assumption is supported strongly by our experimental data and *ab initio* calculations. As visible in Fig. 4, the $P(E_T)$ peaks at values of translational energy close to zero, thus mirroring the absence of any exit barrier as predicted by the calculations. Also, if we subtract the collision energy from the $P(E_T)$ energy maximum, we derive an exothermicity of 138–180 kJ mol^{-1}, which conforms with the *ab initio* value computed for this reaction channel. In addition, the decomposing butatriene complex resides in a deep potential energy well; this can well account for the observation of a backward–forward symmetric CM angular distribution. Also the assumption that the second mechanism is attributable to the $C_2(^3\Pi_u)$ reaction gains full support from the comparison of experimental results and theoretical calculations. That the second mechanism leading to C_4H_3 product has about the same reaction exothermicity as the first one, definitely rules out the occurrence of reaction channels leading to other isomers from the $C_2(^1\Sigma_g^+)$ reactions: the reaction enthalpies of the channels leading to **p2**, **p3** and **p4** are very different, see Fig. 6. Also, the peak of the translational energy distribution is quite broad and extends to 46 kJ mol^{-1} for the higher energy experiment, while in the lower energy experiment the peak is close to 0 kJ mol^{-1}. This change in the peak shape can be rationalized if the second channel has a (modest) exit barrier, as appears to be the case for the triplet reaction (see Fig. 7). In conclusion, the $C_2(^3\Pi_u)$ reaction on the triplet surface at high collision energy can well account for all the experimental facts.

4.2 The C_3–C_2H_4 system

The experimental data analysis is still in progress and only some findings are presented. Interestingly, the C_3–H exchange channel which forms products of $m/z = 63$ ($C_5H_3^+$) was found to occur, but was open only at collision energies larger than 40–42 kJ mol^{-1} in strong contrast to the $C_2 + C_2H_4$ reaction (which yields the products $C_4H_3 + H$ at lower collision energies). It is interesting to compare these results with the potential energy surface involved, *cf.* Fig. 8, as derived from our *ab initio* calculations. Similarly to the reaction of dicarbon with ethylene, the tricarbon molecule attacks the π-bond of the olefin to form a three-membered (**i1**) or a five-membered ring **i3** *via* addition of one or two terminal carbon atoms, respectively. Both structures

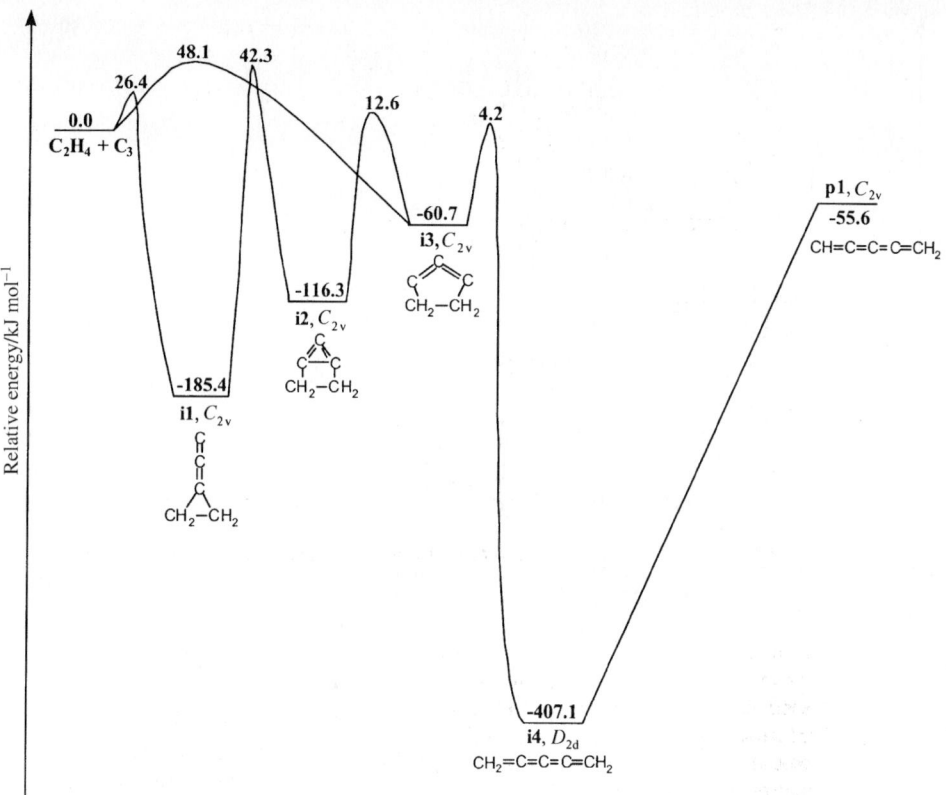

Fig. 8 Potential energy surface involved in the reaction of $C_3(X\,^1\Sigma_g^+)$ with ethylene.

are stabilized by 185.7 and 61 kJ mol^{-1}, respectively, and are connected through a bicyclic C_5H_4 isomer **i2**; the latter resides in a potential energy well of 116.5 kJ mol^{-1}. However, in strong contrast to the C_2 reaction, the C_3 addition involves entrance barriers of 26.4 and 48.2 kJ mol^{-1} to form **i1** and **i3**, respectively, as both processes are formally symmetry forbidden. This symmetry-imposed barrier was verified in previous bulk experiments which derived a value of 26.6 kJ mol^{-1}.[12] Interestingly, TOF and angular distributions measured at $m/z = 63$ and 62 are slightly different and that seems to imply that an H_2 elimination channel leading to C_5H_2 isomer(s) is open. A more accurate analysis of the raw data will shed light on this suggestion.

4.3 The C_2–C_2H_2, C_3–C_2H_2, C_2–C_3H_4 and C_3–C_3H_4 systems

The data analysis and electronic structure calculations of these reactions are still ongoing. We report only some observations here. Crossed beam experiments of C_2 with acetylene and with methylacetylene have resulted in the observation of the C_2 vs. H atom exchange pathway (forming C_4H ($m/z = 49$) and C_5H_3 ($m/z = 63$)) at collision energies as low as 8 kJ mol^{-1} and as high as 39 kJ mol^{-1}. A detailed study of the reaction with partially deuterated CD_3CCH showed that only the D atom is released (signal at $m/z = 65$, $C_5D_2H^+$); no H atom elimination at $m/z = 66$ ($C_5D_3^+$) could be detected. Both C_3 systems show reactive scattering signal of the C_3 vs. H atom exchange yielding C_5H ($m/z = 61$) and C_6H_3 ($m/z = 75$), only when the collision energy was higher than 80 ± 9 kJ mol^{-1} (C_3–C_2H_2) and 45 ± 5 kJ mol^{-1} (C_3–C_3H_4). Like the $C_2 + CH_3CCH$ system, the hydrogen atom was only emitted from the methyl group as in the $C_3 + CH_3CCH$ reaction. These findings, at this stage, suggest the possible reaction pathways sketched in Figs. 9 and 10, but these mechanisms are subject to further investigations.

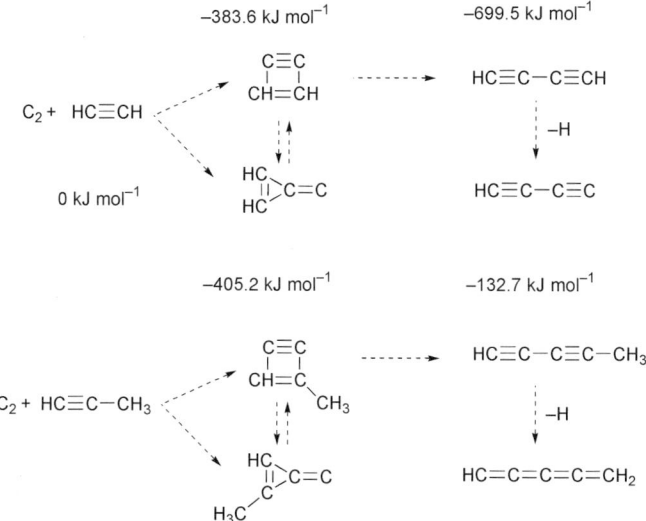

Fig. 9 Schematic pathways involved in the reaction of dicarbon with acetylene and methylacetylene. Energies for the C_2–C_2H_2 system are given with respect to the separated reactants.

4.4 The $C(^3P_j)$–C_2H_2 system

The reaction between ground state carbon atoms, $C(^3P_j)$, and acetylene, $C_2H_2(X\ ^1\Sigma_g^+)$, was studied previously at three collision energies between 8.8 and 45.0 kJ mol^{-1} using the crossed molecular beams technique. Product angular distributions and time-of-flight spectra of C_3H at $m/z = 37$ and 36 were recorded. Reaction dynamics inferred from the experimental data and *ab initio* calculations on the triplet C_3H_2 and doublet C_3H potential energy surface suggest two microchannels initiated by addition of $C(^3P_j)$ either to one acetylenic carbon to form *s-trans*-propenediylidene or to two carbon atoms to yield triplet cyclopropenylidene *via* loose transition states located at their centrifugal barriers. Propenediylidene rotates around its B/C axis and a [2,3]-H-migration leads to propargylene. This is followed by C–H-bond cleavage *via* a symmetric

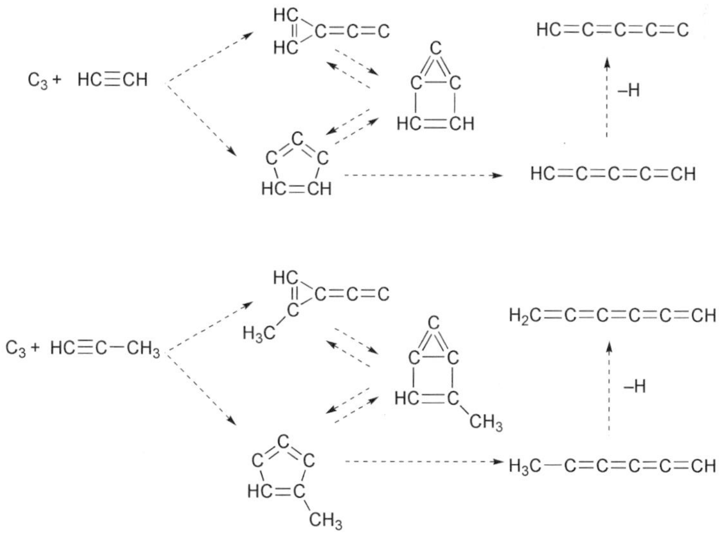

Fig. 10 Schematic pathways involved in the reaction of tricarbon with acetylene and methylacetylene.

exit transition state to give l-C$_3$H(X $^2\Pi$) and H. Direct stripping dynamics contribute to the forward-scattered, second microchannel to yield c-C$_3$H(X ^{2}B$_2$) and H. This contribution is quenched with rising collision energy. TOFs of reactively scattered signal at $m/z = 36$ were found to be superimposable with those recorded at $m/z = 37$ suggesting that signal at $m/z = 36$ originates from cracking of $m/z = 37$ in the ionizer. However, the performance and signal to noise ratio of this experimental set-up was limited by the fact that diffusion pumps and oil lubricated roughing pumps were employed. Casavecchia et al.[39] suggested that C$_3$ might be an additional product of the C(^{3}P$_j$) + C$_2$H$_2$ reaction. However, their experiments employ continuous carbon beams which contain ground state and electronically excited C(^{1}D). Since multiple channel fits to one m/z ratio are not always 'unique', the need to reinvestigate the potential C$_3$($^1\Sigma_g^+$) + H$_2$($^1\Sigma_g^+$) pathway with the improved, 35″ crossed beams machine was stressed.[27]

This experiment was repeated at a collision energy of 16.6 kJ mol^{-1}, under conditions where the beam contains only ground state carbon atoms. TOF spectra were recorded at $m/z = 37$ and $m/z = 36$ at angles between 12° and 72° with respect to the primary beam.[40] Fig. 11 shows the TOF recorded at the center-of-mass angle at 47° at $m/z = 36$; the data accumulation time was 120 min. Two components are clearly visible, i.e. a slow peak which arises from fragmentation of $m/z = 37$, and a fast contribution which could not be fit with the CM functions employed for the forward-convolution of the $m/z = 37$ data. The best fit CM translational energy distributions of the fast contribution is shown in Fig. 12, incorporating products of 36 and 2 u, i.e. the channel C$_3$($^1\Sigma_g^+$) + H$_2$($^1\Sigma_g^+$), in the fitting routine. The TOFs were fit with a symmetric, slightly sideways peaking $T(\theta)$. Within the error limits, acceptable fits were achieved with intensity ratios $I(90°)/I(0°) = 1.3–1.1$; an isotropic distribution yields a slightly worse fit. These findings suggest that the decomposing complex is either long-lived or 'symmetric'. Further, the $P(E_T)$ peaks between 70 and 100 kJ mol^{-1} indicating that the reaction involves a tight exit transition state. If we subtract the collision energy from the high energy cut-off of the $P(E_T)$ within the error limits, the reaction is found to be exothermic by 140–150 kJ mol^{-1}. These data are in good agreement with a reaction energy of 130–135 kJ mol^{-1} calculated from thermodynamical data.[41] The reaction from the triplet manifold to form C$_3$($^1\Sigma_g^+$) and molecular hydrogen is spin-forbidden, and intersystem crossing (ISC) from the singlet manifold is involved. A recent investigation of C$_3$H$_2$ isomers suggested that triplet propargylene, HCCCH, can indeed undergo ISC to the singlet surface,[42] followed by loss of molecular hydrogen from singlet propargylene or—after isomerization—from vinylidenecarbene via exit transition states well above the separated products.[43] This finding of a tight exit transition state correlates nicely with the translational energy distribution peaking well away from zero. Summarizing, we conclude that the C$_3$($^1\Sigma_g^+$) + H$_2$($^1\Sigma_g^+$) pathway is open.

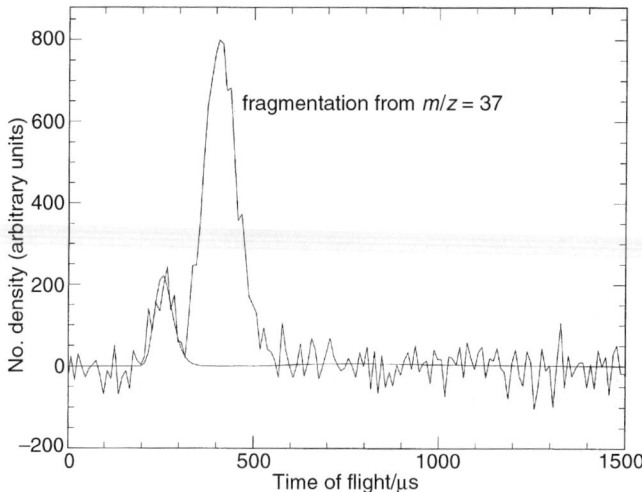

Fig. 11 Time-of-flight data at $m/z = 36$ of the reaction of atomic carbon with acetylene recorded at a laboratory angle of 47°. The smooth line represents the best fit.

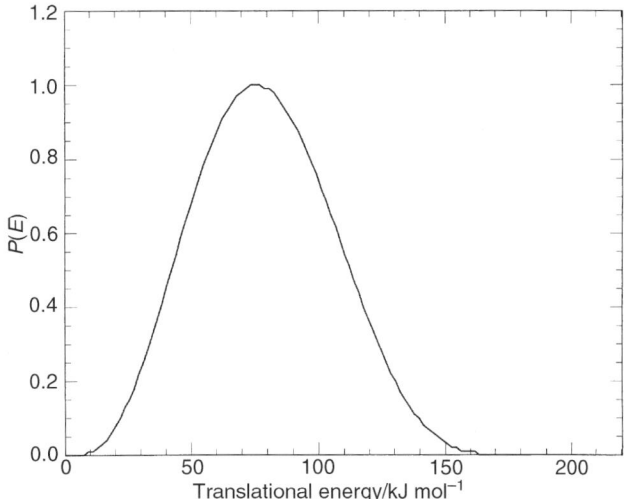

Fig. 12 Center-of-mass translational energy distribution of the reaction $C(^3P_j) + C_2H_2(^1\Sigma_g^+) \rightarrow C_3(^1\Sigma_g^+) + H_2(^1\Sigma_g^+)$.

This finding of an ISC to the singlet surface has far reaching consequences for the $C(^3P_j)$–C_2H_2 system as our previous interpretation focused solely on the triplet surface. Therefore, a more detailed interpretation of the experimental data including ISC is necessary. Further, this finding makes it interesting to investigate similar, substituted triplet propargylene systems, *cf.* Fig. 13. As triplet methylpropargylene and dimethylpropargylene were found to be the decomposing complexes in the reactions of atomic carbon with methylacetylene and dimethylacetylene, respectively, a possible ISC might be followed by CH_4 and C_2H_6 elimination *via* five-membered ring transition

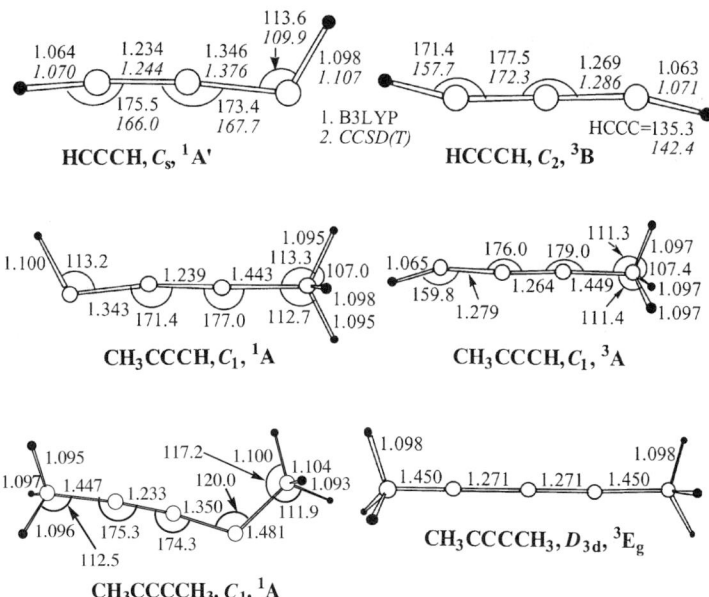

Fig. 13 Triplet (right) and singlet (left) structures of propargylene (top), methylpropargylene (middle) and dimethylpropargylene (bottom).

states from singlet methylpropargylene and dimethylpropargylene. Interestingly, all triplet propargylene structures were found to be lower in energy than the singlet species, and the singlet–triplet splitting was found to be 52.8 kJ mol^{-1} (progargylene), 33.5 kJ mol^{-1} (methylpropargylene), and 38.1 kJ mol^{-1} (dimethylpropargylene). The pathways of a potential methane and ethane elimination are subject to further investigations.

5 Conclusions

Our combined crossed molecular beam experiments and *ab initio* calculations on the reactions of dicarbon and tricarbon molecules with three unsaturated hydrocarbons, acetylene, methylacetylene, and ethylene, are governed by an initial addition of C_2/C_3 to the π bonds. Under single collision condition, highly unsaturated cyclic structures are formed. These intermediates are interconnected *via* various transition states and are suggested to ring open to HCCCCH ($C_2 + C_2H_2$), HCCCCCH ($C_3 + C_2H_2$), HCCCCCH$_3$ ($C_2 + CH_3CCH$), HCCCCCCH$_3$ ($C_3 + CH_3CCH$), H$_2$CCCCH$_2$ ($C_2 + C_2H_4$) and H$_2$CCCCCH$_2$ ($C_3 + C_2H_4$). The chain isomers were found to decompose predominantly by atomic hydrogen loss yielding C_4H, C_5H, HCCCCH$_2$, HCCCCCH$_3$, H$_2$CCCCH and H$_2$CCCCCH. The role of possible cyclic reaction products and potential H$_2$ elimination pathways are currently under investigation. Our preliminary data suggest further that all reactions of $C_2(^1\Sigma_g^+)$ most likely have no entrance barriers and are exothermic. This is in strong contrast to bimolecular reactions of the tricarbon molecule $C_3(^1\Sigma_g^+)$ as these reactions have characteristic thresholds between 40 and 85 kJ mol^{-1}.

These findings have strong implications with respect to combustion processes. First, the explicit identification of the hydrogen replacement by C_2/C_3 demonstrates that hydrogen-deficient hydrocarbon molecules can be formed in combustion environments. We identified at least six distinct reaction products and more than 24 potential reaction intermediates. Although under our experimental single collision conditions the intermediates involved cannot be stabilized *via* a third body reaction, the conditions are more complex in actual combustion processes. Here, number densities are typically in the order of 10^{17} cm^{-3} at temperatures between 1000 and 2000 K, and ternary reactions might stabilized the reaction intermediates. The elevated temperatures open further reaction pathways which, like the elementary reactions of $C_3(^1\Sigma_g^+)$, either are endothermic and/or involve a significant entrance barrier. Although the full data analyses are still continuing, our investigations demonstrated the capability of reactions of small carbon molecules with unsaturated hydrocarbons to synthesize hydrogen-deficient molecules in combustion flames. This is a first step towards building a systematic database of intermediates and reaction products involved in this important reaction class. These should be included in future models of combustion flames as well as PAH and soot formation.

Although this Faraday Discussion focuses on combustion processes, we wish to briefly address the astrophysical implications of this research. In fact, more than once, combustion chemists and astrophysicists have employed similar reaction networks to explain the chemistry of these environments; multi-component models of carbon cluster growth and the correlation with PAH and soot formation are examples.[44] Several isomers of the carbon chain radicals with the general formula C_nH have been detected in the interstellar medium.[45] Therefore, our experimental and theoretical findings can help in explaining the formation of the C_4H and C_5H radicals as detected towards the cold molecular clouds TMC-1 and the carbon star IRC+10126, respectively. To our best knowledge, no ion–molecule reaction network can explain the formation of these isomers satisfactorily, and the crossed beam results present compelling evidence that neutral–neutral reactions might produce C_4H and C_5H *via* reactions of small carbon clusters with acetylene. While the average translation temperature of species in cold molecular clouds is typically 10 K, the temperatures rise up to 4000 K close to the photosphere of carbon stars. Based on these considerations, reactions of $C_3(^1\Sigma_g^+)$ are prohibited in interstellar dark clouds since the average translational energy cannot overcome the experimentally-determined entrance barriers and/or endothermicities. However, these elementary processes are open in the outflows of carbon stars because of the elevated temperature. In contrast, C_2 reactions were found to be exothermic and barrier-less and should occur even at temperatures as low as 10 K. Our preliminary data on the C_2/C_3 reactions with acetylene suggest that the alternating C_nH and C_mH (n = even, m = odd) abundance in interstellar environments might be a result of distinct chemical reactivity and exothermicity; clearly, more experimental data—especially reactions of small carbon clusters with diacetylene and larger clusters with acetylene and diacetylene—are crucial to obtain more detailed, systematic correlations.

Acknowledgements

RIK is indebted to the Deutsche Forschungsgemeinschaft (DFG) for a Habilitation fellowship (IIC1-Ka1081/3-1). The experimental work was supported by Academia Sinica until September 2000. Hereafter, support from University of York, Department of Chemistry, is acknowledged. AMM thanks the National Science Council for support on the C_2–C_2H_4 and C_3–C_2H_4 calculations. FS greatly acknowledges a DAAD HSP III - Doktoranden fellowship (D/99/22963). Work in Erlangen was further supported by the DFG and the Fond der Chemischen Industrie. Work in Georgia was funded by the US Department of Energy. This work was performed within the International Astrophysics Network.

References

1. A. G. Gordon, *The Spectroscopy of Flames*, Wiley, New York, 1974; A. P. Baronavski and J. R. McDonald, *J. Chem. Phys.*, 1977, **66**, 3300; S. Green, *Annu. Rev. Phys. Chem.*, 1981, **32**, 103; D. C. Clary, *Annu. Rev. Phys. Chem.*, 1990, **41**, 61; S. R. Federman and D. L. Lambert, *Astrophys. J.*, 1988, **328**, 777; H. W. Kroto and K. McKay, *Nature*, 1988, **331**, 328.
2. M. R. Combi and U. Fink, *Astrophys. J.*, 1997, **484**, 879; K. S. K. Swamy, *Astrophys. J.*, 1997, **481**, 1004; P. Rousselot, C. Laffont, G. Moreels and J. Clairemidi, *Astron. Astrophys.*, 1998, **335**, 765.
3. A. P. Baranovski and J. R. McDonald, *J. Phys. Chem.*, 1977, **66**, 3300; A. G. Gaydon and H. G. Wolfhard, *Flames*, Chapman and Hall, New York, 1979; W. Weltner and R. J. van Zee, *Chem. Rev.*, 1989, **89**, 1713.
4. R. Gredel, *Astron. Astrophys.*, 1999, **351**, 657; I. A. Crawford, *Mon. Not. R. Astron. Soc.*, 1997, **290**, 41.
5. S. B. Yorka, *Astrophys. J.*, 1983, **88**, 1816.
6. I. A. Crawford and M. J. Barlow, *Mon. Not. R. Astron. Soc.*, 2000, **311**, 370.
7. W. Huggins, *Proc. R. Soc. London*, 1882, **33**, 1.
8. A. E. Douglas, *Astrophys. J.*, 1951, **114**, 446; H. H. Nelson, H. Helvajian, L. Pasternack and J. R. McDonald, *Chem. Phys.*, 1982, **73**, 431.
9. R. G. Norrish, G. Porter and B. A. Thrush, *Proc. R. Soc. London, Ser. A*, 1953, **216**, 165.
10. L. M. Haffner and D. M. Meyer, *Astrophys. J.*, 1995, **453**, 450.
11. Y. T. Lin, R. K. Mishra and S. L. Lee, *J. Phys. Chem. B*, 1999, **103**, 3151; N. Balucani et al., *J. Chem. Phys.*, 1991, **94**, 8611; M. Alagia et al., *J. Chem. Soc., Faraday Trans.*, 1995, **91**, 575; M. Alagia et al., *Faraday Discuss.*, 1999, **113**, 133.
12. W. M. Pitts, L. Pasternack and J. R. McDonald, *Chem. Phys.*, 1982, **68**, 417; L. Pasternack, W. M. Pitts and J. R. McDonald, *Chem. Phys.*, 1981, **57**, 19; V. M. Donnely and L. Pasternack, *Chem. Phys.*, 1979, **39**, 427; L. Pasternack and J. R. McDonald, *Chem. Phys.*, 1979, **43**, 173; H. Reisler, M. Mangir and C. Wittig, *J. Chem. Phys.*, 1979, **71**, 2109; K. H. Becker, B. Donner, H. Geiger, F. Schmidt and P. Wiesen, *Z. Phys. Chem.*, 2000, **214**, 503; M. Martin, *J. Photochem. Photobiol. A*, 1992, **66**, 263.
13. P. S. Skell, L. M. Jackman, S. Ahmed, M. L. McKee and P. B. Shevlin, *J. Am. Chem. Soc.*, 1989, **111**, 4422; G. H. Jeong, K. J. Klabunde, O. G. Pan, G. C. Paul and P. B. Shevlin, *J. Am. Chem. Soc.*, 1989, **111**, 8784.
14. R. I. Kaiser, C. Ochsenfeld, M. Head-Gordon, Y. T. Lee and A. G. Suits, *Science*, 1996, **274**, 1508.
15. R. I. Kaiser, D. Stranges, Y. T. Lee and A. G. Suits, *J. Chem. Phys.*, 1996, **105**, 8721.
16. L. C. L. Huang, H. Y. Lee, A. M. Mebel, S. H. Lin, Y. T. Lee and R. I. Kaiser, *J. Chem. Phys.*, 2000, **113**, 9637.
17. R. I. Kaiser, W. Sun, A. G. Suits and Y. T. Lee, *J. Chem. Phys.*, 1997, **107**, 8713.
18. R. I. Kaiser, Y. T. Lee and A. G. Suits, *J. Chem. Phys.*, 1996, **105**, 8705.
19. R. I. Kaiser, D. Stranges, H. M. Bevsek, Y. T. Lee and A. G. Suits, *J. Chem. Phys.*, 1997, **106**, 4945.
20. N. Balucani, A. M. Mebel, H. Y. Lee, Y. T. Lee and R. I. Kaiser, *J. Chem. Phys.*, 2001, in press.
21. I. Hahndorf, H. Y. Lee, A. M. Mebel, S. H. Lin, Y. T. Lee and R. I. Kaiser, *J. Chem. Phys.*, 2000, **113**, 9622.
22. R. I. Kaiser, A. M. Mebel, A. H. H. Chang, S. H. Lin and Y. T. Lee, *J. Chem. Phys.*, 1999, **110**, 10300.
23. R. I. Kaiser et al., *J. Chem. Phys.*, 1999, **110**, 6091.
24. R. I. Kaiser et al., in *Astrochemistry—From Molecular Clouds to Planetary Systems*, ed. Y. C. Minh and E. F. van Dishoek, Associated Scientific Publishers, San Francisco, 2000, p. 251.
25. P. Casavecchia, *Rep. Progr. Phys.*, 2000, **63**, 355; P. Casavecchia, N. Balucani and G. Volpi, *Annu. Rev. Phys. Chem.*, 1999, **50**, 347.
26. C. Ochsenfeld, R. I. Kaiser, A. G. Suits, Y. T. Lee and M. Head-Gordon, *J. Chem. Phys.*, 1997, **106**, 4141.
27. N. Balucani, O. Asvany, Y. Osamura, L. C. L. Huang, Y. T. Lee and R. I. Kaiser, *Plan. Space Sci.*, 2000, **48**, 447.

28 R. I. Kaiser and A. G. Suits, *Rev. Sci. Instrum.*, 1995, **66**, 5405.
29 R. I. Kaiser, C. Ochsenfeld, D. Stranges, M. Head-Gordon and Y. T. Lee, *Faraday Discuss.*, 1998, **109**, 183.
30 A. M. Mebel, K. Morokuma and M. C. Lin, *J. Chem. Phys.*, 1995, **103**, 7414.
31 G. D. Purvis and R. J. Bartlett, *J. Chem. Phys.*, 1982, **76**, 1910; C. Hampel, K. A. Peterson and H. J. Werner, *Chem. Phys. Lett.*, 1992, **190**, 1; P. J. Knowles, C. Hampel and H. J. Werner, *J. Chem. Phys.*, 1994, **99**, 5219; M. J. O. Deegan and P. J. Knowles, *Chem. Phys. Lett.*, 1994, **227**, 321.
32 A. D. Becke, *J. Chem. Phys.*, 1993, **98**, 5648; C. Lee, W. Yang and R. G. Parr, *Phys. Rev. B*, 1988, **37**, 785.
33 M. J. Frisch *et al.*, *GAUSSIAN 94, Revision E.2*, Gaussian, Inc., Pittsburgh, PA, 1995.
34 *MOLPRO* is a package of *ab initio* programs written by H.-J. Werner and P. J. Knowles, with contributions from J. Almlöf, R. D. Amos, M. J. O. Deegan, S. T. Elbert, C. Hampel, W. Meyer, K. Peterson, R. Pitzer, A. J. Stone, P. R. Taylor and R. Lindh.
35 J. F. Stanton, J. Gauss, J. D. Watts, W. J. Lauderdale and R. J. Bartlett, *ACES-II*, University of Florida, Gainsville, FL.
36 T. Helgaker *et al.*, *DALTON, an ab initio electronic structure program, Release 1.0*, 1997.
37 M. J. Frisch *et al.*, *GAUSSIAN 98, Revision A.7*, Pittsburgh, PA, 1999.
38 J. F. Stanton, J. Gauss, J. D. Watts, M. Nooijen, N. Oliphant, S. A. Perera, P. G. Szaley and R. J. Bartlett, "*ACES II*', Gainesville, FL, 1995.
39 P. Casavecchia, N. Balucani, L. Cartechini, G. Capozza, A. Bergeat and G. G. Volpi, *Faraday Discuss.*, 2001, **119**, (DOI: 10.1039/b102634h).
40 TOF close to the carbon beam showed a significant background from elastically scattered C_3. We redid the $C(^3P_j) + C_2H_4$ reaction as well at a similar collision energy at 18 kJ mol^{-1}, but in this case no H_2 elimination pathway was observed experimentally.
41 *Handbook of Chemistry and Physics*, CRC Press, Boca Raton, 1998.
42 A. M. Mebel, W. M. Jackson, A. H. H. Chang and S. H. Lin, *J. Am. Chem. Soc.*, 1998, **120**, 5741.
43 R. I. Kaiser, A. M. Mebel and Y. T. Lee, *J. Chem. Phys.*, 2001, **114**, 231.
44 G. Pascoli and A. Polleux, *Astron. Astrophys.*, 2000, **359**, 799.
45 M. Ohishi and N. Kaifu, *Faraday Discuss.*, 1998, **109**, 205; M. Guelin, S. Green and P. Thaddeus, *Astrophys. J.*, 1978, **224**, L27; M. Cernicharo, C. Kahane, J. Gomez-Gonzelez and M. Guelin, *Astron. Astrophys.*, 1986, **164**, L1; C. M. Walmsley, P. R. Jewell, L. E. Snyder and G. Winnewisser, *Astron. Astrophys.*, 1984, **134**, L11; J. Cernicharo, M. Guelin, K. M. Menten and C. M. Walmsley, *Astron. Astrophys.*, 1987, **181**, L1; J. Guelin *et al.*, *Astron. Astrophys.*, 1997, **317**, L1; J. Cernicharo and M. Guelin, *Astron. Astrophys.*, 1996, **309**, L27; C. A. Gottlieb, M. C. McCarthy, K. J. Travers and P. J. Thaddeus, *Chem. Phys.*, 1998, **109**, 5433; T. J. Millar and E. Herbst, *Astron. Astrophys.*, 1994, **288**, 561.

Determination of the CH + O₂ product channels

Astrid Bergeat,[a] Teresa Calvo,[a] Françoise Caralp,[a] Jean-Hugues Fillion,[b] Gérard Dorthe[a] and Jean-Christophe Loison*[a]

[a] *Laboratoire de Physico-Chimie Moléculaire, UMR no. 5803, Université Bordeaux I, F-33405 Talence Cedex, France. E-mail: jc@lpcm.u-bordeaux.fr*
[b] *Observatoire de Paris-Meudon, DAMAP, 5 place J. Jansen, F-92195 Meudon Cedex, France*

Received 26th February 2001
First published as an Advance Article on the web 21st September 2001

The multichannel CH + O_2 reaction was studied at room temperature, in a low-pressure fast-flow reactor. CH radical was obtained from the reaction of $CHBr_3$ with potassium atoms. The overall rate constant was determined from the decay of CH with distance, O_2 being introduced in excess. The result, after corrections for axial and radial diffusion, is $k = (3.6 \pm 0.5) \times 10^{-11}$ cm^3 molecule^{-1} s^{-1}. The OH(A $^2\Sigma^+$) chemiluminescence was observed, confirming the existence of the OH + CO channel. The vibrational population distribution of OH(A $^2\Sigma^+$) is 32% in the $v' = 1$ level and 68% in the $v' = 0$ level (± 5%). The relative atomic concentrations were determined by resonance fluorescence in the vacuum ultraviolet. A ratio of 1.4 ± 0.2 was found between the H atom density (H atoms being produced from the H + CO_2 channel and from the HCO dissociation) and the O atom density (O + HCO). *Ab initio* calculations of the transition structures have been performed, associated with statistical estimations. The estimated branching ratios are: O + HCO, 20%; O + H + CO, 30%; H + CO_2, 30%; and CO + OH, 20%.

1. Introduction

The CH radical is one of the most reactive and simplest organic free radicals. It is well known as an important intermediate in various reaction systems in fields from the planetary atmosphere to hydrocarbon combustion.[1] For example, the CH + N_2 reaction producing N + HCN at high temperature is believed to be the source of "prompt" NO in N-containing flames.[2] The CH + NO reaction used to reduce NO emission is believed to play a key role in the overall reburning mechanism.[3] The characteristic OH(A → X) emission observed in hydrocarbon/air flames may arise mainly from the CH + O_2 reaction to produce excited OH. This OH* chemiluminescence is often used to follow and probe the CH radical. However, although numerous previous studies have examined the CH + O_2 reaction kinetics (Table 1), very little is known about possible product channels.[11,15–17] The oxidation of CH is highly exothermic and thus can lead to many thermodynamically accessible product channels with vibrationally or even electronically excited products.

$$CH(X\,^2\Pi) + O_2(X\,^3\Sigma_g^-) \rightarrow H(^2S) + CO_2(X\,^1\Sigma_g^+, A\,^1B_2), \qquad \Delta_rH^\circ_{298} = -8.02 \text{ eV} \qquad (1)$$

$$\rightarrow CO(X\,^1\Sigma^+, a\,^3\Pi) + OH(X\,^2\Pi, A\,^2\Sigma^+), \qquad \Delta_rH^\circ_{298} = -6.92 \text{ eV} \qquad (2)$$

$$\rightarrow O(^3P, ^1D) + HCO(X\,^2A', A\,^2A''), \qquad \Delta_rH^\circ_{298} = -3.16 \text{ eV} \qquad (3)$$

DOI: 10.1039/b101815i

Table 1 Global rate constants of the CH(X $^2\Pi$, $v = 0$) + O_2(X $^3\Sigma_g^-$) reaction, at room temperature

k(CH + O_2) /10^{-11} cm^3 molecule^{-1} s^{-1}	P /Torr	Apparatus[a]	CH radical source[b]	Detection[c]	Ref.
3.3 ± 0.4	30 (Ar)	cell	CH$_3$NH$_2$ IRMPD	OH(A) chem. and CH(A–X) LIF	4
5.9 ± 0.8	100 (Ar)	cell	CHBr$_3$ LP (193 nm)	CH(A–X) LIF	5
5.1 ± 0.5	100 (Ar)	cell	CHBr$_3$ LP (266 nm)	CH(A–X) LIF	6
0.21 ± 0.02	10 (Ar)	cell	CH$_3$OH IRMPD	CH(A–X) LIF	7
8 ± 3	20 (Ar)	cell	CHBr$_3$ LP (266 nm)	OH(A) chem.	8
2.3 ± 0.5	1.8/0.2 (He/Ar)	reactor	CHBr$_3$ + K or Na	CH(A–X) LIF	9
5.1 ± 0.3	2 (Ar)	cell	CH$_2$Br$_2$ or CHClBr$_2$ LP (248 nm)	CH(A–X) LIF	10
3.5 ± 0.3	20 (Ar)	cell	CHBr$_3$ or (CH$_3$)$_2$CO LP (193 nm)	CH(A–X) LIF	11
3.74 ± 0.19 3.62 ± 0.11	30 (Ar/He)	cell CRESU	+ CHBr$_3$ LP (193 nm)	CH(A–X) LIF	12
4.7 ± 0.4	50 (He)	cell	CHBr$_3$ LP (248 nm)	OH(A) chem.	13
3.26 ± 0.20	20 (Ar)	cell	CHBr$_3$ LP (248 nm)	CH(A–X) LIF	14
3.6 ± 0.5	2 (He)	reactor	CHBr$_3$ + K	OH(A) chem. and CH(A–X) LIF	This work

[a] Reactor: fast-flow reactor, CRESU: Laval nozzles. [b] LP: laser photolysis, IRMPD: infrared multiple-photon dissociation (9551 nm). [c] LIF: laser-induced fluorescence, chem.: chemiluminescence.

$$\rightarrow O(^3P) + H(^2S) + CO(X\ ^1\Sigma^+), \qquad \Delta_r H^{\circ}_{298} = -2.50\ \text{eV} \quad (4)$$

$$\rightarrow O(^3P) + HOC, \qquad \Delta_r H^{\circ}_{298} = -1.48\ \text{eV} \quad (5)$$

$\Delta_r H^{\circ}_{298}$ denotes the enthalpy change for the reaction at 298 K when products are formed in their ground electronic states. As the HOC isomerization barrier[18] is 2.95 eV above HCO(X ^{2}A′) and thus 0.21 eV below the total available energy of the reactants, the 2 channels HCO + O and HOC + O will not be distinguished. Furthermore, owing to the exoergicity of channel (3) and the easy dissociation of HCO, this channel should partially turn into O + H + CO. Moreover, the OH + CO channel could also lead to O + H + CO through the predissociation of OH(A $^2\Sigma^+$).

The first observation of the CH + O_2 products was the CO and CO_2 laser emissions found by Lin et al.[15] in the flash photolysis of O_2/CHBr$_3$/(SF$_6$ or Ar or He) mixtures. However, no product branching ratios or internal energy state distributions were measured. The only quantitative estimations of product branching ratios were performed on channel (2) yielding OH + CO. Grebe et al.[16] in the system C_2H_2/O/H, have estimated that the CH + O_2 → CO + OH* rate constant is $k = 8 \times 10^{-14}$ cm^3 molecule^{-1} s^{-1} at 298 K and 2 Torr. However, in their discharge flow, the overall rate constant of the CH + O_2 reaction was 1.7×10^{-11} cm^3 molecule^{-1} s^{-1}, 2 times slower than the recently reported experimental values (Table 1). Yet their rate constant for the channel yielding excited OH is in good agreement with that determined by Porter et al.[17] in C_2H_2 or CH_4 flames (CH + O_2 → CO + OH*, $k = 1 \times 10^{-13}$ cm^3 molecule^{-1} s^{-1}). So the branching ratio of the channel yielding excited OH is about 0.2–0.5%. The branching ratio of the CO + OH(X $^2\Pi$, $v = 0$) channel has been estimated at around 20% by Okada et al.,[11] probing OH and CH radicals by laser-induced fluorescence (LIF). The experiment was performed in a cell at room temperature and at a total pressure of 20 Torr Ar, the CH radicals being produced by laser photolysis of CHBr$_3$ and (CH$_3$)$_2$CO. At the same time, they performed some ab initio calculations to determine the topology of the potential energy surface. No significant barrier in the entrance was found and the barrier heights for all the transition states were much lower than the energies of the reactants. These calculations are consistent with experimental studies (no pressure dependence between 5 and 350 Torr[11,13] and no temperature dependence between 2200 and 3500 K[19,20] and 297 and 720 K[6,12] or a very small one according to Taatjes et al.,[13] and an important increase at low temperature[12]). The reaction intermediate HCOO which leads to HCO + O, and HCO$_2$ were found but the structures and the energies were not calculated in this study. However, they propose to use the well-known potential energy surface[21] of the CO + OH reaction which leads to H + CO_2 through the HCO_2 intermediate. As the exact topology of the potential energy surface is not known, any conclusion about the main channel may be given.

To determine the product branching ratio, we have performed a study of the CH + O_2 reaction in a low-pressure fast-flow reactor at room temperature. A clean source of CH radical was provided by the reaction of potassium atoms with bromoform; i.e., CH radical in the vibronic level X $^2\Pi$, $v = 0$. The decay of CH probed by LIF, O_2 being introduced in excess, allowed verification of the overall rate constant. Our main interest, however, was to improve, as far as possible, the product branching ratio determination. The branching ratio for the channel yielding CO + OH could not be determined as, owing to the exoergicities, many rovibronic levels, were populated and only a few of them could be probed by LIF. Under our experimental conditions, thermalization might be completed over the rotational levels but not over the vibrational levels and electronic states. The lack of vibronic thermalization prevented recovery of the relative concentrations of diatomic or triatomic species from the probed levels. However, chemiluminescence allowed some insight into the dynamics of the radiative products and of course allowed us to check the occurrence of these pathways. Actually, relative branching ratios could be determined for the reaction channels leading to O or H atoms, using the VUV atomic resonance fluorescence. In order to complete and confirm the branching ratios, theoretical calculations were performed. As the OH + CO → H + CO_2 potential energy surface is now well known,[21] emphasis was put on the calculations for the entrance channel, the CH + O_2 → O + HCO channel and the pathways leading to the HCO_2 intermediate of the OH + CO ↔ H + CO_2 reaction.

2. Experimental

The experimental set-up has been described in detail previously,[22,23] and only a summary is given here. The set-up consists of a fast-flow reactor, i.e. a 36 mm inner tube with four optical ports for detection. The CH radicals are produced in an "injector" which slides in the reactor. At the end of the injector, the CH radicals are mixed with the oxygen flow. Then, the distance between the end of the injector and the observation windows is directly proportional to the reaction time (the gas speed is 26.5 m s^{-1}, at room temperature, 2 Torr, the buffer gas being He with purity ⩾99.995%). The distance (d) between the window detection and the injector nozzle aperture could be varied over the range 0–100 mm with 0.5 mm precision. The pressure was measured by a capacitance manometer (Barocel 0–10 Torr) and the flow rates were adjusted by thermal mass flow meters (Tylan). Before each experiment, the vacuum and the leak-plus-outgassing rate were checked with a Pirani gauge (respectively <0.05 mTorr and <10 mTorr min^{-1}). The oxygen was used directly from the cylinder without further purification (99.995% Air liquid). The CH radicals are produced by the reaction of potassium vaporized in a microfurnace, with bromoform.[22] The temperature of the microfurnace and the flow ratio of bromoform are adjusted to give total abstraction of the bromine atoms in the injector and to avoid the secondary reaction CH + CHBr. This was checked by probing both CH and CHBr by LIF. The only problem with this source was the possibility of the CH + CH reaction in the injector when the bromoform concentration was high. To detect the atoms and determine the atomic branching ratios, high CH concentrations were needed. In this case, the source produced CH radicals, KBr molecules, and a small amount of C_2H and C_2 [22b] by the CH + CH reaction.

The detection systems are an ARFS apparatus, a diagnostic laser system (dye laser pumped by a ND-YAG laser, Quantel YG 581C, and a Hamamatsu R106 photomultiplier with interference filters), and two chemiluminescence collectors (Jobin-Yvon HRS2 monochromator for the range 195–850 nm and an ARC VM502 monochromator for the vacuum UV emissions). As atom concentrations in the reactor were too low for absorption measurements, H and O atoms were probed by their resonance fluorescence at 121.57 and 130.35 nm, respectively. The ARFS apparatus consists of a microwave discharge lamp powered by 125 W and 2450 MHz (EMS, Microtron 200 microwave generator) and the vacuum UV monochromator. The flow in the lamp (40 sccm, 2 Torr) consisted of He (AirGaz 99.9995% purity) carrying a known premixed $N_2/H_2/O_2$ mixture to obtain optimal emission intensities from excited N, H and O atoms. The relative concentrations in the lamp were typically 12 ppm H_2, 6 ppm N_2 and 61 ppm O_2. Atomic emission from the lamp and fluorescence from atomic products were collected through a MgF$_2$ window and a LiF biconvex lens and into a ARC VM502 monochromator with a Al + MgF$_2$ grating blazed at 120 nm and a Hamamatsu R1459 solar-blind photomultiplier tube. The vacuum in the monochromator was kept below 10^{-5} mbar by a turbomolecular pump (Turbovac 50, Leybold) backed by a mechanical pump (Trivac D1,6B,Leybold). The atomic excitation line intensities of the lamp were checked at the beginning and at the end of each experimental. To ensure the linear dependence of the atomic fluorescence *vs.* the lamp emission intensity, we verified that the lamp emission atomic lines were non-reversed and that the sub-level electronic populations of the excited atoms in the lamp were proportional to their statistical weights.[22] As the atomic concentration in the reactor was very low, the fluorescence signal was divided by the lamp emission intensity to obtain the relative atomic concentrations in the reactor.

3. Experimental results

3.1. Global rate constant

The global rate constant was calculated by following the CH LIF signal, O$_2$ being introduced in excess. After the "mixing" time, the CH exponential decay constant must vary linearly with the oxygen concentration (pseudo-first-order conditions). This is the case only after corrections for radial and axial diffusion (Fig. 1). These important corrections have been validated by study of the CH + NO reaction whose rate constant is close to the gas kinetic limit. The CH + NO rate constant, found with the same correction procedure, is in very good agreement with the published values.[22a] As this reaction has a rate constant 5 times larger than for CH + O$_2$, diffusion corrections were more important than in the present study. The corrected values of the pseudo-first-order rate constant *vs.* the oxygen concentration give an overall rate constant of $k_{CH+O_2} = (3.6 \pm 0.5) \times 10^{-11}$ cm^3 molecule^{-1} s^{-1}. This value is in agreement with recently reported values (Table 1).

3.2. Product branching ratios

The atomic products have been probed by resonance fluorescence, after first checking that atomic absorption was small. Under this condition, the fluorescence signal divided by the emission intensity of the lamp is proportional to the atomic concentration multiplied by f_A/δ_A, f_A being the oscillator strength[24] and δ_A the Doppler broadening ($T = 300$ K).[25] Then, the relative concentrations of atoms in the reactor could easily be determined.

A typical VUV atomic fluorescence is shown in Fig. 2(b). One can see the fluorescences of H(^{2}P^0 → ^{2}S) and O(^{3}S^0 → ^{3}P). The atomic concentrations in the reactor had the same behaviour as the CH concentration when the oven temperature or the bromoform flow was increased, O$_2$

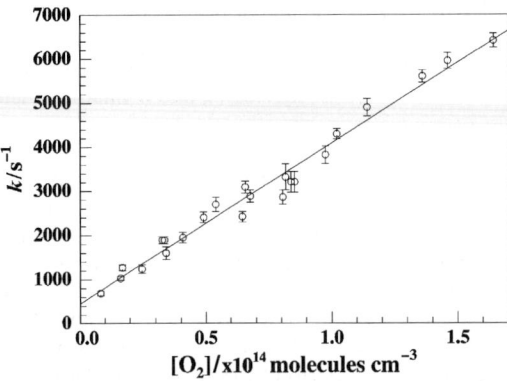

Fig. 1 Evolution of the pseudo-first-order rate constant *vs.* the oxygen concentration. A linear dependence is obtained after corrections for radial and axial diffusion.

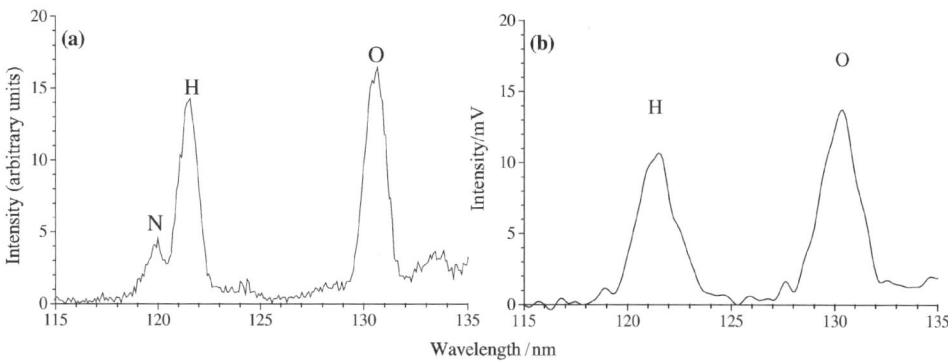

Fig. 2 (a) VUV lamp emission spectrum. (b) Atomic resonance fluorescence spectrum produced by the CH + O_2 reaction.

being in excess. Moreover, the evolution *vs.* the distance (directly proportional in a flow-reactor to the reaction time) was the evolution of primary products. The O and H atoms are thus primary products of the CH + O_2 reaction. The only source of H density overestimation in the product branching ratio was the contribution of the CH + CH → H + C_2H reaction.[22b] Without oxygen, the H fluorescence intensity started from zero at the exit of the "injector". This means that nearly all the H atoms produced by the CH + CH reaction in the injector are removed by the metallic wall of the injector. The H atom contribution from CH + CH reaction could thus only be by production in the reactor.[22] When a large excess of oxygen is introduced this secondary reaction is minimized, but it is still not negligible. In order to estimate the contribution of H production by the CH + CH reaction, different experiments were performed. First, with an excess of oxygen, the H production from the CH + CH reaction was followed by integration of the C_2H* chemiluminescence intensity (CH + CH → C_2H* + H). No difference in the kinetics of the H production from CH + CH or CH + O_2 reactions could be seen. It is thus very important to estimate the H density contribution of the secondary reaction CH + CH for all distances or reaction times. As the H atom production rate of the CH + CH reaction is directly proportional to the C_2H* chemiluminescence, integration of the chemiluminescence signal is proportional to the H production. Comparison of this integrated value with and without O_2 gives the ratio of H production by the CH + CH reaction with and without O_2. Typically, the chemiluminescence signal was reduced by 2.1 when O_2 was added (this ratio depends on the [O_2] added, and all the measurements of the product branching ratios were carried out with a constant excess value of [O_2] to give this constant ratio of 2.1). The H production from the CH + CH reaction in the presence of O_2 is thus equal to 1/2.1 of the H production when no O_2 is added, this last signal being directly accessible. Typically the H production of the CH + CH reaction when O_2 is added is around 10% of the H production of the CH + O_2. The evolution of the ratio [H]/[O] *vs.* the bromoform concentration is stabilized to a value of 1.4 for small $CHBr_3$ concentrations where the CH + CH is negligible. We conclude here that for the overall rate constant measurement, based on LIF measurement of CH, the CH concentration is very low, and the CH + CH reaction is thus negligible, and thus *the branching ratio between the two atoms H(^{2}S)/O(^{3}P) is (1.4 ± 0.2)*.

The CH + O_2 reaction produces H(^{2}S) and O(^{3}P) atoms. We have never been able to detect the presence of the very reactive O(^{1}D) atom. Moreover, in the lowest symmetry C_s, the CH(X $^2\Pi$) + O_2(X $^3\Sigma^-$) reactants correlate with the ground-state products but not with O(^{1}D) + HCO(X ^{2}A'). In order to give an idea of the absolute branching ratios, we have compared the atomic fluorescences produced by the two reactions CH + O_2 and CH + NO under the same pseudo-first-order approximation conditions. The atomic production is 1.2 times more important in the CH + O_2 reaction than in the CH + NO reaction. In the CH + NO study, only the relative atomic branching ratios have been determined, but the agreement between theory and experiment allow us to propose an absolute distribution.[22a] Combining this distribution with the atomic production, and comparing the two reactions, this means that the CH + O_2 reaction should produce about 110% of atoms. We check this result on comparing the H production from CH + O_2

versus the CH + CH$_4$ and CH + H$_2$ reactions. This result is not very surprising as the available energies for the channels O + HCO and OH + CO are higher than the dissociation barriers of HCO and OH. As the product branching ratio for the CO + OH(A $^2\Sigma^+$) channel is between 0.2 and 0.5%,[17,18] this "extra" atomic production comes essentially from the dissociation of HCO into H + CO. To summarise, $\tau_1 + \tau_3 + 2\tau_4 = 1.1 \pm 0.2$ and $(\tau_1 + \tau_4)/(\tau_3 + \tau_4) = 1.4 \pm 0.2$, with τ_1, τ_3, τ_4 the branching ratios of the channels H + CO$_2$ (1), O + HCO (3) and O + H + CO (4).

The absolute branching ratios could not be determined for the OH + CO channel; we have to probe all the rovibronic levels of OH or CO, since under our experimental conditions, the thermalization was completed over rotational levels, but not over vibrational levels and electronic states. However, in order to have an idea of the absolute branching ratios, HCO was probed by LIF (B ^{2}A′ ← X ^{2}A′).[26] The intensity of the signal was compared with that obtained from the O + C$_2$H$_4$ reaction, O atoms being produced by microwave discharge in an O$_2$/He mixture. The O + C$_2$H$_4$ reaction has an overall rate constant[27a] of $k = 7.1 \times 10^{-13}$ cm^3 molecule^{-1} s^{-1} whereas O$_2$ + C$_2$H$_4$ reaction does not occur ($k \ll 10^{-20}$ cm^3 molecule^{-1} s^{-1}). The branching ratios are the following: H + CH$_2$CHO, 40% and HCO + CH$_3$, 60%. The HCO + CH$_3$ channel is exothermic by 1.21 eV with very little dissociation of HCO.[27b] We also compared the LIF signal of the two reactions, O + C$_2$H$_4$ and CH + O$_2$. With an estimation of the CH and O concentrations (the CH concentration is estimated by the simulation of its production in the oven,[22] and the O concentration by using our lamp in absorption), we can assume then that the HCO branching ratio produced by the CH + O$_2$ reaction is between 8 and 30%. This is in agreement with the other results. The branching ratios expressed as a function of the OH + CO branching ratio τ_2 are: H + CO$_2$ (1), $\tau_1 = 0.5 - \tau_2$; OH + CO (2), τ_2; O + HCO (3), $\tau_3 = 0.4 - \tau_2$; O + H + CO (4), $\tau_4 = 0.1 + \tau_2$. Thus, τ_2 is 0–40%, τ_1 10–50%, τ_3 0–40% in agreement with the HCO LIF detection, and τ_4 10–50%.

3.3. Internal distribution of energy

The oxidation of CH is highly exothermic and thus can lead to many vibrationally and electronically excited products. The major chemiluminescent product is the well-known OH(A $^2\Sigma^+ \to$ X $^2\Pi$) in the ultraviolet range (Fig. 3(a)). As we can see in Fig. 1, analysis of the OH chemiluminescence decays gives the same overall rate constant as analysis of the CH LIF decays. The excited OH is thus really a primary product of the CH + O$_2$ reaction. According to the correlation diagram in the lowest symmetry C_s, this could only be possible if there is an intersystem crossing between two surfaces (our source only produces CH in its ground vibronic state). The first quadruplet surface leads to the products CO(a $^3\Pi$) + OH(X $^2\Pi$). The second doublet surface correlates the excited reactants CH(a $^4\Sigma^-$) + O$_2$(X $^3\Sigma_g^-$) with the products OH(A $^2\Sigma^1$) + CO(X $^1\Sigma^+$). To have an idea of the dynamics, the rovibrational distribution was determined by simulating the chemiluminescence spectrum. There is good agreement between the experimental spectrum and the simulation (Fig. 3(a) and (b)). The truncated singular value decomposition method, imposing a statistical distribution over the e and f levels, gave a vibrational population distribution of OH(A $^2\Sigma^+$): 32% in the $v' = 1$ level and 68% in the $v' = 0$ level (± 5%) with very hot rotational distributions (Fig. 3(c) and (d)). Usually, under the same conditions, the rotational thermalization was total. Moreover, the rovibrational distributions did not change significantly over the first ms of reaction time and over the pressure range 0.8–5.0 Torr. As the OH* rotational constant B_e is 17.36 cm^{-1},[28] the gap between the high rotational levels is so large that He collisions cannot induce deactivation. This distribution is thus very close to the nascent distribution and intersystem crossing leads to population of the high rotational levels of OH(A $^2\Sigma^+$). However, this inter-system crossing between the two surfaces, leading respectively to OH* + CO and OH + CO*, does not seem to be of a conical intersection. One can see the CO(a $^3\Pi \to$ X $^1\Sigma^+$) chemilumiscence of the vibrational levels up to $v' = 4$. This is consistent with the total available energy to the CH + O$_2$ reaction at 300 K. It was verified that the CO chemiluminescence signal was mainly from CH reaction and not from the C$_2$ radicals produced by CH + CH reaction in the injector. The C$_2$ + O$_2$ reaction, exothermic by $\Delta\varepsilon_0 = -10.78 \pm 0.04$ eV, produces a lot of CO chemiluminescence but only a very small amount of CO(a $^3\Pi$): this was checked by producing C$_2$ radicals using the C$_2$Cl$_4$ + 4K system. Moreover, the rate constant[29] estimated to be 3×10^{-12} cm^3 molecule^{-1} s^{-1} is too small to interfere with the CH + O$_2$ product determination. In any

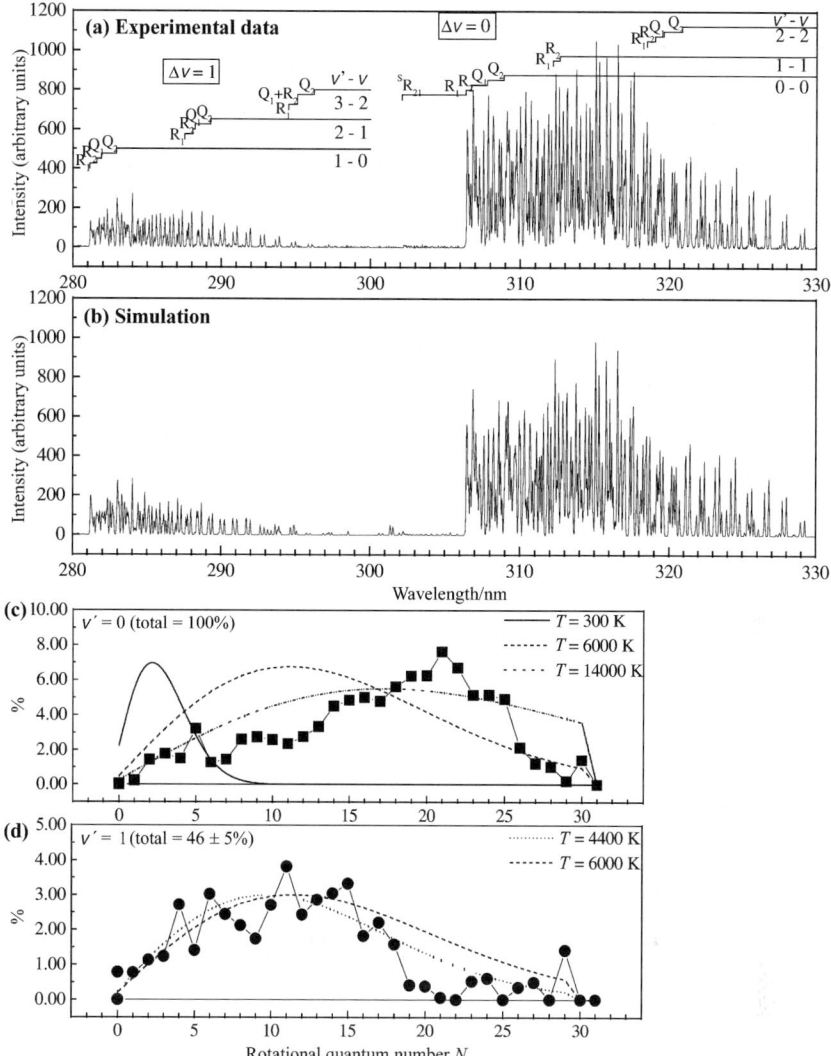

Fig. 3 (a) Chemiluminescence spectrum of OH(A $^2\Sigma^+ \to$ X $^2\Pi$) produced by the CH + O_2 reaction (only the weak branch heads are indicated). (b) Simulation (with a 0.07 nm width of the peak at half height and statistical distribution over the e and f levels) in order to determine the rovibrational distribution. (c) Relative rotational distributions of OH(A $^2\Sigma^+$) in the vibrational level $v' = 0$. (d) As (c) in $v' = 1$. Comparisons of the experimental data (squares and circles) with Boltzmann distributions at different temperatures.

case, we could not confirm that CO* is only produced by the CH + O_2 reaction as its lifetime is very long (between 2.0 and 500 ms[30]). Thus one has to take into account the possible reaction CH + O, the possibly significant role of the secondary reaction CO(a $^3\Pi$) + CO(a $^a\Pi$) → CO(A $^1\Pi$) + CO(X $^1\Sigma^+$), and relaxation induced by He collisions. The CO(a $^3\Pi$) + CO(a $^3\Pi$) and C_2 + O_2 reactions might also explain the CO(A $^1\Pi$) chemiluminescence between 130 and 260 nm.

The Meinel bands of OH have not been detected.[31] This indicates that the high vibrational levels of the OH ground state are not strongly populated or are undetectable with our experimental sensitivity. We have not detected the HCO(A ^{2}A″) chemiluminescence, but HCO(A ^{2}A″) is a predissociated state and the reactants in their ground states do not correlate with HCO(A ^{2}A″) + O in C_s symmetry.

4. Theoretical study

Even if the rate constant of the $CH + O_2$ reaction had been determined over the range 15–3500 K, neither the pathways leading to the different products, nor the nature of the intermediates involved are known. Early quantum chemical studies on [C,H,O,O,] isomers were mainly focused on the $OH + CO \rightarrow H + CO_2$ reaction,[21] except for the qualitative *ab initio* study on the $CH + O_2$ reaction by Okada *et al.*[11] We have therefore performed new *ab initio* calculations associated with RRKM estimation to identify the key intermediates and the reaction pathways and thereby to gain insight into the reaction kinetics.

4.1. Quantum chemical calculations

Radical reactions are difficult to study theoretically because the accurate description of the low energy barriers they could present requires a very high level of theory. To obtain reliable results on this kind of surface with open shells, it is necessary to use a method where the dynamical correlation energy is well described. The complete active space second-order perturbation theory (CASPT2) has been shown to be a reliable and computationally quite cheap approach for calculation of the dynamical correlation energy.[32] We first performed restricted Hartree–Fock (RHF) and complete active space self-consistent field (CASSCF) calculations. The CASSCF functions were then used in the CASPT2 calculations. For the CASPT2 calculations, the geometry was optimised only at the CASSCF level. All calculations were performed using MOLPRO 2000[33] with the double-zeta cc-pVDZ basis set from Dunning.[34] For the CASSCF and CASPT2 calculations, the 1s orbitals of the carbon and the two oxygen atoms were always kept doubly occupied as they do not participate in the bond breaking and bond forming which occurs along the reaction path. The 2s orbitals of the two oxygen atoms are also, in general, kept doubly occupied because they play a minor role (this effect was checked at the CASSCF level). Therefore, for the $CH + O_2$ system, the active space was built with 13 electrons distributed on 11 orbitals.

Two different entrance channels are expected: addition of C to one of the O atoms leading to HCOO, or insertion of C into the O_2 bond leading to HCO_2. We discuss firstly the features of the potential energy curves when the CH radical approaches the diatomic O_2. As CH is in a $^2\Pi$ state and O_2 in a $^3\Sigma^-$ state, there are 4 possible combinations, giving two doublet state and two quadruplet states. The $C_{\infty v}$ (linear approach of C toward the O_2 axis) and C_{2v} (perpendicular approach of C toward the centre of O_2) approaches present a barrier in the entrance. The C_s approach gives no, or a negligible (less than 50 cm^{-1}, which has no significance at our level of calculation), barrier for the $^2A'$, $^2A''$ and $^4A'$ surfaces, CH being almost parallel to the O_2 axis (with almost no difference for the *cis* and the *trans* configurations). However, this approach gives a barrier for the $^4A''$ surface. For the C_1 approach, there is no barrier for the surfaces associated with the first two $^2A'$ states and with one of the $^4A'$ states, where the CH is almost perpendicular to the triangle formed by COO. The second $^4A'$ state has a barrier in the entrance channel of more than 1 eV.

Quadruplet surface. For the symmetries C_s or C_1, there is only one surface without any barrier; the $^4A'$ state for C_s and the first $^4A'$ for C_1. The only allowed spin exit channel leads to the HCO (A $^2A''$) + O(^3P) products. The minimum energy path leads directly from the reactants to the products, without an intermediate state. This result seems reasonable as the first $^4HCO_2$ state was found to be localized 1.6 eV above the exit channel. Nevertheless, it should be noticed that this first $^4HCO_2$ state presents a minimum in the symmetries C_{2v}, C_s and C_1, and thus is not a dissociated state, but presents a barrier toward the dissociation to HCO(A $^2A''$) + O(^3P).

Doublet surfaces. In C_s symmetry, the topology of the two doublet surfaces $^2A'$ and $^3A''$ at the entrance was found to be flat, from CASPT2 calculations performed with the geometries calculated at the CASCF level. Then the minimum energy path leads directly to HCO(A $^2A''$) + O(^3P) for the two surfaces without any conventional transition state, in disagreement with the preliminary results of Okada *et al.*[11] Only a van der Waals type complex (HCOO) in the exit channel (HCO + O) was found. On the $^2A'$ and $^2A''$ surfaces in C_s symmetry, the isomerization, from any point on the HCOO reaction path to HCO_2 proceeds with a small barrier, lower than the energy of the reactants.

If we allow all angles to be varied, then in C_1 symmetry, the minimum energy path of the first doublet surface, corresponding to the $^2A'$ or $^2A''$ path of C_s symmetry, leads directly to the formation of HCO_2 without any barrier in the entrance valley calculated at CASSCF or CASPT2 levels. Thus the C_1 calculation shows that the C_s pathways leading to $HCO(A\,^2A'') + O(^3P)$ is not the minimum energy path for the entrance channel and we should thus consider that the two first doublet surfaces of the system lead to the intermediate $^2HCO_2$. For this first intermediate $^2HCO_2$ system, the fundamental state is a $^2A'$ state of C_s symmetry, the excited $^2A''$ state being localized 0.38 eV above. As the minimum energy entrance channel is for C_1 symmetry and as these two intermediate states are close in energy, they will be scrambled along the reaction pathway and we should consider that the flow of reactants is all going to the first $^2HCO_2$ state.

Fig. 4 displays a schematic picture of the potential energy surface (PES) correlating with the CH and O_2 ground states.

To reiterate, the $^4A'$ surface leads to the formation of $HCO(A\,^2A'') + O(^3P)$ without a barrier in the entrance valley. The doublet surfaces lead, in a first step, to the production of the $^2HCO_2^*$ radical without a barrier in the entrance valleys. Then the subsequent evolution of $^2HCO_2$ can be considered using the potential energy surface developed by Schatz et al.[21] It should be noted that the geometry obtained in this work for $^2HCO_2$ is in good agreement with that of Schatz et al.[21] This energized adduct can dissociate into $^2H + \,^1CO_2$ or, in a minor way, into $^3O + \,^2HCO$. Furthermore, isomerization to ^2HOCO is possible. When formed from activated $^2HCO_2^*$, the ^2HOCO isomer will be energetically excited and will dissociate into $OH + CO$ and $H + CO_2$.

To estimate the contributions of the different exit channels from $^2HCO_2^*$ on the doublet surface, we could only perform statistical calculations, the trajectory calculation on the complete *ab initio* surface for this kind of system being beyond computational possibility. However, due to the large excess energy of $^2HCO_2^*$ (8.3 eV), the RRKM unimolecular lifetimes of the various processes are about 10^{-14} s, too short to apply statistical theory, particularly for the assumption of a rapid intramolecular transfer leading to energy randomisation among internal degrees. Nevertheless, RRKM statistical evaluations were performed to obtain an approximate limiting case. The first result indicates that the lifetime of the energized $^2HCO_2^*$ will be very short and microcanonical rate constants of the various processes indicate reaction rates that are several orders of magnitude faster than the collision frequency of *ca.* 10^8 s^{-1} for a pressure of 1–2 Torr of He at 298 K. Thus there should be no pressure effect on the rate constant, this result being valid even if there is no randomisation among internal degrees.

The relative importance of the three unimolecular reactions of $^2HCO_2^*$ was evaluated from the ratios of the corresponding microcanonical rate constants at the entrance energy $E_e = 8.3$ eV.

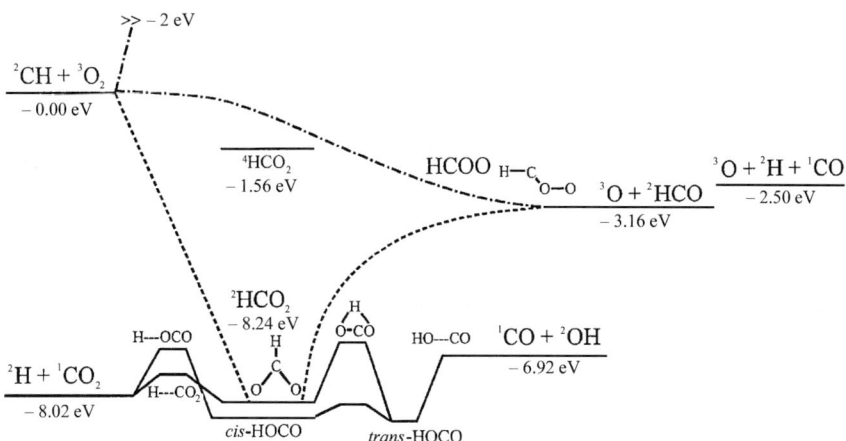

Fig. 4 Potential energy diagram of the CH + O_2 reaction: (– – –) doublet surfaces and (– · –) quadruplet surfaces. The channels yielding OH + CO and H + CO_2 correspond to the potential energy surface (———) of the CO + OH reaction calculated by Schatz et al.[21a]

Thus, from RRKM theory, the HCO + O channel is negligible, and the branching ratios estimated are about 80% for H + CO_2 and 20% for OH + CO in the dissociation of $^2HCO_2^*$. The large difference between the H + CO_2 channel and OH + CO channel is due to the nature of the transition state for these two channels; the OH + CO channel presents a tighter transition state with a higher barrier height than for the H + CO_2 channel.

5. Discussion

The temperature and pressure dependences of the overall rate constant, as the isotopic effect, agree with this potential energy surface topology. Indeed, there is no pressure dependence between 5 and 350 Torr.[13,11] The pressure independence of the rate constants implies only that most of the complexes formed dissociate to products or to the original reactants before being stabilized. There is no temperature dependence between 2200 and 3500 K[19,20] and between 297 and 720 K[6,12] or only a very small one according to Taatjes et al.,[13] and a strong increase at low temperature.[12] Moreover, there is no isotopic effect, other than those due to the difference in masses.[13,14] All these points are consistent with the potential energy surface topology: no barrier in the entrance, no high barrier along the topology of the surface and a lifetime of the adduct HCO_2 too short to allow collisional stabilization.

We have considered that the quadruplet and doublet surfaces with no barrier in the entrance channel are populated following the spin degeneracy; each quadruplet state has the same population as the total of the doublet states. As one of the quadruplet states has a barrier in the entrance channel, it will not participate in the formation of the products. Moreover, as the barrierless quadruplet surface leads only to the formation of $HCO(A\,^2A'') + O(^3P)$ and the doublet surfaces lead almost only to HCO_2^*, we estimate that near 50% produce HCO + O and 50% HCO_2^*. As the dissociation barrier of HCO (+0.910 eV) is much smaller than the available energy for the channel yielding O + HCO (+3.16 eV), a part of the HCO produced will dissociate into H + CO, this fraction depending upon the internal energy distribution of HCO produced by the CH + O_2 reaction. Nevertheless, we should consider that the O + H + CO channel comes almost exclusively from the dissociation of HCO and not from the dissociation of OH or CO_2. Our theoretical estimation of 50% for the O + HCO (3) and O + H + CO (4) channels is fully consistent with our experimental results. Then the main question is still the evaluation of the OH + CO (2) channel branching ratio. The RRKM estimation gives a ratio of 10% and that of Okada et al.,[11] 20%. Even if the CH + O_2 reaction is a limiting case for RRKM calculations, our experimental and ab initio results validate the Okada et al. result for the CO + OH branching ratio, showing that this channel is a minor one. Combining all the results, we anticipate that the actual distribution should not be very far from the following values, the error for each channel being estimated at about 10%. H + CO_2 (1): $\tau_1 = 0.30$; OH + CO (2): $\tau_2 = 0.20$; O + HCO (3): $\tau_3 = 0.20$; O + H + CO (4): $\tau_4 = 0.30$.

Acknowledgements

We thank J.-H. Fillion of the Observatoire de Meudon (France) for providing the $OH(A\,^2\Sigma^+)$ population distribution, and P. Jimeno for help in the determination of the potential energy surface.

References

1. W. A. Sanders and M. C. Lin, in *Chemical Kinetics of Small Organic Radicals*, ed. Z. B. Alfassi, CRC Press, Boca Raton, FL, 1986, vol. III.
2. J. A. Miller and C. T. Bowman, *Prog. Energy Combust. Sci.*, 1989, **15**, 287.
3. W. S. Lanier, J. A. Mulholland and J. T. Beard, *Proc. Int. Combust. Soc.*, 1988, **21**, 1171.
4. I. Messing, C. M. Sadowski and S. V. Filseth, *Chem. Phys. Lett.*, 1979, **66**, 95.
5. J. E. Butler, J. W. Fleming, L. P. Goss and M. C. Lin, *Chem. Phys.*, 1981, **56**, 355.
6. M. R. Berman, J. W. Fleming, A. B. Harvey and M. C. Lin, in *19th Symposium (Int.) on Combustion*, The Combustion Institute, Pittsburgh, 1982, p. 73.
7. J. A. Duncanson and W. A. Guillory, *J. Chem. Phys.*, 1983, **78**, 4958.
8. D. A. Lichtin, M. R. Berman and M. C. Lin, *Chem. Phys. Lett.*, 1984, **108**, 18.

9   S. M. Anderson, A. Freedman and C. E. Kolb, *J. Phys. Chem.*, 1987, **91**, 6272.
10 K. H. Becker, B. Engelhardt, P. Wiesen and K. D. Bayes, *Chem. Phys. Lett.*, 1989, **154**, 342.
11 S. Okada, K. Yamasaki, H. Matsui, K. Saito and K. Okada, *Bull. Chem. Soc. Jpn.*, 1993, **66**, 1004.
12 P. Bocherel, L. B. Herbert, B. R. Rowe, I. R. Sims, I. W. M. Smith and D. Travers, *J. Phys. Chem.*, 1996, **100**, 3063.
13 C. A. Taatjes, *J. Phys. Chem.*, 1996, **100**, 17840.
14 C. Mehlmann, M. J. Frost, D. E. Heard, B. J. Orr and P. F. Nelson, *J. Chem. Soc., Faraday Trans.*, 1996, **92**, 2335.
15 M. C. Lin, *J. Chem. Phys.*, 1974, **61**, 1835.
16 J. Grebe and K. H. Homann, *Ber. Bunsen-Ges. Phys. Chem.*, 1982, **86**, 581.
17 R. P. Porter, A. H. Clark, W. E. Kaskan and W. E. Browne, *11th Symposium (Int.) on Combustion*, The Combustion Institute, Pittsburgh, 1967, p. 907.
18 L. C. Geiger, G. C. Schatz and L. B. Harding, *Chem. Phys. Lett.*, 1985, **114**, 520.
19 M. Röhrig, E. L. Petersen, D. F. Davidson, R. K. Hanson and C. T. Bowman, *Int. J. Chem. Kinet.*, 1997, **29**, 781.
20 M. W. Markus, P. Roth and T. Just, *Int. J. Chem. Kinet.*, 1996, **28**, 171.
21 (*a*) G. C. Schatz, M. S. Fitzcharles and L. B. Harding, *Faraday Discuss. Chem. Soc.*, 1987, **84**, 359; (*b*) K. Kudla, G. C. Schatz and A. F. Wagner, *J. Chem. Phys.*, 1991, **95**, 1635.
22 (*a*) A. Bergeat, T. Calvo, N. Daugey, J.-C. Loison and G. Dorthe, *J. Phys. Chem. A*, 1998, **102**, 8124; (*b*) A. Bergeat, T. Calvo, G. Dorthe and J.-C. Loison, *J. Phys. Chem. A*, 1999, **103**, 6360.
23 (*a*) N. Daugey, A. Bergeat, A. Schuck, P. Caubet and G. Dorthe, *Chem. Phys.*, 1997, **87**, 222; (*b*) N. Daugey, A. Bergeat, P. Caubet, E. Cecarrelli, A. Schuck and G. Dorthe, *News Lett. Anal. Astron. Spectrosc.*, 1995, **22**, 10.
24 W. L. Wiese, J. R. Fuhr and T. M. Deters, *J. Phys. Chem. Ref. Data*, 1996, Monograph 7.
25 K. P. Lynch, T. C. Schwab and J. V. Michael, *Int. J. Chem. Kinet.*, 1976, **8**, 651.
26 A. D. Sappey and D. R. Crosley, *J. Chem. Phys.*, 1990, **93**, 7601.
27 (*a*) D. L. Baulch, C. J. Colbos, R. A. Cox, P. Frank, G. Hayman, Th. Just, J. A. Kerr, T. Murrells, M. J. Pilling, J. Troe, R. W. Walker and J. Warnatz, *J. Phys. Chem. Ref. Data*, 1994, **23**, 847; (*b*) U. C. Sridharan and F. Kaufman, *Chem. Phys. Lett.*, 1983, **102**, 45.
28 G. Herzberg, *Molecular Spectra and Molecular Structure I. Spectra of Diatomic Molecules*, Van Nostrand, Princeton, NJ, 2nd edn., 1950.
29 (*a*) M. Mangir, H. Reisler and C. Wittig, *J. Chem. Phys.*, 1980, **73**, 829; (*b*) H. Reisler, M. Mangir and C. Wittig, *J. Chem. Phys.*, 1980, **71**, 2109; (*c*) L. Pasterneck and J. R. McDonald, *Chem. Phys.*, 1979, **43**, 173; (*d*) V. M. Donnelly and L. Pasternack, *Chem. Phys.*, 1979, **39**, 427.
30 R. H. Bonham and M. Fink, *J. Chem. Phys.*, 1971, **55**, 4118.
31 G. E. Streit and H. S. Johnston, *J. Chem. Phys.*, 1976, **64**, 95.
32 (*a*) J. Villà, A. Gonzalez-Lafont, J. M. Lluch, J. C. Corchado and J. Espinosa-Garcia, *J. Chem. Phys.*, 1997, **107**, 7266; (*b*) A. Rauk, D. Yu, P. Borowski and B. Ross, *Chem. Phys.*, 1995, **197**, 73; (*c*) B. J. Person, B. O. Ross and M. Simonson, *Chem. Phys. Lett.*, 1995, **234**, 382.
33 *MOLPRO 2000* is a package of *ab initio* programs written by H.-J. Werner and P. J. Knowles.
34 T. H. Dunning, *J. Chem. Phys.*, 1989, **90**, 1007.

A theoretical analysis of the reaction between propargyl and molecular oxygen

David K. Hahn,[†,a] Stephen J. Klippenstein*[b] and James A. Miller*[b]

[a] *Chemistry Department, Case Western Reserve University, Cleveland, OH, 44106-7078, USA*
[b] *Combustion Research Facility, Sandia National Laboratories, Livermore, CA, 94551-0969, USA*

Received 8th March 2001
First published as an Advance Article on the web 17th October 2001

The temperature- and pressure-dependent kinetics of the reaction between propargyl and molecular oxygen have been studied with a combination of electronic structure theory, transition state theory, and the time-dependent master equation. The stationary points on the potential energy surface were located with B3LYP density functional theory. Approximate QCISD(T,Full)/6-311++G(3df,2pd) energies were obtained at these stationary points. At low temperatures the reaction is dominated by addition to the CH_2 side of the propargyl radical followed by stabilization. However, addition to the CH side, which is followed by one of various possible internal rearrangements, becomes the dominant process at higher temperatures. These internal rearrangements involve a splitting of the O_2 bond *via* the formation of 3-, 4- or 5-membered rings, with the apparent products being CH_2CO + HCO. Rearrangement *via* the 3-membered ring is found to dominate the kinetics. Rearrangement from the CH_2 addition product, *via* a 4-membered ring, would yield H_2CO + HCCO, but the barrier to this rearrangement is too high to be kinetically significant. Other possible products require H transfers and, as a result, appear to be kinetically irrelevant. Modest variations in the energetics of a few key stationary points (most notably the entrance barrier heights) yield kinetic results that are in good agreement with the experimental results of Slagle and Gutman (I. R. Slagle and D. Gutman, *Proc. Combust. Inst.*, 1986, **21**, 875) and of Atkinson and Hudgens (D. B. Atkinson and J. W. Hudgens, *J. Phys. Chem. A*, 1999, **103**, 4242).

Introduction

Resonantly stabilized free radicals (RSFR's) are generally thought to play an important role in the formation of aromatic compounds, polycyclic aromatic compounds (PAH), and soot during the combustion of hydrocarbon fuels.[1–16] The property of RSFR's that distinguishes them from ordinary free radicals is that the unpaired electron in an RSFR is delocalized, *i.e.* it is spread out over two or more sites in the radical, resulting in at least two resonant electronic structures of comparable importance. A consequence of this delocalization of the unpaired electron is that RSFR's are more stable than ordinary radicals, generally having lower enthalpies of formation than one might

[†] Present address: Chemistry Department, University of Nevada, Reno, NV, 89557, USA.

DOI: 10.1039/b102240g

expect from a single Kekulé structure. Because of this increased stability, RSFR's are better able to survive combustion environments than are ordinary radicals, and their concentrations can grow to quite large values in flames. With such large concentrations, the reactions of RSFR's among themselves become attractive ways to build higher hydrocarbons in flames.[1,2,13,14]

From a combustion chemistry point of view, the most important resonantly stabilized free radical is propargyl (C_3H_3); it is also the simplest, possessing two resonant Kekulé structures,

$$\underset{H}{\overset{H}{\diagdown}}\ddot{C}-C\equiv C-H \quad \longleftrightarrow \quad \underset{H}{\overset{H}{\diagdown}}C=C=\overset{\cdot}{\underset{H}{C}},$$

the first of which is dominant. Its "resonance energy" is frequently cited as being approximately 8 or 9 kcal mol^{-1}. In flames, propargyl normally is formed either by hydrogen abstraction from allene or propyne or through the reaction of singlet methylene with acetylene,[13]

$$^1CH_2 + C_2H_2 \rightarrow C_3H_3 + H. \tag{R1}$$

The importance of propargyl in combustion stems from its ability to react with itself and form cyclic species,

$$C_3H_3 + C_3H_3 \rightarrow \text{Phenyl} + H$$

$$\rightarrow \text{Benzene}$$

$$\rightarrow \text{Fulvene}. \tag{R2}$$

Reaction (R2) is frequently cited as the dominant "ring-forming" reaction in flames of aliphatic fuels.[3,4,6–9,11,13] The most important competing steps are the reactions of propargyl with molecular oxygen and hydrogen atom,[11–13]

$$C_3H_3 + O_2 \rightarrow \text{products} \tag{R3}$$

$$C_3H_3 + H \rightarrow C_3H_2 + H_2, \tag{R4}$$

where C_3H_2 is normally taken to be triplet propargylene. Recently Blitz et al.[17] have studied reaction (R1) using master-equation methods, and Miller and Klippenstein.[18] have studied (R2) using related techniques. Klippenstein and Harding[19] studied the $C_3H_3 + H$ addition reaction, but they did not consider the bimolecular channel. The present investigation continues the theme of trying to understand propargyl chemistry in combustion by studying reaction (R3) theoretically in some detail.

Perhaps the most important property of resonantly stabilized free radicals is that they form relatively weak bonds with stable molecules, most notably with molecular oxygen.[1,13,15] For example, if one considers other typical hydrocarbon radicals found in flames, such as CH_3, C_2H, C_2H_3 and C_2H_5, one finds that their bond energies with O_2 are 29,[20] 49,[21] 45[22,23] and 34 kcal mol^{-1},[24–26] respectively. These values are to be contrasted with the much smaller value for the C_3H_3–O_2 bond energy of 18 kcal mol^{-1} deduced by Slagle and Gutman in their experimental investigation of the $C_3H_3 + O_2$ reaction, a value that is confirmed in the present work. The resonantly stabilized allyl radical, C_3H_5, has a similarly small bond energy with O_2 of approximately 20 kcal mol^{-1}. Such weakly bound addition complexes are not easily stabilized by collisions, particularly at high temperature, nor do they readily support the rearrangement that may be required to form oxidized bimolecular products. It is this property of RSFR's that is most responsible for their ability to reach high concentrations in flames.

In the present work we characterize the important features of the $C_3H_3 + O_2$ potential using QCISD(T), with MP2 basis set corrections, and DFT-B3LYP methods. From this information we calculate microcanonical (RRKM) rate coefficients and solve various forms of the master equation (ME), principally the time-dependent, multiple-well form, to extract information about the thermal rate coefficient and product distribution as a function of temperature and pressure. We pay particular attention to comparing our results with the pioneering experiments of Slagle and Gutman[15] and the more recent work of Atkinson and Hudgens.[16]

Theory

A. Quantum chemistry

The geometric structures and vibrational frequencies for all stationary points considered here were obtained *via* density functional theory employing the Becke-3 Lee–Yang–Parr (B3LYP) functional[27] and the 6-311+G(d,p) basis set.[28] Intrinsic reaction coordinate calculations were performed for each saddlepoint in order to verify the connection with the local minima. In each instance, both $^2A'$ and $^2A''$ states were considered, and when imaginary frequencies were obtained, the structures were allowed to relax to C_1 symmetry. The correlation between the bonding in these $^2A'$, $^2A''$ and ^2A states and the lowest singlet and triplet states of O_2 follows in much the same way as described in the detailed work of Schaefer and coworkers,[25,26] and so in the interest of brevity, it is not described here. The GAUSSIAN 98 quantum chemistry software[29] was employed in each of the present quantum-chemical simulations.

Higher level energies were obtained for each stationary point *via* a combination of quadratic configuration interaction calculations [QCISD(T)][30] and second order Møller–Plesset perturbation theory (MP2).[28] In particular, approximate QCISD(T,Full)/6-311++G(3df,2pd) energies [E_{HL}] are obtained *via* the relation

$$E_{HL} = E_{QCI,base} + \Delta_{Basis} + \Delta_{Core} \tag{1}$$

where the base energy is

$$E_{QCI,base} = E[QCISD(T)/6\text{-}311\text{++}G(d,p)]. \tag{2}$$

The term

$$\Delta_{Basis} = E[MP2/6\text{-}311\text{++}G(3df,2pd)] - E[MP2/6\text{-}311\text{++}G(d,p)] \tag{3}$$

corrects for limitations in the 6-311++G(d,p) basis set, while

$$\Delta_{Core} = E[MP2(Full)/6\text{-}311\text{++}G(d,p)] - E[MP2/6\text{-}311\text{++}G(d,p)]$$

provides an MP2 correction for core electron correlation. The present core–valence correlation correction for the stationary point energies of the complexes relative to the $C_3H_3 + O_2$ reactants is generally about -0.3 kcal mol^{-1}. In contrast, the basis set correction is both much larger and more variable, with a typical correction being -5 kcal mol^{-1}.

For comparison purposes, standard Gaussian-3 (G3)[31] and multi-coefficient (MC-)G3[32,33] energies have also been obtained for $C_3H_3 + O_2$ [but again employing the B3LYP/6-311+G(d,p) geometries and harmonic frequencies]. In these two methods QCISD(T)/6-31G(d) calculations provide a base energy that is supplemented with a series of MP4 and MP2 energies with larger basis sets, with the MC-G3 method differing from the G3 method primarily in the use of a parametrized sum of the different contributions. The largest basis set in the G3 and MC-G3 methods is essentially the 6-311++G(2df,2p) basis, with some additional core functions in the G3 calculation. For saddlepoints there are some ambiguities in the application of these two methods related to the extent of electron pairing and numbers of bonds. However, these ambiguities result in uncertainties of 0.3 kcal mol^{-1} or less, and so are of minor significance here.

The use of larger basis sets, particularly at the QCISD(T) level, and fewer separability assumptions for the present HL method should be an improvement over the G3 and MC-G3 methods, while the use of the MP4 estimates for some components of the basis set corrections in the G3 and MC-G3 methods provides some improvement over the present HL method. The root-mean-squared errors in the dissociation energies for a set of test molecules are 1.24 and 1.10 kcal mol^{-1} for the G3 and MC-G3 methods, respectively.[32] It seems reasonable to assume similar sorts of errors for the energies of the wells from the present 'HL' method. However, for all these methods the saddle point energies are more uncertain, due in large part to the generally increased multi-reference character of the wavefunctions.

B. The master equation

In previous work[34,35] we have described in detail our methods for calculating partition functions, sums and densities of states, and microcanonical rate coefficients. Consequently, we shall not repeat that discussion here. However, it is worthwhile to discuss various forms of the master equation and our methods for solving them, although some of this discussion can be found elsewhere.[18,35,36]

The one-dimensional, time-dependent, multiple-well master equation. For our purposes the multiple-well master equation in one dimension can be expressed as a set of M coupled integro-differential equations, where M is the number of wells,

$$\frac{dn_i(E)}{dt} = Z \int_{E_{0i}}^{\infty} P_i(E, E') n_i(E') dE' - Z n_i(E) - \sum_{j \neq i}^{M} k_{ji}(E) n_i(E)$$

$$+ \sum_{j \neq i}^{M} k_{ij}(E) n_j(E) - k_{di}(E) n_i(E) + K_{eqi} k_{di}(E) F_i(E) n_R n_m - k_{pi}(E) n_i(E) \quad (i = 1, \ldots, M). \quad (4)$$

In these equations, t is the time, Z is the collision number per unit time; $n_i(E)dE$ is the number density of molecules or complexes in well i with energy between E and $E + dE$; E_{0i} is the ground state energy of well i; $P_i(E,E')$ is the probability that a molecule in well i with energy between E' and $E' + dE'$ will be transferred by collision to a state with energy between E and $E + dE$; $k_{ij}(E)$ is the unimolecular rate coefficient for isomerization from well j to well i; $k_{di}(E)$ and $k_{pi}(E)$ are the dissociation rate coefficients from well i to reactants ($C_3H_3 + O_2$) and bimolecular products, respectively; n_R and n_m are the number densities of the deficient (C_3H_3) and excess (O_2) reactants, respectively; and K_{eqi} is the equilibrium constant for the addition reaction, $C_3H_3 + O_2 \leftrightarrow$ well i. The function $F_i(E)$ is the equilibrium energy distribution in well i at temperature T,

$$F_i(E) = \rho_i(E) e^{-\beta E} / Q_i(T), \quad (5)$$

where $Q_i(T)$ is the vibrational–rotational partition function for the ith well, $\rho_i(E)$ is the corresponding density of states, $\beta = (k_B T)^{-1}$, and k_B is Boltzmann's constant. In eqn. (4) we have allowed for only one set of bimolecular products, although at least two are possible. By a much simpler analysis it is shown below that only one is significant; consequently we neglect the others in solving the full master equation.

For future reference, it should be noted that the term in eqn. (4) containing $F_i(E)$ is more naturally written as $k_{ai} n_R n_m \rho_{Rm}(E) e^{-\beta E} / Q_{Rm}$, where k_{ai} is the association rate coefficient into well i, Q_{Rm} is the reactant partition function (including relative translational motion), and $\rho_{Rm}(E)$ is the corresponding density of states. The form chosen in eqn. (4) comes from the detailed balance condition, assuming that the reactants are maintained in thermal equilibrium. It has the advantage of restricting the dependence of eqn. (4) on the reactant properties to the calculation of equilibrium constants.

In the present work nitrogen (N_2) is taken to be the bath gas in all cases, and we restrict our attention to the situation most commonly encountered in chemical kinetics experiments, *i.e.*

$$n_{N_2} \gg n_m \gg n_R \quad (6)$$

The condition (6) implies that $n_m \approx$ constant and thus renders the master equation linear. It is necessary to add to the master equation only an equation for n_R,

$$\frac{dn_R}{dt} = \sum_{i=I, II} \int_{E_{0i}}^{\infty} k_{di}(E) n_i(E) dE - n_R n_m \sum_{i=I, II} K_{eqi} \int_{E_{0i}}^{\infty} k_{di}(E) F_i(E) dE. \quad (7)$$

Eqn. (4) and (7) assume that the reactants are maintained in thermal equilibrium throughout the course of the reaction, a condition that is largely guaranteed by the inequalities (6).

To solve eqn. (4) and (7) it is best to rewrite them in a vector form where the "transition matrix" is Hermitian.[35] Defining $x_R(t) = n_R(t)/n_R(0)$, $x_i(E, t) = n_i(E, t)/n_R(0)$ and $y_i(E, t) = x_i(E, t)/f_i(E)$, where $f_i^2(E) = F_i(E) Q_i(T)$, and approximating the integrals in the equations as discrete sums with a constant energy spacing δE, one obtains (after considerable algebra) the desired form of the equations (in Dirac notation),

$$\frac{d}{dt} | w(t) \rangle = G | w(t) \rangle, \quad (8)$$

where $|w(t)\rangle$ is the vector,

$$|w(t)\rangle \to \left[y_I(E_{0I}), \ldots y_I(E_l), \ldots y_i(E_{0i}), \ldots y_i(E_l), \ldots, \left(\frac{n_m}{Q_{Rm}\delta E}\right)^{1/2} x_R \right]^T, \qquad (9)$$

G is real and symmetric, and E_l is the energy of the lth gridpoint. The solution vector $|w(t)\rangle$ clearly has $1 + \sum_{i=1}^{M} N_i$ components, where N_i is the number of grid points in the ith well.

From the solution vector, $|w(t)\rangle$, one can extract information about the relative populations $x_R(t)$, $X_i(t)$ and $x_p(t)$, where

$$X_i(t) = \int_{E_{0i}}^{\infty} x_i(E, t)dE \qquad i = I, \ldots, M \qquad (10)$$

and

$$x_p(t) = 1 - x_R(t) - \sum_{i=I}^{M} X_i(t) = \frac{n_{CH_2CO}(t)}{n_R(0)} = \frac{n_{HCO}(t)}{n_R(0)}; \qquad (11)$$

$x_p(t)$ is the fraction of the initial reactant concentration, $n_R(0)$, that has formed bimolecular products at time t. Eqn. (11) is a consequence of our assumption that the bimolecular products constitute an "infinite sink" and that the only bimolecular products are $CH_2CO + HCO$, justified below. It is useful in interpreting our solutions to compute a time-dependent rate coefficient $k(T, p, t)$,

$$k(T, p, t) = -\frac{1}{n_m x_R(t)} \frac{dx_R(t)}{dt}. \qquad (12)$$

If $x_R(t)$ decays to zero exponentially in time, $k(T, p, t) = k(T, p)$ is constant in time, and one may reasonably say that a "good" rate coefficient exists. In cases where the reactant decay is not a simple exponential, it is frequently useful to characterize the rate using eqn. (12) at the time when $x_R(t)$ has dropped to a value of $\frac{1}{2}$. Similarly, to avoid any ambiguity in the product distributions, the branching fractions are defined as

$$\alpha_i = X_i(\tau) \qquad i = I, \ldots, M \qquad \text{and} \qquad \alpha_{bi} = x_p(\tau), \qquad (13)$$

where τ is the time when $x_R(\tau)$ has dropped to a value $x_R(\tau) \approx 0.01$.

As in our previous work on the recombination of propargyl radicals, we use two different methods to solve eqn. (8). The first method is the one we prefer, i.e. expansion in terms of the eigenvectors and eigenvalues of G, i.e.

$$|w(t)\rangle = \sum_{j=1}^{N_I + \cdots N_M + 1} e^{\lambda_j t} |g_j\rangle\langle g_j|w(0)\rangle, \qquad (14)$$

where λ_j and $|g_j\rangle$ are the aforementioned $1 + \sum_{i=I}^{M} N_i$ (negative) eigenvalues and eigenvectors. One may fruitfully think of eqn. (14) as an expansion in the "normal modes of relaxation" of G.[37] The slow modes describe "chemical reactions," and under certain conditions they provide explicit information about particular thermal isomerization and dissociation/recombination reactions. Typically there is one chemically significant eigenvalue/eigenvector pair for each transition state in the problem, consistent with the notion that a transition state is a "bottleneck" for reaction in configuration space. Such is the case in the present investigation. In the discussion below we interpret our results in terms of these eigenpairs. Also, eqn. (12) suggests that a good rate coefficient exists only when $x_R(t)$ is governed by a single eigenpair. In such cases, $x_R(t)$ falls and the products rise with the same time constant. We use the DSYEV routine from LAPACK[38] to diagonalize G and thus express $|w(t)\rangle$ in the form given by eqn. (14).

In problems with deep wells at low temperature, it frequently occurs that some of the eigenvalues are so small in magnitude that they cannot be computed accurately by the diagonalization routine. In some instances, this causes no difficulty because these eigenvalues are incidental to the problem, but in other cases they can contaminate the solution badly. In these latter cases we solve eqn. (8) directly using the stiff ODE integrator VODE.[39] This method is more robust than the eigenvector expansion method, but it is somewhat more time consuming and provides less physical insight.

The collisionless (zero-pressure) limit. In certain limiting cases it is not difficult to solve the two-dimensional master equation. One of those cases is the collisionless limit. Our method is an extension of the method first developed by Miller et al.[40] If one takes $Z = 0$ in eqn. (4) and considers $n_i(E)$ to be a function of both E and J, i.e. $n_i(E, J)$, where $n_i(E, J)$ dE is the number density for well i with energy between E and $E + dE$ and with angular momentum quantum number J, eqn. (4) can be written in the simple vector form,

$$\frac{d|n(E, J)\rangle}{dt} = -K(E, J)|n(E, J)\rangle + n_R n_m |a(E, J)\rangle, \quad (15)$$

where $K(E, J)$ is a matrix of isomerization and dissociation rate coefficients and $|a(E, J)\rangle$ is a vector of association rate coefficients multiplied by a reactant Boltzmann factor. Applying the steady-state approximation to eqn. (15) and solving for $|n(E, J)\rangle$, one obtains

$$|n(E, J)\rangle = K^{-1}(E, J)|a(E, J)\rangle n_R n_m, \quad (16)$$

where $K^{-1}(E, J)$ is the inverse matrix of $K(E, J)$. The rate of formation of bimolecular products can be described by the vector equation,

$$\frac{d|P(E, J)\rangle}{dt} = D(E, J)|n(E, J)\rangle, \quad (17)$$

where the components of $|P(E, J)\rangle$ are the number densities per unit energy of the various possible sets of bimolecular products, and $D(E, J)$ is the matrix of rate coefficients describing dissociation to these products from the various wells. Substituting eqn. (16) into (17), one has

$$\frac{d|P(E, J)\rangle}{dt} = D(E, J)K^{-1}(E, J)|a(E, J)\rangle n_R n_m. \quad (18)$$

Integrating over E and summing over J, one can identify a vector of thermal rate coefficients. As indicated above, $|a(E, J)\rangle$ can be written as

$$|a(E, J)\rangle = |b(E, J)\rangle \rho_{Rm}(E, J) e^{-\beta E}/Q_{Rm}(T), \quad (19)$$

where $\rho_{Rm}(E, J)$ is the appropriate reactant density of states and $|b(E, J)\rangle$ contains only the association rate coefficients. Using this information one can write the vector of bimolecular rate coefficients as

$$|k_0(T)\rangle = \frac{1}{Q_{Rm}(T)} \sum_J (2J + 1) \int_0^\infty D(E, J)K^{-1}(E, J)|b(E, J)\rangle \rho_{Rm}(E, J) e^{-\beta E} dE. \quad (20)$$

As noted by Miller et al.,[40] when one substitutes the appropriate RRKM rate coefficients into eqn. (20) all the densities of states cancel, and one is left with the result,

$$|k_0(T)\rangle = \frac{1}{hQ_{Rm}(T)} \sum_J (2J + 1) \int_0^\infty N_D(E, J) N_K^{-1}(E, J)|N_b(E, J)\rangle e^{-\beta E} dE, \quad (21)$$

where h is Planck's constant, and N_D, N_K^{-1} and $|N_b\rangle$ are related to D, K^{-1} and $|b\rangle$ in that the former contain only the sums of states $N^\pm(E, J)$ that appear in the numerators of the corresponding RRKM rate coefficient expressions, $k(E, J) = [N^\pm(E, J)]/[h\rho(E, J)]$. The vector $|k_0(T)\rangle$ has as its components the thermal rate coefficients for the bimolecular product channels. In practice we recognize at the outset that we do not need the densities of states and work only with N_D, N_K and $|N_b\rangle$.

Evaluating eqn. (21) is relatively straightforward as long as one takes care to avoid singularities in N_K. Discussing this in any detail would take us too far afield, but it can be done by considering the "connectivity" of the wells among themselves and with reactants and products, eliminating rows and columns in N_D, N_K and $|N_b\rangle$ at E, J combinations where linear dependences occur.

We call the theoretical treatment just described, *i.e.* with both E and J conserved, microcanonical, J-conservative theory, μJT. One can go through the same analysis without conserving J; we call such a treatment microcanonical theory, μT. Full multiple-well ME rate coefficients reduce to the μT result as $Z \to 0$.

One-well, one-product dissociation. Smith and Gilbert[41] have provided a very useful method for solving the two-dimensional master equation for simple one-well problems by reducing the two-dimensional ME to an approximately equivalent one-dimension problem. Our method is closely associated with theirs. Smith and Gilbert take the independent variables in the two-dimensional master equation to be ε and J, where ε is the energy in the active degrees of freedom. The key to the method is that the energy transfer probability function $P(\varepsilon, J; \varepsilon', J')$ is assumed to have the form

$$P(\varepsilon, J; \varepsilon', J') = P(\varepsilon, \varepsilon')\Phi(\varepsilon, J), \qquad (22)$$

where $\Phi(\varepsilon, J)$ is a function only of the state of the molecule *after* the collision. In their analysis, Smith and Gilbert use the function,

$$\Phi(\varepsilon, J) = (2J + 1)e^{-\beta E_J}/Q_J, \qquad (23)$$

where E_J and Q_J are the rotational energy (*i.e.* that associated with J) and rotational partition function for the molecule, respectively. This model assumes the final state J distribution is Boltzmann for each and every initial ε', J' combination individually. Such an assumption may not be very physically realistic.

The derivation of our model is formally the same as that of Smith and Gilbert. However, we formulate the problem in terms of E and J directly, rather than ε and J, and take the energy transfer function to be

$$P(E, J; E', J') = P(E, E')\varphi(E, J) \qquad (24)$$

where $\varphi(E, J)$ is assumed to be

$$\varphi(E, J) = (2J + 1)\rho(E, J)/\rho(E), \qquad (25)$$

with $\rho(E) = \sum_J (2J + 1)\rho(E, J)$. The density of states $\rho(E, J)$ is such that $\rho(E, J)$ dE is the number of states with energy between E and $E + $ dE and with angular momentum quantum number equal to J. Eqn. (24) and (25) imply that rotational energy is transferred just like vibrational energy and that the J distribution after collision is proportional to the volume of phase space available at any particular E, J combination. We believe that this model is more realistic than the one described in the previous paragraph. We shall discuss both these models (and others), as well as various forms for $P(E, E')$, in a forthcoming paper.[42]

In all the calculations reported here, a simple exponential-down model was used for $P(E, E')$. The computer code VARIFLEX[43] was used to perform all the chemical kinetics calculations.

Results and discussion

A. Energetics and pathways

A schematic diagram of the HL calculated mechanism for the $C_3H_3 + O_2$ reaction is provided in Fig. 1, corrected as discussed below to reflect experimental rate coefficients and equilibrium constants. The corresponding stationary point energies are summarized in Table 1 for each of the quantum-chemical methods considered. All energies mentioned in the paper are measured relative to reactants and include zero-point corrections. In the kinetics section, the saddle points are presumed to be the equilibrium geometry for the transition state dividing surfaces and so are labeled ts here. Those structures for which G3 and MC-G3 results are absent correspond to stationary points for which the predicted kinetics of the $C_3H_3 + O_2$ reaction has relatively little dependence on the calculated energetics.

The geometric structures for each of these stationary points are illustrated in Fig. 2. Various structures related to simple torsional motions from these minima have also been explored at the

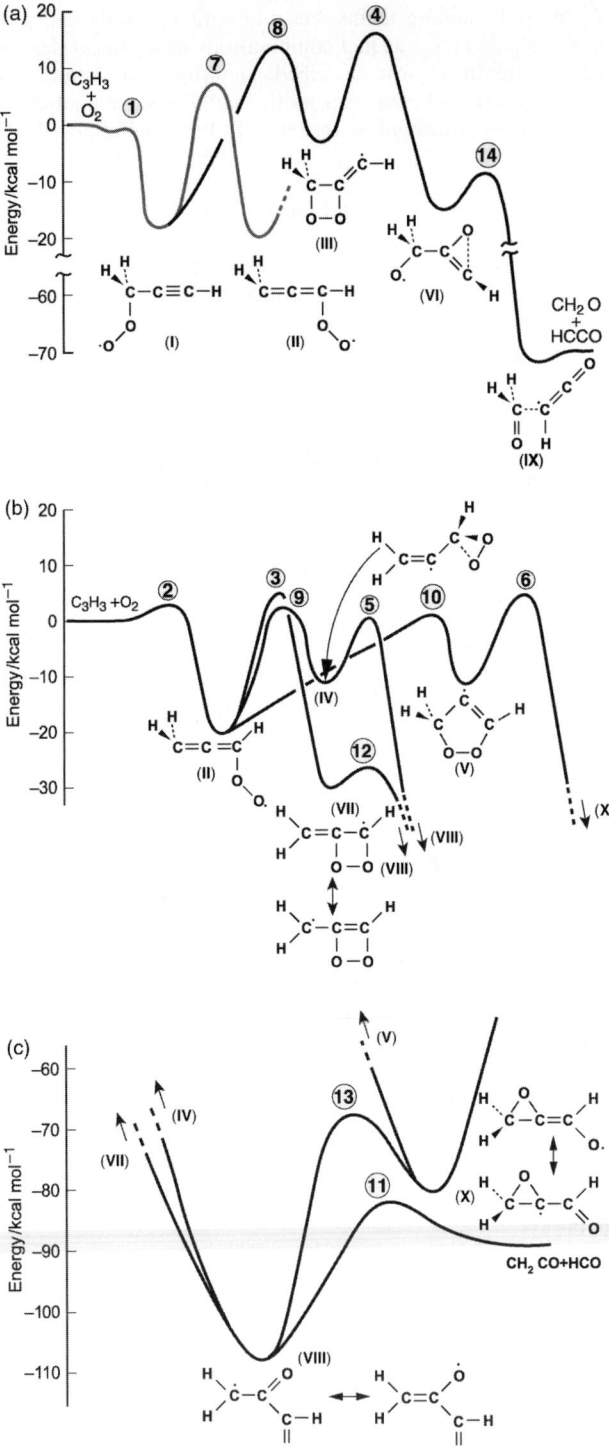

Fig. 1 Reaction coordinate diagram for the $C_3H_3 + O_2$ reaction.

Table 1 Energetics of stationary points for $C_3H_3 + O_2$[a]

Species	State	B3LYP[b]	G3	HL	MCG3	Expt	$\langle S^2 \rangle$[c]	A_{max}
I; CH_2 add'n.	$^2A''$	−11.2	−17.2	−19.2	−18.5	−18.2	0.75(0.76)	0.15
II; CH add'n.	$^2A''$	−16.3	−19.4	−20.1	−22.4		0.76(0.90)	0.14
III; CCOO ring; CH_2 side	$^2A'$	9.5	−0.4	−2.8	−2.3		0.76(0.92)	0.09
IV; COO ring; CH side	$^2A'$	1.7	−8.5	−11.2	−10.5		0.76(0.96)	0.10
V; CCOOO ring	^2A	2.4	−8.0	−10.8	−9.9		0.76(0.98)	0.10
VI; CH_2OCOCH	^2A	−7.3		−15.3			0.76(0.76)	0.10
VII; CCOO ring; CH side	$^2A''$	−19.6		−30.5			0.77(0.95)	0.11
VIII; CH_2COCHO	$^2A''$	−98.9		−107.7			0.77(1.02)	0.11
IX; CH_2O···CHCO	$^2A''$	−66.6		−72.7			0.76(0.82)	0.06
X; COC ring; CH_2 side	$^2A''$	−71.6		−80.1			0.76(0.90)	0.11
ts1(R↔I)	$^2A''$	2.5	4.9	3.7	2.6	−0.2	1.27(1.87)	0.12
ts2(R↔II)	$^2A''$	4.3	8.2	7.1	5.5	∼3	1.23(1.95)	0.12
ts3(II↔VII)	^2A	6.7	7.5	5.9	1.9		0.76(1.39)	0.18
ts4(III↔VI)	^2A	21.2	19.4	18.0	16.9		1.02(1.91)	0.16
ts5(IV↔VIII)	$^2A''$	11.5	4.0	1.5	−0.8		0.86(0.97)	0.25
ts6(V↔X)	^2A	9.9	7.8	7.7	4.7		1.12(1.96)	0.24
ts7(I↔II)	$^2A''$	5.8	9.8	8.9	5.1		0.82(1.83)	0.11
ts8(I↔III)	^2A	22.2	17.2	15.1	13.6		0.78(1.11)	0.16
ts9(II↔IV)	^2A	9.2	5.4	3.6	1.6		0.77(1.13)	0.11
ts10(II↔V)	$^2A''$	8.6	5.2	2.2	−0.8		0.76(0.84)	0.16
ts11(VIII↔P1)	^2A	−76.4		−81.9			0.76(0.81)	0.05
ts12(VII↔VIII)	^2A	−16.0		−26.9			0.78(0.95)	0.10
ts13(VIII↔X)	^2A	−60.5		−68.0			0.76(0.93)	0.21
ts14(VI↔IX)	^2A	−5.8		−8.3			0.99(1.50)	0.21
P1 = HCO + CH_2CO		−83.5		−89.3			0.75(0.77)	0.05
P2 = H_2CO + HCCO		−66.1		−71.4			0.76(0.82)	0.06
$C_2H_3 + CO_2$		−102.0		−109.9			0.76(0.95)	0.010
CH_3CO + CO		−105.1		−117.1			0.75(0.76)	0.039
CH_2CHO + CO		−99.8		−110.4			0.77(0.93)	0.099
CH_2CCHO + O		16.9		14.4			0.77(1.07)	0.068

[a] All energies are relative to $C_3H_3 + O_2$ and include B3LYP/6-311+G(d,p) estimated zero-point corrections.
[b] B3LYP/6-311+G(d,p) energy. [c] The first number is the average spin squared for the B3LYP/6-311+G(d,p) calculation and the number in parentheses is the value for the MP2/6-311++G(d,p) calculation.

B3LYP/6-311+G** level. The latter information is used in conjunction with the vibrational frequencies in generating Fourier series representations of the torsional potentials.

Also provided in Table 1 are the average spin squared, $\langle S^2 \rangle$, and the maximum amplitude in the QCISD(T) analysis, A_{max}, both of which provide some indication of the multireference character of each state, and thereby of the relative uncertainties in the calculated energies. For the wells, the spin contaminations are quite minor, and one expects the higher-level energy estimates to be fairly accurate. Indeed, the HL, G3 and MC-G3 estimates for the well depths are reasonably similar, with the maximum difference being 3.1 kcal mol^{-1}. In contrast, these well depths are significantly underestimated at the B3LYP level, with a typical underestimate being 10 kcal mol^{-1}.

For the saddle point structures, there is more variation in the G3, MC-G3 and HL estimates, while the B3LYP values appear to be less of an overestimate. The extreme spin contamination for ts1, ts2, ts4, ts6 and ts14 suggests some need for caution in interpreting the energies for these geometries. The spin contamination for a number of the other saddle points is also large enough to be of some concern.

The first step in the mechanism illustrated in Fig. 1 involves addition of the oxygen molecule to either of the radical sites from the two resonance structures for propargyl. At all levels of theory considered here there is a well-defined saddlepoint with a significant barrier height for each of these additions. The presence of a barrier for these additions is the result of the loss of the resonance stabilization of both C_3H_3 and O_2 upon formation of the addition complex. This loss of resonance stabilization in C_3H_3 also results in greatly reduced well depths for the two $C_3H_3O_2$

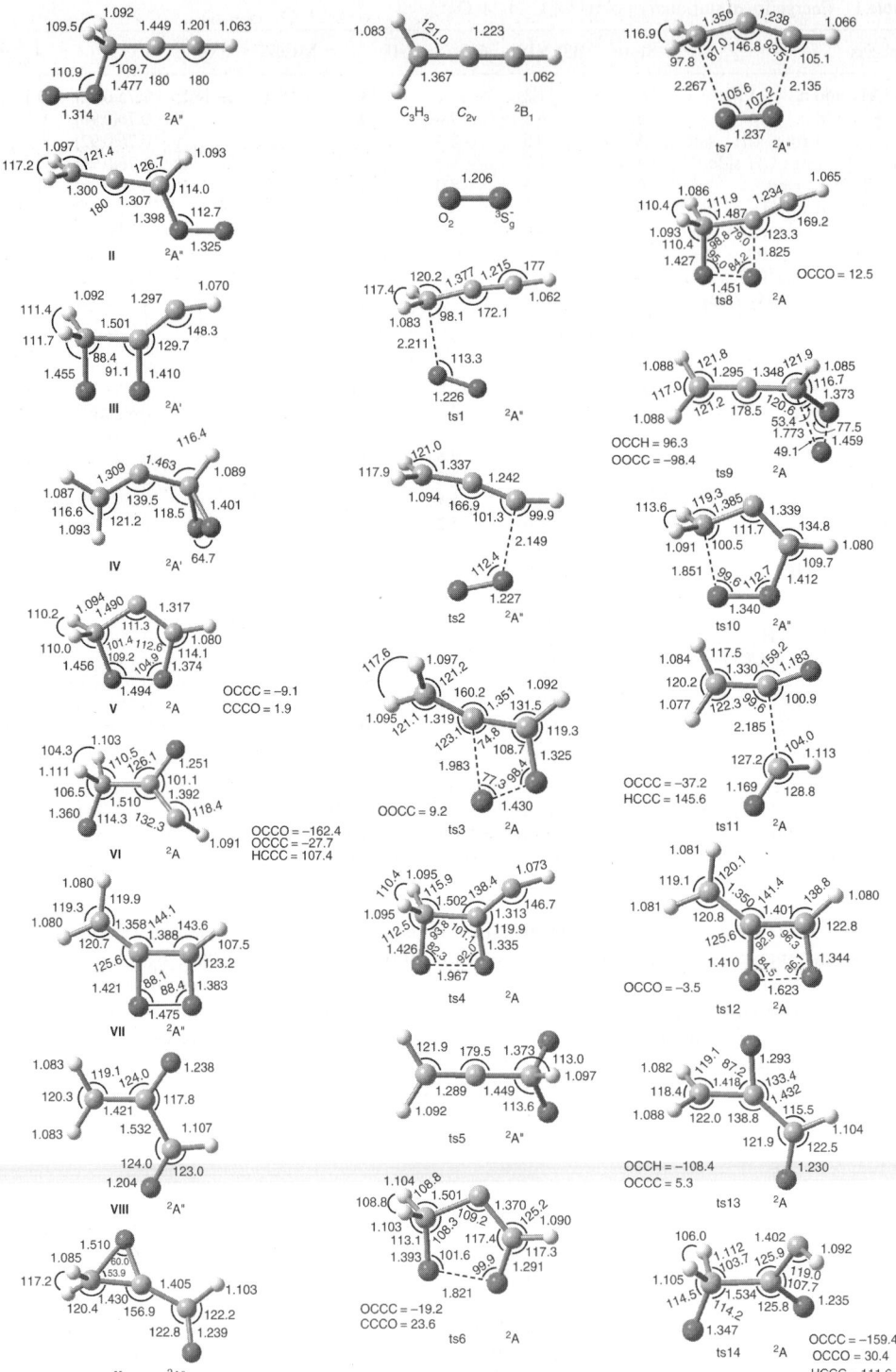

Fig. 2 Geometric structures of stationary points on $C_3H_3 + O_2$ potential energy surface.

addition complexes (~19 kcal mol^{-1} each) and the related $C_3H_5O_2$ complex (20 kcal mol^{-1}), as compared with those for the HCCOO, C_2H_3OO and C_2H_5OO complexes (48, 45 and 34 kcal mol^{-1}, respectively, each from the present HL calculations). The 3 kcal mol^{-1} lower value for the addition barrier on the CH_2 side of C_3H_3 than on the CH side correlates with the dominance of the corresponding resonance structure in the propargyl radical.

A plot illustrating the B3LYP/6-31G* calculated dependence of the optimal interaction energy on the CO separation is provided in Fig. 3 for the addition of O_2 to HCC, C_2H_3, C_2H_5, C_3H_5 and C_3H_3. Interestingly, even though the spin contaminations are comparable in each case, allyl radical, which is the only other resonantly stabilized radical in this set, is also the only other radical that is predicted to have a saddlepoint for the addition. To verify further the presence of such saddlepoints, we have reoptimized the ts1 geometry at the CCSD(T)/6-31G(d) level. An HL calculation at this geometry yields a barrier height that is essentially the same as that calculated at the B3LYP/6-311+G(d,p) geometry. It is also worth noting that the observed high-pressure addition rate constants are substantially smaller for the two resonantly stabilized radicals.[15,16] Regardless, given the high spin-contamination at these entrance saddlepoints, it would be worthwhile, in future calculations, to perform a multi-reference based analysis of these entrance saddlepoints.

Related calculations for unsaturated hydrocarbon radicals have indicated that further reaction generally involves the splitting of the OO bond.[21–23] For both C_2H_3OO and HCCOO this splitting proceeds via the formation of either a 3-membered –COO– ring or a 4-membered –COOC– ring, with the 3-membered ring providing the dominant pathway. For C_3H_3OO a 5-membered –CCCOO– ring may also play a role.

We now focus on the available pathways for reaction from HCCCH$_2$OO (**I**). The formation of the 3-membered ring from **I** does not occur since the CH_2 carbon already has its full valence shell occupied with single bonds to separate moieties. The 4-membered ring (**III**) has a large barrier to both its formation from **I** and to the splitting of the OO bond to form **VI**, and thus it is expected to be kinetically insignificant. The height of this barrier relative to the bottom of well **I** is roughly equivalent to the analogous energy difference for HCC + O_2 and for C_2H_3 + O_2. Thus, it is just the absence of resonance stabilization in **I** and **III** that makes this 4-membered ring route highly endothermic.

It would seem that the 5-membered –CCCOO- ring (**V**) could be formed directly from **I**. However, attempts to locate the saddlepoint for this process instead led to saddlepoints for either the initial addition to form **I**, or for the direct transfer of the O_2 group from the CH_2 side to the CH side of C_3H_3 (*i.e.* from **I** to **II** via ts7). Apparently, the barrier to forming the 5-membered ring

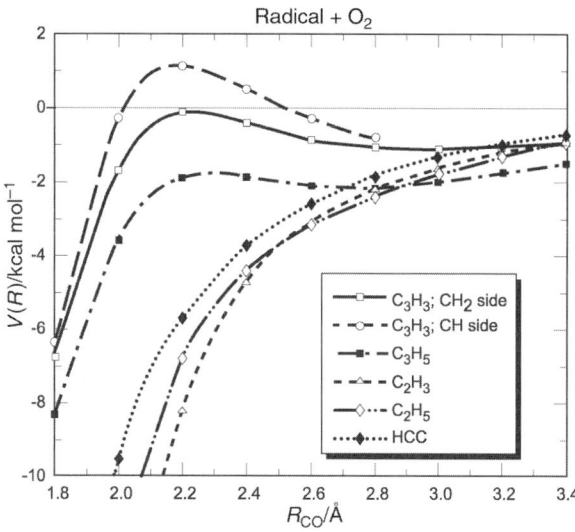

Fig. 3 Reaction path energy for the addition of several flame radicals to O_2.

from **I** is higher than that for either of these two processes, likely due to the energy penalty involved in bending the CC triple bond. The additional fact that the barrier from **I** to **II** is higher than those for forming **I** and **II** from reactants, suggests that the formation of well **I** is essentially a dead end, with no further reaction occurring from it. To verify this a few secondary possibilities for reaction are considered next.

With saturated alkyl radicals the decompositions initially involve transfer of either the O_2 group to abstract an H atom and form HO_2 plus an alkene, or an H atom to the O_2 group, to form a COOH moiety within a C centered radical.[25,26] The H transfer route can produce either HO_2 plus an alkene again, or OH plus an aldehyde or ketone. For $C_3H_3 + O_2$, the formation of HO_2 is accompanied by the production of C_3H_2, which is highly endothermic. However, one possible complex formed *via* an H transfer route (HCCCHOOH) has two resonance structures and as a result is energetically allowed, with a well depth of 8.7 kcal mol^{-1}. Furthermore, the OH plus HCCCHO products arising from this complex are estimated from HL calculations to be 45.7 kcal mol^{-1} exothermic. However, HL calculations estimate the barriers (relative to reactants) to formation of HCCCHOOH from **I** and **II** to be 22 and 27 kcal mol^{-1}, respectively. The barrier heights relative to wells **I** and **II** are quite similar to the analogous energies in C_2H_5OO. Thus, the large positive value for these barriers (relative to reactants) are again simply reflective of the shallowness of wells **I** and **II**.

In $C_2H_3 + O_2$ and $HCC + O_2$, the initially formed wells may also decompose *via* the exothermic loss of an O atom. In both cases the resulting molecular radical is resonance stabilized. For well **I**, the resulting radical is not resonantly stabilized and so the loss of an O atom is highly endothermic [at the B3LYP/6-31G(d) level it is calculated to be 48 kcal mol^{-1}]. For well **II**, the CH_2CCHO product is resonance stabilized. However, HL calculations suggest it is still 14 kcal mol^{-1} endothermic. This increased endothermicity, relative to that for the corresponding products in the C_2H_3 reaction, is again indicative of the resonance stabilization of the C_3H_3 reactants, with the actual OO bond dissociation energies being roughly equivalent. Simple H loss channels from both **I** and **II** also appear to be highly endothermic, as do the transition states for H transfer from these wells. In summary, the usual paths for producing bimolecular products all lead to barriers of at least 14 kcal mol^{-1} for pathways proceeding directly from well **I**.

In contrast with well **I**, the formation of the three-, four- and five-membered rings from well **II** each proceeds with a barrier predicted to be lower than that for the entrance channel saddlepoint. The formation of the five-membered ring has the lowest saddlepoint at 2.2 kcal mol^{-1}, likely due to the diminished ring strain. However, the subsequent splitting of the OO bond, which ultimately yields the low energy species **X**, has the largest barrier of 7.7 kcal mol^{-1}. Further conversion from **X** to **VIII** is facile. It is perhaps worth noting that the 5-membered ring itself has two nearly degenerate electronic states in C_s symmetry (the $^2A'$ state is at -10.9 kcal mol^{-1}, while the $^2A''$ state is at -5.0 kcal mol^{-1}), and the HL calculations actually suggest that the $^2A'$ state is lower than the symmetry-broken ^2A state.

For HCCOO and C_2H_3OO, the formation of the three-membered COO rings provide the dominant pathways for reaction, with saddlepoints at -29.0 and -21.4 kcal mol^{-1} respectively. Due once again to the resonance stabilization of the C_3H_3 radical, the saddlepoint for formation of the three-membered ring (**IV**) from **II** lies significantly higher at 3.6 kcal mol^{-1}. However, this value is still below that for the entrance channel, and the saddlepoint for the subsequent conversion to the diketone **VIII** is even lower at 1.5 kcal mol^{-1}. Thus, based solely on energetics, one expects this channel to provide the dominant pathway for reaction.

Interestingly, the saddlepoint at 5.9 kcal mol^{-1} for formation of the four-membered CCOO ring (**VII**) from **II**, is only 7 kcal mol^{-1} higher than the corresponding saddlepoints in HCCOO and C_2H_3OO. Apparently, the resonance stabilization of the four membered ring species, **VII**, offsets at the saddlepoint some of the energy cost relating to the initial loss of resonance stabilization from C_3H_3. As a result, the four-membered ring pathway lies only slightly higher than the three-membered ring pathway and might be expected to be competitive with it. The splitting of the OO bond from **VII** again produces the diketone species **VIII**, now with a saddlepoint (ts12) well below the reactants.

Each of the ring-forming pathways from well **II** ultimately leads to the diketone species **VIII**. The lowest energy decomposition of this species is expected to produce $HCO + CH_2CO$. However, it is worth noting that there are a large number of other energetically accessible products. For example, $C_2H_3 + CO_2$, $CH_3CO + CO$ and $CH_2CHO + CO$ (at -110, -118 and -110 kcal mol^{-1}, respectively) are each more exothermic than are the HCO + CH$_2$CO products. However, their formation (either from **VIII** or from other wells such as **IV**) requires an H transfer that is expected to have a substantially greater barrier than that for some other available pathway. Similarly, a number of other products (both unimolecular and bimolecular) are accessible *via* O atom insertion to form ether (and even di-ether) complexes, and it is not clear (*cf.* ts10 for C_2H_3OO in ref. 9) that the barriers for such processes would be larger than that for dissociation of well **VIII**, for example. However, for well **VIII**, the initially formed ether would still be expected ultimately to yield HCO + CH$_2$CO products. Another interesting species is an $-H_2CC(O)CHO-$ ring, which is resonantly stabilized and (at the B3LYP/6-311+G** level) is only 21 kcal mol^{-1} higher than **VIII**. This species might yield either HCO + CH$_2$CO or H$_2$CO + HCCO upon dissociation. However, the transition state for formation of this species from **VIII** is unlikely to be below that for dissociation of **VIII**. In summary, although a more complete exploration of the lower energy pathways is necessary to be certain, it does appear likely that HCO + CH$_2$CO will be the only significant bimolecular products from the $C_3H_3 + O_2$ reaction.

B. Chemical kinetics calculations

Our rate coefficient calculations are all based on the potential energy surface (PES) depicted in Fig. 1. This differs from the HL results discussed above in the following ways:

1. The $C_3H_3-O_2$ bond energy for well **I** has been reduced from 19.2 kcal mol^{-1} to 18.2 kcal mol^{-1} in order to predict accurately the equilibrium constants measured by Slagle and Gutman.[15] The threshold energy for ts1, E_{01}, has been reduced from 3.7 kcal mol^{-1} to -0.2 kcal mol^{-1} in order to obtain good agreement with the high-pressure rate coefficients of Atkinson and Hudgens. A similar adjustment was made to E_{02} in the belief that the HL method would make similar errors in the two barrier heights. This latter change also helped the agreement of the theoretical results with the high temperature experiments of Slagle and Gutman.

2. The threshold energies E_{0i}, $i = 3, \ldots, 10$, were all reduced by 1 kcal mol^{-1} to improve agreement with experiment. They were all reduced by the same amount, because we believe the HL method should result in similar errors in all of them. Other stationary point energies were left unchanged, because we believed that they were inconsequential to the kinetics, a suspicion that was later confirmed.

Fig. 1 suggests a useful simplification to the kinetics analysis. The transition states ts3, ts4, ts5 and ts6 are essentially "points of no return," *i.e.* any complex that passes through these transition states moving to the right on the diagram inevitably goes to bimolecular products. Making this assumption *a priori* converts a 10-well problem to a 5-well problem, substantially reducing the demand on our computing resources. To check the accuracy of this approximation we did a number of calculations, covering the range of temperatures and pressures of interest (including the zero pressure limit), both with and without this assumption. There were no significant differences in the results. In fact, what differences there were are most likely due to the larger value of δE required for the 10-well problem ($\delta E = 150$ cm^{-1} as opposed to $\delta E = 75$ cm^{-1} for the 5-well case). Most of the results presented below come from solving the 5-well master equation.

Fig. 4 is an Arrhenius plot of the "limit" rate coefficients: $k_{I\infty}(T)$ and $k_{II\infty}(T)$, the high-pressure limit rate coefficients for formation of $C_3H_3O_2(I)$ and $C_3H_3O_2(II)$, respectively (the roman numerals in parentheses refer to the well numbers) and $k_0^{(1)}(T)$ and $k_0^{(2)}(T)$, the zero-pressure limit rate coefficients for formation of CH$_2$O + HCCO and CH$_2$CO + HCO, respectively. There are some important conclusions to be drawn and observations to be made from Fig. 4. First, the CH$_2$O + HCCO channel is never significant; $k_0(T) = k_0^{(1)}(T) + k_0^{(2)}(T) \approx k_0^{(2)}(T)$ for all temperatures, confirming the assertion made in the Theory section that only one bimolecular product channel is significant. In fact, $k_0^{(1)}(T)$ is never as much as 1% of $k_0(T)$ for temperatures below 2000 K. Second, the μJT and μT results for $k_0^{(1)}(T)$ and $k_0^{(2)}(T)$ are virtually identical, indicating that angular momentum conservation is not important for predicting these rate coefficients. Third, $k_{I\infty}(T)$ and $k_{II\infty}(T)$ are, strictly speaking, rate coefficients for formation of $C_3H_3O_2^*(I)$ and $C_3H_3O_2^*(II)$ complexes, respectively. Because there is never any appreciable conversion of (**I**) to (**II**) through ts7, the ratio $k_0^{(1)}(T)/k_{I\infty}(T)$ and $k_0^{(2)}(T)/k_{II\infty}(T)$ are the probabilities that the complexes, once formed, will go on to form bimolecular products. $C_3H_3O_2^*(II)$ is converted to CH$_2$CO + HCO approximately

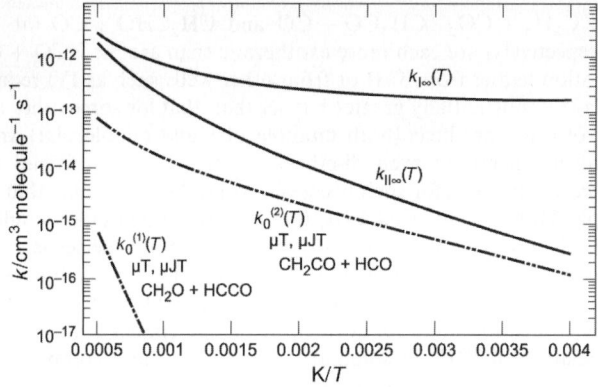

Fig. 4 High-pressure limit and zero-pressure limit rate coefficients for $C_3H_3 + O_2 \to$ products.

1/3 of the time at low T, a value that is reduced to about 1/20 at 2000 K. By contrast, the probability of $C_3H_3O_2^*(I)$ producing $CH_2O + HCCO$ is always less than 10^{-3} for $T \leq 2000$ K, a consequence of the large energy barriers at ts4 and ts8. Well (I) is a dead end to reaction as long as no stabilization occurs.

Fig. 5 is an overview of the kinetic behavior one might observe in an experiment designed to cover a broad range of temperatures and pressures, p. All the results shown, except the limiting rate coefficients $k_0(T)$ and $k_{1\infty}(T)$, were calculated with the 5-well, time-dependent master equation model. At low temperatures, i.e. for $T < 380$ K, the rate coefficient $k(T, p)$ is pressure dependent with $C_3H_3O_2(I)$ the only product, a result that one might have anticipated from Fig. 4. In the temperature range, $380 < T/K < 430$, the reaction $C_3H_3 + O_2 \leftrightarrow C_3H_3O_2(I)$ equilibrates, making it difficult to measure a rate coefficient. In fact, Slagle and Gutman were unable to measure rate coefficients in this region, although they were able to measure K_{eqI}, a valuable contribution. Although we shall continue to use it, the term "equilibrates" is not a very good one. What is meant is that reactants begin to be as important as products at equilibrium. Emerging from the high-temperature end of this equilibration region is a reaction whose rate coefficient $k(T, p) = k_0(T)$ independent of pressure and may be written as the elementary step, $C_3H_3 + O_2 \to CH_2CO + HCO$. This behavior is exactly that observed by Slagle and Gutman. In the paragraphs that follow we discuss each of these regimes in some detail.

As mentioned above, Slagle and Gutman measured equilibrium constants for the reaction $C_3H_3 + O_2 \leftrightarrow C_3H_3O_2(I)$. We compare our theoretical values for $K_{eqI}(T)$ with their results in Fig. 6. The HL bond energy, $D_0 = 19.2$ kcal mol^{-1}, gives equilibrium constants that are too high, but reducing D_0 to 18.2 kcal mol^{-1} produces excellent agreement with the experiments. The bond

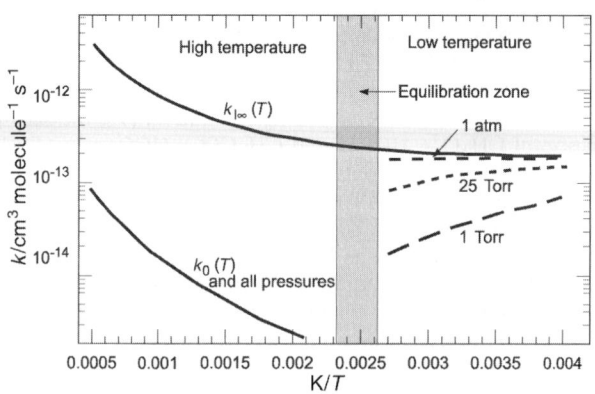

Fig. 5 Arrhenius plot of the thermal rate coefficient for $C_3H_3 + O_2 \to$ products.

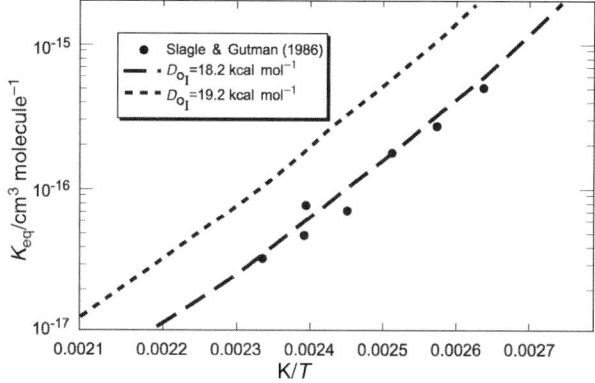

Fig. 6 Equilibrium constants for $C_3H_3 + O_2 \leftrightarrow C_3H_3O_2(I)$.

energy is important in the kinetics analysis, because it largely determines the lifetime of $C_3H_3O_2^*(I)$ complexes and thus plays a major role in describing the pressure dependence of $k(T, p)$ in the low-temperature region. It also determines the boundary between the low-temperature regime and the equilibration zone.

Fig. 7 is a fall-off plot of $k(T, p)$ at room temperature, $T = 295$ K, i.e. a plot of the rate coefficient as a function of pressure. Since, as discussed above, the multiple-well ME indicates that the reaction is a simple one-well dissociation/recombination reaction under these conditions, we have calculated the rate coefficient from the two-dimensional master equation (discussed in the Theory section), as well as from the one-dimensional ME. The value of $E_{01} = -0.2$ kcal mol^{-1} was determined by adjusting this energy to give good agreement with the higher pressure experiments shown in the plot. A value of $\langle \Delta E_d \rangle = 500$ cm^{-1} was determined by matching the 2-d ME solutions to the experimental fall-off behavior. The 1-d solution using the same value of $\langle \Delta E_d \rangle$ as for the 2-d case differs more and more from the 2-d solution as the pressure is reduced, as one might expect. All the calculations described above and below were performed with this same value of $\langle \Delta E_d \rangle$.

Fig. 8 compares our predictions for $k(T, p)$ with the experiments of Slagle and Gutman in the low-temperature regime. The experiments were performed at pressures of $p \approx 2$ Torr and $p \approx 0.7$ Torr; the calculations used exactly these values. The 2-d master equation solutions are in excellent agreement with the experiments. The 1-d solutions are approximately 25% to 35% higher. All in all, the theory provides a very good quantitative description of the chemical kinetics in the low-temperature regime.

The most interesting kinetic behavior occurs in the equilibration zone of Fig. 5. Fig. 9 is an Arrhenius plot of several rate coefficients, calculated by various methods, that allow us to interpret

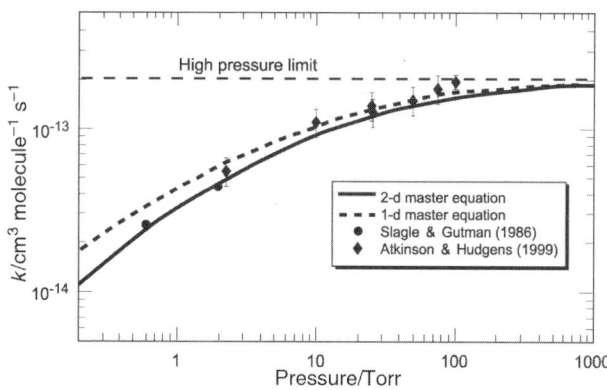

Fig. 7 Comparison of theory with experiment for pressure dependence of $k(T, p)$ at 295 K.

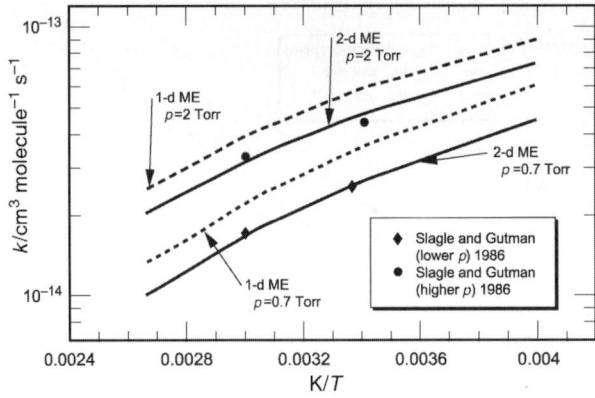

Fig. 8 Comparison of theory with experiment for $k(T, p)$ in the low-temperature regime, $T < 380$ K.

this behavior. The calculations were all performed for conditions that are very close to those of Slagle and Gutman's experiments, i.e. $p = 2.5$ Torr with a partial pressure of O_2 of 0.2 Torr. Both the 1-d and 2-d solutions of the one-well problem are shown for $k_1(T, p)$, the rate coefficient for $C_3H_3 + O_2 \leftrightarrow C_3H_3O_2(I)$. They differ by only 20% or so under these conditions. If $x_R(t)$ is governed by a single eigenpair of G (λ_i and $|g_i\rangle$), and if $x_R(t)$ decays to zero with the time constant $-\lambda_i$, the measured rate coefficient is equal to

$$k(T, p) = -\lambda_i/n_m. \qquad (26)$$

One can see that this is the case up to the beginning of the equilibration zone, i.e.

$$k(T, p) = k_1(T, p) = -\lambda_6/n_m, \qquad (27)$$

where λ_6 is the sixth largest eigenvalue (algebraically) of G at room temperature. In other reactions, the deviation of $-\lambda_i/n_m$ from $k(T, p)$ has been indicative of very complicated behavior.[18,24,35,36] However, in the present case it is simply a result of the equilibration of the stabilization reaction, i.e. $x_R(t)$ approaches a non-zero asymptote. One can still extract $k_1(T, p)$ from $-\lambda_6$ with knowledge of the equilibrium constant,

$$k_1 = \frac{-\lambda_6 K_{eq}}{1 + n_m K_{eq}}. \qquad (28)$$

Eqn. (28) and a companion expression for the reverse rate coefficient should be familiar from the theory of isomerization (first-order) reactions if n_m is taken equal to unity.[37,44] However, in the

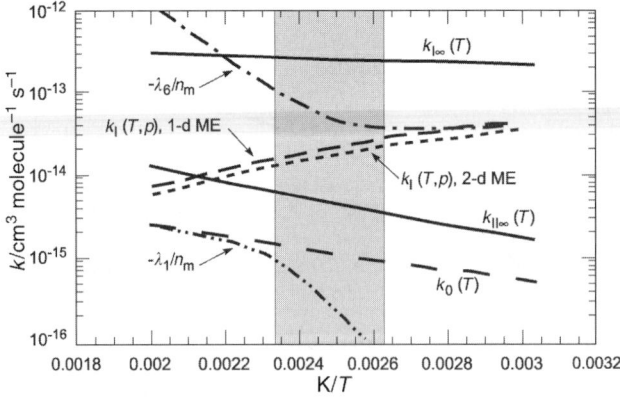

Fig. 9 Rate coefficients, as indicated, in the equilibration zone.

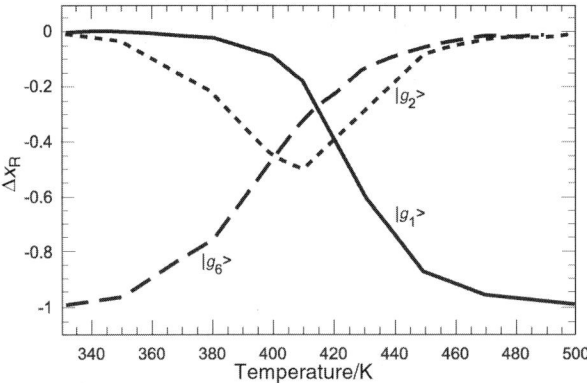

Fig. 10 Contributions of eigenvectors $|g_j\rangle$ to Δx_R.

present case, the product $n_m k_l(T, p)$ is the effective, pseudo-first-order rate coefficient, and $n_m K_{eq}$ is the effective first-order equilibrium constant. Thus n_m appears in eqn. (28). If one uses eqn. (28) to calculate $k_l(T, p)$, the results are the same as those plotted for the 1-d, 1-well problem in Fig. 9. It is important to emphasize that the reaction, $C_3H_3 + O_2 \leftrightarrow C_3H_3O_2(I)$, retains its integrity as an elementary step over the entire range of conditions considered here. Its rate coefficient is unobservable at high temperature only because the reaction equilibrates. Emerging from the equilibration zone, the measurable rate coefficient, $k(T, p) = k_0(T)$, gradually becomes equal to $-\lambda_1/n_m$, another good rate coefficient. Understanding how this occurs requires a more careful consideration of the eigenvalues and eigenvectors of G.

It is convenient to label the eigenvalues of G by their values at room temperature, $T = 295$ K. Typically there is one chemically significant eigenpair for each transition state. These correspond to the slow modes of relaxation of the entire system. Because our mathematical model does not allow the bimolecular products to recombine, all the eigenvalues are negative. Let us define λ_1 to be the algebraically largest eigenvalue (the least negative one) at $T = 295$ K, λ_2 to be the next largest one, and so on. The corresponding eigenvectors are $|g_1\rangle, |g_2\rangle$, etc. If we had allowed the bimolecular products to recombine, there would be another eigenvalue, $\lambda_0 = 0$, corresponding to complete thermal and chemical equilibrium. Although there are a number of crossings of the chemically significant eigenvalue curves as T increases, we maintain the 295 K labeling throughout our discussion. In fact, we shall be concerned with only $\lambda_1, \lambda_2, \lambda_6$ and their corresponding eigenvectors.

If one thinks of each term in eqn. (14) as propagating forward in time independently from $t = 0$ to $t = \infty$, it is not difficult to envision each chemically significant eigenpair as a "chemical reaction." Each eigenpair has the conservation property,[45,46]

$$\left(\Delta x_R + \sum_{i=1}^{M} \Delta X_i + \Delta x_p\right)_j = 0, \tag{29}$$

where the subscript j stands for the jth eigenpair, and Δ indicates the change in population that accompanies the evolution of that eigenpair from $t = 0$ to $t = \infty$. Frequently it happens, particularly at low temperature, that only one term in eqn. (29) is negative (the reactant) and only one is positive (the product). Under such conditions, that eigenpair truly describes an elementary chemical reaction as one normally thinks of it, i.e. an isomerization or dissociation/recombination reaction, and its forward and reverse rate coefficients can be determined from expressions analogous to eqn. (28). However, at higher temperatures, "mixing" of these eigenvectors can occur, resulting in bi-exponential reactant decays and jumping of rate coefficients from one eigenvalue curve to another on an Arrhenius plot.[18,36] There is another conservation property that is rigorously satisfied because the bimolecular products represent an infinite sink,

$$\sum_j (\Delta X_i)_j = 0, \quad i = 1, \ldots, M \tag{30}$$

i.e. there is no long term accumulation of population in any of the wells. Eqn. (30) should be closely satisfied in any event, because equilibrium heavily favors the bimolecular products.

Armed with this information, consider Fig. 10–14, which must be viewed together to get a complete picture of what occurs in the equilibration zone. Fig. 10–13 are plots of $(\Delta x_R)_j$, $(\Delta X_I)_j$, $(\Delta X_{II})_j$ and $(\Delta x_p)_j$, respectively, for $j = 1, 2$ and 6. Fig. 14 is a plot of the corresponding eigen-

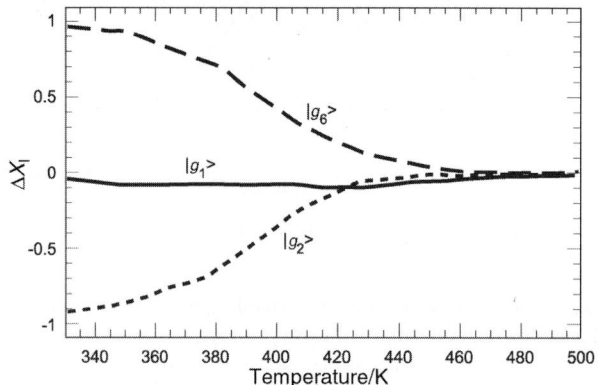

Fig. 11 Contributions of eigenvectors $|g_j\rangle$ to ΔX_I.

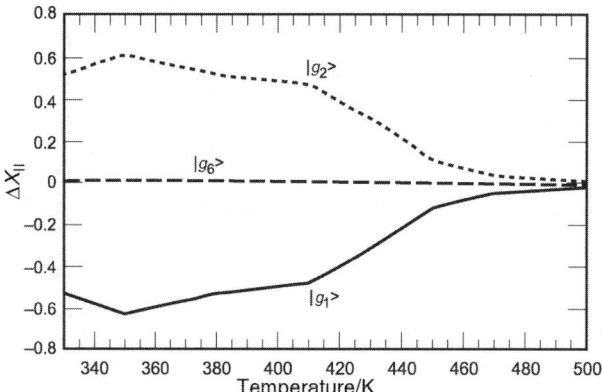

Fig. 12 Contributions of eigenvectors $|g_j\rangle$ to ΔX_{II}.

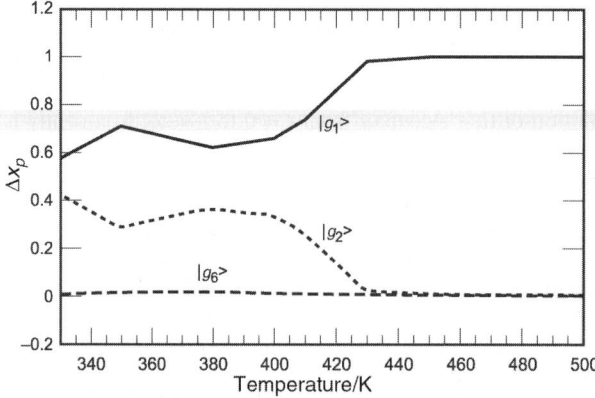

Fig. 13 Contributions of eigenvectors $|g_j\rangle$ to Δx_p.

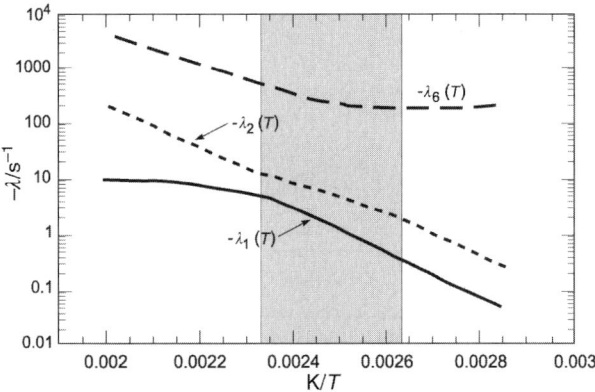

Fig. 14 Eigenvalues of G as a function of T through the equilibration zone.

values. At low temperature, C_3H_3 is removed completely by $|g_6\rangle$ and converted to $C_3H_3O_2(I)$ on a relatively fast timescale (see Fig. 14). In fact, $(\Delta x_R)_6 = -(\Delta X_I)_6$ under all conditions, indicating, as noted above, that $|g_6\rangle$ describes the elementary step, $C_3H_3 + O_2 \leftrightarrow C_3H_3O_2(I)$ at all temperatures. At low temperature, $|g_2\rangle$ describes a process that converts $C_3H_3O_2(I)$ to $C_3H_3O_2(II)$ and $CH_2CO + HCO$ on a slower timescale, and finally $|g_1\rangle$ converts the $C_3H_2O_2(II)$ formed through $|g_2\rangle$ to $CH_2CO + HCO$ on an even slower scale. Under these conditions, $|g_1\rangle$ describes a simple thermal dissociation reaction. As is normally the case, one can associate each eigenvalue with a transition state.[18,36] In this case, λ_6 is associated with ts1, λ_2 is associated with ts2 and λ_1 is associated with ts9. From this information, one can conclude that the removal of $C_3H_3O_2(I)$ actually occurs as the sequence $C_3H_3O_2(I) \to C_3H_3 + O_2 \to C_3H_3O_2(II)$ and $CH_2CO + HCO$, because the barrier at ts7 precludes direct (I)–(II) isomerization.

The situation becomes more complicated in the equilibration zone. Initially, the $C_3H_3O_2(I)$ stabilization reaction equilibrates to smaller and smaller values of $C_3H_3O_2(I)$, and $|g_2\rangle$ converts both $C_3H_3 + O_2$ and $C_3H_3O_2(I)$ to $C_3H_3O_2(II)$ and $CH_2CO + HCO$, i.e.

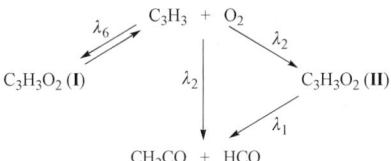

However, as temperature increases through the equilibration zone, two things occur simultaneously. Equilibrium no longer allows any significant conversion of propargyl to $C_3H_3O_2(I)$, and well **II** reaches its "stabilization limit",[18,36] no longer allowing any $C_3H_3O_2(II)$ to be formed. This causes a mixing of $|g_1\rangle$ and $|g_2\rangle$, through which more and more propargyl is converted directly to $CH_2CO + HCO$ by $|g_1\rangle$ as the temperature increases. What comes out of the equilibration zone at high temperature is an elementary reaction, $C_3H_3 + O_2 \to CH_2CO + HCO$, whose rate coefficient is $-\lambda_1/n_m$, which is determined principally by ts9.

We used the term "stabilization limit" in the last paragraph. This concept is discussed extensively in our previous work,[18,36] so we shall not belabor the point here. However, the stabilization limit is reached for a well at a temperature beyond which it is no longer possible to stabilize complexes in that well, *no matter how high one makes the pressure*. Such a condition is reached when reactivating collisions become almost as important as deactivating ones at the isomerization or dissociation energy threshold.

Fig. 14 shows that $-\lambda_6(T)$ is always at least an order of magnitude greater than either $-\lambda_2(T)$ or $-\lambda_1(T)$, indicating that $C_3H_3 + O_2 \leftrightarrow C_3H_3O_2(I)$ always equilibrates much faster than any other chemical changes occur. Consequently, the equilibrium constant measurements made by Slagle and Gutman are unlikely to have been contaminated by any of the secondary chemistry described by our master equation.

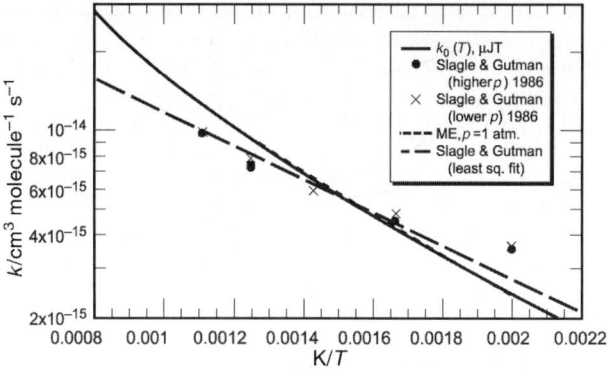

Fig. 15 Comparison of high-temperature rate coefficient predictions with experiments of Slagle and Gutman.[15]

Fig. 15 compares our predictions of the high-temperature rate coefficient, equal to $k_0(T)$, with the experimental results of Slagle and Gutman. The agreement is good, although the theory shows a stronger temperature dependence than the experiments. With the potential energy surface shown in Fig. 1, the reaction goes primarily through the 3-membered ring, $ts9 \to \textbf{IV} \to ts5 \to CH_2CO + HCO$. Reaction through the 4-membered ring contributes 13% at 1000 K and 16% at 2000 K; the fraction is less at lower temperatures. Similarly, reaction through the 5-membered ring contributes only 2% at 1000 K and 3% at 2000 K. Reaction via $\textbf{(I)} \to \textbf{(II)}$ isomerization never contributes as much as 1% to the reaction, i.e. $C_3H_3 + O_2 \to ts1 \to \textbf{(I)} \to ts7 \to \textbf{(II)} \to \cdots$ is never competitive with $C_3H_3 + O_2 \to ts2 \to \textbf{(II)} \to \cdots$ Of course all these conclusions are conditional on at least the relative barrier heights in Fig. 1 being accurate. Reducing E_{03} or E_{06} by as little as 2 or 3 kcal mol^{-1} relative to E_{09} could change the picture.

The theoretical rate coefficient in Fig. 15 can be represented by the modified Arrhenius expression,

$$k_0(T) = 2.83 \times 10^{-19} T^{1.7} \exp(-1500/RT) \text{ cm}^3 \text{ molecule}^{-1} \text{ s}^{-1}, \tag{31}$$

for $500 < T/K < 2000$. Eqn. (31) results in larger values of $k_0(T)$ at high temperature than does the Slagle–Gutman expression. Although the truth may lie somewhere in between, we recommend eqn. (31) for modeling—the Arrhenius expression given by Slagle and Gutman probably does not extrapolate well to 2000 K.

Concluding remarks

In the present investigation, we have analyzed in some detail the reaction between propargyl and molecular oxygen. Electronic structure theory (DFT-B3LYP and QCISD(T) methods) provided information about stationary points of the potential; RRKM theory was used to calculate microcanical rate coefficients, and various forms of the master equation were solved to extract information about thermal rate coefficients and product distributions. The electronic structure calculations themselves show two important features associated with the addition of resonantly stabilized free radicals, such as C_3H_3, to O_2:

1. The bond energy is relatively small whether the CH_2 end or the CH end of propargyl adds to the oxygen, $D_o < 20$ kcal mol^{-1}. The addition of allyl (C_3H_5), another RSFR, to O_2 shows a similarly small bond energy, whereas bond energies for non-resonantly stabilized flame radicals are typically between 30 and 50 kcal mol^{-1}.

2. The resonantly stabilized radicals have potential energy barriers for their addition to O_2, whereas the ordinary radicals do not.

Both of these PES properties act to make resonantly stabilized free radicals less reactive in flames than ordinary radicals.

With relatively modest modifications to important features of the PES, the chemical kinetics analysis was able to account quantitatively for the experimental results of Slagle and Gutman[15] and Atkinson and Hudgens.[16] In agreement with experiment, the master equation calculations show that the $C_3H_3 + O_2$ reaction is dominated by simple addition at low temperatures, $C_3H_3 + O_2 \leftrightarrow C_3H_3O_2(I)$, and by the reaction, $C_3H_3 + O_2 \to CH_2CO + HCO$, at high temperatures. The rate coefficient for the latter is independent of pressure even though it takes place over multiple potential wells. It is initiated by attack of the CH end of propargyl to O_2. The high temperature rate coefficient is reasonably well represented by the expression, $k(T) = 2.83 \times 10^{-19} T^{1.7} \exp(-1500/RT)$ cm^3 molecule^{-1} s^{-1}, in the temperature range, $500 < T/K < 2000$ K.

From a theoretical perspective, the most interesting feature of the reaction is the "equilibration zone." In the temperature range, $380 < T/K < 430$, equilibrium increasingly favors the reactants of the $C_3H_3 + O_2 \leftrightarrow C_3H_3O_2(I)$ reaction, which is described over a wide range of temperatures and pressures by the eigenpair of G, $|g_6\rangle$ and λ_6. At lower temperatures, the eigenvectors $|g_2\rangle$ and $|g_1\rangle$ describe the processes,

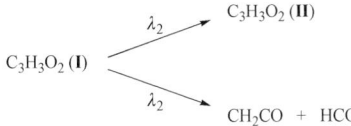

and

$$C_3H_3O_2\,(II) \xrightarrow{\lambda_1} CH_2CO + HCO \quad ,$$

respectively, which occur on a time scale much slower than $-1/\lambda_6$. In the equilibration zone, these eigenvectors begin to "couple" strongly, largely because of the equilibration of $C_3H_3 + O_2 \leftrightarrow C_3H_3O_2(I)$ and the passage of the stabilization limit of $C_3H_3O_2(II)$. Beyond 430 K, the observable reaction can be described by the elementary step, $C_3H_3 + O_2 \to CH_2CO + HCO$, with a rate coefficient equal to $-\lambda_1/n_m$, where λ_1 is associated with ts9 of Fig. 1.

Note added in proof

After completing this manuscript we investigated further the possibility of $C_3H_3O_2^*(II)$ dissociating into $C_3H_3O + O$. Our best estimate is that this channel contributes about 30% to the reaction at 2000 K, dropping off to less than 2% at 1000 K. For most applications it is probably negligible.

Acknowledgement

This work was supported by the Division of Chemical Sciences, Geosciences, and Biosciences, the Office of Basic Energy Sciences, United States Department of Energy.

References

1 J. A. Miller, *Proc. Combust. Inst.*, 1996, **20**, 461.
2 H. Richter and J. B. Howard, *Prog. Energy Combust. Sci.*, 2000, **26**, 565.
3 A. D'Anna, A. Violi and A. D'Allessio, *Combust. Flame*, 2000, **121**, 418.
4 A. D'Anna and A. Violi, *Proc. Combust. Inst.*, 1998, **27**, 425.
5 T. Faravelli, A. Goldaniga and E. Ranzi, *Proc. Combust. Inst.*, 1998, **27**, 1489.
6 P. Lindstedt, *Proc. Combust. Inst.*, 1998, **27**, 269.
7 C. J. Pope and J. A. Miller, *Proc. Combust. Inst.*, 2000, **28**, 1519.
8 M. J. Castaldi, N. M. Marinov, C. F. Melius, J. Hwang, S. M. Senkan, W. J. Pitz and C. K. Westbrook, *Proc. Combust. Inst.*, 1996, **26**, 693.
9 N. M. Marinov, W. J. Pitz, C. K. Westbrook, M. J. Castaldi and S. M. Senkan, *Combust. Sci. Technol.*, 1996, **116/117**, 211.
10 C. F. Melius, M. E. Colvin, N. M. Marinov, W. J. Pitz and S. M. Senkan, *Proc. Combust. Inst.*, 1996, **26**, 685.
11 J. A. Miller, J. V. Volponi and J.-F. Pauwels, *Combust. Flame*, 1996, **105**, 451.
12 J.-F. Pauwels, J. V. Volponi and J. A. Miller, *Combust. Sci. Technol.*, 1995, **110–111**, 249.
13 J. A. Miller and C. F. Melius, *Combust. Flame*, 1992, **91**, 21.
14 C. F. Melius, J. A. Miller and E. M. Evleth, *Proc. Combust. Inst.*, 1992, **24**, 621.

15 I. R Slagle and D. Gutman, *Proc. Combust. Inst.*, 1986, **21**, 875.
16 D. B. Atkinson and J. W. Hudgens, *J. Phys. Chem. A*, 1999, **103**, 4242.
17 M. A. Blitz, M. S. Beasley, M. J. Pilling and S. H. Robertson, *Phys. Chem. Chem. Phys.*, 2000, **2**, 805.
18 J. A. Miller and S. J. Klippenstein, *J. Phys. Chem., A*, 2001, **105**, 7254.
19 S. J. Klippenstein and L. B. Harding, *Proc. Combust. Inst.*, 2000, **28**, 1503.
20 S. P. Walch, *Chem. Phys. Lett.*, 1993, **215**, 81.
21 R. Sumathi, J. Peeters and M. T. Nguyen, *Chem. Phys. Lett.*, 1998, **287**, 109.
22 B. K. Carpenter, *J. Phys. Chem.*, 1995, **99**, 9801.
23 A. M. Mebel, E. W. G. Diau, M. C. Lin and K. Morokuma, *J. Am. Chem. Soc.*, 1996, **118**, 9759.
24 J. A. Miller, S. J. Klippenstein and S. H. Robertson, *Proc. Combust. Inst.*, 2000, **28**, 1479.
25 X. S. Ignatyev, Y. Xie, W. D. Allen and H. F. Schaefer III, *J. Chem. Phys.*, 1997, **107**, 141.
26 J. C. Rienstra-Kiracofe, W. D. Allen and H. F. Schaefer III, *J. Phys. Chem. A*, 2000, **104**, 9823.
27 A. D. Becke, *J. Chem. Phys.*, 1993, **98**, 5648.
28 W. J. Hehre, L. Radom, J. A. Pople and P. von R. Schleyer, *Ab Initio Molecular Orbital Theory*, Wiley, New York, 1987.
29 *GAUSSIAN98*, M. J. Frisch, G. W. Trucks, H. B. Schlegel, G. E. Scuseria, M. A. Robb, J. R. Cheeseman, V. G. Zakrzewski, J. A. Montgomery, Jr., R. E. Stratmann, J. C. Burant, S. Dapprich, J. M. Millam, A. D. Daniels, K. N. Kudin, M. C. Strain, O. Farkas, J. Tomasi, V. Barone, M. Cossi, R. Cammi, B. Mennucci, C. Pomelli, C. Adamo, S. Clifford, J. Ochterski, G. A. Petersson, P. Y. Ayala, Q. Cui, K. Morokuma, D. K. Malick, A. D. Rabuck, K. Raghavachari, J. B. Foresman, J. Cioslowski, J. V. Ortiz, B. B. Stefanov, G. Liu, A. Liashenko, P. Piskorz, I. Komaromi, R. Gomperts, R. L. Martin, D. J. Fox, T. Keith, M. A. Al-Laham, C. Y. Peng, A. Nanayakkara, C. Gonzalez, M. Challacombe, P. M. W. Gill, B. Johnson, W. Chen, M. W. Wong, J. L. Andres, C. Gonzalez, M. Head-Gordon, E. S. Replogle and J. A. Pople, Gaussian, Inc., Pittsburgh PA, 1998.
30 J. A. Pople, M. Head-Gordon and K. Raghavachari, *J. Chem. Phys.*, 1987, **87**, 5968.
31 L. A. Curtiss, K. Raghavachari, P. C. Redfern, V. Rassolov and J. A. Pople, *J. Chem. Phys.*, 1998, **109**, 7764.
32 P. L. Fast, M. L. Sanchez and D. G. Truhlar, *Chem. Phys. Lett.*, 1999, **306**, 407.
33 P. L. Fast and D. G. Truhlar, *J. Phys. Chem. A*, 1999, **103**, 3802.
34 J. A.Miller and S. J. Klippenstein, *J. Phys. Chem. A*, 2000, **104**, 2061.
35 J. A. Miller, S. J. Klippenstein and S. H. Robertson, *J. Phys. Chem. A*, 2000, **104**, 7525; also J. A. Miller, S. J. Klippenstein and S. H. Robertson, *J. Phys. Chem. A*, 2000, **104**, 9806 (correction).
36 J. A. Miller and S. J. Klippenstein, *Int. J. Chem. Kinet.*, 2001, in press.
37 B. Widom, *Science*, 1965, **148**, 1555.
38 E. Anderson, Z. Bai, C. Bishof, J. Demmel, J. Dongarra, J. DuCroz, A. Greenbaum, S. Hammerling, A. McKenney, S. Ostronchov and D. Sorensen, *LAPACK Users' Guide*, SIAM, Philadelphia, PA, 1992.
39 P. N. Brown, G. D. Byrne and A. C. Hindmarsh, *SIAM J. Sci. Stat. Comput.*, 1989, **10**, 1038.
40 J. A. Miller, C. Parrish and N. J. Brown, *J. Phys. Chem.*, 1986, **90**, 3339.
41 S. C. Smith and R. G. Gilbert, *Int. J. Chem. Kinet.*, 1988, **20**, 367.
42 J. A. Miller, S. J. Klippenstein and C. Raffy, in preparation.
43 S. J. Klippenstein, A. F. Wagner, R. C. Dunbar, D. M. Wardlaw, S. H. Robertson and J. A. Miller, *Variflex Version 1.08m*, 2000.
44 B. Widom, *J. Chem. Phys.*, 1971, **55**, 44.
45 E. W. Montroll and K. E. Shuler, *Adv. Chem. Phys.*, 1958, **1**, 361.
46 R. K. Boyd, *J. Chem. Phys.*, 1974, **60**, 1214.

Infrared frequency-modulation probing of product formation in alkyl + O_2 reactions

Part IV.† Reactions of propyl and butyl radicals with O_2‡

John D. DeSain,[a] Craig A. Taatjes,*[a] James A. Miller,[a] Stephen J. Klippenstein[a] and David K. Hahn§[b]

[a] *Combustion Research Facility, Mail Stop 9055, Sandia National Laboratories, Livermore, California, 94551-0969, USA*
[b] *Department of Chemistry, Case Western University, Cleveland, OH, 44106-7078, USA*

Received 8th March 2001
First published as an Advance Article on the web 19th October 2001

The time-resolved production of HO_2 in the Cl-initiated oxidation of iso- and n-butane is measured using continuous-wave (CW) infrared frequency modulation spectroscopy between 298 and 693 K. The yield of HO_2 is determined relative to the $Cl_2/CH_3OH/O_2$ system. As in studies of smaller alkanes, the branching fraction to HO_2 + alkene in butyl + O_2 displays a dramatic rise with increasing temperature between about 550 and 700 K (the "transition region") which is accompanied by a qualitative change in the time behavior of the HO_2 production. At low temperatures the HO_2 is formed promptly; a second, slower production of HO_2 is responsible for the bulk of the increased yield in the transition temperature region. In contrast to reactions of smaller alkyl radicals with O_2, the total HO_2 yield in the butyl radical reactions appears to remain significantly below 1 up to 700 K, implying a significant role for OH-producing channels. The slower HO_2 production in butane oxidation displays an apparent activation energy similar to that measured for smaller alkyl + O_2 reactions, suggesting that the energetics of the HO_2 elimination transition state are similar for a broad range of R + O_2 systems. A combination of QCISD(T) based characterizations of the propyl and butyl + O_2 potential energy surfaces and master equation based characterization of the propyl + O_2 kinetics provide the framework for explanation of the experimentally observed HO_2 production in Cl-initiated propane and butane oxidation. These calculations suggest that the HO_2 elimination channel is similar in all reaction systems, and that hydroperoxyalkyl (QOOH) species produced by internal H-atom abstraction in RO_2 can provide a path to OH formation. However, the QOOH formed by the energetically favorable 1,5 isomerization (*via* a six-membered ring transition state) generally experiences significant barriers (relative to the radical + O_2 reactants) to the production of an oxetane + OH. In contrast, the barriers to forming OH + an oxirane or an oxolane, *via* 1,4 or 1,6 isomerizations, respectively, are generally below reactants.

---

† For Part III see ref. 12.
‡ Electronic Supplementary Information available. See http://www.rsc.org/suppdata/fd/b1/b102237g/
§ Present address: Department of Chemistry, University of Nevada, Reno, NV, 89557, USA.

DOI: 10.1039/b102237g

Introduction

The reactions of alkyl radicals with molecular oxygen are central to the mechanism of low and moderate temperature hydrocarbon oxidation. The behavior of alkyl + O_2 reactions in this temperature region is largely governed by the kinetics of the intermediate alkylperoxy (RO_2) adduct. At low temperature, stabilization of the RO_2 radical is the dominant reactive channel. As the temperature increases other channels grow in importance: thermal dissociation back to reactants, elimination of HO_2, and (especially for larger alkyl radicals) isomerization to hydroperoxyalkyl (QOOH) radicals, followed by dissociation to form OH or HO_2. Hydroperoxyalkyl radicals have long been recognized as crucial intermediates in the formation of oxygen atom heterocycles in hydrocarbon oxidation, and their dissociation or reaction with O_2 has been proposed as an important chain propagation step producing OH radicals.[1-4] An increase in the relative importance of the isomerization with increasing alkyl radical chain length has been determined from initial product formation measurements. Since direct information is largely unavailable, rates for isomerization have often been rationalized based on estimates of ring strain in the cyclic transition state for H transfer. The changing roles of isomerization, dissociation, and subsequent reactions of these RO_2 species are key to modeling the mechanism of pre-ignition chemistry, autoignition, and engine knock.[5-9]

Recently we have begun a series of measurements of HO_2 formation in reactions of alkyl radicals with O_2. Studies of the reactions of ethyl, propyl, and cyclopentyl radicals with O_2 have previously been reported.[10-12] The detailed chemistry of the alkylperoxy radicals formed in the reaction of alkyl radicals with O_2 rapidly becomes complicated as the size of the alkyl radical and the number of possible isomerization reactions increases. Nonetheless, many important mechanistic features are common to nearly all R + O_2 reactions: the formation of the alkylperoxy radical, the elimination of HO_2 to form the conjugate alkene, internal hydrogen abstractions to form QOOH, and dissociation of QOOH species to form OH or HO_2. Correlations among different R + O_2 reaction systems with these shared features have been a key part of the extensive product formation studies that underlie our present understanding of alkylperoxy and hydroperoxyalkyl chemistry.[1,2,4,13] In investigating reactions of increasing complexity the characterization of each reaction provides a basis for interpreting the next and it is hoped that comparison among the reactions can be used to gain a general representation of R + O_2 reactions. The reaction of ethyl with O_2 has been the most thoroughly investigated R + O_2 reaction both experimentally and theoretically,[2,3,10,14-31] and has been proposed as a prototype for the entire series of R + O_2 reactions. However, it is not clear to what degree information gained from the simplest R + O_2 reactions can be applied and extended to larger systems. The strategy of this paper is to employ theoretical characterization of propyl and butyl radical reactions with O_2 to model new experimental measurements of HO_2 production in butane oxidation and to investigate general characteristics of R + O_2 reactions.

This work presents new experimental results on the Cl-initiated oxidation of iso- and n-butane, which proceeds *via* the reaction of butyl radicals with O_2. This reaction can result in a number of different products:

$$C_4H_9 + O_2 \xrightarrow{[M]} C_4H_9O_2 \tag{1a}$$

$$\longrightarrow C_4H_8 + HO_2 \tag{1b}$$

$$\longrightarrow C_4H_8O + OH \tag{1c}$$

The butylperoxy radicals formed in reaction (1a) can be intermediates in reactions (1b) and (1c), and can also undergo isomerization reactions of the form $C_4H_9O_2 \leftrightarrow C_4H_8OOH$. The experiments monitor the formation of HO_2 in reaction (1b), following pulsed photolytic initiation, using near-infrared frequency modulation (FM) spectroscopy. The interpretation of these butane oxidation experiments is more complicated than those for smaller alkanes because of thermal dissociation of the butyl radicals and the greater contribution of reaction (1c).

Analysis of the experimental results draws on parallels between the butyl + O_2 reactions and previously investigated ethyl and propyl + O_2 reactions. Kaiser and coworkers have measured alkene yields as a function of temperature and pressure for the ethyl and propyl + O_2 reactions.[19,20,23,32–34] Both reactions show an inverse dependence of the alkene yield on pressure and display a steep rise in the HO_2 + alkene yield with temperature above ~500 K. This increase is attributed to the onset of alkylperoxy radical dissociation. Measurements of Cl-initiated oxidation of ethane and propane, using the same method as the present measurements,[10,11] showed that this increase in HO_2 yield was associated with the emergence of a "delayed" component to the HO_2 production. In both of these reactions the HO_2 yield reaches ~100% by 683 K. The upper limits placed on the formation of OH in the Cl-initiated oxidation experiments are consistent with the initial formation of the coincident oxygenated products (oxiranes and oxetane) in ethane and propane oxidation at higher temperatures.[2,4,14,35] The experimental data on butane oxidation indicates a larger production of oxygen heterocycles in isobutyl, n-butyl and sec-butyl + O_2 reactions than for propyl + O_2 or ethyl + O_2. The reaction of tert-butyl with O_2 is however, thought to lead almost exclusively to isobutene + HO_2.[2]

We also present a complete QCISD(T)-based quantum chemical characterization of the stationary points for the interaction of O_2 with each of the butyl and propyl isomers. Various aspects of the potential energy surface for the propyl and butyl + O_2 reaction systems have been explored in earlier theoretical studies. For example, Pritchard and coworkers have studied both the isomerizations from RO_2 to QOOH, and the dissociation of the QOOH species to OH with density functional theory.[36,37] More quantitative energetic estimates were obtained by Chen and Bozzelli in their CBS-q study of the stationary points in the tert-butyl and isobutyl + O_2 reaction systems[38] and by Seinfeld and coworkers in a CBS-q study of a more limited set of stationary points in the n-butyl and sec-butyl + O_2 systems.[39] Related studies of Bozzelli and coworkers have also provided energies for some of the stationary points in the propyl and ethyl + O_2 systems.[40,41] Various other studies have provided more limited, and generally less accurate, results for the propyl and/or butyl + O_2 systems.[42–45]

The present work extends these prior studies with a consistent treatment of the ethyl, propyl, and butyl + O_2 reactions, employing a methodology that is closely analogous to that which was found to yield quantitatively meaningful results for the C_2H_5 + O_2 system.[29,31] For the most part, the present results are in reasonably good agreement with the prior CBS-q studies.[38,39] The present quantum chemical characterization of the propyl + O_2 reaction is employed in time-dependent master equation simulations of its kinetics, allowing detailed comparisons with related experimental data.[1,11,46] The propyl and butyl reactions share many common mechanistic features, which allows for the prediction of the HO_2 production in the butyl reactions from the results of the propyl radical simulations. Such predictions are found to be in reasonable agreement with the experimental measurements.

Experiment

Measurement of Cl-initiated butane oxidation

The reactions of butyl radicals with O_2 are investigated using the laser photolysis–CW infrared frequency-modulation method employed previously for other R + O_2 reactions.[10–12] The Cl is generated by photolysis of Cl_2 at 355 nm, and butyl radicals are generated by subsequent Cl abstraction from iso- or n-butane, which produces a mixture of butyl radical isomers. The C_4H_9 radical then reacts with O_2 to produce the HO_2 radical:

$$Cl_2 \xrightarrow{h\nu(355\ nm)} 2Cl \qquad (2)$$

$$Cl + n\text{-}C_4H_{10} \rightarrow (n\text{-}C_4H_9, sec\text{-}C_4H_9) + HCl \qquad (3a)$$

$$Cl + iso\text{-}C_4H_{10} \rightarrow (iso\text{-}C_4H_9, tert\text{-}C_4H_9) + HCl \qquad (3b)$$

$$C_4H_9 + O_2 \rightarrow \text{products} \qquad (1)$$

The O_2 concentration is kept at least 30 times greater than the Cl_2 concentration in order to minimize the effects of the competing reaction of butyl radicals with Cl_2.

The progress of reaction (1b) is monitored by two-tone frequency modulation (FM) spectroscopy on the overtone of the O–H stretch in HO_2 near 1.5 µm[47] using a tunable diode laser. The diode laser output is passed 17 times through the reaction region by means of a Herriott-type multipass cell. The quartz reactor is 1.3 m long, with CaF_2 windows, and the gold-coated spherical mirrors of the Herriott cell are located outside the flow cell. The photolysis beam, a 5 ns pulse from a Nd : YAG laser at 355 nm, travels down the central axis of the reactor. The IR probe beam follows the off-axis path of the Herriott arrangement,[48] tracing a circular pattern on the outer edge of the Herriott mirrors and a smaller circle in the center of the cell. The IR probe and the UV photolysis beam overlap only in the center of the flow cell, where the temperature is more readily controlled.[49] In these experiments the effective pathlength (*i.e.*, total overlapping path of the photolysis and probe) is approximately 9 m.

The relative yield of the HO_2 radical produced by reaction (1) is measured by comparing the observed HO_2 FM signal with that from the corresponding reaction of $CH_2OH + O_2$. This reaction is assumed to produce a 100% yield of HO_2 over the temperature range of these experiments.[50] The CH_2OH is produced by Cl abstraction of hydrogen atom from methanol.

$$CH_3OH + Cl \rightarrow CH_2OH + HCl, \quad 100\% \tag{4}$$

$$CH_2OH + O_2 \rightarrow CH_2O + HO_2, \quad 100\% \tag{5}$$

To relate the observed quantities to characteristics of the reaction, corrections must in general be made for the removal reactions of HO_2, as well as for certain side reactions, as discussed below. Typical gas concentrations are $[O_2] = 6.6 \times 10^{16}$ cm^{-3}, $[Cl_2] = 2.1 \times 10^{15}$ cm^{-3}, and 8.0×10^{15} cm^{-3} of either butane or methanol. Helium is added to maintain a total gas density of 8.5×10^{17} cm^{-3}.

Data analysis

The qualitative behavior of the HO_2 signal from $C_4H_9 + O_2$ is similar to that observed in previous studies of C_2H_5, c-C_5H_9 and C_3H_7 reactions with O_2.[10–12] The HO_2 observed below 500 K appears rapidly after the photolysis pulse, with a rise time faster than the experimental resolution of ~ 10 μs, (limited by the low-pass filter used in the two-tone FM method). Above 500 K a second slower HO_2 signal is observed. Fig. 1 shows the HO_2 signals from the reactions of a mixture of iso- and *tert*-C_4H_9 (produced by Cl + iso-C_4H_{10}) and of CH_2OH with O_2 at 648 K. In Fig. 1 the HO_2 signal from the butyl + O_2 reaction shows two general components: a small initial sharp rise and a second much slower rise. Determining the production of HO_2 *via* the delayed mechanism requires correction for the ongoing removal of HO_2 by self-reaction and by reactions with other radical species.

Correction for the self-reaction of HO_2 uses only information inherent to the observed HO_2 signals from $Cl_2/CH_3OH/O_2$ and $Cl_2/C_4H_{10}/O_2$, and requires no assumed rate coefficients. The decay of the HO_2 signal from the $CH_2OH + O_2$ reference reaction is dominated by $HO_2 + HO_2$ recombination,

$$HO_2 + HO_2 \rightarrow products. \tag{6}$$

The time profile of the HO_2 signal from the reference reaction is given by

$$I_{ref}(t) = \alpha[HO_2]_t = \frac{\alpha[HO_2]_0}{1 + 2k_6 t[HO_2]_0}, \tag{7}$$

where α is a detection constant relating the observed FM signal to HO_2 concentration. A plot of the inverse of the reference HO_2 signal *vs.* time gives a line with slope $2k_6/\alpha$. Since the temperature-dependent linestrength of the probe transition is unknown, the absolute value of k_6 remains undetermined in these experiments; however, the analysis requires only the phenomenological rate coefficient $2k_6/\alpha$. As in previous work, the time-dependent FM signal, $I(t)$, from HO_2 produced in reaction (1) can be written using a formal solution to the kinetic equation for HO_2 production:

$$I(t) = \alpha \int_0^t R_{production}(t')dt' - \frac{2k_6}{\alpha} \int_0^t I(t')^2 dt' - \alpha \int_0^t R_{removal}(t')dt'. \tag{8}$$

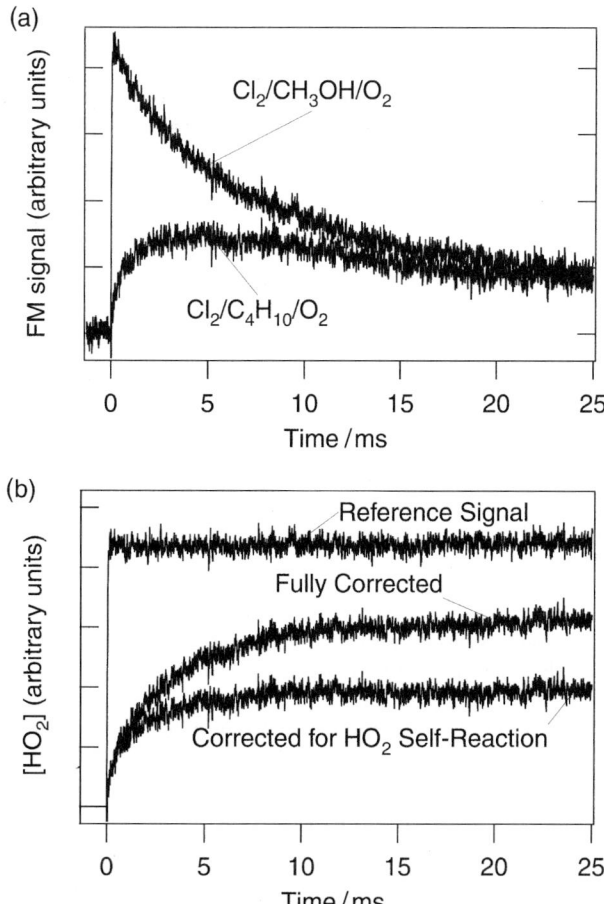

Fig. 1 Time-resolved HO_2 signal from Cl-atom initiated oxidation of isobutane at 648 K. (a) The experimental data traces for the reference $Cl_2/CH_3OH/O_2$ and the experimental $Cl_2/iso-C_4H_{10}/O_2$ system. (b) The signals after data analysis. The largest amplitude trace is the HO_2 production from the reference reaction, after removing the second-order decay. The apparent rate coefficient for this reaction, $2k_6/\alpha$, is used to correct the HO_2 signal from the isobutane system for the ongoing removal of HO_2 by self-reaction, as shown by the lowest amplitude trace. Modeling the removal of HO_2 by reaction with other radical species, principally $C_4H_9O_2$ under these conditions, results in the middle trace as the best estimate of the HO_2 production in the Cl-initiated butane oxidation. (Colour version available as ESI.†)

where $R_{production}$ is the effective time-dependent rate of HO_2 production and $R_{removal}$ is the effective time-dependent rate for removal of HO_2 by processes besides self-reaction. This removal rate consists of HO_2 reactions with other radicals X in the system, so $R_{removal} = k_x[X]_t[HO_2]_t$. Determining the time-resolved production of HO_2 from the Cl-initiated butane oxidation, denoted $R_{production}$, is the aim of the measurement. The integrated profiles method[51,52] can be applied to this formal solution, along with the available time-resolved relative HO_2 concentrations, to correct for known rate processes. Were HO_2 recombination the only significant loss mechanism, then $R_{removal}$ would be equal to zero and all parameters in eqn. (8) would be available directly from the experiment. This assumption produces a lower bound to the actual HO_2 production rate, since additional corrections for HO_2 signal lost by other mechanisms ($R_{removal}$) will increase the amplitude of the final "corrected" time profile.

Whereas the HO_2 self-reaction can be accounted for using only the measured data, correction for removal of HO_2 by reaction with other species requires additional modeling and assumed rate coefficients. In this context a distinction should be made between the extraction of the production rate for HO_2 and the interpretation of that production rate. Extraction of the effective total HO_2 production simply requires estimating a value, or range of possible values, for $R_{removal}$, but interpretation of the production in terms of elementary reactions may require knowledge of the nature of the removal as well as production processes. In previous investigations, these two tasks were linked, since experimental conditions could be maintained so that the $RO_2 + HO_2$ reaction dominated $R_{removal}$, i.e.,

$$HO_2 + C_4H_9O_2 \rightarrow products, \qquad (9)$$

and $R_{removal} \approx k_9[C_4H_9O_2]_t[HO_2]_t$. Unfortunately for the present investigation, none of the $C_4H_9O_2 + HO_2$ reaction rate coefficients have been measured. The competing removal of the RO_2 by self-reaction,

$$RO_2 + RO_2 \rightarrow products. \qquad (10)$$

must also be modeled, and of the required $C_4H_9O_2 + C_4H_9O_2$ reactions only the tert-$C_4H_9O_2$ + tert-$C_4H_9O_2$ reaction has been measured. Further, in other alkyl + O_2 experiments, an initial value for the alkylperoxy radical was obtained by assuming that immediately after the fast establishment of the steady-state one has $[RO_2] \approx [R]_0 - [HO_2]$, i.e., all butyl radicals react with O_2 to produce either HO_2 or $C_4H_9O_2$. This assumption is good only if the steady-state for reaction (1a) favors the products, if there is no significant competing removal of either C_4H_9 or $C_4H_9O_2$, and if the contribution of reaction (1c) can be ignored. However, the equilibrium constants for reactions (1a) have not been measured, except for tert-butyl + O_2.[53] The iso- and n-butyl radicals can be expected to undergo thermal decomposition at the highest temperatures of the present study. Finally, the contribution of OH producing channels is likely to be significant at high temperature,[2,54] rendering the analysis used in previous work problematic.

Nonetheless, at these higher temperatures whatever unobserved radicals are present must still react with HO_2 and with each other, and in the absence of significant chain branching the total radical concentration will remain less than $[Cl]_0$. Therefore, extraction of a phenomenological total HO_2 production remains formally similar, although a rationalization in terms of elementary reactions will be different. The HO_2 signal corrected only for self-reaction remains a lower bound to the total HO_2 production in the entire oxidation mechanism; however, direct identification of this production with a single elementary reaction may be impossible.

A qualitative modeling of the removal of HO_2 by other radicals has been carried out using the method of previous experiments. Table 1 lists available kinetic data for several $RO_2 + HO_2$ reactions. A rate coefficient equal to that for the c-$C_5H_9O_2 + HO_2$ reaction was chosen to represent $C_4H_9O_2 + HO_2$. We further assume that all isomers of the butylperoxyl radical react very similarly with HO_2 and with themselves, and assume a $C_4H_9O_2 + C_4H_9O_2$ rate coefficient from extrapolation of lower temperature measurements of the tert-$C_4H_9O_2$ self-reaction. Estimation of $R_{removal}$ in these systems clearly involves a number of assumptions. The uncertainties introduced by these assumptions depend on the importance of $R_{removal}$, and are discussed in more detail in the

Table 1 Reaction rate coefficients for $X + HO_2$, $X + X$ reactions ($X = R, RO_2$)

X	k_{X+X}/cm^3 molecule^{-1} s^{-1}	k_{X+HO_2}/cm^3 molecule^{-1} s^{-1}
CH_3O_2	$9.1 \times 10^{-14}\ e^{420\ K/T}$ (298–700 K)[71]	$4.1 \times 10^{-13}\ e^{790\ K/T}$ (298–700 K)[71]
$C_2H_5O_2$	6.4×10^{-14} (250–450 K)[72]	$2.7 \times 10^{-13}\ e^{1000\ K/T}$ (200–500 K)[72]
$CH_2=CHCH_2O_2$	$5.4 \times 10^{-14}\ e^{760\ K/T}$ (286–394 K)[73]	5.6×10^{-12} (393–426 K)[73]
$(CH_3)_3CCH_2O_2$	$1.7 \times 10^{-15}\ e^{1961\ K/T}$ (248–373 K)[74]	$1.43 \times 10^{-13}\ e^{1380\ K/T}$ (248–365 K)[75]
$C_5H_9O_2$	$2.9 \times 10^{-13}\ e^{-555\ K/T}$ (243–373 K)[74]	$3.2 \times 10^{-13}\ e^{1150\ K/T}$ (214–539 K)[76]
$C_6H_{11}O_2$	$7.7 \times 10^{-14}\ e^{-184\ K/T}$ (253–373 K)[77]	$2.6 \times 10^{-13}\ e^{1245\ K/T}$ (248–364 K)[78]
$C_6H_5CH_2O_2$	$2.76 \times 10^{-14}\ e^{1680\ K/T}$ (273–450 K)[79]	$3.75 \times 10^{-13}\ e^{980\ K/T}$ (273–450 K)[79]
C_2H_5	2×10^{-11} (300–1200 K)[80]	1×10^{-12} (300–2500 K)[81]
		5×10^{-11} (250–1200 K)[24]
$CH_3\cdot CHCH_3$	$1.65 \times 10^{-11}\ (T/298)^{-0.7}$ (300–2500 K)[82]	4×10^{-11} (300–2500 K)[82]
$CH_3CH_2\cdot CH_2$	2×10^{-11} (300–2500 K)[82]	4×10^{-11} (300–2500 K)[82]
$(CH_3)_3\cdot C$	$5.5 \times 10^{-12}\ (T/298)^{-1.5}$ (300–2500 K)[83]	3×10^{-11} (300–2500 K)[83]

Results section below. The kinetic equations are cast as equations using the observed signals:

$$A \equiv \alpha[X]_t = I_{ref}(0) - \alpha \int_0^t R_{production}(t')dt' - \frac{2k_x}{\alpha}\int_0^t (\alpha[X]_{t'})^2 dt' - \frac{k_y}{\alpha}\int_0^t \alpha[X]I(t')dt' \quad (11)$$

$$B \equiv \alpha \int_0^t R_{production}(t')dt' = I(t) + \frac{2k_6}{\alpha}\int_0^t I(t')^2 dt' + \frac{k_y}{\alpha}\int_0^t \alpha[X]_{t'} I(t')dt' \quad (12)$$

Initially $A_{(0)}$ is assumed to be 0, and then successive approximations to the quantities A and B can be calculated:

$$B_{(n)}(t) = I(t) + \frac{2k_6}{\alpha}\int_0^t I(t')^2 dt' + \frac{k_y}{\alpha}\int_0^t A_{(n-1)} I(t')dt' \quad (13)$$

$$A_{(n)} = (I_{ref}(0) - B_{(n)}) - \frac{2k_x}{\alpha}\int_0^t (I_{ref}(0) - B_{(n)})^2 dt' - \frac{k_y}{\alpha}\int_0^t (I_{ref}(0) - B_{(n)})I(t')dt' \quad (14)$$

Iteration of these equations converges to a solution for B that represents the production of HO_2 from reaction (1) that gives rise to the observed signal under the conditions of the model. For $T \lesssim 660$ K, we identify [X] with [$C_4H_9O_2$], k_x with k_{10} and k_y with k_9 under the conditions of our experiments. The temperature of 660 K is chosen based on estimates of K_{eq} using the quantum chemistry described below and a harmonic-oscillator calculation of the partition functions for the butylperoxyl radicals (which should somewhat underestimate the equilibrium constant). Combining the estimated K_{eq} and literature determinations of the thermal dissociation rates for butyl radicals[55–58] suggests that near 660 K $k_{dissociation}$ becomes comparable to $K_{eq}[O_2]$ divided by the measured effective time constant for HO_2 production. At higher temperatures a correction for reactions of HO_2 with other radicals, principally alkyl radicals, may be more realistic. This amounts to modeling a different set of unobserved radicals with estimated rate coefficients, and the procedure is identical, since radical–radical reactions remain the dominant mechanism for total radical removal. However, this modeling is clearly subject to very large uncertainties and must be regarded as merely qualitative. Literature estimates of several R + HO_2 and R + R reactions are also listed in Table 1 for reference.

Calculations

Quantum chemistry

The key stationary points on the ethyl, propyl and butyl + O_2 potential energy surfaces have been characterized with B3LYP density functional theory employing the 6-31G* basis set. Higher-level energies are obtained for each stationary point *via* a combination of quadratic configuration interaction calculations (QCISD(T)) with a small basis set and second order Møller–Plesset perturbation theory (MP2) with a larger basis set. The higher-level (HL) estimates, corresponding to approximate QCISD(T)/large energies, are then obtained from the relation:

$$E_{HL} = E(QCISD(T)/small) + E(MP2/large) - E(MP2/small) \quad (15)$$

For the consistent set of energies (E_{HL1}) the 6-31G(d) basis is employed for the smaller basis set, and the 6-311++G(2df,2pd) basis is employed for the larger basis. For the propyl + O_2 reaction, an additional set of higher level energies (E_{HL2}) are obtained by employing the 6-311G(d,p) basis as the smaller basis, and the 6-311++G(3df,2pd) basis as the larger basis. A complete investigation of the torsional modes is carried out for the propyl + O_2 isomers, but only a limited treatment has been performed for the butyl + O_2 systems; other torsional states may be slightly (*e.g.*, several tenths of a kcal mol^{-1}) lower in energy. Each of these quantum chemical simulations is performed with the GAUSSIAN 98 quantum chemistry software.[59]

Kinetics

The master equation simulations have been performed as described in our previous work.[60–62] For channels with a well-defined saddlepoint the microcanonical rate coefficients have been evaluated on the basis of conventional transition state theory. The various torsional modes have been treated as one-dimensional hindered rotors using quantum-chemical evaluations of all the minima to generate a Fourier representation of the potential surface. Such torsional potentials were then employed in Pitzer–Gwinn based evaluations of the partition functions, and sums and densities of states.[63] The remaining modes were treated with rigid-rotor harmonic oscillator assumptions.

For the barrierless entrance channel, variable reaction coordinate transition state theory was employed.[64,65] The requisite transitional mode potential was based on B3LYP/6-31G(d) evaluations of the interaction energies along the minimum energy path. The force field for the transitional bending modes consisted of sums of sinusoidally hindered rotors, with an exponential decay of the force constants with separation. The relative magnitude of the force constants for different modes was adjusted to reproduce the calculated values at the equilibrium C_3H_7OO geometry. The absolute magnitude and decay of such force constants was adjusted to reproduce experimentally observed high-pressure rate constants.[46,66]

The one-dimensional time-dependent master equation was solved *via* diagonalization of the transition matrix. The time-dependent populations of reactants, and unimolecular and bimolecular products, were determined directly from the eigenvalues and eigenvectors. An exponential down model, with a ΔE_{down} value of 350 cm^{-1}, was employed for the energy transfer process. The VARIFLEX software was used in these master equation simulations.[67]

Results

Butane oxidation experiments

The yields of HO_2 extracted from the data analysis are displayed as a function of temperature in Fig. 2 and summarized in Table 2. At room temperature very little HO_2 is observed, and an upper limit of 0.01 is placed on the HO_2 yield at 298 K. As is the case with previous investigations of $R + O_2$ reactions, the total HO_2 yield displays a marked increase near 600 K, rising from ~10% to ≥80% between 573 and 663 K. This behavior is explicable in terms of a coupled mechanism for HO_2 formation, in which stabilization of the alkylperoxyl radical competes with elimination to form HO_2 + alkene. In the transition temperature range where the yield of the HO_2 + alkene product increases sharply, the thermal decomposition of RO_2, the dominant product at lower temperatures, begins to become larger than the reactive removal of the RO_2 species.

Fig. 2 shows both raw yields and "corrected" yields after the qualitative correction for other removal reactions of HO_2. The modeling of the $C_4H_9O_2$ reactions has the greatest effect on the

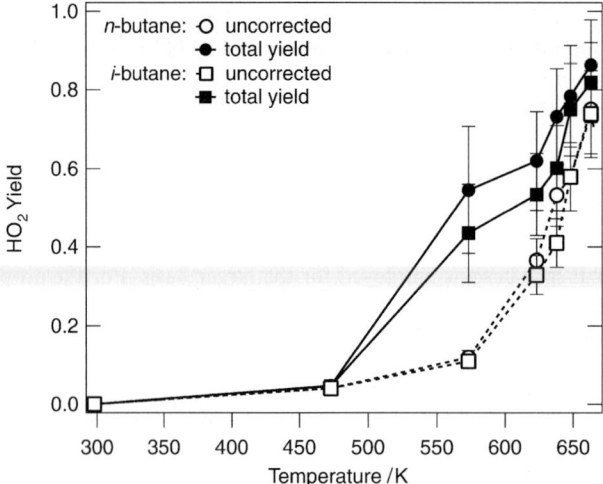

Fig. 2 Effective total yields of HO_2 in the Cl-initiated oxidation of butanes as a function of temperature. The circles represent the yields from n-butane, and the squares yields from isobutane. Open symbols (○, □) represent Φ_{raw}, the yields after accounting only for HO_2 self-reaction. The filled (●, ■) symbols are total yields Φ_{total} after modeling for butylperoxyl reactions.

Table 2 Yields and effective rate constants for HO_2 formation in Cl-initiated butane oxidation[a]

	Φ_{raw}		Φ_{total}		$(1/\tau)/s^{-1}$	
T/K	n-butane	isobutane	n-butane	isobutane	n-butane	isobutane
298	⩽0.01	⩽0.01	⩽0.01	⩽0.01	—	—
473	0.04	0.04	0.04 ± 0.02	0.05 ± 0.04	—	—
573	0.10	0.11	0.55 ± 0.16	0.44 ± 0.13	43 ± 40	63 ± 50
623	0.39	0.33	0.62 ± 0.12	0.54 ± 0.10	210 ± 60	130 ± 80
638	0.53	0.41	0.73 ± 0.12	0.60 ± 0.11	350 ± 60	270 ± 140
648	0.59	0.58	0.78 ± 0.12	0.75 ± 0.12	470 ± 40	360 ± 70
663	0.76	0.74	0.86 ± 0.11	0.82 ± 0.10	760 ± 90	440 ± 65
673	$(0.75)^b$	$(0.57)^b$	$(0.86 ± 0.11)^b$	$(0.75 ± 0.12)^b$	980 ± 150	590 ± 90
683	$(0.88)^b$	$(0.59)^b$	$(0.92 ± 0.11)^b$	$(0.77 ± 0.12)^b$	1350 ± 200	770 ± 120
693	$(0.81)^b$	$(0.80)^b$	$(0.85 ± 0.09)^b$	$(0.83 ± 0.09)^b$	1460 ± 300, 1720 ± 350[c]	1120 ± 220, 800 ± 160[c]

[a] Error estimates are ±2σ and include estimates of systematic uncertainty, as discussed in the text. The statistical uncertainty in the raw yields is ±15%. [b] Yield measurements at these temperatures are subject to additional uncertainties because of competing butyl dissociation and OH production. See text for details. [c] Measurements taken at $[O_2] = 6.6 \times 10^{16}$ and 6.0×10^{17} cm^{-3}, respectively.

yield determinations at moderate temperature, where the difference between the "raw" yield and the total yield is greatest. Varying the assumed values for both k_9 and k_{10} by a factor of two causes a relative change in the extracted yield of ~5%, which is less than the statistical uncertainty in the yield determinations. However, measurements that rely heavily on modeling the reactions of the unobserved $C_4H_9O_2$ radical must be considered more uncertain. The error estimates given in Table 2 and shown in Fig. 2 represent the sum of statistical uncertainties and the effects of a factor of two variation in both k_x and k_y, plus an additional uncertainty equal to 25% of the total correction for $R_{removal}$. Within this estimate of the experimental uncertainty, there is no evidence of a difference in HO_2 yield between the two butane isomers.

Above ~660 K, the identification of $R_{removal}$ with the $C_4H_9O_2 + HO_2$ reaction becomes dubious, as discussed above, although the extraction of a phenomenological yield remains formally the same. It must again be emphasized that at these higher temperatures, this total effective HO_2 yield in the Cl-initiated oxidation cannot be regarded as the branching fraction of any elementary reaction step. The raw and "corrected" yields at these temperatures, using the same assumed k_x and k_y, are given in Table 2 but are not shown in Fig. 2, since they provide only qualitative information. At these temperatures the contribution of channel (1c) may be significant,[1,2,54] presumably occurring via RO_2 isomerization,

$$C_4H_9OO \rightarrow C_4H_8OOH \rightarrow OH + C_4H_8O \tag{16}$$

The OH product will regenerate butyl radicals:

$$C_4H_{10} + OH \rightarrow C_4H_9 + H_2O \tag{17}$$

The sequence of reactions (17) and (1b) transforms OH into HO_2, so the total observed HO_2 yield can be larger than the branching fraction (k_{1b}/k_1) whenever OH production is significant. Other removal reactions of OH (e.g., with HO_2, RO_2 or alkyl radicals) can compete with reaction (17) and prevent the observed HO_2 yield from reaching 100%.

The yields of HO_2 in the butyl radical reactions show qualitative similarities with the previously reported ethyl, propyl and cyclopentyl radical reactions. However, unlike those reactions, the "raw" HO_2 yields Φ_{raw} (corrected only for HO_2 self-reaction) in the Cl-initiated butane oxidation process do not reach unity, (even at 693 K). Assuming that the total radical concentration is ⩽$[Cl]_0$, an HO_2 yield of 1 implies [X] = 0, the $R_{removal}$ term in eqn. (12) becomes negligible, and Φ_{raw} must reach 1. The lower observed yields for butane oxidation are experimental evidence of significant contribution from non HO_2-producing channels such as OH + C_4H_8O.

In previous investigations of the reactions of ethyl, propyl and cyclopentyl radicals with O_2, the production of HO_2 in the transition temperature region has displayed two timescales. The total yield could therefore be separated into a "prompt" and a "delayed" component. The timescale of the prompt rise in the HO_2 signal is not resolved in these experiments. The delayed rise is approximately exponential, with oxidation of isobutane showing larger deviations from single exponential behavior than n-butane. In all the $R + O_2$ reactions measured to date, the large increase in HO_2 which occurs in the transition region is attributable to the delayed component. In the present experiments, the amplitude of the faster initial rise in the HO_2, the prompt yield, is difficult to determine. The qualitative prompt yield, roughly estimated from extrapolation back to $t = 0$ of an exponential fit to the slower rise in HO_2, rises far more slowly with temperature than does the total yield. Like the smaller alkyl radical + O_2 reactions previously studied, the increase in alkene + HO_2 yield for butyl + O_2 in the transition temperature region arises from delayed HO_2 formation.

The fit of the slower component of the HO_2 signal to an exponential gives a time constant τ for the delayed production of HO_2. The inverse of the time constant is plotted as a function of temperature in Fig. 3. The butyl radicals derived from the different butanes appear to show slightly different behavior. The uncertainties arising from the treatment of the $C_4H_9O_2$ reactions are greatest at the low temperatures, where the extracted slow risetime is nearly completely determined by the treatment of the butylperoxy radical reactions. At 573 K the extracted time constant is roughly inversely proportional to the assumed $C_4H_9O_2 + HO_2$ rate coefficient. At higher temperatures, the extracted time constant is dominated by the rise of the observed signal. At higher temperatures, any correction for $C_4H_9O_2 + HO_2$ becomes less important as the dissociation of $C_4H_9O_2$ increases. The error estimates shown in Fig. 3 and listed in Table 2 are a sum of statistical uncertainties and the effects of a factor of two variation in the assumed values for both k_x and k_y.

Above ~ 660 K correction for $R_{removal}$, *i.e.*, that all butyl radicals react with O_2 to produce either HO_2 or $C_4H_9O_2$, may no longer be valid. Reactions of HO_2 with whatever other radicals are present must still occur, but the prediction of their reaction rate coefficients is not feasible. However, at the highest temperatures of this study the phenomenological time constant is nearly insensitive to the model for $R_{removal}$. At 673 K, a factor of two increase in the assumed values for k_x and k_y changes the time constant by only a few percent, and the total correction for $R_{removal}$ changes the time constant by only $\sim 10\%$ at 693 K. The phenomenological HO_2 production rates for $T > 663$ K, calculated using the same method, are shown as the open symbols in Fig. 3. At

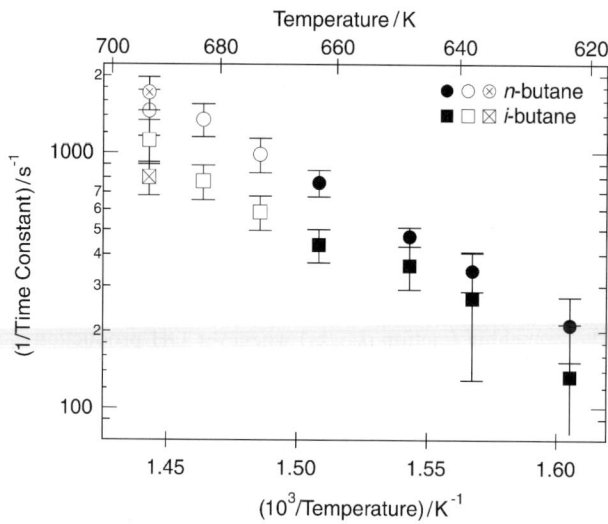

Fig. 3 Effective rate coefficients for the delayed production of HO_2 in the Cl-initiated oxidation of isobutane (squares) and n-butane (circles). The measurements at $T \geqslant 673$ K are shown as open symbols to indicate that the identification of $R_{removal}$ with $C_4H_9O_2 + HO_2$ reactions is uncertain at these temperatures, and that cycling reactions of CH_3 and OH radicals must also be considered. All measurements are taken with an O_2 concentration of 6.6×10^{16} cm^{-3} except for the hatched symbols (⊗, ⊠), where $[O_2] = 6.0 \times 10^{17}$ cm^{-3}.

693 K, the highest temperature investigated experimentally, a dependence of the HO_2 production rate constant on $[O_2]$ is measured. The apparent yield increases slightly for both butane isomers and the time constants (shown in Fig. 4) change by $\leqslant 30\%$ as the O_2 concentration is increased by a factor of 10. This substantiates the proposition that dissociation competes with RO_2 formation at these temperatures.

Interpretation of the trend with O_2 in the present system must account for cycling reactions of the CH_3 or C_2H_5 product of butyl radical dissociation,

$$CH_3 + Cl_2 \rightarrow CH_3Cl + Cl \tag{18}$$

$$C_2H_5 + Cl_2 \rightarrow C_2H_5Cl + Cl \tag{19}$$

$$C_4H_{10} + Cl \rightarrow C_4H_9 + HCl. \tag{3}$$

The butyl radicals produced from the Cl + butane reactions are a mix of isomers, with Cl + n-butane producing a mixture of n-butyl and sec-butyl radicals and the Cl + isobutane reaction producing an mixture of isobutyl and tert-butyl radicals. The dissociation and cycling reactions (3, 18, 19) serve to increase the relative concentration of the more stable isomer (tert-butyl for isobutane oxidation and sec-butyl for n-butane oxidation). The measured HO_2 production rate constant is related to the rate-limiting step for HO_2 production, which can be altered by the contribution of different isomers.

Quantum chemistry and master equation calculations for propyl + O_2

In order to provide a foundation for discussion of the butane oxidation, we first consider a detailed theoretical characterization of the simpler propyl + O_2 system. The $C_3H_7 + O_2$ reactions undergo the same types of isomerization and elimination steps as do the larger butyl + O_2 reactions. High-level calculations and master-equation simulations are more feasible to perform on the smaller system. The features of the potential energy surface (E_{HL2}) for the iso- and n-propyl reactions with O_2 are summarized in Table 3. Addition of O_2 to either isomer is barrierless; the calculated well depth for isopropyl·O_2 is slightly larger than for n-propyl·O_2. In both isomers the concerted pathway for HO_2 elimination via the five-membered ring transition state, which we denote as TS_{elim}, is energetically favored over isomerization via a five-membered ring to form a QOOH radical. However, these transition states, for formation of either CH_3CHCH_2OOH or $CH_3CH(OOH)CH_2$, are both at negative energies relative to the reactants (Table 3), and both

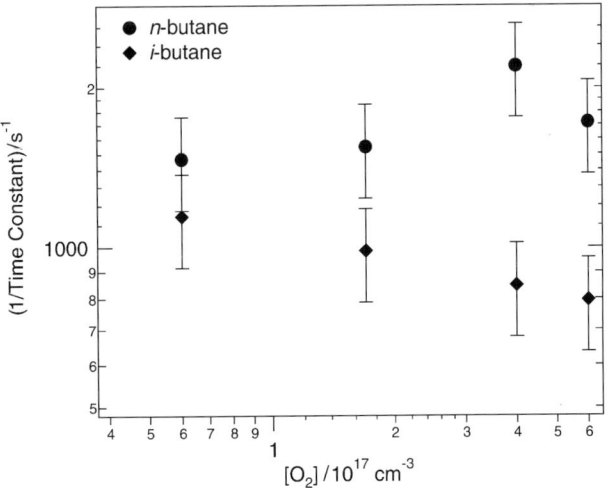

Fig. 4 Dependence of effective HO_2 production rate coefficients on O_2 concentration at 698 K. At this temperature, the rate constant for dissociation of n-butyl and isobutyl radicals is $>10^4$ s^{-1}. The effective rate constant at lower $[O_2]$ is therefore weighted towards the more stable isomer (sec-butyl for n-butane oxidation and tert-butyl for isobutane oxidation).

Table 3 Stationary point energies on the propyl + O_2 potential energy surfaces[a]

R	RO_2	TS_{elim}[b]	TS_{isom}[b]	QOOH		TS_{OH}	TS_{HO_2}
n-C_3H_7	−34.9	−5.2(1,4s)	−2.6(1,4s)	−21.6 CH_3CHCH_2OOH		−6.5	−3.4
			−11.2(1,5p)	−19.8 $CH_2CH_2CH_2OOH$		3.6	21.9
iso-C_3H_7	−36.8	−7.0(1,4p)	−1.4(1,4p)	−20.2 $CH_2CH(OOH)CH_3$		−4.9	−1.9

[a] Results from HL2 calculations in kcal mol^{-1} relative to R + O_2, including a B3LYP/6-31G* calculated zero-point energy correction. [b] The notation 1,n refers to an (n + 1)-membered ring transition state, and the letter refers to transfer of a *primary*, *secondary* or *tertiary* hydrogen atom. The TS_{elim} is a concerted elimination transition state, not strictly speaking an H transfer, but the same designation is used to indicate the site involved.

have accessible pathways to OH and HO_2 product channels, listed as TS_{OH} and TS_{HO_2} in Table 3. The contribution of these isomerizations can therefore be expected to be significant, especially at higher temperatures. The (1,5p) isomerization *via* a six-membered ring in n-propyl·O_2 to form ˙$CH_2CH_2CH_2OOH$ is at a much lower energy, as may be expected from lowered ring strain in the transition state. (The notation 1,n refers to an (n + 1)-membered ring transition state, and the letter refers to transfer of a *primary*, *secondary* or *tertiary* hydrogen atom.) However, the ˙$CH_2CH_2CH_2OOH$ radical which is formed has a sizeable barrier to formation of bimolecular products (OH + oxetane) and will be unstable with respect to isomerization back to $CH_3CH_2CH_2OO˙$. The calculated equilibrium constant for the $CH_3CH_2CH_2OO˙ \leftrightarrow$ ˙$CH_2CH_2CH_2OOH$ vastly favors the RO_2 form, with K_{eq} of more than 100 even at 800 K, a temperature above the stabilization limit for the n-propyl + O_2 reaction.

Minor adjustment (1–2 kcal mol^{-1}) of the calculated potential energies allows the time-resolved master equation calculations to fit the available experimental data for the propyl + O_2 reactions. Specifically, the well depth for n-propyl was reduced by 1 kcal mol^{-1}, the transition state for HO_2 elimination was raised by 1 kcal mol^{-1} in n-propyl·O_2 and by 2 kcal mol^{-1} for isopropyl·O_2, the transition state for $CH_3CH_2CH_2OO \rightarrow CH_3CHCH_2OOH$ was lowered by 1 kcal mol^{-1} and that for $CH_3CH(OO)CH_3 \rightarrow CH_3CH(OOH)CH_2$ was lowered by 0.5 kcal mol^{-1}. The published measurements of alkene yields, HO_2 production rate constants, total rate coefficients, and equilibrium constants are consistently fit using the master-equation analysis with this adjusted quantum chemistry.[2,33,34,46,53,66] As an example, Fig. 5 compares the results of the master-equation calculations to the measured rate coefficients for n-propyl + O_2 from Slagle *et al.*[46] The

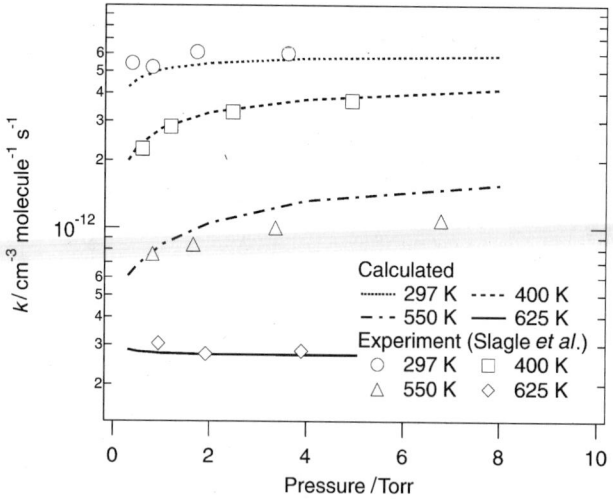

Fig. 5 Comparison of master-equation calculations with measurements of total rate coefficients for n-propyl reaction with O_2.

calculations give a biexponential time dependence for the n-propyl radical disappearance at the higher two temperatures (550 and 635 K). For purposes of comparison, an effective rate constant was estimated from the time at which the calculated population had decayed to 10% of its starting value. The predicted pressure and temperature dependence agrees well with the experimental values. Fig. 6 and Table 4 show the master-equation results for the effective rate coefficient for HO_2 production as a function of temperature. Individual rate coefficient determinations from the experiments of DeSain et al.[11] are shown for comparison. In order to compare the master equation results with the HO_2 production measurements of ref. 11 it is necessary to average the results from iso- and n-propyl + O_2, weighted by their relative production in the Cl + propane initiation reaction. Based on an extrapolation of the temperature dependent branching fraction measurements of Tschuikow-Roux et al.[68] and the room temperature branching measurement of Tyndall et al.[69] we use a 0.55 : 0.45 ratio of n-propyl to isopropyl. Fig. 7 shows the predicted time-dependent HO_2 signal for Cl-initiated oxidation of propane at 668 K from the master equation analysis, along with an experimental data trace. The amplitude and time behavior of the experimental HO_2 signal is reasonably well-modeled by the average of the n-propyl and isopropyl + O_2 calculations.

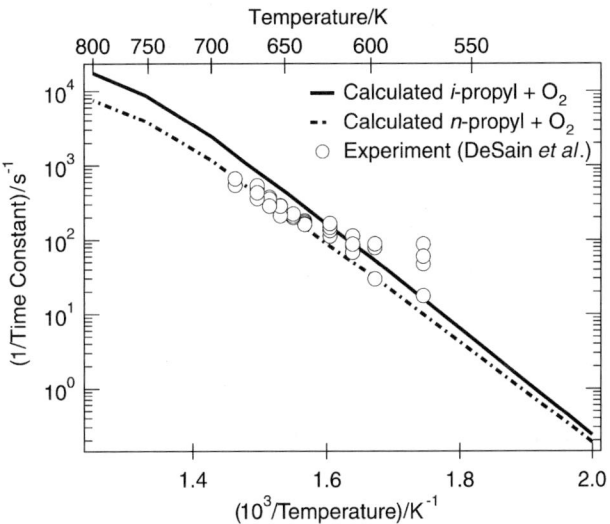

Fig. 6 Master-equation calculations of the effective rate coefficient for HO_2 production in the reactions of propyl radical isomers with O_2. The experimental measurements of ref. 11 are for an approximately 55 : 45 n- : iso isomeric mix. As in the present experiments, the uncertainties in the rate coefficient determinations increase dramatically at low temperature, and additional systematic uncertainty should be added to the statistical uncertainty evident from the scatter in the individual determinations.

Table 4 Calculated effective HO_2 production rate constants $(1/\tau)$ in s^{-1} for the reactions of propyl with O_2

T/K	$1/\tau$(isopropyl)	$1/\tau$(n-propyl)	$1/\tau$(experiment)[11]
500	0.24	0.19	
550	4.7	3.1	39
600	55	32	66 (598 K)
625	160	89	140 (623 K)
650	430	230	255 (653 K)
675	1100	530	602 (683 K)
700	2400	1100	
750	8700	3900	
800	18000	7600	

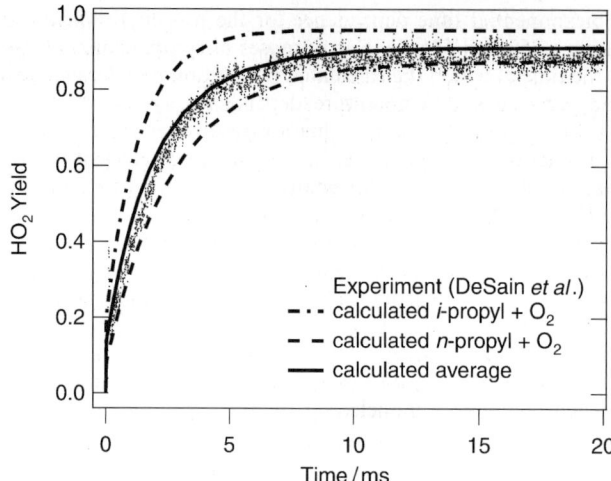

Fig. 7 Predicted time behavior of HO_2 production in the Cl-atom initiated oxidation of propane at 668 K from the master equation calculations. The experimental data are taken from the experiments of ref. 11. (Colour version available as ESI.†)

The predicted effective rate coefficients for delayed HO_2 production from the iso- and n-propyl + O_2 reactions for $P_{O_2} = 4.5$ Torr can be fit by an Arrhenius form between 300 and 750 K, yielding $k_{i\text{-}C_3H_7} = 5.9 \times 10^{12}\ e^{-15300/T}$ and $k_{n\text{-}C_3H_7} = 8.7 \times 10^{11}\ e^{-14500/T}$. The fitted activation energy for the HO_2 production is slightly less than the energy difference between the bottom of the RO_2 well and the transition state for HO_2 elimination, ΔE_{elim}, which is 31.8 kcal mol^{-1} in the isopropyl + O_2 master-equation calculations (adjusted from the 29.8 kcal mol^{-1} QCISD(T)-based value) and 29.7 kcal mol^{-1} for the n-propyl + O_2 calculations.

Calculations of butyl + O_2 potentials

The potential energy surfaces for the reactions of butyl radical isomers with O_2 are qualitatively similar to those observed in the reactions of smaller alkyl radicals with O_2. The main features of the potential energy surfaces are summarized in Table 5, calculated at a slightly lower level than the propyl + O_2 reactions discussed above. The transition states for HO_2 elimination are consistently 3–7 kcal mol^{-1} below the reactant energy. The energies of the transition states for the competing isomerizations to form QOOH species decrease with increasing size of the ring formed in the transition state. The internal abstraction of a hydrogen atom from the carbon to which the

Table 5 Stationary point energies on the butyl + O_2 potential energy surfaces[a]

R	RO_2	TS_{elim}[b]	TS_{isom}[b]	QOOH	TS_{OH}	TS_{HO_2}
n-C_4H_9	−33.2	−3.2(1,4s)	0.2(1,4s)	−19.5 $CH_3CH_2CHCH_2OOH$	−5.2	−1.9
			−10.9(1,5s)	−19.4 $CH_3CHCH_2CH_2OOH$	10.9	47.8[c]
			−9.3(1,6p)	−16.3 $CH_2CH_2CH_2CH_2OOH$	−2.3	52.2[c]
sec-C_4H_9	−35.2	−4.5(1,4p)	1.8(1,4p)	−18.2 $CH_3CH_2CH(OOH)CH_2$	−3.8	−1.1
		−4.3(1,4s)	−2.4(1,4s)	−21.4 $CH_3CHCH(OOH)CH_3$	−8.7	−5.4
			−10.6(1,5p)	−19.5 $CH_2CH_2CH(OOH)CH_3$	2.7	30.7[c]
iso-C_4H_9	−33.0	−3.1(1,4t)	−3.3(1,4t)	−22.2 $(CH_3)_2CCH_2OOH$	−9.9	−1.0
			−8.7(1,5p)	−16.8 $CH_2CH(CH_3)CH_2OOH$	3.9	26.1[c]
tert-C_4H_9	−36.7	−6.5(1,4p)	1.2(1,4p)	−19.0 $(CH_3)_2(CH_2)COOH$	−7.2	−2.9

[a] Results from HL1 calculations in kcal mol^{-1} relative to R + O_2, including a B3LYP/6-31G* calculated zero-point energy correction. [b] The notation 1,n refers to an (n + 1)-membered ring transition state, and the letter refers to transfer of a *primary*, *secondary* or *tertiary* hydrogen atom. The TS_{elim} is a concerted elimination transition state, not strictly speaking an H transfer, but the same designation is used to indicate the site involved. [c] B3LYP/6-31G* result.

O_2 is bonded *via* a four-membered ring correlates with a low energy product channel, OH + aldehyde or ketone, but the transition state energies are too high for this to be a kinetically significant pathway except at much higher temperatures, and are not reported here. Isomerization *via* a five-membered transition state is near or slightly above the transition state for HO_2 elimination, depending on the nature of the hydrogen being abstracted. Larger ring transition states for isomerization are significantly lower than TS_{elim}. The QOOH species formed by isomerization have relatively shallow wells and could rapidly isomerize back to the RO_2 form in the absence of a competitive route to bimolecular products. In particular, 1,5 isomerization *via* a six-membered ring transition state in n-butyl·O_2, *sec*-butyl·O_2 or isobutyl·O_2 produces a QOOH species which has a significant barrier to production of OH and a substituted oxetane. The calculated transition states for $CH_3CHCH_2CH_2OOH$ and $CH_2CH_2CH(OOH)CH_3$ to produce 2-methyloxetane + OH are 10.9 and 2.7 kcal mol^{-1} above the R + O_2 reactants, respectively, and that for $CH_2CH(CH_3)CH_2OOH$ to form 3-methyloxetane is 3.9 kcal mol^{-1} above reactants. These species will likely rapidly establish equilibrium with the corresponding RO_2 form. In contrast, the $CH_2(OOH)CH_2CH_2CH_2$ radical, formed by isomerization of the n-butyl·O_2 radical *via* a (1,6p) seven-membered ring transition state, has a transition state to form oxolane (tetrahydrofuran, THF) + OH which lies 2.3 kcal mol^{-1} below the energy of the reactants. The QOOH species formed from 1,4 isomerizations have transition states for OH + oxirane production below the isomerization transition state, making their formation essentially irreversible. It seems likely that the main contribution to OH production in the reactions of propyl and butyl radicals with O_2 will be through the 1,4 and 1,6 isomerizations.

Discussion

The initially formed RO_2 radical in the reactions of butyl or propyl radicals with O_2 can isomerize by internal hydrogen atom abstraction to produce a number of QOOH isomers, which can dissociate by C–O or O–O bond fission to produce HO_2 or OH. This isomerization must compete with the direct elimination of HO_2 from the RO_2 well. As the size and complexity of the alkyl radical increases beyond C_4 species, the further increase in the number of possible pathways rapidly makes detailed computational characterization difficult. However, the important transition state structures are already present in small RO_2 radicals, raising the hope that analysis of their reactions can serve as a prototype for larger alkyl + O_2 systems. This work uses a detailed theoretical investigation of the propyl + O_2 system as a framework for predicting the behavior of the Cl-initiated butane oxidation studies.

The master-equation results for the reactions of the propyl radical isomers with O_2 produce good agreement with the available rate coefficients, branching fractions and equilibrium constants. A similar detailed master-equation simulation for butyl + O_2 is beyond the scope of the present paper, which intends to explore the applicability to larger R + O_2 systems of models based on small alkyl radical reactions. A simplistic extension of the propyl + O_2 master equation results to analogous processes in the oxidation of butanes yields a qualitative model for the present experimental results and provides a test for general mechanistic conclusions about R + O_2 systems. Table 6 gives a comparison of calculated energies for stationary states of ethyl, propyl, and butyl + O_2 reactions calculated at the same level of theory (HL1 ~ QCISD(T)/6-311++G(2df,2pd)). The calculated energies at this level tend to be ~1–3 kcal mol^{-1} too high, based on comparisons with higher level calculations (Table 3) and detailed comparisons with experiment for ethyl and propyl + O_2.[29] The well depths for the addition of O_2 to form an alkylperoxyl radical differ among the alkyl radicals investigated by several kcal mol^{-1}; however, the calculated well depth relative to the HO_2 elimination transition state is remarkably consistent, varying by only about 1 kcal mol^{-1} among ethyl-, propyl- and butyl-O_2 systems. The calculated kinetic behavior of the propyl + O_2 reactions are dominated by the five-member ring transition states for HO_2 elimination. The simplest approximation to the contribution of similar processes in the butyl + O_2 reactions is to simply apply the fitted Arrhenius form for the calculated HO_2 production in iso- and n-propyl + O_2 to reactions *via* comparable transition states in butyl + O_2, taking into account differences in the calculated energies.

Using this simplified picture, the HO_2 production from n-butyl + O_2 should have approximately the same *A*-factor as the n-propyl reaction; that from *sec*-butyl should be approximately

Table 6 Consistent energies for key stationary points in alkyl oxidation reactions[a]

R	RO_2	TS_{elim}[b]	ΔE_{elim}	TS_{isom}[b] 1,4p	1,4s	1,4t	1,5p	1,5s	1,6p
C_2H_5	−32.5	−1.4(1,4p)	31.2	5.7	—	—	—	—	—
n-C_3H_7	−33.2	−2.7(1,4s)	30.5	—	0.4	—	−8.0	—	—
iso-C_3H_7	−35.0	−4.3(1,4p)	30.7	1.6	—	—	—	—	—
n-C_4H_9	−33.2	−3.2(1,4s)	30.0	—	0.2	—	—	−10.9	−9.3
sec-C_4H_9	−35.2	−4.5(1,4p)	30.7	1.8	—	—	−10.6	—	—
		−4.3(1,4s)	30.9	—	—	−2.4	—	—	—
iso-C_4H_9	−33.0	−3.1(1,4t)	29.9	—	—	—	−3.3	−8.7	—
tert-C_4H_9	−36.7	−6.5(1,4p)	30.2	1.2	—	—	—	—	—

[a] Results from HL1 calculations in kcal mol^{-1} relative to $R + O_2$, including a B3LYP/6-31G* calculated zero-point energy correction. [b] The notation 1,n refers to an $(n + 1)$-membered ring transition state, and the letter refers to transfer of a *primary*, *secondary* or *tertiary* hydrogen atom. The TS_{elim} is a concerted elimination transition state, not strictly speaking an H transfer, but the same designation is used to indicate the site involved.

that for n-propyl (from the CH_2 group) plus 1/2 that for isopropyl (from the CH_3). Similarly, the A-factor for isobutyl + O_2 can be approximated as 1/2 that for n-propyl + O_2 (because there is only one available H) and *tert*-butyl + O_2 as 3/2 isopropyl + O_2. The activation energies can be simply estimated by adjusting the fitted master equation values by the difference between the calculated ΔE_{elim} ($TS_{elim} - RO_2$) for the relevant propyl and butyl radical transition states (given in Table 6). The results of this simplistic model, as shown in Fig. 8, are remarkably close to the experimentally measured values for the effective HO_2 production rate coefficients. This near-agreement suggests that the concerted elimination also dominates HO_2 production in the butyl + O_2 reactions, and indicates that general models for at least some aspects of $R + O_2$ reactions may be possible from detailed consideration of smaller prototype reactions.

The estimated values for the HO_2 production rate coefficients in the n- and *sec*-butyl + O_2 reactions are somewhat smaller than the experimentally measured values from n-butane oxidation. However, an increased role for the isomerization pathway would increase the effective rate constant for delayed HO_2 production, and is corroborated by the evidence for significant non-

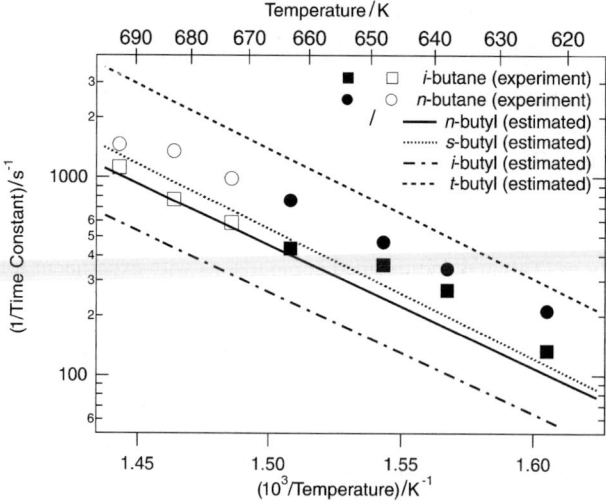

Fig. 8 Comparison of experimental effective rate constants for HO_2 production in Cl-atom initiated butane oxidation with estimated rate constants based on a simple extrapolation from master equation analysis of iso- and n-propyl + O_2 reactions.

HO_2 producing channels (presumably reaction (1c)) in butane oxidation. The calculated transition state for the 1,4s isomerization of sec-butyl·O_2 to $CH_3CHCH(OOH)CH_3$ is relatively close in energy to TS_{elim}. Moreover, the estimate based on the propyl + O_2 master-equation calculations can provide information only on transition states analogous to those which determine the HO_2 production in propyl + O_2. In particular, consider the 1,6p isomerization in n-butyl + O_2, which produces $CH_2CH_2CH_2CH_2OOH$, and has a transition state to produce OH + THF below the energy of the butyl + O_2 reactants. This process can be expected to contribute to the removal of n-butyl·O_2, and hence alter the observed overall time constants for delayed HO_2 formation, but it of course has no analogy in propyl + O_2.

The estimated rate coefficients for HO_2 production in the isobutyl and tert-butyl + O_2 reactions differ by more than a factor of five. The experimental data for isobutane oxidation has been fit using a single exponential to gain an effective rate constant, but the isobutane data does show a departure from single exponential behavior. The behavior of the effective rate coefficient with O_2 concentration at 693 K is also consistent with a significantly smaller rate constant for the less stable isomer. Unfortunately the signal-to-noise at early time in the present experiments does not permit reliable extraction of two exponentials from all of the i-butane oxidation experiments. Nevertheless, in several cases signal levels are high enough to permit such an analysis; Fig. 9 shows a semi-logarithmic plot of the experimental corrected HO_2 production for isobutane oxidation at 648 K (also shown in Fig. 1). The dashed line shows the single exponential fit used to derive the effective rate coefficient in Table 2 and Fig. 4. The solid line is a double exponential fit. The two effective rate constants from this double exponential fit are 265 and 2813 s^{-1}, larger than the estimated isobutyl + O_2 and tert butyl + O_2 rate coefficients. As for the n-butane case, an increase in the contribution from the QOOH pathway in isobutyl + O_2 is expected to result in larger effective rate constants than may be predicted from a simple analogy with propyl + O_2.

In mechanisms constructed for modeling hydrocarbon oxidation, the 1,5 isomerization via a six-membered ring transition state is often especially important, since the ring strain is not large, resulting in smaller activation energy, and the entropy loss is not too great, resulting in relatively large A-factors.[5,6,8,9] However, the present calculations suggest that this isomerization may be reversible for many alkyl·O_2 systems, since a sizeable barrier exists to formation of an oxetane + OH. Effective activation energies for isomerization reactions have been derived based on a comprehensive analysis of the initial products of alkane addition to H_2/O_2 mixtures,[1,2] and from direct measurements of OH production in the reaction of neo-pentyl radicals with O_2.[70] The analysis of these experiments assumes irreversibility of the isomerization reaction $RO_2 \rightarrow QOOH$; the present calculations in $C_3H_7O_2$ and $C_4H_9O_2$ suggest that this condition does not hold for 1,5 (and perhaps 1,6) isomerizations. Correlation of the observed effective activation energies for

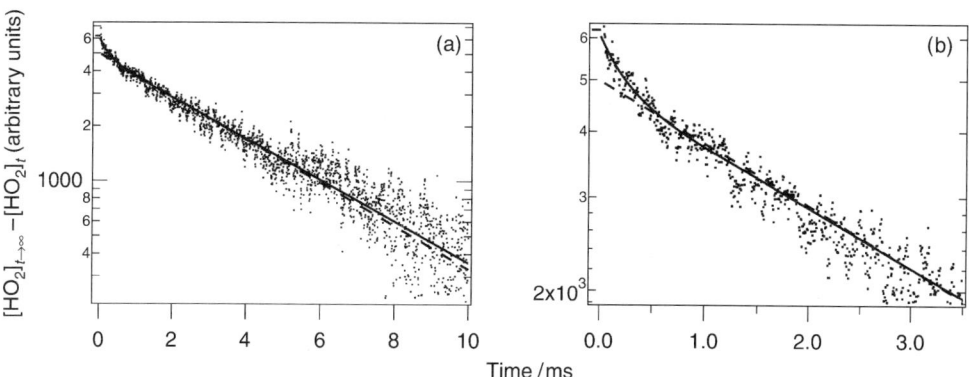

Fig. 9 (a) Semi-logarithmic plot of the corrected HO_2 production from Cl-initiated oxidation of isobutane at 648 K (dots, data also shown in Fig. 1), with a single exponential (dashed line) and double exponential (solid line) fit to the delayed production. (b) Expanded view of the early portion of the HO_2 signal, showing the biexponential behavior. The departure from a single exponential is interpreted as evidence for different effective HO_2 production rate constants for the reaction with O_2 of the two isomers (isobutyl and tert-butyl) formed in the Cl + isobutane initiation step. (Colour version available as ESI.†)

product formation with elementary isomerization steps and hence with specific features of the potential energy surface may need to be reconsidered.

Investigation of the details of OH production in alkyl + O_2 reactions is critical for modeling chain propagation in low temperature hydrocarbon oxidation. The potential energy surfaces calculated here imply that the OH production is likely to be dominated by products of 1,4 and 1,6 or higher isomerizations. The master equation modeling of the propyl + O_2 reactions produces reasonable agreement with literature measurements of branching fractions to cyclic ether products, as well as with LIF measurements of OH production in Cl-initiated alkane oxidation experiments.[1,2,54] Further work is underway to more thoroughly characterize the OH-producing channels in these and larger R + O_2 reactions.

Conclusions

The HO_2 production in the Cl-atom initiated oxidation of isobutane and n-butane have been experimentally measured between 297 and 693 K. The HO_2 produced is attributed to the reactions of butyl radicals with O_2. A significant fraction of the butyl radicals do not produce HO_2 even at the highest temperatures investigated, in accord with end product measurements of significant O-heterocycle yields in butane oxidation. A mechanism based on master equation modeling of the propyl + O_2 reactions, coupled with quantum chemical characterizations of stationary points on the butyl + O_2 potential energy surfaces, produces reasonable agreement with the observed time behavior of the HO_2 production. The present work suggests that the 1,4 pathways continue to dominate HO_2 production in the butyl + O_2 reactions. Finally, the calculations suggest that the $RO_2 \leftrightarrow QOOH$ isomerizations *via* 1,5 and 1,6 H transfers may be reversible.

Acknowledgements

This work is supported by the Division of Chemical Sciences, Geosciences, and Biosciences, the Office of Basic Energy Sciences, the U.S. Department of Energy.

References

1. R. W. Walker, in *Research in Chemical Kinetics*, ed. R. Compton and G. Hancock, Elsevier, Amsterdam, 1995, vol. 3, p. 1.
2. R. W. Walker and C. Morley, in *Low-Temperature Combustion and Autoignition*, ed. M. J. Pilling, Elsevier, Amsterdam, 1997, p. 1.
3. S. H. Robertson, P. W. Seakins and M. J. Pilling, in *Low-Temperature Combustion and Autoignition*, ed. M. J. Pilling, Elsevier, Amsterdam, 1997, p. 125.
4. R. T. Pollard, in *Gas Phase Combustion*, ed. C. H. Bamford and C. F. H. Tipper, Elsevier, Amsterdam, 1977, p. 249.
5. W. J. Pitz and C. K. Westbrook, *Combust. Flame*, 1986, **63**, 113.
6. E. Ranzi, A. Sogaro, P. Gaffuri, G. Pennati, C. K. Westbrook and W. J. Pitz, *Combust. Flame*, 1994, **99**, 201.
7. J. F. Griffiths, *Prog. Energy Combust. Sci.*, 1995, **21**, 25.
8. H. J. Curran, P. Gaffuri, W. J. Pitz and C. K. Westbrook, *Combust. Flame*, 1998, **114**, 149.
9. V. Warth, N. Stef, P. A. Glaude, F. Battin-Leclerc, G. Scacchi and G. M. Côme, *Combust. Flame*, 1998, **114**, 81.
10. E. P. Clifford, J. T. Farrell, J. D. DeSain and C. A. Taatjes, *J. Phys. Chem. A*, 2000, **104**, 11549.
11. J. D. DeSain, E. P. Clifford and C. A. Taatjes, *J. Phys. Chem. A*, 2001, **105**, 3205.
12. J. D. DeSain and C. A. Taatjes, *J. Phys. Chem. A*, 2001, **105**, 6646.
13. R. R. Baker, R. R. Baldwin and R. W. Walker, *J. Chem. Soc., Faraday Trans. 1*, 1975, **71**, 756.
14. R. R. Baldwin, I. A. Pickering and R. W. Walker, *J. Chem. Soc., Faraday Trans. 1*, 1980, **76**, 2374.
15. I. C. Plumb and K. R. Ryan, *Int. J. Chem. Kinet.*, 1981, **13**, 1011.
16. I. R. Slagle, E. Ratajczak and D. Gutman, *J. Phys. Chem.*, 1986, **90**, 402.
17. K. G. McAdam and R. A. Walker, *J. Chem. Soc., Faraday Trans. 2*, 1987, **83**, 1509.
18. A. F. Wagner, I. R. Slagle, D. Sarzynski and D. Gutman, *J. Phys. Chem.*, 1990, **94**, 1853.
19. E. W. Kaiser, I. M. Lorkovic and T. J. Wallington, *J. Phys. Chem.*, 1990, **94**, 3352.
20. E. W. Kaiser, T. J. Wallington and J. M. Andino, *Chem. Phys. Lett.*, 1990, **168**, 309.
21. R. R. Baldwin, D. R. Stout and R. W. Walker, *J. Chem. Soc., Faraday Trans.*, 1991, **87**, 2147.
22. O. Dobis and S. W. Benson, *J. Am. Chem. Soc.*, 1993, **115**, 8798.
23. E. W. Kaiser, *J. Phys. Chem.*, 1995, **99**, 707.

24 J. W. Bozzelli and A. M. Dean, *J. Phys. Chem.*, 1990, **94**, 3313.
25 G. E. Quelch, M. M. Gallo and H. F. Schaefer III, *J. Am. Chem. Soc.*, 1992, **114**, 8239.
26 G. E. Quelch, M. M. Gallo, M. Shen, Y. Xie, H. F. Schaefer III and D. Moncrief, *J. Am. Chem. Soc.*, 1994, **116**, 4953.
27 I. S. Ignatyev, Y. Xie, W. D. Allen and H. F. Schaefer III, *J. Chem. Phys.*, 1997, **107**, 141.
28 P. K. Venkatesh, A. M. Dean, M. H. Cohen and R. W. Carr, *J. Chem. Phys.*, 1999, **111**, 8313.
29 J. A. Miller, S. J. Klippenstein and S. H. Robertson, *Proc. Combust. Inst.*, 2000, **28**, 1479.
30 J. C. Rienstra-Kiracofe, W. D. Allen and H. F. Schaefer III, *J. Phys. Chem A*, 2000, **104**, 9823.
31 J. A. Miller and S. J. Klippenstein, *Int. J. Chem. Kinet.*, The reaction between ethyl and molecular oxygen II: Further analysis, in press.
32 E. W. Kaiser, L. Rimai and T. J. Wallington, *J. Phys. Chem.*, 1989, **93**, 4094.
33 E. W. Kaiser and T. J. Wallington, *J. Phys. Chem.*, 1996, **100**, 18770.
34 E. W. Kaiser, *J. Phys. Chem. A*, 1998, **102**, 5903.
35 R. R. Baker, R. R. Baldwin and R. W. Walker, *Proc. Combust. Inst.*, 1971, **13**, 291.
36 W. T. Chan, H. O. Pritchard and I. P. Hamilton, *Phys. Chem. Chem. Phys.*, 1999, **1**, 3715.
37 W. T. Chan, I. P. Hamilton and H. O. Pritchard, *J. Chem. Soc., Faraday Trans.*, 1998, **94**, 2303.
38 C. J. Chen and J. W. Bozzelli, *J. Phys. Chem. A*, 1999, **103**, 9731.
39 T. P. W. Jungkamp, J. N. Smith and J. H. Seinfield, *J. Phys. Chem. A*, 1997, **101**, 4392.
40 C.-J. Chen and J. W. Bozzelli, *J. Phys. Chem. A*, 2000, **104**, 4997.
41 T. H. Lay and J. W. Bozzelli, *J. Phys. Chem. A*, 1997, **101**, 9505.
42 P. N. Skancke, *Int. J. Quantum Chem.*, 1992, **41**, 591.
43 R. Mereau, M. T. Rayez, F. Caralp and J. C. Rayez, *Phys. Chem. Chem. Phys.*, 2000, **2**, 1919.
44 M. Kranenburg, M. V. Ciriano, A. Cherkasov and P. Mulder, *J. Phys. Chem. A*, 2000, **104**, 915.
45 I. Garcia-Cruz, M. E. Ruiz-Santoyo, J. R. Alvarez-Idaboy and A. Vivier-Bunge, *J. Comput. Chem.*, 1999, **20**, 845.
46 I. R. Slagle, J.-Y. Park and D. Gutman, *Proc. Combust. Inst.*, 1984, **20**, 733.
47 H. E. Hunziker and H. R. Wendt, *J. Chem. Phys.*, 1974, **60**, 4622.
48 D. Herriott, H. Kogelnik and R. Kompfner, *Appl. Opt.*, 1964, **3**, 523.
49 J. S. Pilgrim, R. T. Jennings and C. A. Taatjes, *Rev. Sci. Instrum.*, 1997, **68**, 1875.
50 W. B. DeMore, S. P. Sander, D. M. Golden, R. F. Hampson, M. J. Kurylo, C. J. Howard, A. R. Ravishankara, C. E. Kolb and M. J. Molina, *Chemical Kinetics and Photochemical Data for Use in Stratospheric Modeling*, Jet Propulsion Laboratory, Pasadena, CA, 1997.
51 K. Yamasaki, A. Watanabe, T. Kakuda and I. Tokue, *Int. J. Chem. Kinet.*, 1998, **30**, 47.
52 K. Yamasaki and A. Watanabe, *Bull. Chem. Soc. Jpn.*, 1997, **70**, 89.
53 V. D. Knyazev and I. R. Slagle, *J. Phys. Chem. A*, 1998, **102**, 1770.
54 J. D. DeSain and C. A. Taatjes, in preparation.
55 V. D. Knyazev, I. A. Dubinsky, I. R. Slagle and D. Gutman, *J. Phys. Chem.*, 1994, **98**, 11099.
56 V. D. Knyazev, I. A. Dubinsky, I. R. Slagle and D. Gutman, *J. Phys. Chem.*, 1994, **98**, 5279.
57 V. D. Knyazev and I. R. Slagle, *J. Phys. Chem.*, 1996, **100**, 5318.
58 Á Bencsura, V. D. Knyazev, S.-B. Xing, I. R. Slagle and D. Gutman, *Proc. Combust. Inst.*, 1992, **24**, 629.
59 M. J. Frisch, G. W. Trucks, H. B. Schlegel, G. E. Scuseria, M. A. Robb, J. R. Cheeseman, V. G. Zakrzewski, J. A. Montgomery, Jr., R. E. Stratmann, J. C. Burant, S. Dapprich, J. M. Millam, A. D. Daniels, K. N. Kudin, M. C. Strain, O. Farkas, J. Tomasi, V. Barone, M. Cossi, R. Cammi, B. Mennucci, C. Pomelli, C. Adamo, S. Clifford, J. Ochterski, G. A. Petersson, P. Y. Ayala, Q. Cui, K. Morokuma, D. K. Malick, A. D. Rabuck, K. Raghavachari, J. B. Foresman, J. Cioslowski, J. V. Ortiz, B. B. Stefanov, G. Liu, A. Liashenko, P. Piskorz, I. Komaromi, R. Gomperts, R. L. Martin, D. J. Fox, T. Keith, M. A. Al-Laham, C. Y. Peng, A. Nanayakkara, C. Gonzalez, M. Challacombe, P. M. W. Gill, B. Johnson, W. Chen, M. W. Wong, J. L. Andres, C. Gonzalez, M. Head-Gordon, E. S. Replogle and J. A. Pople, GAUSSIAN 98, Gaussian, Inc., Pittsburgh, PA, 1998.
60 J. A. Miller, S. J. Klippenstein and S. H. Robertson, *J. Phys. Chem. A*, 2000, **104**, 7525.
61 J. A. Miller, S. J. Klippenstein and S. H. Robertson, *J. Phys. Chem. A*, 2000, **104**, 9806.
62 J. A. Miller and S. J. Klippenstein, *J. Phys. Chem. A*, 2000, **104**, 2061.
63 K. S. Pitzer and W. D. Gwinn, *J. Chem. Phys.*, 1942, **10**, 428.
64 S. J. Klippenstein, *J. Chem. Phys.*, 1992, **96**, 367.
65 S. J. Klippenstein, *J. Phys. Chem.*, 1994, **98**, 11459.
66 R. P. Ruiz and K. D. Bayes, *J. Phys. Chem.*, 1984, **88**, 2592.
67 S. J. Klippenstein, A. F. Wagner, R. C. Dunbar, D. M. Wardlaw, S. H. Robertson and J. A. Miller, VARIFLEX; 1.08m ed., 2001.
68 E. Tschuikow-Roux, T. Yano and J. Niedzielski, *J. Chem. Phys.*, 1985, **82**, 65.
69 G. S. Tyndall, J. J. Orlando, T. J. Wallington, M. Dill and E. W. Kaiser, *Int. J. Chem. Kinet.*, 1997, **29**, 43.
70 K. J. Hughes, P. A. Halford-Maw, M. J. Pilling and T. Turanyi, *Proc. Combust. Inst.*, 1992, **24**, 645.
71 D. L. Baulch, C. J. Cobos, R. A. Cox, P. Frank, G. Hayman, T. Just, J. A. Kerr, T. Murrells, M. J. Pilling, J. Troe, R. W. Walker and J. Warnatz, *J. Phys. Chem. Ref. Data*, 1994, **23**, 847.
72 R. Atkinson, D. L. Baulch, R. A. Cox, R. F. Hampson, Jr., J. A. Kerr, M. J. Rossi and J. Troe, *J. Phys. Chem. Ref. Data*, 1997, **26**, 521.

73 A. A. Boyd, B. Nozière and R. Lesclaux, *J. Chem. Soc., Faraday Trans.*, 1996, **92**, 201.
74 P. D. Lightfoot, R. A. Cox, J. N. Crowley, M. Destriau, G. D. Hayman, M. E. Jenkin, G. K. Moortgat and F. Zabel, *Atmos. Environ., Part A*, 1992, **26**, 1805.
75 D. M. Rowley, R. Lesclaux, P. D. Lightfoot, K. Hughes, M. D. Hurley, S. Rudy and T. J. Wallington, *J. Phys. Chem.*, 1992, **96**, 7043.
76 M. A. Crawford, J. J. Szente, M. M. Maricq and J. S. Francisco, *J. Phys. Chem. A*, 1997, **101**, 5337.
77 D. M. Rowley, P. D. Lightfoot, R. Lesclaux and T. J. Wallington, *J. Chem. Soc., Faraday Trans.*, 1992, **88**, 1369.
78 D. M. Rowley, R. Lesclaux, P. D. Lightfoot, B. Nozière, T. J. Wallington and M. D. Hurley, *J. Phys. Chem.*, 1992, **96**, 4889.
79 B. Nozière, R. Lesclaux, M. D. Hurley, M. A. Dearth and T. J. Wallington, *J. Phys. Chem.*, 1994, **98**, 2864.
80 D. L. Baulch, C. J. Cobos, R. A. Cox, C. Esser, P. Frank, T. Just, J. A. Kerr, M. J. Pilling, J. Troe, R. W. Walker and J. Warnatz, *J. Phys. Chem. Ref. Data*, 1992, **21**, 411.
81 W. Tsang and R. F. Hampson, *J. Phys. Chem. Ref. Data*, 1986, **15**, 1087.
82 W. Tsang, *J. Phys. Chem. Ref. Data*, 1988, **17**, 887.
83 W. Tsang, *J. Phys. Chem. Ref. Data*, 1990, **19**, 1.

General Discussion

Prof. Hippler opened the discussion of the Introductory Lecture: (1) You showed a rectangular temperature profile during combustion in an engine. Would not this lead to a rectangular NO-profile both from thermal and prompt NO?

(2) How well can it be decided that the contribution of prompt NO to total NO is low, since the rate constant for $CH + N_2$ shows a strong non-Arrhenius behaviour and experiments under combustion conditions around 1000 K are lacking?

Prof. Wolfrum responded: (1) The rectangular temperature profile represents a 'snapshot' through the growing flame. Therefore, in the middle of the combustion chamber, where the flame started at the spark plug, hot gases have been present for the longest time. This temporal history is responsible for the increase of thermal NO towards the center of the combustion chamber.

(2) We are watching here an unsteady, growing flame. Residence times within the flame front are short compared to the time where thermal NO can be formed in the hot post-flame gases. The calculation checks if an additional contribution from NO formed directly in the flame front is necessary to explain the experimentally found profiles. Therefore, we did a phenomenological approach modeling prompt NO formation as an instantaneous process at the current flame front position. Flame conditions were assumed to be identical for all volume experiments when the flame front is crossing. Therefore, no temperature effects had to be included in the calculation. For details please refer to ref. 1.

1 C. Schulz, J. Wolfrum and V. Sick, *Twenty-seventh Symp. (Int.) Combust.*, The Combustion Institute, Pittsburgh, 1998, p. 2077.

Prof. Lin asked: We have recently shown by a high-level *ab initio* calculation that the prompt NO reaction, $CH + N_2$, produces the spin-allowed HNCN and H + NCN products,[1] instead of the commonly assumed, spin-forbidden HCN + N. Have you considered the new mechanism in your modeling?

1 L. V. Moskaleva and M. C. Lin, *Proc. Combust. Inst.*, 2000, **28**, 2393.

Prof. Wolfrum responded: As mentioned in the response to Hippler's question, no detailed information on rate constants was included in the simulation calculation presented here.

Dr Klippenstein opened the discussion of Prof. Casavecchia's paper: I was intrigued by your observation of what appears to be a highly non-statistical product branching in the $N(^2D) + CH_4$ reaction. To further investigate this possibility I have performed some direct dynamics simulations with the forces directly determined from B3LYP/6-31G* evaluations. Such simulations nicely complement statistical theories in that they are most feasible and most applicable to reactions occurring on a short time scale (*e.g.*, less than 1 ps) where statistical simulations are of dubious validity.

For the present simulations the initial conditions were chosen to correspond to a fixed C to N separation of 4.0 a_0, which roughly corresponds with the separation at the saddlepoint for the insertion of the N atom into a CH bond. The internal coordinates and conjugate momenta for the vibrational modes of the CH_4 fragment were chosen quasiclassically with a random vibrational phase and the harmonic oscillator normal mode zero-point energy in each mode. The remaining coordinates and conjugate momenta were chosen randomly, as described in detail in our related work for the $CH_3 + O$ reaction.[1] While this choice of initial conditions does not exactly mimic the experimental conditions, it should still provide qualitatively meaningful predictions for the branching ratios.

Simulations were performed for a total angular momentum J of 25 at two different energies: 12.6 and 63 kcal mol^{-1}, in excess of the reactant's zero-point energy. A time step of 0.5 fs was employed for each of the 47 and 33 trajectories completed for these two energies. The results of these simulations are given here in Table 1.

Qualitatively these results are in good agreement with your experimental observations. In particular, there is a larger than expected branching to the higher energy $CH_3N + H$ channel, and this branching increases with increasing energy. Furthermore, the numerical values for the relative branching are of a similar magnitude to the experimental observations. The simulations also predict minor branching to other unexplored product channels such as $CH_3 + NH$.

As you suggest in your paper, the increased production of the $CH_3N + H$ channel over likely statistical theory expectations appears to be related to a correlation between the insertion and dissociation processes. Indeed, the average timescale for the H atom loss is much shorter for those trajectories that produce CH_3N (50 fs being fairly typical). Furthermore, in many instances the $CH_3N + H$ producing trajectories appear to proceed directly from insertion to bimolecular products.

The question I have for you is whether one can reconcile this direct short time nature of the trajectories that produce $CH_3N + H$ with the observed anisotropy in the centre-of-mass (CM) angular distributions?

1 T. P. Marcy, R. R. Díaz, D. Heard, S. R. Leone, L. B. Harding and S. J. Klippenstein, *J. Phys. Chem. A*, 2001, **105**, 8361.

Prof. Casavecchia responded: I am very glad to learn that direct dynamics simulations have become feasible for a polyatomic reaction such as $N(^2D) + CH_4$. Your results are indeed very interesting; in particular I notice that there really is a good qualitative agreement between your theoretical predictions and the results of our crossed molecular beam investigations as a function of collision energy, *i.e.*, the branching to the less exoergic $CH_3N + H$ channel is found also theoretically to be higher than expected on statistical grounds and to increase with increasing energy. Your theoretical work contributes considerably to the understanding of the dynamics of this multichannel reaction. Specifically, your finding that $CH_3N + H$ formation is prompt, following the insertion of the $N(^2D)$ atom into the C–H bond of methane, can help to rationalize the shape of the experimental center-of-mass angular distribution. In fact, an isotropic angular distribution, as that found for the CH_3N channel in our study, is usually attributed (at least for simple A + BC reactions) to a reaction micromechanism implying a long-lived-complex formation, *i.e.* a CH_3NH complex whose lifetime is significantly longer than its rotational period. This would be in line with the deep potential well of CH_3NH. However, an isotropic angular distribution can also have other

Table 1 Results of simulations performed for total angular momentum $J = 25$ at two different energies in excess of the reactants zero point energy

Product	Branching	
	$E = 12.6$	$E = 63.0$
$CH_2NH + H$	0.55	0.34
$CH_3N + H$	0.27	0.41
$NH + CH_3$	0.11	0.13
$HCNH + H_2$	0.07	0.03
$H_2CN + H_2$	0	0.06
$NH_2 + CH_2$	0	0.03

interpretations, especially for 'insertion' reactions. An isotropic angular distribution could also arise from a combination of two dynamically different 'direct' micromechanisms, one leading to backward scattering (arising from small impact parameter collisions) and the other leading to forward scattering (arising from large impact parameter collisions). But perhaps, more likely than that, the formalism derived for simple, limiting cases, is not valid here since, once the N atom has inserted into one of the four equivalent CH bonds, the geometry of the intermediate changes dramatically, and the memory of the initial direction of the reagents is lost. As a result, the intermediate is likely to fragment isotropically in space. The time scale of a 'direct' mechanisms is sub-picosecond, i.e., of the time-scale of molecular vibrations. The theoretical results of Klippenstein suggest that this is indeed the case. Interestingly, the fact that the $P(E'_T)$ distribution for the CH_3N + H forming channel corresponds to a large fraction (more than 40%) of total available energy released in translation, indicates that the energy release is not statistical and since no potential barrier is theoretically predicted to exist in the exit channel,[1] this is consistent with a CH_3NH intermediate which dissociates promptly, on a short time scale, before energy randomization can take place, i.e. the reaction proceeds via a 'direct' mechanism. In conclusion, although CH_3NH is very stable with respect to CH_3N + H products, the mechanism of formation of CH_3N appears to be via a direct insertion of N into the CH bond followed by prompt NH bond cleavage. The fact that the CM angular distribution is isotropic (backward–forward symmetric) is still consistent with the short time nature of the reactive trajectories, as found in your theoretical investigation (see above).

1   Y. Kurosaki, T. Takayanagi, K. Sato and S. Tsunashima, *J. Phys. Chem. A*, 1998, **102**, 254.

Dr Kaiser said: The almost isotropic (flat) angular distribution in the case of the CH_4 + $N(^2D)$ reaction could have an alternative explanation. For example, it might be a superposition of two direct reaction mechanisms: mechanism I is dictated by relatively large impact parameters leading to a 'forward' scattered contribution; mechanism II is dominated by small impact parameters giving rise to a 'backward' scattered component, i.e. possibly via a transition state including a penta-coordinated carbon atom (Fig. 1). These transition states are well known in 'classical' organic chemistry; their contribution to the PES and hence the reactive scattering signal of CH_3N and should be investigated computationally.

Prof. Casavecchia replied: Although the superposition of two direct reaction mechanisms may be an alternative explanation, this is not the only other possible rationalization of an isotropic angular distribution (see my response to Dr Klippenstein previously). Actually, since we find the

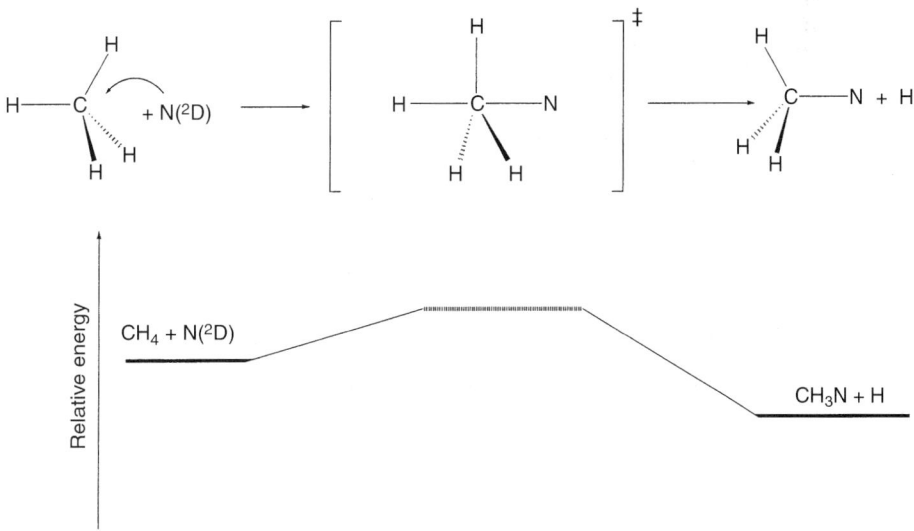

Fig. 1   The CH_4 + $N(^2D) \rightleftharpoons CH_3N$ + H reaction.

CM angular distribution of CH_3N to be isotropic in a wide range of collision energies (from 21.3 to 63.2 kJ mol^{-1}), it is unlikely that two different micromechanisms, one resulting from small impact parameter collisions and the other from large impact parameter collisions, give rise to a constant amount of backward and forward scattering, respectively, as a function of collision energy; in fact, one would expect an increasing forward scattering with increasing E_c, something which is not found experimentally. In addition, although transition states with a penta-coordinated carbon atom exist and may play some role also in this reaction, we note that detailed *ab initio* molecular orbital calculations of the potential energy surfaces for the $N(^2D) + CH_4$ reaction found that $N(^2D)$ 'inserts' into the C–H bond of methane.[1] Of course, a theoretical treatment of the dynamics using classical trajectory calculations on the global potential energy surface of this reaction, when a global CH_4N surface will become available, is desirable to fully characterize the reaction dynamics.

1   Y. Kurosaki, T. Takayanagi, K. Sato and S. Tsunashima, *J. Phys. Chem. A*, 1998, **102**, 254.

Prof. Golden asked: Can the difference in branching ratio of CH_3N and CH_2NH be due to symmetry and other factors that contribute to the *A*-factors (density of states at the transition states)?

Prof. Casavecchia answered: This is an interesting comment. Such factors could also be important. A detailed theoretical investigation of this reaction is however desirable to explore what is the role of energy and symmetry factors *vs.* dynamics. I notice that previous statistical (RRKM) considerations by Takayanagi and coworkers[1] did not find branching to the higher energy $CH_3N + H$ channel to be important, while the direct dynamics treatment of Klippenstein provides quite opposite results. Clearly, dynamics is at play in this multichannel reaction.

1   Y. Kurosaki, T. Takayanagi, K. Sato and S. Tsunashima, *J. Phys. Chem. A*, 1998, **102**, 254.

Prof. Kohse-Höinghaus said: I would like to comment on the observation of hydrogen-deficient compounds in low-pressure flames. We have seen C_4H_2 and C_6H_2 in propene–oxygen, pentene–oxygen and acetylene–oxygen flames at 50 mbar, in particular these compounds (and C_8H_2) have been observed in fuel-rich acetylene flames used for diamond deposition.[1] While their dependence on flame stoichiometry may be quite well predicted using the combustion mechanism by Miller and Melius[2] with a few addenda,[3] the C_4H_2 concentration is underpredicted by about a factor of 10 (ref. 4) and the C_6H_2 concentration by about two orders of magnitude.[5]

1   A. G. Löwe, A. T. Hartlieb, J. Brand, B. Atakan and K. Kohse-Höinghaus, *Combust. Flame*, 1999, **118**, 37.
2   J. A. Miller and C. F. Melius, *Combust. Flame*, 1992, **91**, 21.
3   J. A. Miller, personal communication to B. Atakan, 1999.
4   K. Kohse-Höinghaus, A. Löwe and B. Atakan, *Thin Solid Films*, 2000, **368**, 185.
5   A. Löwe, PhD thesis, University of Bielefeld, Germany 1999.

Prof. I. W. M. Smith commented: I share with Prof. Casavecchia and Dr Kaiser, a strong interest in the reaction between $C(^3P)$ atoms and C_2H_2. The experiments reported here by Prof. Casavecchia confirm the importance of the 'spin-forbidden' channel to $C_3(X\,^1\Sigma_g^+) + H_2$ which appears to be competitive with the 'spin-allowed' channel to $C_3H + H$. As pointed out in the paper by Casavecchia, and co-workers, the reaction to $C_3(X\,^1\Sigma_g^+) + H_2$ is likely made possible by the long lifetime of the strongly bound triplet propargylene intermediate allowing intersystem crossing to the singlet surface and subsequent separation to $C_3(X\,^1\Sigma_g^+) + H_2$. There would seem to be an opportunity for theory of two kinds. First, the lifetime of the triplet state is likely to be strongly dependent on energies of the triplet products: $l\text{-}C_3H + H$ and $c\text{-}C_3H + H$, and any barriers along the paths leading to these products. Can theory define these energies more closely than has been done so far? I note that the values quoted in the paper (from ref. 63) show these reactions to be very slightly exothermic, whereas the calculations of Guadagnini *et al.* (ref. 58) appear to suggest that they are very slightly endothermic. Second, very exact RRKM calculations on this system would be valuable. Such calculations should help not only to establish the lifetimes of the triplet propargylene and also the branching ratios, both between $C_3(X\,^1\Sigma_g^1) + H_2$ and $C_3H + H$, and between $l\text{-}C_3H + H$ and $c\text{-}C_3H + H$, but also how those branching ratios depend on temperature over a wide range. This information would be useful in models of both combustion and interstellar cloud chemistry.

Prof. Casavecchia responded: Yes, I agree with Prof. I. W. M. Smith that there is a need for a determination more accurate than the existing ones of the energetics of the various channels of the $C(^3P) + C_2H_2$ reaction, as well as of exact RRKM calculations in order to provide estimates of branching ratios as a function of temperature. From an experimental point of view, we plan to study this reaction both at lower and higher collision energies than that ($E_c = 29.3$ kJ mol^{-1}) reported in our paper[1] in order to explore how the branching ratio between $C_3H + H$ and $C_3 + H_2$ formation varies with collision energy. From experiments at very low E_c (4 kJ mol^{-1} or less) we also hope to be able to explore the branching ratio between linear C_3H and cyclic C_3H formation. Accurate theoretical values of the energetics of the corresponding reaction channels would be valuable in the data analysis.

1 P. Casavecchia, N. Balucani, L. Cartechini, G. Capozza, A. Bergeat and G. G. Volpi, *Faraday Discuss.*, 2001, **119**, 27.

Dr Bergeat opened the discussion of Dr Kaiser's paper: You have reported a new experiment which allows you to detect the $C_3 + H_2$ channel of the $C + C_2H_2$ reaction. How have you increased your sensitivity? Have you any idea or estimation of the branching ratio of the $C_3 + H_2$ and $C_3H + H$ channels?

In our fast-flow reactor, at 300 K and 1 Torr, we have found a branching ratio less than 0.53 ± 0.04 for the $H + C_3H$ channel, by comparison with the H production by the $C + H_2S$ reaction.[1,2]

(1) May you agree with this estimation?
(2) May you estimate the sensitivity of your experiments for the other C, C_2 and C_3 + hydrocarbon reactions?

1 A. Bergeat and J. C. Loison, *Phys. Chem. Chem. Phys.*, 2001, **3**, 2038.
2 N. Galland, F. Caralp, M. T. Rayez, Y. Hannachi, J. C. Loison, G. Dorthe and A. Bergeat, *J. Phys. Chem. A*, in press.

Dr Kaiser responded: (1) The very first study of the C–C_2H_2 system was performed with the old setup of the 35″ machine (1994–1997) employing diffusion pumps and oil lubricated roughing pumps. Therefore, the background at $m/z = 37$ and 36 was very high. Further, a significant background at $m/z = 36$ came from elastically scattered C_3 which was in the primary beam. Last but not least, the filament in the electron impact ionizer consisted of a tungsten wire coated with carbon; this contributed to the background at $m/z = 36$ as well. The new setup (after the machine was moved from Berkeley to Taiwan) had significant improvements. The filament was replaced by thoriated iridium. In addition, the detector system was equipped with a closed cycle helium refrigerator cold head (4.5 K) which lowered the pressure to $<8 \times 10^{-13}$ mbar. Second, no diffusion pumps were used; only magnetically suspended turbo molecular pumps for the main chamber as well as both source chambers were utilized, and the main chamber was backed with an oil-free scroll pump. Third, all cables inside the machine were replaced by Teflon-coated wires which had a very low out gassing rate. Fourth, a second 4.5 K cold shield was placed between the interaction region and the chopper wheel. All these four improvements reduced the background at $m/z = 36$ significantly (the only background arises from electronic background which can be eliminated). Finally, the carbon source was modified: the atomic carbon number density was enhanced, and—at the same time—the contribution of C_3 to the elastically scattered background was minimized. These factors together allow a detection of the $C_3 + H_2$ channel which could not have been observed before.

(2) At a collision energy of 16.6 kJ mol^{-1}, it is about 30–40%. However, it is not feasible to compare this branching ratio to ones obtained from kinetic studies. As you pointed out in your paper, your experiments were not performed under single collision conditions. So numerous processes can complicate the true 'binary picture' of a collision as present in crossed beam setups. These are, for example: (a) third body reactions and hence a stabilization of long lived reaction intermediates, (b) collision induced inter-system crossing. So one always has to be very careful to compare these bulk studies with crossed beam data.

(3) As you know, the reactions of atomic, ground state carbon have rate constants in the order of 10^{-10} cm^3 s^{-1}. On the other hand, kinetic data suggest that reactions of tricarbon with unsaturated hydrocarbons are very slow (about 10^{-12} cm^3 s^{-1} and smaller); please refer to the literature cited in our paper presented at this meeting. Based on our experiments, the reaction of tricarbon with acetylene to form atomic hydrogen and CCCCCH is endothermic by about 80 ± 9 kJ mol^{-1}. Therefore, if the scattering experiments are performed at a collision energy of, let's say, 100 kJ mol^{-1}, this channel is open. But the scattering signal of the heavy fragment is confined to a very narrow angular range due to the low amount of energy which can channel into the translational degrees of freedom. So we can say that in principle, once the collision energy is larger than the endothermicity of a channel, this one can be detected in crossed experiments—even if the rate constants from bulk experiments (for example at room temperature) indicate that the reactions are very slow.

Prof. Casavecchia said: With reference to the paper by Kaiser et al., I have a few questions regarding the results for the reaction $C(^3P) + C_2H_2$.

(1) You studied this reaction a few years ago and published the results in several articles.[1–4] Although you never showed TOF spectra at $m/z = 36$ in refs. 1–4, you always stated that the TOF and angular distributions at $m/z = 36$ were identical to those measured at $m/z = 37$, and this result lead you to the conclusion that $C_3H + H$ formation is the only reaction pathway taking place in the $C(^3P) + C_2H_2$ reaction, while formation of $C_3 + H_2$, a spin-forbidden channel, was clearly not occurring. However, I notice that the signal-to-noise of the TOF spectra at $m/z = 37$ reported in the previous articles[1–4] is similar to that exhibited in the spectrum at $m/z = 36$ shown in Fig. 11 of your paper.[5] Since the detector background is expected to be lower at $m/z = 36$ than at $m/z = 37$ (because of the ^{37}Cl isotope), I would have expected you to see the fast peak at $m/z = 36$, attributable to the $C_3 + H_2$ channel, also in the previous experiments. Then the first question is: why didn't you see the two peak structure, as that shown in Fig. 11, in the previous[1–4] measurements?

(2) In your pulsed C beams, obtained by laser ablation of graphite,[6] you have reported the presence of a large concentration of C_2 and C_3, with C_3 being 2.5 times more abundant than C when Ne is used as seeding gas, and 25% of C when He is the seeding gas.[3,6] As a matter of fact, using these beams you can study the reactive scattering of C_2 and C_3, which is the subject of your paper at this meeting. Since we know that C_3 does not react with C_2H_2 at $E_c = 16.6$ kJ mol^{-1}, having a threshold at $E = 80$ kJ mol^{-1},[5] the second question is then: Why don't you see elastically scattered C_3 from the C_2H_2 beam in the TOF spectrum at $m/z = 36$ at $E_c = 16.6$ kJ mol^{-1}? Elastically scattered C_3 should appear as a strong fast peak in the spectrum. I am very puzzled by the absence of the C_3 elastic peak in your spectrum of Fig. 11. What is the explanation? You report in your paper that you have measured TOF spectra from 12 to 72 degrees: can you show us the complete set of data, which are required to derive the CM product translational energy distribution shown in Fig. 12? Furthermore, when you say (footnote 40 in ref. 5) that close to the beam you have an elastic background, what do you mean? You are measuring TOF spectra and if you have elastically scattered particles, these should appear in the spectrum as a peak!

(3) You say that the results at $m/z = 36$ are different from those at $m/z = 37$. Can you show us the corresponding laboratory angular distributions at $m/z = 37$ and $m/z = 36$ that you have used to obtain the results shown in Fig. 12.?

(4) The authors say that by subtracting the collision energy from the high energy cut-off of the $P(E_T)$ (shown in Fig. 12), the reaction is found to be exothermic by 140–150 kJ mol^{-1}, and that this value is in good agreement with a reaction energy of 130–135 kJ mol^{-1} calculated from thermodynamical data (ref. 41). However, if one takes the thermodynamical data, which are rather accurate, one actually finds that the exoergicity of the reaction $C(^3P) + C_2H_2 \rightarrow C_3 + H_2$ is 100–105 kJ mol^{-1}, a value much lower than the value of 140–150 kJ mol^{-1} derived from your experiment. How do the authors explain this significant difference?

Finally, we would like to emphasize that our experiments on the $C(^3P) + C_2H_2$ reaction reported at this meeting,[7] although carried out using C beams containing both $C(^3P)$ and $C(^1D)$, allow a clean determination of the dynamics of the two competing reaction pathways leading to $C_3H + H$ and $C_3 + H_2$ formation. The reason is due (a) to the fact that our continuous C beams do not contain any detectable amount of C_3 which, elastically scattered, would represent a serious interference in the measurements of reactive signal at $m/z = 36$, and (b) to the high signal-to-noise of the experiment which permitted very clearly differences between the angular distribution at $m/z = 37$ and $m/z = 36$ (see Fig. 11a of ref. 7). Once the data at $m/z = 37$, corresponding to the $C_3H + H$ channel, are analyzed with great sensitivity to derive the CM angular and translational energy distributions for $C_3H + H$ formation from $C(^3P)$ and $C(^1D)$, the data at $m/z = 36$, which are partly derived from fragmentation of the C_3H product detected at $m/z = 37$, can also be analyzed in detail by adding to the components derived from the $m/z = 37$ data analysis, a new component corresponding to the new reaction channel contributing to the $m/z = 36$ signal, that is the $C_3 + H_2$ channel. We have derived a ratio of cross sections $\sigma(C_3 + H_2)/[\sigma(C_3 + H_2) + \sigma(C_3H + H)]$ of 0.3 at $E_c = 29.3$ kJ mol^{-1}. What is the ratio derived at $E_c = 16.6$ kJ mol^{-1} in your study? It would be interesting to know it in order to see how the above branching ratio varies with collision energy.

1 R. I. Kaiser, Y. T. Lee and A. G. Suits, *J. Chem. Phys.*, 1995, **103**, 10395.
2 R. I. Kaiser, C. Ochsenfeld, D. Stranges, M. Head-Gordon, Y. T. Lee and A. G. Suits, *Science*, 1996, **274**, 1508.
3 R. I. Kaiser, C. Ochsenfeld, M. Head-Gordon, Y. T. Lee and A. G. Suits, *J. Chem. Phys.*, 1997, **106**, 1729.
4 R. I. Kaiser, C. Ochsenfeld, D. Stranges, M. Head-Gordon and Y. T. Lee, *Faraday Discuss.*, 1998, **109**, 183.
5 R. I. Kaiser, T. N. Le, T. L. Nguyen, A. M. Mebel, N. Balucani, Y. T. Lee, F. Stahl, P. v. R. Schleyer and H. F. Schaefer III, *Faraday Discuss.*, 2001, **119**, 51.
6 R. I. Kaiser and A. G. Suits, *Rev. Sci. Instrum.*, 1995, **66**, 5405.
7 P. Casavecchia, N. Balucani, L. Cartechini, G. Capozza, A. Bergeat and G. G. Volpi, *Faraday Discuss.*, 2001, **119**, 27.

Dr Kaiser responded: The identity of the TOFs which was stated in the previous articles refers to the *reactively* scattered product. As you know, this does not refer to the background from inelastically scattered tricarbon molecules. There was a significant contribution of $m/z = 36$ from elastically scattered tricarbon in the very first study of this reaction back in Berkeley. For further details, please refer to my answer to Dr Bergeat's question. However, it should be really stressed that there is—in the case of multiple fits—always space for flexibility. Check for instance the data on the backward scattered channel in the $C-C_2H_4$ system. This backward scattering is very difficult to explain. If this pathway really exists, it should rather be forward–backward symmetric or—as the collision energy increases—forward-scattered.

Dr Bergeat said: The $C + C_2H_2$ reaction leads to $H + C_3H$ and $H_2 + C_3$ channels. The last channel is spin forbidden and the adduct lying on the path leading to $H + C_3H$ is quite long-lived, allowing time for inter-system crossing (ISC) (calculations were performed by your group[1,2] and others[3,4]). Our results (less than 53% of H atom produced by the reaction) and the estimation of Casavecchia *et al.* (presented here) under single collision conditions confirm that the ISC is facilitated. However, the C + methylacetylene reaction leads to the formation of around 79% of H atoms. Your previous experiment and the experiment of Casavecchia showed that there is no H_2 elimination. In your article, you mentioned the possibility of an ISC to lead to the formation of CH_4. As the exit channels leading to $H + C_3H_3$ are strongly exothermic (-177, -132 and -54 kJ mol^{-1}), while in the case of $C + C_2H_2$, the H elimination is nearly athermic (-1.5 and -8.6 kJ mol^{-1}), do you think that the ISC could be efficient? The experiment of Casavecchia shows that no H_2 is formed. Is it connected with the small ISC probability or with the barrier on the potential energy surface and the possibility of CH_4 production?

1 C. Ochsenfeld, R. I. Kaiser, Y. T. Lee, A. G. Suits and M. Head-Gordon, *J. Chem. Phys.*, 1997, **106**, 4141.
2 A. M. Mebel, W. M. Jackson, A. H. H. Chang and S. H. Lin, *J. Am. Chem. Soc.*, 1998, **120**, 5751.
3 J. Takahashi and K. Yamashsita, *J. Chem. Phys.*, 1996, **104**, 6613.
4 R. Guadagnini, G. C. Schatz and S. P. Walch, *J. Phys. Chem. A*, 1998, **102**, 5837.

Dr Kaiser replied: You cannot compare the H_2 loss channel in the HCCCH intermediate ($C-C_2H_2$ system) with the H_2 loss pathway from the $HCCCCH_3$ intermediate ($C-CH_3CCH$ system)! The H_2 channel in the acetylene reaction is similar to the CH_4 loss pathway in the methylacetylene reaction. No one expects H_2 loss from $HCCCCH_3$; but a CH_4 elimination might be feasible—if ISC takes place.

Dr Klippenstein said: A BAC-MP4 theoretical study of the decomposition of vinylacetylene (CH_2CHCCH) by Melius et al. indicated that the H loss channel actually lies significantly higher in energy than both the H_2 + HCCCCH and HCCH + H_2CC: channels.[1] Thus, it is somewhat surprising that you conclude that there are no H_2 products in the C_2 + C_2H_4 reaction. This could be due to a looser bottleneck for the H atom loss channel in combination with the high energy of the chemical activation process you have studied. However, I suspect that statistical theories would still predict a significant fraction (e.g. 10%) in the other channels. Can you comment on what is the maximum branching to the H_2 channel that is allowed by your experiments.

1   C. F. Melius, J. A. Miller and C. M. Evleth, *Twenty-fourth Symp. (Int.) Combust.*, The Combustion Institute, Pittsburgh, 1992, p. 621.

Dr Kaiser replied: The data on C_2–C_2H_4 presented here are the first data on this system obtained using the crossed beam technique. We mentioned that a detailed data analysis is 'in progress'; we have about 30 Mbyte additional data on C_2–C_3 reactions—at higher signal-to-noise ratio as presented here; unfortunately, they have not been analyzed yet. So stay tuned to find out what these data will tell.

Dr Kaiser addressed Prof. Casavecchia: Recent crossed beam reactions of $C(^1D)$ with unsaturated hydrocarbons acetylene, methylacetylene, propylene and ethylene[1,2] at high collision energies of 48–104 kJ mol^{-1} showed that they follow direct reaction dynamics. Further, all systems are dominated by a $C(^1D)$ vs. H exchange pathway to form C_3H, n-C_4H_3, C_4H_5 isomers and C_3H_3 (propargyl). Based on these results it is surprising that Casavecchia et al. could not detect any reactive scattering signal from $C(^1D)$ with C_2H_4. Although their collision energy is lower than 48 kJ mol^{-1}, electronic structure calculations predict[2] that at least an H atom elimination pathway should be observable. Since Casavecchia et al. performed their experiments with continuous beams of similar velocity, speed ratio and composition, $C(^1D)$ should be present in each beam. Is there any explanation why no reactive scattering signal was observed in the $C(^1D)$–C_2H_4 system?

1   R. I. Kaiser, A. M. Mebel and Y. T. Lee, *J. Chem. Phys.*, 2001, **114**, 231.
2   R. I. Kaiser, T. L. Nguyen, A. M. Mebel and Y. T. Lee, *J. Chem. Phys.*, submitted June 2001.

Prof. Casavecchia responded: Kaiser et al.[1] have recently reported a study of the reaction $C(^1D)$ + C_2H_2 at high collision energies (E_c = 45 kJ mol^{-1} and 109 kJ mol^{-1}) using pulsed C beams generated by laser ablation of graphite. In their study at the lowest E_c, the C beam appears to contain both $C(^3P)$ and $C(^1D)$, while at the highest E_c only $C(^1D)$.[1] We find it surprising that in that study[1] the C beam corresponding to E_c = 109 kJ mol^{-1} is estimated to contain exclusively $C(^1D)$: some $C(^3P)$ should always be present on the basis of simple electronic partition function arguments. In fact, the experimental data at E_c = 109 kJ mol^{-1} (see Fig. 4 of ref. 1) do not really permit one to exclude some contribution to C_3H + H formation also from the $C(^3P)$ reaction. We have studied the same reaction at E_c = 29.3 kJ mol^{-1} using a continuous beam of C atoms obtained by radio-frequency discharge in a dilute mixture of CO_2 in He and from the analysis presented here,[2] the ratio of concentration of $C(^1D)/C(^3P)$ in the beam appears to be much smaller than in the experiment of Kaiser et al.[1] at E_c = 45 kJ mol^{-1} (compare Fig. 11(a) of ref. 2 with Fig. 3 of ref. 1; these figures show the relative contribution of $C(^3P)$ and $C(^1D)$ reactions to the same C_3H product detected at m/z = 37).

In his comment, Kaiser says that in a similar study on the reaction $C(^3P, {}^1D)$ + C_2H_4 at E_c = 48–104 kJ mol^{-1} similar results on the $C(^1D)$ reaction were obtained.[3] Unfortunately, the work is not published yet and we don't know what is the relative signal attributable to the $C(^1D)$ and $C(^3P)$ reactions at E_c = 48 and 104 kJ mol^{-1}. Presumably, at the lowest E_c the contribution of $C(^1D)$ is again considerably larger than that of $C(^3P)$, while at the highest E_c all the reactive signal is attributed to $C(^1D)$, as in the case of the $C(^3P, {}^1D)$ + C_2H_2 reaction. We have studied the C + C_2H_4 reaction at E_c = 8.8, 15.9 and 29.7 kJ mol^{-1}. Since the E_c = 29.7 kJ mol^{-1} data (see Fig. 11(c) in ref. 2) have been obtained by using a beam similar (i.e., CO_2 seeded in He) to that used for the $C(^3P, {}^1D)$ + C_2H_2 experiment (see Fig. 11(a) in ref. 2), we would have expected some contribution also from $C(^1D)$ to the C_3H_3 signal. However, the m/z = 38 and m/z = 39 angular and TOF distributions could be well fit without invoking any ^1D contribution (see Fig. 11(c) in ref. 2).

Since the best-fit of the angular distribution measured at $m/z = 38$ (and characterized by rather small error bars) is slightly underestimating the last two experimental points at small angles (see Fig. 11(c) in ref. 2), after we submitted the *Faraday Discussion* paper presented here we studied the same reaction at the higher E_c of 37.8 kJ mol^{-1} by using the same C beam while accelerating the C_2H_4 beam by the seeded beam technique. The experimental laboratory angular distribution at $m/z = 38$ is shown in Fig. 2. As can be seen, because of the different kinematics, we are now able to observe the complete fall-off of the angular distribution also at small angles and, clearly, the small angle intensity cannot be accounted for by invoking only the reaction channels leading to $H_2CCCH + H$ and $H_3CCC + H$ from the reaction of $C(^3P)$, as was the case at the lower E_c of 8.8–29.7 kJ mol^{-1}: a small, additional contribution is now needed to fit the data at small angles. In fact, the dashed line in Fig. 2 corresponds to a contribution from the reaction $C(^1D) + C_2H_4 \rightarrow C_3H_3(\text{propargyl}) + H$, in a relative amount, with respect to the $C(^3P)$ contributions (dotted and dashed-dotted lines), which is comparable to what we have observed in the $C + C_2H_2$ reaction at the slightly lower E_c of 29.3 kJ mol^{-1}. These results may suggest that the relative cross section $^1D/^3P$ is somewhat different in the reactions with C_2H_2 and C_2H_4, and/or that the $C(^1D)$ concentration in the experiment with C_2H_4 was lower than in that with C_2H_2, something plausible because the radio-frequency power conditions and backing pressure were not exactly the same in the two experiments. Another possible reason for a small $C(^1D)$ contribution to the $C_3H_3 + H$ channel is that the $C(^1D)$ reaction can also lead to other, competitive product channels, as for instance to $C_2H_2(X\,^1\Sigma_g^+) + CH_2$, which has an exoergicity comparable to the $C_3H_3 + H$ channel. Indeed, photodissociation of ground state singlet C_3H_4 (allene) at 157 nm (corresponding to a total available energy similar to that of our experiments) has shown that the C–C bond cleavage channel is sizeable (relative yields for the H, H_2, and CH_2 formation processes were determined to be 1 : 0.15 : 0.27).[4] Since in the experiment at $E_c = 29.7$ kJ mol^{-1} we did not observe the complete fall-off of the lab angular distribution at small angles, and the CM angular distribution from the $C(^1D)$ reaction is expected to be significantly forward peaked (as at $E_c = 37.8$ kJ mol^{-1}), a small contribution of $C(^1D)$ also to the results at $E_c = 29.7$ mol^{-1} cannot be ruled out; however, this contribution would be quite small. On the whole, the results seem to suggest that the relative contribution of $C(^1D)$ increases with E_c with respect to that of $C(^3P)$. This is supported by the fact that the angular distribution at $E_c = 8.8$ kJ mol^{-1} can very well be fitted without invoking any $C(^1D)$ contribution.[5] It should be noted that we were able to observe the fall-off to zero both at small and large angles in the angular distribution at this low E_c,[5] and its width would be quite sensitive to the extra available energy (121.9 kJ mol^{-1}) carried by the

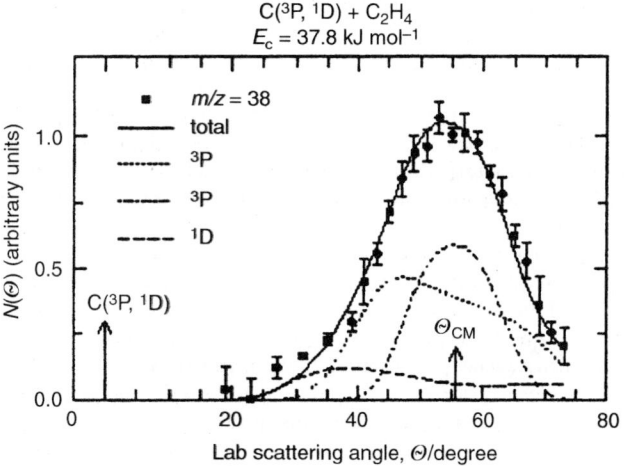

Fig. 2 Product ($m/z = 38$) laboratory angular distribution from the $C(^3P, ^1D) + C_2H_4$ reaction at $E_c = 37.8$ kJ mol^{-1}. The solid line is the total best-fit, dotted and dotted-dashed lines are the contributions (dotted: propargyl formation; dotted-dashed: cyclopropenyl and propyn-1-yl formation) from the $C(^3P)$ reaction, and dashed line is the contribution to propagyl formation from the $C(^1D)$ reaction.

electronically excited $C(^1D)$ atom. Also note that the C beam used for the 8.8 kJ mol^{-1} experiment is obtained by seeding CO_2 in Ne and has been shown, from studies of the $C(^1D) + H_2$ reaction,[6] to contain both $C(^3P)$ and $C(^1D)$.

So, we do not see appreciable $C(^1D)$ contribution to the H-displacement channel at $E_c = 29.7$ kJ mol^{-1}, and we start to see a small amount of it at $E_c = 37.8$ kJ mol^{-1}, because our C beam contains little $C(^1D)$ (relative to $C(^3P)$) with respect to the pulsed beam used by Kaiser et al. at $E_c = 48-104$ kJ mol^{-1}.

Finally, we note that even invoking a small $C(^1D)$ contribution in order to fit the data at $E_c = 37.8$ kJ mol^{-1}, the small angle intensity cannot be entirely reproduced (see Fig. 2). This suggests that the H_2 elimination channel may also have some importance in this reaction; in fact, it can be noted that the data in Fig. 2 have been recorded at $m/z = 38$ ($C_3H_2^+$) since the signal-to-noise ratio was higher than at $m/z = 39$ ($C_3H_3^+$). The experimental data shown in Fig. 2 have been obtained by using a very dilute (2%) mixture of C_2H_4 in He, which reduced significantly the signal level given a $m/z = 39$ angular distribution with large error bars. We plan to repeat the experiment with a more concentrated mixture of C_2H_4 in He and plan to compare the $m/z = 39$ and $m/z = 38$ angular distributions to verify whether they are identical (in this case the $m/z = 38$ signal would come only from fragmentation of C_3H_3 in the detector ionizer, corresponding only to the H-displacement channel) or somewhat different at small angles (in this case the $m/z = 38$ signal would partially come from a dynamically different channel, that corresponding to H_2 elimination and leading to $C_3H_2 + H_2$). In regard to the last point, we note that photodissociation of singlet C_3H_4 at 157 nm has been found to lead to both H and H_2 elimination with a ratio of 1 : 0.15,[4] and H_2 elimination from a chemically activated C_3H_4 complex may well occur also.

All this indicates that the dynamics of the $C(^3P, ^1D) + C_2H_4$ reaction are rather complex and accurate data at several collision energies are crucial to disentangle them in detail.

1 R. I. Kaiser, A. M. Mebel and Y. T. Lee, J. Chem. Phys., 2001, **114**, 231.
2 P. Casavecchia, N. Balucani, L. Cartechini, G. Capozza, A. Bergeat and G. G. Volpi, Faraday Discuss., 2001, **119**, 27.
3 R. I. Kaiser, T. L. Nguyen, A. M. Mebel and Y. T. Lee, J. Chem. Phys., submitted.
4 S. Harich, Y. T. Lee and X. Yang, Phys. Chem. Chem. Phys., 2000, **2**, 1187.
5 L. Cartechini, G. Capozza, P. Casavecchia, G. G. Volpi, W. Geppert, C. Naulin and M. Costes, in preparation.
6 A. Bergeat. L. Cartechini, N. Balucani, G. Capozza, L. F. Phillips, P. Casavecchia, G. G. Volpi, L. Bonnet and J.-C. Rayez, Chem. Phys. Lett., 2000, **327**, 197.

Prof. Plane addressed Dr Kaiser and Prof. Casavecchia: Both of these molecular beam techniques require a high flux of radicals in one of the crossed beams. These are made either by high-powered microwave dissociation (Casavecchia) or by laser ablation of a solid target (Kaiser). Presumably highly excited metastable states are produced by both techniques. What is the evidence that they do not play a significant role in forming the observed reaction products?

Prof. Casavecchia responded: In general, we generate radical beams by high-power, high-pressure radio-frequency discharge.[1-4] In the case of nitrogen and carbon atoms, as well as in the case of oxygen atoms, both ground state ($N(^4S)$, $C(^3P)$ and $O(^3P)$) and excited state metastable ($N(^2D)$, $C(^1D)$ and $O(^1D)$) atoms are contained in the beam. However, the nitrogen atom reactions discussed in our paper[5] are endoergic for $N(^4S)$ while exoergic for $N(^2D)$, so we are only observing products from the excited state atom reactions. In contrast, in the case of C atom reactions, those discussed here[5] are exoergic for both $C(^3P)$ and $C(^1D)$; in particular, the reactions with unsaturated hydrocarbons are known to be fast both for $C(^3P)$ and $C(^1D)$. In this case, we observe contributions to formation of the same reaction product from both $C(^3P)$ and $C(^1D)$. As discussed in our paper,[5] by measuring accurately product angular and velocity distributions and relying on energy and momentum conservation, we are able to separate the distinct contributions of $C(^3P)$ and $C(^1D)$ to the formation of a given product. This in the past was also done successfully for O atom reactions, such as $O(^3P, ^1D) + H_2S$.[1,6] For instance, in the reaction $C + CH_3CCH$, the $C(^3P)$ and $C(^1D)$ contributions to formation of $C_4H_3 + H$ are clearly visible in both the laboratory angular and TOF distributions (see Fig. 11(b) of ref. 5), and the CM product angular and translational energy distributions have been derived for both $C(^3P)$ and $C(^1D)$ reactions (see Fig. 12(b) in ref. 5 for the $P(E'_T)$). In conclusion, excited atoms, if present in the beam, do also contribute, in general, to forming the observed product. The extent of the contribution depends on the relative concentration and relative reactive cross section of excited and ground state atoms; this may vary from system to system. Finally, the presence of excited species in these discharge generated beams, although experimentally may in some cases represent a complication when one wishes to study the ground state reactions, offers the exciting opportunity of exploring the reaction dynamics of electronically excited (metastable) atoms, such as $C(^1D)$.

But perhaps you were wondering about the possible role of even more highly excited metastable states, such as $O(^1S)$, $N(^2P)$ and $C(^1S)$. Although these more highly excited states may actually be present in our beams, though in lower concentration than $O(^1D)$, $N(^2D)$ and $C(^1D)$, we have never found evidence of their contribution. Please note that because of the much higher energies involved, their contribution to the observed products would be noticeable in the width of the product angular distribution as well as in the product velocity distributions. This is not surprising since it is well known that $O(^1S)$, $N(^2P)$ and $C(^1S)$, despite their much higher energy content, are much less reactive than $O(^1D)$, $N(^2D)$ and $C(^1D)$, respectively. The reason being that those very excited reactants do not correlate adiabatically with ground state products (for instance, see ref. 1 for O atoms).

1 M. Alagia, N. Balucani, P. Casavecchia, D. Stranges and G. G. Volpi, *J. Chem. Soc., Faraday Trans.*, 1995, **91**, 575.
2 M. Alagia, V. Aquilanti, D. Ascenzi, N. Balucani, D. Cappelletti, L. Cartechini, P. Casavecchia, F. Pirani, G. Sanchini and G. G. Volpi, *Isr. J. Chem.*, 1997, **37**, 329.
3 A. Bergeat, L. Cartechini, N. Balucani, G. Capozza, L. F. Phillips, P. Casavecchia, G. G. Volpi, L. Bonnet and J.-C. Rayez, *Chem. Phys. Lett.*, 2000, **327**, 197.
4 P. Casavecchia, *Rep. Prog. Phys.*, 2000, **63**, 355.
5 P. Casavecchia, N. Balucani, L. Cartechini, G. Capozza, A. Bergeat and G. G. Volpi, *Faraday Discuss.*, 2001, **119**, 27.
6 N. Balucani, L. Beneventi, P. Casavecchia, D. Stranges and G. G. Volpi, *J. Chem. Phys.*, 1991, **94**, 8611; P. Casavecchia, N. Balucani, M. Alagia, L. Cartechini and G. G. Volpi, *Acc. Chem. Res.*, 1999, **32**, 503.

Dr Kaiser responded to Prof. Plane: Yes, laser ablation of solid targets will certainly produce metastable and even long-lived, highly excited Rydberg states. However, the great advantage of pulsed beams compared to continuous sources is that these excited states can be eliminated. Note that excited states are only present in the fast part of the pulsed beam. For example, $C(^1D)$ is present only if the carbon beam is faster than about 3000 m s^{-1}. Likewise, vibrationally excited CN radicals are only present at velocities of more than about 2000 m s^{-1}. However, a chopper wheel located after the ablation zone and after the skimmer can select a part of the pulsed beam in which no metastable species are present. Note that in some cases we can select a fast pulse of the beam to study the reactions of electronically excited species, such as carbon atoms.

Dr Whitaker communicated to Prof. Casavecchia and Dr Kaiser: In the conclusion of their paper Casavecchia *et al.* stress that their recent achievements in the study of elementary reactions by crossed molecular beam (CMB) methods has been made possible by developments in continuous supersonic beam sources and yet Dr Kaiser in his presentation of the equally impressive results obtained for the reactions of di- and tri-carbon with unsaturated hydrocarbons was at pains to point out the importance of pulsed molecular beam sources. Although clearly it is difficult to produce C_2 and C_3 in a continuous beam source for excited atomic reactants. My question to both Prof. Casavecchia and Dr Kaiser is, what are the advantages and disadvantages of each of these sources?

Prof. Casavecchia communicated in response: Continuous and pulsed beam sources have both advantages and disadvantages, depending on the application.[1] Usually, one wishes to use pulsed beams in conjunction with pulsed laser beam generation and/or pulsed laser detection schemes, for obvious reasons. In contrast, in crossed beam experiments with electron-impact ionization mass spectrometric detection, because of the continuous nature of the detection scheme, one wishes, in general, to use continuous beams, for duty cycle reasons, and this has always been done since the development of the technique.[1] However, pulsed beams can also be successfully used in reactive scattering with electron-impact ionization mass spectrometric detection, as the pioneering work of Gentry and coworkers and Lee and coworkers on the $D + H_2$ reaction demonstrated.[2]

More recently, a great deal of reaction dynamics studies have been carried out in Lee's group[3–6] using pulsed beams of C, CN and O(^{1}D), and in Davis's group[7] using pulsed beams of a variety of transition metal atoms. The main difference, from a practical point of view, between using continuous or pulsed beams in this type of experiments is that with continuous beams one measures directly the product angular distribution by modulating with a tuning fork chopper one of the two beams for background subtraction. Since typical counting times are 50–100 s per angle, by taking 4–5 angular scans one obtains readily angular distributions over a fine grid of angles with high accuracy and small error bars (typically, 1–3%). Then, product velocity distributions are measured by the TOF technique at selected laboratory angles (typical counting times being 20–120 min depending on signal intensity). Since the area of each TOF spectrum at a given lab angle corresponds to the intensity of the angular distribution at that angle, the various TOF spectra can be normalized to an accurate angular distribution. In contrast, in experiments with pulsed beams, one measures only TOF spectra at selected angles (at least, in the applications seen so far), and then derives the product angular distribution by integrating the TOF spectra at each angle. Since the measurement of a TOF spectrum requires a considerable amount of time (from the order of 5–10 min to hours, depending on signal intensity) and 4–5 scans are needed to derive an angular distribution, during this extended period of time fluctuations of experimental conditions may occur. Although this can be taken into account (*i.e.*, by time normalization), the resulting angular distribution is often measured on a sparce grid of angles and is characterized by sizeable error bars (typically 10%).

Of course, there can be valid reasons, and often advantages, for using pulsed rather than continuous beams in CMB experiments with mass spectrometric detection. For instance, photolysis is usually a clean way of generating atoms and radicals, and there may not be other ways to generate these species efficiently in a continuous fashion. For example, generation of O(^{1}D) by 157 nm laser photolysis of O_2 is very efficient and clean and this has permitted beautiful work in recent years by Yang and coworkers.[6] One great advantage of using pulsed beams is the very low gas consumption: this permits one to carry out studies with isotopically marked pure reagents that are not usually possible using continuous beams. Pulsed laser ablation is perhaps the only way to generate beams of transition metals.[7] Generation of C atoms by pulsed laser ablation is well established since the development of an efficient beam source by Costes *et al.*;[8] this has permitted one to study in recent years the dynamics of a large variety of C atom reactions with unsaturated hydrocarbons.[3,4] However, the error bars of the angular distributions derived for these reactions are often significantly larger than those we have obtained using our newly developed continuous beams of C atoms (compare for instance the results on $C + C_2H_2$ and $C + C_2H_4$ presented here by us[9] and the corresponding results obtained using pulsed C beams[4]). In addition, measurements of angular distributions with continuous beams are typically much faster. On another note, pulsed beams of C atoms generated by laser ablation of graphite are heavily contaminated by C_2 and C_3 species, and perhaps higher C_n clusters.[3,4] However, while this may represent a complication in studies of some C atom reactions, it offers the opportunity of studying the reactive scattering of C_2 and C_3, as reported here by Kaiser *et al.*[10]

In conclusion, there are not in general strict prescriptions on the use of pulsed or continuous beams in CMB experiments with electron-impact ionization mass spectrometric detection. Both types of beams can be used with various degree of success, resolution and accuracy.

1 P. Casavecchia, *Rep. Prog. Phys.*, 2000, **63**, 355, and references therein.
2 S. A. Buntin, C. F. Giese and W. R. Gentry, *J. Chem. Phys.*, 1987, **87**, 1443; R. E. Continetti, B. Balko and Y. T. Lee, *J. Chem. Phys.*, 1990, **93**, 5719.
3 R. I. Kaiser, C. Ochsenfeld, D. Stranges, M. Head-Gordon and Y. T. Lee, *Faraday Discuss.*, 1998, **109**, 183, and references therein.
4 R. I. Kaiser, C. Ochsenfeld, M. Head-Gordon, Y. T. Lee and A. G. Suits, *J. Chem. Phys.*, 1997, **106**, 1729; R. I. Kaiser, Y. T. Lee and A. G. Suits, *J. Chem. Phys.*, 1996, **105**, 8705; R. I. Kaiser, A. M. Mebel and Y. T. Lee, *J. Chem. Phys.*, 2001, **114**, 231.
5 N. Balucani, O. Asvany, Y. Osamura, L. C. L. Huang, Y. T. Lee and R. I. Kaiser, *Planet. Space Sci.*, 2000, **48**, 447, and references therein.
6 J. Shu, J. J. Lin, C. C. Wang, Y. T. Lee, X. Yang, T. M. Nguyen and A. M. Mebel, *J. Chem. Phys.*, 2001, **115**, 7, and references therein.
7 P. A. Willis, H. U. Stauffer, R. Z. Hinrichs and H. F. Davis, *J. Phys. Chem. A*, 1999, **103**, 3706; P. A. Willis, H. U. Stauffer, R. Z. Hinrichs and H. F. Davis, *Rev. Sci. Instrum.*, 1999, **70**, 2606.

8   M. Costes, C. Naulin, G. Dorthe, G. Daleau, J. Joussot-Dubien, C. Lalaude, M. Vinckert, A. Destor, C. Vaucamps and G. Nouchi, *J. Phys. E*, 1989, **22**, 1017.
9   P. Casavecchia, N. Balucani, L. Cartechini, G. Capozza, A. Bergeat and G. G. Volpi, *Faraday Discuss.*, 2001, **119**, 27.
10 R. I. Kaiser, T. N. Le, T. L. Nguyen, A. M. Mebel, N. Balucani, Y. T. Lee, F. Stahl, P. v. R. Schleyer and H. F. Schaefer III, *Faraday Discuss.*, 2001, **119**, 51.

Dr Kaiser communicated in response to Dr Whitaker: The advantages of pulsed beams are as follows.

(1) Pulsed beams are very versatile, and can produce almost any reactive species. Carbon containing radicals for instance C_2D, C_2H_3, C_3H_3 and C_6H_5 which have been easily produced in pulsed beams *in situ* (C_2D), *via* photolysis (C_2H_3 and C_3H_3) and flash pyrolysis (C_6H_5) are extremely difficult to produce in continuous (CW) beams *via* discharge techniques.

(2) Coupling a chopper wheel with a pulsed source allows one to select a part of the pulse and hence get rid of highly excited metastable species in the beam. This is difficult in continuous beams as ground state and excited species might coexist.

(3) The *in situ* production of the cyano and ethynyl radicals *via* pulsed laser ablation of graphite seeding the ablated species in nitrogen or deuterium carrier gas allows us to produce an intense beam of hydrocarbon radicals. This method can be extended very easily to use nitrogen as a carrier gas of ablated boron atoms. This produces a supersonic beam of BN radicals, which are very difficult to make with continuous beams. Likewise, the carbon rod in the pulsed source can be replaced by any material in a rod form—even metals. CW beams might have great difficulties in producing supersonic beams of refractory metal atoms.

(4) Due to the excessive gas load, CW beams require a significant pumping setup and hence large capital investment. The gas load of pulsed beams is lower, and the costs are relatively moderate—especially if the beam sources are going to be operated under completely hydrocarbon-free conditions. Likewise, pulsed valves allow one to perform reactions with expensive reagents, for example deuterated molecules. Except HD of D_2, these experiments might be not fundable for CW sources.

(5) Further developments of the detection schemes in crossed beam setups will certainly involve pulsed laser techniques. Therefore, pulsed sources can always be coupled with pulsed laser detection/production techniques. However, if quasi-continuous, tunable VUV light from, for example, the Advanced Light Source will be used to photoionize the reactively scattered products, CW sources provide a larger duty cycle compared to pulsed valves.

Prof. I. W. M. Smith said: My question is addressed to Dr Kaiser and to others in the audience of a theoretical bent. It concerns the relative reactivities of $C_2(X\,^1\Sigma_g^+)$ and $C_3(X\,^1\Sigma_g^+)$ to alkenes and alkynes. The experiments and calculations of Kaiser and co-workers, along with those of others, demonstrate that $C_2(X\,^1\Sigma_g^+)$ adds to simple unsaturated hydrocarbons over a surface without a potential energy barrier. On the other hand, the addition $C_3(X\,^1\Sigma_g^+)$ to the same species is impeded by a substantial barrier. Is there any relatively simple theoretical reason for this large difference in behaviour? The question is not only important in relation to understanding these particular reactions, but also it might help us make sensible guesses at the rate coefficients for other reactions of alkenes and alkynes.

Dr Kaiser responded: Regarding the C_3 reactions, our paper said 'that signal was observed if the collision energy is higher than ...'. This means (a) there is an entrance barrier, (b) the reaction to form the observed channel is endothermic or (c) there is an entrance barrier and the reaction to form the observed channel is endothermic. One possibility to understand the different reactivities of dicarbon *vs.* tricarbon might (and it is only a suggestion!) be based on the molecular orbitals in C_2 and C_3, respectively. Both carbon clusters have a $^1\Sigma_g^+$ electronic ground state. However, dicarbon has an empty, energetically low lying $3\sigma_g^+$ orbital (LUMO, lowest unoccupied molecular orbital) which could be an 'acceptor' for the π electrons of the hydrocarbon. On the other hand, the LUMO of tricarbon is energetically very high, and therefore, might make it difficult to act as an 'acceptor' for the π electrons of the hydrocarbons (this could result in a lower reactivity). I agree, this explanation is only very qualitative.

Dr Mebel added: The reason why C_2 is more reactive toward unsaturated hydrocarbons than C_3 is the following. C_2 has a low lying excited electronic state; the singlet–triplet energy gap in C_2 is less than 2 kcal mol^{-1}. On the contrary in C_3 the energy difference between the ground and first excited electronic states is large. The presence of low-lying excited state usually results in higher reactivity. In this view, one can expect that C_n species with even n should be more reactive than those with odd n.

Prof. Pilling commented: The papers of Casavecchia and Kaiser have shown the determination of channel yields under single collision conditions, with excellent resolution. For applications to combustion, we need to be able to convert this information into rate coefficients, but this is difficult without considerable theoretical analysis. It would be of considerable value to be able to use molecular beam data, and results from reaction dynamics experiments and *ab initio* calculations also, in evaluations of reaction rate data for combustion and atmospheric applications. At present this is not routinely possible, but some way needs to be found to provide the theoretical input into the evaluation process, in order to use a wider range of both experimental and theoretical information.

Prof. S. C. Smith responded: The great value of the crossed-beam data is that it provides 'cleaner' data against which to validate theory, which can then be used to work through *via* calculations to rate constants. Checking theory against thermal data always involves additional uncertainties, particularly in relation to the energy transfer parameters. Crossed-beam data under collisionless conditions provides an important alternative source of validation of parameters which are then used in the theory to go on and calculate rate constants.

Dr Klippenstein said: Ion chemists have been using beam data in theory and modeling of reaction kinetics for many years.

Prof. Casavecchia said: With reference to Prof. Pilling's comment, I fully agree that ways need to be found to better exploit molecular beam data in kinetic evaluations. Unfortunately, this aspect is not trivial because beam experiments are usually carried out under well defined relative collision energy conditions and with (at least partial) quantum state control. At present, the main contribution of our kind of experiments to kinetic evaluations stems from the fact that we can identify unambiguously the primary reaction products and also determine their relative importance under the specific conditions of the experiment. Then the ratio of integral cross sections at a given relative velocity can be approximated to the ratio of rate constants at the temperature corresponding to that average relative velocity.

Prof. Troe commented: In order to make these scattering experiments more useful to kineticists, it would be helpful to have at least a semiquantitative estimate of the absolute cross-sections. The difficulty in obtaining these are well known. Nevertheless, even some tentative information would be helpful.

Dr Kaiser responded: As you know, it is very difficult and time consuming to measure absolute cross sections with currently existing crossed beam setups. However, with significant capital investment and laser based detection/calibration techniques it will certainly be feasible to provide absolute cross sections in modified crossed beam setups in the future.

Prof. Casavecchia responded to Prof. Troe: I agree with Prof. Troe on the usefulness of absolute cross sections for kinetics use. Unfortunately, this is one of the most difficult quantities to measure in scattering experiments with neutral particles. Still, the procedure of using the small angle elastic scattering to calibrate the absolute reactive scale, as it has been used in the past, at least for simple A + BC reactions,[1] may be extended also to polyatomic reactions, if an estimate of the long range van der Waals interaction can be obtained (for instance, from theory or semiemirical rules). The uncertainty of this procedure was typically of a factor of two.[1] Despite the additional complication of heavy fragmentation under electron impact for polyatomic species, it should be possible, with some effort, to derive semiquantitative estimates of integral reactive cross sections also for polyatomic reactions, as those discussed here.

1   C. H. Becker, P. Casavecchia, P. W. Tiedemann, J. J. Valentini and Y. T. Lee, *J. Chem. Phys.*, 1980, **73**, 2833, and references therein.

Dr Seakins† commented and opened the discussion of Dr Loison's paper: We have studied the kinetics and products of the $CH + O_2$ reaction for some time. There is much in the interesting results of Bergeat *et al.* that we would agree with, but there are some significant differences.

There is agreement in two respects. (a) The overall rate coefficient $(4.40 \pm 0.52) \times 10^{-11}$ cm^3 $molecule^{-1}$ s^{-1}. (b) There is significant H atom production.

We note three differences. (a) We observe ~90% H atom production. (b) Questions about the 20% yield of OH (although this is partially based on the work of Okada *et al.*, ref. 11 of the paper presented here by Loison). (c) Questions about the significant production of HCO.

Methodology: Bromoform ($CHBr_3$) is photolysed at 248 nm and concentrations of either CH or H are observed by laser induced fluorescence, the latter at 121.6 nm. Fig. 3 (presented here) shows that the kinetics of H atom production (from methane) and CH removal are well correlated, *i.e.* that the H is being produced in the same reaction as CH is removed. This observation is important as the photolysis of bromoform can lead to a variety of other reactive products.

We use a calibration reaction $CH + CH_4 \rightarrow C_2H_4 + H$ to avoid the need for measuring absolute concentrations. The H atom branching ratio is obtained by comparing the final H atom signal strengths (Fig. 4a).

The photolysis of bromoform also has an H producing channel. We account for this by removing any CH with nitrogen ($CH + N_2 \rightarrow HCN_2$) to give us a baseline signal.

In order to make a final quantitative comparison we need to account for the differing signal strengths in the presence of oxygen or methane. We calibrate the system by photolytically generating identical [H] from H_2S photolysis in the presence of either oxygen or methane. (Fig. 4b).

Results: After accounting for both prompt H atom formation and calibrating for both methane and oxygen we measure a branching ratio of H atoms from the $CH + O_2$ reaction of $90 \pm 15\%$. Our current experiments cannot differentiate between H atoms produced from the $H + CO_2$ or

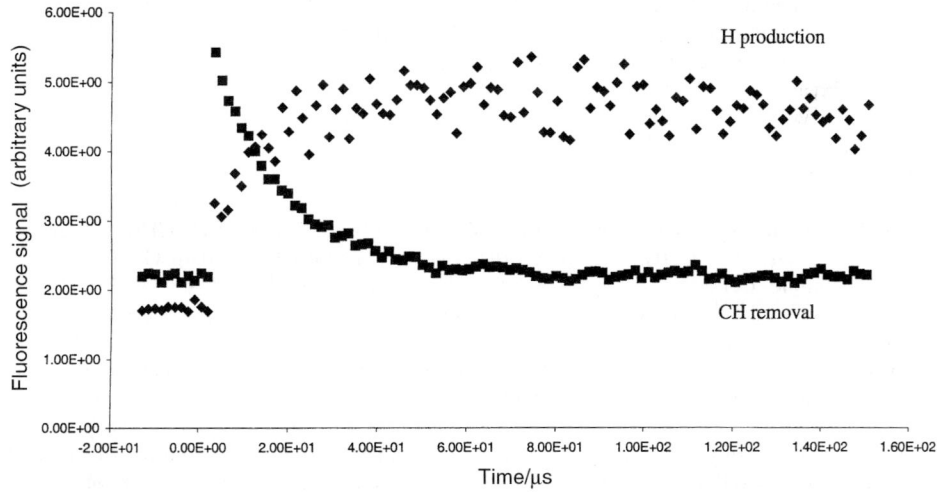

Fig. 3 Correlation of CH removal and H atom production from the $CH + CH_4$ reaction.

† Also Dr M. Blitz (University of Leeds), Dr H. Qian (University of Leeds) and Mr K. McKee (University of Leeds).

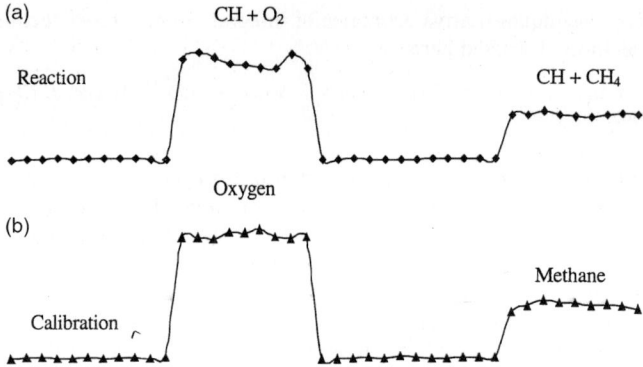

Fig. 4 (a) H atom signals obtained from CH + O_2 and CH + CH_4. The signal from the CH + CH_4 reaction is lower due to absorption of 121.6 nm radiation by methane. (b) Calibration signals for identical [H] in the presence of oxygen and methane.

H + O + CO channel and therefore any comparisons must be with the sum of channels 1 and 4 of Loison's paper, presented here.

Issues and questions: (a) Our current results are incompatible with the 60% yields of channels 1 and 4 reported in Loison's paper.

(b) Could the authors discuss the possibility of C_2H reactions influencing their measurements? Any C_2H produced in the injector or in the reactor by CH + CH → C_2H + H could influence yields via $C_2H + O_2$ → HCCO + O, this could lower the H : O ratio. A description of corrections for H atoms produced in the reactor is given in section 3.2 of their paper but no reference is made to corrections for the C_2H co-product.

(c) Relative atom yields put onto an absolute basis by comparison with the complex CH + NO reaction. How much uncertainty does this introduce?

(d) The observation of a significant HCO yield is interesting considering the reaction exothermicity and the weak H–CO bond. Does the partial stabilization of HCO under the 2 Torr conditions of the experiment give any information about this channel of the reaction (partition of energy to HCO, lifetime of HCO, ΔE_d)?

(e) Although it is not possible to probe the complete rovibrational manifold of the OH product, do you have any idea about the vibrational distribution?

The experiments discussed in Loison's paper and in this contribution are complex. Error bars are significant and therefore the discrepancies may not be that significant. The two techniques are complementary in approach, a combination of the results may offer the possibility of unravelling the product distribution of this complex and interesting reaction.

Dr Loison responded: During the photolysis of $CHBr_3$ products other than CH(X $^2\Pi$) are produced such as CH(a $^4\Sigma^-$) CHBr, ..., these other products being reactive with O_2 and not with CH_4 (for example CH(a $^4\Sigma^-$) + O_2: $k = 2.6 \times 10^{-11}$ molecules^{-1} cm^3 s^{-1} and CH(a $^4\Sigma^-$) + CH_4: $k < 0.07 \times 10^{-11}$ molecules^{-1} cm^3 s^{-1} (ref. 1)) they could be responsible for an extra H production in the $CHBr_3/O_2/h\nu = 248$ nm system vs. $CHBr_3/CH_4/h\nu = 248$ nm. Then I think that your result of 90 ± 15% of H atoms production is greatly influenced by these secondary reactions.

C_2H is indeed produced in the micro-furnace and in the fast flow reactor with a maximum ratio of 15% vs. the CH production. As $C_2H + O_2$ could give H and O atoms, and as the global rate constant of the $C_2H + O_2$ is equal to 3.0×10^{-11} molecules^{-1} cm^3 s^{-1}, we studied this reaction, producing C_2H by C_2HCl_3 + 3K. We found, in preliminary results, that about 35% of H atoms and 25% of O atoms are produced in the $C_2H + O_2$ reaction. Then this reaction could contribute less than 6% of H atoms and 4% of O atoms in the CH + O_2 reaction studies in our experimental conditions and then is included in the systematic error.

We compare the H atom production of the CH + O_2 reaction not only with CH + NO reaction, but also with CH + CH_4 and CH + H_2 reactions (this correction has been made in the article).

HCO has been probed by LIF (laser induced fluorescence) using the B(0,0,1)–X(0,0,0) transition near 251.5 nm, and no vibrational hot band has been detected. Nevertheless, no systematic study on the vibrational distribution HCO has been made, and therefore we don't know the energy partition precisely. The branching ratio of 8–30% on HCO production has been found with the supposition that the HCO probed was relaxed at 300 K. The lifetime of HCO above its dissociation barrier is very short (less than 1 ps) but HCO could have a very long lifetime. In our experiment we probe only HCO(0,0,0) relaxed at 300 K and in this state it's lifetime is much more than the time scale of the fast flow reactor which is typically 1 ms. Then we found that 60% of the HCO is dissociated in H + CO, and 40% is stabilized by collisions at 300 K. Of course at combustion temperature, the part of HCO dissociated will be higher.

The CO + OH channel is exothermic by 6.92 eV and only the $v'' = 0$ and a part of $v'' = 1$ could be probed by LIF due to predissociation of the excited state and we didn't try to probe OH by LIF. The only information on the OH energy distribution is the absence of the Meinel bands (ref. 31 of our paper) which are overtone transitions in fundamental state of OH in the 550–850 nm range, indicating that the high vibrational levels ($v'' = 4$–9) of OH (X $^2\Pi$) are not strongly populated.

Even if this reaction seems simple, it appears quite complex to study and we agree with Dr Seakins that complementary approaches will help provide a good description of the CH + O_2 reaction.

1   J. Phys. Chem., 1992, **96**, 5685.

Dr Bergeat said: Your results on the CH + O_2 reaction are based on the comparison with the CH + CH_4 reaction. Can you confirm that the CH + CH_4 reaction leads only to H + C_2H_4 ? The CH production is performed by photolysis of $CHBr_3$. What are the other products of the photolysis? The source was used by other groups to study C, CBr, CHBr .... Have you any estimation of the secondary reaction contributions (like CHBr + CH_4, CBr + CH_4 ...)? Have you any idea of the quartet CH production and contribution in your experiment?

Dr Seakins replied: As far as we are aware H + C_2H_4 is the sole channel for the reaction. The negative temperature dependence of the overall kinetics of the CH + CH_4 reaction rules out the endothermic CH_2 + CH_3 reaction. One other possible set of exothermic products would be C_2H_3 + H_2, however, this would require either a three or four centered elimination reaction from an ethyl intermediate and we do not believe that this would compete with the simpler H atom elimination. Strictly speaking our reported yields are referenced to the H atom yield from the CH + CH_4 reaction, and whilst we believe this to be 100%, the possibility of other minor channels does exist.

Bromoform is not an ideal source for photolysis, as there will be other photolysis products, but we believe it is the best available for our technique. Our rationale for ignoring the effects of such species is two-fold. Firstly the kinetics of CH removal and H atom production are well correlated. Halogenated species such as CHBr or CBr are likely to react on a slower timescale than the more reactive CH radical. Secondly, the most likely elimination channel from the formation of any halogenated intermediate product would be the elimination of a bromine atom, rather than an H atom. Thus for both kinetic and mechanistic reasons, we believe that secondary reactions of bromoform co-photolysis products will not affect our H atom yield.

We have not looked at the effects of quartet CH on H atom production and this is something that we should address. Once again, however, the correlation of H atom production and CH(X) removal seems to indicate that the dominant source of the observed H atom production is the CH(X $^2\Pi$) radical.

Dr Klippenstein commented: The CH + O_2 system is again amenable to direct dynamics simulations due to the great exothermicity of the various product channels. In these simulations the classical equations of motion for the nuclei are solved *via* numerical propagation with the time dependent forces directly determined from *ab initio* quantum chemical simulations. The B3LYP/6-31G* method, which generally provides at least a qualitatively meaningful description of the global energetics, was again employed for the direct determination of the forces.

For this system, I have focused on the question of the branching between the various product channels for the dynamics on the doublet electronic state. The procedures employed are essentially identical to those described in my comment on the $CH_4 + N(^2D)$ reaction. 48 trajectories were propagated, with a typical trajectory lasting for 100–200 fs. The trajectories were initiated at a CH to O_2 center-of-mass separation of 4.0 a_0 and were terminated when any atom–atom separation exceeded 10.0 a_0.

For $J = 25$, at an energy of 12.6 kcal mol^{-1} above the zero-point energy of the reactants, these simulations predict a branching of 0.42 : 0.25 : 0.19 : 0.15 for the CO + OH : HCO + O : CO_2 + H : CO + O + H channels. Interestingly, there appears to be a highly non-statistical branching to the much less exothermic HCO + O channel. Correspondingly, it is not necessary to incorporate a significant contribution from the quartet state to explain your observed population in the HCO + O channel.

However, one should note that these trajectory predictions are for the nascent distribution. In many instances the HCO molecules are produced with an internal energy that exceeds the dissociation threshold. As a result, longer timescale observations would find less HCO and more CO + O + H. By the same token, we also observe that even some of the OH fragments have an internal energy that exceeds its dissociation threshold and some of the CO_2 fragments might also. Thus, a direct correlation with your experimental results should also consider the internal energy distribution and ultimate fate of the nascent products.

Dr Loison replied: The direct dynamics simulations made by Dr Klippenstein are very interesting. The main difference with our work is the fact that the doublet surface could lead to the HCO + O channel, for 25% on one doublet surface and thus between 6 and 12% on the total of the reaction. This is not a major change and should be confirmed. The most surprising result is the ratio of OH + CO vs. H + CO_2. This may be due to the fact that at 4.0 Å the long range part of the CH + O_2 potential could not be described by the B3LYP/6-31G* method. I think that these calculations should take into account the two doublet surfaces with precise description of the long range part. The calculation of energy distribution of the nascent products should also be very interesting.

Prof. Kohse-Höinghaus asked: (1) I wonder how clean your source of CH really is—would reactions including K + O_2 or O with bromoform affect your chemistry?

(2) The overall reaction rate of CH + O_2 in your paper is found to be independent of temperature at high temperatures (*i.e.* approximately 2000 K). It is intriguing to see that Desgroux *et al.*, in their paper to be presented at this meeting, discuss their CH measurements in a low pressure flame, in agreement with previous work, with respect to a potential temperature dependence of the CH + O_2 reaction in this temperature range around 2000 K. Would this finding indicate a conflict with your investigation?

Dr Loison replied: We do not make any investigation at high temperature and extrapolation of measurements to temperatures above 1000 K should be done with great care. The overall reaction rate of CH + O_2 is equal to 5.5×10^{-11} molecules^{-1} cm^3 s^{-1} (ref. 1) in the 300–2000 K range, but the direct measurements are only in the 15–700 K temperature range. Two recent combustion experiments[2,3] find a rate constant of $1.2–1.6 \times 10^{-10}$ molecules^{-1} cm^3 s^{-1} near 2000 K. That could be due to the crossing of the first $^4A''$ surface (repulsive in C_s symmetry) and the second $^4A'$ surface (coming from CH(a $^4\Sigma^-$) + $O_2(^3\Sigma^-)$) this crossing being avoided in C_1 symmetry leading to an entrance channel with a quite high barrier, and thus this channel will be open only at high temperature.

1   J. Phys. Chem. Ref. Data, 1992, **21**, 411.
2   M. Rohrig, E. L. Petersen, D. F. Davidson, R. K. Hanson and C. T. Bowman, Int. J. Chem. Kinet., 1997, **29**, 781.
3   M. W. Markus, P. Roth and T. Just, Int. J. Chem. Kinet., 1996, **28**, 171.

Prof. Plane asked: The method of using atomic K to strip Br atoms successively from $CHBr_3$, thereby producing CH radicals, is very elegant. One possible problem may arise from having to mix $CHBr_3$ with an excess of K. The remaining K will enter the flow tube and recombine with O_2 to produce KO_2. Does this play a significant role in the observed kinetics?

Dr Loison replied: In our source, K–bromoform and O_2 are in separated parts. We produce CH in a microfurnace with only $CHBr_3$, K and He with an excess of K (typically 1 mTorr in the microfurnace). Then the only K reactions in the microfurnace are $CHBr_3 + 3K \rightarrow CH + 3KBr$. Some of the K could escape from the microfurnace (less than 0.1 mTorr), with CH radicals, but the only possible reaction of K is adduct production such as $K + O_2 + M \rightarrow KO_2 + M$, $K + O + M \rightarrow KO + M$, $K + CH + M \rightarrow KCH + M$ reaction with rate constant equal to 8×10^{-30} cm^6 $molecule^{-2}$ s^{-1} for the $K + O_2 + N_2$ and then at 1.4 Torr (4.6×10^{16} molecule cm^{-3}) we estimate the global rate constant of this reaction with He as a bath gas to be 2.0×10^{-13} $molecule^{-1}$ cm^3 s^{-1} and thus negligible under our conditions for the consumption of any species (O_2, O, H or CH) and the small amount of KO_2 adduct produced (less than 0.003 mTorr) will not perturb the reaction (for example the $KO_2 + H \rightarrow HO_2 + K$ or $KO_2 + O \rightarrow O_3 + K$ are absolutely negligible as KO_2 is at least 300 times less concentrated than O_2, even if the global rate constant of this reaction is equal to the $CH + O_2$ one.

Dr Miller commented: An important finding of this paper is the theoretical result that the reaction can occur on both the doublet and quartet surfaces *with no energy barrier*. I wonder if this might be the reason that the shock tube experiments (ref. 19 and 20 of the paper) yield rate constants that are much higher than those of the low-temperature experiments (ref. 6, 12 and 13 of the paper). Perhaps the reaction takes place predominantly on the doublet surface at low T, with both surfaces contributing at high T.

Prof. Pilling opened the discussion of Dr Miller's paper: Miller and coworkers have discussed the application of eigenvector, eigenvalue analysis to master equations. The following comments refer to the process of relating the eigenvalues to the rate coefficients for the chemical system under study.

For reactions involving a single reactive process, such as dissociation, the unimolecular rate coefficient can be equated to the modulus of the eigenvalue of smallest magnitude, λ_1, so that $k_{uni} = -\lambda_1$. The eigenvalues of larger magnitude are related to collisional relaxation processes and refer to the relaxation of the system to a steady state population distribution over the energy states. Problems arise at high temperatures, especially for molecules with small dissociation energies, because λ_1 becomes so large that it is comparable in magnitude with the larger eigenvalues and the separation of chemical and collisional timescales is no longer possible: the system does not relax to a steady state distribution on a time short compared with dissociation and the reaction rate coefficient cannot simply be equated to $-\lambda_1$.[1] This problem also applies to more complex multi-well systems and has been discussed by Tsang *et al.*[2] in relation to alkyl radical isomerisation and dissociation.

For reactions with several wells and reaction channels, and with the eigenvalues related to chemical reaction well separated from those associated with collisional relaxation, a matrix G_{chem} can be factored out whose eigenvalues are related to the rate coefficients for the component chemical reactions. This relationship can be direct, with the modulus of the eigenvalue equal to the first order rate coefficient, or may be more complicated. The question arises as to what affects this relationship and how easily can the rate coefficients be related to the eigenvalues.

For the purposes of this comment, it is easiest to consider a more general situation of a complex reaction whose kinetics are defined by the coupled rate equations $(dc/dt) = G_{chem} c$ where c is a vector of concentrations of chemical species and G_{chem} is a matrix of first order (or pseudo first order) rate coefficients connecting the species; we order c according to the lifetimes of the species, with the shortest-lived occurring first. The diagonal elements of the matrix are minus the sum of the first order rate coefficients for removal of the species. The lower triangular section of G_{chem} contains the rate coefficients for processes that 'feed forward' from shorter-lived to longer-lived species, while the upper section relates to 'feed back' processes from longer- to shorter-lived species. If all the elements in the upper triangular section of G_{chem} are zero, *i.e.* if there is no feed back, the moduli of the eigenvalues of the matrix are identical to the diagonal elements and the relationship between the eigenvalues and the rate coefficients is a simple one. This situation remains true for species with only zero elements in the upper section of the appropriate row of G_{chem} and is approximately true if the species responsible for the feed back are *much* longer lived than the species under consideration.[3] In all other situations, the relationship is more complex. This situation arises in two of the posters we have presented, on alkyl radical isomerization and dissociation,[4] and on $H + SO_2$.[5] It is important that the problem is analysed to define chemical networks that can describe the system and to provide the associated rate coefficients. We have done this through a classical kinetic analysis of the coupled rate equations,[4,5] but this is not the only, or even the most direct, way. The component rate coefficients may depend on both temperature and pressure and must be represented in a form that can be used readily by combustion modelers—presumably a Troe or modified Troe format.

1 H. O. Pritchard, *The quantum theory of unimolecular reactions*, Cambridge University Press, 1984.
2 W. Tsang, V. Bedanov, M. R. Zachariah, *Ber. Bunsen-Ges. Phys. Chem.*, 1997, **101**, 491.
3 N. Bell, M. J. Pilling and A. S. Tomlin, *J. Phys. Chem.*, submitted.
4 L. D. Jones, L. C. Jitariu, I. H. Hillier, S. H. Robertson and M. J. Pilling, poster presented at this meeting.
5 M. A. Blitz, K. J. Hughes and M. J. Pilling, poster presented at this meeting.

Dr Miller responded: I agree, more or less, with everything you say. We have discussed in several places how to extract rate coefficients from an eigenvector/eigenvalue analysis (*e.g.* ref. 18 and 36 of our paper, the paper itself and my comment following the paper by Frankcombe and Smith). In ref. 36 we extracted pressure and temperature dependent rate constants in the Troe format for the three reactions,

$$C_2H_5 + O_2 \rightarrow C_2H_5O_2,$$

$$C_2H_5 + O_2 \rightarrow C_2H_4 + HO_2,$$

and

$$C_2H_5O_2 \rightarrow C_2H_4 + HO_2.$$

With the corresponding equilibrium constants, these three rate constants completely describe our master equation results to reasonable accuracy.

Although I have wavered on this point at times, I believe that, as long as Δx_R is -1 in eqn. (29) of the paper for some eigenvector of *G*, that eigenvector describes an elementary reaction (perhaps with several product channels), and the associated eigenvalue can be used to extract the corresponding phenomenological rate constants. Such an identification is legitimate, in my opinion, even though the products indicated by eqn. (29) may require a complex to pass over several potential wells and suffer many collisions over each of those wells.

Prof. Troe said: For kinetics modellers it would be helpful not only to have energy profiles from the given calculations but also rotational constants (or structures) and frequencies at the stationary points. In this way, the provided rate calculations could be modified, adapted or corrected whenever new information or better methods become available. Likewise centrifugal barriers $E_0(J)$ should be given explicitly for the addition processes such that proper falloff curves for the addition processes including anharmonicity effects could be calculated.

Dr Miller replied: We would be happy to provide rotational constants and vibrational frequencies upon request. In fact, we could simply provide the VARIFLEX input files. However, centrifugal barriers are determined implicitly and used internal to the code; they are not part of the input.

Prof. Golden said: The question of how evaluators should represent data for modelers, prompts me to point out that evaluation of data needs considerably more support. This difficult task must be undertaken by researchers with experimental experience. It seems a colossal waste of resources to fill journals and the web with unevaluated data. Researchers themselves and funding agencies must put a higher value on these activities.

Prof. Troe addressed Dr Miller: (1) You show that there are shallow entrance wells in some addition reactions of O_2. Are these kinetically relevant? Are the determining activated complexes the inner maxima?

(2) Why do the 2-d and 1-d master equation calculations of the addition reaction in the low pressure limit give such different results? Could you provide the limiting termolecular rate constants?

Dr Miller replied: (1) The inner maximum is the bottleneck.

(2) Although the 2-d master equation predicts non-equilibrium J distributions, I believe the major effect is simply one of conserving angular momentum between the molecule and the transition state, i.e. a certain amount of '$BJ(J + 1)$' energy is tied up in conserving angular momentum and cannot be used to break the bond. Remember that our formulation is in terms of E and J, where E is the *total energy* of the molecule, and not in terms of ε and J, where ε is the energy in the active degrees of freedom. In the latter formulation, one would expect the 2-d model to give larger rate constants than the 1-d model, because the effective rotational constant is smaller at the transition state than for the molecule, thus freeing some 'rotational energy' to be used to break the bond.

In the present problem k_0 (2-d) at $T = 295$ K is 2.97×10^{-30} cm^6 molecule^{-2} s^{-1} and k_0 (1-d) is 1.12×10^{-29} cm^6 molecule^{-2} s^{-1}. The difference is a factor of about 3.8.

Prof. S. C. Smith said: (a) There is a perturbational extension of the Smith–Gilbert rotationally averaged master equation which deals with weak-collisional relaxation of the angular momentum. This is described in ref. 1 in the context of ion–molecule association modelling.

(b) Have you explored the possibility of applying the Smith–Gilbert rotationally averaged master equation for multi-well systems?

1 S. C. Smith, M. J. McEwan and R. G. Gilbert, *J. Chem. Phys.*, 1989, **90**, 1630.

Dr Miller responded: (a) I am aware of your paper, but I prefer our variant of the original Smith–Gilbert method. By taking E and J as the independent variables in the master equation, rather than ε and J, we are able to take into account weak-collisional relaxation of the angular momentum approximately with a method that is as simple mathematically as the Smith–Gilbert strong-J model.

(b) If either the Smith–Gilbert rotationally averaged master equation or our variant of it is used unaltered in a problem where reaction must be allowed to take place in both directions, detailed balance is not satisfied and significant errors can result. This situation occurs because extra terms appear in the equations when one sums the 2-d master equation over J. However, it should be possible to solve the resulting equations in this case by an iterative procedure, but we have not yet attempted to do this.

Prof. Lin said: The energy transfer step-size you employed for N_2, $\langle \Delta E_{\text{down}} \rangle = 500$ cm^{-1}, seems rather large to me. In our recent modeling of the OH + CO reaction, we obtained a value of 250 cm^{-1} for N_2.[1] A similar value (270 cm^{-1}) has been used by Gilbert and coworkers[2] for the decomposition of C_4H_9Br.

1 R. S. Zhu, E. W. G. Diau, M. C. Lin and A. M. Mebel, *J. Phys. Chem.*, accepted.
2 T. C. Brown, K. King and R. G. Gilbert, *Int. J. Chem. Kinet.*, 1988, **20**, 549.

Dr Miller answered: We deduced $\langle \Delta E_d \rangle = 500$ cm^{-1} from modeling the fall-off curve of Atkinson and Hudgens (ref. 16 of the paper) at 295 K using the 2-d master equation and assuming a single exponential-down model for the energy transfer function; we also used Lennard-Jones potentials to calculate Z, the collision rate. In a similar analysis of $H + O_2 + N_2 \rightarrow HO_2 + N_2$ at room temperature, we deduced a value of $\langle \Delta E_d \rangle = 300$ cm^{-1}. For $H + NO + N_2 \rightarrow HNO + N_2$ we deduced $\langle \Delta E_d \rangle \approx 500$ cm^{-1}. However, in both these cases the difference between 300 cm^{-1} and 500 cm^{-1} was less than 20% in the rate constant. Moreover, I do not think it unreasonable that different molecules could transfer different quantities of energy in collisions with N_2, particularly if larger molecules are concluded to transfer more energy than smaller ones. The values of $\langle \Delta E_d \rangle$ that we (and everyone else) deduce from modeling dissociation/recombination reactions apply only to energies of the molecule near the dissociation limit. The value of $\langle \Delta E_d \rangle$ probably depends on just how high the dissociation limit is above the ground state. My point is that $\langle \Delta E_d \rangle$ is not a fundamental constant for a given collider gas.

Your question brings to mind a more practical issue. Typically, one deduces $\langle \Delta E_d \rangle$ from an RRKM/master equation model using a single exponential-down energy transfer function, a harmonic-oscillator/rigid-rotor model to calculate the density of states, and Lennard-Jones potentials to calculate Z. All three of these assumptions could be questionable under certain conditions. Including anharmonicities will cause the $\langle \Delta E_d \rangle$ values inferred to be slightly smaller, perhaps more than slightly in some cases. Clearly, $\langle \Delta E_d \rangle$ depends somewhat on the form of $P(E,E')$, although such a dependence may be more important for chemically activated or photo-activated systems than for thermally activated ones. Also, Z may not be accurately described by Z_{LJ}. Classical trajectory calculations[1,2] seem to indicate that Z should be about 25% or more larger for ordinary colliders and that it may depend on E and J as well as T. Experiments[3] clearly show that Z should be *much larger* if water is the collider. A slightly more subtle effect, commonly omitted from analyses, is that $\langle \Delta E_d \rangle$ may depend on J[1,4,5] as well as E and T.

1 N. J. Brown and J. A. Miller, *J. Chem. Phys.*, 1984, **80**, 5568.
2 G. Lendvay and G. C. Schatz, in *Advances in Chemical Kinetics and Dynamics. Vol. 2: Vibrational Energy Transfer Involving Large and Small Molecules*, ed. J. R. Barker, JAI Press, Greenwich, CT, 1995, p. 481.
3 K.-J. Hsu, J. L. Durant and F. Kaufman, *J. Phys. Chem.*, 1987, **91**, 1895; K.-J. Hsu, S. M. Anderson, J. L. Durant and F. Kaufman, *J. Phys. Chem.*, 1989, **93**, 1018.
4 A. Gelb, *J. Phys. Chem.*, 1985, **89**, 4189.
5 S. Nordholm and H. W. Schranz, in *Advances in Chemical Kinetics and Dynamics. Vol. 2: Vibrational Energy Transfer Involving Large and Small Molecules*, ed. J. R. Barker, JAI Press, Greenwich, CT, 1995, p. 245.

Prof. Wolfrum opened the discussion of Dr Taatjes' paper: Did you observe the formation of heterocyclic compounds due to isomerization steps in your experiments?

Dr Taatjes responded: We are unable to directly observe the formation of heterocycles in our experiments. However the amount of OH suggested by the non-unity yield of HO_2 in our experiments, and implied by separate OH LIF measurements in our laboratories, is broadly consistent with the heterocycle branching fractions deduced from literature measurements of the initial products of butyl + O_2 reactions.[1]

1 See, *e.g.*, R. W. Walker and C. Morley, in *Low-Temperature Combustion and Autoignition*, ed. M. J. Pilling, *Comprehensive Chemical Kinetics*, Elsevier, Amsterdam, 1997, vol. 35, p. 1–124, and extensive references therein.

Prof. Griffiths asked: (1) In order to understand the chain branching processes that occur during alkane oxidation at low temperatures in a quantitative way, it is important to establish the extent to which RO_2 isomerises to QOOH and then dissociates to generate OH radicals, or acquire another O_2 to create diperoxy species. Are you able to deduce this information from your fraction Φ (Table 2) that leads to HO_2 radicals?

(2) From your data are you able to quantify the kinetic origins of the much greater reactivity of n-butane relative to that of isobutane leading to spontaneous ignition?

Dr Taatjes answered: The HO_2 yields measured in our experiments are an effective yield for the overall oxidation. They reflect a competition between HO_2 elimination from RO_2 and other removal processes for RO_2 and are not necessarily directly interpretable as the branching fraction of an elementary reaction. The rise in HO_2 yields with temperature in the transition region represents the onset of thermal instability of the RO_2 species. Further, the production of OH, which occurs to a significant extent in the butyl + O_2 reactions, tends to produce HO_2 eventually because of the regeneration of butyl radicals *via* the OH + butane reaction. Nevertheless, the fact that the yield of HO_2 remains significantly below 100% even at the highest temperatures of this study implies a contribution from other bimolecular channels, *i.e.*, isomerization to QOOH, followed by dissociation to OH + heterocycle (or by reaction with O_2, also producing OH). A qualitative, though naïve, estimate of the branching to these channels is given by $(1 - \Phi_{total})$ at temperatures above the transition region, which would suggest OH fractions of approximately

15–20% near 700 K. However, because the recycling reaction of OH with butane will tend to inflate the HO_2 yield, the true branching fraction will be larger than this estimate. For comparison, initial product measurements imply ~35% branching to OH-producing channels from n-butyl and *sec*-butyl reactions with O_2 at 753 K.[1,2] A positive activation energy for branching to oxirane + OH relative to $C_2H_4 + HO_2$ has been observed in the ethyl + O_2 reaction,[3] and might also be expected for the analogous channels in the butyl + O_2 reactions. Combined modeling of time-resolved OH measurements and our HO_2 signals from Cl-initiated butane oxidation may permit a more precise branching determination.

As to the different ignition behavior of the two butane isomers, the present experiments using Cl-initiated oxidation provide only qualitative suggestions. The individual butyl isomers may not be formed by Cl-initiation in the same ratio as occurs in spontaneous ignition, so direct quantitative comparison between the two systems can be difficult. For example, the combustion behavior of isobutane may have a larger contribution from the *tert*-butyl radical (which produces almost exclusively alkene + HO_2 in its reaction with O_2) than does our Cl-initiated oxidation. However, we certainly hope that our calculations, coupled with validation on test systems such as Cl-initiated oxidation, can contribute to improved modeling of ignition and related phenomena.

1 R. R. Baker, R. R. Baldwin, A. R. Fuller and R. W. Walker, *J. Chem. Soc., Faraday Trans. 1*, 1975, **71**, 736.
2 R. R. Baker, R. R. Baldwin and R. W. Walker, *J. Chem. Soc., Faraday Trans. 1*, 1975, **71**, 756.
3 R. R. Baldwin, I. A. Pickering and R. W. Walker, *J. Chem. Soc., Faraday Trans. 1*, 1980, **76**, 2374.

Prof. Golden asked: Can there be any contribution from the simple abstraction process?

Dr Miller responded: We have calculated the abstraction rate constant for alkyl + O_2 reactions only for ethyl + O_2. In this case the abstraction has a barrier of ~18 kcal mol^{-1}, and the rate constant is about a factor of ten smaller than that for the $C_2H_5 + O_2 \rightleftharpoons C_2H_5O_2^* \rightarrow C_2H_4 + HO_2$ mechanism at 2000 K. The abstraction is even less significant at lower temperatures.

Dr Seakins communicated to Dr Taatjes: One complication in your analysis of experimental data is the generation of multiple butyl and propyl fragments during the initial chlorine atom abstraction. What effect does this have on the quantitative data that can be obtained from the experimental results?

One possible method of eliminating this would be to use the less reactive, and hence more selective bromine atom to initiate the reaction. For example at 621 K, Br atoms react with n-butane and isobutane to give 96 and 98% of *sec*-butyl or *tert*-butyl respectively.[1] The disadvantage is that the rate coefficients are significantly slower than for Cl atom abstraction, but if your signals are not affected by hydrocarbons, then you can drive the pseudo-first-order rate coefficient to comparable levels by increasing the butane concentrations. The high activation energies of bromine atom reactions will however limit the temperature range of study. Bromine atoms react in a similar manner to chlorine with methanol[2] so you would still in principle use the same calibration reaction.

1 P. W. Seakins, M. J. Pilling, J. T. Niiranen, D. Gutman and L. N. Krasnoperov, *J. Phys. Chem.*, 1992, **96**, 9847.
2 S. Dobe, T. Berces, T. Turanyi, F. Marta, J. Grussdorf, F. Temps and H. Gg. Wagner, *J. Phys. Chem.*, 1996, **100**, 19864.

Dr Taatjes communicated in response: The production of multiple isomers is a complication for detailed quantitative modelling. The derivation of structural effects on isomerization and elimination pathways in RO_2 would be greatly facilitated by measurements on individual isomers. In this connection the use of Br initiation is an excellent suggestion. We had briefly considered the use of bromine atoms but were daunted by the relatively low rate coefficients for Br + hydrocarbon reactions. However, as you point out, the Br-initiated scheme would become more useful precisely in the elevated temperature range where the important kinetic phenomena of QOOH formation and HO_2 elimination occur. We will certainly consider this method in our future studies.

Rotational effects in broadening factors of fall-off curves of unimolecular dissociation reactions

Jürgen Troe* and Vladimir G. Ushakov[†]

Institut für Physikalische Chemie, Universität Göttingen, Tammannstrasse 6, D-37077 Göttingen, Germany

Received 1st March 2001
First published as an Advance Article on the web 15th October 2001

Strong collision fall-off curves of unimolecular dissociation and the reverse recombination reactions are calculated by using the statistical adiabatic channel/classical trajectory model (SACM/CT). This formalism properly accounts for angular momentum coupling of transitional modes with overall rotation. Calculations are made for linear molecules dissociating into linear fragments and atoms with randomly chosen properties of the transitional modes and for isotropic as well as anisotropic potentials. Analytical representations of center broadening factors as a function of molecular parameters are given. A comparison between fall-off curves from rigid activated complex RRKM theory, from the present loose activated complex SACM/CT model, and from CT calculations on an *ab initio* potential is made for the $HO_2 \to H + O_2$ system. It is shown that, besides rotational effects, energy-dependent anharmonicities of the density of states also influence the shape of the fall-off curves in this system.

1. Introduction

Unimolecular dissociation and the reverse recombination reactions are important constituents of combustion mechanisms. The adequate representation of their temperature-, pressure- and bath gas-dependence, therefore, is of considerable basic as well as practical importance. The problem has a series of aspects:

(i) One may want to have an analytical representation of the fall-off curves which as precisely as possible reproduces the results of a detailed theoretical modelling. A series of articles with this aim have appeared (see *e.g.* ref. 1–10) which, with increasing complexity of the proposed expressions, arrive at an increasingly precise representation of the theoretical modelling results. However, one may argue that a complete theoretical modelling using a sufficiently fast computing algorithm is always superior (see *e.g.* ref. 11 and 12) and that accurate theoretical modelling at present is not possible anyway, because too many of the molecular parameters and processes involved are not yet characterized sufficiently well.

(ii) One may want to have analytical representations which, with the sacrifice of some precision, are of the simplest possible form such that they demonstrate the effects of varying molecular parameters and processes in a transparent way. Fall-off expressions of this type have *e.g.* been proposed in ref. 1–6. Apart from the limiting low pressure (k_0) and high pressure (k_∞) rate constants (pseudo-first order for dissociation or pseudo-second order for recombination), these

[†] Present address: Institute of Problems of Chemical Physics, Russian Academy of Sciences, 142432 Chernogolovka, Russia.

DOI: 10.1039/b101964n

expressions relate characteristic parameters of the fall-off curves, such as their broadening factors $F(k_0/k_\infty)$ (see ref. 1–3), to a small number of "global parameters" (such as an "effective number of oscillators S", an "effective barrier height ratio B", and a weak collision efficiency β_c). However, one may argue that a multi-parameter problem of the considered type can never be reduced to a representation with as limited a number of parameters as proposed.

(iii) One may need fall-off expressions of the simplest possible form such that the rate constants can conveniently be tabulated and introduced into the modelling of large reaction systems, see *e.g.* ref. 13–15. Simple broadening factors may also be required as fitting expressions to represent experimental results. In this case experimental limitations (scatter of the data and inaccessibility of the limiting rate constants) and theoretical uncertainties may render useless a more elaborate treatment. However, one should be aware of the consequences of an oversimplified approach. Employing inadequate center broadening factors may lead to incorrect extrapolations to the limiting rate constants while the center parts of the fall-off curves may nevertheless be well represented.

Which aspect is most important, depends on the situation and also on "the philosophy" of the user. It appears important to emphasize that at present there is no single optimum approach. An RRKM master equation treatment has its limitations: conventional rigid activated complex RRKM theory does not treat the bond fission properly; variational RRKM, SACM or classical trajectory (CT) treatments are complicated and in fall-off calculations have only rarely been used in full detail; adequate potential energy surfaces for bond fission are available only in selected cases; lifetime fluctuations of dissociating molecules have been considered most rarely; rovibrational energy transfer is hardly understood at all and only rarely included in two-dimensional master equation treatments. It, therefore, appears appropriate to continue model investigations, identifying the influence of various factors on the shape of fall-off curves (see also ref. 16–19). It is "our philosophy" that this helps to assess the reliability of the employed fall-off expressions. After having calculated strong collision fall-off curves for the $HO_2 \Leftrightarrow H + O_2$ system on an *ab initio* potential,[20] in the present work we do more systematic calculations with model valence potentials. We focus attention on the influence of molecular rotations on strong collision fall-off curves which requires a proper handling of the contributions from "transitional modes". In addition, going beyond the work of ref. 2 we provide a more systematic investigation of the influence of characteristic molecular parameters on broadening factors of fall-off curves. A comparison with fall-off curves based on *ab initio* potentials or on conventional rigid activated complex RRKM theory will also be given for the $HO_2 \to H + O_2$ system.

2. Calculation of strong collision broadening factors

Following the policy of ref. 1–3 and 21, doubly-reduced fall-off curves k/k_∞ as a function of k_0/k_∞ with broadening factors $F(k_0/k_\infty)$ defined by

$$k/k_\infty = [(k_0/k_\infty)/(1 + k_0/k_\infty)]F(k_0/k_\infty) \qquad (1)$$

are employed (k = rate coefficient, being pseudo-first order for dissociation or pseudo-second order for recombination). Neglecting weak collision effects and assuming that the specific rate constants $k(E,J)$ of the unimolecular reaction can be expressed by statistical rate theory in the form

$$k(E, J) = W(E, J)/h\rho(E, J) \qquad (2)$$

as a function of the energy E and the angular momentum (quantum number J), the strong collision broadening factors are given by

$$F^{SC}(k_0/k_\infty) = (1 + x) \int \sum_J (2J + 1)[F_\rho F_W/(xF_\rho + F_W)]\exp(-E/kT)\mathrm{d}(E/kT). \qquad (3)$$

Here

$$x = k_0/k_\infty \qquad (4)$$

is a measure of the bath gas concentration [M], because $k_0 \propto$ [M], and

$$F_\rho = \frac{\rho(E, J)}{\int \sum_J (2J + 1)\rho(E, J)\exp(-E/kT)\mathrm{d}(E/kT)} \tag{5}$$

$$F_W = \frac{W(E, J)}{\int \sum_J (2J + 1)W(E, J)\exp(-E/kT)\mathrm{d}(E/kT)}. \tag{6}$$

$W(E, J)$ denotes the total number of open channels: it is calculated by convolution of the number of open channels of the transitional modes $W_{tr}(E, J)$ and of the conserved modes $W_{cons}(E, J)$ via

$$W(E, J) = \int W_{tr}(E - E_i, J)[\partial W_{cons}(E_i, J/\partial E]\mathrm{d}E_i; \tag{7}$$

$\rho(E, J)$ is the rovibrational density of states.

In the present work we focus attention on a dissociation/recombination model "linear molecule ⇔ linear fragment + atom" where the interaction potential is of "anisotropic standard valence form".[22] In center-of-mass coordinates this potential is of Morse-character in the inter-fragment distance, and of cos-character in the two transitional bending modes, with distance-dependent transitional mode frequencies corresponding to a ratio of $\alpha/\beta = 0.5$ (α = looseness parameter,[23] β = Morse parameter). Using this potential we have done extensive statistical adiabatic channel/classical trajectory (SACM/CT) calculations, studying the dependence of high pressure limiting rate constants on molecular parameters.[22] We use the corresponding (E, J)-specific results[24] for the determination of $W_{tr}(E, J)$ in the present work. In the spirit of the SACM/CT treatment, the conserved modes are identified with the internal modes of the fragment (here taken as oscillators), for which numbers of states $W_{cons}(E, J)$ are determined by accurate state counting. Correspondingly, we have the partition functions

$$Q = \int \sum_J (2J + 1)W(E, J)\exp(-E/kT)\mathrm{d}(E/kT) = Q_{cons}^{s-3} Q_{tr} \tag{8}$$

with

$$Q_{cons}^{s-3} = \prod_{i=1}^{s-3}[1 - \exp(-\hbar\omega_i/kT)]^{-1} \tag{9}$$

for the conserved modes, and

$$Q_{tr} = \int \sum_J (2J + 1)W_{tr}(E, J)\exp(-E/kT)\mathrm{d}(E/kT) \tag{10}$$

for the transitional modes. The partition functions define[2] an effective number of oscillators

$$S_k = 1 - T^{-1}\,\mathrm{d}\ln Q/\mathrm{d}(1/T) = 1 + S_{cons} + S_{tr} \tag{11}$$

with

$$S_{cons} = \sum_{i=1}^{s-3} (\hbar\omega_i/kT)[\exp(\hbar\omega_i/kT) - 1]^{-1} \tag{12}$$

and

$$S_{tr} = -T^{-1}[\mathrm{d}\ln Q_{tr}/\mathrm{d}(1/T)]. \tag{13}$$

In the present work, the frequencies of the fragment have been randomly chosen from a range between ω_{min} and ω_{max} with $\omega_{max}/\omega_{min} = 10$ and with average quanta $\langle\hbar\omega_i\rangle$ from the range $D/30$ to $D/3$ (D = Morse dissociation energy).

The density of states $\rho(E, J)$ is expressed in Whitten–Rabinovitch form

$$\rho(E, J) \propto (E_0 + E - B_{tot} J(J + 1) + a_{WR} E_z)^{s-1} \tag{14}$$

where the energy E is counted from the ground state of the separated atom + fragment, B_{tot} is the rotational constant (in energy units) of the linear molecule, a_{WR} is the energy-dependent Whitten–Rabinovitch function, s is the number of oscillators of the molecule, and E_Z is the zeropoint energy of the molecule. The rotational constant B_{tot} of the linear molecule is related to the rotational constant of the fragment B through

$$B_{tot} = B/(1 + 2M_e r_e^2) \qquad (15)$$

where r_e defines a reduced distance between atom and fragment in the molecule through $r_e = \beta R_e$ (R_e = equilibrium bond length of the dissociating bond in center of mass (c.o.m) coordinates, β = Morse parameter) and M_e defines an effective mass of the system through $M_e = 2BD/\hbar^2\omega_{str}^2$ (ω_{str} corresponds to the stretching vibration of the dissociating bond chosen randomly such as indicated above). M_e and r_e are important quantities determining the centrifugal barriers of the system. The ratio B/D was chosen randomly in the range 10^{-7} to 10^{-4}. The bond energy D, which in this case is identical with the reaction threshold energy E_0, determines the parameter

$$B_K = [(S_K - 1)/(s - 1)][E_0 + a_{WR}(E_0)E_Z]/kT = [(S_K - 1)/(s - 1)]B' \qquad (16)$$

which, besides S_K, governs the center broadening factors[2]

$$F^{SC}_{cent} = F^{SC}(k_0/k_\infty = 1). \qquad (17)$$

Fig. 1–3 show a representative set of broadening factors $F^{SC}(k_0/k_\infty)$ calculated in the described way. For the smallest extent of broadening ($F^{SC} \to 1$), they approach the simple symmetric form initially proposed in ref. 1

$$\log F^{SC} \approx \frac{\log F^{SC}_{cent}}{1 + (\log x/N^{SC})^2} \qquad (18)$$

with $x = k_0/k_\infty$ and

$$N^{SC} \approx 0.75 - 1.27 \log F^{SC}_{cent} \qquad (19)$$

With increasing extent of broadening, asymmetries arise, with the minima shifting either to the left or to the right of the center $k_0/k_\infty = 1$ and with different widths of the curves at either side of the center. We have tested the expressions for the full fall-off curves proposed in ref. 1–10 in order to

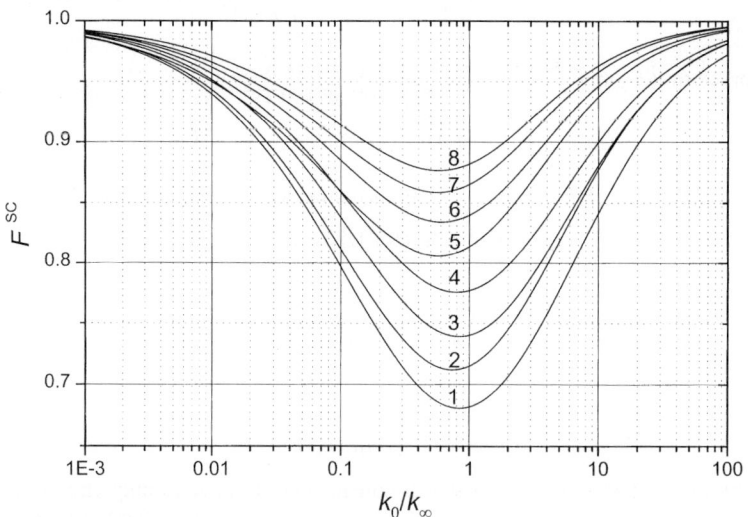

Fig. 1 Strong collision broadening factors $F_{SC}(k_0/k_\infty)$ from SACM/CT calculations for $s = 4$ (molecular parameters $\{S'_K, B'_K\}$ from eqn. (32)–(34) = {2.22, 41.8}, {2.51, 13.5}, {2.3, 19.1}, {2.07, 61.5}, {1.79, 16.6}, {1.77, 31.6}, {1.81, 6.91} and {2.00, 5.37} for curves 1, ..., 8, respectively; calculations for isotropic potential, i.e. phase space theory (PST)).

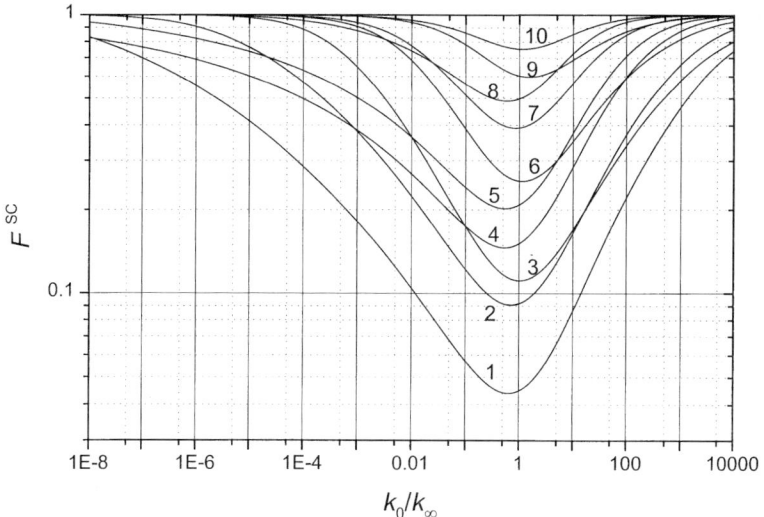

Fig. 2 As Fig. 1, for $s = 22$ (molecular parameters $\{S'_K, B'_K\} = \{16.2, 24.7\}, \{13.8, 18.3\}, \{10.5, 26.9\}, \{18.3, 15.3\}, \{16.9, 12.9\}, \{6.8, 23.3\}, \{8.10, 6.50\}, \{12.4, 6.44\}, \{3.91, 7.44\}$ and $\{3.22, 2.44\}$ for curves 1, ..., 10, respectively).

reproduce the asymmetries. None of these expressions was able to reproduce the full variety of curves shown in Fig. 1–3. Part of the problem lies in the fact that fall-off curves from rigid activated complex RRKM theory, from which most of the proposed expressions were derived differ in their shape to some extent from the present results, see below. In addition, the energy dependence of anharmonicities of the densities of states may play a role, see below. Nevertheless, the proposed expressions are useful for fitting purposes if they are not of too complicated form. Going beyond eqn. (18), in the present work, we found that a simple four-parameter relation

$$\ln F^{SC} \approx -c_1 x^{c_2}/(1 + c_3 x^{c_4}) \tag{20}$$

fits most of the curves surprisingly well. Representative values of the parameters $c_1, \ldots, c_4$ are

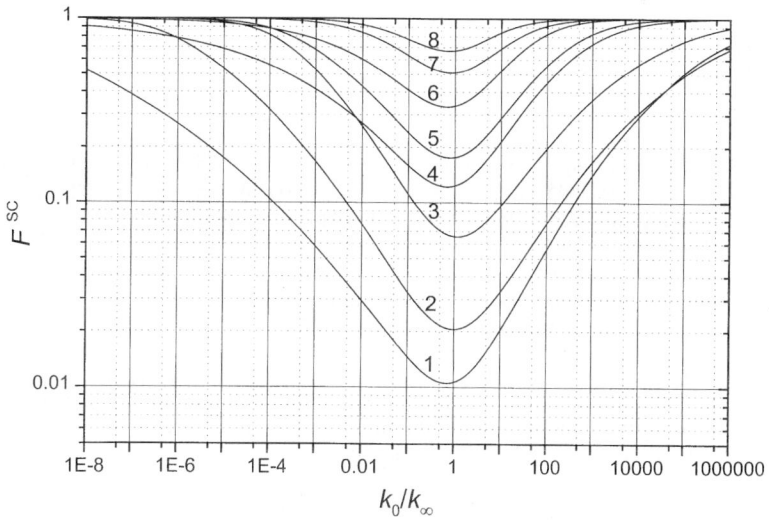

Fig. 3 As Fig. 1, for $s = 61$ (molecular parameters $\{S'_K, B'_K\} = \{29.4, 31.9\}, \{19.8, 34.4\}, \{15.5, 23.1\}, \{47.2, 23.6\}, \{16.3, 12.1\}, \{40.2, 15.9\}, \{53.2, 10.9\}, \{53.2, 10.9\}$ and $\{51.1, 8.43\}$ for curves 1, ..., 8, respectively).

given in Section 4. We, therefore, adopt the following strategy: first we relate the center broadening factor $F_{\text{cent}}^{\text{SC}}$ to the parameters S_K and B_K; in addition we use eqn. (18) and (19) as a first approximation for $F^{\text{SC}}(x)$. Second, eqn. (20) is used as an empirical fitting expression. The parameters $c_1, \ldots, c_4$ may also be extracted from the graphical representations given in Section 4. Alternatively, conventional RRKM fall-off curves are scaled to account for transitional modes and rotational effects, see Section 4.

3. Analysis of center broadening factors

By analyzing a limited number of fall-off curves from rigid activated complex RRKM calculations, as well as from classical Kassel integrals, in ref. 2 a simple relation between the two parameters S_K and B_K from eqn. (11) and (16) and the center broadening factors $F_{\text{cent}}^{\text{SC}}$ was derived. For the Kassel integrals,

$$\log F_{\text{cent}}^{\text{SC}} \approx -(1.06 \log S_K)^2/(1 + C_1 S_K^{c_2}) \tag{21}$$

with

$$C_1 = 0.10 \exp(2.5 B_K^{-1} - 0.22 B_K - 6 \times 10^{-10} B_K^6) \tag{22}$$

$$C_2 = 1.9 + 4.6 + 10^{-5} B_K^{2.8} \tag{23}$$

was found to give results of about $\pm 15\%$ accuracy over the range $0 \leqslant B_K \leqslant 36$ and $1 \leqslant S_K \leqslant 30$. In the present work, at first, we have extended the range of calculations to $0 \leqslant B_K \leqslant 100$ and $1 \leqslant S_K \leqslant 100$. Accepting a more complicated analytical expression, a more precise fit over the extended range of B_K and S_K was derived. We obtained

$$\ln F_{\text{cent}}^{\text{SC}} \approx -c_1 X^{c_2 + c_3 X}/(1 + c_4 X^{c_5})^{c_6} \tag{24}$$

with $X = \ln S_K$ and the parameters $c_1, \ldots, c_6$ expressed as a function of B_K through

$$c_1 = a_1(1 - \exp(-a_2 B_K^{a_3})) + a_4(\exp(a_5 B_K^{a_6}) - 1), \tag{25}$$

$$c_2 = a_7 - a_8 B_K - a_9 \exp(-a_{10} B_K^{a_{11}}), \tag{26}$$

$$c_3 = -a_{12} + a_{13} B_K, \tag{27}$$

$$c_4 = \exp[a_{14} + a_{15}(B_K - a_{17}) + \sqrt{a_{16}^2(B_K - a_{17})^2 + a_{18}^2} + a_{19} \exp(-a_{20} B_K)], \tag{28}$$

$$c_5 = a_{21} + a_{22}(B_K - a_{24}) - \sqrt{a_{23}^2(B_K - a_{24})^2 + a_{25}^2}, \tag{29}$$

$$c_6 = \exp[-a_{26} + a_{27} \exp(-a_{28} B_K) + \exp(a_{29} B_K - a_{30})]. \tag{30}$$

The parameters $a_1, \ldots, a_{30}$ have the values $a_1 = 0.4143$, $a_2 = 0.3316$, $a_3 = 0.8$, $a_4 = 0.002$, $a_5 = 0.0285$, $a_6 = 1.0$, $a_7 = 2.9115$, $a_8 = 0.008\,94$, $a_9 = 1.3856$, $a_{10} = 0.075\,65$, $a_{11} = 0.8$, $a_{12} = 0.096\,64$, $a_{13} = 0.002\,411$, $a_{14} = -35.8695$, $a_{15} = -0.5342$, $a_{16} = 0.6940$, $a_{17} = 16.2702$, $a_{18} = 23.6629$, $a_{19} = 1.6784$, $a_{20} = 0.9286$, $a_{21} = 26.2406$, $a_{22} = 0.4148$, $a_{23} = 0.5596$, $a_{24} = 11.9012$, $a_{25} = 16.4441$, $a_{26} = 1.4854$, $a_{27} = 1.3574$, $a_{28} = 0.087\,94$, $a_{29} = 0.021\,96$, $a_{30} = 1.3087$.

Switching from Kassel integrals to rigid activated complex RRKM calculations, it was shown in ref. 2 that the expressions (21)–(23) obtained from classical Kassel integrals could still be used if S_K and B_K were replaced by S_T and B_T with $S_T = S_K$ and

$$B_T = [(S_K - 1)/(s - 1)]^{0.6} B_K = [(S_K - 1)/(s - 1)]^{1.6}[E_0 + a_{WR}(E_0)E_Z]/kT. \tag{31}$$

In our present work we investigated to what extent eqn. (21)–(23) and eqn. (24)–(30) can be further used for a representation of our full fall-off calculations, including two transitional modes and rotational effects such as described in Section 2. Obviously, the parameters B_T and S_T have to be changed in order to accommodate transitional modes and rotational effects. The new parameters to be used with eqn. (24)–(30) are denoted by S_K' and B_K' whereas the new parameters to be used with eqn. (21)–(23) are denoted by S_T' and B_T'. At first, we identified S_K' with S_K from eqn. (11)–(13).

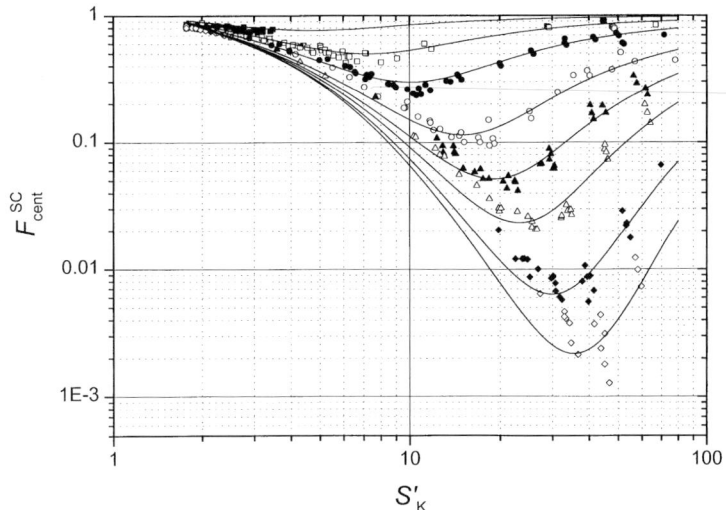

Fig. 4 Strong collision center broadening factors F_{cent}^{SC} from SACM/CT calculations as a function of the molecular parameters S'_K and B'_K from eqn. (32)–(34) (full lines: analytical approximation from eqn. (24)–(30); symbols: SACM/CT results for $B'_K = 2$–3 (■), 5–6 (□), 8–10 (●), 15–16 (○), 20–22 (▲), 25–28 (△), 33–38 (◆) and 38–48 (◇) for isotropic potential (PST)).

However, it turned out that better results were derived by the simpler choice of

$$S'_K = 1 + S_{cons} + 0.77. \qquad (32)$$

On the other hand, the best choice of B'_K was

$$B'_K = B'[(S'_K - a)/(s_{cons} + 1.77 - a)]^{1.31 + 5.14s^{-0.73}} \qquad (33)$$

where s is the total number of oscillators of the adduct, $s_{cons} = s - 3$ is the number of conserved

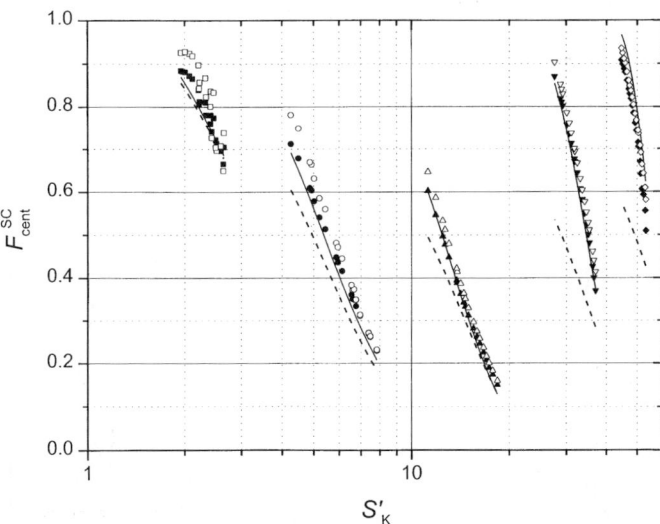

Fig. 5 As Fig. 4; SACM/CT calculations for $B' = [E_0 + a_{WR}(E_0)E_Z]/kT = 20$ (filled symbols: isotropic potential (PST); open symbols: anisotropic potential with $\varepsilon = 10$ from eqn. (36); full lines: analytical approximation from eqn. (24)–(30); dashed lines: analytical approximation from eqn. (21)–(23), (32), (35); (■) $s = 4$, (●) $s = 10$, (▲) $s = 22$, (▼) $s = 43$, (◆) $s = 61$).

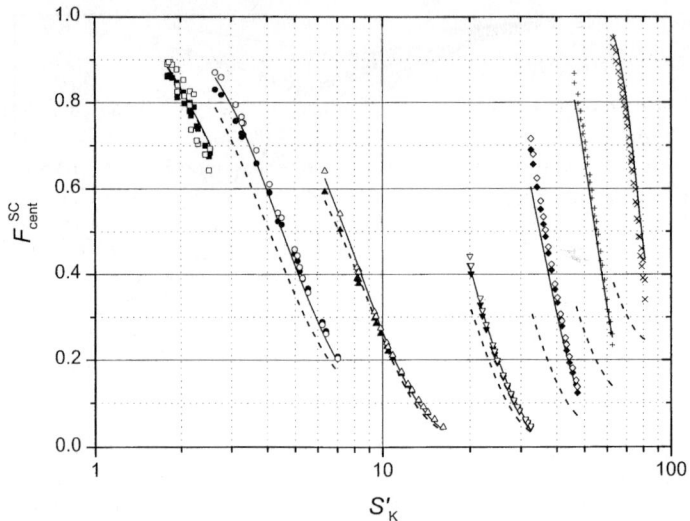

Fig. 6 As Fig. 5, for $B' = 40$ (symbols as in Fig. 5, in addition (+) $s = 79$ and (×) $s = 100$ for PST and $\varepsilon = 10$).

oscillators, S_{cons} in eqn. (32) is given by eqn. (12), B' is defined by eqn. (16), and a is defined by

$$a = 0.5 + (s - 5.2)(6.47B'/s)/[\exp(4.27B'/s) - 1]. \tag{34}$$

For moderate broadening (at $S'_K \leqslant 5$, where $F_{\text{cent}}^{\text{SC}} \gtrsim 0.6$), $F_{\text{cent}}^{\text{SC}}$ becomes nearly independent of B'_K, see ref. 2. In this range, eqn. (32) indicates that the two transitional modes, because of the presence of centrifugal barriers and angular momentum coupling, contribute less to S'_K than two free rotors, for which 0.77 would be replaced by unity. For more extensive broadening, the relative contribution from the transitional modes becomes increasingly less important.

In addition to eqn. (24)–(30) with eqn. (32)–(34), we tested the simpler eqn. (21)–(23). We found

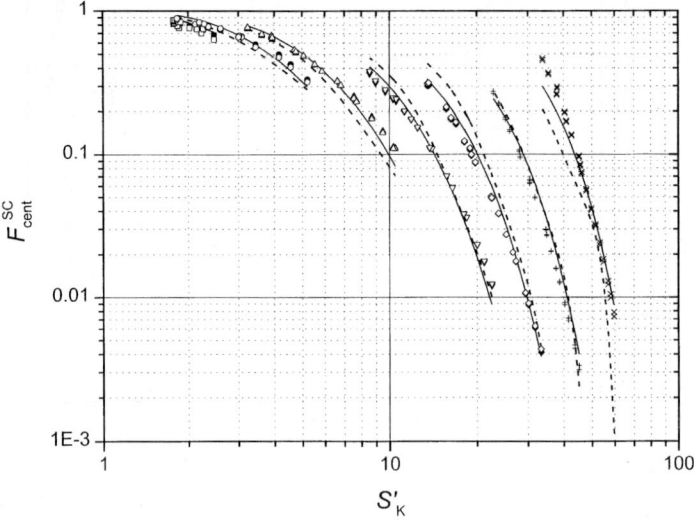

Fig. 7 As Fig. 6, for $B' = 100$.

that $S_T' = S_K'$ and

$$B_T' = B'[(S_K' - 1)/(s_{cons} + 1.77 - 1)]^{1.6} \quad (35)$$

gave optimum results which, however, were of a quality inferior to that of eqn. (24)–(30) and (32)–(34).

In Fig. 4–7 we illustrate our calculated F_{cent}^{SC} and their analytical representation by eqn. (21)–(35). Confirming our earlier conclusion (see Fig. 4 from ref. 2), F_{cent}^{SC} essentially only depends on two parameters, chosen to be S_K' and B_K' in Fig. 4. There is some scatter of the calculated points around the full lines from eqn. (24)–(30) which is due to the range of B_K' values combined into one curve.

Fig. 5–7 provide an enlargened view of F_{cent}^{SC} over the range $s = 4$–100 and $B' = 20$–100. Calculations treating the transitional modes without anisotropy of the potential (phase space theory, filled symbols) and with a very anisotropic potential ($\varepsilon = 10$, open symbols) with

$$\varepsilon = [\hbar\omega_{bend}]^2/4BD \quad (36)$$

see ref. 22, are included. Eqn. (24)–(30) and (32)–(34) (full lines) over wide ranges of conditions work very well; the simpler eqn. (21)–(23) and (35) (dashed lines) are less precise but still provide satisfactory results for medium ranges of conditions. One observes that the given equations slightly overestimate the extent of broadening for small s and small B'. In this range one also notices separate contributions from centrifugal barriers (PST) and from anisotropy ($\varepsilon = 10$). One should, however, note that the fall-off calculations here, i.e. at the corresponding very large temperatures, become relatively unprecise. At larger B', these differences disappear and the general agreement between F_{cent}^{SC} from the full calculation and from eqn. (24)–(30) and (32)–(34) becomes very satisfactory.

4. Shapes of fall-off curves

After having analyzed the magnitude of broadening factors through the center broadening factors in Section 3, in the following we look at asymmetries which are not accounted for by eqn. (18) and (19). Instead of following the approach of ref. 2, we use eqn. (20). Representative sets of parameters $c_1, \ldots, c_4$ obtained by fitting of our calculated fall-off curves to eqn. (20) are illustrated in Fig. 8–10. We did not succeed in providing an analytical relation between these fitting parameters and the molecular parameters S_K', s and B'. However, Fig. 8–10 provide a sufficiently extensive guide to the prediction of asymmetries of broadening factors.

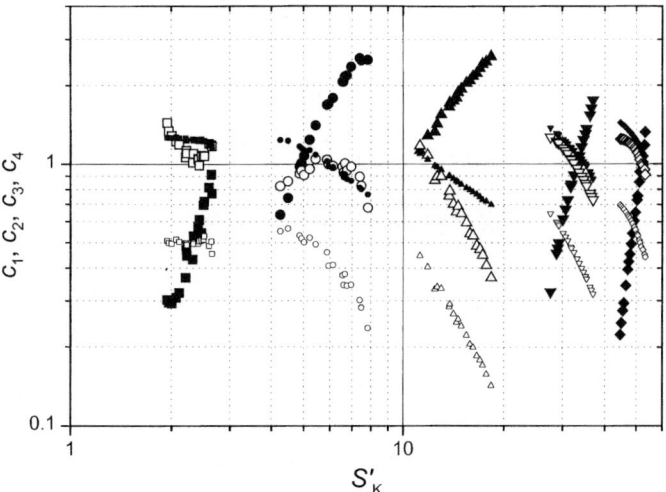

Fig. 8 Shape parameters $c_1, \ldots, c_4$ of strong collision broadening factors in eqn. (20) as a function of molecular parameters ($B' = 20$; large filled symbols: c_1; small open symbols: c_2; large open symbols: c_3; small filled symbols: c_4; (■) $s = 4$, (○) $s = 10$, (△) $s = 22$, (▽) $s = 43$, (◇) $s = 61$; PST).

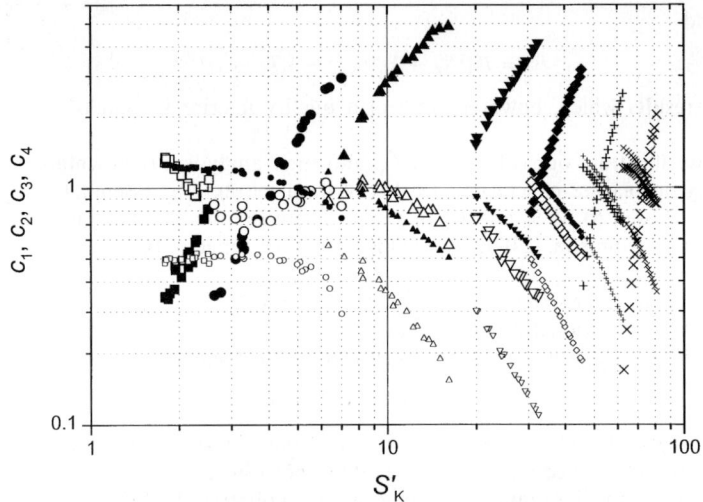

Fig. 9 As Fig. 8, for $B' = 40$ (in addition (+) $s = 79$ and (×) $s = 100$ with ordering as for the curves with $s = 4$–61).

A different access to asymmetries of broadening factors may consist in the following strategy: one calculates broadening factors from classical Kassel integrals or from rigid activated complex RRKM theory with parameters which are close to those of the real system. These calculations are numerically very quickly done. One then compares the resulting center broadening factors with those obtained from the analytical expressions given in Section 3 (see Fig. 4–7) and scales the broadening factors from Kassel integrals or from RRKM calculations. Examples for this procedure are illustrated in Fig. 11. We haven chosen four SACM/CT calculations represented by full lines in the figure. We have then replaced the two transitional modes by low frequency harmonic oscillators and done rigid activated complex RRKM calculations leading to the dashed lines in the figure. We have finally scaled these RRKM results such that the same F_{cent}^{SC} as from the SACM/CT calculations are obtained, leading to the dotted lines in the figure. One notices good agreement between the SACM/CT and scaled RRKM results for broad fall-off curves where the relative

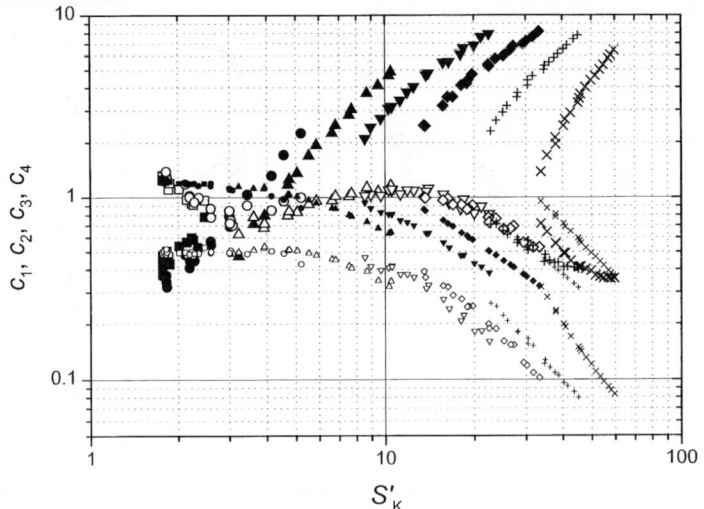

Fig. 10 As Fig. 9, for $B' = 100$.

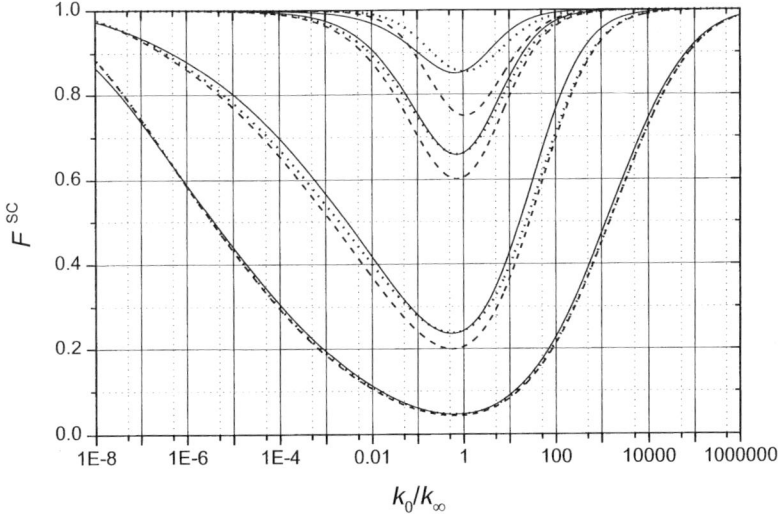

Fig. 11 Comparison of SACM/CT (full lines), rigid activated complex RRKM (dashed lines) and scaled RRKM (dotted lines) broadening factors ($\{s, B', S'_K, B'_K\} = \{4, 20, 2.15, 7.20\}$, $\{43, 20, 32.0, 8.2\}$, $\{22, 20, 16.2, 11.8\}$ and $\{22, 40, 15.9, 23.9\}$ for groups of three curves from top to bottom, see text).

contribution from transitional modes is small. However, SACM/CT fall-off curves with decreasing broadening become increasingly narrower than rigid activated complex RRKM curves. Furthermore, for small s, SACM/CT curves are more asymmetric, being shifted towards smaller k_0/k_∞.

We finally compare broadening factors of the present model calculations employing a Morse/standard anisotropy potential with our recent fall-off calculations for the $HO_2 \Leftrightarrow H + O_2$ system on an *ab initio* potential surface.[20] The *ab initio* potential from ref. 25 is of more complicated form than the present model potential; furthermore, the HO_2 calculations included energy-dependent anharmonicities of the density of states $\rho(E, J)$.[25–27] Fig. 12 shows broadening factors at four temperatures. One observes that the detailed calculations from ref. 20 led to a pronounced

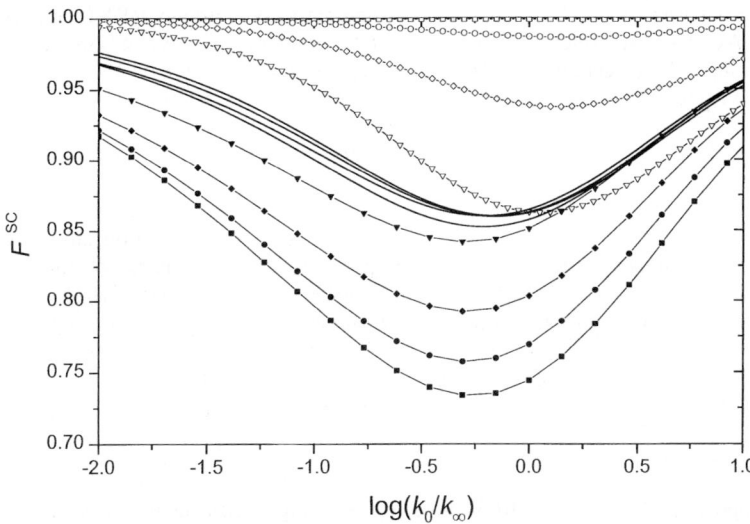

Fig. 12 Strong collision broadening factors for the $HO_2 \Leftrightarrow H + O_2$ system (($\blacksquare$) $T = 310$ K, ($\bullet$) $T = 580$ K, ($\blacklozenge$) $T = 1080$ K, ($\blacktriangle$) $T = 2000$ K; curves with filled symbols: SACM/CT results from ref. 20 with *ab initio* potential and energy-dependent anharmonic densities of states from ref. 25; curves with open symbols: rigid activated complex RRKM results, see text; curves without symbols: SACM/CT results for model potential of present work and harmonic densities of states with $T = 310, 580, 1080$ and 2000 K from bottom to top at the left side of the diagram; see text).

decrease of the extent of broadening with increasing temperature, with F_{cent}^{SC} increasing from about 0.74 at 310 K to 0.85 at 2000 K. The present SACM/CT calculations, treating HO_2 as a linear molecular system, produced broadening factors of similar asymmetry but with nearly temperature-independent center broadening factors near $F_{cent}^{SC} \approx 0.86$. We have investigated the origin of this difference, finding that the temperature dependence of the results from ref. 20 is mostly due to the energy dependence of the anharmonicity correction of the vibrational density of states while the difference in absolute values of F_{cent}^{SC} is mostly due to the use of the true reduced mass of $H + O_2$ instead of the randomly chosen larger value of the present treatment. For comparison, we also include rigid activated complex RRKM calculations, for which the transitional mode of a nonlinear HO_2 is represented by the HO_2 bending oscillator. Clearly this mode is too rigid, produces increasing broadening with increasing temperature and leads to too symmetric fall-off curves. Scaling of the RRKM results would not improve the situation. This confirms the trends illustrated in Fig. 11.

5. Conclusions

The present systematic calculations of fall-off curves for bond fission reactions of linear molecules dissociating into linear fragments and atoms confirmed the conclusion from ref. 2 that center broadening factors F_{cent}^{SC}, in spite of the multiparameter nature of the system, can be represented satisfactorily in terms of two molecular parameters (S'_K and B'_K from eqn. (32)–(34) or S'_T and B'_T from eqn. (32) and (35)). We were able to provide an analytical expression for F_{cent}^{SC} as a function of these two parameters (see eqn. (24)–(30) or eqn. (21)–(23)). On the other hand, we could only graphically relate the shifts and asymmetries of the broadening factors F^{SC} (k_0/k_∞) to the mentioned molecular parameters.

Our SACM/CT calculations account for angular momentum coupling of the two transitional modes with the overall rotation of the system. We could illustrate the influence of centrifugal barriers and of anisotropy of the potential on the fall-off curves. It was shown that the contribution of the transitional modes to the effective number of oscillators in the fall-off curve on average was about 0.385 per transitional mode, see eqn. (32). We presume that this result, derived from a "linear molecule ⇔ linear fragment + atom" system, can be generalized up to the most complex "nonlinear molecule ⇔ nonlinear fragment + nonlinear fragment" system. (For a simplified treatment of the $HNO_3 \rightarrow HO + NO_2^-$ system, see ref. 28.)

We showed that shapes of fall-off curves from rigid activated complex RRKM theory and from the present SACM/CT treatment do not agree in the finer details, even if agreement in F_{cent}^{SC} is forced by scaling of the center broadening factors. This observation is most pronounced for systems with a small number of oscillators s; with increasing s, the relative influence of transitional modes becomes increasingly less important. Because of the differences in the shapes of the fall-off curves from RRKM and SACM/CT treatments, it appears premature to employ multiparameter fits of RRKM fall-off curves beyond simple analytical expressions such as eqn. (18)–(20). In reality, more complicated ab initio potentials of individual molecular systems as well as energy-dependent anharmonicities of the densities of states lead to individual shapes of fall-off curves which differ from the "standard shapes" of the present SACM/CT treatment to some extent. In addition, weak collision effects[3,21] influence the shapes of fall-off curves in a, so far, not well documented way. With improved knowledge about rovibrational collisional energy transfer, more systematic master equation treatments become possible. One may hope that the resulting weak collision broadening factors can then also be approximated by simple analytical expressions as a function of two molecular parameters (such as S'_K and B'_K) in addition to the weak collision efficiency β_c.[1,3,21]

Acknowledgements

Financial support of this work by the Deutsche Forschungsgemeinschaft (SFB 357 "Molekulare Mechanismen unimolekularer Prozesse") is gratefully acknowledged.

References

1 J. Troe, *J. Phys. Chem.*, 1979, **83**, 114.
2 J. Troe, *Ber. Bunsen-Ges. Phys. Chem.*, 1983, **87**, 161.

3   R. G. Gilbert, K. Luther and J. Troe, *Ber. Bunsen-Ges. Phys. Chem.*, 1983, **87**, 169.
4   I. Oref, *J. Phys. Chem.*, 1989, **93**, 3465.
5   Z. Pawlowska and I. Oref, *J. Phys. Chem.*, 1990, **94**, 567.
6   Z. Pawlowska, W. C. Gardiner and I. Oref, *J. Phys. Chem.*, 1993, **97**, 5024.
7   H. Wang and M. Frenklach, *Chem. Phys. Lett.*, 1993, **205**, 271.
8   A. Kazakov, H. Wang and M. Frenklach, *J. Phys. Chem.*, 1994, **98**, 10598.
9   O. Prezhdo, *J. Phys. Chem.*, 1995, **99**, 8633.
10 P. K. Venkatesh, *J. Phys. Chem. A*, 2000, **104**, 280.
11 J. S. Poole and R. G. Gilbert, *Int. J. Chem. Kinet.*, 1994, **26**, 273.
12 S. C. Smith and T. J. Frankcombe, *Faraday Discuss.*, 2001, **119**, (DOI: 10.1039/b102562g).
13 R. Atkinson, D. L. Baulch, R. A. Cox, R. F. Hampson, J. A. Kerr, M. J. Rossi and J. Troe, *J. Phys. Chem. Ref. Data*, 2000, **29**, 167.
14 D. L. Baulch, C. J. Cobos, R. A. Cox, P. Frank, G. Hayman, Th. Just, J. A. Kerr, T. Murrells, M. J. Pilling, J. Troe, R. W. Walker and J. Troe, *J. Phys. Chem. Ref. Data*, 1994, **23**, 847.
15 S. P. Sander, R. R. Friedl, W. B. DeMore, D. M. Golden, M. J. Kurylo, R. F. Hampson, R. E. Huie, G. K. Moortgat, A. R. Ravishankara, C. E. Kolb and M. J. Molina, *JPL Publication 00-3, March, 2000*, http://jpldataeval.jpl.nasa.gov/.
16 H. O. Pritchard and S. R. Vatsya, *J. Phys. Chem.*, 1992, **96**, 172.
17 V. K. Knyazev, *J. Phys. Chem.*, 1995, **99**, 14738.
18 V. D. Knyazev and I. R. Slagle, *J. Phys. Chem.*, 1996, **100**, 16899.
19 C. Nyeland, *Z. Phys. Chem.*, 2000, **214**, 1329.
20 J. Troe, *28th Symp. (Int.) on Combustion*, The Combustion Institute, Pittsburgh, 2000, p. 1463.
21 J. Troe, *Ber. Bunsen-Ges. Phys. Chem.*, 1974, **78**, 478.
22 A. I. Maergoiz, E. E. Nikitin, J. Troe and V. G. Ushakov, *J. Chem. Phys.*, 1998, **108**, 5265.
23 M. Quack and J. Troe, *Ber. Bunsen-Ges. Phys. Chem.*, 1974, **78**, 240.
24 A. I. Maergoiz, E. E. Nikitin, J. Troe and V. G. Ushakov, to be submitted.
25 L. B. Harding, J. Troe and V. G. Ushakov, *Phys. Chem. Chem. Phys.*, 2000, **2**, 631.
26 L. B. Harding, A. I. Maergoiz, J. Troe and V. G. Ushakov, *J. Chem. Phys.*, 2000, **113**, 11019.
27 J. Troe, *Chem. Phys.*, 1995, **190**, 381.
28 J. Troe, *Int. J. Chem. Kinet.*, 2001, in press.

Time-dependent master equation simulation of complex elementary reactions in combustion: Application to the reaction of $^1CH_2$ with C_2H_2 from 300–2000 K

Terry J. Frankcombe and Sean C. Smith

Department of Chemistry, University of Queensland, Brisbane, Qld, 4072, Australia

Received 19th March 2001
First published as an Advance Article on the web 7th September 2001

Computational simulations of the title reaction are presented, covering a temperature range from 300 to 2000 K. At lower temperatures we find that initial formation of the cyclopropene complex by addition of methylene to acetylene is irreversible, as is the stabilisation process *via* collisional energy transfer. Product branching between propargyl and the stable isomers is predicted at 300 K as a function of pressure for the first time. At intermediate temperatures (1200 K), complex temporal evolution involving multiple steady states begins to emerge. At high temperatures (2000 K) the timescale for subsequent unimolecular decay of thermalized intermediates begins to impinge on the timescale for reaction of methylene, such that the rate of formation of propargyl product does not admit a simple analysis in terms of a single time-independent rate constant until the methylene supply becomes depleted. Likewise, at the elevated temperatures the thermalized intermediates cannot be regarded as irreversible product channels. Our solution algorithm involves spectral propagation of a symmetrised version of the discretized master equation matrix, and is implemented in a high precision environment which makes hitherto unachievable low-temperature modelling a reality.

I Introduction

An active area of research is the kinetics of the formation of the propargyl radical C_3H_3 in flames. The formation of C_3H_3 is believed to be a significant step in the formation of simple aromatic hydrocarbons and thus polycyclic aromatic hydrocarbons and soot.[1–8] The major route to C_3H_3 proposed by Miller and Melius[1] involves insertion of singlet methylene into acetylene to form C_3H_4, which isomerises before decomposing to C_3H_3. In experiments investigating the kinetics, any $^1CH_2$ formed is destroyed through a combination of collisionally induced intersystem crossing to the triplet methylene state and reaction with acetylene to form C_3H_4. The rate constant for the reaction of $^1CH_2$ with C_2H_2 to form C_3H_4 and eventually C_3H_3 has been measured experimentally by several different methods over the past 20 years. A good summary is given by Blitz *et al.*[9] While it appears well-established that a single time-independent bimolecular rate constant is appropriate for the disappearance of $^1CH_2$ under pulsed conditions, once energised C_3H_4 has been formed the competition between collisional stabilisation to thermalized C_3H_4 isomers and dissociation to form the C_3H_3 radical remains largely unquantified. While experimental data fail to provide this information the most practical approach to determining the dynamics of the full reaction is by numerical simulation. This paper extends previous efforts to model this reaction with a master equation (ME) formulation.[9–11]

DOI: 10.1039/b102562g

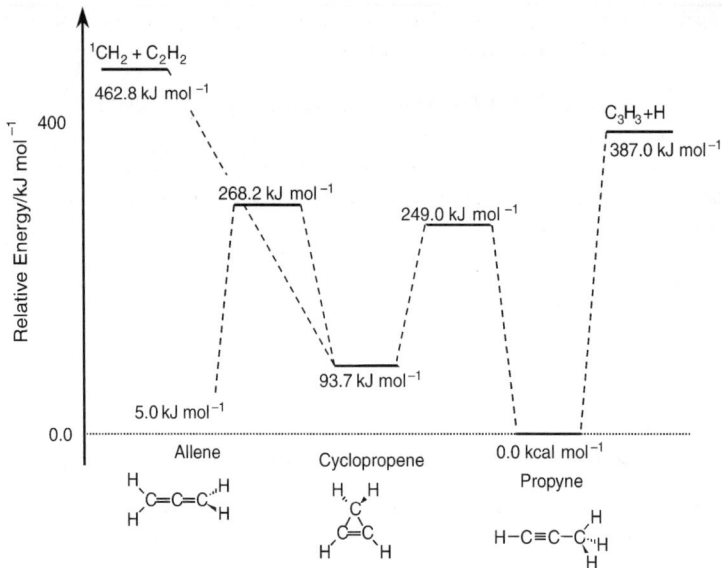

Fig. 1 Schematic potential energy surface of the reaction $^1CH_2 + C_2H_2 \rightarrow C_3H_3 + H$ via isomerizing C_3H_4.

The potential energy surface (PES) used to model the overall reaction

$$^1CH_2 + C_2H_2 \rightleftharpoons C_3H_4 \rightarrow C_3H_3 + H \qquad (1)$$

is shown in Fig. 1. There is disagreement in the literature regarding the heats of formation of the species present. While most authors agree to within around 5 kJ mol^{-1} for the heats of formation of the stable isomers, estimates of the heats of formation of the isomerisation transition states vary over a range of 40 kJ mol^{-1}. Following the previous modelling of this reaction,[9–11] relevant energies were taken as those quoted by Karni et al.[12]

The PES shown in Fig. 1 utilises only the lowest energy paths for the reaction. Other possible reaction paths exist, such as direct isomerisation from allene to propyne and reaction paths between cyclopropene and the other two isomers through higher energy transition states.[13,14] While the neglect of these higher energy paths slightly underestimates the rates of the conversion processes, it is likely that this effect is smaller than that of the existing uncertainties in this PES.

Experiments investigating this system are typically conducted at room temperatures and low pressure. The calculations of Gates et al.[10] and those of Frankcombe et al.[11] modelled the system at the reasonably high pressure of 10^4 Torr and at temperatures of 1400 and 2000 K respectively. Frankcombe et al. found that the method used was stable to a temperature of around 800 K at the pressure investigated, below which numerical instabilities rendered the results unreliable. Blitz et al.[9] modelled the system over a range of pressures down to 1 Torr. By implementing the model in quadruple precision (doubling the precision of the calculations from standard double precision) the model was able to be used down to 600 K before the results became unreliable. These numerical difficulties at low temperatures are well known in MEs describing unimolecular reactions. Recently we[15] have shown that in the case of the unimolecular decomposition of ethane at 300 K numerical difficulties can be circumvented by increasing the precision in which the calculations are performed beyond quadruple precision. In this work a similar approach has been taken, implementing the ME model in a numerical environment in which an arbitrary precision level can be set—well beyond the approximately 32 decimal digits available through quadruple precision.

II Master equation model

The ME formulation for unimolecular reactions is well known.[16–18] For more complex isomerising reactions it is less common, despite being first applied in the 1960s.[19] Even so, it is reasonably well established in the literature,[9–11,19–25] so only a brief description will be given here.

The unimolecular energy grained ME describes the evolution of the reactant molecule population through a set of coupled differential equations of the form

$$\frac{d\rho_i}{dt} = \omega \sum_j P_{ij}\rho_j - \omega\rho_i - k_i\rho_i \qquad (2)$$

where ρ_i represents the population of the reactant in the ith energy grain, P_{ij} is the conditional probability that a collision event involving a molecule in the jth energy grain leaves the molecule in the ith energy grain, k_i is the microscopic reaction rate for the ith energy grain and ω is the collision frequency. The three terms on the right of eqn. (2) describe gain and loss of population through collisional energy transfer (CET) and loss through unimolecular reaction, respectively.

For an isomerising system eqn. (2) is generalised to

$$\frac{d\rho_i}{dt} = \omega \sum_j P_{ij}\rho_j - \omega\rho_i - \rho_i \sum_j k_j^{(L,i)} + \sum_j k_j^{(G,i)}\rho_j \qquad (3)$$

Here the populations ρ_i have been extended to include all of the energy grains from all of the isomers. The sum over the index j is over all energy grains, and each sum is independent. This formulation means that significant portions of the quantities P_{ij}, $k_j^{(L,i)}$ and $k_j^{(G,i)}$ are zero.

The first two terms on the right-hand side of eqn. (3) remain unchanged from eqn. (2). The next term is analogous to the third term in eqn. (2) and describes the reactive loss of population from state i. In this term the sum is over all reactive loss channels for state i and each term $k_j^{(L,i)}$ gives the microscopic rate for the channel. In an isomerising system there will be a nonzero $k_j^{(L,i)}$ for each isomerisation that is possible from state i. There will also be $k_j^{(L,i)}$ terms for each unimolecular dissociation, such as the formation of $C_3H_3 + H$ from propyne or $^1CH_2 + C_2H_2$ from cyclopropene in the system shown in Fig. 1. The final term describes the gain in population from isomerisation of other isomers and from other formation processes such as the formation of cyclopropene from $^1CH_2 + C_2H_2$. There will be a nonzero population gain rate constant $k_j^{(G,i)}$ for each channel, similar to the loss case. If the system is being described in a conservative manner there will be one nonzero $k_j^{(L,i)}$ loss term in the system of equations for each nonzero $k_j^{(G,i)}$ gain term.

If the set of energy grain populations ρ_i are formed into a vector $\boldsymbol{\rho}$ the system of coupled equations can be expressed as

$$\frac{d\boldsymbol{\rho}}{dt} = \boldsymbol{A}\boldsymbol{\rho} \qquad (4)$$

where the matrix $\boldsymbol{A}$ is constructed from all the components of eqn. (3) other than the ρ_i terms. As described in detail in ref. 10 and 11, ordering the elements in $\boldsymbol{\rho}$ so that the energy grains for each isomer are grouped together and in order of ascending energy within each isomer yields a well-structured matrix with dense blocks on the main diagonal and diagonal blocks off the diagonal.

Bimolecular reactions can be incorporated into the system if the reaction is performed under pseudo-first-order conditions. Pseudo-first-order conditions allow the rate of change of the reactant not in excess to be expressed in a linear fashion as an element of the product $\boldsymbol{A}\boldsymbol{\rho}$. The $^1CH_2 + C_2H_2$ and $C_3H_3 + H$ channels can be handled in this way by adding a single population element ρ_i to the end of the state space $\boldsymbol{\rho}$ for each, with the population element describing the population of the species not in excess. The inclusion of each linearised bimolecular channel adds one nonzero row and column to the matrix $\boldsymbol{A}$ describing the rates of reaction to and from the energy grained isomers. As the final dissociation to form C_3H_3 is essentially irreversible, an alternative is to include only a grain for the $^1CH_2 + C_2H_2$ channel and treat the system nonconservatively. The treatment of these bimolecular terms has been discussed in detail in Frankcombe et al.[11]

The collisional and reactive processes for an equilibrating system must obey detailed balance.[17] As a consequence the matrix $\boldsymbol{A}$ can be symmetrised by the transform

$$\boldsymbol{B} = \boldsymbol{SAS}^{-1} \qquad (5)$$

where the diagonal matrix $\boldsymbol{S}$ is defined from the Boltzmann distribution $\boldsymbol{f}$ (corresponding to the ordering used in $\boldsymbol{\rho}$) by $S_{ii} = f_i^{-1/2}$. When the matrix is symmetrised the population vector $\boldsymbol{\rho}$ is redefined, with each element being scaled by the square root of the corresponding element of f. Generally speaking, decompositions of symmetric matrices such as B are faster and more stable than of asymmetric matrices such as A. Symmetrisation is standard procedure for the unimolecular ME and extends naturally to the isomerisation ME in most cases.[11]

By expanding the exponential operator in terms of the eigenvalues and eigenvectors of A (so that $Ax^{(i)} = \lambda_i x^{(i)}$) or B one obtains as the solution to systems of the form of eqn. (4)

$$\rho(t) = \sum_i \alpha_i \exp(\lambda_i t) x^{(i)} \qquad (6)$$

where the α_i terms are determined so that

$$\rho(0) = \sum_i \alpha_i x^{(i)} \qquad (7)$$

The ME problem then becomes an eigenproblem, with energy-resolved populations being analytically available once the eigenpairs have been determined.

Eqn. (6) provides analytically a representation $\rho(t)$ of the population profile at all times t. This representation is either the raw population (if the asymmetric eigenproblem $Ax = \lambda x$ has been solved) or the raw population scaled by the square root of the Boltzmann population (if the matrix was symmetrised to B before the eigenproblem was solved). If the symmetrised version of the problem was solved then the raw population can quickly be gleaned by reversing the Boltzmann scaling. Once the raw population profile has been attained for the desired time the total population of any species is given by the sum of the population grains belonging to the species in question.

The calculation of the rate matrix and the solution of the eigenproblem was implemented using a version of the MPFUN multi-precision package of Bailey.[26] This package allows specification of virtually any numerical precision level. Calculations were typically performed maintaining 50 decimal digits of precision, though calculations at 40, 70 and 100 digits were performed in some cases to test the precision level required. The eigenpairs of the matrix were calculated with adaptations of the EISPACK routine "rs" and the LAPACK routine "dsyev".

Finding a set of α_i to satisfy eqn. (7) is normally achieved through the projection theorem so that $\alpha_i = \langle \rho(0), x^{(i)} \rangle$. This relies on the set of eigenvectors forming not only a complete basis for $\mathbb{R}^n$ (where n is the number of energy grains used for ρ), but also being orthonormal. This is the case when the matrix has been symmetrised, and this is one of the properties that has lead to the popularity of the symmetric approach. When the asymmetric matrix is diagonalized the eigenvector basis is no longer orthonormal with respect to the usual dot product. However an inner product weighted by the diagonal matrix S^2 yields orthonormality[11] so eqn. (7) is satisfied by $\alpha_i = \langle \rho(0), x^{(i)} \rangle_{S^2}$ in this case.

III Modeling

Several quantities need to be defined to fully specify the ME described in eqn. (3). In this work the popular exponential down model was used for the CET kernel. When discretized this model gives

$$P_{ij} = c_i \exp(-(E_j - E_i)/\alpha) \qquad (8)$$

for $E_i < E_j$, where E_i is the energy describing the ith energy grain, α is approximately equal to $\langle \Delta E \rangle_{\text{down}}$, the average energy transfered in a collision resulting in a lower energy, and c_i is a normalisation constant to ensure $\sum_i \Delta E_i P_{ij} = 1$. Detailed balance is used to determine P_{ij} for transitions to higher energy prior to normalisation.

The Lennard-Jones collision frequency was used for ω. Equally spaced energy grains were used with a grain size of 2.39 kJ mol^{-1} (200 cm^{-1}). 250 energy grains above the zero-point of the lowest energy isomer (propyne) were used, covering an energy range up to almost 600 kJ mol^{-1}. Unless explicitly varied the α parameter for the CET model was taken as 2.27 kJ mol^{-1} (190 cm^{-1}). A single α parameter was used for all three C_3H_4 isomers in all cases.

The rate matrix was built, symmetrised and diagonalized for a range of temperatures from 300 to 2000 K and a range of pressures for 1 to 10^4 Torr. For each pressure and temperature time-

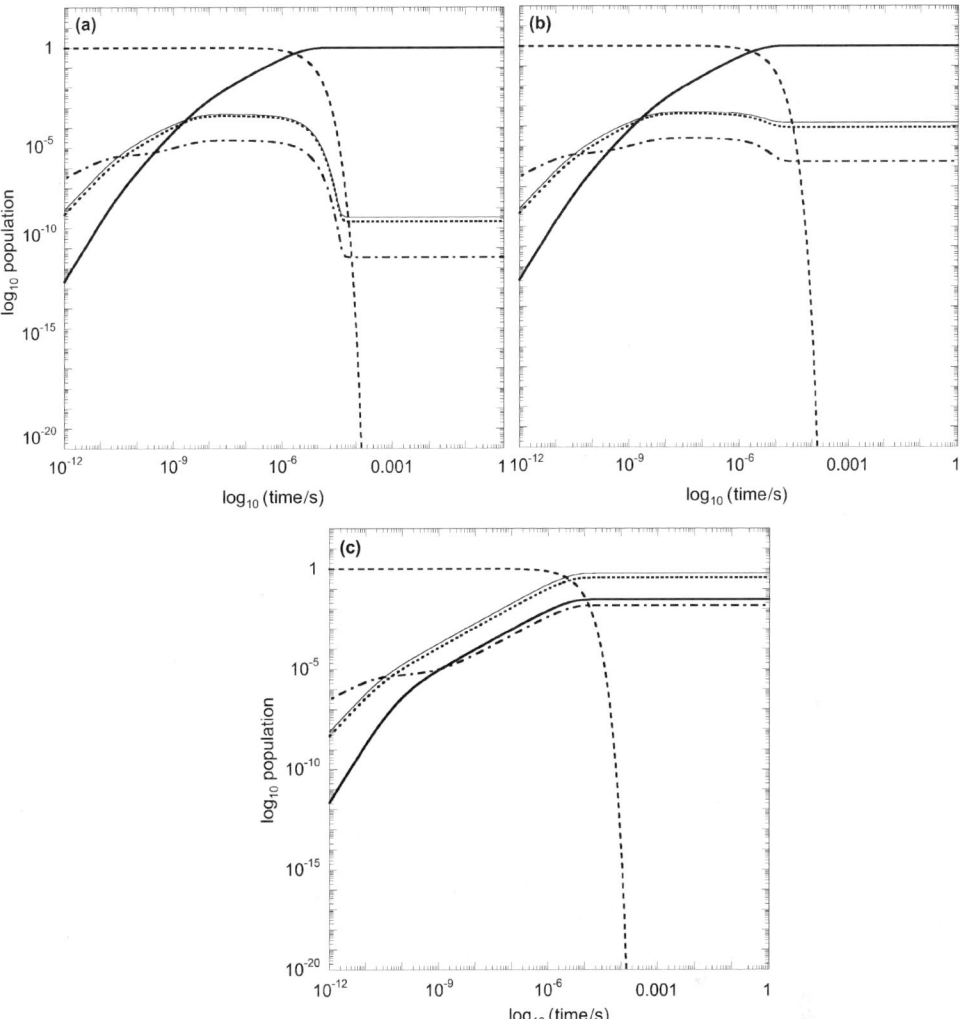

Fig. 2 Time-dependent population profiles for $^1CH_2$ (– – –), C_3H_4 isomers allene (· · ·), cyclopropene (· – ·) and propyne (——, fine) and C_3H_3 (——, thick) at 300 K and 1 Torr (a), 10 Torr (b) and 10^4 Torr (c). Calculated with 50 digit arithmetic.

dependent population profiles were calculated using eqn. (6) and summing the relevant grain populations (with the projection coefficients α_i being determined using the projection theorem, as discussed at the end of the previous section) for an initial population being solely composed of $^1CH_2$. A range of numerical precision was used, up to 100 digits. Generally 50 digits was sufficient precision to yield converged total populations at all reasonable times. Initially the system was modelled both conservatively, including an explicit grain for the product state, and non-conservatively, without the product state grain and using conservation of population to calculate the C_3H_3 population. To include the product state grain in a symmetrisable manner requires the linearisation of this channel by introducing an excess of hydrogen atoms to achieve pseudo-first-order conditions.[11] Under low pressure conditions it was found that the recombination of the C_3H_3 being formed in the reaction with the excess hydrogen atoms (which were in excess by a factor of 10^{10}) contributed significantly to the population of the C_3H_4 isomers. Indeed, from medium times the population of the C_3H_4 isomers increased linearly with time. At the lowest pressures the linear increase from back reaction intruded into the time-frame of the dynamics of

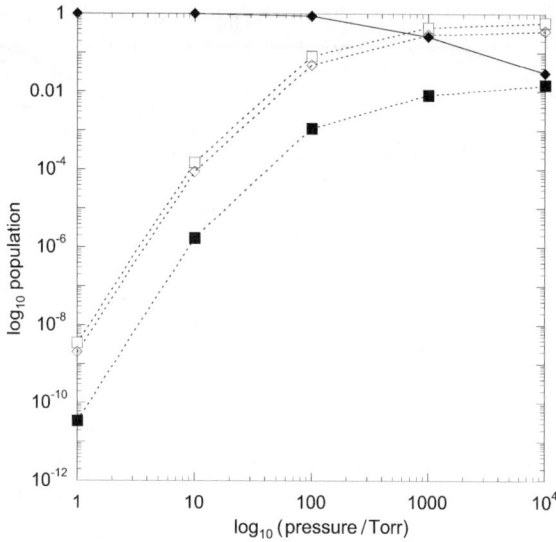

Fig. 3 Long time populations (time of order 1 s) of C_3H_3 (-◆-) and C_3H_4 isomers allene (··◇··), cyclopropene (··■··) and propyne (··□··) as a function of pressure at 300 K.

the forward reaction determining the degree of stabilisation of the C_3H_4 isomers. At shorter times the populations calculated with the conservative and nonconservative approach agreed, as did the two approaches at higher pressures, in accord with the earlier work.[11] As the linearisation of the product channel is not realistic and the reaction is essentially irreversible the nonconservative approach was used for the data reported here.

In the remainder of this section a range of temperature regimes will be discussed. It is worth noting that all the features that are present in the high temperature case are also present in the lower temperature cases, but occur at far longer timescales. These features will generally not be mentioned until the timescale at which they act (which generally gets shorter with increasing temperature) becomes close to the timescales of other processes being discussed.

1 Modelling at 300 K

Fig. 2 shows calculated population profiles for the five species modelled at 300 K and 1, 10 and 10^4 Torr. Profiles for pressures between 10 and 10^4 Torr are intermediate between those shown. At times shorter than those shown the C_3H_4 isomer and C_3H_3 populations maintain a linear dependence on time. At short times the populations may deviate from the expected profile through numerical error associated with insufficient precision. These features are not present in populations calculated in 70 or 100 digit algebra except at extremely short times.

The short time population profiles shown in Fig. 2 are independent of pressure. This population corresponds to C_3H_4 being formed at high energy (greater than 460 kJ mol^{-1}), isomerising to propyne and dissociating to C_3H_3 before CET (the only pressure-dependent component) could significantly alter the energy-dependent population distribution. All populations remaining in the C_3H_4 isomer states at long times have been deactivated to energies below the reaction threshold by CET. This C_3H_4 population, though in a reasonably steady state, is not in thermal equilibrium (see modelling at 1200 K below) and undergoes very slow unimolecular decay to C_3H_3 + H. The system was additionally modelled without CET by removing the first two terms from eqn. (3), leaving only three diagonals and the "arrowhead" of the matrix nonzero. This matrix could be compressed to 406 grains by removing inaccessible energy grains from the state space. When this matrix was diagonalized and used to propagate an initial $^1CH_2$ population the population profiles produced were identical to the parts of the profiles shown in Fig. 2(a) at times shorter than 3×10^{-5} s, just before the C_3H_4 isomer populations stabilised to the long time steady state. In the absence of CET the C_3H_4 isomer populations continued to fall dramatically as the reaction continued to form the C_3H_3 product. In the 10 Torr case (Fig. 2(b)) the populations deviate slightly from the no CET and 1 Torr cases from around 10^{-8} s to around 10^{-5} s of simulation time, at which point the populations diverge.

It is clear from Fig. 2 that the population that remains in the C_3H_4 isomers is captured by CET at intermediate times (around 10^{-8} to 10^{-5} s) *en route* to forming C_3H_3. As one would expect from such a process the amount of population caught in such a way increases with the pressure of the system. The branching ratios between the various products of the $^1CH_2 + C_2H_2$ reaction are therefore reasonably strongly pressure dependent. This pressure dependence is shown in Fig. 3.

The long time channel ratios, being dependent on the rate of CET, are also very sensitive to the value of α used in the CET kernel, eqn. (8). At 300 K and 1 Torr the long time populations of the C_3H_4 isomers showed a near exponential increase with increasing α, increasing by around four orders of magnitude when α was increased from 2.27 to 4.78 kJ mol^{-1} (190 to 400 cm^{-1}). However as the populations were still of the order of 10^{-4} or less there was very little impact on the final population of propargyl. At higher pressures the population dependence of the C_3H_4 isomers and propargyl is considerably altered. At 1000 Torr the dependence on α was much weaker for the C_3H_4 isomers, a two-fold increase in α producing at most a 40% increase in C_3H_4 (20% for the higher concentration isomers), as shown in Fig. 4. However as the C_3H_4 isomers accounted for nearly 75% of the long time population at the weaker CET this increase in stabilisation reduced the long time C_3H_3 population by 60%.

The population profile of the purely reactive (as opposed to CET driven) component is dependent on the excess population of the input channel reaction partner (C_2H_2 in this case). Most of the calculations reported here used an excess C_2H_2 population a factor of 10^{15} greater than the $^1CH_2$ population. A lower excess of C_2H_2 leads to lower steady state concentrations at intermediate times. The long time population ratios remain independent of the excess population.

2 Modelling at 1200 K

Fig. 5 shows the population profiles calculated for 1200 K at pressures of 1, 100 and 10^4 Torr. It is worth noting that the pressures for which data are shown in Fig. 2(b) and 5(b) are different, and have been chosen to highlight salient features. The behaviour of the populations shown in Fig. 5 is largely similar to the behaviour shown in Fig. 2 for the system at 300 K. There is a purely reactive profile of $^1CH_2$ reacting with the excess C_2H_2 to form intermediate C_3H_4 before dissociating to form C_3H_3. At medium times the C_3H_4 isomers achieve a steady state with the rate of gain

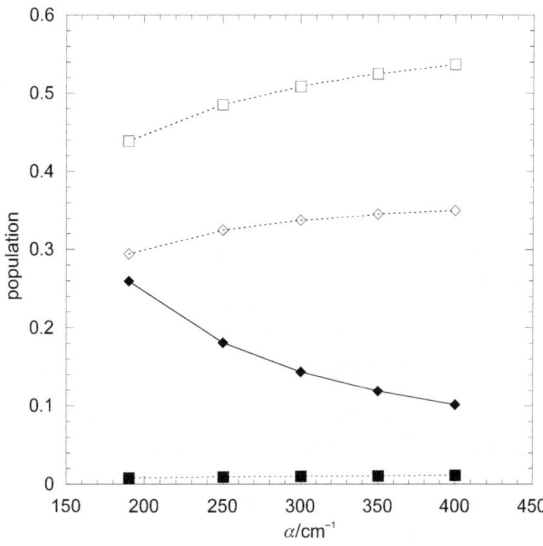

Fig. 4 Long time populations (time of order 1 s) of C_3H_3 (–◆–) and C_3H_4 isomers allene (··◇··), cyclopropene (··■··) and propyne (··□··) as a function of CET parameter α (approximately $\langle \Delta E \rangle_{down}$) at 300 K and 1000 Torr.

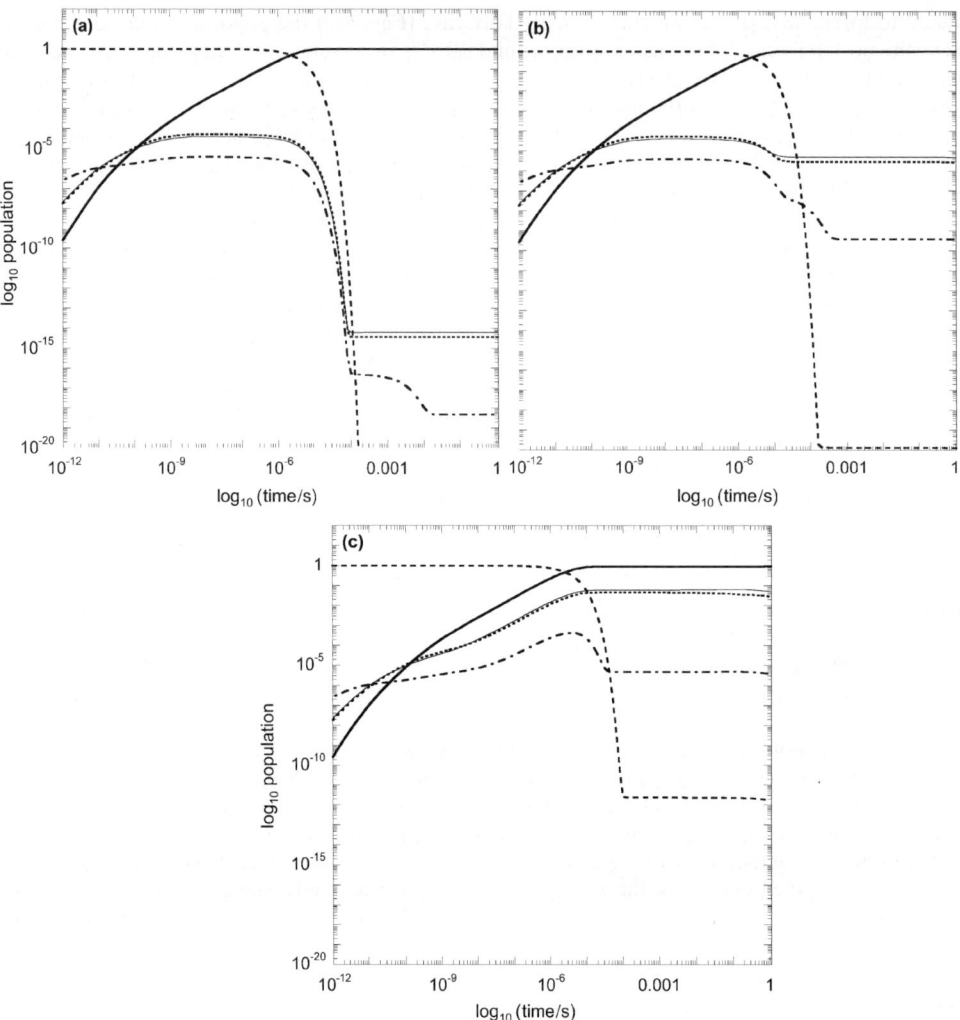

Fig. 5 Time-dependent population profiles for $^1CH_2$ (---), C_3H_4 isomers allene (···), cyclopropene (-·-·) and propyne (——, fine) and C_3H_3 (——, thick) at 1200 K and 1 Torr (a), 100 Torr (b) and 10^4 Torr (c). Calculated with 50 digit arithmetic.

through reaction of $^1CH_2$ equalling the rate of loss to C_3H_3. Once the $^1CH_2$ supply becomes depleted (at times in the vicinity of 10^{-5} s) the available energised population in the isomers continued to form C_3H_3 via isomerisation to propyne.

Population remaining as C_3H_4 after the incoming flux has dropped off due to depletion of $^1CH_2$ at long times is again that which has been stabilised by CET. Two features of this longer time population are evident in Fig. 5 that were not evident in Fig. 2. The first is that for a pressure of 10^4 Torr (Fig. 5(c)) the population of the $^1CH_2$ reactant reaches a steady state at long times. This is actually present in the 300 K case but the steady state $^1CH_2$ population is much lower than the lower range of the plots in Fig. 2. It is readily explained by the reverse reaction from the C_3H_4 isomer cyclopropene to $^1CH_2 + C_2H_2$. At higher temperatures the rate of this reverse reaction increases, as expected from a unimolecular reaction, leading to a higher equilibrium concentration. This 'equilibrium' being a steady state relies on the unimolecular decay of the C_3H_4 isomers to $C_3H_3 + H$ being very slow so that the cyclopropene population is essentially constant at these times.

The second feature present in Fig. 5 that is not present in Fig. 2 lies in the population profile of the cyclopropene isomer. In Fig. 5(a) in particular two plateaux are evident in the long time population of cyclopropene following the steep decline associated with depletion of $^1CH_2$. This is indicative of the isomer populations not being in thermal equilibrium once the influx of population from $^1CH_2 + C_2H_2$ has ceased. This is evident from examining either the rate constant as a function of time of the isomer interconversion processes (the rates of which are given by appropriate sums over the last two terms on the right of eqn. (3)) or the energy resolved population profiles at times within the two plateaux. The populations at times within the shorter time plateau are clearly depleted at high energy in cyclopropene and augmented at high energy in allene and propyne. Cyclopropene undergoes slow unimolecular reaction to allene and propyne once the cyclopropene population ceases to be artificially inflated by incoming flux from $^1CH_2 + C_2H_2$. This redistribution process still occurs at lower temperatures but at a much lower rate due to the lower temperature—the decay of the population to the long time steady state occurring at times greater than that shown in Fig. 2. The increase in the population of the allene and propyne due to the slow redistribution of population from cyclopropene is minimal due to the smallness of the cyclopropene population compared to that of allene and propyne. The time-frame of the equilibration of the isomer populations is clearly pressure dependent—depending as it does on CET to bring the cyclopropene population into equilibrium. At elevated pressures, such as in Fig. 5(c), the equilibration of the isomer populations occurs fast enough for the change in cyclopropene population due to redistribution to be competing effectively with the collisional stabilisation of cyclopropene. Hence the cyclopropene population at 1200 K and 10^4 Torr falls to an equilibrium steady state without reaching an initial non-equilibrium steady state, indicated by the single hump in the population at times 10^{-6}–10^{-4} s in Fig. 5(c).

The pressure dependence of the long time populations of the C_3H_4 isomers is qualitatively similar to that shown in Fig. 3 for 300 K. At the higher temperature the population remaining as C_3H_4 at long time was greatly reduced, as indicated by Fig. 5. The long time residual $^1CH_2$ population, still very small despite being orders of magnitude larger than at 300 K, exhibited a similar qualitative pressure dependence to the C_3H_4 isomers. The smaller residual C_3H_4 population at 1200 K was reflected in the much weaker pressure dependence of the final C_3H_3 population, reduced from unity by a mere 0.4% at 1000 Torr and 10% at 10^4 Torr.

At 1200 K and 1 Torr the dependence of the final C_3H_4 population on the value for α used in the CET model is stronger than in the 300 K, 1 Torr case. However the impact on the production of C_3H_3 is negligible due to the smaller population of C_3H_4 isomers. At 1000 Torr a similar trend is observed with an increased sensitivity to α for the C_3H_4 populations: doubling α increases the C_3H_4 populations by around a factor of 10. The lower C_3H_4 population means that such an increase yields a decrease of only 3% of the C_3H_3 formation at long times.

3 Modelling at 2000 K

Fig. 6 shows the population profiles calculated for 2000 K at pressures of 1, 100 and 1000 Torr. Once more these pressures are different from those selected for Fig. 2 and 5. All the features highlighted in Fig. 2 and 5 are present, such as a purely reactive, pressure-independent short time population profile and C_3H_4 isomer population steady state, CET-driven and hence pressure-dependent capture of C_3H_4 population, and slow isomeric redistribution of cyclopropene (visible in Fig. 6(a) between 10^{-4} and 10^{-3} s). The major new feature shown in Fig. 6 is the decay of the C_3H_4 isomer populations for long time. Similarly to the cyclopropene population redistribution, this does occur at the lower temperatures but at much longer times than shown in Fig. 2 and 5. The C_3H_4 population decay is explained simply by unimolecular decay of the propyne isomer to $C_3H_3 + H$ and maintenance of the equilibrium between the three C_3H_4 isomers and $^1CH_2$ leading to an overall slow decrease in C_3H_4 population.

The long time decay of propyne to $C_3H_3 + H$ is both temperature- and pressure-dependent, as one would expect from a weakly chemically activated unimolecular decay. At 1 Torr (Fig. 6(a)) this decay is slow enough to readily define a decaying equilibrium of reacting species around 10^{-3} s of simulation time. Increasing the pressure to 100 Torr (Fig. 6(b)) increases the rate of the decay so that it is fast enough for the populations never to stabilise once the population from the purely reactive processes begins to fall, though a definite approach to equilibrium is evident by the

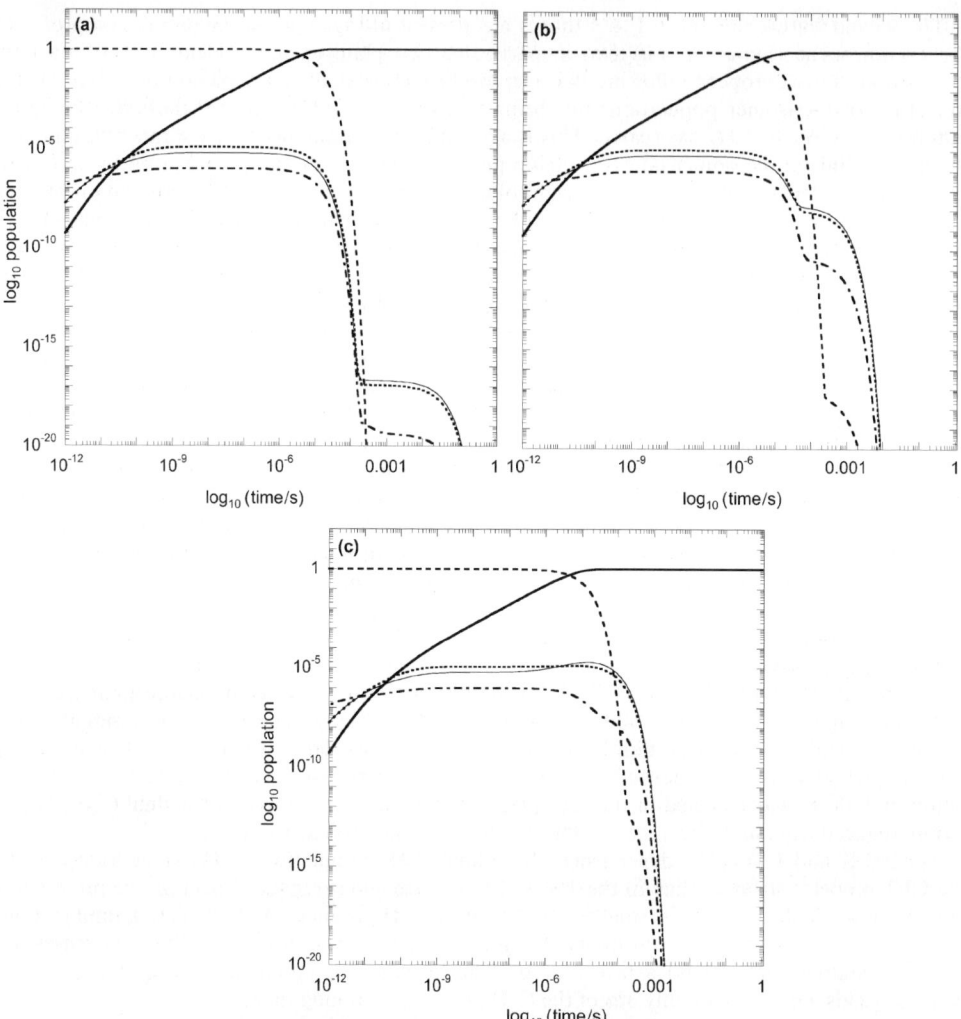

Fig. 6 Time-dependent population profiles for $^1CH_2$ (---), C_3H_4 isomers allene (···), cyclopropene (–·–·) and propyne (——, fine) and C_3H_3 (——, thick) at 2000 K and 1 Torr (a), 100 Torr (b) and 1000 Torr (c). Calculated with 50 digit arithmetic.

sudden decrease in the rates of depopulation at times of the order of 10^{-4} s. Further increasing the pressure to 1000 Torr (Fig. 6(c)) increases the rate of the decay of the isomers to such a degree that no estimate of the population of the C_3H_4 isomers by collisional stabilisation can be made. Once the purely reactive steady state populations are perturbed by stabilised C_3H_4 population the $^1CH_2$ and C_3H_4 populations never again approach a steady state.

IV Discussion and conclusion

The microscopic rate constants determined by Blitz et al.[9] for the $^1CH_2 + C_2H_2 \rightleftharpoons C_3H_4$ reaction were calculated by inverse Laplace transform (ILT) of measured temperature-dependent rates of disappearance of $^1CH_2$. Implicit in this approach is the assumption that the measured rate (after other mechanisms for disappearance of $^1CH_2$ have been accounted for) is indicative of the forward rate only, with no significant repopulation due to the reverse reaction. The ability to model the reaction with the microscopic rate constants so derived at the temperatures the experiments were conducted at means that the assumption of negligible back reaction can be tested explicitly. Blitz et al. measured the rate of decay of $^1CH_2$ over the temperature range 205–773 K at pressures around 1 Torr. In this work calculations were performed at 300, 600 and 900 K. The rate of back reaction from cyclopropene was calculated (as an appropriate sum over the last term on the right of eqn. (3)) and compared with the total rate of disappearance of $^1CH_2$. At 300 K the total rate of disappearance exceeded the rate of repopulation by a factor of 6×10^4, indicating that the assumption of no back reaction is very good at this temperature. At 600 and 900 K the rates of disappearance exceeded the rates of back reaction by factors of 4000 and 800 respectively. Whilst the back reaction becomes more significant at higher temperatures, at the temperatures measured by Blitz et al. the negligible back reaction assumption appears to hold.

The microscopic rate constants for the propyne $\rightleftharpoons C_3H_3 + H$ reaction were determined by Blitz et al.[9] in a similar manner to the $^1CH_2 + C_2H_2$ reaction. In the absence of appropriate experimental data the rate constant was assumed to be the same as the rate constant for the reaction $C_3H_5 + H$. The room temperature value was taken from Hanning-Lee and Pilling[27] and was assumed to be independent of temperature. The $C_3H_3 + H$ reaction was modelled with a conservative, symmetrized matrix and the forward and reverse fluxes compared in a similar manner to the $^1CH_2 + C_2H_2$ reaction described above. At 100 and 1000 Torr (spanning Hanning-Lee and Pilling's experiment which covered 98–400 Torr) the total rate of disappearance of C_3H_3 exceeded the rate of the propyne decomposition by factors of 1000 and 8500 respectively. Hence the ILT-derived microscopic rate constants for both the $^1CH_2 + C_2H_2$ and $C_3H_3 + H$ channels are self-consistent.

The discussion in the previous section indicates that in this reaction there are essentially four competing processes:

1. The shortest timescale process is the CET-, and hence pressure-independent, purely reactive conversion from $^1CH_2$ to energised C_3H_4, isomerisation and decomposition to form C_3H_3. After an initial very short induction period during which the populations of the C_3H_4 isomers build up, there is a steady state with the conversion efficiency of $^1CH_2$ to C_3H_3 being 100%. When the population of $^1CH_2$ drops the remaining C_3H_4 population passes through propyne to form C_3H_3 irreversibly.

2. The collisional stabilization of energetic C_3H_4 formed in process 1. This is the mechanism primarily influencing the channel yield ratio for the production of C_3H_4 or C_3H_3. This mechanism typically becomes significant during the steady state phase of the purely reactive evolution, though is likely to become important earlier at higher pressures, particularly at low temperatures.

3. The thermal equilibration of the C_3H_4 population formed by process 2 (and indeed any remaining $^1CH_2$).

4. The slow decomposition of stabilized C_3H_4 population to $C_3H_3 + H$.

If τ_1 through τ_4 are the characteristic timescales of the processes 1 through 4 above, then typically $\tau_1 < \tau_2 < \tau_3 < \tau_4$. All of these timescales bar τ_1 are both temperature- and pressure-dependent, becoming shorter at higher temperatures and pressures. At high temperatures and pressures $\tau_2 \sim \tau_3 \sim \tau_4$, meaning that long-lasting populations of C_3H_4 are not formed, as noted by Blitz et al.[9] While channel fractions cannot be given in these conditions it is worth noting that at these long times the $^1CH_2$ population becomes depleted under pulsed conditions. Thermalised C_3H_4 isomers irreversibly decay to C_3H_3 (via isomerisation to propyne) with well-defined time-independent rate constants.

Throughout the calculations reported in this work 50 digit arithmetic was used. This was not strictly required for the higher temperatures, but was used for consistency. The high precision approach is very much a "brute force" approach to the problem, quite contrary to the sophisticated approach taken in Frankcombe et al.[11] While this extended precision approach has not been applied to this type of problem before it is a natural extension of the quadruple precision approach taken by Blitz et al.[9] While accessing low temperature dynamics has proven in the past to be impossible with normal precision levels for the current system,[9,11] others have had success with systems at first glance similar to this. De Avillez Pereira et al.[24] achieved success for the reaction of methyl with OH, which proceeds through a single deep well and then on to multiple bimolecular product channels. They used steady state methods to access the long time behaviour by incorporating an absorbing boundary placed some way below the stabilization threshold for

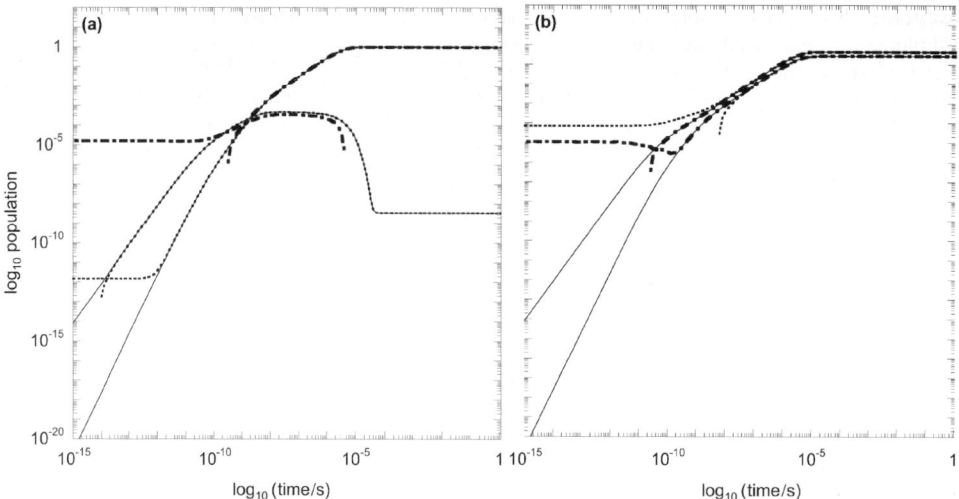

Fig. 7 Time-dependent population profiles for propyne and C_3H_3 at 300 K and 1 Torr (a) and 1000 Torr (b) calculated in 100 digit arithmetic (——) and in quadruple precision using the routines "rs" from EISPACK (· · ·) and "dsyev" from LAPACK (– · –).

methanol. The rationale here is that, at suitably low temperatures, the process of stabilization is essentially irreversible because of the negligibly small rate constant for subsequent unimolecular decay of the thermalized molecules. Miller et al.[23] achieved apparently accurate results for long time populations using a similar spectral method to this work. The $C_2H_3 + C_2H_2$ reaction which they studied moves through a system of isomerising wells sufficiently shallow (ca. 200 kJ mol^{-1}) to allow solution using standard eigensolvers, even at temperatures as low as 300 K. As is demonstrated by the discussions above, the title reaction of this study exhibits both irreversible and reversible stabilization, depending on the temperature. Hence, one could potentially use the steady state approach of De Avillez Pereira et al. at low temperatures and the direct approach of Miller et al. at high temperatures. The power of the approach we have demonstrated in this work is that it handles both limits and the transition in between rigorously.

The ME for the present system was solved with standard eigensolvers in quadruple precision in parallel with the high precision calculations. The quadruple precision eigendecomposition was calculated with both the LAPACK routine "dsyev" and the older EISPACK routine "rs". There was a significant difference between the results calculated with the two routines, with the older rs routine being faster and often more robust. Population profiles calculated using the 50 digit methods and the two quadruple precision methods are shown in Fig. 7 for the 300 K case at 1 and 1000 Torr. In general the results agree reasonably well for reasonable times. However the quadruple precision results suffer from reliability problems. A good example of this is the propyne population calculated with the quadruple precision dsyev routine, shown in Fig. 7(a). At no stage does the method yield a population for this isomer in accord with the more accurate 50 digit result, and indeed produces a negative population for long times. Not shown in Fig. 7(a) is the population profile calculated for allene, which shows similar disagreement with the accurate high precision population but which does not become negative. While that population profile may appear to agree qualitatively with the observed trends for the population with respect to temperature and pressure, the long time population is in error by several orders of magnitude. Such an error would not be easy to detect without reference calculations at higher precision.

To date the experimental measurements of the rate of reaction of the present system have focused on the disappearance of $^1CH_2$. This input channel is not sensitive to the kinetics of the remainder of the C_3H_3 formation path. New experimental investigations are required to study the other side of the reaction, the formation of C_3H_3, in order to verify this modelling and the predicted strong pressure dependence of the long time C_3H_3 (and consequently hydrogen atom) population at low temperatures.

Acknowledgements

We thank Dr Struan Robertson and Dr Kevin Gates for ongoing communication in relation to this work. The high precision code used in this work was based on code originally provided by Drs Robertson and Gates. Funding by the Australian Research Council Large Grant Scheme is also gratefully acknowledged (Grant number A10027155). TJF supported financially by an Australian Postgraduate Award.

References

1 J. A. Miller and C. F. Melius, *Combust. Flame*, 1992, **91**, 21.
2 J. D. Adamson, C. L. Morter, J. D. DeSain, G. P. Glass and R. F. Curl, *J. Chem. Phys.*, 1996, **100**, 2125.
3 H. Wang and M. Frenklach, *Combust. Flame*, 1997, **110**, 173.
4 H. Böhm, K. Kohse-Höinghaus, F. Lacas, C. Rolon, N. Darabiha and S. Candel, *Combust. Flame*, 2001, **124**, 127.
5 C. H. Wu and R. D. Kern, *J. Phys. Chem.*, 1987, **91**, 6291.
6 R. D. Kern, H. J. Singh and C. H. Wu, *Int. J. Chem. Kinet.*, 1988, **20**, 731.
7 U. Alkemade and K. H. Homann, *Z. Phys. Chem.* (*Neue Folge*), 1989, **161**, 19.
8 R. D. Kern, H. Chen, J. H. Kiefer and P. S. Mudipalli, *Combust. Flame*, 1995, **100**, 177.
9 M. Blitz, M. S. Beasley, M. J. Pilling and S. H. Robertson, *Phys. Chem. Chem. Phys.*, 2000, **2**, 805.
10 K. E. Gates, S. H. Robertson, S. C. Smith, M. J. Pilling, M. S. Beasley and K. J. Maschhoff, *J. Phys. Chem. A*, 1997, **101**, 5765.
11 T. J. Frankcombe, S. C. Smith, K. E. Gates and S. H. Robertson, *Phys. Chem. Chem. Phys.*, 2000, **2**, 793.
12 M. Karni, I. Oref, S. Barzilai-Gilboa and A. Lifshitz, *J. Phys. Chem.*, 1988, **92**, 6924.
13 M. Yoshimine, J. Pacansky and N. Honjou, *J. Am. Chem. Soc.*, 1989, **111**, 4198.
14 R. Kakkar and B. S. Padhi, *Indian J. Chem. B*, 1999, **38**, 1262.
15 T. J. Frankcombe and S. C. Smith, *Comput. Phys. Commun.*, in press.
16 R. G. Gilbert and S. C. Smith, *Theory of Unimolecular and Recombination Reactions*, Blackwell Scientific, Oxford, 1990.
17 S. Nordholm and H. W. Schranz, in *Advances in Chemical Kinetics and Dynamics*, ed. J. R. Barker, JAI, Connecticut, 1995, vol. 2A, p. 245.
18 K. A. Holbrook, M. J. Pilling and S. H. Robertson, *Unimolecular Reactions*, Wiley, Chichester, 2nd edn., 1996.
19 N. Snider, *J. Chem. Phys.*, 1964, **42**, 548.
20 M. Quack, *Ber. Bunsen-Ges. Phys. Chem.*, 1984, **88**, 94.
21 N. J. B. Green, P. J. Marchant, M. J. Perona, M. J. Pilling and S. H. Robertson, *J. Chem. Phys.*, 1992, **96**, 5896.
22 P. K. Venkatesh, A. M. Dean, M. H. Cohen and R. W. Carr, *J. Chem. Phys.*, 1997, **107**, 8904.
23 J. A. Miller, S. J. Klippenstein and S. H. Robertson, *J. Phys. Chem. A*, 2000, **104**, 9806.
24 R. De Avillez Pereira, D. L. Baulch, M. J. Pilling, S. H. Robertson and G. Zeng, *J. Phys. Chem. A*, 1997, **101**, 9681.
25 W. Tsang, V. Bedanov and M. R. Zachariah, *Ber. Bunsen-Ges. Phys. Chem.*, 1997, **101**, 491.
26 D. H. Bailey, *ACM Trans. Math. Software*, 1995, **21**, 379.
27 M. A. Hanning-Lee and M. J. Pilling, *Int. J. Chem. Kinet.*, 1992, **24**, 271.

Use of quantum methods for a consistent approach to combustion modelling: Hydrocarbon bond dissociation energies

Juan P. Senosiain,[a] Joseph H. Han,[b] Charles B. Musgrave[ab] and David M. Golden†[c]

[a] Department of Materials Science & Engineering, Stanford University, CA 94305, USA
[b] Department of Chemical Engineering, Stanford University, CA 94305, USA
[c] Department of Mechanical Engineering, Stanford University, CA 94305, USA.
E-mail: golden@navier.stanford.edu

Received 3rd April 2001
First published as an Advance Article on the web 6th November 2001

An attempt has been made to use modern quantum methods to codify the data base concerning bond dissociation energies in hydrocarbons. Calculations have been performed using two hybrid DFT methods, the well-known B3LYP formalism and a newly developed alternative named KMLYP. CBS-Q has also been employed where possible. The combination of experimental uncertainty and theoretical limitations is less than completely satisfactory. However, within uncertainties that translate to a factor of two at 1500 K, many transferable quantities are elucidated. A hybrid method has been developed for the correction of DFT calculations using group additivity. Given that the philosophy behind this work is the understanding that all data bases must be optimised for specific applications, so that avoidance of large errors is more important than absolute precision, the results appear to be quite useful. We are particularly encouraged by the performance of the KMLYP method, given its ease of application to molecules of practical interest.

Introduction

Understanding of any complex process, such as combustion or atmospheric chemistry, requires a model. The full understanding of combustion means a model that couples transport phenomena and chemistry. The chemical part of the model is represented by differential equations relating to "elementary" chemical processes parametrised by rate constants dependent on the temperature, pressure and nature of the bath gas. The model must be able to access rate constants in both the possible directions for the chemistry. These forward and reverse rate constants are related *via* the equilibrium constant and thus the thermochemistry of each step. Large numbers of the chemical species are reactive intermediates such as free radicals, whose thermochemistry has been the subject of many years of experiment and theory. Many of these properties are archivally tabulated.[1–7] Unfortunately, these tabulations are often not in agreement with each other. In addition, they often take values from disparate sources.

The general philosophy that we bring to this effort is embodied in the study and natural gas combustion mechanism known as GRI-Mech.[8] The essence of this approach requires the estab-

† Also associated with Molecular Physics Laboratory, SRI International, Menlo Park, CA 94025, USA.

lishment of a chemical mechanism and the assignment of rate constants to the steps of that mechanism. Sensitivity analysis is then applied to establish on which particular sets of individual rate constants (or thermochemical data) a specific experiment depends. Hopefully sufficient numbers of experiments can be found, or performed, to represent an important segment of the mechanism as a whole. At this point the formalism of statistical design of experiments is used to optimise the rate constants (or thermochemical data), within their reported error bounds, creating the final mechanism. Hopefully there will be other experiments, not used in the optimisation, that can be used as validation tests.

The above procedure requires that the original rate data (and thermochemical values) be sufficiently accurate that the optimisation does not explore false minima. *Thus, a self-consistent reasonably accurate data set is a more important goal than precise values for only a small subset of these data.*

With this philosophy as the driver, and given the extensive reliance on modern quantum methods in the recent literature, we have examined through comparison of experiments with the use of some methods that can be applied to reasonably sized species, the heats of formation (bond dissociation energies) of a large base of hydrocarbon radicals up to seven carbons. We point out inconsistencies in some accepted values as well as transportable quantities that can be used in extrapolation to other species. Interestingly we find both some discrepancies between theory and experiment and that some widely held beliefs about relative bond dissociation energies may be misplaced.

There are already many theoretical studies that examine thermochemical properties of free radicals. Typically, these address a few molecules per study, often using different quantum methods. Here the attempt is to consolidate a large amount of experimental information while developing a self-consistent database. This is in keeping with the emphasis above on self-consistent data sets that will lend themselves to optimisation.

For practical engineering applications an improvement of group additivity values for free radicals[9] would be a welcome advance. Although group contributions may be computed from the results of application of quantum methods, the DFT methods are themselves too inaccurate and the CBS-Q method is computationally intensive. We present a hybrid method that bridges the gap.

Experimental data

There are several well-known compilations of experimental data that relate to heats of formation of hydrocarbon radicals (bond dissociation energies). Experimental methods were reviewed by one of us several years ago[5] and by Berkowitz *et al.* more recently.[6] These remain, albeit in improved versions, the mainstays of the experimental techniques. A problem is that as good as experimental techniques may be, there are variations among them. Furthermore, there is a combination of older and newer data cited in most compilations. The newer data often replace values obtained earlier. The most obvious example is the replacement of values obtained with halogenation techniques[10–16] with values some 4 to 15 kJ mol^{-1} higher.

In most halogenation studies, the Arrhenius parameters for the reaction $X + RH \rightarrow R + HX$ (where X is a halogen and R an alkyl radical), are well-determined at temperatures near 500 K. In these studies it was argued that the activation energy for the reverse reaction is 4.2 ± 4.2 kJ mol^{-1} for X = I and perhaps somewhat higher for X = Br. (Most of the halogenation studies were iodination studies). The current replacement of halogenation values is due to the consensus that reactions of radicals with HI and HBr, actually have negative activation energies of about the same magnitude.[17,18] However, if current compilations are examined (*e.g.* the CRC Handbook[19]), one finds that values of some of the smaller hydrocarbon radicals have been raised, whereas values for some of the larger radicals, whose values remain unchanged, are taken from references that relied on the halogenation technique. In addition, corrections are not so straightforward as there are instances where halogenation experiments give the same result as those determined by other methods more widely believed to be correct.

Attempts, using theoretical methods, to show the origin of the negative activation energies for radical reactions with HBr and HI, have been at best marginally successful. For example, the potential energy surface for $Br + C_2H_6$ computed at QCISD/6-311G(d,p) shows an adduct lying about 6 kJ mol^{-1} below the transition state. However when the basis set is expanded to 6-

311+G(3df,2p) the transition state disappears.[20] Clearly the large number of electrons and effects such as spin–orbit coupling make it difficult to compute a potential energy surface for halogenation reactions of interest with resolutions better than 4.2 kJ mol^{-1}. In this study, we address the bond dissociation energies and radical heats of formation directly in an attempt to examine the hydrocarbon data base consistently.

Calculation methods

Density functional (DFT) theory and a complete basis set (CBS) compound method (CBS-Q)[21] were employed to calculate bond dissociation energies of a series of hydrocarbons.

The CBS-Q energy is calculated from the results of several computations. First, the geometry is optimised at the second-order Møller–Plesset perturbation theory (MP2) with the 6-31G(d) basis set. Then a frequency calculation is done at this geometry at a Hartree–Fock/6-31G* level, and scaled by a factor of 0.918 44 to obtain the zero-point energy and thermal corrections. A base SCF energy and an MP2 energy are obtained with the 6-311+G(3d2f,2df,2p) basis set. Extrapolation to the CBS limit is done by means of an expansion in pair natural orbitals. Higher order corrections are done at the MP4(SDQ)/6-31+G(d(f),d, p) and QCISD(T)/6-31+G* levels of theory. Additional corrections are added to account for core correlation, spin contamination and size consistency.

Two distinct hybrid DFT methods were used: Becke's 3-parameter exchange-correlation functional (B3LYP[22]) and a new hybrid DFT method.[23] Geometries and harmonic frequencies for the set of hydrocarbons and their radicals were calculated using unrestricted (U)B3LYP with Pople's 6-311G(2d,2p) basis set. Zero-point energies and thermal corrections to 298.15 K were done using unscaled harmonic frequencies at this basis set and level of theory. Subsequent single point calculations were done using a larger basis set: 6-311 + (3df,2p) with B3LYP and the new method.

New DFT method

This method (referred to as "KMLYP" throughout this text) uses a combination of exchange and correlation functionals to minimise the errors for the hydrogen and fluorine atoms, while preserving the correct asymptotic values. The exchange functional is 44.3% Slater exchange and 55.7% exact (Hartree–Fock) exchange. It also uses a correlation functional consisting of 44.8% Lee–Yang–Parr and 55.2% Vosko–Wilk–Nusair correlation functionals.

This method can be specified in the GAUSSIAN98[24] program using the following keywords in the route section:

\# UBLYP/6-311+G(3df,2p)
\# iop(5/45 = 10 000 557) iop(5/46 = 00 000 443) iop(5/47 = 04 481 000.)

Additional corrections due to spin contamination and lone pair correlation were added in the following way:

$$E_{KMLYP} = E'_{KMLYP} + A(N_\alpha - N_\beta) + BN_{lp}$$

Where E_{KMLYP} and E'_{KMLYP} are the corrected and uncorrected KMLYP energies, respectively, N_α, N_β the number of α and β electrons, N_{lp} the number of lone electron pairs or π-bonds, and A and B are empirical constants determined by minimizing the error in reproducing the G2 molecular set. The former is -0.00258 E_h and the latter -0.01053 and -0.01231 E_h for atoms and molecules respectively. This correction scheme is similar to that used by the CBS and Gaussian compound methods.[21,25]

Bond dissociation energies were calculated directly as:

$$D^0_{298}(H-R) = \Delta_f H°(R) + \Delta_f H°(H) - \Delta_f H°(RH)$$

Unless otherwise noted, all energies include zero-point corrections and thermal corrections to 298.15 K.

C–H Bond dissociation energies (BDEs)

Homolytic bond dissociations are a challenge for theoretical methods. These involve a change in the net number of paired electrons, thus the electron correlation energy (~ 1 eV per electron pair) has to be accounted for appropriately. We have studied the C–H bond dissociations of the hydrocarbon species shown in Table 1 using the quantum chemistry methods mentioned above. We are aware of other studies using higher-level methods (for example, see ref. 26–30). *Our purpose is to cover a large subset of species where we would not always be able to make isodesmic corrections, and form a consistent body of bond dissociation energies.*

The accuracy and reliability of CBS-Q for calculating the heats of formation of small molecules has been tested by Petersson et al.,[31] who showed it to be at least as good as G2's and superior to

Table 1 Calculated and experimental bond dissociation energies (D_{298}) and their deviations

	D_{298}/kJ mol^{-1}				Deviations from experiment/kJ mol^{-1}		
	B3LYP	KMLYP	CBS-Q	Experiment	B3LYP	KMLYP	CBS-Q
C1							
H–CH$_3$	421.1	427.4	439.5	438.9[1]	−17.8	−11.5	0.6
C2							
H–C$_2$H$_5$	411.6	419.3	425.5	420.7[4]	−9.1	−1.4	4.9
H–CH=CH$_2$	465.0	476.6	462.0	464.7[4]	0.3	11.9	−2.6
H–C≡CH	563.3	572.9	557.6	556.1[4]	7.2	16.8	1.5
C3							
H–C$_3$H$_7$	413.0	424.0	426.9	422.7[4]	−9.7	1.3	4.2
H–CH(CH$_3$)$_2$	394.4	405.4	414.9	416.8[4]	−22.4	−11.4	−1.9
H–CH$_2$CH=CH$_2$	350.9	359.5	361.3	361.9[6]	−11.0	−2.4	−0.6
H–CH=CH$_2$CH$_3$	458.2	468.9	471.7				
H–C(CH$_3$)=CH$_2$	436.3	447.7	452.1				
H–CH$_2$C≡CH	364.1	377.2	380.1	339.0[4]	−9.9	3.2	6.1
H–C≡CCH3	562.2	570.2	566.1				
H–[CHCH$_2$CH$_2$][a]	441.3	452.6	458.5	444.8[5]	−3.5	7.8	13.7
H–[CHCH=CH]	406.5	413.8	421.3	379.1[5]	27.4	34.7	42.2
C4							
H–C$_4$H$_9$	409.8	418.0	427.6	425.4[51]	−15.6	−7.4	2.2
H–CH(CH$_3$)C$_2$H$_5$	393.9	404.0	413.0	411.2[51]	−17.3	−7.2	1.8
H–C(CH$_3$)$_3$	383.9	396.4	409.1	404.3[51]	−20.4	−7.9	4.8
H–CH$_2$CH(CH$_3$)$_2$	413.9	423.0	427.6	425.2[51]	−11.3	−2.2	2.4
H–CH=CHC$_2$H$_5$	413.8	422.2	471.2				
H–CH$_2$CH$_2$CH=CH$_2$	415.4	424.6	427.9				
H–CH(CH$_3$)CH=CH$_2$	333.9	343.3	347.9				
H–CH$_2$CH=CHCH$_3$	348.1	357.3	360.5	358.2[5]	−10.1	−0.9	2.3
H–C(=CH$_2$)CH$_2$CH$_3$	447.3	459.4					
H–C(=CH$_2$)CH=CH$_2$	402.1	414.2	408.8				
H–CH=CHCH=CH$_2$	462.5	471.7	468.3				
H–CH$_2$C(CH$_3$)=CH$_2$	357.8	367.2	364.5	358.2[52]	−0.4	9.0	6.3
H–CH$_2$C=CCH$_3$	359.9	373.7	355.1	364.8[5]	−4.9	8.9	−9.7
H–CH$_2$CH$_2$C≡CH	421.1	429.7	427.5				
H–[CHCH(CH$_3$)CH$_2$]	425.2	439.0					
H–CH$_2$–[CHCH$_2$CH$_2$]	399.1	411.3	413.8	407.5[13]	−8.4	3.8	6.3
H–[CHCH$_2$CH$_2$CH$_2$]	401.3	411.3	419.0	403.8[10,44]	−2.5	7.5	15.2
H–[CHCH=CHCH$_2$]	366.2	373.8	376.2				
H–[C=CHCH$_2$CH$_2$]	459.0	472.7	467.5				
H–[C=CHCH=CH]	460.5	472.8	474.0				
C5							
H–C$_5$H$_{11}$	413.9	415.9	428.3				
H–CH(CH$_3$)C$_3$H$_7$	396.6	398.9					
H–CH(C$_2$H$_5$)$_2$	394.6	404.9					
H–C(CH$_3$)$_2$C$_2$H$_5$	386.6	400.1		404.0[53,54]	−17.4	−3.9	
H–C(=CH$_2$)CH$_2$CH$_2$CH$_3$	439.0	450.8					
H–C(C$_2$H$_5$)CH=CH$_2$	338.2	348.6					
H–C(=CH$_2$)CH=CHCH$_3$	383.0	402.9					
H–CH$_2$CH=CHCH=CH$_2$	325.6	334.5	333.5	347.0[40,41,55]	−21.4	−12.5	−13.5
H–CH(CH=CH$_2$)$_2$	289.9	296.2		318.0[40,41,55]	−28.1	−21.7	
H–CH$_2$C(CH$_3$)$_3$	415.8	425.9		418.0[56]	−2.2	7.9	
H–CH(CH$_3$)C≡CCH$_3$	344.0	358.2		365.3[5]	−21.3	−7.1	
H–C(CH$_3$)$_2$C≡CH	335.4	350.2		338.9[5]	−3.5	11.3	
H–C(CH$_3$)$_2$CH=CH$_2$	324.9	334.2		323.0[5]	1.9	11.2	
H–[CHCH$_2$CH$_2$CH$_2$CH$_2$]	386.9	395.1	406.5	403.5[57]	−19.6	−11.4	0.0
H–[CHCH=CHCH$_2$CH$_2$]	334.1	336.0	347.1	344.3[5]	−10.2	−8.3	2.8
H–[CHCH$_2$CH=CHCH$_2$]	389.2	389.6	406.4				
H–[CHCH=CHCH=CH]	329.9	344.4		346.7[19]	−16.8	−2.3	

Table 1 Continued

	D_{298}/kJ mol^{-1}				Deviations from experiment/kJ mol^{-1}		
	B3LYP	KMLYP	CBS-Q	Experiment	B3LYP	KMLYP	CBS-Q
C6							
H-Hexa-1,3-diene-5-yl	308.9	318.0					
H-Cyclohexyl	397.9	409.0		399.6[6]	−1.7	9.4	
H-Cyclohexene-1-yl	446.3	459.3					
H-Cyclohexene-3-yl	333.1	342.8					
H-Cyclohexene-4-yl	396.7	407.5					
H-Cyclohexa-1,3-diene-5-yl	296.7	305.5		305.0[5]	−8.3	0.5	
H-Cyclohexa-1,4-diene-3-yl	296.6	308.8		305.4	−8.8	3.4	
H-Phenyl	462.3	474.0		473.1[6]	−10.8	0.9	
C7							
H-Benzyl	361.7	371.8		375.7[6]	−14.0	−3.9	

[a] Cyclic groups shown in brackets.

that of B3LYP/6-311+G(3df,2p). There are however some fairly large deviations from experiment for bond dissociations leading to cyclopropenyl (61 kJ mol^{-1}), cyclobutyl (15 kJ mol^{-1}), cyclopropyl (14 kJ mol^{-1}) and pentadienyl (14 kJ mol^{-1}) radicals. (The experimental values for these species are not beyond question!) In general terms, CBS-Q systematically overpredicts the C–H bond strengths of cyclanes by ~14 kJ mol^{-1}. It is worth noting that the computational cost of this method is much higher than the DFT methods, so its applicability to larger hydrocarbons is quite limited. While the DFT methods usually scale as approximately N^3 (where N is the number of basis functions), the QCISD(T) part of the CBS-Q method scales as N^7.

The errors in the results obtained with B3LYP are too large to be considered chemically accurate. The largest disagreement of the BDEs calculated with this method, compared to experiments, are for those leading to: cyclopropenyl (27 kJ mol^{-1}), pentadienyl‡ (28 kJ mol^{-1} from penta-1,4-diene, 21 kJ mol^{-1} from penta-1,3-diene), isopropyl (22 kJ mol^{-1}), pent-2-yne-4-yl (21 kJ mol^{-1}), tert-butyl (20 kJ mol^{-1}), cyclopentane (20 kJ mol^{-1}), methane (18 kJ mol^{-1}), 2-methylbut-2-yl (17 kJ mol^{-1}), vinyl (17 kJ mol^{-1}) and cyclopentadienyl (17 kJ mol^{-1}). The BDEs of all other species calculated are within 14 kJ mol^{-1} of the cited experimental values. In most cases, B3LYP underestimates the BDEs. This may be due to the tendency of the local density approximation (LDA) to excessively delocalise charge density, resulting in lower energy semi-occupied molecular orbitals (SOMO) and thus favouring radicals over closed-shell species. Although spin contamination is usually not an issue for this method, DiLabio and co-workers[32] reported that better results are obtained by using a restricted open-shell mode for the radicals, (RO)B3LYP, and adjusting the heat of formation of the hydrogen atom. The latter is equivalent to a rigid increase of BDEs by ~6 kJ mol^{-1}, thereby reducing the underbinding effect. It should be noted that the 1σ error-bars for available experimental numbers are about 8 kJ mol^{-1} and can sometimes be as large as 16 kJ mol^{-1}.

The BDEs calculated with the new KMLYP DFT method, present considerably smaller deviations than B3LYP. The accuracy is close to that obtained by CBS-Q, yet it is a significantly less computationally demanding method. With the exception of BDEs leading to cyclopropenyl (35 kJ mol^{-1}), ethynyl (17 kJ mol^{-1}) and pentadienyl (22 and 13 kJ mol^{-1} from penta-1,4-diene and penta-1,3-diene, respectively), all other values calculated with this method deviate from experiment by less than 12 kJ mol^{-1}, roughly a factor of four in a Boltzmann term at 1000 K.

It has been established[33] that energy converges with respect to basis sets faster with DFT methods than with more sophisticated treatments of electron correlation such as QCISD(T) or CCSD(T). Thus, the accuracy of B3LYP is not expected to improve significantly when the basis set is increased. Scaling the frequencies by an empirical correction causes a mild increase in the bond strength due to changes in zero-point energy (ZPE) and heat capacity. For instance, multiplying the B3LYP frequencies by an empirical factor[2]§ reduced the ZPE corrections of n-propyl,

‡ The experimental numbers for this radical are uncertain (vide infra).
§ A factor of 0.9804 was used, as suggested by Wong.[34]

isopropyl, allyl and cyclopropyl bonds by only 0.77, 0.80, 0.70 and 0.75 kJ mol^{-1}, respectively. Thermal corrections were reduced by 0.77, 0.79, 0.70 and 0.70 kJ mol^{-1}, respectively. In fact, if frequency calculations and geometry optimisations are done with the 6-311G(d,p) basis set instead, the differences are minimal and much CPU-time is saved.

Since much of the error involved in quantum calculations is systematic, by studying a series of similar molecules and comparing the results we can suggest if a given experimental measurement should be repeated. Such is the case for the cyclopropenyl radical, where the experimental BDE seems too low (by about 42 kJ mol^{-1}) compared to the three theoretical methods used in this study. In this case the disagreement is larger for the higher level method (CBS-Q), and the magnitude of this is more than twice that of the next largest error. Another theoretical study[35] using the G2 method also predicts higher $\Delta_f H_{298}$ values for the cyclopropenyl radical. To our knowledge, the only experimental number for this bond energy comes from Defrees and co-workers,[36] who used ion-cyclotron double-resonance spectroscopy combined with gas-phase basicity measurements. It is possible that ring-opening during the ionisation led to an erroneously low $\Delta_f H$ of the radical.[37]

Radical stabilisation

The transferability of group contributions in calculating the heats of formation of radicals as well as closed-shell compounds was noted by Benson and co-workers in the 1960s.[9] In this approach, properties such as heat of formation are primarily associated with the heavy atoms and their nearest neighbours. Thus, a C–H bond in a hydrocarbon can be understood in terms of the cardinality of the radical (whether it is a primary, secondary or tertiary bond). Contributions from non-nearest neighbour effects, such as resonance stabilisation, changes in ring strain energy, polarisable groups, steric effects and branching (roughly in order of decreasing importance), can be added.

Cardinality and hyperconjugation

Inspecting the bond energies for (non-stabilized) primary, secondary and tertiary bonds, we find that the experimental values are roughly constant, about 422, 414 and 404 kJ mol^{-1} respectively. The values calculated for these quantities (excluding CH_4) are within 10 kJ mol^{-1} for a given method and bond cardinality, the variations being due primarily to the different polarisabilities of adjacent substituents.

In this observation, methane is singled out as an exception because it lacks a polarisable C–C bond to stabilise the radical, so its BDE is about 12 kJ mol^{-1} stronger than other primary bonds. The increased stabilisation of the larger alkyls can be understood as a hyperconjugation of the SOMO perpendicular to the C–C bond, with the sp^3 hybridised carbon (see ref. 38). It should be noted that the DFT methods have relatively large errors in this bond energy (−17.8 and −11.5 kJ mol^{-1} for B3LYP and KMLYP, respectively) compared to CBS-Q (0.6 kJ mol^{-1}).

Allyl and propargyl-type resonance stabilisation

The presence of an unsaturated bond neighbouring the carbon atom involved in the homolytic bond cleavage enables the radical to delocalise over a large space. We calculated the magnitude of these effects, using the following operational definition of the radical stabilisation energy (RSE)

$$RSE = D^o_{298}(R'-H) - D^o_{298}(R-H)$$

where R is the radical in question and R' the corresponding saturated radical. These results are shown in Table 2. In general, allyl-type stabilisation (i.e. the presence of an α double-bond) in the radical is about 60 ± 5 kJ mol^{-1}, suggesting that this is a transferable quantity within those limits.

Here there is an example of a situation suggesting an error in the experimental value. Accepting ±5 kJ mol^{-1} as generally good agreement, we see that for first three entries in Table 2, H–(CH$_2$CH=CH$_2$), H–(CH$_2$CH=CHCH$_3$) and H–(CH$_2$C(CH$_3$)=CH$_2$), the allyl stabilisation energy given by theory and experiment conforms to the value above. Theory would suggest the same value for all other allyl entries in Table 2. However, the experimental value for H–(C(CH$_3$)$_2$CH=CH$_2$) is 81 kJ mol^{-1} and would seem to be too high. (BDEs can be uncertain by ~13 kJ mol^{-1}, but the differences between these can be uncertain by up to 25 kJ mol^{-1}).

Table 2 Radical stabilisation energies of several unsaturated hydrocarbons[a]

Stabilisation group	Radical stabilisation energy/kJ mol^{-1}			
	B3LYP	KMLYP	CBS-Q	Experimental[b]
Allyl				
H–CH$_2$CH=CH$_2$	62.2	64.4	65.6	60.8
H–CH$_2$CH=CHCH$_3$	61.8	60.6	67.1	67.2
H–CH(CH$_3$)CH=CH$_2$	63.0	63.6	65.2	
H–CH$_2$C(CH$_3$)=CH$_2$	56.2	55.8	63.1	67.0
H–C(CH$_3$)$_2$CH=CH$_2$	61.7	65.9		81.0[c]
H–C(C$_2$H$_5$)CH=CH$_2$	56.5	56.3		
Propargyl				
H–CH$_2$C≡CH	49.0	46.8	46.8	48.7
H–CH$_2$C≡CCH$_3$	52.2	46.5	72.4	60.6
H–CH(CH$_3$)C≡CH	48.2	44.5	46.9	
H–CH(CH$_3$)C≡CCH$_3$	52.6	40.7		
H–C(CH$_3$)$_2$C≡CH	51.2	49.8		65.1
Conjugated allyl				
H–CH$_2$CH=CHCH=CH$_2$	88.3	81.5	94.7	78.4
H–CH(CH=CH$_2$)$_2$	104.7	108.6		
Cyclic allyl				
H–[CHCH=CH]	34.8	38.8	18.2	65.7
H–[CHCH=CHCH$_2$]	35.1	37.5	42.9	
H–[CHCH=CHCH$_2$CH$_2$]	52.8	59.1	59.4	62.2
H–Cyclohexene-3-yl	64.8	66.2		
Conjugated cyclic allyl				
H–[CHCH=CHCH=CH]	57.0	50.7		59.8
H–Cyclohexa-1,3-diene-5-yl	101.2	103.4		94.6
H–Cyclohexa-1,4-diene-3-yl	101.3	100.2		94.2

[a] Radical stabilisation energies were defined as the difference in bond energies of the stabilised species minus the corresponding saturated hydrocarbon. [b] Experimental values taken from Table 1. [c] Ref. 5.

The propargyl-type stabilisation (resonance with an α triple bond) is generally smaller then the allyl-type. For instance, the experimental value for propargyl stabilisation is 12 kJ mol^{-1} lower than that of allyl. Note that the experimental value for the BDE leading to 3-methyl-but-1-ene-3-yl (cited above) results in a stabilisation energy 16 kJ mol^{-1} higher than that of 3-methyl-but-1-yne-3-yl. Although both types of stabilisation involve the delocalisation of 3 electrons in a π orbital formed by 3 p-type atomic orbitals, the sp^1 hybridisation results in a higher energy due to the extra orthogonalisation constraint. Another factor contributing to the difference in stabilities is the shortened C–C bond, as suggested by Tsang.[39]

Conjugation

It is well-known that 1,3-conjugation of double bonds leads to extra stability due to conjugation. The conjugation energies were calculated from the isodesmic reactions shown in Table 3. While the experimental conjugation energy of the two π-bonds in buta-1,3-diene and penta-1,3-diene¶ is about 17 kJ mol^{-1}, the quantum methods estimate a decreased conjugation energy in hexa-1,3-diene, probably due to the different chemical nature of the terminal and chain π-bonds. When three linear π-bonds are conjugated (hexa-1,3,5-triene), this energy is predicted to be about twice that of two conjugated bonds.

An interesting test of the transferability of these stabilisation energy concepts comes from consideration of penta-1,3-diene and penta-1,4-diene, both of which lead to the same radical. The stabilisation energies for the penta-1,4-diene secondary C–H bond is the difference between this bond strength and the secondary C–H bond in pentane. The same radical is formed when the

---

¶ This number is confirmed by group additivity calculations.

Table 3 Conjugation energies in linear and cyclic alkenes[a]

Reactants	Products	Conjugation energy/kJ mol^{-1}			
		B3LYP	KMLYP	CBS-Q	Experiment[b]
Linear					
Buta-1,3-diene	Buta-1,3-diene + butane → 2 But-1-ene	18.2	16.3	15.9	17.0
Penta-1,3-diene	Penta-1,3-diene + pentane → Pent-1-ene + pent-2-ene	17.5	22.3	34.9	16.6
Hexa-1,3-diene	Hexa-1,3-diene + hexane → Hex-1-ene + hex-3-ene	15.1	10.8		
Hexa-1,3,5-triene	Hexa-1,3,5-triene + 2 hexane → Hex-3-ene + 2 hex-1-ene	37.8	29.8		
Cyclic					
Cyclobutadiene[c]	c-Butadiene + c-butane → 2 c-Butene	−146.9	−158.1	−134.3	−191.6
Cyclopentadiene	c-Pentadiene + c-pentane → 2 c-Pentene	11.9	18.7	14.9	8.6
Cyclohexadiene	1,3 c-Hexadiene + c-hexane → 2 c-Hexene	6.5	3.8		9.9
Benzene	Benzene + 2 c-hexane → 3 c-Hexene	−157.3	−159.7		−150.4

[a] Conjugation energies were defined as the difference in hydrogenation energies of the conjugated polyene and that of the single alkene. [b] Experimental values taken from Table 1. [c] Negative conjugation energies reflect antiaromatic effects.

primary C–H bond in penta-1,3-diene is broken. However, the difference between this bond strength and a primary bond in pentane also includes the conjugation energy in penta-1,3-diene. Thus the DFT values and the experimental values of the bond dissociation energies (see Table 1) will yield a value for the conjugation energy. The values so obtained are 21,∥ 19 and 21 kJ mol^{-1} from experiment, B3LYP and KMLYP respectively. The conjugation energy value computed directly from the difference in heats of hydrogenation of penta-1,3-diene, pent-1-ene and pent-2-ene is about 17, 18 and 22 kJ mol^{-1} for experiment, B3LYP and KMLYP respectively. These numbers are in good agreement, well within the experimental uncertainty.

The stabilisation energy of the pentadienyl radical has been estimated to be 90.8 ± 4 kJ mol^{-1} by Frey and Krantz,[40] who studied the isomerisation of trans-1-cyclopropylbuta-1,3-diene. Similarly, Egger and Jola[41] measured the 3-cis/trans isomerisation rate of hepta-1,trans-3,trans-5-triene and concluded that hepta-1,3-dienyl was stabilised by an extra 77.4 kJ mol^{-1}. The uncertainty in this quantity is given by McMillen and Golden (op. cit.) as ±13 kJ mol^{-1}.

To compare the conjugation stabilisations in cyclic conjugated hydrocarbons, care should be taken to account for ring strain and aromaticity effects. The dramatic effects observed in the cases of cyclobutadiene and benzene are due to antiaromatic and aromatic behaviour (these are discussed in the next section). In the cases of c-pentadiene and c-hexa-1,3-diene the calculated conjugation energies are considerably lower than those of their linear counterparts, even though ring strains are small and there is no antiaromatic behaviour.

Ring strain and aromaticity

The thermodynamic stability of cyclic hydrocarbons is noticeably different from their aliphatic counterparts due to ring strain and aromaticity. When a conjugated system is allowed to delocalise in a ring, strong stabilization effects (aromaticity) are observed when the number of π-electrons obeys Hückel's condition ($4n + 2$). Since aromaticity and ring strain are defined with respect to a hypothetical reference, there is no unique way of determining them.[9] In the light of this, we define ring strains via homodesmotic reactions, that is, reactions where the products and reactants have the same groups. We have studied the ring strains ($\Delta_{rs}^{(n)}$) in a series of hydrocarbons, where n is the number of sp^2 carbons in the ring, as hydrogen atoms are progressively removed from the ring. Calculated and experimental $\Delta_{rs}^{(n)}$, along with the corresponding homodesmotic reactions, are shown in Table 4. The values of ring corrections in Table 4 can also be obtained from group contributions rather than the use of individual molecules.

The origin of ring strain in cyclopropane and cyclobutane is different than in larger rings. The strain energy of cyclopropane and cyclobutane is a result of the reduced orbital overlaps as the bonds are bent away from their tetrahedral value (this can sometimes be alleviated by out-of-plane bending). In larger cyclanes, on the other hand, the strain is mostly due to repulsions between neighbouring hydrogens.

As we remove hydrogens from cyclopropane, the ring strain energy sharply increases at a rate of about 80 kJ mol^{-1} per sp^2 carbon introduced. In cyclobutane this strain energy increase is about 16 kJ mol^{-1} per sp^2 carbon for the first three carbons. The ring is dramatically destabilised when a fourth sp^2 carbon is introduced due to the antiaromatic character of cyclobutadiene. Deniz et al.[42] determined a formation enthalpy of cyclobutadiene of 477 ± 46 kJ mol^{-1} using photoacoustic calorimetry. They estimated an antiaromatic destabilisation of 201 kJ mol^{-1}, relative to a reference with isolated double bonds. If we assume a linear increment of 16 kJ mol^{-1} of the strain energy, we obtain an aromaticity of ∼170, 162 and 140 kJ mol^{-1} for B3LYP, KMLYP and CBS-Q methods, respectively. If we use the isodesmic reaction:

$$2\,\square \longrightarrow \square + \square$$

and add the conjugation stabilisation energies (from Table 2), we obtain antiaromatic destabilisations of 165, 174 and 150 kJ mol^{-1} for these three methods respectively, and 209(±46) kJ mol^{-1} for the experimental value.

∥ The experimental number is derived using the primary and secondary BDEs in butane.

Table 4 Calculated ring strain energies (kJ mol^{-1}) and corresponding homodesmotic reactions

# sp^2	Reactants		Products	B3LYP	KMLYP	CBS-Q	Exp
C3 0	3 Propane	→	c-Propane + 3 ethane	93.7	127.4	117.6	115.8
1	Penta-3-yl	→	c-Propyl + ethane	143.8	174.3		
2	Penta-1,4-diene	→	c-Propene + ethene	214.0	249.2		224.8
3	Pentadienyl	→	c-Propenyl + ethene	330.6	366.8	354.0	
C4 0	4 Propane	→	c-Butane + 4 ethane	92.3	123.6	111.5	111.8
1	Penta-3-yl + propane	→	c-Butyl + 2 ethane	102.3	129.2		
2	2 But-1-ene	→	c-Butene + ethene + ethane	144.7	47.8	139.2	126.1
3	But-1-ene-3-yl + but-1-ene-4-yl	→	c-Butenyl + ethyl + ethene	148.7	169.1	158.6	
4	2 Butadiene	→	c-Butadiene + 2 ethene	326.2	355.0	315.7	364.3
C5 0	5 Propane	→	c-Pentane + 5 ethane	15.7	38.0	24.9	27.8
1	Penta-3-yl + 2 propane	→	c-Pentyl + 3 ethane	11.3	27.3		
2	2 But-1-ene + propane	→	c-Pentene + ethene + 2 ethane	17.8	32.0	25.5	26.3
3	Pent-1-ene-3-yl + but-1-ene	→	c-But-2ene-yl + ethene + ethane	17.6	21.8		
4	Penta-1,4-diene + butadiene	→	c-Pentadiene + 2 ethene	19.9	22.8		28.7
5	Pentadienyl + butadiene	→	c-Pentadienyl + 2 ethene	60.0	70.9		
C6 0	6 Propane	→	c-Hexane + 6 ethane	-6.9	9.0		1.9
1	Penta-3-yl + 3propane	→	c-Hexyl + 4 ethane	-0.2	12.3		
2	2 But-1-ene + 2 propane	→	c-Hexene + ethene + 3 ethane	0.5	10.1		7.2
3	Penta-2-ene-yl + pen-1-tene	→	c-Hexa-2ene-yl + ethene + ethane	3.1	9.0		
4	2 Penta-1,4-diene	→	1,4-c-Hexadiene + 2 ethene	-5.9	2.4		-2.9
5	Pentadienyl + 1,4-pentadiene	→	1,4-c-Hexadienyl + 2 ethene	0.8	14.9		
6	3 Butadiene	→	Benzene + 3 ethene	-95.3	-98.1		-86.0

Ring strain energy in five-membered hydrocarbons shows the opposite trend, decreasing slightly as more hydrogens are removed. This could be due to the reduced hydrogen repulsions, which favour the planarisation of the ring and result in bond angles closer to their tetrahedral value. However the ring strain sharply increases upon introducing the fifth sp^2 carbon, possibly due to the Jahn–Teller distortion of the cyclopentadienyl radical. Cycloalkanes larger than cyclopropane have several low-energy conformations. In the case of six-membered rings, the ring strain energy is close to zero and the variations are probably due to the intrinsic errors of the methods. Interestingly, the values of $\Delta_{rs}^{(4)}$ calculated *via* cyclohexa-1,3-diene and cyclohexa-1,4-diene have opposite signs. If ring strain is neglected, DFT methods predict an aromatic stabilisation of benzene of ~ 96 kJ mol^{-1}, in good agreement with the 88 kJ mol^{-1}, estimated by Deniz.[42] Ring strain energy is expected to increase slightly for larger cycloalkanes due to the larger number of hydrogen interactions. For example, Cox[43] reports ring strain energies of 115, 111, 27, 0 and 26 kJ mol^{-1} for cycloalkanes with three to seven carbons, respectively.

Since ring strain increases sharply in three and four-membered rings, the C–H bond energies of these rings are expected to be larger than a linear secondary carbon, provided there are no allyl or aromatic stabilisations. The fact that the reported experimental BDE of H-cyclobutane (404 kJ mol^{-1}, ref. 10 and 44) is lower than that of H-isopropyl is an indication that the former BDE may be in error. It was determined from bromination and iodination experiments that assumed a positive activation energy of 4 ± 4 kJ mol^{-1} for c-butyl + HX → c-butane + X reaction.[10,14] KMLYP and CBS-Q bond energies also suggest this number is 7–16 kJ mol^{-1} too low.

Although the reactions listed in Table 4 are homodesmotic, the bonding environment changes considerably upon ring formation due to cyclic delocalisation of the electron density. For this reason the performance of both DFT methods is less than desirable, with B3LYP predicting ring strains substantially lower than those obtained from experimental heats of formation, and KMLYP generally over-predicting them. Ring strains obtained with CBS-Q, on the other hand, are in good agreement with experimental values.

Group additivity

The concept of group additivity has been used with great success by Benson and others[9] to predict thermodynamic properties of families of compounds. It relies on the fact that a thermodynamic property Φ can be approximated by a sum of subunits referred to as groups, assigned to each non-hydrogen atom:

$$\Phi_i = \sum_j n_{ij} g_j$$

where the group g_i appears n_{ij} times in the compound i. Group assignments take into account atom-type and bond order of a given atom as well as its neighbours. In the case of hydrocarbons, all the methyl groups are accounted for by the same group, regardless of the bonding situation of the carbon it is bound to, thus C–(Cd)(H)$_3$ is accounted for as C–C(H)$_3$.

Heats of formation

We have attempted to compute group additivity parameters for the heats of formation of radicals for the molecules and radicals listed in Table 1. The heat of formation of a hydrocarbon R can be obtained from the atomisation energy and the formation energies of the atoms in the gaseous state:

$$\Delta_f H^\circ_{calc}(R) = E(R) + \Delta_{298}(R) + n_c[\Delta_f H^\circ_{exp}(C) - E(C) - \Delta_{298}(C)]$$
$$+ n_H[\Delta_f H^\circ_{exp}(H) - E(H) - \Delta_{298}(H)]$$

where n_C and n_H are the number of carbon and hydrogen atoms in R, respectively, and Δ_{298} is the difference between the enthalpy at 298 K and the energy at 0 K, due to zero-point energy and heat capacity effects. The heats of formation of the atoms are taken from experiment.[19] However this procedure alone yields heats of formation with errors too large to be chemically useful, even when

Table 5 Group additivity correction parameters, $\{g_i\}$/kJ mol^{-1}

Group	B3LYP-GA	MLYP-GA
C–(C)(H)$_3$	−3.7	30.8
C–(C)$_2$(H)$_2$	−12.2	31.3
C–(C)$_3$(H)	−25.8	29.3
C–(C)$_4$	−43.0	25.6
Cd–(H)$_2$	−1.0	16.6
Cd–(C)(H)	−7.3	20.6
Cd–(C)$_2$	−17.6	18.8
Cd–(Cd)(H)	−7.3	17.5
C–(Cd)(C)(H)$_2$	−12.5	26.6
C–(Cd)(C)$_2$ (H)	−12.6	13.2
C–(Cd)$_2$ (H)$_2$	−14.5	24.8
Ct–(H)	−3.8	−1.9
Ct–(C)	−4.9	4.3
C–(Ct)(C)(H)$_2$	−11.4	30.5
C–(Ct)(C)$_2$ (H)	−27.0	25.1
C˙–(C)(H)$_2$	8.6	34.8
C˙–(C)$_2$ (H)	6.8	40.5
C˙–(C)$_3$	−9.3	33.2
C–(C˙)(C)(H)$_2$	−13.6	29.4
C–(C˙)(C)$_2$ (H)	−27.2	26.6
C–(C˙)(C)$_3$	−53.2	13.7
Cd–(C˙)(H)	2.5	33.0
Cd–(C˙)(C)	−18.9	23.0
C˙–(Cd)(H)$_2$	−2.0	17.7
C˙–(Cd)(C)(H)	−8.8	19.8
C˙–(Cd)(C)$_2$	−36.8	2.8
C˙–(Cd)$_2$ (H)	−5.9	21.8
Ct–(C˙)	−12.4	−9.9
C˙–Ct(H)$_2$	−11.9	39.7
C˙–Ct(C)$_2$	−16.0	28.0

it is used with accurate methods such as CCSD(T). This accumulation of error is caused by the radically different bonding environment surrounding the atoms in a molecule, and those of the isolated atoms.

Although DFT methods are very useful to map the local shape of the potential energy surface (PES), the total electronic energies contain intrinsic errors large enough that the heats of formation obtained directly using the above expression are unphysical. Nonetheless, we calculated the heats of formation for a set of 25 molecules and 25 radicals, in two different ways using the formula given above. In the first method we used uncorrected DFT energies and thermal corrections, and in the second we used equivalent atom enthalpies of hydrogen and carbon atoms by fitting the heats of formation of linear pentane and hexane. As expected, the heats of formation obtained in these two ways were in poor agreement with experiment. Interestingly enough, however, we found that heats of formation calculated this way can be cast into group additivity parameters and reproduce the computed data set with rms error of $\sim$1.3 and 1.9 kJ mol^{-1}, for B3LYP and KMLYP respectively. Note that even though the group parameters calculated with uncorrected DFT enthalpies and atom equivalents are different, the errors caused by group additivity are identical since the atom equivalents are superseded by group corrections.

DFT and group additivity scheme for predicting heats of formation

Given the difficulties for predicting heats of formation with *ab initio* methods, whenever these are needed, chemists frequently use isodesmic reactions or apply empirical corrections. The former usually give more reliable results, but their use is limited to cases where there is known data for a similar reaction.

Table 6 Heats of formation and deviations from experiment of CBS-Q, B3LYP-GA and KMLYP-GA

Species	$\Delta_f H^o$			Exp	Deviations from experiment		
	CBS-Q	B3LYP-GA	KMLYP-GA		CBS-Q	B3LYP-GA	KMLYP-GA
Hex-1-ene (E)		−41.0	−42.7	−42.1		1.1	−0.6
Hexa-1,3,5-triene (E)[a]		173.7	130.2	167.8		−8.6	−2.6
c-Propane	60.0	35.0	64.7	53.3	6.7	−18.3	11.4
c-Propene	292.0	266.5	302.0	278.7	13.4	−12.2	23.3
c-Butadiene	441.2	436.6	468.3	477.0	−35.8	−40.4	−8.7
c-Butane	34.8	14.0	39.9	28.5	6.3	−14.5	11.5
c-Butene	170.8	151.1	171.9	156.9	13.9	−5.8	15.0
c-Pentadiene	145.0	134.9	114.7	138.9	6.1	−11.3	−6.8
c-Pentane	−71.1	−82.3	−66.6	−76.4	5.3	−5.9	9.8
c-Pentene	44.4	29.2	39.2	35.6	8.9	−6.4	3.6
c-Hexa-1,3-diene		103.0	108.5	104.6		−1.6	4.0
c-Hexa-1,4-diene		99.2	107.9	104.8		−5.6	3.2
c-Hexane		−124.3	−116.5	−123.1		−1.2	6.6
c-Hexene		−7.7	−3.6	−4.3		−3.4	0.7
Methyl-c-propane	30.7	1.2	33.6	24.3	6.4	−23.1	9.4
Pent-1-yl	64.8	60.1	53.6	59.1	5.8	1.0	−5.5
Pent-2-yl		49.4	41.8	44.8		4.6	−3.0
Pent-3-yl		46.0	45.9	46.9		−0.9	−1.0
c-Propyl	300.4	274.5	304.7	339.8	−39.3	−65.3	−35.1
c-Butyl	235.8	213.4	238.7	214.2	21.6	−0.8	24.4
c-Propyl-carbinyl	226.5	167.3	215.3	213.8	12.7	−46.5	1.5
c-Pentyl	117.5	102.9	115.9	105.0	12.5	−2.1	10.9

[a] Calculated using the 6-311G(d,p) basis set.

An example of empirical corrections are the so-called "high level corrections" used in compound methods such as G2 or CBS-Q. These typically involve only two parameters: an extensive term that scales with the number of electrons and a term that compensates for the number of unpaired electrons, but their success is limited to computationally-intensive methods that include an advanced treatment of electron correlation. An example is the work of Marsi et al.[45]

Several correction schemes have also been developed specifically for DFT energies with varying degrees of success. For example, those using atom equivalents,[46] bond equivalents,[47] bond density functions[48] and group equivalents.[49,50]

With engineering applicability in mind, we have developed a set of group equivalents to correct the heats of formation obtained with B3LYP and KMLYP methods. This was done by doing a through-the-origin regression of:

$$\Delta_i = \Delta_f H_{298}^{exp}(i) - \Delta_f H_{298}^{DFT}(i) = \sum_j n_{ij} g_j$$

where Δ_i is the difference between the experimental and DFT formation enthalpies of component i in a set of 19 radicals and 25 closed-shell molecules. Since the group equivalents account for the differences between experiment and DFT, the heats of formation of a specific compound are given by:

$$\Delta_f H_{298}^{calc}(i) = \Delta_f H_{298}^{DFT}(i) + \sum_j n_{ij} g_j$$

These DFT energies corrected with group additivity are referred to as B3LYP-GA and KMLYP-GA in this paper. The group equivalents obtained in this way are listed in Table 5.

The hope is that the group parameters will correct the incomplete electron correlation of DFT, while the DFT calculation is general enough to accurately estimate non-local interactions. The usefulness of these empirical corrections depends, of course, on their ability to predict experimental values for compounds other than those used for the fit. We tested this method on a set of 22 species, most of them are cyclic hydrocarbons (7 of them radicals) where we expect non-local contributions to be large. The results are shown in Table 6. As expected, the KMLYP-GA methods gave heats of formation for hex-1-ene and hexa-1,3,5-triene within 6 kJ mol^{-1} of the experimental value, but errors were larger in structures with large ring strains. The KMLYP-GA method showed the best performance overall, and had the largest deviations in the case of cyclopentadiene and cyclobutyl.

Concluding remarks

We have attempted to create a consistent data base for hydrocarbon bond dissociation energies. At 1500 K, an uncertainty of 9 kJ mol^{-1} will yield a factor of two uncertainty in an equilibrium constant. The RMS deviations for the whole series of BDEs (excluding cyclopropenyl) are 13.6, 8.4 and 6.6 kJ mol^{-1} for B3LYP, KMLYP and CBS-Q, respectively. The chemical accuracy of B3LYP is insufficient even for high temperature modelling applications. However the two alternative methods used in this study show RMS deviations that are close to the reported experimental error bars. The CBS-Q method is difficult to apply to larger species, but the applications of KMLYP-GA in this field are promising, considering its accuracy and computational efficiency.

Factors leading to radical stabilisation have been studied by performing quantum chemistry calculations of the C–H bond dissociation. These stabilising contributions are not easily decoupled from each other, so simple additive calibrations generally cannot be used to estimate the BDEs without loss in accuracy.

The series of homologous reactions used to estimate the stabilising effects present in different hydrocarbons is often internally inconsistent. Since the origin of experimental measurements used is quite diverse (both in terms of reactions, analytic techniques and experimental conditions) the experimental errors are expected to be much more uncorrelated than those of theoretical calculations. Thus a self-consistent body of theoretical BDEs may be less susceptible to error propagation and can be used effectively to signal those experiments that are worth revisiting.

Ring strain energies were obtained for three to six-membered rings with varying numbers of sp^2 carbons *via* homodesmotic reactions. We found B3LYP seriously underestimates ring strain energy, while the results obtained with CBS-Q are within the error limits of experiments.

A scheme for obtaining accurate heats of formation from DFT calculations and group additivity equivalents is presented. This method relies on the fact that contributions to the electronic energy are mostly local, and can be parametrized with the group additivity protocol. The non-local parts are then obtained through the DFT calculation. In a test set of molecules with strong non-local effects (cyclic hydrocarbons) the group-additivity corrected KMLYP method performed best.

Acknowledgements

We are grateful to Jeung-Ku Kang for useful discussions concerning the unpublished KMLYP method.

Appendix

Vertical C–H bond strength

Branching affects the BDEs mainly through changes in hyperconjugation. In order to estimate the contribution of this effect, energies were calculated for a number of these radicals at the same geometries as the closed-shell molecule, but with the hydrogen atom under consideration removed. We hypothesized that this 'vertical C–H bond strength' would be transferable for a

Table A1 'Vertical' BDEs[a] and energy changes (kJ mol^{-1}) due to geometry relaxation

Bond	'vertical' D_e(R–H)	ΔE_{relax}
C2		
H–C$_2$H$_5$	475.2	30.6
H–CH=CH$_2$	497.7	13.1
C3		
H–C$_3$H$_7$	475.2	30.2
H–CH(CH$_3$)$_2$	458.0	30.8
H–CH$_2$CH=CH$_2$	452.8	72.0
H–CHC≡CH	448.7	52.8
H–[CHCH$_2$CH$_2$]	489.0	15.7
C4		
H–C(CH$_3$)$_3$	445.8	28.9
H–CH(CH$_3$)CH=CH$_2$	455.3	67.3
H–CH$_2$CH=CHCH$_3$	497.4	119.5
H–C(=CH$_2$)CH=CH$_2$	488.7	54.1
H–[CHCH$_2$CH$_2$CH$_2$]	465.4	30.9
H–[CHCH=CHCH$_2$]	453.5	56.9
H–[C=CHCH=CH]	498.1	7.8
C5		
H–CH$_2$CH=CHCH=CH$_2$	443.6	88.7
H–CH=CHCH$_2$CH=CH$_2$		
H–[CHCH$_2$CH$_2$CH$_2$CH$_2$]	472.4	52.6
H–[CHCH=CHCH$_2$CH$_2$]	436.2	71.3
H–[CHCH=CHCH=CH]	441.3	78.4
C6		
H–Phenyl	498.1	7.5

[a] Radical calculated at the geometry of the molecule (see Appendix).

given type of bond, provided that hyperconjugation before geometry relaxation was small and ring strain in the molecule similar to that in the radical at the 'frozen geometry'. However, this does not seem to be the case and the calculated B3LYP/6-311G(2d,2p) energy changes due to geometry relaxations are extremely large (~ 120 kJ mol^{-1} for primary bonds). These "vertical BDEs" are listed in Table A1 for the reader's interest.

References

1. M. W. J. Chase, *J. Phys. Chem. Ref. Data*, 1998, Monograph 9.
2. M. Frenkel, K. N. Marsh, R. C. Wilhooit, G. J. Kabo and G. N. Roganov, *Thermodynamics of Organic Compounds in the Gas Phase*, Thermodynamics Research Center, College Station, TX, 1994.
3. NIST, Chemistry Webbook, 2000. http://webbook.nist.gov/chemistry/
4. W. Tsang, in *Energetics of Organic Free Radicals*, ed. J. A. Martinho Simoes, A. Greenberg and J. F. Liebman, Blackie Academic and Professional, London, 1996, p. 22.
5. D. F. McMillen and D. M. Golden, *Annu. Rev. Phys. Chem.*, 1982, **33**, 493.
6. J. Berkowitz, G. B. Ellison and D. Gutman, *J. Phys. Chem.*, 1994, **98**, 2744.
7. J. D. Cox, D. D. Wagman and V. A. Medvedev, *CODATA Key Values for Thermodynamics*, Hemisphere, New York, 1989.
8. G. P. Smith, D. M. Golden, Frenklach, N. W. Moriarty, B. Eiteneer, M. Goldenberg, C. T. Bowman, R. K. Hanson, S. Song, W. C. J. Gardiner, V. V. Lissianski and Z. Qin, GRI-Mech, 2000. http://www.me.berkeley.edu/griimech/
9. S. W. Benson, *Thermochemical Kinetics*, Wiley, New York, 1968.
10. K. C. Ferguson and E. Whittle, *J. Chem. Soc., Faraday Trans.*, 1971, **67**, 2618.
11. S. Furuyama, D. M. Golden and S. W. Benson, *Int. J. Chem. Kinet.*, 1970, **2**, 93.
12. S. Furuyama, D. M. Golden and S. W. Benson, *Int. J. Chem. Kinet.*, 1971, **3**, 237.
13. D. F. McMillen, D. M. Golden and S. W. Benson, *Int. J. Chem. Kinet.*, 1971, **3**, 359.
14. D. F. McMillen, D. M. Golden and S. W. Benson, *Int. J. Chem. Kinet.*, 1972, **4**, 487.
15. O. Kondo, R. M. Marshall and S. W. Benson, *Int. J. Chem. Kinet.*, 1988, **20**, 297.
16. D. M. Golden and S. W. Benson, *Chem. Rev.*, 1969, **69**, 125.
17. J. M. Nicovich, C. A. Vandijk, K. D. Kreutter and P. H. Wine, *J. Phys. Chem.*, 1991, **95**, 9890.
18. M. J. Pilling, *Pure Appl. Chem.*, 1992, **64**, 1473.
19. J. A. Kerr and D. W. Stocker, in *Handbook of Chemistry and Physics*, 81st edn., The Chemical Rubber Company, Boca Raton, FL, 2001.
20. P. Marshall, personal communication, 1999.
21. M. R. Nyden and G. A. Petersson, *J. Chem. Phys.*, 1981, **75**, 1843.
22. A. D. Becke, *J. Chem. Phys.*, 1993, **98**, 5648.
23. J. K. Kang and C. B. Musgrave, *J. Chem. Phys.*, 2001, in press.
24. M. J. Frisch, G. W. Trucks, H. B. Schlegel, G. E. Scuseria, M. A. Robb, J. R. Cheeseman, V. G. Zakrzewski, J. Montgomery, R. E. Stratmann, J. C. Burant, S. Dapprich, J. M. Millam, A. D. Daniels, K. N. Kudin, M. C. Strain, O. Farkas, J. Tomasi, V. Barone, M. Cossi, R. Cammi, B. Mennucci, C. Pomelli, C. Adamo, S. Clifford, J. Ochterski, G. A. Petersson, P. Y. Ayala, Q. Cui, K. Morokuma, D. K. Malick, A. D. Rabuck, K. Raghavachari, J. B. Foresman, J. Cioslowski, J. V. Ortiz, A. G. Baboul, B. B. Stefanov, G. Liu, A. Liashenko, P. Piskorz, I. Komaromi, R. Gomperts, R. L. Martin, D. J. Fox, T. Keith, M. A. Al-Laham, C. Y. Peng, A. Nanayakkara, C. Gonzalez, M. Challacombe, P. M. W. Gill, B. Johnson, W. Chen, M. W. Wong, J. L. Andres, C. Gonzalez, M. Head-Gordon, E. S. Replogle and J. A. Pople, *GAUSSIAN 98, Revision A.7*, Gaussian Inc., Pittsburgh, PA, 1998.
25. L. A. Curtiss, K. Raghavachari, G. W. Trucks and J. A. Pople, *J. Chem. Phys.*, 1991, **94**, 7221.
26. D. Feller and D. A. Dixon, *J. Phys. Chem. A*, 2000, **104**, 3048.
27. C. W. Bauschlicher and H. Partridge, *Chem. Phys. Lett.*, 1995, **239**, 246.
28. W. Klopper, K. L. Bak, P. Jorgensen, J. Olsen and T. Helgaker, *J. Phys. B*, 1999, **32**, R103.
29. P. Marshall, *J. Phys. Chem.*, 1999, **103**, 4560.
30. J. M. L. Martin and G. deOliveira, *J. Chem. Phys.*, 1999, **111**, 1843.
31. G. A. Petersson, D. K. Malick, W. G. Wilson, J. W. Ochterski, J. A. Montgomery and M. J. Frisch, *J. Chem. Phys.*, 1998, **109**, 10570.
32. G. A. DiLabio and D. A. Pratt, *J. Phys. Chem. A*, 2000, **104**, 1938.
33. C. W. Bauschlicher and H. Partridge, *Chem. Phys. Lett.*, 1995, **240**, 533.
34. M. W. Wong, *Chem. Phys. Lett.*, 1996, **256**, 391.
35. M. N. Glukhovtsev, S. Laiter and A. Pross, *J. Phys. Chem.*, 1996, **100**, 17801.
36. D. J. Defrees, R. T. McIver and W. J. Hehre, *J. Am. Chem. Soc.*, 1980, **102**, 3334.
37. S. D. Han, M. C. Hare and S. R. Kass, *Int. J. Mass Spectrom.*, 2000, **201**, 101.
38. K. U. Ingold and J. S. Wright, *J. Chem. Educ.*, 2000, **77**, 1062.
39. W. Tsang, *Int. J. Chem. Kinet.*, 1970, **2**, 23.
40. H. M. Frey and A. Krantz, *J. Chem. Soc. A*, 1969, 1159.
41. K. W. Egger and M. Jola, *Int. J. Chem. Kinet.*, 1970, **2**, 265.

42 A. A. Deniz, K. S. Peters and G. J. Snyder, *Science*, 1999, **286**, 1119.
43 J. D. Cox and G. Pilchner, in *Thermochemistry of Organic and Organometallic Compounds*, Academic Press, London, 1970.
44 M. H. Baghalvayjooee and S. W. Benson, *J. Am. Chem. Soc.*, 1979, **101**, 2838.
45 I. Marsi, B. Viskolcz and L. Seres, *J. Phys. Chem. A*, 2000, **104**, 4497.
46 D. Habibollahzadeh, M. E. Grice, M. C. Concha, J. S. Murray and P. Politzer, *J. Comput. Chem.*, 1995, **16**, 654.
47 M. R. Ibrahim and P. V. Schleyer, *J. Comput. Chem.*, 1985, **6**, 157.
48 J. Cioslowski, G. H. Liu and P. Piskorz, *J. Phys. Chem. A*, 1998, **102**, 9890.
49 N. L. Allinger, K. Sakakibara and J. Labanowski, *J. Phys. Chem.*, 1995, **99**, 9603.
50 K. B. Wiberg, *J. Comput. Chem.*, 1984, **5**, 197.
51 J. A. Seetula and I. R. Slagle, *J. Chem. Soc., Faraday Trans.*, 1997, **93**, 1709.
52 A. B. Trenwith and S. P. Wrigley, *J. Chem. Soc., Faraday Trans. 1*, 1977, **73**, 817.
53 J. B. Pedley and J. Rylance, *Sussex-NPL Computer Analysed Thermochemical Data; Organic and Organometallic Compounds*, University of Sussex, 1977.
54 S. E. Stein, in *SRD Thermochemical Database, 25. NIST Structures and Properties Database and Estimation Program*, US Department of Commerce, 1992.
55 A. B. Trenwith, *J. Chem. Soc., Faraday Trans. 1*, 1980, **76**, 266.
56 C. W. Larson, E. A. Hardwidge and B. S. Rabinovich, *J. Chem. Phys.*, 1969, **2**, 265.
57 A. L. Castelhano and D. Griller, *J. Am. Chem. Soc.*, 1982, **104**, 3655.

Low-energy paths for the unimolecular decomposition of CH_3OH: A G2M/statistical theory study

W. S. Xia,[a] R. S. Zhu,[a] M. C. Lin*[a] and A. M. Mebel[b]

[a] *Department of Chemistry, Emory University, Atlanta, GA, 30322, USA.*
   E-mail: chemmcl@emory.edu
[b] *Institute of Atomic and Molecular Sciences, Taipei 166, Taiwan*

Received 5th March 2001
First published as an Advance Article on the web 31st October 2001

The potential energy surface (PES) of the CH_3OH system has been characterized by *ab initio* molecular orbital theory calculations at the G2M level of theory. The mechanisms for the decomposition of CH_3OH and the related bimolecular reactions, $CH_3 + OH$ and $^1CH_2 + H_2O$, have been elucidated. The rate constants for these processes have been calculated using variational RRKM theory and compared with available experimental data. The total decomposition rate constants of CH_3OH at the high- and low-pressure limits can be represented by $k^\infty = 1.56 \times 10^{16} \exp(-44\,310/T)$ s^{-1} and $k^0_{Ar} = 1.60 \times 10^{36} T^{-12.2} \exp(-48\,140/T)$ cm^3 molecule^{-1} s^{-1}, respectively, covering the temperature range 1000–3000 K, in reasonable agreement with the experimental values. Our results indicate that the product branching ratios are strongly pressure dependent, with the production of $CH_3 + OH$ and $^1CH_2 + H_2O$ dominant under high ($P > 10^3$ Torr) and low ($P < 1$ atm) pressures, respectively. For the bimolecular reaction of CH_3 and OH, the total rate constant and the yields of $^1CH_2 + H_2O$ and $H_2 + HCOH$ at lower pressures ($P < 5$ Torr) could be reasonably accounted for by the theory. For the reaction of $^1CH_2$ with H_2O, both the yield of $CH_3 + OH$ and the total rate constant could also be satisfactorily predicted theoretically. The production of $^3CH_2 + H_2O$ by the singlet to triplet surface crossing, predicted to occur at 4.3 kcal mol^{-1} above the $H_2C\cdots OH_2$ van der Waals complex (which lies 82.7 kcal mol^{-1} above CH_3OH), was neglected in our calculations.

I. Introduction

Methanol is an important alternate fuel, which can be employed in internal combustion engines. The thermal decomposition of CH_3OH, carried out primarily by shock-heating has been investigated by many research groups.[1–11] The rate constant for the initial decomposition was obtained invariably by kinetic modeling of the complex chemistry assuming primarily the C–O splitting process producing CH_3 and OH:

$$CH_3OH \rightarrow CH_3 + OH \quad (1)$$

In a study carried out with H atom and OH radical detection behind incident shock waves, Dombrowsky *et al.*[9] reported that CH_3OH decomposition produces OH with about 80% yield and less than 16% and 4% of $^1CH_2 + H_2O$ and $H + CH_3O/CH_2OH$, respectively, by the following reactions:

DOI: 10.1039/b102057i

$$CH_3OH \rightarrow {}^1CH_2 + H_2O \quad (2)$$

$$\rightarrow H + CH_2OH/CH_3O \quad (3)$$

In addition to these initiation processes, H_2 molecular elimination producing the three isomers of formaldehyde, cis-HCOH, trans-HCOH and CH_2O, may also be accessible energetically under combustion conditions:[12–16]

$$CH_3OH \rightarrow H_2 + CH_2O \quad (4)$$

$$\rightarrow H_2 + cis\text{-HCOH} \quad (5)$$

$$\rightarrow H_2 + trans\text{-HCOH} \quad (6)$$

In a series of experiments, Lee and co-workers recently studied the dynamics of fragmentation of chemically activated CH_3OH produced by the $O(^1D) + CH_4$ reaction[17] with 135 kcal mol^{-1} of internal energy. They indeed reported the observation of H_2 + CH_2O/HCOH in addition to CH_3 + OH and H + CH_3O/CH_2OH product pairs, with the methyl formation channel being the dominant one. In this bimolecular reaction, a large fraction of CH_3 may be formed by the direct abstraction process, which is indicated by forward scattering of OH product.[17] The HCOH diradical was reported to be generated in the thermal reaction of CH_3 and OH by Just and co-workers under low-pressure conditions.[18]

The kinetics for the recombination of CH_3 with OH, which is directly relevant to the major fragmentation step mentioned above, has also been investigated by many research groups.[18–26] These results, primarily determined at lower temperatures than those carried out in pyrolytic experiments with shock tubes, should be valuable for characterization of the barrierless recombination process.

In this study, we investigate the thermal decomposition of CH_3OH in its ground electronic state focusing on the aforementioned low-energy paths, which may play a major role in the combustion reaction. The predicted results will be compared with existing kinetic data for the decomposition reaction and the recombination of CH_3 with OH.

II. Computational methods

1. *Ab initio* calculations

The geometries of the reactant, intermediates, transition states, and products of the CH_3OH decomposition reaction were optimized at the B3LYP/6-311G(d,p) (*i.e.*, Becke's three-parameter nonlocal exchange functional[27–29] with the nonlocal correlation functional of Lee et al.[30] The energies of all species were calculated by the G2M[31] method, which uses a series of calculations with B3LYP/6-311G(d,p) optimized geometries to approximate the CCSD(T)/6-311+G(3df,2p) level of theory, including a "higher level correction (HLC)" based on the number of paired and unpaired electrons. The total G2M energy with zero-point energy (ZPE) correction is calculated as follows:[31]

$$E[(G2M(CC2)] = E_{bas} + \Delta E(+) + \Delta E(2df) + \Delta E(CC) + \Delta' + \Delta E(HLC, CC2) + ZPE.$$

$$E_{bas} = E[PMP4/6\text{-}311G(d,p)]$$

$$\Delta E(+) = E[PMP4/6\text{-}311+G(d,p)] - E_{bas}$$

$$\Delta E(2df) = E[PMP4/6\text{-}311G(2df,p)] - E_{bas}$$

$$\Delta E(CC) = E[CCSD(T)/6\text{-}311G(d,p)] - E_{bas}$$

$$\Delta' = E[UMP2/6\text{-}311+G(3df,2p)] - E[UMP2/6\text{-}311G(2df,p)]$$

$$- E[UMP2/6\text{-}311+G(d,p)] + E[UMP2/6\text{-}311(d,p)]$$

$$\Delta E(HLC, CC2) = -5.78n_\beta - 0.19n_\alpha \text{ in units of } 10^{-3} E_h$$

where n_α and n_β are the numbers of valence electrons, $n_\alpha \geq n_\beta$. All calculations were carried out with the GAUSSIAN 98[32] and MOLPRO 96[33] programs.

2. RRKM calculations

The rate constant and product branching ratios were computed with a microcanonical variational RRKM (Variflex[34]) code which solves the master equation[35–37] involving multi-step vibrational energy transfers for the excited intermediate ($CH_3OH^\dagger$). The *ab initio* PES calculated at the G2M level, to be discussed in the next section, was used in the calculation.

Similar to our previous calculation[38–41] with the Variflex code, the component rates were evaluated at the *E,J*-resolved level. The pressure dependence was treated by 1-D master equation calculations using the Boltzmann probability of the complex for the *J* distribution. The master equation was solved by an inversion based approach.[35,36] In order to achieve convergence in the integration over the energy range, energy grain size varied from 30 to 100 cm^{-1}, depending on the third body; the grain size used provides numerically converged results for all temperatures studied with the energy spanning range from 15 000 cm^{-1} below to 64 900 cm^{-1} above the threshold. The total angular momentum *J* covered the range from 1 to 241 in steps of 10 for the *E,J*-resolved calculation. For the barrierless transition states, the Varshni potential[42]

$$V(R) = D_e\{1 - \alpha \exp[-\beta(R^2 - R_0^2)]\}^2 - D_e$$

was used to represent the potential energy along the individual reaction coordinate. In the above equation, D_e is the bond energy excluding zero-point vibrational energies and $\alpha = R_0/R$, where R is the reaction coordinate (*i.e.* the distance between the two bonding atoms, in the present case C–O) and R_0 is the equilibrium value of R. For the tight transition states, the numbers of states were evaluated according to the rigid-rotor harmonic-oscillator assumption.

III. Results and discussion

The optimized geometries of the reactants, intermediates, transition states and products are shown in Fig. 1; the potential energy diagram obtained at the G2M level is presented in Fig. 2; the total and relative energies are compiled in Table 1 and the vibrational frequencies and moments of inertia of all species used in RRKM calculations are summarized in Table 2. As illustrated in Fig. 2, the decomposition of CH_3OH can occur *via* several product channels as alluded to before.

Table 1 The energetics (in kcal mol^{-1}) of species optimized at the B3LYP/6-311G(d,p) level [I: 6-311G(d,p); Ia: 6-311+G(3df,2p)]

Species	B3LYP/I	ZPE	MP2/I	MP2/Ia	CCSD(T)/I	G2M(cc2)
CH_3OH	0.0^a	0.0^b	0.0^c	0.0^d	0.0^e	0.0^f
$CH_3 + OH$	85.4	−8.2	97.8	95.2	84.1	91.9
$H + CH_2OH$	91.5	−8.7	90.7	92.1	91.2	96.2
$H_2CO + H_2$	16.9	−9.1	16.7	19.0	15.9	18.1
c-CHOH + H_2	74.6	−9.6	77.8	80.0	72.7	74.8
t-CHOH + H_2	69.9	−9.0	72.3	75.3	67.5	70.5
$^1CH_2 + H_2O$	95.5	−8.4	97.0	96.6	90.6	90.3
$^3CH_2 + H_2O$	83.4	−7.9	79.6	82.0	78.6	80.7
vdw	80.8	−2.7	84.5	86.1	81.1	82.7
TS1	80.9	−4.6	85.8	87.6	82.4	84.1
TS2	87.2	−9.1	98.1	102.4	91.8	98.6
TS3a	83.7	−6.4	91.4	92.6	86.8	88.1
TS3b	81.1	−6.0	88.4	89.9	84.0	85.5
TS4	88.5	−5.8	93.3	90.4	93.2	90.6
MSX	87.3	−7.7	91.1	93.1	85.7	87.0

^{a} −115.706277 au. ^{b} 0.051117 au. ^{c} −115.385188 au. ^{d} −115.462626 au. ^{e} −115.417442 au. ^{f} −115.539263 au.

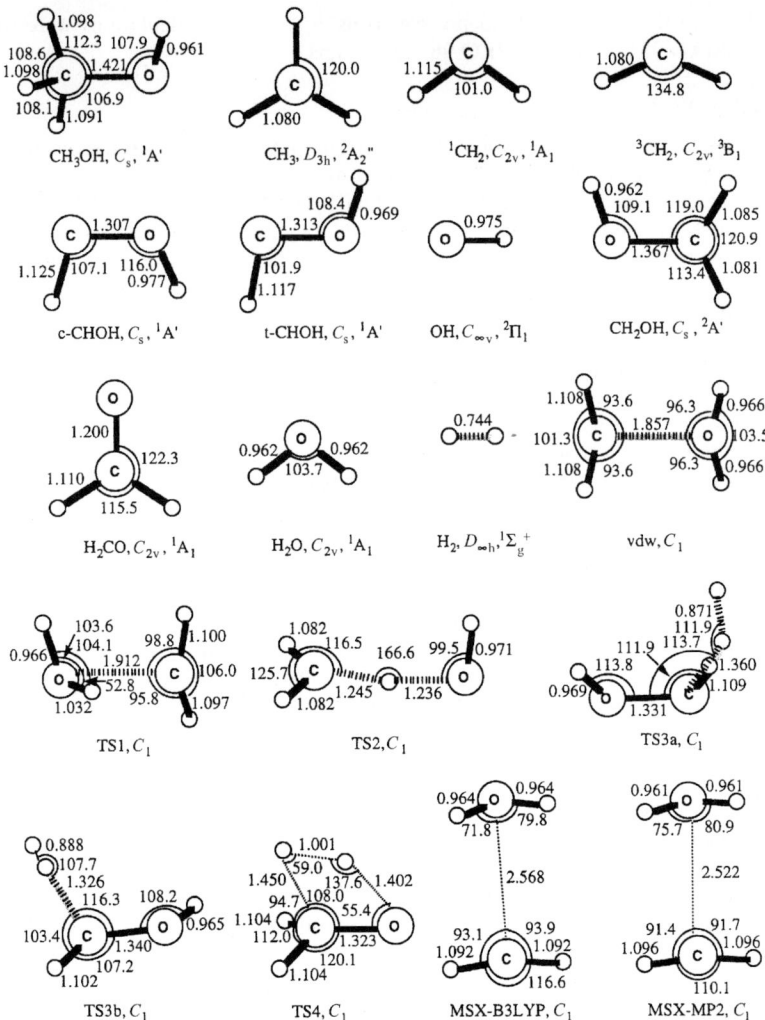

Fig. 1 The optimized geometries of the reactants, intermediates, transition states and products computed at the B3LYP/6-311G(d,p) level.

1. Potential energy surface of the system

The association reaction of CH_3 and OH (channel (1)) by a barrierless process forms CH_3OH with C_s symmetry, the association energy is predicted to be 91.9 kcal mol^{-1} at the G2M level, which is slightly higher than the experimental value, 90.2 kcal mol^{-1}.[43] In channel (2), one of the H atoms in the CH_3 group migrates to the OH group to form a van der Waals complex, vdw, via TS1. The C–O bond length in TS1 is about 0.50 and 0.06 Å longer than that in CH_3OH and the vdw complex, respectively. The barrier height at TS1 is 84.1 kcal mol^{-1} and the complex is 82.7 kcal mol^{-1} above CH_3OH; the results agree with those of 82.3 and 80.7 kcal mol^{-1} calculated by Walch[15] and Harding et al.[14] using CASSCF (CCI)/DZP and RMP4/6-31G (d,p) methods, respectively. The vdw complex dissociates barrierlessly without a well-defined TS to produce products $^1CH_2 + H_2O$ with an endothermicity of 90.3 kcal mol^{-1} above CH_3OH at the G2M level of theory; this value is in good agreement with the experimental result, 89.6 kcal mol^{-1},[43,44] that from the Sandia database,[45] 89.5 kcal mol^{-1} and the value predicted by Walch,[15] 88.8 kcal mol^{-1}. These values are lower than Harding's older result, 94.9 kcal mol^{-1}.[14] Pilling and co-

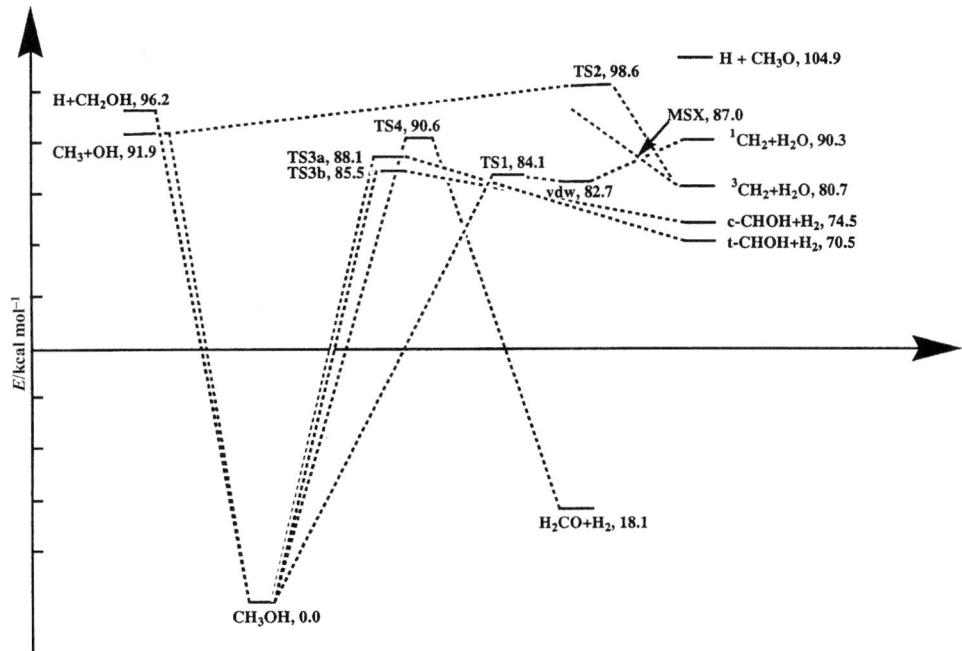

Fig. 2 Schematic energy diagram of the CH_3–OH system computed at the G2M level.

workers[13] estimated the overall enthalpy change from $CH_3 + OH$ to $^1CH_2 + H_2O$ to be endothermic by 0.4 ± 0.5 kcal mol^{-1}. The decomposition by channel (3) is a barrierless process, the calculated energy of $H + CH_2OH$ is 96.2 kcal mol^{-1} higher than that of CH_3OH, which is again in excellent agreement with 96.2 kcal mol^{-1} from the Sandia database.[45] In channel (4), the elimination of H_2 forming H_2CO and H_2 via a four-member ring transition state, TS4, with a barrier height of 90.6 kcal mol^{-1}, which is slightly lower than the values of 91.9 and 96.5 kcal mol^{-1} reported in ref. 15 and 14, respectively. The 18.1 kcal mol^{-1} endothermicity for the production of CH_2O and H_2 is in good agreement with the experimental value, 18.7 kcal mol^{-1} [43] and the value of 18.1 kcal mol^{-1} calculated by Walch.[15] Channels (5) and (6) correspond to the H_2 elimination from the CH_3 group to form cis- and trans-CHOH through TS3a and TS3b, which lie 88.1 and 85.5 kcal mol^{-1} above CH_3OH, respectively; these barriers may be compared with those reported by Walch[15] and Harding et al.[14] of 85.0 and 90.2 kcal mol^{-1} for H_2 + HCOH.

According to the known kinetics of the $^1CH_2 + H_2O$ reactions studied by Wagner and coworkers,[46–49] the $^3CH_2 + H_2O$ products can, in principle, be formed by surface crossing from the singlet to the triplet surface in the vicinity of the vdw complex. The result of our search indeed located a seam-of-crossing (MSX) between the singlet and triplet PES with the configuration, $H_2C\cdots OH_2$, which lies 4.3 kcal mol^{-1} above the complex (see Fig. 2); the structures of MSX calculated at B3LYP/6-311G (d,p) and MP2/6-311G (d,p) levels are similar (see Fig. 1) and the C–O bond length in the MSX is 2.568 and 2.522 Å at the above levels, respectively. The $^3CH_2 + H_2O$ products can also be formed by the direct abstraction reaction via a triplet surface. The barrier for the reaction was predicted to be 6.7 kcal mol^{-1}, which is close to the 6.0 kcal mol^{-1} obtained by Wilson and Balint-Kurti[50] using the CASSCF/cc-pvdz//MP2/cc-pvdz method. According to the result of our preliminary calculation, the triplet surface can cross over to the singlet surface near the $^1CH_2 + H_2O$ plateau with the $H_2C\cdots HOH$ configuration, corresponding to the abstraction process.

The search for minima on the seam of crossing (MSX) between singlet and triplet PES of CH_3OH has been performed using the program written by Cui, Dunn and Morokuma.[51] This approach uses the Lagrange multiplier method and the wavefunctions for intersecting electronic states are calculated separately. The intersection can be investigated at various levels of theory

Table 2 Frequencies and moments of inertia for TSs, reactants, products, and van der Waals complex with the geometry optimized at the B3LYP/6-311G(d,p) level

Species	I_i/au	ν_j/cm^{-1}
HO	0.00000, 3.21921, 3.21921	3703
CH$_3$	6.29682, 6.29682, 12.59365	501, 1402, 1402, 3108, 3287, 3287
CH$_3$OH	14.08662, 72.97294, 75.60766	331, 1053, 1085, 1167, 1379, 1489, 1492, 1508, 2975, 3017, 3103, 3839
CH$_2$OH	9.37199, 60.42662, 69.11064	441, 586, 1062, 1207, 1367, 1488, 3111, 3256, 3840
H$_2$CO	6.34252, 46.23039, 52.57291	1203, 1270, 1539, 1826, 2869, 2918
H$_2$	0.00000, 0.99655, 0.99655	4419
c-CHOH	6.45499, 49.54586, 56.00084	1023, 1210, 1334, 1477, 2721, 3518
t-CHOH	6.26219, 49.45364, 55.71582	1109, 1223, 1330, 1519, 2829, 3718
^{1}CH$_2$	3.09966, 5.32602, 8.42568	1405, 2884, 2946
^{3}CH$_2$	1.06310, 7.15159, 8.21469	1056, 3118, 3359
H$_2$O	2.25747, 4.11907, 6.37654	1640, 3810, 3906
vdw	14.47378, 105.49810, 109.86914	137, 360, 611, 645, 1055, 1138, 1421, 1659, 2915, 2974 3785, 3860
TS1	13.53666, 109.78968, 113.40013	627(i), 350, 447, 713, 944, 1077, 1413, 1556, 2796, 3006 3090, 3807
TS2	9.95043, 181.48108, 184.87358	1220(i), 139, 309, 447, 516, 775, 1089, 1238, 1335, 3131 3302, 3758
TS3a	19.50989, 72.59660, 78.55656	803(i), 579, 667, 929, 1069, 1247, 1263, 1436, 1547, 2621 2915, 3682
TS3b	18.65191, 72.44493, 78.04774	899(i), 582, 677, 982, 1088, 1198, 1314, 1458, 1610, 2515 3011, 3778
TS4	17.52319, 65.04985, 70.50916	2202(i), 886, 905, 1145, 1180, 1264, 1447, 1509, 1915 2275, 2918, 2950
^{1}msx(dft)	14.28553, 188.21551, 194.44229	264(i), 167(i), 154, 343, 501, 575, 1221, 1638, 3020, 3193 3805, 3886
^{3}msx(dft)	14.28553, 188.21551, 194.44229	121(i), 184, 260, 271, 381, 622, 1313, 1656, 3046, 3186 3795, 3882
^{1}msx(mp2)	14.38278, 181.99122, 187.13620	197(i), 120(i), 319, 335, 516, 615, 1363, 1676, 3075, 3209 3882, 3975
^{3}msx(mp2)	14.38278, 181.99122, 187.13620	197(i), 235, 284, 297, 378, 738, 1414, 1691, 3095, 3199 3874, 3972

and the method can work within any *ab initio* or density functional approximation if the energies and gradients for the two states are provided. The computer program is written as an external script that takes the energies and gradients for the current geometry from the *ab initio* program (GAUSSIAN) and generates a next geometry for which the calculations of energies and gradients are run again. The iterations continue until a MSX is located and then the single point energy of MSX was calculated at the G2M level.

2. Rate constant calculations

As described before, the rate constants for the unimolecular decomposition of CH$_3$OH and its related bimolecular processes have been computed with the Variflex code of Klippenstein et al.[34] On account of the absence of well-defined transition states for the fragmentation reactions producing CH$_3$ + OH and ^{1}CH$_2$ + H$_2$O, their association potential functions were computed variationally to cover the range of C–O separations from 2.1 to 3.5 Å for the former and from 2.0 to 7.0 Å for the latter. The computed potential energies could be fitted to the Varshni function with the parameters of $\beta = 0.446$ and 0.178 Å^{-2} for the former and the latter process, respectively. The L-J parameters of CH$_3$OH, Ar, He, N$_2$ and SF$_6$ are taken from the literature.[52,53]

A. Unimolecular decomposition of CH$_3$OH. On the basis of the PES and the mechanism employed in kinetic modeling for the CH$_3$ + OH reaction by Pilling and co-workers,[13] the unimolecular decomposition of CH$_3$OH may occur by several accessible low-energy paths:

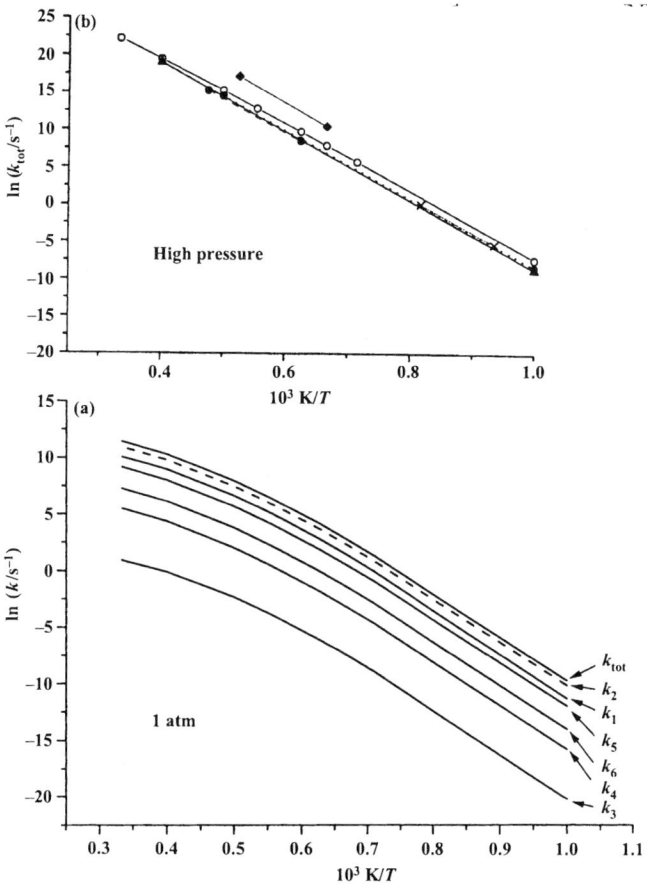

Fig. 3 (a) Predicted CH$_3$OH decomposition individual and total rate constants at 1 atm as functions of temperature. (b) Comparison of predicted and experimental total rate constants at high pressure: (○) this work, (▲) ref. 11, (●) ref. 16, (⊕) ref. 6, (◆) ref. 5, (×) ref. 2.

$$CH_3OH \rightarrow CH_3 + OH \quad (1)$$
$$\rightarrow {}^1CH_2 + H_2O \quad (2)$$
$$\rightarrow H + CH_2OH \quad (3)$$
$$\rightarrow H_2 + CH_2O \quad (4)$$
$$\rightarrow H_2 + cis\text{-HCOH} \quad (5)$$
$$\rightarrow H_2 + trans\text{-HCOH} \quad (6)$$

In addition, the $^3CH_2 + H_2O$ products may be formed by surface crossing at MSX, which is predicted to be located at 87 kcal mol^{-1} above CH$_3$OH or 4.3 kcal mol^{-1} above the H$_2$C···OH$_2$ complex. For the $^1CH_2 + H_2O$ formation, because of the presence of the molecular complex and the possibility of surface crossing, we assessed the location of the rate-limiting dividing surface at TS1, MSX (assuming unit crossing probability) or the variational TS beyond the molecular complex. The result of the test with the Variflex code coupling all 6 product channels indicates that TS1 controls the production of CH$_2$ at either electronic state above 500 K (*vide infra*). Accordingly, in all subsequent calculations for the decomposition of CH$_3$OH at high temperatures, TS1 was employed as the sole transition state for the production of H$_2$O.

The results of calculations by Variflex gave the total rate constants in Ar at the high- and low-pressure limits: $k_{tot}^\infty = 1.56 \times 10^{16}$ exp($-44310/T$) s^{-1} and $k^0 = 1.60 \times 10^{36}$ $T^{-12.2}$

Table 3 The formulae for the individual and total rate constants predicted at 1, 10 and 50 atm Ar in units of s^{-1}

k_i/s^{-1}	1 atm	10 atm	50 atm
k_1 $CH_3OH \rightarrow CH_3 + OH$	$1.86 \times 10^{62} T^{-14.3} \exp(-55709/T)$	$7.41 \times 10^{62} T^{-14.1} \exp(-53861/T)$	$2.72 \times 10^{63} T^{-14.0} \exp(-57859/T)$
k_2 $CH_3OH \rightarrow {}^1CH_2 + H_2O$	$2.68 \times 10^{59} T^{-13.5} \exp(-53862/T)$	$3.24 \times 10^{59} T^{-13.3} \exp(-548171/T)$	$4.03 \times 10^{59} T^{-13.2} \exp(-55585/T)$
k_3 $CH_3OH \rightarrow H + CH_2OH$	$3.78 \times 10^{61} T^{-15.3} \exp(-56891/T)$	$4.40 \times 10^{62} T^{-14.8} \exp(-58029/T)$	$3.75 \times 10^{63} T^{-14.6} \exp(-59239/T)$
k_4 $CH_3OH \rightarrow H_2CO + H_2$	$1.53 \times 10^{58} T^{-13.8} \exp(-54614/T)$	$4.66 \times 10^{58} T^{-13.6} \exp(-55758/T)$	$1.68 \times 10^{59} T^{-13.6} \exp(-56777/T)$
k_5 $CH_3OH \rightarrow c\text{-}CHOH + H_2$	$7.36 \times 10^{58} T^{-13.5} \exp(-54013/T)$	$1.16 \times 10^{57} T^{-13.4} \exp(-55026/T)$	$1.75 \times 10^{59} T^{-13.3} \exp(-55855/T)$
k_6 $CH_3OH \rightarrow t\text{-}CHOH + H_2$	$1.08 \times 10^{59} T^{-13.8} \exp(-54658/T)$	$2.82 \times 10^{59} T^{-13.6} \exp(-55762/T)$	$9.18 \times 10^{59} T^{-13.6} \exp(-56758/T)$
k_{tot}	$2.65 \times 10^{60} T^{-13.7} \exp(-54323/T)$	$4.32 \times 10^{61} T^{-13.7} \exp(-55929/T)$	$3.67 \times 10^{62} T^{-13.7} \exp(-57192/T)$

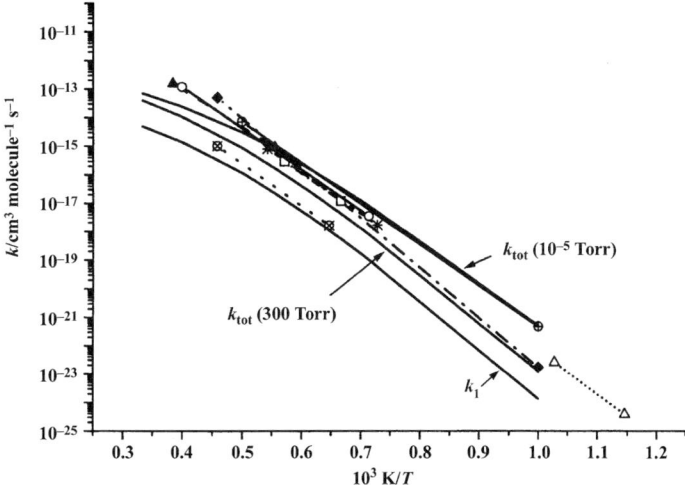

Fig. 4 The comparison of the second order decomposition rate constants between predicted and experimental values. Solid lines are the predicted values and symbols are experimental results: (▲) ref. 7, (●) ref. 1, (◆) ref. 4, (∗) ref. 8, (△) ref. 3, (□) ref. 9, (○) ref. 10, (⊕) ref. 11.

$\exp(-48140/T)$ cm^3 molecule^{-1} s^{-1} covering the temperature range 1000–3000 K. The downward energy transfer stepsize was assumed to be 400 cm^{-1} for Ar. It is worth noting that under high-pressure conditions, the decomposition reaction is dominated by the production of CH$_3$ and OH ($E_0 = 91$ kcal mol^{-1}), whereas under low-pressure conditions the decomposition reaction, which is controlled by the upward collisional energy transfer, produces predominantly the products with lower energy barriers, such as ^{1}CH$_2$ + H$_2$O and H$_2$ + cis/trans-HCOH. The individual rate constants predicted at 1, 10 and 50 atm are summarized in Table 3, and those for 1 atm are also presented graphically in Fig. 3. These results clearly show that product yields for the various channels are strongly influenced by the pressure and temperature of the system. Dombrowsky *et*

Table 4 Comparison of the predicted and measured total rate constants for CH$_3$ + OH at various pressures and temperatures[a]

T/K	M	P/Torr	k_p
290	He	∞	2.2(0.82[b])
473	He	∞	1.7(0.56[b])
700	He	∞	1.4(0.41[b])
298	He	∞	2.2(1.7[c], 0.35[d])
		0	6.9(20[c], 25[d])
298	He	∞	2.2(0.73[e])
294	Ar	750	1.2(0.94[f])
1200	Ar	760	0.17(0.18[g])
296	N$_2$	760	1.8(0.93[h])
200	SF$_6$	∞	2.8(2.5[i])
298	SF$_6$	∞	2.2(2.6[i])
		0	2.8(72[i])
400	SF$_6$	∞	1.9(2.6[i])

[a] k_p is given in units of 10^{-10} cm^3 molecule^{-1} s^{-1} for the second-order and 10^{-28} cm^6 molecule^{-2} s^{-1} for the third-order rate coefficient. Measured values are given in parentheses. [b] Ref. 13. [c] Ref. 24. [d] Ref. 23. [e] Ref. 46. [f] Ref. 21. [g] Ref. 20. [h] Ref. 19. [i] Ref. 25.

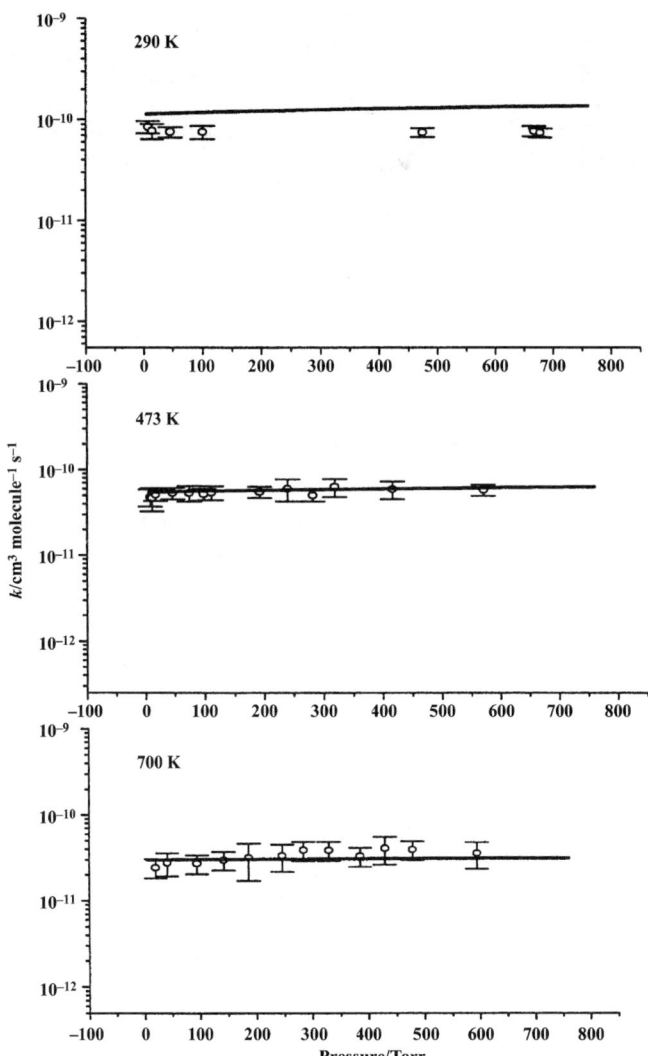

Fig. 5 Predicted rate constants for $CH_3 + OH \rightarrow$ products in comparison with the experimental data at different temperatures. Solid lines are the predicted values and symbols are the results from ref. 13.

al.[9] reported that at pressures between 96 and 686 Torr, the thermal decomposition of CH_3OH in the temperature range 1400–2200 K produced $CH_3 + OH$ with about 80% yields and less than 16% and 4% $^1CH_2 + H_2O$ and $H + CH_2OH/CH_3O$, respectively. This product distribution is consistent with our high-pressure limit value, but is inconsistent with our medium pressure (0.5–1 atm) result.

Fig. 4 presents the predicted second-order rate coefficients for all product formation (k_{tot}) as well as for $CH_3 + OH$ production (k_1), which has been assumed in most previous data analyses. These results are compared with shock-tube data typically obtained at 0.5 atm pressure with Ar as diluent. The predicted values lie within the scatter of the existing data. It should be noted that the second-order rate coefficients evaluated with the 300 Torr and very low-pressure data differ noticeably below 1500 K, clearly suggesting that at 300 Torr, the decomposition reaction is still not fully in the second-order region. In addition, k_1 is noticeably lower than k_{tot}.

B. Bimolecular $CH_3 + OH$ reactions. The bimolecular reaction of CH_3 with OH may occur by direct abstraction and association/fragmentation mechanisms. The direct abstraction process

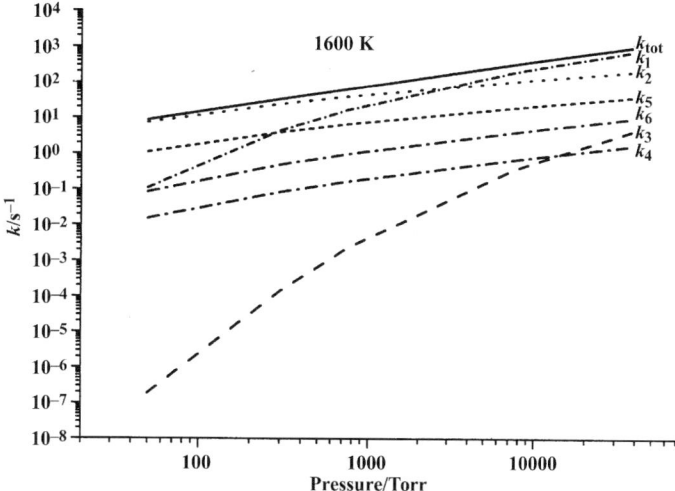

Fig. 6 The pressure-dependent individual and total rate constants for the decomposition of CH_3OH at 1600 K.

taking place by a triplet surface producing $^3CH_2 + H_2O$ via TS2 was predicted to have a well-defined barrier, which lies 6.7 kcal mol^{-1} above the reactants. The TST calculation by Variflex gives rise to the following non-linear Arrhenius expression:

$$k_7 = 2.0 \times 10^{-22} \times T^{3.39} \exp(-1412/T) \text{ cm}^3 \text{ molecule}^{-1} \text{ s}^{-1}$$

covering the range 300–3000 K. The predicted reaction barrier, 6.7 kcal mol^{-1}, agrees reasonably with that predicted by Wilson and Balint-Kurti,[50] 6.0 kcal mol^{-1}, at the CASSCF/cc-pvdz//MP2/cc-pvdz level of theory as alluded to before.

The association reaction of CH_3 with OH has been studied by many investigators.[18–26] Under high-pressure conditions, the reaction occurs primarily by the recombination/stabilization process producing CH_3OH. Under low-pressure conditions, the reaction may yield CH_3OH fragmentation products such as $^1CH_2 + H_2O$[13,46] and $H_2 + HCOH/CH_2O$.[12,18] In Table 4, we compare the predicted total rate constants with the values measured with different inert gases (He, Ar, N$_2$ and SF$_6$, with their $\langle \Delta E \rangle_{down}$ taken to be 40, 400, 210 and 1000 cm^{-1} respectively). The agreement between theory and experiment is seen to be better than a factor of two for most cases. With SF$_6$ as the diluent, the most efficient third-body studied, the agreement between the predicted

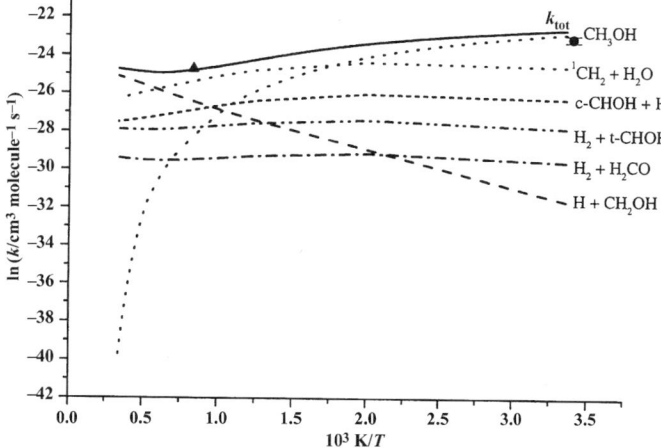

Fig. 7 Bimolecular reaction rate constants for $CH_3 + OH \rightarrow$ products in 760 Torr Ar covering the range 300–3000 K. Curves are the predicted results. Symbols are from experiments: (▲) ref. 20, (●) ref. 21.

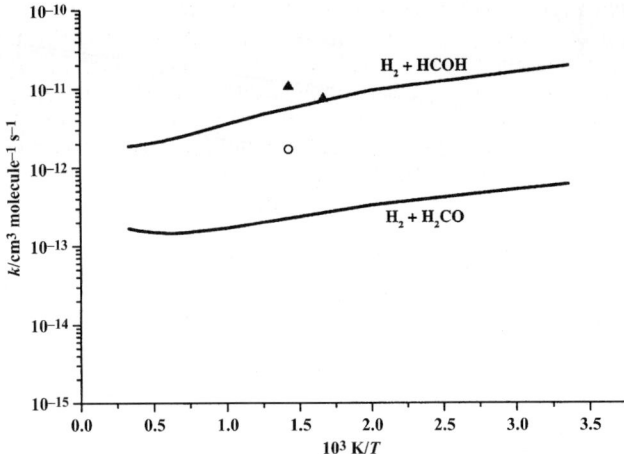

Fig. 8 Comparison of the rate constants for $CH_3 + OH \rightarrow H_2 + HCOH$ and $H_2 + H_2CO$ channels with experimental results from ref. 12.

high-pressure limits and the extrapolated data of Fagerstrom et al.[25] is excellent for the 200–400 K temperature range. However, the predicted low-pressure limit is about a factor of 30 smaller than the extrapolated value, suggesting that the extrapolation to the low-pressure limit might not be adequate. Bott and Cohen[20] measured the total rate constant for the $CH_3 + OH$ reaction by shock-heating in Ar to 1200 K, they reported the value, 1.8×10^{-11} cm^3 molecule^{-1} s^{-1}, which agrees closely with our result, 1.7×10^{-11} cm^3 molecule^{-1} s^{-1}. With He as diluent, Pilling and co-workers[13] measured the recombination rate as a function of pressure; their results reported for

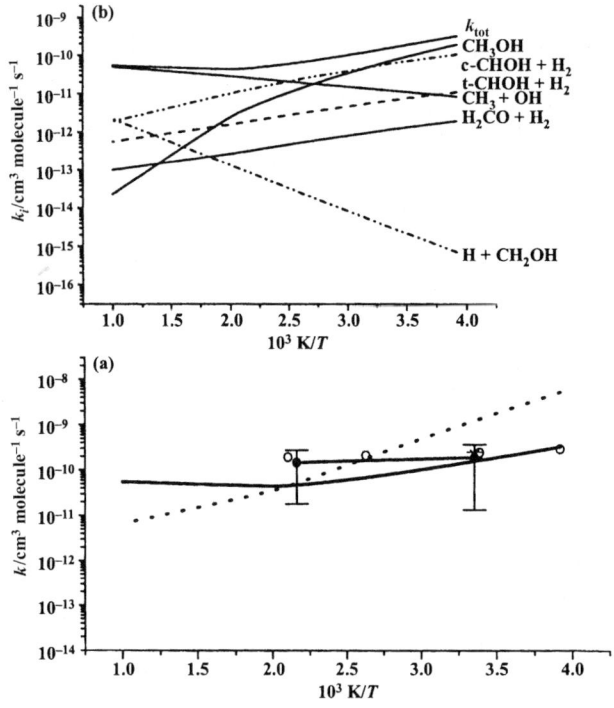

Fig. 9 (a) Comparison of the rate constants for $^1CH_2 + OH \rightarrow$ products with experimental results at 5 Torr in Ar. The solid line is the result assuming $^1CH_2 + OH \rightarrow H_2C\cdots OH_2$ and the dotted line assuming TS1 being the controlling barrier, respectively. Symbols are from experiments: (×) ref. 47, (○) ref. 49, (●) ref 55. (b) The branching ratios for all channels mentioned in the text.

the three temperatures are compared in Fig. 5 with the predicted values using $\langle \Delta E \rangle_{down} = 40$ cm^{-1}. The observed data could be reasonably accounted for theoretically. To fit these experimental data, Pilling and co-workers[13] had to use a rather large step-size, $\langle \Delta E \rangle_{down} = 230$ cm^{-1}.

Because of the strong C–O bond in CH_3OH, the reaction of CH_3 with OH producing various products must be strongly pressure dependent. Fig. 6 presents the branching rate constants for all products produced by the decomposition of the chemically activated CH_3OH at 1600 K and Fig. 7 presents the effect of temperature at 760 Torr Ar pressure. The predicted total rate constants for $CH_3 + OH$ at room temperature and 1200 K are seen to agree reasonably with experimental values.

At low pressures where collisional deactivation of excited CH_3OH becomes less competitive, both dehydration and dehydrogenation processes producing $^1CH_2 + H_2O$ and $H_2 + HCOH/CH_2O$, respectively, become measurable. Wagner and co-workers[46] studied the formation of $^1CH_2 + H_2O$ from $CH_3 + OH$ at 298 K and 1.3 Torr He. The yield of $^1CH_2 + H_2O$ was determined to be 89%, which agrees closely with our predicted values of 88% and 81% assuming TS1 and the variational TS leading to the formation of the products as the rate-limiting step, respectively. In this calculation, the effect of surface crossing from the singlet to the triplet state, which cannot be readily accounted for at present, was fully ignored.

The formation of $H_2 + HCOH$ and CH_2O from $CH_3 + OH$ was studied by Humpfer et al.[12] at 600 and 700 K under low-pressure conditions. Their results are compared with the predicted values in Fig. 8. For the HCOH product, which includes both cis- and trans-isomers, our theoretical result agrees reasonably with the experimental values, although with a negative, instead of positive temperature dependence as reported. The negative temperature dependence is expected theoretically because both TS3a and TS3b lie below the $CH_3 + OH$ reactants. For the $H_2 + CH_2O$ channel, the yield of CH_2O was found to be considerably higher than the predicted value.

3. **The $^1CH_2 + H_2O$ reaction.** The kinetics of the $^1CH_2 + H_2O$ reaction has been studied by Wagner, King and their collaborators[46–48,54,55] by laser induced fluorescence. Wagner et al.[46–48] also determined the yields of $CH_3 + OH$ from the reaction. Fig. 9 compares the predicted total rate constant with experimentally determined results, again excluding the electronic quenching contribution. The theoretical values calculated under low-pressure conditions ($P \leq 5$ Torr), with either TS1 or the variational TS for the decomposition of the $H_2C \cdots OH_2$ complex as the rate-limiting step, agree satisfactorily with the experimentally results. In addition, the yields of $CH_3 + OH$ from this reaction were predicted to be strongly temperature dependent, 16% at 295 K, 44% at 380 K and 69% at 475 K; they compare reasonably with Wagner's room temperature results, $50 \pm 20\%$[48] and 50–100%.[47]

IV. Conclusions

The kinetics and mechanisms for the thermal decomposition of CH_3OH and the related bimolecular reactions, $CH_3 + OH$ and $^1CH_2 + H_2O$, have been investigated by high-level molecular orbital (G2M) and variational RRKM calculations over a wide range of reaction conditions. Under high-pressure ($P > 10^3$ Torr) and high-temperature ($T > 1000$ K) combustion conditions, the decomposition of CH_3OH produces primarily the $CH_3 + OH$ products as has been assumed in most earlier kinetic modeling of the reaction in shock waves. At lower pressures ($P < 1$ atm), the formation of $^1CH_2 + H_2O$ was found to be dominant, assuming the reaction to be controlled either by the transition state leading to the formation of the $H_2C \cdots OH$ van der Waals complex or by the variational TS leading to the formation of the $^1CH_2 + H_2O$ products.

For the $CH_3 + OH$ reaction, which occurs primarily by the association/decomposition process via vibrationally excited CH_3OH, the predicted total rate constant and the product yields of $^1CH_2 + H_2O$ and $H_2 + HCOH$ agree reasonably with exppherimental data. Similarly, the total rate constant and the yield of $CH_3 + OH$ from $^1CH_2 + H_2O$ could be reasonably accounted for theoretically. The $CH_3 + OH$ reaction by the direct abstraction mechanism over the triplet surface producing $^3CH_2 + H_2O$, which has a predicted barrier of 6.7 kcal mol^{-1}, was concluded to be unimportant although it may play a role in the production of $^3CH_2$ under combustion conditions.

Acknowledgements

This work is sponsored partially by the Basic Energy Science, Department of Energy under grant No. DE-FG02-97-ER14784 (to RZ) and partially by the Caltech Multidisciplinary University Research Initiative under ONR grant No. N00014-95-1-1388 (to MCL).

References

1. C. T. Bowman, *Combust. Flame*, 1975, **25**, 343.
2. D. Aronowitz, D. W. Naegeli and I. Glassman, *J. Phys. Chem.*, 1977, **81**, 2555.
3. M. Cathonnet, J.–C. Boettner and H. James, *J. Chim. Phys.*, 1979, **76**, 183.
4. C. K. Westbrook and F. L. Dryer, *Combust. Sci. Technol.*, 1979, **20**, 125.
5. T. Tsuboi, M. Katoh, S. Kikuchi and K. Hashimoto, *Jpn. J. Appl. Phys.*, 1981, **20**, 985.
6. K. Spindler and H. Gg. Wagner, *Ber. Bunsen-Ges. Phys. Chem.*, 1982, **86**, 2.
7. P. H. Cribb, J. E. Dove and S. Yamazaki, *Symp. (Int.) Combust.*, [*Proc.*], 1985, **20**, 779; P. H. Cribb, J. E. Dove and S. Yamazaki, *Combust. Flame*, 1992, **88**, 169.
8. Y. Hidaka, T. Oki and H. Kawano, *J. Phys. Chem.*, 1989, **93**, 7134.
9. Ch. Dombrowsky, A. Hoffmann, M. Klatt and H. Gg. Wagner, *Ber. Bunsen-Ges. Phys. Chem.*, 1991, **95**, 1685.
10. T. Koike, M. Kudo, I. Maeda and H. Yamada, *Int. J. Chem. Kinet.*, 2000, **32**, 1.
11. W. Tsang, *J. Phys. Chem. Ref. Data*, 1987, **16**, 471.
12. R. Humpfer, H. Oser and H. H. Grotheer, *Int. J. Chem. Kinet.*, 1995, **27**, 577.
13. R. De. Avillez Pereira, D. L. Baulch, M. J. Pilling, S. H. Robertson and G. Zeng, *J. Phys. Chem. A*, 1997, **101**, 9681.
14. L. B. Harding, H. B. Schlegel, R. Krishnan and J. A. Pople, *J. Phys. Chem.*, 1980, **84**, 3394.
15. S. P. Walch, *J. Chem. Phys.*, 1993, **98**, 3163.
16. D. L. Baulch, C. J. Cobos, R. A. Cox, P. Frank, G. Hayman, Th. Just, J. A. Kerr, T. Murrells, M. J. Pilling, J. Troe, R. W. Walker and J. Warnatz, *J. Phys. Chem. Ref. Data*, 1994, **23**, 847.
17. J. J. Lin, J. Shu, Y. T. Lee and X. Yang, *J. Chem. Phys.*, 2000, **113**, 5287.
18. H. Oser, N. D. Stothard, R. Humpfer, H. H. Grotheer and T. Just, *Symp. (Int.) Combust.* [*Proc.*], 1992, **24**, 597.
19. T. J. Sworski, C. J. Hochanadel and P. J. Ogren, *J. Phys. Chem.*, 1980, **84**, 129.
20. J. F. Bott and N. Cohen, *Int. J. Chem. Kinet.*, 1991, **23**, 1071.
21. C. Anastasi, S. Beverton, T. Ellermann and P. Pagsberg, *J. Chem. Soc., Faraday Trans.*, 1991, **87**, 2325.
22. K. J. Hughes, A. R. Pereira and M. J. Pilling, *Ber. Bunsen-Ges. Phys. Chem.*, 1992, **96**, 1352.
23. H. Oser, N. D. Stothard, R. Humpfer and H. H. Grotheer, *J. Phys. Chem.*, 1992, **96**, 5359.
24. K. A. Bhashkaran, P. Frank and Th. Just, *Proc. Int. Symp. Shock Tubes Waves*, 1980, **2**, 503.
25. K. Fagerstrom, A. Lund, G. Mahmoud, J. T. Jodkowski and E. Ratajczak, *Chem. Phys. Lett.*, 1993, **204**, 226.
26. W. Tsang and R. F. Hampson, *J. Phys. Chem. Ref. Data*, 1986, **15**, 1087.
27. A. D. Becke, *J. Chem. Phys.*, 1993, **98**, 5648.
28. A. D. Becke, *J. Chem. Phys.*, 1992, **96**, 2155.
29. A. D. Becke, *J. Chem. Phys.*, 1992, **97**, 9173.
30. C. Lee, W. Yang and R. G. Parr, *Phys. Rev. B*, 1988, **37**, 785.
31. A. M. Mebel, K. Morokuma and M. C. Lin, *J. Chem. Phys.*, 1995, **103**, 7414.
32. M. J. Frisch, G. W. Trucks, H. B. Schlegel, P. M. W. Gill, B. G. Johnson, M. A. Robb, J. R. Cheeseman, T. Keith, G. A. Petersson, J. A. Montgomery, K. Raghavachari, M. A. Al-Laham, V. G. Zakrzewski, J. V. Ortiz, J. B. Foresman, J. Cioslowski, B. B. Stefanov, A. Nanayakkara, M. Challacombe, C. Y. Peng, P. Y. Ayala, W. Chen, M. W. Wong, J. L. Andres, E. S. Replogle, R. Gomperts, R. L. Martin, D. J. Fox, J. S. Binkley, D. J. Defrees, J. Baker, J. P. Stewart, M. Head-Gordon, C. Gonzalez and J. A. Pople, *GAUSSIAN 98, REVISION A.1*, Gaussian, Inc., Pittsburgh PA, 1998.
33. *MOLPRO* is a package of *ab initio* programs written by H.-J. Werner and P. J. Knowles, with contributions from J. Almlöf, R. D. Amos, A. Berning, D. L. Cooper, M. J. O. Deegan, A. J. Dobbyn, F. Eckert, S. T. Elbert, C. Hampel, R. Lindh, A. W. Lloyd, W. Meyer, A. Nicklass, K. Peterson, R. Pitzer, A. J. Stone, P. R. Taylor, M. E. Mura, P. Pulay, M. Schütz, H. Stoll and T. Thorsteinsson.
34. S. J. Klippenstein, A. F. Wagner, R. C. Dunbar, D. M. Wardlaw and S. H. Robertson, *VARIFLEX: VERSION 1.00*, 1999.
35. R. G. Gilbert and S. C. Smith, *Theory of Unimolecular and Recombination Reactions*, Blackwell Scientific, Carlton, Australia, 1990.
36. K. A. Holbrook, M. J. Pilling and S. H. Robertson, *Unimolecular Reactions*, Wiley, New York, 1996.
37. J. Troe, *J. Chem. Phys.*, 1977, **66**, 6745.
38. R. S. Zhu and M. C. Lin, *J. Phys. Chem. A*, 2000, **104**, 10807.
39. W. S. Xia and M. C. Lin, *PhysChemComm*, 2000, 13.
40. W. S. Xia and M. C. Lin, *J. Chem. Phys.*, 2001, **114**, 4522.

41. R. S. Zhu and M. C. Lin, *J. Phys. Chem. A*, 2001, **105**, 243.
42. V. P. Varshni, *Rev. Mod. Phys.*, 1957, **29**, 664.
43. M. W. Chase, Jr., *J. Phys. Chem. Ref. Data*, 1998, Monograph, No. 9.
44. P. R. Bunker, P. Jensen, W. P. Kraemer and R. Beardsworth, *J. Chem. Phys.*, 1985, **85**, 3724.
45. R. J. Lee, F. M. Ripley and J. A. Miller, *Chemkin Thermodynamic Data Base. SANDIA Report 87-8215 UC4*, Sandia National Laboratories, Livermore, CA, 1987.
46. R. Deters, M. Otting, H. G. Wagner, F. Temps, B. Laszlo, S. Dobe and T. Berces, *Ber. Bunsen-Ges. Phys. Chem.*, 1998, **102**, 58.
47. W. Hack, H. G. Wagner and A. Wilms, *Ber. Bunsen-Ges. Phys. Chem.*, 1988, **92**, 620.
48. H. H. Carstensen and H. G. Wagner, *Ber.-Max-Planck-Inst. Stroemungsforsch*, 1993, **9**, 31.
49. H. H. Carstensen and H. G. Wagner, *Ber. Bunsen-Ges. Phys. Chem.*, 1995, **92**, 1539.
50. C. Wilson and G. G. Balint-Kurti, *J. Phys. Chem. A*, 1998, **102**, 1625.
51. Some details of the algorithm are described in K. Dunn and K. Morokuma, *J. Chem. Phys.*, 1995, **102**, 4904 and Q. Cui, Ph.D. Thesis, Emory University, Atlanta, 1998. The program uses the HONDO 8.4 package by M. Dupuis.
52. H. Hippler, J. Troe and H. J. Wendelken, *J. Chem. Phys.*, 1983, **78**, 6709.
53. F. M. Mourits and F. H. A. Rummens, *Can. J. Chem.*, 1977, **55**, 3007.
54. G. J. Gutsche, W. D. Lawrance, W. S. Staker and K. D. King, *J. Phys. Chem.*, 1995, **99**, 11867.
55. F. Hayes, W. D. Lawrance, W. S. Staker and K. D. King, *J. Phys. Chem.*, 1996, **100**, 11314.

A direct transition state theory based analysis of the branching in $NH_2 + NO$

De-Cai Fang,[a] Lawrence B. Harding,*[a] Stephen J. Klippenstein*[b] and James A. Miller*[b]

[a] *Chemistry Division, Argonne National Laboratory, Argonne, IL, 60439, USA*
[b] *Combustion Research Facility, Sandia National Laboratories, Livermore, CA, 94551-0969, USA*

Received 8th March 2001
First published as an Advance Article on the web 11th October 2001

A combination of high-level quantum-chemical simulations and sophisticated transition state theory analyses is employed in a study of the temperature dependence of the $N_2H + OH \to HNNOH$ recombination reaction. The implications for the branching between $N_2H + OH$ and $N_2 + H_2O$ in the $NH_2 + NO$ reaction are also explored. The transition state partition function for the $N_2H + OH$ recombination reaction is evaluated with a direct implementation of variable reaction coordinate (VRC) transition state theory (TST). The orientation dependent interaction energies are directly determined at the CAS + 1 + 2/cc-pvdz level. Corrections for basis set limitations are obtained *via* calculations along the *cis* and *trans* minimum energy paths employing an ~aug-pvtz basis set. The calculated rate constant for the $N_2H + OH \to HNNOH$ recombination is found to decrease significantly with increasing temperature, in agreement with the predictions of our earlier theoretical study. Conventional transition state theory analyses, employing new coupled cluster estimates for the vibrational frequencies and energies at the saddlepoints along the $NH_2 + NO$ reaction pathway, are coupled with the VRC-TST analyses for the $N_2H + OH$ channels to provide estimates for the branching in the $NH_2 + NO$ reaction. Modest variations in the exothermicity of the reaction (1–2 kcal mol^{-1}), and in a few of the saddlepoint energies (2–4 kcal mol^{-1}), yield TST based predictions for the branching fraction that are in satisfactory agreement with related experimental results. The unmodified results are in reasonable agreement for higher temperatures, but predict too low a branching ratio near room temperature, as well as too steep an initial rise.

I. Introduction

The $NH_2 + NO$ reaction is of key importance in the thermal De-NO_x process.[1–6] The initial step in this reaction is a radical–radical recombination to form NH_2NO. A 1,3-H atom transfer then yields a chemically activated HNNOH species, which is initially formed with a *trans* configuration for the HNNO torsion. This *trans* species can dissociate to form $N_2H + OH$, or isomerize to a *cis* HNNO configuration. The *cis* species may dissociate to any one of at least three different

DOI: 10.1039/b102235k

bimolecular product channels:

$$NH_2 + NO \rightarrow NH_2NO^* \rightarrow HNNOH(trans)^* \rightarrow N_2H + OH \quad (a)$$
$$HNNOH(trans)^* \rightarrow HNNOH(cis)^* \rightarrow N_2H + OH \quad (a')$$
$$\rightarrow N_2 + H_2O \quad (b)$$
$$\rightarrow N_2O + H_2 \quad (c)$$

A schematic diagram of the reaction path energies is provided in Fig. 1. Due to the small molecular size and the shallowness of the potential wells, there is essentially no stabilization of the unimolecular species at ordinary pressures. Note the presence of cis and trans conformers also for the NNOH torsion. However, in contrast with that for the HNNO torsion, the barrier separating these pairs of conformers is low enough that they are best considered as a single species. As illustrated in Fig. 1, there is also a saddlepoint (labeled ts7) directly connecting H_2NNO with $N_2O + H_2$, but this saddlepoint lies too high in energy to be kinetically significant.

The N_2H species is metastable, dissociating on a microsecond or shorter timescale into $N_2 + H$. Importantly, it is only this channel that is chain branching, whereas the other two are chain terminating. As a result, the branching among these channels is of key importance and has been examined in a large number of experimental studies.[7–19] There is considerable scatter in the experimental observations for this branching fraction, due perhaps to the generally indirect nature of the observations. However, in recent years it has become clear that the branching ratio $\alpha \equiv (k_a + k_{a'})/k_T$, where k_T is the total rate coefficient, rises from about 0.1 at room temperature up to about 0.6 or higher at 2000 K. Furthermore, the detailed temperature dependence appears to have an unusual sigmoidal shape, with the peak rate of rise at about 1100 K.

Only a few theoretical studies of this branching ratio have been reported.[20–22] The recent study of Diau and Smith generated and employed high-level ab initio data[23] (building on the work of earlier electronic structure studies[24–29]) for the channels with well-defined saddlepoints, and clearly demonstrated that the branching ratio α should rise significantly with increasing temperature.[21] Furthermore, for the first time, a satisfactory theoretical framework was employed for the "barrierless" channels (i.e., those with Morse-like reaction path potentials) corresponding to the initial $NH_2 + NO$ addition channel and the $N_2H + OH$ forming channels. For these barrierless channels, the absence of sufficient quantum-chemical data led to their use of empirical model potentials. Notably, regardless of the model employed, their predicted rate rose much too rapidly with temperature, for temperatures between 300 and 1000 K.

In an extension and elaboration of Diau and Smith's work, we constructed a model that reproduced the experimentally observed temperature dependence for the branching ratio α.[22] Due to the great exothermicity of ts5 relative to ts6 and ts8, essentially all of the cis isomers dissociate via

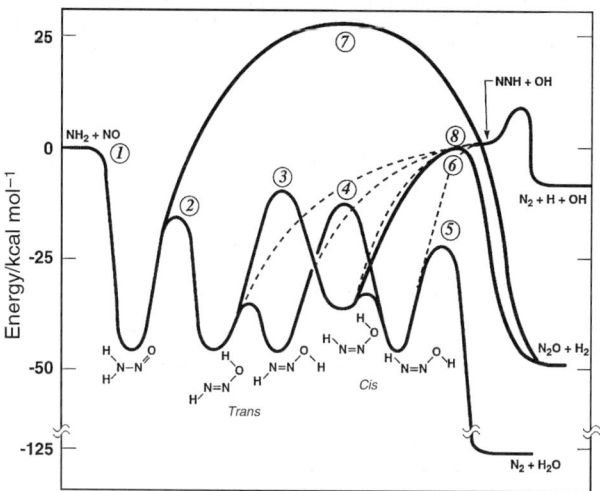

Fig. 1 Reaction coordinate diagram for the $NH_2 + NO$ reaction.

ts5 to $N_2 + H_2O$. As a result, the branching is primarily determined by a competition between the dissociation of the *trans* conformer to $N_2H + OH$ (*via* ts8) and the isomerization from the *trans* to the *cis* HNNO conformer of HNNOH (*via* the ts3/ts4 pair). Correspondingly, the details of the potential employed for the interaction between N_2H and OH (*i.e.*, in ts8) are key determining factors in the predictions for α.

The incorporation of a potential form for the $N_2H + OH$ addition that yielded (i) a fairly deep minimum in its temperature dependence, and (ii) a fairly small rate constant, was required to obtain good agreement with the experimentally observed temperature dependence for α. Furthermore, the best agreement was obtained by allowing the *trans* addition rate initially (*i.e.*, at low temperatures) to be smaller than that for the *cis* addition side and then gradually increase up to the same value as that for the *cis* side, as the temperature was raised from 1100 K. In effect, the latter variation, incorporated *via* the parameter $\gamma \equiv k_{trans}/(k_{trans} + k_{cis})$ in ref. 22, allowed for the use of an even smaller rate constant for the *trans* addition, and for it to have an even more well-defined minimum.

In this work, *a priori* theoretical estimates are obtained for the rate coefficients for the *cis* and *trans* addition of OH to N_2H. The direct VRC-TST formalism is employed in these calculations.[30,31] The orientation-dependent interaction energies are obtained *via* multi-reference configuration interaction calculations. The computed transition state partition functions for the $N_2H + OH$ addition channels are implemented in zero-pressure rate coefficient calculations to provide *a priori* theoretical predictions for the branching ratio α. The latter analysis incorporates hindered rotor treatments for the NNOH modes in the ts3/ts4 pair and in ts5, using newly determined torsional potentials. New coupled-cluster quantum-chemical simulations of the stationary point energies, and vibrational frequencies along the $NH_2 + NO$ reaction pathway are also employed in this analysis.

The methodology and results from the present quantum chemical simulations are presented in Section II. Then, in Section III, a brief summary is given of the present implementation of the direct VRC-TST methodology. Some discussion of the overall kinetic methodology is also provided therein. The resulting predictions for the temperature dependence of the $N_2H + OH$ addition rates, and the $NH_2 + NO$ branching ratio are presented and discussed in Section IV. Some concluding remarks are made in Section V.

II. Potential surface calculations

A. Methodology

Three kinds of electronic structure theory were used: single-reference, coupled-cluster (CC) theory, multi-reference configuration interaction (MRCI) theory, and B3LYP density functional theory. Coupled cluster calculations are generally very accurate for regions of the potential surface that are described well by a single reference configuration. However, single reference configuration methods, such as CC, generally do not yield reliable results for bond cleavage reactions where multi-configuration reference wavefunctions are needed. In this work CC calculations were used to characterize the energies and frequencies of the stationary points (minima and transition states), while MRCI calculations were used for the direct statistical calculations and reduced dimensional surfaces describing the dissociation of the HNNOH adduct into $N_2H + OH$. The B3LYP density functional calculations were used to explore the NNOH torsional potentials in the ts3/ts4 pair and in ts5.

Both the MRCI and CC electronic structure calculations were carried out using the MOLPRO package of codes,[32] while the B3LYP calculations were performed with the GAUSSIAN98 software package.[33] The basis sets used in all calculations were the correlation-consistent basis sets of Dunning.[34-36] The CC calculations employ the open shell, spin unrestricted, coupled cluster theory, CCSD(T), of Knowles *et al.*[37] Geometry optimizations and normal mode frequencies were evaluated using both the cc-pvdz and the cc-pvtz basis sets. For some stationary points, single point calculations using the larger aug-cc-pvtz basis set (at the cc-pvtz geometries) were also done. The aug-cc-pvtz basis set was employed in the B3LYP torsional potential calculations.

All of the MRCI calculations are based on orbitals from state-averaged, complete active space, self-consistent field (CASSCF) calculations and are for only the singlet states.[38] The CASSCF reference wavefunction consists of four active electrons and three active orbitals (the N_2H radical

Table 1 CCSD(T) stationary point energies

Species	cc-pvdz[a]	cc-pvtz[a]	aug-cc-pvtz[b]	G2M[c]
$NH_2 + NO$	0.0	0.0	0.0	0.0
$N_2 + H_2O$	−114.7	−120.5	−121.4	−124.3
$N_2H + OH$	+4.5	+3.3	+2.5	3.6
H_2NNO	−36.4	−41.7		−46.7
$HOHN_2$	−16.4	−23.0		−28.0
HNNOH(c,c)	−32.5	−36.8	−37.2	−40.2
HNNOH(c,t)	−38.0	−42.4	−43.1	−46.3
HNNOH(t,c)	−38.1	−42.8	−43.5	−46.6
HNNOH(t,t)	−35.8	−41.3	−42.7	−46.0
H_2NNO → HNNOH(t,c); ts2	−7.8	−11.6	−12.1	−14.8
HNNOH(t,t) → HNNOH(c,t); ts4	+2.6	−5.1	−7.8	−10.9
HNNOH(t,c) → HNNOH(c,c); ts3	+4.4	−2.5	−4.4	−8.1
HNNOH(t,t) → HNNOH(t,c)	−28.1	−33.4		−37.8
HNNOH(c,c) → HNNOH(c,t)	−29.1	−34.0		−37.4
HNNOH(c,t) → $N_2 + H_2O$; ts5	−15.2	−19.4	−22.1	−24.9

[a] All values include CCSD(T)/cc-pvtz calculated zero-point energy corrections. [b] Single point calculation using the CCSD(T)/cc-pvtz geometries. [c] G2M calculations from ref. 23.

orbital and two oxygen p orbitals). The two lowest roots of this wavefunction were averaged with equal weights. This is the minimum necessary to describe correctly the two degenerate singlet states of the $N_2H + OH$ asymptote. The CASSCF orbitals were used in internally-contracted, multi-reference configuration interaction (CAS + 1 + 2) calculations.[39] The majority of CAS + 1 + 2 calculations employ the cc-pvdz basis set. Both size consistency and basis set corrections were added to the CAS + 1 + 2/cc-pvdz energies in a manner described below.

As noted above, the MRCI calculations focus on the dissociation of the HNNOH adduct, HNNOH → $N_2H + OH$. Since the reverse recombination reaction is barrierless, the dynamical bottlenecks for this process are expected to occur at relatively large distances and will be most influenced by changes in the energy and transitional modes as a function of the distance between the OH and N_2H fragments. Changes involving the internal degrees of freedom of the two fragments, OH and N_2H, are expected to be much less important. Consequently, in these calculations, all of the internal degrees of freedom of the OH and N_2H fragments were kept frozen at the asymptotic values.

To estimate the error associated with the use of the relatively small cc-pvdz basis set and the truncation of the CI at double excitations, additional calculations were carried out using a larger ∼aug-pvtz basis set and a multi-reference Davidson correction[40] (for higher order excitations). The ∼aug-pvtz basis is the same as the Dunning aug-cc-pvtz basis set except that the diffuse nitrogen and oxygen f functions and the diffuse hydrogen d functions were eliminated to make the calculations more tractable. In previous studies of radical–radical recombination reactions we have found one-dimensional basis set corrections (based on the difference between large and small basis set calculations along the minimum energy path, MEP) to be sufficient. For this reaction test calculations showed the correction to be much larger for HON angles near zero than for larger HON angles. Since the HON angle of the MEP varies greatly, from near zero at large distances to ∼100° at shorter distances, the use of a one-dimensional basis set correction that depends only on the distance, not the HON angle, was deemed to be unacceptable.

The HON angle dependence of the basis set correction was evaluated using the following procedure. Eight one-dimensional potential curves (as a function of R_{NO}) were evaluated with the ∼aug-pvtz basis set. For all eight curves the NNO angle was fixed at 114° (approximately the angle of the minimum energy path for addition). In four of the curves the HNNO dihedral angle was fixed at 0° (cis), in the other four it was fixed at 180° (trans). In one pair of cis and trans curves the HON angle was fixed at 0°, in a second pair this angle was fixed at 180°. In the final four curves (two cis and two trans) the HON angle was fixed at 90° and the HONN dihedral angle was set to either 0° or 180°. The same curves were evaluated with the smaller, cc-pvdz basis set and the differences, ΔE, were fit to two functions (one for cis HNNO and one for trans) of the following form,

$$\Delta E(R_{NO}, \theta, \phi) = a(R_{NO}) + b(R_{NO})\cos\theta + c(R_{NO})\sin\theta + d(R_{NO})\sin\theta\cos\phi,$$

where θ is the HON angle and ϕ is the NNOH dihedral angle. These functions were then used to shift the cc-pvdz cis and trans energies. Note that the correction is assumed to be independent of the NNO angle. This assumption introduces small discontinuities in the potential for NNO angles of 0° and 180°. These discontinuities are not expected to affect the kinetics predictions significantly, since the reactive flux is small in these regions.

B. Results

The CCSD(T) energies and normal mode frequencies are given in Tables 1 and 2, respectively. For comparison purposes the G2M results from Diau and Smith are also recorded therein.[23] The CCSD(T)/aug-cc-pvtz energies are in reasonable agreement with earlier calculations. For example, the energies of the various N_2H_2O minima and transition states are typically ~ 3 kcal mol^{-1} higher, relative to $NH_2 + NO$, than the G2M calculations of Diau and Smith.[23] This difference in energy is primarily due to the inclusion of a 3.5 kcal mol^{-1} "higher level correction" in the G2M calculations. The CCSD(T)/cc-pvtz normal mode frequencies reported in Table 2 are employed in the kinetics calculations without refinement.

As noted above, the MRCI calculations focus on the HNNOH $\leftrightarrow$ NNH + OH dissociation–recombination reaction. There are four planar rotamers of HNNOH, (cis,cis), (cis,trans), (trans,cis) and (trans,trans), where in this (a,b) notation, a refers to the HNNO dihedral angle and b refers to the NNOH dihedral angle. As noted in the introduction, barriers for rotation about the ON bond are quite small (~ 10 kcal mol^{-1}) while barriers for rotation about the NN bond are larger (~ 40 kcal mol^{-1}), consistent with the single bond vs. double bond character of these two bonds. The CCSD(T) calculations predict that the barriers to rotation about the NO bond lie more than 30 kcal mol^{-1} below the reactants, $NH_2 + NO$, while barriers to rotation about the NN bond lie only 4 to 8 kcal mol^{-1} below the reactants.

Fig. 2 shows a two-dimensional contour plot of the ground state OH + NNH interaction potential using a molecule fixed, Cartesian coordinate system. The origin of the coordinate system is the central nitrogen atom and the Y axis coincides with the NN bond axis. Each point on this plot corresponds to a position for the oxygen atom in this coordinate system. To define the geometry uniquely, two more angles are needed to specify the orientation of the OH bond axis. In this particular plot the NOH angle and the NNOH dihedral angle are both fixed at 90°. Four attractive channels are visible in this plot. The two on the right side of the plot lead to different rotamers of the HNNOH molecule. The upper left channel corresponds to a barrierless hydrogen

Table 2 CCSD(T)/cc-pvtz normal mode frequencies

Species	Frequencies/cm^{-1}
NH_2	3457, 3364, 1558
NO	1903
N_2	2346
H_2O	3947, 3842, 1669
OH	3719
N_2H	2891, 1809, 1112
H_2NNO	3621, 3499, 1635, 1523, 1249, 1079, 763, 629, 486
$HOHN_2$	3762, 3322, 1871, 1406, 1180, 818, 692, 425, 400
HNNOH(c,c)	3619, 3173, 1626, 1445, 1303, 953, 896, 674, 467
HNNOH(c,t)	3820, 3273, 1664, 1408, 1315, 1008, 843, 628, 450
HNNOH(t,c)	3625, 3425, 1597, 1446, 1381, 952, 849, 665, 649
HNNOH(t,t)	3826, 3409, 1634, 1453, 1342, 971, 919, 658, 491
$H_2NNO \rightarrow$ HNNOH(t,c); ts2	3538, 2126, 1451, 1348, 1183, 1169, 969, 579, 1881i
HNNOH(t,t) $\rightarrow$ HNNOH(c,t); ts4	3870, 3806, 1812, 1224, 651, 468, 462, 311, 1288i
HNNOH(c,c) $\rightarrow$ HNNOH(t,c); ts3	3861, 3681, 1753, 1284, 710, 706, 546, 509, 1365i
HNNOH(t,t) $\rightarrow$ HNNOH(t,c)	3795, 3361, 1591, 1418, 1313, 950, 840, 660, 543i
HNNOH(c,c) $\rightarrow$ HNNOH(c,t)	3734, 3121, 1567, 1390, 1285, 975, 770, 619, 419i
HNNOH(c,t) $\rightarrow N_2 + H_2O$; ts5	3709, 2179, 1850, 1125, 1005, 786, 537, 133, 1184i

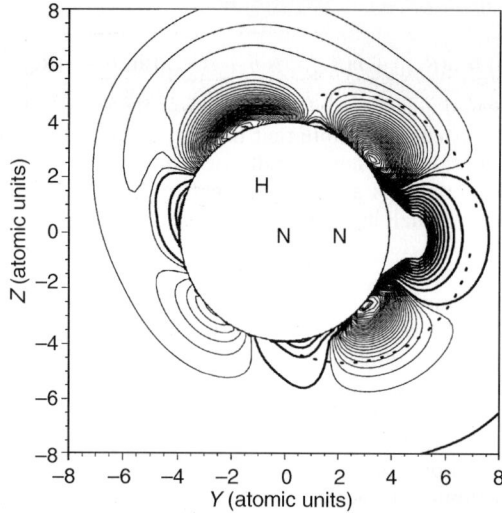

Fig. 2 Two dimensional contour plot of the NNH + OH interaction potential. The heavy solid contours correspond to zero (NNH + OH) and positive energy. The lighter solid contours correspond to negative energies. The contour increment for positive contours is 5 kcal mol^{-1}. For negative contours, the increment is 1 kcal mol^{-1}. See text for an explanation of the coordinate system used. Transition state dividing surfaces (for 800 K) are shown with dotted lines.

abstraction forming $N_2 + H_2O$, and the lower left channel leads to the $HOHN_2$ complex. Superimposed on this plot are transition state dividing surfaces that will be discussed in the next section.

This abstraction channel may also be reached from the *cis* HNNOH well *via* the transfer of the OH group through the saddlepoint near the top-middle portion of the plot. The latter saddlepoint corresponds to that labeled ts5 in Fig. 1. The minimum on the bottom left corresponds to an HOHNN complex. Note the saddlepoint (at the middle left) connecting this $HOHN_2$ complex with the abstraction channel is significantly below reactants. Here we focus on the addition to form the *cis* and *trans* HNNOH wells.

Fig. 3 shows one-dimensional potential curves for the OH + NNH addition forming each of the

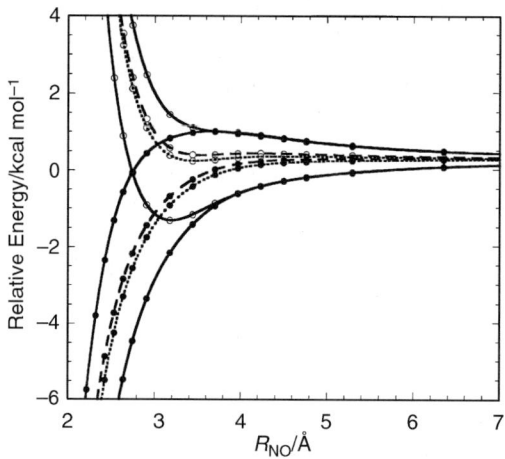

Fig. 3 One dimensional, CAS + 1 + 2/cc-pvdz, potentials for the planar dissociation of HNNOH. The internal degrees of freedom of the NNH and OH fragments are fixed as in Fig. 2. The NNO angle is fixed at 110° and the NOH angle is fixed at 90°. Solid symbols are for the ground state, open symbols for the lowest excited state. The two lowest solid curves (at larger R) are for the (*cis,trans*) orientation. The highest solid curves are for the (*cis,cis*) orientation. The dashed curves are for the (*trans,cis*) orientation and the dot curves are for the (*trans,trans*) orientation.

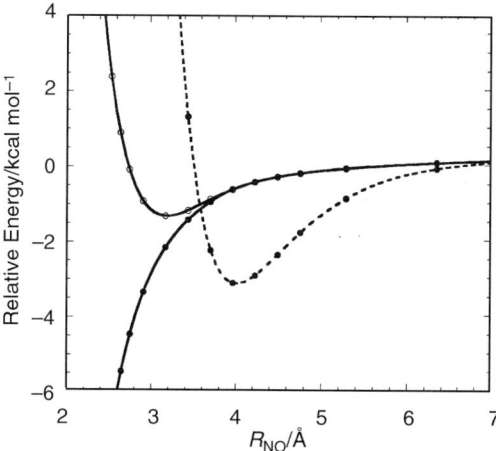

Fig. 4 One dimensional, CAS + 1 + 2/cc-pvdz, potentials for the planar dissociation of HNNOH. The internal degrees of freedom of the NNH and OH fragments are fixed as in Fig. 2. Solid symbols refer to the ground state, open symbols to the lowest excited state. The solid curves are for the most favorable covalent (*cis,trans*) orientation (same as shown in Fig. 3). The dashed curve is for the most favorable hydrogen bonding orientation (NOH angle = 0° and the NNO angle = 40°).

four, covalently bonded, rotamers of HNNOH. From this plot one can see that the (*cis,trans*) orientation is significantly more attractive at all distances than the other three orientations. The (*cis,cis*) potential is the least attractive, predicted to have a barrier to addition of ~ 1 kcal mol^{-1} when constrained in this orientation. The average of the (*cis,trans*) and (*cis,cis*) ground state potentials is quite similar to the average of the (*trans,trans*) and (*trans,cis*) potentials. In the direct statistics calculations discussed below, the contributions from the HNNO torsions (torsion *a*) are separated into *cis* and *trans* channels *via* restrictions on the range of torsional angles in the configurational integrals ($-90°$ to $90°$ for *cis* and $90°$ to $270°$ for *trans*), while the contributions from the NNOH torsion (torsion *b*) are integrated over the full range.

From the plot in Fig. 3 one can also see that the excited singlet state curves are nearly degenerate with the ground state curves into an NO separation of ~ 3.5 Å. At shorter distances the excited state curves rapidly become repulsive. This near-degeneracy of the two singlet states at large separations suggests that the excited state may make a significant contribution to the reactive flux at low temperatures where the transition state generally lies at large separations.

Fig. 4 compares a hydrogen bonding orientation (HON angle of 0°) with the most favorable, (*cis,trans*), covalent orientation. From this plot it is seen that hydrogen bonding leads to a fairly strongly bound (~ 3 kcal mol^{-1}) complex having a significantly longer-range interaction than the covalently bonded complexes. This strong interaction for the H-bonded orientation has important effects on the predicted association rate constant as discussed below.

III. Kinetics

A. Methodology

VRC-TST is employed in the present analysis of the reactive flux for the $N_2H + OH$ addition reaction.[41,42] The additions to form *cis* HNNO and *trans* HNNO forms of HNNOH are considered separately. The strong repulsion in the middle right region of the contour plot in Fig. 2 suggests such a separation should be valid. The basis of the VRC-TST procedure involves an assumed separation into the internal vibrational modes of the two reacting fragments (N_2H and OH), and the remaining rotational and orbital modes. The former modes, termed the conserved modes, are treated *via* direct summation over the quantized rigid-rotor harmonic oscillator energy levels. Here, these modes are assumed to be invariant along the reaction path and the asymptotic N_2H and OH values are employed. The rotational and orbital modes, termed the transitional modes, are evaluated *via* Monte Carlo integration of the classical phase space integral representation for the number of states.

The evaluation of the transitional mode phase space integrals, for the relative orientational motions of the two reacting fragments, requires the sampling of the potential energy for a large number of configurations in five-dimensional configuration space. Here, the requisite potential values are directly determined *via* CAS + 1 + 2/cc-pvdz calculations, as described in the quantum chemical section. This direct statistical approach[30,31] bypasses the difficult and time-consuming step of generating an accurate analytical representation of a five-dimensional potential energy surface.

In the VRC-TST approach a range of definitions for the reaction coordinate are considered. Each definition yields a family of transition state dividing surfaces, corresponding to different reaction coordinate values within the given definition. According to the kinetic variational principle, minimization of the calculated reactive flux with respect to these reaction coordinate definitions and values yields an optimal estimate for the reaction rate constant. Here, each reaction coordinate is defined in terms of the distance between the O atom and a given pivot point in the N_2H fragment.

Due to the computational requirements of the direct sampling procedure only a limited number of reaction coordinate definitions are considered here. At low temperatures the transition state lies at quite large separations, where the maximum attractiveness is not too large. At large separations, the optimal choice for each pivot point is generally the corresponding fragments center-of-mass, due to the absence of kinetic coupling terms for this choice. N_2H center-of-mass to O atom separations ranging from 5 to 6.5 Å were considered for the *trans* addition and from 5 to 7 Å for the *cis* addition, both on a 0.5 Å grid. Ideally, one would also choose the OH fixed point to be at its center-of-mass for such dividing surfaces. However, the O atom is close enough to the OH center-of-mass that such deviations should be irrelevant.

At higher temperature, the transition state moves in to quite short separations, and the terminal N atom provides a better approximation to the optimum N_2H pivot point. Thus, for both addition paths, a bond length reaction coordinate, defined by a fixed distance between the O atom and the terminal N atom was also considered. For the *trans* addition, separations ranging from 2.0 to 3.5 Å on a 0.25 Å grid and from 3.5 to 6.0 Å on a 0.5 Å grid were considered. The separations considered for the *cis* addition differed only in extending up to 6.5 Å.

The two-dimensional contour plot, shown in Fig. 2, also illustrates the optimum fixed bond length dividing surfaces for both the *cis* and *trans* additions at 800 K. As expected, these dividing surfaces follow the potential contours near the minimum energy path, and then move sharply in to the repulsive region of the potential at greater deviations from the minimum energy path. It is possible that other fixed points for the N_2H fragment, such as locating one along the N_2H radical orbital, would yield even lower estimates for the rate constant. However, simple inspection of the contour plots suggests that such further optimizations are unlikely to yield a significant reduction in the rate constant, and so computational expediency led us to consider only the bond length and center-of-mass reaction coordinates. For the OH radical, one might also consider a fixed point located along the radical orbital. However, our current VRC-TST program does not properly treat the symmetry aspects of a fixed point of the diatomic axis, and so such variations were not considered.

The number of configurations sampled in the direct evaluations of the transitional mode partition function were chosen to yield statistical error bars, from the Monte Carlo integration, of 10% or less. The anisotropy in the potential is greater at shorter separations and for the *trans* addition side. Thus, these regions required the maximum sampling of 2000 configurations. In contrast, only 500 configurations were sampled for the center-of-mass reaction coordinates. Altogether, the energies were directly determined for a total of about 50 000 configurations.

The determination of the individual *trans* and *cis* addition rate constants requires an approximate division of the total reactive flux into that contributing separately to the two addition channels. A plane through the midpoint of the NN bond and perpendicular to both the NNH plane and the NN bond was taken to represent the separation between the two HNNOH forming addition channels, and the abstraction and weak HOHNN complex channels. The plane passing through the NN bond and again perpendicular to the NNH plane was taken to separate the *cis* and *trans* addition channels. The presence of a strongly repulsive region at the middle-right portion of the contour plot suggests that the separation between the *cis* and *trans* addition is valid. Similarly, at least for short enough separations, the repulsive region between the *trans* addition channel and the HOHNN channel indicates that this separation is reasonably valid. Unfortunately, the region between the *cis* channel and the abstraction channel (at the top-middle of Fig. 2) is purely attractive and so the division of flux is more approximate here. The particular separation plane chosen here passes through the apparent region of the transition state for the abstraction channel and so provides a reasonable first approximation to the proper dividing surface.

The two-dimensional basis set correction described in Section II includes both positive and negative corrections to the directly determined energies. Its inclusion results in a reduction of the estimated rate constant by only 20% at most for the *trans* channel and about 10% at most for the *cis* channel. The magnitude of these reductions peaks at about 600 K, and then gradually decreases with increasing temperature. At the highest temperatures considered here (2200 K), the basis set correction yields the more typical result of an increase by a few %.

A similar two-dimensional Davidson correction for higher order excitations has also been implemented. This correction is much more significant, yielding an increase in the predicted rate constant by anywhere from 20 to 50%, with the maximal increase for a temperature of about 800 K. All the results presented below include both the basis set and Davidson corrections.

A plot of the calculated dependence of the transitional mode contribution to the number of available states, $N_{E,J}^{\text{transitional}}$, on the NO separation is provided in Fig. 5 for a variety of energies and $J = 3$ in the *trans* addition channel. Plots for other J values are similar. There are clearly two separate minima in this plot, with the outer minimum dominating at low energies and the inner minimum dominating at higher energies. The inner minimum first arises where the covalent bonding starts to be the dominant attractive interaction. As the bonding interactions get much stronger at even shorter separations the number of available states rises very rapidly. The outer minimum lies well to the right of the H-bonding minimum.

A statistical treatment of the crossing probabilities at each encounter of the inner and outer transition states yields an effective transition state number of states given by

$$\frac{1}{N_{\text{eff}}} = \frac{1}{N_{\text{inner}}} + \frac{1}{N_{\text{outer}}} - \frac{1}{N_{\text{max}}}, \tag{1}$$

where N_{max} is the maximum in the number of available states for separations between the two transition states.[43–46] For the *trans* addition, the inner transition state was taken to correspond to separations ranging from 2.0 to 2.75 Å, and the outer transition state was considered to be at 4.5 Å or greater. The maximum in between was evaluated from the grid points between 3.0 and 4.0 Å. A similar separation of grid points into inner and outer minima and central maximum was made for the *cis* addition.

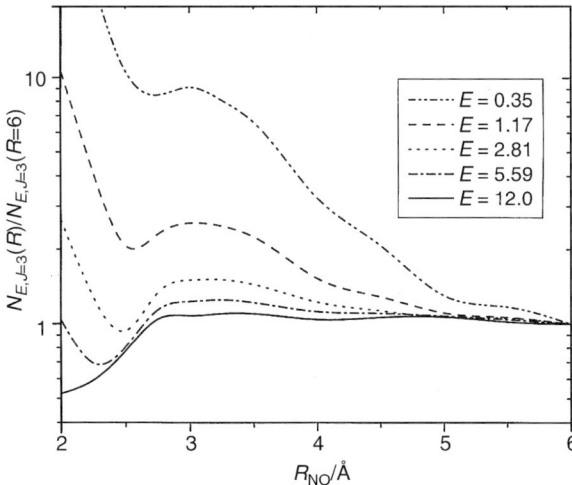

Fig. 5 Plot of the normalized transitional mode number of available states, $N_{E,J}(R)$, vs. NO separation, R, for $J = 3$ and a range of energies in kcal mol^{-1}. All values are normalized to unity at $R_{\text{NO}} = 6$ Å.

Both the direct statistical analysis and the subsequent kinetic modeling of the overall NH_2 + NO reaction were performed with a slightly modified version of the VARIFLEX software package.[47] The procedure for obtaining the zero pressure limit rate constants, and thereby branching fractions, from the transition state theory analyses for the individual channels is based on the original work of Miller et al.[48] as described in ref. 49. The transition state theory analyses employ the CCSD(T)/aug-cc-pvtz energies and CCSD(T)/cc-pvtz normal mode vibrational frequencies reported in Tables 1 and 2, respectively. The transition state for the barrierless NH_2 + NO entrance channel is treated as in our prior theoretical study.[22] The HONN torsional mode in transition states 3/4 and transition state 5, is treated as a one-dimensional hindered rotor employing the Pitzer–Gwinn expression for the rotational constants and Fourier fits to the B3LYP/aug-cc-pvtz calculated torsional potential.[22,50]

B. Results

i. HN_2 + OH rate constant. The kinetics of the N_2H + OH addition is complicated by the presence of multiple, nearly degenerate, electronic states. As in other radical–radical reactions, the combination of the OH and N_2H doublets generates both singlet and triplet states. At large separations the singlet and triplet states are nearly degenerate, while at shorter separations the bonding in the singlet state leads to a large splitting between the two. Furthermore, the $^2\Pi_{1/2}$ state of the OH radical, which differs from the ground $^2\Pi_{3/2}$ state by the spin–orbit splitting of 143 cm^{-1}, results in an additional set of singlet and triplet states. The proper treatment of all these nearly degenerate electronic states requires some knowledge of their transition rates. In the absence of this knowledge, the possible effect of the excited states can be estimated by assuming rapid or zero transition rates.

Various a priori predictions for the temperature dependence of the sum of the *trans* and *cis* OH plus N_2H addition rates are plotted in Fig. 6. The dashed line in this plot corresponds to a calculation in which the transitions between the two singlet states are assumed to be rapid enough to maintain a statistical distribution of the vibrational states. In contrast, the upper solid line is based on an assumed zero transition rate and includes just the contribution from the ground singlet state. For both of these calculations there is assumed to be no contribution from the triplet states. The inclusion of the excited singlet state is seen to yield only a modest increase in the

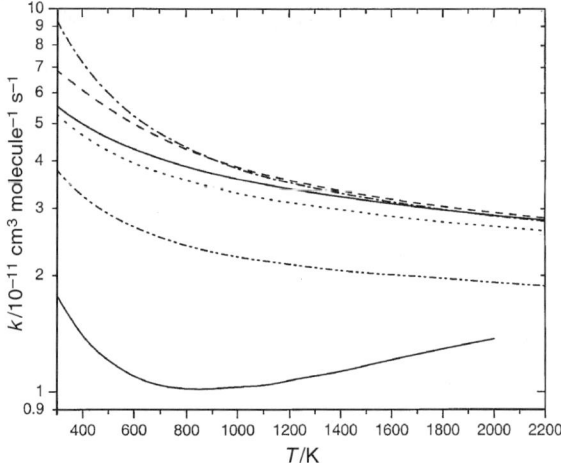

Fig. 6 Plots of the sum of the *cis* and *trans* addition rates *vs.* temperature for a number of different treatments of the electronic states and the multiple minima in $N_{E,J}(R)$. The upper solid line results from employing the overall minimum in $N_{E,J}(R)$ and considering only the ground singlet state. The dashed line differs from the solid line by including the contribution from the first excited singlet state. The dotted line employs the effective treatment of the two minima [*cf.* eqn. (1)], again considering only the ground singlet state. The dash-dot line employs the effective number of states treatment and includes a degeneracy of four at long-range. The dash-dot-dot line also employs an effective treatment for only the ground singlet state, but with N_{max} set to infinity. The lower solid line corresponds to the results of our prior theoretical study.

predicted rate constant. The smallness of this correction is due in part to the spin–orbit splitting of the OH radical removing the degeneracy of the two singlet states. This splitting is included here as a constant shift in the energy applied to the MRCI calculated excited singlet state since spin–orbit effects are ignored in the MRCI analysis.

The dotted line corresponds to a calculation in which the effect of the two separate transition states is accounted for *via* eqn. (1), with only the ground singlet state included. For the upper solid curve the transition state partition functions were instead evaluated in terms of the overall minimum. The effective number of states treatment yields a modest reduction in the predicted rates over a very wide range of temperatures. The dash-dot-dot line is also obtained from an effective number of states calculation for the ground singlet state, but with N_{max} set to infinity. This calculation corresponds to the maximal statistical effect of two transition states acting in series and has been found to provide a good representation of the observations in the dissociation of CH_2CO.[51] The large range of temperature over which the two effective number of states calculations yield a reduction correlates with the large range in energy for which the inner and outer minima in the number of states are similar.

A crude estimate for the possible contribution of the triplet states to the reaction was obtained by assuming that the ground singlet and ground triplet states are degenerate for separations of 3.0 Å or greater, and that the singlet to triplet transition rate is rapid enough to maintain a statistical distribution. Implementing this triplet state contribution into the effective number of states based calculation yields the dashed-dotted line. The difference between the dotted and the dashed-dotted line suggests that the triplet state may make a significant contribution at lower temperatures.

Each of the calculated rate constant curves shows a significant decrease with increasing temperature. Furthermore, this decrease is qualitatively similar to that predicted in our prior study (*cf.* the lower solid line). The increase in the predicted rate constant at low temperatures is likely a result of the relatively strong long-range attractions of the H-bonded geometries. The predicted rate constants are all substantially greater than that predicted in our prior investigation. That for the effective number of states calculation with N_{max} set to infinity is closest, being about a factor of two greater on average. The latter calculation provides what is likely the best estimate to the temperature dependence of the association rate and will be considered as our reference case in the following discussion.

The temperature dependence of the parameter γ, which defines the branching into the *trans* channel for the $N_2H + OH$ addition, is plotted in Fig. 7 for the reference case. The results for γ indicate that the *trans* addition rate constant has a sharper decrease with temperature than does the total rate constant, and is thus even more like the predictions of ref. 22. Actually, from room temperature to 2000 K the *trans* addition rate constant decreases by a factor of 2.1. The difference between the repulsive nature of the interactions for the bottom-middle portion of the contour plot

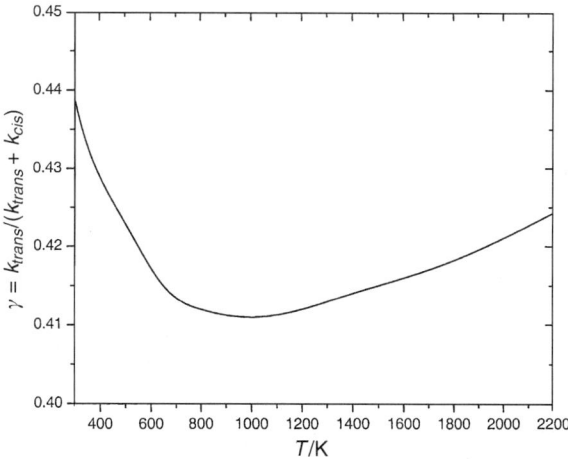

Fig. 7 Plot of the temperature dependence of the parameter $\gamma \equiv k_{trans}/(k_{trans} + k_{cis})$ defining the branching into the *trans* channel for the $N_2H + OH$ addition rate constant.

(*cf.* Fig. 2), and the attractive nature of the interactions at the top-middle portion, provides a likely explanation for the finding that the *trans* addition rate is less than the *cis* addition rate.

ii. Branching fraction. The calculated temperature dependence for the total rate constant in the $NH_2 + NO$ reaction is plotted together with the corresponding experimental data[10,11,13,17,18,20,52–56] in Fig. 8. The unadjusted results, represented with the solid line, are seen to be at the low end of the experimental data. This rate constant depends primarily on the properties of ts2. Decreasing the barrier height for this transition state by 2 kcal mol^{-1} yields the dashed curve, which provides a satisfactory representation of the experimental data. As discussed below, also decreasing the channel (a) endothermicity to 0.9 kcal mol^{-1} and decreasing the barrier height for the ts3/4 pair by 4 kcal mol^{-1} yields our best agreement with experimental results for the branching fraction. The dotted curve illustrates the calculated total rate constant for this set of revisions, and is seen again to provide a satisfactory representation of the experimental data. Each of these calculations employed the reference treatment for the number of states in the $HN_2 + OH$ channel, but are only weakly dependent on such results.

The current predictions for the temperature dependence of the branching fraction α are illustrated in Fig. 9, again together with the corresponding experimental data,[7–19] as well as the model of Miller and Glarborg. The solid line employs the reference treatment for the number of available states. The dashed line employs the effective number of states based on the full evaluation of eqn. (1) considering only the ground singlet state. The dotted line differs from the latter calculation by including the approximate triplet contribution at large separation. For each of these calculations, the energy for ts2 was reduced by 2 kcal mol^{-1} from the CCSD(T) values of Table 1, in keeping with the adjustment required to reproduce the experimental data for the total rate constant. We note however, that the calculated branching fraction is only weakly dependent on this energy.

Overall, the agreement between the theoretical predictions and the experimental data is quite good, especially that for the reference case treatment of the $N_2H + OH$ association. Interestingly, each of the predicted curves illustrated in Fig. 9 shows the desired sigmoidal behaviour, but is too low near room temperature and then rises too rapidly for about the first 500 K. The underprediction at low temperatures is strongly dependent on the endothermicity of the $N_2H + OH$ channel, as is the overestimate of the initial rise in the rate constant. For instructive purposes we now consider what variations in the energies are required to reduce or remove even these minor

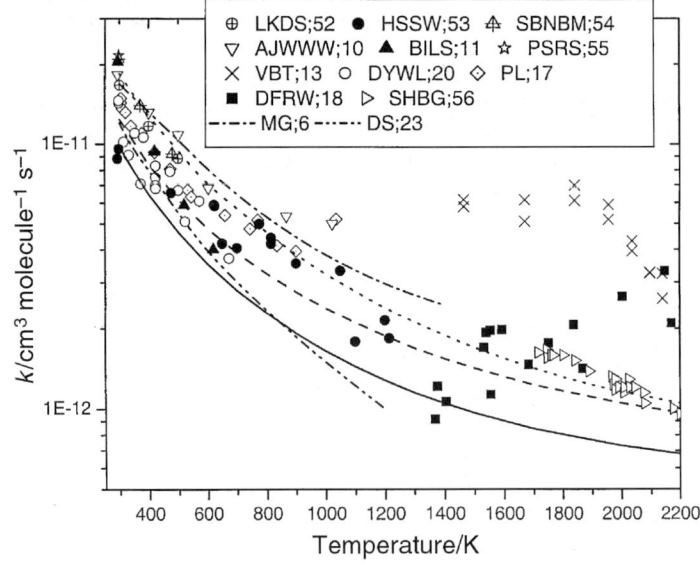

Fig. 8 Comparison of our theoretical predictions for $k_T(T)$ with experiment. The solid, the dashed, and the dotted lines denote the present theoretical predictions. The solid line corresponds to the unadjusted results, for the dashed line the ts2 value was decreased by 2 kcal mol^{-1}, and for the dotted line the $N_2H + OH$ reaction endothermicity and ts5 were also reduced by 2 and 4 kcal mol^{-1}, respectively.

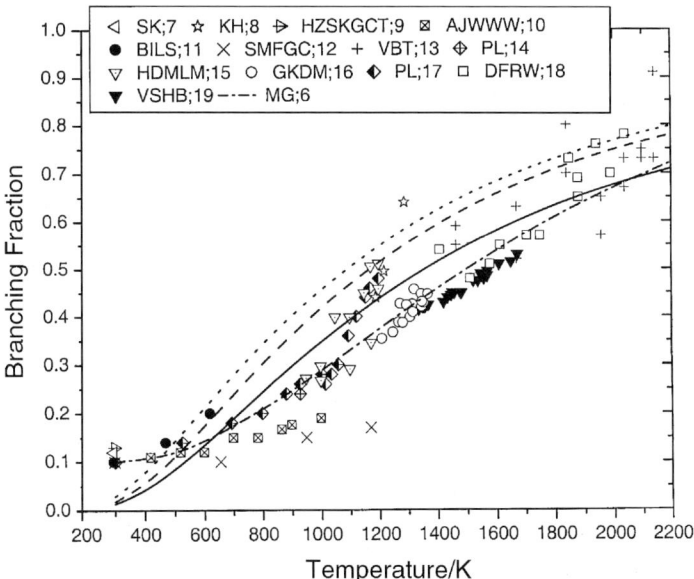

Fig. 9 Comparison of our theoretical predictions for $\alpha(T)$ with experiment. The solid, the dashed, and the dotted lines denote the present unadjusted theoretical predictions. The solid line represents the reference case results, the dashed line represents the results from the full implementation of eqn. (1) for the singlet state only, and the dotted line includes an approximate contribution from the triplet state.

discrepancies in the predictions. In reality, such discrepancies may be due to other effects such as anharmonicity or nonstatistical effects. Alternatively, they may simply be indicative of the range of uncertainty in the experiments.

In our prior study we constrained the theoretical predictions to reproduce the experimental branching fraction of 0.10 at room temperature. Lowering the $N_2H + OH$ endothermicity to 1.2 kcal mol^{-1} yields a branching fraction for the reference case that also reproduces the room-temperature experimental result, as illustrated by the solid line in Fig. 10. However, this result clearly has an overly rapid initial rise in the predicted branching fraction. Further reductions in the endothermicity yield rate constants that are too large at room temperature. However, the reduction of the barrier height for the ts3/4 pair can be used to compensate for this increase in low-temperature branching fraction. The dashed line illustrates the results obtained when the endothermicity is set to 0.8 kcal mol^{-1} and the barrier height for the ts3/4 pair is reduced by 2 kcal mol^{-1} to reproduce again the room-temperature branching fraction. Meanwhile, the dotted line corresponds to an endothermicity of 0.3 kcal mol^{-1} and a reduction of the ts3/4 energy by 4 kcal mol^{-1}.

The constraint of reproducing the rate data at room temperature exactly has been relaxed in generating the solid line illustrated in Fig. 11. Also plotted therein are the results from previous theoretical models for the branching fraction.[20,21] This curve again employs a reduction of the ts3/4 energy by 4 kcal mol^{-1}, but now adjusts the endothermicity to 0.9 kcal mol^{-1}, which are both close to the limit of reasonable errors in the quantum chemical estimates. The agreement between this theoretical prediction and the experimental data for 400 K and higher is remarkable.

The underestimate at 300 K, and the difficulty in obtaining a branching fraction that has such a strongly sigmoidal temperature dependence, is suggestive of some experimental errors in the room-temperature measurements. However, there are, of course, still a number of possible shortcomings in the theoretical model. For example, the combination of a decrease in the predicted *trans* $HN_2 + OH$ association rate with a further decrease in the $NH_2 + NO \rightarrow HN_2 + OH$ reaction endothermicity should remove this discrepancy. Alternatively, an increase in the transition state number of states for the ts3/ts4 pair (due to anharmonic vibrational effects), when coupled once again with a decrease in the reaction endothermicity, might also remove this discrepancy. A much stronger decrease in the $N_2H + OH$ recombination rate constant with increasing tem-

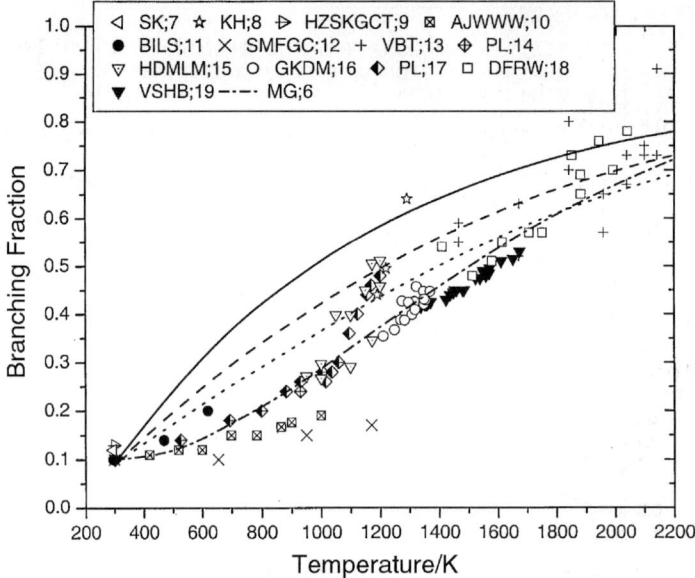

Fig. 10 Comparison of our theoretical predictions for α(T) with experiment. The solid, the dashed, and the dotted lines denote the present theoretical predictions. The solid line represents the reference case with the N$_2$H + OH reaction endothermicity reduced to 1.2 kcal mol^{-1}. For the dashed line the endothermicity is reduced to 0.8 kcal mol^{-1} and the ts3/4 barrier height is reduced by 2 kcal mol^{-1}. For the dotted line the endothermicity is reduced to 0.3 kcal mol^{-1} and the ts3/4 barrier height is reduced by 4 kcal mol^{-1}.

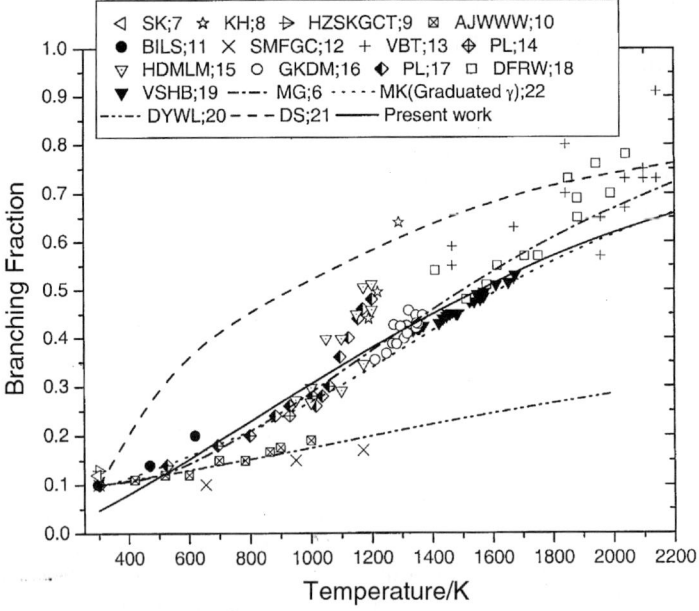

Fig. 11 Comparison of our theoretical predictions for α(T) with experiment and with prior theoretical results. The solid lines represents the reference case with the N$_2$H + OH reaction endothermicity reduced to 0.9 kcal mol^{-1} and the ts3/4 barrier height reduced by 4 kcal mol^{-1}.

perature might also improve the agreement. In any case, the results plotted in Fig. 11 demonstrate that reasonable adjustments in the potential can yield a temperature-dependent branching fraction that is largely consistent with experiment.

The importance of the $NH_2 + NO \rightarrow HN_2 + OH$ reaction endothermicity to the modeling led us to consider two further aspects of its calculation. In particular, we have also performed CCSD(T) calculations with the aug-cc-pvqz basis, which yields a reduction in the endothermicity from that for the aug-cc-pvtz basis of only 0.1 kcal mol^{-1}. We have also performed anharmonic vibrational analyses at the CCSD(T)/aug-cc-pvtz level, which actually increases the endothermicity by 0.2 kcal mol^{-1}. Thus, it appears unlikely that the calculated endothermicity is in error by more than 1–2 kcal mol^{-1}.

IV. Concluding remarks

In a previous investigation[22] we were able to describe the rate coefficient and branching fraction α of the $NH_2 + NO$ reaction,

$$NH_2 + NO \rightarrow N_2H + OH \qquad (a)$$
$$\rightarrow N_2 + H_2O, \qquad (b)$$

where $\alpha = k_a/(k_a + k_b)$, with some degree of accuracy. However, the primary conclusion of that work was that $\alpha(T)$ depends critically on the potential governing the fragmentation of various HNNOH isomers into $N_2H + OH$. In the present work we have described that potential in some detail using high-level electronic structure theory.

It is useful and instructive to couch our results in terms of association rate coefficients for the $N_2H + OH$ reaction. In agreement with the conclusion from previous work, this rate coefficient drops rapidly with increasing temperature. However, it does not have a relative minimum at $T \approx 900$ K, as did our earlier model. The function $\gamma(T)$, where $\gamma = k_{trans}/(k_{trans} + k_{cis})$ is the fraction of the association rate coefficient that forms the *trans* HNNOH isomer, first drops and then rises with increased temperature. The earlier work anticipated the high-temperature rise, but not the fall in γ with increased T at lower temperatures.

The drop in the $N_2H + OH$ rate coefficient is a key aspect of reproducing the observed temperature dependence of the branching fraction. With relatively modest changes in selected internal rearrangement barriers we are able to predict $\alpha(T)$ over a wide range of temperatures. As in our previous study, the endothermicity of reaction (a) turns out to be approximately zero (± 1 kcal mol^{-1}) when it is adjusted to give optimum agreement with experiment for $\alpha(T)$. The value of 0.9 kcal mol^{-1} for our best model, amounts to adjusting the *ab initio* electronic structure theory down by 1.5 kcal mol^{-1}. The greater adjustment in the ts3/4 barrier height, down by 4 kcal mol^{-1}, is also reasonable given the greater uncertainty in predicting saddlepoint energies and the difficulty in predicting absolute energies for processes involving the pairing of electrons.

Acknowledgement

This work was supported by the U. S. Department of Energy, Office of Basic Energy Sciences, Division of Chemical Sciences, Geosciences and Biosciences. The DOE Contract Number for the work at Argonne is W-31-109-ENG-38.

References

1 R. K. Lyon, *U.S. Pat.*, 1975, 3 900 554.
2 J. A. Miller, M. C. Branch and R. J. Kee, *Combust. Flame*, 1981, **43**, 81.
3 J. A. Miller and C. T. Bowman, *Prog. Energy Combust. Sci.*, 1981, **15**, 287.
4 J. A. Miller and P. Glarborg, in *Gas Phase Chemical Reaction Systems: Experiments and Models 100 years after Max Bodenstein*, ed. J. Wolfrum, H.-R. Volpp, R. Rannacher and J. Warnatz, Springer, Berlin, 1996, p. 318.
5 J. A. Miller, *Proc. Combust. Inst.*, 1996, **26**, 461.
6 J. A. Miller and P. Glarborg, *Int. J. Chem. Kinet.*, 1999, **31**, 757.
7 J. A. Silver and C. E. Kolb, *J. Phys. Chem.*, 1982, **86**, 3249.
8 M. A. Kimball-Linne and R. K. Hanson, *Combust. Flame*, 1986, **64**, 377.
9 J. L. Hall, D. Zeitz, J. W. Stephens, J. V. V. Kasper, G. P. Glass, R. F. Curl and F. K. Tittel, *J. Phys. Chem.*, 1986, **90**, 2501.

10 B. Atakan, A. Jacobs, M. Wahl, R. Weller and J. Wolfrum, *Chem. Phys. Lett.*, 1989, **155**, 609.
11 V. P. Bulatov, A. A. Ioffee, V. A. Lozovsky and O. M. Sarkisov, *Chem. Phys. Lett.*, 1989, **161**, 141.
12 J. W. Stephens, C. L. Morter, S. K. Farhat, G. P. Glass and R. F. Curl, *J. Phys. Chem.*, 1993, **97**, 8944.
13 J. Vandooren, J. Bian and P. J. van Tiggelen, *Combust. Flame*, 1994, **98**, 402.
14 J. Park and M. C. Lin, *J. Phys. Chem.*, 1996, **100**, 3317.
15 M. J. Halbgewachs, E. W.-G. Diau, A. M. Mebel, M. C. Lin and C. F. Melius, *Proc. Combust. Inst.*, 1996, **26**, 2109.
16 P. Glarborg, P. Kristensen, K. Dam-Johansen and J. A. Miller, *J. Phys. Chem. A*, 1997, **101**, 3741.
17 J. Park and M. C. Lin, *J. Phys. Chem. A*, 1997, **101**, 5.
18 J. Deppe, G. Friedrichs, H.-J. Römming and H. Gg. Wagner, *Phys. Chem. Chem. Phys.*, 1999, **1**, 427.
19 M. Votsmeier, S. Song, R. K. Hanson and C. T. Bowman, *J. Phys. Chem. A*, 1999, **103**, 1566.
20 E. W.-G. Diau, F. Yu, M. A. G. Wagner and M. C. Lin, *J. Phys. Chem.*, 1994, **98**, 4034.
21 E. W.-G. Diau and S. C. Smith, *J. Phys. Chem.*, 1996, **100**, 12349.
22 J. A. Miller and S. J. Klippenstein, *J. Phys. Chem. A*, 2000, **104**, 2061.
23 E. W.-G. Diau and S. C. Smith, *J. Chem. Phys.*, 1997, **106**, 9236.
24 H. Abou-Rachid, C. Pouchan and M. Chaillet, *Chem. Phys.*, 1984, **90**, 243.
25 C. F. Melius and J.S. Binkley, *Proc. Combust. Inst.*, 1985, **20**, 575.
26 J. A. Harrison, R. G. A. MacLagan and A. R. Whyte, *J. Phys. Chem.*, 1987, **91**, 6683.
27 S. P. Walch, *J. Chem. Phys.*, 1993, **99**, 5295.
28 M. Wolf, D. L. Yang and J. L. Durant, *J. Photochem. Photobiol. A*, 1994, **80**, 85.
29 X. Duan and P. Page, *J. Mol. Struct. (Theochem)*, 1995, **333**, 233.
30 S. J. Klippenstein, A. L. L. East and W. D. Allen, *J. Chem. Phys.*, 1996, **105**, 118.
31 S. J. Klippenstein and L. B. Harding, *J. Phys. Chem. A*, 1999, **103**, 9388.
32 MOLPRO is a package of *ab initio* programs written by H.-J. Werner and P. J. Knowles with contributions from J. Almlof, R. D. Amos, A. Berning, D. L. Cooper, M. J. O. Deegan, A. J. Dobbyn, F. Eckert, S. T. Elbert, C. Hampel, R. Lindh, A. W. Lloyd, W. Meyer, A. Nicklass, K. Peterson, R. Pitzer, A. J. Stone, P. R. Taylor, M. E. Mura, P. Pulay, M. Schutz, H. Stoll and T. Thorsteinsson.
33 GAUSSIAN98, M. J. Frisch, G. W. Trucks, H. B. Schlegel, G. E. Scuseria, M. A. Robb, J. R. Cheeseman, V. G. Zakrzewski, J. A. Montgomery, Jr., R. E. Stratmann, J. C. Burant, S. Dapprich, J. M. Millam, A. D. Daniels, K. N. Kudin, M. C. Strain, O. Farkas, J. Tomasi, V. Barone, M. Cossi, R. Cammi, B. Mennucci, C. Pomelli, C. Adamo, S. Clifford, J. Ochterski, G. A. Petersson, P. Y. Ayala, Q. Cui, K. Morokuma, D. K. Malick, A. D. Rabuck, K. Raghavachari, J. B. Foresman, J. Cioslowski, J. V. Ortiz, A. G. Baboul, B. B. Stefanov, G. Liu, A. Liashenko, P. Piskorz, I. Komaromi, R. Gomperts, R. L. Martin, D. J. Fox, T. Keith, M. A. Al-Laham, C. Y. Peng, A. Nanayakkara, C. Gonzalez, M. Challacombe, P. M. W. Gill, B. Johnson, W. Chen, M. W. Wong, J. L. Andres, M. Head-Gordon, E. S. Replogle and J. A. Pople, Gaussian, Inc., Pittsburgh PA, 1998.
34 T. H. Dunning, Jr., *J. Chem. Phys.*, 1989, **90**, 1007.
35 R. A. Kendall, T. H. Dunning, Jr. and R. J. Harrison, *J. Chem. Phys.*, 1992, **96**, 6796.
36 D. E. Woon and T. H. Dunning, Jr., *J. Chem. Phys.*, 1993, **98**, 1358.
37 P. J. Knowles, C. Hampel and H.-J. Werner, *J. Chem. Phys.*, 1993, **99**, 5219.
38 H.-J. Werner and P. J. Knowles, *J. Chem. Phys.*, 1985, **82**, 5053; P. J. Knowles and H.-J. Werner, *Chem. Phys. Lett.*, 1985, **115**, 259.
39 H.-J. Werner and P. J. Knowles, *J. Chem. Phys.*, 1988, **89**, 5803; P. J. Knowles and H.-J. Werner, *Chem. Phys. Lett.*, 1988, **145**, 514.
40 S. R. Langhoff and E. R. Davidson, *Int. J. Quantum Chem.*, 1974, **8**, 61; D. W. Silver and E. R. Davidson, *Chem. Phys. Lett.*, 1978, **52**, 403.
41 S. J. Klippenstein, *J. Chem. Phys.*, 1992, **96**, 367.
42 S. J. Klippenstein, *J. Phys. Chem.*, 1994, **98**, 11459.
43 J. O. Hirschfelder and E. Wigner, *J. Chem. Phys.*, 1939, **7**, 616.
44 W. H. Miller, *J. Chem. Phys.*, 1976, **65**, 2216.
45 W. J. Chesnavich, L. Bass, T. Su and M. T. Bowers, *J. Chem. Phys.*, 1981, **74**, 2228.
46 S. N. Rai and D. G. Truhlar, *J. Chem. Phys.*, 1983, **79**, 6046.
47 S. J. Klippenstein, A. F. Wagner, R. C. Dunbar, D. M. Wardlaw, S. H. Robertson and J. A. Miller, *VARIFLEX Version 1.07mt* (December, 2000) and *Version 0.31* (July 1999).
48 J. A. Miller, C. Parrish and N. J. Brown, *J. Phys. Chem.*, 1986, **90**, 3339.
49 D. K. Hahn, S. J. Klippenstein and J. A. Miller, *Faraday Discuss.*, 2001, (DOI: 10.1039/b102240g).
50 K. S. Pitzer and W. D. Gwinn, *J. Chem. Phys.*, 1942, **10**, 428.
51 E. A. Wade, A. Mellinger, M. A. Hall and C. B. Moore, *J. Phys. Chem. A*, 1997, **101**, 6568.
52 R. Lesclaux, P. V. Khé, P. DeZauzier and J. C. Soulignac, *Chem. Phys. Lett.*, 1975, **35**, 493.
53 H. Hack, H. Schacke, M. Schroter and H. Gg. Wagner, *Proc. Combust. Inst.*, 1979, **17**, 505.
54 L. J. Stief, W. D. Brobst, D. F. Nava, R. P. Borkowski and J. V. Michael, *J. Chem. Soc., Faraday Trans. 2*, 1982, **78**, 1391.
55 P. Pagsberg, B. Sztuba, E. Ratajczak and A. Sillesen, *Acta Chem. Scand.*, 1991, **45**, 329.
56 S. Song, R. K. Hanson, C. T. Bowman and D. M. Golden, *Proc. Combust. Inst.*, 2000, **28**, 2403.

An experimental and theoretical study of the product distribution of the reaction CH_2 ($\tilde{X}\,^3B_1$) + NO

Mustapha Fikri, Stefan Meyer, Jan Roggenbuck and Friedrich Temps*

Institut für Physikalische Chemie, Christian-Albrechts-Universität zu Kiel, Olshausenstr. 40, D-24098 Kiel, Germany. E-mail: temps@phc.uni-kiel.de

Received 19th March 2001
First published as an Advance Article on the web 2nd November 2001

Measurements of the product branching ratios of the reaction CH_2 ($\tilde{X}\,^3B_1$) + NO (1) are presented together with calculations of the thermal rate constant and branching ratios using unimolecular rate theory. The reaction was investigated experimentally at room temperature using FTIR spectroscopy. The yields of the main products HCNO and HCN were found to be $\Gamma_{HCNO} = 0.89 \pm 0.06$, $\Gamma_{HCN} = 0.11 \pm 0.06$. Other minor products could be rationalized by numerical simulations of the reaction system taking into account possible consecutive reactions. The potential energy surface for the reaction was characterized by quantum chemical calculations using *ab initio* and density functional methods. The proposed reaction pathways connecting reactants to products were explored by multi-channel unimolecular rate theory calculations to determine the CH_2 ($\tilde{X}$) + NO capture rate constant and the rate constants for the different product channels as a function of temperature. The calculated capture rate constant of $k = 2.3 \times 10^{13}$ cm^3 mol^{-1} s^{-1} is in good agreement with experimental values at room temperature. Collisional stabilization of the initial H_2CNO recombination complex was predicted to be negligible up to pressures of >1 bar. For ambient pressures and temperatures up to 2000 K, HCNO + H were calculated as the dominating products, with $\Gamma_{HCNO} \approx 0.94$ in agreement with the experiments. The channel to HCN + OH was calculated with $0.015 \leqslant \Gamma_{HCN} \leqslant 0.05$, only slightly below the experimental value.

1. Introduction

The reaction of methylene (CH_2) with nitric oxide (NO) is a major step in the so-called NO_x-reburn process for the reduction of NO in hydrocarbon combustion.[1] CH_2 can react in the triplet ground electronic state ($\tilde{X}\,^3B_1$, hereafter referred to as $^3CH_2$) or in the low-lying singlet first excited electronic state ($\tilde{a}\,^1A_1$, hereafter denoted as $^1CH_2$),

$$^3CH_2 + NO \rightarrow \text{products}, \qquad (1)$$

$$^1CH_2 + NO \rightarrow \text{products}. \qquad (2)$$

The singlet is only 37.7 kJ mol^{-1} above the triplet.[2,3] The reactions are highly exothermic and may, in principle, give a large number of different products (Fig. 1). However, despite several studies,[4–11] little is known with certainty about their major channels and about the underlying reaction mechanism.

DOI: 10.1039/b102563p

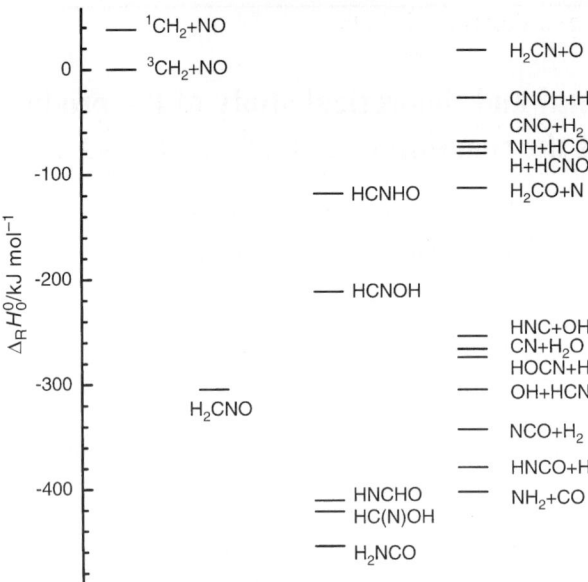

Fig. 1 Possible exothermic reaction channels of reaction (1).

Considering the $^3CH_2$ reaction (1), the overall rate constant at room temperature has been fairly well determined from two direct measurements, carried out by Seidler et al.[6] and by Darwin et al.,[7] who found values of $k_1 = 2.2 \times 10^{13}$ and 1.75×10^{13} cm^3 mol^{-1} s^{-1}, respectively. Atakan et al.[8] also reported on the formation of OH radicals in the reaction. In a preceding study, by Grußdorf et al.,[9] we proposed the two reaction channels

$$^3CH_2 + NO \rightarrow HCNO + H, \quad (1.1)$$

$$\rightarrow HCN + OH. \quad (1.2)$$

The products HCNO and HCN were identified using FTIR spectroscopy. However, the experimental conditions had not been systematically varied and effects of side reactions and the influence of $^1CH_2$ had not been fully clarified. Subsequently, the formation of H atoms and OH radicals was investigated in shock tube experiments by Bauerle et al.[10] Very recently, Kong and co-workers published an FTIR emission study in which they suggested a number of further product channels.[11] They were unable, however, to determine quantitative yields.

The $^1CH_2$ + NO reaction (2) is known to be very fast ($k_2 = 1.5 \times 10^{14}$ cm^3 mol^{-1} s^{-1}).[12–14] However, as for $^3CH_2$, the products and, in addition, the role of the collisional intersystem crossing channel $^1CH_2 + NO \rightarrow {}^3CH_2 + NO$ are not clear.[14]

Quantum chemical calculations for the $^3CH_2$ reaction performed in this laboratory[15] and by the groups of Bacskay[16–18] and Kong[11] at different levels of theory have provided some information on the potential energy surface (PES) and the reaction pathways connecting the reactants to products. According to these calculations, there are several competing reaction channels, involving comparable potential energy barriers. Under these conditions, predictions for the reaction channels from the theoretical calculations alone are difficult.

The aim of this paper is to improve our understanding of reaction (1) on the grounds of more detailed experimental and theoretical work. We present new FTIR results on the reaction products obtained by photolysis of mixtures of H_2CCO and NO in He or Ar as bath gases at different total pressures with systematic variations of the NO pressures. In addition, we describe unimolecular rate calculations for the product branching ratios as a function of temperature, based on the ab initio data.[15–18]

2. Experimental and theoretical methods

2.1. Experimental

The experimental set-up has been described before.[9,19,20] Measurements were carried out under stationary conditions in a cylindrical quartz photoreactor of 20 cm length and 5 cm i.d. equipped with NaCl windows. The cell was located inside a Bruker IFS 66v FTIR spectrometer and connected to an all-glass gas handling system which could be evacuated with a turbomolecular pump. Reaction mixtures containing H_2CCO and NO in He or Ar as inert bath gases were photolyzed using a 200 W Hg lamp imaged into the reaction cell to irradiate its entire volume as evenly as possible through a 313 nm interference filter. The photolysis light was blocked after 5, 10, 20, 40 and 60 min of photolysis to take FTIR spectra of the reacting gas mixtures, using a nominal resolution of 0.25 cm^{-1}. A spectrum before starting the photolysis was subtracted as background trace to generate difference spectra which show the depletion of the reactants by negative peaks and reaction products by positive peaks. Reference spectra of pure samples of the reactants and products in calibrated mixtures with the bath gases were used for converting integrated FTIR signals to partial pressures.

All gases were of the highest commercially available purities: Ar (99.9999%), He (99.9999%), NO (99.5%), CO (99.995%), CO_2 (99.995%), N_2O (99.999%), C_2H_4 (99.5%) and C_2H_2 (99.6%). Ar, He, CO and CO_2 were taken as supplied. NO, N_2O, C_2H_4 and C_2H_2 were purified by repeated fractional distillation until no impurities were detectable in the FTIR spectra. H_2CCO was prepared by pyrolysis of acetic acid anhydride adopting the method of Bak et al.[21] HCNO was synthesized by pyrolysis of 3-phenyl-4-oxoiminoisoxazol-5-(4H)-one.[22] Reference samples of HCN were prepared by heating mixtures of NaCN with stearic acid.

2.2. Quantum chemical and multi-channel unimolecular rate calculations

Quantum chemical calculations to elucidate the CH_2 + NO PES were carried out using the GAUSSIAN 94 program suite.[23] Structures, energies and vibrational frequencies of intermediates and transition states were determined at the HF/6-31G(d), MP2(full)/6-311G(d,p) and B3LYP/6-311G(d,p) levels of theory and, for the energies, also at the G2 level. Transition states were checked by intrinsic reaction coordinate analyses.

The output was used in multiple-well multi-channel statistical unimolecular rate theory calculations to obtain thermal rate constants for the main product channels.[24–26] Reactions with distinctive potential energy barriers (tight transition states) were modeled using the Rice–Ramsperger–Kassel–Marcus (RRKM) model.[24,25] Barrier-less reactions (loose transition states) were treated using the simplified statistical adiabatic channel model (SACM) of Quack and Troe.[27–30]

The central quantities calculated were the microcanonical specific rate constants

$$k(E, J) = \frac{\sigma}{\sigma^{\ddagger}} \frac{W^{\ddagger}(E - E_{0,J}, J)}{h\rho(E, J)} \quad (I)$$

of the vibrationally highly excited species formed in the course of the reaction as a function of energy E and angular momentum J. $W^{\ddagger}(E - E_{0,J}, J)$ is the number of open reaction channels at the energy E counted from the vibrational ground state of the reactant, $E_{0,J}$ is the potential energy barrier height for the reaction step of interest, $\rho(E, J)$ is the density of states of the respective reactant, σ and $\sigma^{\ddagger}$ are symmetry factors and h is Planck's constant. The numbers and densities of states were counted with the Beyer–Swinehart algorithm[31,32] with a grain size of 5 cm^{-1}. Disappearing oscillators in the SACM calculations were taken into account by interpolation formulas.[29,30] K-rotors in the symmetric top approximation were considered as active degrees of freedom, and anharmonicity corrections were applied as recommended.[29,30] The anisotropy parameters α/β in the SACM calculations were close to the "standard value" of $\alpha/\beta = 0.50$ (see below).

Collisional deactivation of the "vibrationally hot" H_2CNO^* intermediate from the $^3CH_2$ + NO association reaction was taken into account by an effective collision frequency $\omega = \gamma_c Z_{LJ}[M]$, where Z_{LJ} is the Lennard-Jones collision frequency, [M] the bath gas concentration and γ_c the collision efficiency.[33,34] Thermal rate constants for the product channels were then derived with the usual steady-state assumption for the vibrationally highly excited intermediates by integrating the $k_i(E, J)$ over the E, J distributions.[24,25]

3. Results

3.1. Experimental data

3.1.1. Observed products. Measurements were carried out at room temperature by photolyzing H_2CCO at $\lambda \approx 313$ nm using reaction mixtures with partial pressures of $p(H_2CCO) = 5$ mbar and $0 \leq p(NO)/\text{mbar} \leq 9$ with overall pressures of $p = 100$ mbar (M = He) or 570 mbar (M = Ar). The experimental conditions were chosen as a compromise between maximization of the reactions of CH_2 in the singlet or in the triplet state and the reduced FTIR detection sensitivities due to increasing pressure broadening at higher pressures. Due to the small changes in the H_2CCO and NO at their high pressures and correspondingly strong absorptions, the absolute depletions of these compounds was usually not determined to better than about ± 15–20%. At the lower experimental pressure, the product yields with respect to the H_2CCO consumed were relatively small because of consecutive reactions competing with reactions (1) and (2). The analysis in the following therefore concentrates on the measurements at the higher experimental pressure.

An FTIR spectrum showing the reaction products in the presence of a large excess of NO is depicted in Fig. 2. The observed compounds were H_2CCO, NO, CO, C_2H_4, HCNO, HCN, N_2O, CO_2, H_2CO and HONO.[9] The only products without added NO were CO, C_2H_4 and C_2H_2.

The results are compiled in Table 1. As can be seen, CO and HCNO are the main products, HCN, N_2O, CO_2 and H_2O are minor products. The NO depletion approximately equals the formation of the CO. However, the H_2CCO consumption is somewhat higher. This is true especially at $p = 100$ mbar (M = He). H_2CO and HONO were detected only in trace amounts.[9] The yields of CO, HCNO and HCN with respect to H_2CCO from the measurements at $p = 570$ mbar (M = Ar) are plotted vs. the NO pressure in Fig. 3. The observed product ratios are $p(HCNO)/-\Delta p(H_2CCO) = (0.48 \pm 0.15)$ and $p(HCN)/-\Delta p(H_2CCO) = (0.08 \pm 0.04)$, $p(HCNO)/-\Delta p(NO) = (0.70 \pm 0.10)$ and $p(HCN)/-\Delta p(NO) = (0.11 \pm 0.03)$, $p(HCNO)/p(CO) = (0.70 \pm 0.10)$ and $p(HCN)/p(CO) = (0.11 \pm 0.03)$.

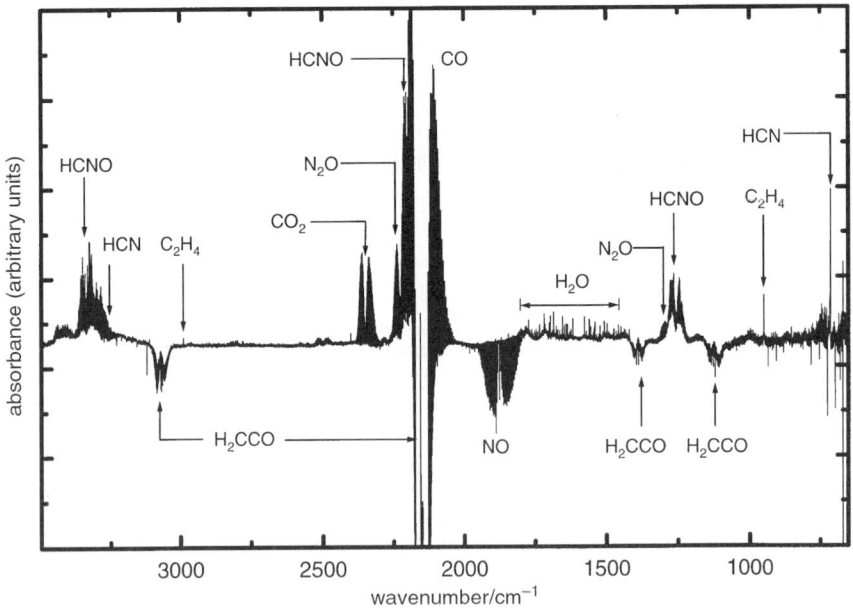

Fig. 2 Measured FTIR product spectrum with $p(CH_2CO) = 5.0$ mbar, $p(NO) = 2.97$ mbar, $p(\text{overall}) = 570$ mbar, M = Ar, photolysis time: $\Delta t = 60$ min.

Table 1 Experimental product yields (partial pressures, all in mbar) after 60 min photolysis ($p_0(H_2CCO) = 5.0$ mbar)[a]

p(overall)[a]	p_0(NO)	Δp(NO)	$\Delta p(H_2CCO)$	p(CO)	$p(C_2H_4)$	p(HCNO)	p(HCN)	$p(N_2O)$	$p(CO_2)$	$p(H_2O)$
100 (M = He)	0.00	—	—	0.043	0.019	0.000	0.0000	—	—	—
	0.30	—	0.077	0.048	0.011	0.007	0.0015	—	—	—
	0.77	—	0.047	0.049	0.012	0.006	0.0012	—	—	—
	2.80	—	0.044	0.050	0.011	0.015	0.0022	—	—	—
	3.17	—	0.068	0.043	0.009	0.013	0.0016	—	—	—
⟨av.⟩[b]	—	—	0.047	⟨0.048⟩	⟨0.011⟩	⟨0.011⟩	⟨0.0016⟩			
570 (M = Ar)	0.00	—	⟨0.052⟩	0.210	0.066	0.000	0.000	0.000	0.000	0.000
	0.93	0.20	0.254	0.200	0.026	0.099	0.011	0.010	0.006	0.002
	2.25	0.17	0.257	0.175	0.019	0.112	0.021	0.015	0.006	0.006
	2.61	0.15	0.274	0.173	0.018	0.102	0.022	0.016	0.007	0.009
	2.97	0.16	0.261	0.164	0.016	0.125	0.020	0.018	0.006	0.007
	3.36	0.19	0.250	0.165	0.017	0.115	0.021	0.018	0.006	0.005
	3.68	0.16	0.224	0.166	0.017	0.129	0.020	0.019	0.006	0.014
	4.03	0.18	0.227	0.185	0.016	0.122	0.022	0.021	0.006	0.010
	6.73	—	0.282	0.152	0.011	0.128	0.018	0.018	0.006	0.005
	8.75	—	0.232	0.141	0.009	0.137	0.017	0.015	0.006	0.005
⟨av.⟩[b]	—	⟨0.17⟩	0.211	⟨0.169⟩	⟨0.016⟩	⟨0.119⟩	⟨0.019⟩	⟨0.017⟩	⟨0.006⟩	⟨0.007⟩
			⟨0.246⟩							

[a] The measurements at p(overall) = 100 mbar (M = He) and 570 mbar (M = Ar) were carried out with different photolysis light filters so that the observed product amounts cannot be directly compared. A dash indicates that a value was not measured. [b] Averaged values (with added NO only).

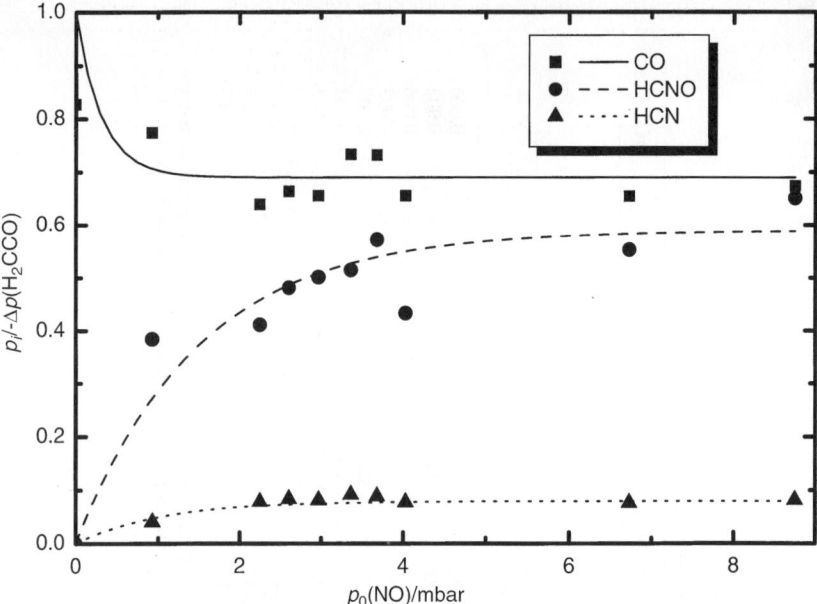

Fig. 3 Plot of the experimental product yields with respect to the observed H_2CCO depletion *vs.* NO pressure ($\Delta p(\text{product})/-\Delta p(CH_2CO)$ *vs.* $p(NO)$). $p(\text{overall}) = 570$ mbar, M = Ar, photolysis time: $\Delta t = 60$ min.

3.1.2. Reaction mechanism. The reactions which have to be considered in order to rationalize the experimental data are compiled in Table 2. The H_2CCO photodissociation proceeds according to[35,36]

$$H_2CCO + h\nu \rightarrow {}^1CH_2 + CO, \tag{3.1}$$

$$\rightarrow {}^3CH_2 + CO. \tag{3.2}$$

Under the conditions used, ${}^1CH_2 + CO$ are the dominating products.[36] Without NO, the ${}^1CH_2$ reacts with H_2CCO or it undergoes collisional deactivation to the triplet ground state,[13,14,37–39]

$$^1CH_2 + H_2CCO \rightarrow C_2H_4 + CO, \tag{4.1}$$

$$\rightarrow {}^3CH_2 + H_2CCO, \tag{4.2}$$

$$^1CH_2 + Ar \rightarrow {}^3CH_2 + Ar, \tag{5}$$

$$^1CH_2 + He \rightarrow {}^3CH_2 + He. \tag{6}$$

The reaction

$$^3CH_2 + H_2CCO \rightarrow \text{products} \tag{7}$$

is too slow to play a significant role.[40] The ${}^3CH_2$ combination reaction leads to small amounts of C_2H_2,[14]

$$^3CH_2 + {}^3CH_2 \rightarrow C_2H_2 + 2H, \tag{8.1}$$

$$\rightarrow C_2H_2 + H_2. \tag{8.2}$$

Side reactions such as[41–43]

$$H + H_2CCO \rightarrow CH_3 + CO, \tag{9}$$

$$CH_3 + CH_3 + M \rightarrow C_2H_6 + M, \tag{10}$$

$$H + CH_3 + M \rightarrow CH_4 + M, \tag{11}$$

$$^3CH_2 + CH_3 \rightarrow H + C_2H_4 \tag{12}$$

Table 2 Compilation of the elementary reactions in the detailed reaction mechanism

No.	Reaction	Rate constant[a] /cm^3 mol^{-1} s^{-1}	Ref.
1.1	$^3CH_2 + NO \rightarrow HCNO + H$	2.2×10^{13}	This work (see text)
1.2	$^3CH_2 + NO \rightarrow HCN + OH$	2.2×10^{12}	This work (see text)
2.1	$^1CH_2 + NO \rightarrow HCNO + H$	1.0×10^{14}	12, 13, 14 (see text)
2.2	$^1CH_2 + NO \rightarrow \rightarrow HCN + OH$	8.0×10^{12}	12, 13, 14 (see text)
2.3	$^1CH_2 + NO \rightarrow {}^3CH_2 + NO$	3.7×10^{13}	12, 13, 14 (see text)
3.1	$H_2CCO + h\nu \rightarrow {}^1CH_2 + CO$	8.0×10^{-6} (1.7×10^{-6})	This work (see text)
3.2	$H_2CCO + h\nu \rightarrow {}^3CH_2 + CO$		
4.1	$^1CH_2 + H_2CCO \rightarrow CO + C_2H_4$	1.2×10^{14}	14, 37
4.2	$^1CH_2 + H_2CCO \rightarrow {}^3CH_2 + H_2CCO$	3.0×10^{13}	14, 37
5	$^1CH_2 + Ar \rightarrow {}^3CH_2 + Ar$	3.2×10^{12}	37
6	$^1CH_2 + He \rightarrow {}^3CH_2 + He$	1.8×10^{12}	37
7	$^3CH_2 + CH_2CO \rightarrow products$	1.0×10^9	40
8.1	$^3CH_2 + {}^3CH_2 \rightarrow C_2H_2 + 2H$	2.0×10^{13}	14
8.2	$^3CH_2 + {}^3CH_2 \rightarrow C_2H_2 + H_2$	1.8×10^{14}	14
9	$H + H_2CCO \rightarrow CO + CH_3$	3.6×10^{10}	41
10	$CH_3 + CH_3 + M \rightarrow C_2H_6 + M$	3.6×10^{13}	42
11	$H + CH_3 + M \rightarrow CH_4 + M$	1.2×10^{14}	42
12	$^3CH_2 + CH_3 \rightarrow H + C_2H_4$	6.5×10^{13}	43
13	$H + NO + M \rightarrow HNO + M$	3.2×10^{11} (5.6×10^{10})	44
14	$OH + NO + M \rightarrow HONO + M$	3.6×10^{12} (6.3×10^{11})	42
15.1	$OH + H_2CCO \rightarrow CH_2OH + CO$	5.0×10^{12}	46, 47
15.2	$OH + H_2CCO \rightarrow CH_3 + CO_2$	2.0×10^{12}	46, 47
16.1	$HNO + HNO \rightarrow H_2O + N_2O$	2.7×10^9	48, 49, 50
16.2	$HNO + HNO + M \rightarrow (HNO)_2 + M$	5.0×10^8	48, 49, 50
17	$CH_2OH + NO \rightarrow H_2CO + HNO$	1.5×10^{13}	51
18	$CH_3 + NO + M \rightarrow CH_3NO + M$	3.9×10^{12}	52
19	$HCNO + H_2CCO \rightarrow products$	2.6×10^2	This work (see text)

[a] Second-order rate constants (cm^3 mol^{-1} s^{-1}) for reactions in the fall-off range are for $T = 298$ K at $p = 570$ mbar with M = Ar (values in parentheses for $p = 100$ mbar with M = He).

can be neglected considering the low steady-state radical concentrations in the experiments. The H$_2$CCO photodissociation rates in the experiments (k_3^{eff}) are thus easily determined from the observed yields of CO or C$_2$H$_4$ + C$_2$H$_2$.

In the presence of NO, at the concentrations used, $^3CH_2$ is rapidly scavenged by the NO. Radical–radical reactions other than with the NO are negligible. The observed HCNO and HCN can thus be attributed to the reaction channels

$$^3CH_2 + NO \rightarrow HCNO + H, \tag{1.1}$$

$$\rightarrow HCN + OH. \tag{1.2}$$

At low bath gas concentrations, some $^1CH_2$ can also react with the NO. In the absence of further information, since the same products (HCNO and HCN) were observed at the two experimental pressures, we assume the same reaction channels, i.e.

$$^1CH_2 + NO \rightarrow HCNO + H, \tag{2.1}$$

$$\rightarrow HCN + OH. \tag{2.2}$$

For the collisional deactivation,

$$^1CH_2 + NO \rightarrow {}^3CH_2 + NO, \tag{2.3}$$

we adopted a value of $k_{2.3}/k_2 = 0.25$, as found for many other colliders.[39] No evidence was seen for significant yields of other products from reactions (1) and (2).

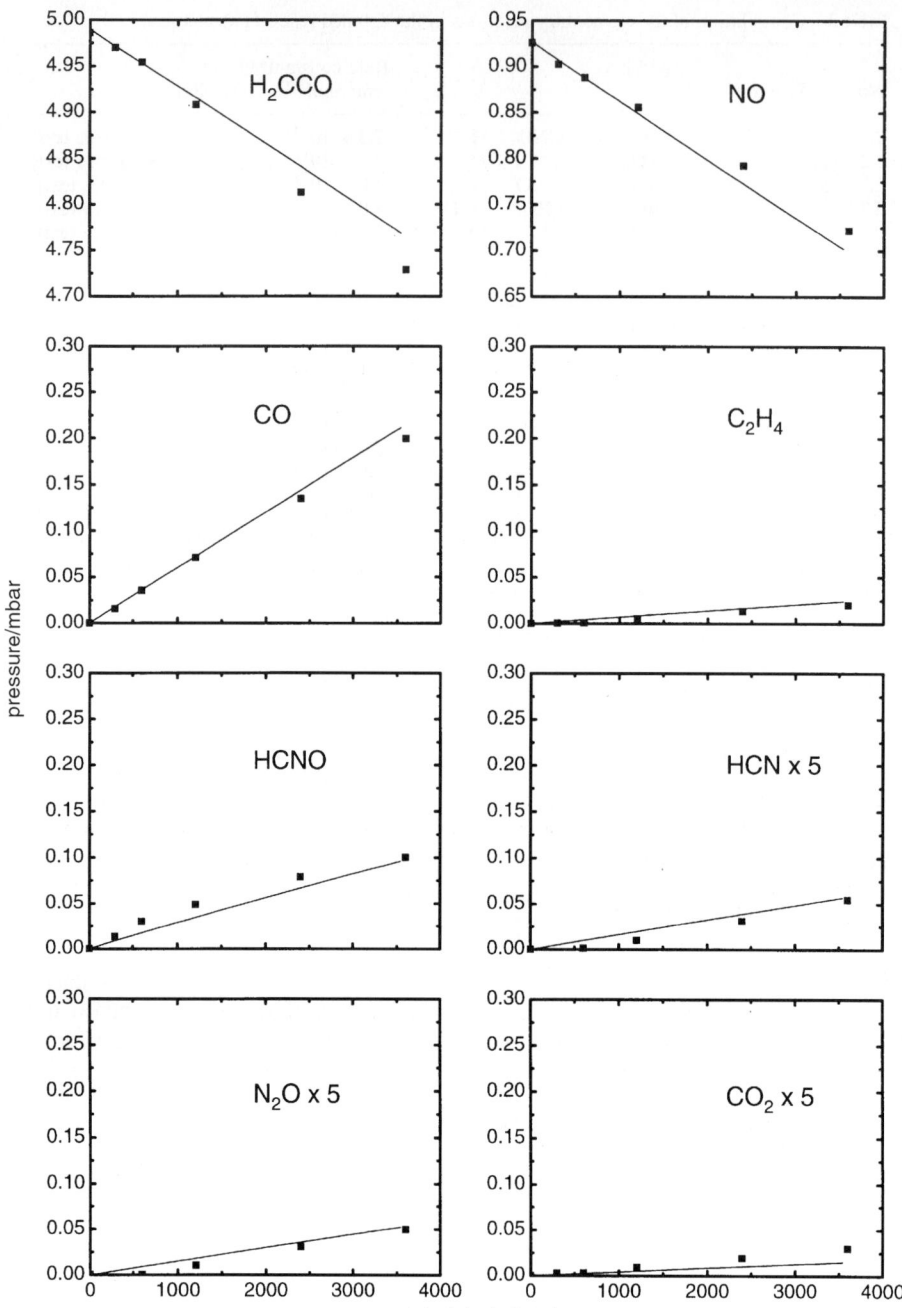

Fig. 4 Measured reactant and product partial pressures as a function of photolysis time. Experimental conditions: $p(H_2CCO) = 5.0$ mbar, $p(NO) = 0.93$ mbar, $p(\text{overall}) = 570$ mbar, $M = Ar$.

The fate of the H atoms and OH radicals from (1) and (2) is determined by their reactions with the NO[44,45] and H_2CCO[46,47] according to (9) and

$$H + NO + M \rightarrow HNO + M, \tag{13}$$

$$OH + NO + M \rightarrow HONO + M, \quad (14)$$

$$OH + H_2CCO \rightarrow CH_2OH + CO, \quad (15.1)$$

$$\rightarrow CH_3 + CO_2. \quad (15.2)$$

Reactions (13) and (14) are in the fall-off regime under the conditions used. The HNO leads to the observed N_2O and H_2O,[48–50]

$$HNO + HNO \rightarrow N_2O + H_2O, \quad (16.1)$$

$$HNO + HNO + M \rightarrow (HNO)_2 + M. \quad (16.2)$$

The CH_3 and CH_2OH radicals are subject to reactions[51,52]

$$CH_2OH + NO \rightarrow H_2CO + HNO, \quad (17)$$

$$CH_3 + NO + M \rightarrow CH_3NO + M. \quad (18)$$

Reaction (17) explains the observed traces of H_2CO. CH_3NO, which can rearrange[53] to CH_2NOH, could not be detected.

Eventually, a slow decay of the HCNO and corresponding consumption of unphotolyzed H_2CCO was observed when the products were monitored for a longer time after the photolysis. This "dark reaction", which appeared to be pseudo-first-order and may be due to heterogeneous processes, could be formally described by reaction (19).

$$HCNO + H_2CCO \rightarrow products. \quad (19)$$

A rate constant k_{19} was determined from the apparent HCNO decay.

3.1.3 Branching ratios for HCNO and HCN.

Fig. 4 shows profiles of the reactants and products as a function of photolysis time for a typical experimental run. At the low H_2CCO consumptions ($\leqslant 5\%$), the H_2CCO and NO decay and CO formation profiles are practically linear. However, reaction (19) leads to a curvature of the HCNO profile. The values of the branching ratios $k_{1.1}/k_1$ and $k_{1.2}/k_1$ were therefore determined from the product formation rates at short reaction times. The experimental data are listed in Table 3. Taking those data and assuming that products channels other than those to HCNO and HCN can be neglected within the experimental

Table 3 Initial formation rates of CO, HCNO and HCN in the photolysis experiments with respect to the H_2CCO depletion rates

mbar	mbar	$\left(\dfrac{d[NO]}{d[H_2CCO]}\right)$	$\left(\dfrac{d[CO]}{-d[H_2CCO]}\right)$	$\left(\dfrac{d[HCNO]}{-d[H_2CCO]}\right)_{t\rightarrow 0}$	$\left(\dfrac{d[HCN]}{-d[H_2CCO]}\right)_{t\rightarrow 0}$
100 (M = He)	0.00	—	0.60	0.00	0.000
	0.30	—	1.03	0.28	0.033
	0.77	—	1.09	0.43	0.026
	2.80	—	0.73	0.36	0.032
	3.17	—	0.91	0.62	0.034
⟨av.⟩[a]	—	—	⟨0.94⟩	⟨0.42⟩	⟨0.030⟩
570 (M = Ar)	0.00	0.00	0.83	0.00	0.00
	0.93	0.77	0.78	0.61	0.05
	2.25	0.63	0.64	0.55	0.08
	2.61	0.58	0.67	0.48	0.08
	2.97	0.64	0.65	0.65	0.08
	3.36	0.85	0.73	0.56	0.09
	3.68	0.71	0.73	0.75	0.09
	4.03	0.64	0.65	0.71	0.08
	6.73	—	0.65	0.60	0.08
	8.75	—	0.67	0.83	0.08
⟨av.⟩[a]	—	⟨0.69⟩	⟨0.69⟩	⟨0.64⟩	⟨0.08⟩

[a] Average values (data with added NO only).

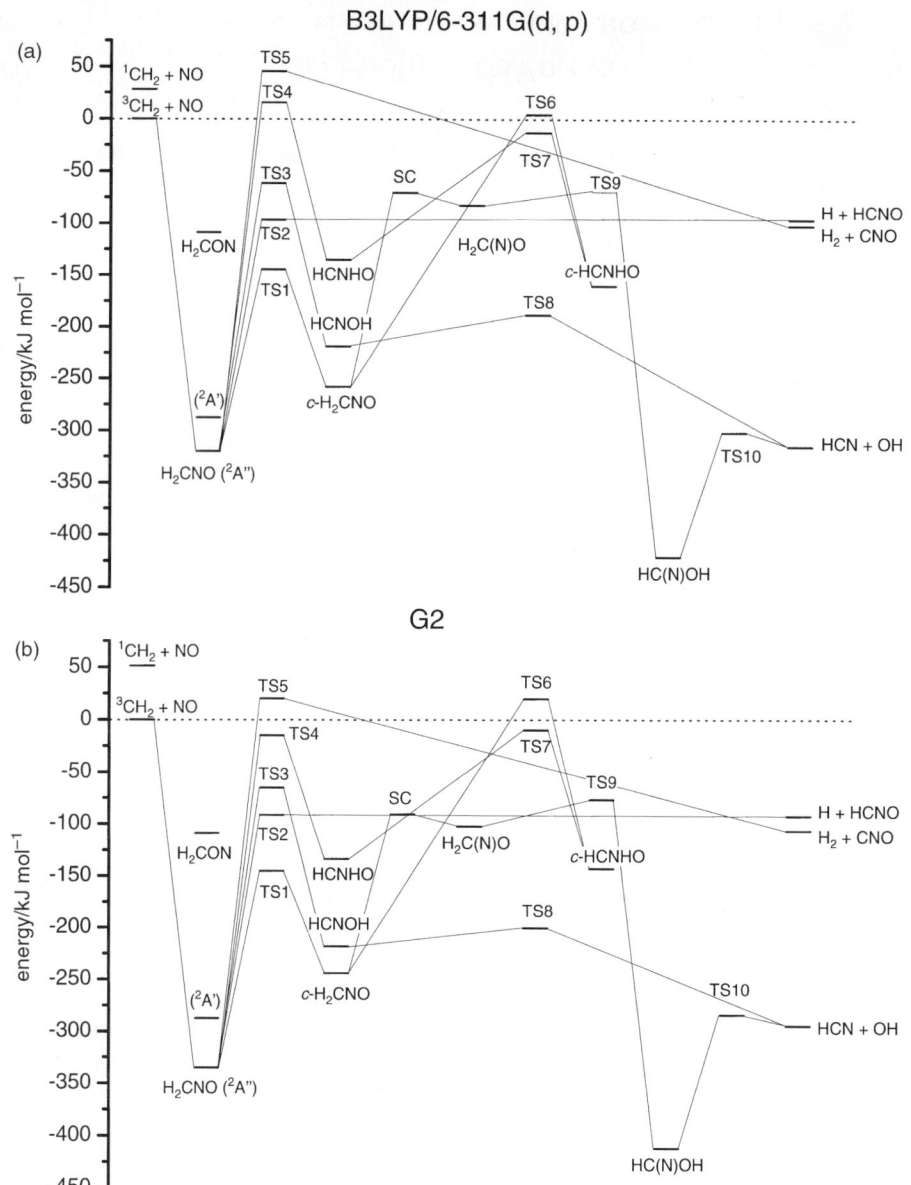

Fig. 5 Energy diagram of the $CH_2 + NO$ reaction system (a) at the B3LYP/6-311G(d,p) and (b) at the G2 levels of *ab initio* theory.

detection limits, one can write

$$\frac{k_{1.1}}{k_1} = \left(\frac{d[HCNO]}{d[HCNO] + d[HCN]}\right)_{t \to 0}, \quad (II)$$

$$\frac{k_{1.2}}{k_1} = \left(\frac{d[HCN]}{d[HCNO] + d[HCN]}\right)_{t \to 0}. \quad (III)$$

The results from the data for $p = 570$ mbar (M = Ar) are $k_{1.1}/k_1 = (0.89 \pm 0.06)$, $k_{1.2}/k_1 = (0.11 \pm 0.06)$, with error limits containing statistical as well as estimated systematic uncertainties.

3.1.4 Numerical simulations.
The direct experimental results were checked by numerical simulations of the measured profiles based on the reaction system and rate constants in Table 2. Calculated time profiles with the above values for $k_{1.1}/k_1$ and $k_{1.2}/k_1$ are given in Fig. 4. A similar level of agreement between the calculated and experimentally measured profiles was found at both experimental pressures.

3.2. Multi-channel unimolecular rate calculations

3.2.1 Potential energy hypersurface.
The potential energy surfaces for the $^3CH_2 + NO$ reaction system calculated at the B3LYP/6-311G(d,p) and at the G2 levels of theory are displayed in Fig. 5(a) and (b). Within the respective error limits, the energies found with the two methods are in satisfactory agreement. The results which we communicated in some detail elsewhere[15] were confirmed by the subsequent work of Shapley and Bacskay.[17,18]

The transition state TS2 (see Fig. 5) for the elimination of an H atom from the H_2CNO complex was the only problematic one. At the G2 level, its energy was calculated slightly below that of the H + HCNO products. The B3LYP/6-311G(d,p) calculations gave a well defined transition state, with a barrier height with respect to H + HCNO of ≈ 100 cm^{-1}, but the H–HCNO bond length came out unusually long ($r = 2.67$ Å). On chemical grounds one can expect a small barrier for the addition of an H atom to a double bond as in HCNO. The B3LYP calculations also resulted in better frequencies for the HCNO fragment.[15] The rate constant calculations described in the following were therefore performed taking the B3LYP/6-311G(d,p) structures and energies.

3.2.2 Detailed reaction scheme.
A reaction scheme extracted from the PES leads is shown in Fig. 6. The first step is the formation of the intermediate H_2CNO^*, reactions ($-1a/1a$). The asterisks here and in the following denote the respective vibrational excitation energies of the different species. In the present work, we considered only H_2CNO^* in its $^2A''$ ground electronic state for reaction entrance channel. We assumed further that only the H_2CNO^* intermediate could be collisionally deactivated (ω).

Four product channels were predicted apart from (ω). One is the direct dissociation reaction (1g) via transition state TS2 to H + HCNO. Two channels were found leading to HCN + OH. The first involves a cyclic intermediate (c-H_2CNO^{**}) reached from the initial H_2CNO^* complex in reaction (1b) via TS1, a potential surface crossing (SC), and isomerization via TS9 to HC(N)OH***. Details of this channel have been taken from Shapley and Bacskay.[17,18] The transformation from c-H_2CNO^{**} to HC(N)OH*** was treated for simplicity as a one-step process (1c), assuming the SC to be fast. This seems justified by the relatively low well of the $H_2C(N)O$ between the SC and TS9 (Fig. 5). HC(N)OH*** then gives HCN + OH in reaction (1d) via TS10. The second pathway (1e) proceeds via TS3 to HCNOH**** which dissociates reaction (1f) to HCN + OH via TS8. The fourth product channel is the formation of H_2 + CNO in reaction (1h). Other intermediate structures which were found on the PES did not seem to lead to energetically accessible products.

3.2.3 Microcanonical specific unimolecular rate constants.
Microcanonical specific rate constants $k_i(E, J)$ for the reaction steps were computed taking the data in Tables 4 and 5. The results are shown in Fig. 7–9, together with normalized energy distribution functions $F_{ss}(E, J)$ of the respective intermediates for two temperatures ($T = 300$ and 1200 K). All data are for an overall

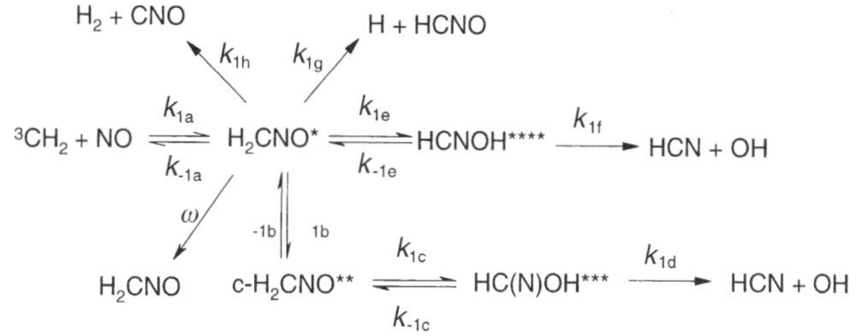

Fig. 6 Reaction scheme used for the kinetic modeling.

Table 4 Energies including zero-point energy relative to the $^3CH_2$ + NO entrance channel, rotational constants and vibrational frequencies for the intermediates and transition states in the reaction scheme at the B3LYP/6-311G(d,p) level for the unimolecular rate calculations

Species	Energy^a/cm^{-1}	Rotational constants/cm^{-1}	Vibrational frequencies/cm^{-1}
$^3CH_2$ + NO	0	56.55, 8.419, 7.328//1.712	1057, 3118, 3358//1989
$^1CH_2$ + NO		19.49, 11.30, 7.144//1.712	1405, 2883, 2945//1989
H$_2$CNO	−28020	4.277, 0.393, 0.359	460, 699, 815, 1130, 1251, 1489, 1723, 3107, 3254
HCNOH	−18260	3.570, 0.390, 0.352	423, 484, 681, 826, 1040, 1375, 1725, 3176, 3660
c-H$_2$CNO	−20400	1.054, 0.876, 0.531	866, 880, 990, 1141, 1160, 1316, 1529, 3077, 3177
H$_2$C(N)O	−8560	2.010, 0.381, 0.339	423, 441, 906, 980, 1102, 1231, 1257, 2825, 2839
HC(N)OH	−34450	2.613, 0.403, 0.349	516, 556, 927, 1096, 1218, 1351, 1702, 3088, 3765
TS1: H$_2$CNO → c-H$_2$CNO	−12140	1.427, 0.612, 0.457	967i, 823, 901, 1086, 1097, 1319, 1500, 3071, 3209
TS2: H$_2$CNO → H + HCNO	−7660	2.642, 0.355, 0.312	128i, 104, 263, 336, 570, 571, 1305, 2321, 3501
TS3: H$_2$CNO → HCNOH	−5460	1.767, 0.524, 0.404	2153i, 350, 635, 832, 923, 983, 1712, 1927, 3201
TS5: H$_2$CNO → H$_2$ + CNO	+1680	4.395, 0.369, 0.345	1191i, 192, 408, 489, 846, 1119, 1208, 1720, 3184
TS8: HCNOH → OH + HCN	−16760	2.893, 0.341, 0.305	710i, 238, 376, 596, 663, 1124, 1818, 3366, 3710
TS9: H$_2$C(N)O → HC(N)OH	−6440	2.468, 0.379, 0.347	1337i, 435, 508, 1043, 1066, 1230, 1374, 2478, 2927
TS10: HC(N)OH → OH + HCN	−24680	1.722, 0.303, 0.258	361i, 221, 275, 733, 785, 823, 2062, 3417, 3727
H + HCNO (linear)	−7740	0.384	244, 244, 570, 570, 1301, 2335, 3514
(nonlinear)	−7850	1158.4, 0.384, 0.384	248, 565, 596, 1302, 2333, 3516
OH + HCN	−24660	18.69//1.495	3701//786, 786, 2201, 3459
H$_2$ + CNO	−8880	60.41//0.415	4419//321, 321, 416, 416, 1201, 1934

^{a} See ref. 15 for the comparable G2 energies.

Table 5 Parameters used in the SACM calculation for the reactions $H_2CNO \to {}^3CH_2 + NO$ and $H_2CNO \to H + HCNO$

H_2CNO	$A, B, C/cm^{-1}$	4.277, 0.393, 0.359
	v_i/cm^{-1}	460, 699, 815, 1130, 1251, 1489, 1723, 3107, 3254
${}^3CH_2$	$A, B, C/cm^{-1}$	56.55, 8.419, 7.873
	v_i/cm^{-1}	1057, 3118, 3358
NO	B/cm^{-1}	1.729 (linear)
	v/cm^{-1}	1989
$H_2CNO \to {}^3CH_2 + NO$	D_{C-N}/cm^{-1}	30 230
	$\beta/\text{Å}^{-1}$	3.73
	α/β	0.48
	Quasi-triatomic model:	$M_{CH_2} = 14, M_N = 14, M_O = 16$ (u)
		$r_{CN} = 1.28$ Å, $r_{NO} = 1.22$ Å, $\delta_{CNO} = 135.1°$
	Adiabatic correlations:	1723 → RC
		4.277 → 1.712
		460 → 1.712
		699 → 7.873
		815 → 8.419
		1130 → 56.55
		1251 → 1057
		1489 → 1989
		3107 → 3118
		3107 → 3358
HCNO	B/cm^{-1}	0.384 (linear)
	v_i/cm^{-1}	244, 244, 570, 570, 1301, 2335, 3514
$H_2CNO \to H + HCNO$	D_{C-H}/cm^{-1}	22 940
	$\beta/\text{Å}^{-1}$	2.59
	α/β	0.50
	Quasi-triatomic model:	$M_H = 1, M_{CH} = 13, M_{NO} = 30$ (u)
		$r_{CH} = 1.08$ Å, $r_{CN} = 1.28$ Å, $\delta_{HCN} = 120.6°$
	Adiabatic correlations:	3254 → RC
		4.277 → 0.384
		460 → 0.384
		699 → 244
		815 → 244
		1130 → 570
		1251 → 570
		1489 → 1301
		1723 → 2335
		3107 → 3514

angular momentum value of $\langle J \rangle = 13$ which corresponds to the average J for room temperature from a collisional capture model[54,55] for the entrance channel. This value increases to $\langle J \rangle \approx 40$ at $T = 2000$ K. Calculations of $k_i(E, J)$ were run up to $J = 200$ to ensure convergence of the thermally averaged rate constants below.

Fig. 7 shows the specific rate constants of the unimolecular reactions of the H_2CNO^* complex. At the energies of interest, above the ${}^3CH_2 + NO$ entrance level, the dissociation to $H + HCNO$, reaction (1g) is by far the fastest channel. A tight TS2 configuration was assumed and treated by an RRKM calculation. An SACM calculation for a loose TS2 was run as a check. As expected, the predicted RRKM rate constant came out somewhat smaller than the SACM value which should be an upper limit. On the other hand, the formation of the cyclic intermediate $c\text{-}H_2CNO^{**}$, reaction (1b) is at least ten times slower at the excitation energies of interest, although it has a much lower threshold energy. The next step is the isomerization of the H_2CNO^* to $HCNOH^{****}$, reaction (1e). The rate constants $k_{-1a}(E, J)$ for the reverse decomposition of the H_2CNO^* to ${}^3CH_2 + NO$ were treated by an SACM calculation with the parameters given in Table 5 and a value for the anisotropy ratio of $\alpha/\beta = 0.48$ (determined below). As seen, $k_{-1a}(E, J)$ increases quickly with E because of the loose transition state and comes into play at higher energies. In contrast, the formation of $H_2 + CNO$ via reaction (1h) seems to be negligible, at least at lower excitation energies.

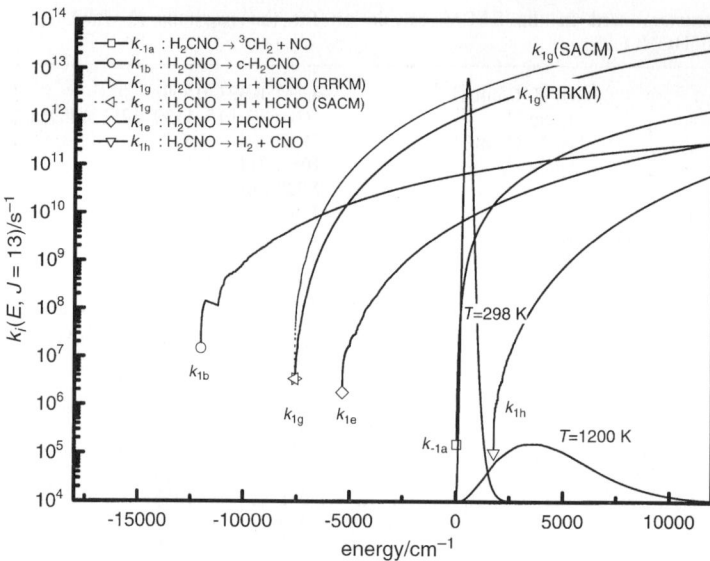

Fig. 7 Calculated microcanonical rate constants $k_i(E, J)$ for the unimolecular reaction steps of the H_2CNO^* complex for $J = 13$ and steady state energy distribution functions of the H_2CNO.

Fig. 8 shows the specific rate constants for the consecutive reactions leading to HCN + OH *via* the cyclic intermediate c-H_2CNO^{**}, reaction (1b). The next step, reaction (1c) gives HC(N)OH (see above). However, the diagram shows clearly that the ring opening, reaction ($-$1b) back to H_2CNO is much faster than reaction (1c). This reduces the yield of HCN + OH from this pathway. The last step in the sequence, dissociation to HCN + OH, reaction (1d), is very fast and not rate determining.

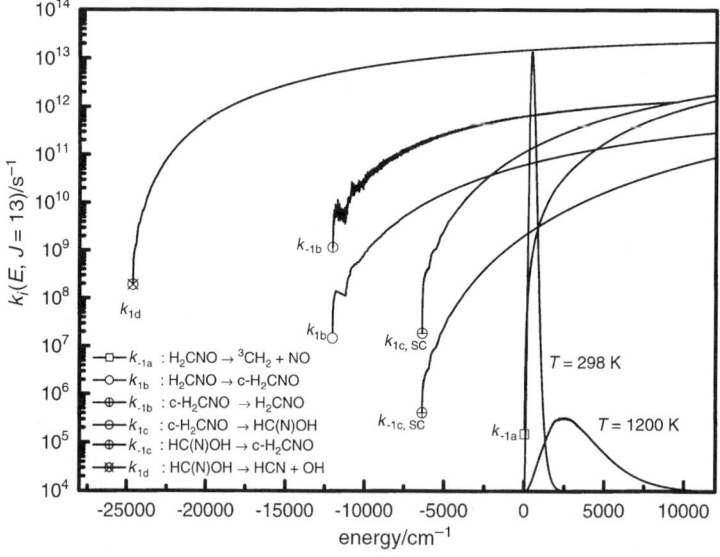

Fig. 8 Calculated microcanonical rate constants $k_i(E, J)$ for the unimolecular reaction steps of the H_2CNO^* complex and steady state energy distribution function of the intermediate $HC(N)OH^{***}$.

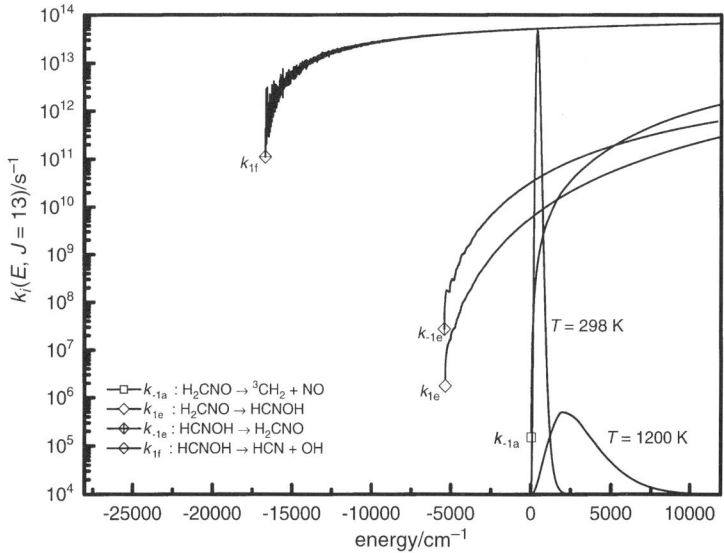

Fig. 9 Calculated microcanonical rate constants $k_i(E, J)$ for the unimolecular reaction steps leading to HCN + OH *via* HCNOH****.

The specific rate constants for the second pathway to HCN + OH are depicted in Fig. 9. Here, the formation of HCNOH****, reaction (1e) is the rate determining reaction. The following dissociation of the HCNOH**** to HCN + OH, reaction (1f) is again very fast.

3.2.4 Collisional stabilization.

The collisional deactivation of the H_2CNO^* complex is described by an effective stabilization rate (ω). Taking the Lennard-Jones parameters ($\sigma_{Ar} = 3.542$ Å, $(\varepsilon/k_B)_{Ar} = 93.3$ K and $\sigma_{H_2CNO} = 4.4$ Å, $(\varepsilon/k_B)_{H_2CNO} = 258$ K, estimated from HCNO)[56] and estimating an average energy transferred per collision of $\langle \Delta E \rangle = 200$ cm^{-1} (which should be on the high side), one obtains[34] a collision frequency of $Z_{LJ} = 2 \times 10^{14}$ cm^3 mol^{-1} s^{-1} and a collision efficiency $\gamma_c = 0.06$. This leads to an effective stabilization rate of $\omega = 5 \times 10^8$ s^{-1} for $p = 1$ bar (room temperature). Comparing this value with the $k_i(E, J)$ curve for the H atom elimination from the H_2CNO^* complex in Fig. 7, it becomes clear that the reaction is in the low pressure regime under the experimental conditions, *i.e.* collisions cannot play a significant role.

3.2.5 High pressure $^3CH_2 + NO$ capture rate constant.

In the high-pressure regime, the H_2CNO^* complex will be deactivated by collisions. The high-pressure "capture" rate constant is given by[25]

$$k_{capture}(T) = \frac{g(T)}{hQ_{CH_2}Q_{NO}} e^{-E_0/k_B T} \sum_{J=0}^{\infty} (2J+1) \int_0^{\infty} W_1(E, J) e^{-E/k_B T} \, dE. \quad (IV)$$

Here, Q_{CH_2} and Q_{NO} are the partition functions of the reactants excluding the electronic part and the center of mass motion and $g(T)$ is the ratio of the electronic partition functions of the complex and the reactants.

Results for $k_{capture}$ at room temperature are given in Table 6. The value of the anisotropy parameter α/β for the calculation of the number of open channels $W_1(E, J)$ was used to fit the calculated high pressure rate constant to the experimental value[6] of $k_1(296 \text{ K}) = (2.2 \pm 0.5) \times 10^{13}$ cm^3 mol^{-1} s^{-1}. Excellent agreement was obtained with $\alpha/\beta = 0.48$ which is close to the "standard" value of $\alpha/\beta = 0.5$. It has to be kept in mind that the calculated value is an upper limit

Table 6 Calculated $^3CH_2 + NO$ capture rate constants ($k_{capture}$) for different values of the anisotropy parameter α/β (experimental value:[6] $k_1 = 2.2 \times 10^{13}$ cm^3 s^{-1} mol^{-1})

α/β	$k_{capture}$(298 K)/cm^3 s^{-1} mol^{-1}
0.40	7.86×10^{12}
0.45	1.65×10^{13}
0.48	2.34×10^{13}
0.50	2.84×10^{13}
0.55	3.91×10^{13}

if redissociation of the H_2CNO^* is important. The predicted temperature dependence up to $T = 2000$ K was very weak.

3.2.6 Rate constants for the product channels. With the steady state assumption for the intermediates and by averaging over the state distribution functions, one obtains the thermal rate constants $k_j(T)$ for the product channels j in the form

$$k_{HCN+OH_{1b-d}}(T) = \frac{g(T)}{hQ_{CH_2}Q_{NO}} \sum_{J=0}^{\infty} (2J+1) \int_0^\infty \frac{k_{1b}(E,J)k_{1c,SC}(E,J)k_{1d}(E,J)}{A_1(E,J)A_2(E,J)A_3(E,J)} W_1(E,J)e^{-E/k_BT} dE \quad (V)$$

$$k_{HCN+OH_{1e-f}}(T) = \frac{g(T)}{hQ_{CH_2}Q_{NO}} \sum_{J=0}^{\infty} (2J+1) \int_0^\infty \frac{k_{1e}(E,J)k_{1f}(E,J)}{A_1(E,J)A_4(E,J)} W_1(E,J)e^{-E/k_BT} dE \quad (VI)$$

$$k_{HCNO+H}(T) = \frac{g(T)}{hQ_{CH_2}Q_{NO}} \sum_{J=0}^{\infty} (2J+1) \int_0^\infty \frac{k_{1g}(E,J)}{A_1(E,J)} W_1(E,J)e^{-E/k_BT} dE \quad (VII)$$

$$k_{CNO+H_2}(T) = \frac{g(T)}{hQ_{CH_2}Q_{NO}} \sum_{J=0}^{\infty} (2J+1) \int_0^\infty \frac{k_{1h}(E,J)}{A_1(E,J)} W_1(E,J)e^{-E/k_BT} dE \quad (VIII)$$

with

$$A_1(E,J) = k_{-1a}(E,J) + k_{1b}(E,J) + k_{1g}(E,J) + k_{1e}(E,J) + k_{1h}(E,J) + \omega$$
$$- \frac{k_{1e}(E,J)k_{-1e}(E,J)}{A_4(E,J)} - \frac{k_{1b}(E,J)k_{-1b}(E,J)}{A_5(E,J)} \quad (IX)$$

$$A_2(E,J) = k_{-1b}(E,J) + k_{1c,SC}(E,J) \quad (X)$$

$$A_3(E,J) = k_{1d}(E,J) + k_{-1c,SC}(E,J) \quad (XI)$$

$$A_4(E,J) = k_{1f}(E,J) + k_{-1e}(E,J) \quad (XII)$$

$$A_5(E,J) = A_2(E,J) - \frac{k_{1c,SC}(E,J)k_{-1c,SC}(E,J)}{k_{-1c,SC}(E,J) + k_{1d}(E,J)}. \quad (XIII)$$

The microcanonical rate constants $k_i(E,J)$ belong to the different reaction steps, $W_1(E,J)$ represents the number of open channels along the entrance channel, and E_0 is the barrier height of the entrance channel counted from $^3CH_2 + NO$.

The calculated thermal rate constants for the different product channels are plotted vs. temperature in Fig. 10. The pathway leading to H + HCNO can be seen to be the major product channel at all calculated temperatures. Its temperature dependence is very weak. The two HCN + OH channels are much slower. The channel that occurs via the cyclic intermediate decreases slightly with temperature, while the other channel increases, leading to a cancellation of the temperature dependences. Taken together, they are predicted with a yield of $\Gamma_{HCN+OH} \approx 0.015$. The $H_2 + CNO$ channel increases with temperature, but remains well below the other channels.

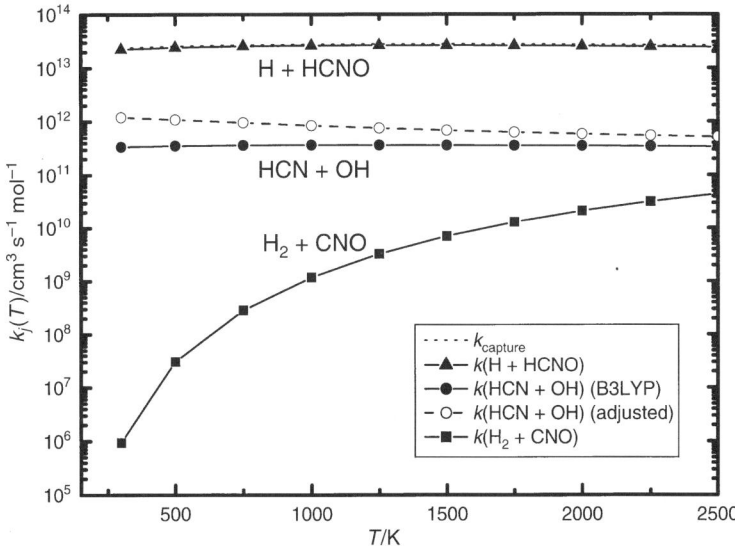

Fig. 10 Calculated thermal rate constants $k(T)$ for the different product channels as a function of temperature.

4. Discussion

4.1. Observed reaction products and comparison with literature data

The experimental results show that the formation of H + HCNO reaction (1.1) accounts for the main product channel of the $^3CH_2$ + NO reaction (1), while OH + HCN, reaction (1.2) are produced in a smaller channel. The branching ratios at room temperature were found to be $k_{1.1}/k_1 = (0.89 \pm 0.06)$, $k_{1.2}/k_1 = (0.11 \pm 0.06)$. This conclusion has been obtained directly from the observed yields of HCNO and HCN. It is supported by the nitrogen balance in the measurements ($\Delta[NO] \approx \Delta[HCNO] + \Delta[HCN] + 2\Delta([N_2O]$ within the experimental error limits) and by the simulations of the reaction system. Thus, the preliminary product study by Grußdorf et al.[9] is confirmed. It is noted, however, that the quoted yields of HCN may be affected by the slow decomposition of the unstable HCNO which can produce some HCN. The above value for the branching ratio for OH + HCN ($k_{1.2}/k_1$) should be an upper limit.

The analysis of the data in Table 3 according to eqn. (II) and (III) did not show a systematic dependence of the product branching ratios on the total bath gas pressure. This indicates that the $^3CH_2$ + NO and $^1CH_2$ + NO reactions (1) and (2) have similar product distributions (apart from the $^1CH_2$ + NO collisional intersystem crossing channel (2.3)). The unlikely alternative would have to be that channel (2.3) accounts for close to 100% of the $^1CH_2$ + NO reaction. The $^1CH_2$ and $^3CH_2$ reactions can be expected to proceed on the same doublet potential energy surface. The small additional internal excitation energy of the H_2CNO^* complex resulting from the electronic excitation energy of the $^1CH_2$ should not make a big difference, considering that the rate determining steps in the different product channels have comparable barrier heights.

The formation of OH radicals and H atoms from the title reaction has also been observed by Atakan et al.[8] and by Bauerle et al.[10] Atakan et al. reported OH + HCN branching ratios of 0.015 ± 0.08 at room temperature, in reasonable agreement with the present data, and 0.63 ± 0.25 at 1025 K. The latter high temperature value does not seem consistent with the present rate constant calculations. In addition, they gave upper limits for NCO + H_2 ($\leqslant 0.15$) and CN + H_2O ($\leqslant 0.10$). Bauerle et al. conducted shock tube experiments. Under the conditions used, the reaction systems were rather complicated.

Kong and co-workers[11] reported FTIR emission measurements of the reaction products. According to their analysis, they observed vibrationally excited CO, OH, HOCN, HCO, as well as NH_2, which they attributed to the reaction channels (1.2) and

$$^3CH_2 + NO \rightarrow H + HOCN, \quad (1.3)$$

$$\rightarrow NH + HCO, \quad (1.4)$$

$$\rightarrow NH_2 + CO. \quad (1.5)$$

There are, however, several problems. First, the IR emission assigned to HOCN persisted for much longer times than the much stronger CO emission, despite the fact that a vibrational relaxation of the polyatomic molecule HOCN should be faster than that of the CO. It is a question, therefore, whether consecutive reactions can be completely ruled out. In our work, an IR band of HOCN at around 2300 cm^{-1} would have been covered by the strong H_2CCO absorption. However, no other IR bands which could be assigned to HOCN could be found. Concerning on the other hand the alleged absence of HCNO, a possible HCNO emission around 2196 cm^{-1} would have been masked in the work of Kong by the very strong CO emission. The emission signals reported at 2472 and 3302 cm^{-1} and assigned to HCO and NH_2, respectively, do indeed coincide with bands of HCNO so that reassignment may be suggested. It is noted, last but not least, that NH_2 from reaction (1.5) would rapidly react with the NO present in large excess according to

$$NH_2 + NO \rightarrow N_2 + H_2O, \quad (20.1)$$

$$\rightarrow N_2 + H + OH. \quad (20.2)$$

N_2 could not be observed in our FTIR experiments, but the OH from reaction (20.2) would lead to detectable products. Hence, there may probably by small contributions of reactions (1.3)–(1.5), but it is unlikely that this would change the conclusion about HCNO + H, reaction (1.1), as the major reaction products.

4.2. Calculated rate constants and branching ratios

The good agreement between the calculated and measured values for the $^3CH_2 + NO$ capture rate constant shows the use of the unimolecular rate calculations and lends support to the product channel calculations.

The calculations showed that H + HCNO are the main products as observed (if collisional stabilization can be neglected). This should not change much with higher temperatures. The two channels forming HCN + OH should give smaller contributions because of the lower rate constants of the rate determining steps. As can be seen from Fig. 7, redissociation to $^3CH_2 + NO$ plays only a side role, even at high energies. The transition state for CNO + H_2 (TS5) was found slightly above the reactants at the B3LYP/6-311G(d,p) and just below the reactants at the G2 level. Nevertheless, considering the other energetically more favorable pathways, this channel may contribute only at high temperatures.

Collisions should not play a significant role in the above experiments and in practical NO_x-reburn conditions unless the pressure is much higher than atmospheric pressure. If necessary, they could be taken into account by a master equation analysis.

The main problem for the rate constant calculations was that the energy barriers of the reaction steps (1c), (1g) and (1e) all lie in a narrow energy range. The accuracy of these barrier heights is therefore of great importance. In particular, the predictions for HCN + OH (Fig. 10) seemed to be on the low side compared to the experiments, but this depends critically on the PES. In order to explore the uncertainties due to the potential barriers, the calculations were repeated with the barrier height for reaction (1g) raised by 1000 cm^{-1} and the barrier heights for reactions (1b) and (1e) lowered by 1000 cm^{-1}. These changes resulted in an increase in the predicted HCN + OH yield to $\Gamma \approx 0.06$ at room temperature (Fig. 10), in better agreement with the experimental values. Nevertheless, HCNO + H remained as the predicted main channel.

Related arguments can be made for other possible reaction channels. In particular, the reaction pathways suggested by Kong and co-workers[11] all involve formation of the cyclic intermediate c-H_2CNO**. As seen from Fig. 7, at the energies of interest, the calculated microcanonical rate constants $k(E, J)$ for this channel (1b) are well below those for production of HCNO + H (1g). Regarding the fate of the c-H_2CNO** intermediate, Kong predicted subsequent reaction pathways which we have not yet taken into account. Nevertheless, the contributions of these pathways to the final product yields should be small because the route via the c-H_2CNO** intermediate cannot compete successfully.

A possibility which remains to be investigated in future work is a reaction via H_2CNO in its $^2A'$ electronic state or the role of a possible H_2CON complex. Both routes may favor formation of HCN + OH, or even other products.

Finally, a comment is in order about angular momentum conservation which is especially important for the simple bond fission reactions and their reverse reactions. The H_2CNO* complex is formed with an overall angular momentum composed of the angular momenta of the reactants $^3CH_2$ and NO and the orbital angular momentum of the reactive collision. At low pressures, the angular momentum of the complex will be conserved during the consecutive unimolecular reactions. Test calculations of the rate constants for the different product channels with estimated average J values of the H_2CNO* of $\langle J \rangle \approx 13$ at room temperature and $\langle J \rangle \approx 40$ at 2000 K gave results which agreed within the uncertainties of the calculations with those obtained by averaging over all J given above.

5. Conclusions

The product distribution of the reaction $^3CH_2$ + NO (1) has been investigated by FTIR measurements and by kinetic modeling by unimolecular rate theory based on quantum chemical PES calculations. The calculated high pressure $^3CH_2$ + NO capture rate constant at room temperature was found to be in good agreement with the experimental value. From the experiments and from the calculations, the reaction channel leading to H + HCNO was found to be the major pathway at all calculated temperatures. The only other important pathway is the production of HCN + OH. The experimental branching ratio at room temperature could be reproduced by minor adjustments to the barrier heights. The data are useful for modeling NO_x-reburn processes in combustion systems.

Acknowledgements

The financial support of this work by the Deutsche Forschungsgemeinschaft and the Fonds der Chemischen Industrie is gratefully acknowledged.

References

1 J. Warnatz, U. Maas and R. W. Dibble, *Combustion*, Springer, Berlin, 1996.
2 A. R. W. McKellar, P. R. Bunker, T. J. Sears, K. M. Evenson, R. J. Saykally and S. R. Langhoff, *J. Chem. Phys.*, 1983, **79**, 5251.
3 P. Jensen and P. R. Bunker, *J. Chem. Phys.*, 1988, **89**, 1327.
4 A. H. Laufer and A. M. Bass, *J. Phys. Chem.*, 1974, **78**, 1344.
5 C. Vinckier and W. J. Debruyn, *J. Phys. Chem.*, 1979, **83**, 2057.
6 V. Seidler, F. Temps, H. Gg. Wagner and M. Wolf, *J. Phys. Chem.*, 1989, **93**, 2070.
7 D. C. Darwin, A. Young, H. S. Johnston and C. B. Moore, *J. Phys. Chem.*, 1989, **93**, 1074.
8 B. Atakan, D. Kocis, J. Wolfrum and P. Nelson, *Proc. Combust. Inst.*, 1992, **24**, 691.
9 J. Grußdorf, F. Temps and H. Gg. Wagner, *Ber. Bunsen-Ges. Phys. Chem.*, 1997, **101**, 134.
10 S. Bauerle, M. Klatt and H. Gg. Wagner, *Ber. Bunsen-Ges. Phys. Chem.*, 1995, **99**, 97.
11 H. Su, F. Kong, B. Chen. M.-B. Huang and Y. Liu, *J. Chem. Phys.*, 2000, **113**, 1885.
12 M. N. R. Ashfold, M. A. Fullstone, G. Hancock and G. W. Ketley, *Chem. Phys.*, 1981, **55**, 245.
13 A. O. Langford, H. Petek and C. B. Moore, *J. Chem. Phys.*, 1983, **78**, 6650.
14 H. H. Carstensen, Dissertation, Max-Planck-Institut für Strömungsforschung, Göttingen, 1995.
15 J. Roggenbuck and F. Temps, *Chem. Phys. Lett.*, 1998, **285**, 422.
16 W. A. Shapley and G. B. Bacskay, *Theor. Chim. Acta*, 1998, **100**, 212.
17 W. A. Shapley and G. B. Bacskay, *J. Phys. Chem. A*, 1999, **103**, 4505.
18 W. A. Shapley and G. B. Bacskay, *J. Phys. Chem. A*, 1999, **103**, 4515.
19 U. Eickhoff and F. Temps, *Phys. Chem. Chem. Phys.*, 1999, **1**, 243.

20 St. Meyer and F. Temps, *Int. J. Chem. Kinet.*, 2000, **32**, 136.
21 B. Bak, D. H. Christensen, J. Christiansen, L. Hansen-Nygaard and J. Rastrup-Andersen, *Spectrochim. Acta*, 1962, **18**, 1421.
22 C. Wentrup, B. Gerecht and H. Briehl, *Angew. Chem.*, 1979, **91**, 503.
23 M. J. Frisch, G. W. Trucks, H. B. Schlegel, P. M. W. Gill, B. G. Johnson, M. A. Robb, J. R. Cheeseman, T. A. Keith, G. A. Petersson, J. A. Montgomery, K. Raghavachary, M. A. Al-Laham, V. G. Zakrzewski, J. V. Ortiz, J. B. Foresman, J. Cioslowski, B. B. Stefanov, A. Nanayakkara, M. Challacombe, C. Y. Peng, P. Y. Ayala, W. Chen, M. W. Wong, J. L. Andres, E. S. Replogle, R. Gomperts, R. L. Martin, D. J. Fox, J. S. Binkley, D. J. Defrees, J. Baker, J. P. Stewart, M. Head-Gordon, C. Gonzalez and J. A. Pople, *GAUSSIAN 94*, Gaussian, Inc., Pittsburgh, 1995.
24 W. Forst, *Theory of Unimolecular Reactions*, Academic Press, New York, 1973.
25 R. G. Gilbert and S. C. Smith, *Theory of Unimolecular and Recombination Reactions*, Blackwell Scientific, Oxford, 1990.
26 J. A. Miller, C. Parrish and N. J. Brown, *J. Phys. Chem.*, 1986, **90**, 3339.
27 M. Quack and J. Troe, *Ber. Bunsen-Ges. Phys. Chem.*, 1974, **78**, 240.
28 L. Brouwer, C. J. Cobos, H.-R. Dübal, F. F. Crim and J. Troe, *J. Chem. Phys.*, 1987, **86**, 6171.
29 J. Troe, *J. Chem. Phys.*, 1981, **75**, 226.
30 J. Troe, *J. Chem. Phys.*, 1983, **79**, 6017.
31 S. E. Stein and B. S. Rabinovitch, *J. Chem. Phys.*, 1973, **58**, 2438.
32 D. C. Astholz, J. Troe and W. Wieters, *J. Chem. Phys.*, 1979, **70**, 5107.
33 J. Troe, *J. Chem. Phys.*, 1977, **66**, 4745.
34 J. Troe, *J. Phys. Chem.*, 1983, **87**, 1800.
35 H. Okabe, *Photochemistry of Small Molecules*, Wiley, New York, 1978.
36 C. G. Morgan, M. Drabbels and A. M. Wodtke, *J. Chem. Phys.*, 1996, **104**, 7460.
37 P. Biggs, G. Hancock, M. R. Heal, D. J. McGarvey and A. D. Parr, *Chem. Phys. Lett.*, 1991, **180**, 533.
38 R. Wagener, *Z. Naturforsch. A*, 1990, **45**, 649.
39 U. Bley and F. Temps, *J. Chem. Phys.*, 1993, **98**, 1058.
40 T. Böhland, F. Temps and H. Gg. Wagner, *Ber. Bunsen-Ges. Phys. Chem.*, 1984, **88**, 455.
41 J. V. Michael, D. F. Nava, W. A. Payne and L. J. Stief, *J. Chem. Phys.*, 1979, **70**, 5222.
42 D. L. Baulch, C. J. Cobos, R. A. Cox, C. Esser, P. Frank, Th. Just, J. A. Kerr, M. J. Pilling, J. Troe, R. W. Walker and J. Waratz, *J. Phys. Chem. Ref. Data*, 1992, **21**, 411.
43 R. Deters, M. Otting, H. Gg. Wagner, F. Temps and S. Dóbé, *Ber. Bunsen-Ges. Phys. Chem.*, 1998, **102**, 978.
44 J. J. Ahumada, J. V. Michael and D. T. Osborne, *J. Chem. Phys.*, 1972, **57**, 3736.
45 Y. Mirokhin, G. Mallard, D. M. Blackslee, F. Wesley, J. Herron, D. Frizzel and R. Hampson, NIST Chemical Kinetics Data Base 17-2Q98, NIST Standard Reference Data, Gaithersburg, MD, 1998.
46 C. Oehlers, F. Temps, H. Gg. Wagner and M. Wolf, *Ber. Bunsen-Ges. Phys. Chem.*, 1992, **96**, 171.
47 J. Grußdorf, J. Nolte, F. Temps and H. Gg. Wagner, *Ber. Bunsen-Ges. Phys. Chem.*, 1994, **98**, 546.
48 A. B. Callear and R. W. Carr, *J. Chem. Soc., Faraday Trans. 2*, 1975, **71**, 1603.
49 M. C. Lin, Y. He and C. F. Melius, *Int. J. Chem. Kinet.*, 1992, **24**, 489.
50 Y. He and M. C. Lin, *Int. J. Chem. Kinet.*, 1992, **24**, 743.
51 P. Pagsberg, J. Munk, C. Anastasi and V. J. Simpson, *J. Phys. Chem.*, 1989, **93**, 5162.
52 J. W. Davies, N. J. B. Green and M. J. Pilling, *J. Chem. Soc., Faraday Trans.*, 1991, **87**, 2317.
53 J. G. Calvert, S. S. Thomas and P. L. Hanst, *J. Am. Chem. Soc.*, 1960, **82**, 1.
54 M. Olzmann, *Ber. Bunsen-Ges. Phys. Chem.*, 1997, **101**, 533.
55 H. Ziemer, S. Dóbé, H. Gg. Wagner, M. Olzmann, B. Viskolcz and F. Temps, *Ber. Bunsen-Ges. Phys. Chem.*, 1998, **102**, 897.
56 N. Marchand, J. C. Rayez and S. C. Smith, *J. Phys. Chem. A*, 1998, **102**, 3358.

Reactions of methyl radicals with propene at temperatures between 750 and 1000 K

Elke Goos,[a] Horst Hippler,*[a] Karlheinz Hoyermann[b] and Bettina Jürges[b]

[a] *Institut für Physikalische Chemie, Universität Karlsruhe, Fritz-Haber-Weg 4, D-76128 Karlsruhe, Germany. E-mail: horst.hippler@chemie.uni-karlsruhe.de*
[b] *Institut für Physikalische Chemie, Universität Göttingen, Tammannstr. 6, D-37077 Göttingen, Germany*

Received 14th March 2001
First published as an Advance Article on the web 7th September 2001

The pyrolysis of propene, initiated by methyl radicals, has been studied in the temperature range 750–1000 K and at a pressure of 0.13 bar in a quasi-wall-free reactor using laser heating by fast vibrational–translational (V–T) energy transfer. This is a convenient method to study homogeneous high-temperature kinetics since the reactor walls remain cold. The radial temperature distribution in the reactor has been investigated by four different methods: a stationary heat balance, optical absorption, pressure rise, and the temperature dependence of the rate of an isomerization reaction. Methyl radicals were produced *via* the fast thermal dissociation of di-*tert*-butyl-peroxide and the products were analysed using GC-MS. The main products of the overall reaction of the model system propene and methyl (C_3H_6 + CH_3) were isopentane (iso-C_5H_{12}) and but-1-ene (1-C_4H_8), whereas allene (C_3H_4), *trans*-but-2-ene (*trans*-2-C_4H_8) and *cis*-but-2-ene (*cis*-2-C_4H_8) were minor components, all showing a strong dependence on temperature. The product distribution and the temperature dependence were analysed by a kinetic model of 61 species and 166 reactions developed for the high-temperature oxidation of butane and the low-temperature oxidation of n-pentane and isopentane. It was necessary to include a few missing reactions and to adjust some rate constants to make the modeling agree with the experimental investigations. This extended mechanism has to be evaluated further in forthcoming experiments.

Introduction

The thermal reactions of hydrocarbons are important in combustion chemistry and petroleum processing. The unsaturated hydrocarbon propene (C_3H_6) is a key intermediate in the combustion of higher alkanes, such as propane, butane, heptane, and iso-octane.[1–3] These hydrocarbons constitute a notable part of practical hydrocarbon fuels. The high-temperature pyrolysis of propene is relevant to the formation of aromatics, possibly *via* allene, propyne, and the propargyl radical.[4–6] An understanding of the thermal decomposition and oxidation of ethene,[7] allene and propene is essential for determining the overall kinetic mechanism of larger fuels and for a better understanding of the high-temperature chemistry of other olefins.

The oxidation of propene has been examined by various studies at high temperatures. Propene/air flame speeds were measured[8,9] and the profiles of stable species and free radicals along a laminar premixed propene–oxygen–argon flame were studied.[10] The oxidation of propene was

DOI: 10.1039/b102411f

also examined behind reflected shock waves[11] and in a jet-stirred reactor.[12,13] Additionally, there have been a number of experimental and computational studies on the oxidation of propene at low and intermediate temperatures in stirred reactors, as reviewed by Wilk *et al.*[14] Recently, Stark and Waddington[15] investigated the oxidation of propene under fuel-rich conditions at 505–549 K and 0.9–4 atm.

A central issue to understanding the pyrolysis of olefins is the explanation of the chemistry of unsaturated radicals, such as ethenyl (C_2H_3) and allyl (a-C_3H_5, CH_2CCH_3), which are formed during the pyrolysis of propene. These unsaturated radicals play a critical role at a much earlier stage in the thermal decomposition of olefins,[16] than in the corresponding reaction of saturated hydrocarbons. Indeed, a lack of knowledge about the rate coefficients for the decomposition of these unsaturated radicals has prevented a comprehensive description of these pyrolyses.

Various studies on propene pyrolysis[17–20] have been reviewed by Hidaka *et al.*,[21] who also studied the thermal decomposition of propene behind reflected shock waves at temperatures between 1200 and 1800 K. The major pyrolysis products were identified as ethene, methane, and ethine. The concentration of propyne was larger than that of allene throughout the experimental range. Hidaka *et al.*[21] also showed, that there were considerable discrepancies in the reported rates for C–H and C–C fission reactions of propene, *i.e.* $C_3H_6 \rightarrow$ a-C_3H_5 + H and $C_3H_6 \rightarrow C_2H_3$ + CH_3. These, however, were the only reaction channels previously considered as initial steps. Hence, the existing measurements using shock-tubes could not be reconciled by just these two initiation reactions. On the basis of thermochemical considerations, they[21] proposed an additional reaction step, $C_3H_6 \rightarrow C_2H_2 + CH_4$, which must be included to obtain reasonable agreement between experiment and modeling. The recently proposed reaction originates from inserting vinylidene (H_2CC) into the C–H bond of CH_4. It was proposed that the latter reaction has an activation energy of 309 kJ mol^{-1}, which is lower by 54.4 and 100.5 kJ mol^{-1}, respectively, than the direct C–H and C–C fission reactions.

In pursuing a converged kinetic mechanism of C_3H_x combustion, Davis *et al.*[22,23] recently undertook a series of studies, which examined the kinetics of the pyrolysis and oxidation of propyne and allene experimentally, as well as theoretically. An experimental and modeling study of the pyrolysis of propyne in a flow reactor and a kinetic modeling of both propyne and allene pyrolysis in shock tubes were reported.[23]

The primary attack on propene by radicals and atoms can proceed by two mechanisms, namely by addition to the double bond and by hydrogen atom abstraction, depending on temperature and pressure. At low temperatures the addition route will prevail, at high temperatures the abstraction route. In the context of hydrocarbon pyrolysis the reactions of CH_3 radicals with C_3H_6 are of main concern, as CH_3 radicals are formed preferentially by the decomposition of larger alkyl radicals. The abstraction route, leading to CH_4 and the radicals propynyl (CH_3CHCH) and/or the resonance stabilized allyl radical a-C_3H_5, has been considered in an experimental study by Roscoe[24] of the overall pyrolysis of propene and in a critical literature survey of mechanisms by Tsang.[25] The addition route, which is of interest in polymerization kinetics, has been investigated at low temperatures (380–441 K) by Miyoshi and Brinton,[26] indicating a predominant CH_3 addition to the terminal carbon atom of propene. The addition of CH_3 radicals to the double bond of C_3H_6 at the terminal or non-terminal position will lead to the formation of the radicals 1-C_4H_9 and iso-C_4H_9, thus entering into the field of the pyrolysis of the saturated hydrocarbons n-C_4H_{10} and iso-C_4H_{10}.

The pyrolytic decomposition of n-butane has received significant attention in the past; Rice[27] first showed its free-radical nature. In 1974 Powers and Corcoran[28] made an effort to discuss the effects of primary and secondary butyl radicals and of secondary reactions. The pyrolysis of n-butane has been reviewed by Corcoran[29] and our experimental and modeling investigation of the methyl-initiated pyrolysis of n-butane was reported recently.[30]

The rationale of the present study is as follows:

(1) By preparing a homogeneous mixture of CH_3 radicals and C_3H_6 at different medium temperatures the competing routes of abstraction and addition can be studied experimentally. Secondary reactions can be controlled by varying the ratio $[CH_3]_0/[C_3H_6]_0$ of the initial concentrations.

(2) The use of the laser powered homogeneous gas heating *via* fast V–T energy transfer avoids heterogeneous pyrolysis at the cold walls of the reactor.

(3) Applying GC-MS analysis after taking samples allows the identification and quantification of the products, especially of the different isomers.

(4) The comprehensive mechanisms for the oxidation and pyrolysis of butane and pentane, derived and applied preferentially for the oxidation of butane and pentane, are used to model our experimental results, thus validating, falsifying, or identifying missing reactions in the mechanism reported so far.

Experimental procedure

The general experimental set-up and the temperature determination have been described previously[30] and are therefore only summarised here. A schematic representation of the experimental arrangement is given in Fig. 1. Before a series of experiments reaction mixtures were prepared and stored in the dark. Their compositions and the experimental conditions are summarized in Tables 1 and 2. The commercial grade chemicals (($CH_3)_3COOC(CH_3)_3$, DTBP: 95%; SF_6: 99.9%; c-C_3H_6: 99.0%; C_3H_6: 99.0%, all Merck; CF_4: 99.8%; Linde; Ar: 99.9998%, Messer-Griesheim) were used without further purification and liquids were carefully degassed before use. After complete mixing of the sample it was placed into a cylindrical cell with KCl windows, which were fixed at the Brewster angle; thus no distortion of the applied laser pulse was observed.

The CO_2-laser (Apollo 560) was operated in its pulse mode at a repetition rate of 1–100 Hz. The laser was stabilised at the P20 line (10.59 µm) in a beam, which can be satisfactorily described by a Gaussian profile.

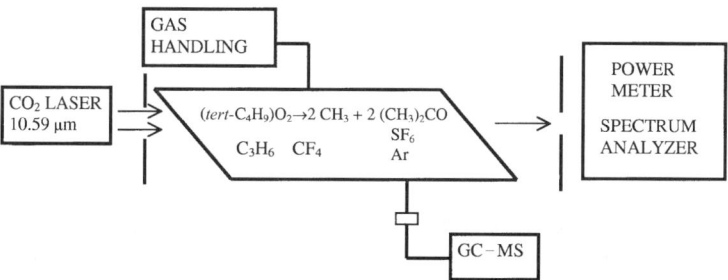

Fig. 1 Schematic representation of the experimental arrangement.

Table 1 Composition of the reaction mixtures, experimental conditions and relative product yields for the reaction $CH_3 + C_3H_6$ for an initial concentration ratio of $[C_3H_6]_0$ to $[DTBP]_0$ of 9. Argon added up to p_{total}; pulse duration: 50 ms; accumulation after 700, 1500 or 9000 pulses

T/K	758	758	845	845	970	970
p_{total}/mbar	133	133	133	133	133	133
$p(SF_6)$/mbar	0.67	0.67	2.22	2.22	4.43	4.43
$p(c-C_3H_6)$/mbar	17.8	17.8	17.8	17.8	17.8	17.8
ME(iso-C_5H_{12})[a]	0.255	0.28	0.255	0.28	0.255	0.28
Educts p/mbar						
DTBP	4.5	4.5	4.5	4.5	4.5	4.5
C_3H_6	40.7	40.7	40.7	40.7	40.7	40.7
Relative product yield (%)						
a-C_3H_4	0.0	0.0	1.0	1.0	13.7	13.9
1-C_4H_8	49.8	50.9	52.7	54.6	59.8	60.7
trans-2-C_4H_8	7.6	7.8	5.7	5.9	5.9	6.0
cis-2-C_4H_8	3.8	3.9	2.4	2.5	3.9	4.0
iso-C_5H_{12}	40.0	37.4	38.2	36.0	16.7	15.4

[a] ME: mass spectroscopic sensitivity.

Table 2 Composition of the reaction mixtures, experimental conditions and relative product yields for the reaction $CH_3 + C_3H_6$ for an initial concentration ratio of $[C_3H_6]_0$ to $[DTBP]_0$ of 100. Argon added up to p_{total}; pulse duration: 50 ms; accumulation after 700, 1500 or 9000 pulses

T/K	758	758	837	837	970	970
p_{total}/mbar	133	133	133	133	133	133
$p(SF_6)$/mbar	0.67	0.67	2.07	2.07	4.43	4.43
$p(c$-$C_3H_6)$/mbar	17.8	17.8	17.8	17.8	17.8	17.8
ME(iso-C_5H_{12})[a]	0.255	0.28	0.255	0.28	0.255	0.28
Educts p/mbar						
DTBP	0.89	0.89	0.89	0.89	0.89	0.89
C_3H_6	89	89	89	89	89	89
Relative product yield (%)						
a-C_3H_4	0.0	0.0	0.0	0.0	8.4	8.5
1-C_4H_8	48.9	50.4	50.1	51.6	57.3	58.2
trans-2-C_4H_8	12.1	12.5	12.5	12.9	12.4	12.6
cis-2-C_4H_8	5.0	5.2	5.4	5.5	5.0	5.1
iso-C_5H_{12}	34.0	31.9	32.0	30.0	16.9	15.6

[a] ME: mass spectroscopic sensitivity.

When a laser pulse of 50 ms duration is applied, the temperature of the given mixture rises due to vibrational to translational energy transfer from vibrationally excited SF_6 to argon. A stationary temperature profile is reached within 1–3 ms with a maximum temperature along the optical axis of the cell. The energy absorbed by the gas mixture was varied with the partial pressure of the SF_6 and the incident laser power, which was monitored continuously during the experiment. During the 50 ms duration of the CO_2 laser pulse the temperature profile in the cell remains constant and decays after the laser pulse within 1–3 ms. Cycles of heating and cooling were obtained by repeated application of laser pulses, giving complete mixing of the heated gas in the middle of the cell with the cold gas near the wall by gas convection. This leads to a completely homogeneous composition between two succeeding laser pulses. The time for ordinary diffusion from the centre to the wall is ~ 30 ms, so that radicals produced in the centre are removed mainly by homogeneous reactions. Convective flow into the non-isothermal region leads to mixing with the gas in the dead space, but no deposit was found on the reactor's walls.

As already described,[30] the radial temperature profiles in the cell were deduced by four different methods: 1. By measurement of the total energy absorbed in the volume and analysis of the heat transfer balance considering the absorbed power and heat conduction. 2. By determination of the radial distribution of the absorbed energy using the temperature-dependent optical absorption coefficient of SF_6. 3. By measurement of the pressure rise in the total volume of the cell and its dependence on the energy absorbed, and 4. by the study of the conversion of cyclopropane to propene via the well established temperature-dependent rate coefficient for the isomerization.

By using a chemical thermometer, e.g. the isomerisation of cyclopropane (c-C_3H_6) to propene (C_3H_6) or the thermal decomposition of di-tert-butyl peroxide (DTBP), the maximum temperature in the cell can be measured. As the rate constants indicate considerable activation energies, measurable conversion occurs only at the highest temperature in the centre of the reaction cell. The maximum temperature of each experiment was checked via the conversion of cyclopropane.

In the examined temperature range of 750 to 1000 K the thermal decomposition of DTBP, the source of CH_3 radicals in this study, proceeds on a millisecond timescale and the CH_3 radicals disappear predominantly by recombining to produce ethane. The rate of the reaction with the hydrocarbon (e.g. propene) can be adjusted by varying the partial pressure of the hydrocarbon. After applying a great number of laser pulses, part of the mixture was sampled to analyse the products of the reaction studied (methyl + propene). The products were identified by gas chromatography via their different retention times, as well as by mass spectroscopy using listed fragmentation spectra, which were taken from the NBS Library Compilation.[31] The concentrations of products were determined by integrating the area of the gas chromatogram by characteristic fragment masses. Adding CF_4, which is not involved in the reaction, allowed consideration of the different mass spectroscopic sensitivities of the products.

Experimental results

The pyrolysis of propene as initiated by methyl radicals was studied at temperatures between 750 and 1000 K and at a pressure of 133 mbar. The relative yields of the main products of the reaction $CH_3 + C_3H_6$ at three temperatures are also listed in Tables 1 and 2. For the experiments summarized in Table 1, the ratio of the initial concentrations of $[C_3H_6]_0$ and $[DTBP]_0$ was 9 and in Table 2 the ratio of $[C_3H_6]_0/[DTBP]_0$ was chosen as 100 to reduce the recombination of methyl radicals to ethane.

The main products of these pyrolyses are 1-C_4H_8 and iso-C_5H_{12}; the minor products are 2-C_4H_8, cis-2-C_4H_8 and C_3H_4. It was not possible to distinguish between allene and propyne. At temperatures equal to or greater than 845 K traces of C_5H_{10} were found. Besides isobutane and C_3H_8, which were impurities in the propene (C_3H_6) used, small amounts of C_2H_4, 1,3-C_4H_6 and C_5H_8 were detected. A quantitative determination of methane and ethane was not possible with the GC columns used, as the analysis was aimed at the detection of higher hydrocarbons up to the aromatics. The alternative values in Tables 1 and 2 for the products were obtained using different mass spectroscopic sensitivities.[31] The uncertainties shown in the yields have no significant influence on the product distribution. The relative yields of products have an uncertainty of $\pm 20\%$, which results from uncertainties of the signal areas and uncertainties in the mass-spectroscopic sensitivities.

The measured temperature distributions indicate an average temperature of 795 K for a pressure of 1.3 mbar of SF_6 and 945 K in the case of 4 mbar of SF_6 with an uncertainty of $\pm 4\%$.

Reaction mechanism and modeling

To describe the reaction sequence, following initiation of the pyrolysis of propene by methyl radicals, a combination of existing detailed chemical kinetic reaction mechanisms has been used, including the high-temperature oxidation of n-butane, the oxidation of n-pentane at low temperatures and the oxidation of isopentane.[32] The rate coefficient for DTBP decomposing to tert-C_4H_9O radicals was taken from the literature,[33,34] and the rate coefficient for the subsequent decomposition[35] of the tert-C_4H_9O radicals to $(CH_3)_2CO$ and CH_3 radicals was derived from their low temperature values at ~ 400 K and extrapolated to our experimental conditions. Furthermore the formation of iso-C_5H_{12} via the reactions of the radicals CH_3, sec-C_4H_9 and iso-C_4H_9 was incorporated. In Table 3 the rate coefficients used for the essential reactions are summarized, in comparison with the literature values, together with the modifications indicated by our modeling. The pressure dependence of some reactions was included and the fall-off behaviour was calculated with k_0, k_∞ and F_{center} through the Troe formula.[36]

Kinetic modeling was done on a personal computer using the CHEMKIN programs.[37] We assumed isothermal conditions and neglected the heating period, since the temperature rise was always very fast compared to the reaction time. We also disregarded the small pressure increase because of the large dead space of the system. The initial concentrations of the reactants di-tert-butyl peroxide and C_3H_6, as well as the actual pressure of argon were used as inputs for the modeling calculations. Adjustment of the initial values of the rate constants for the reactions in the model was guided by a sensitivity analysis and by the reported estimates of uncertainty for the kinetic parameters obtained from the literature. By virtue of the chosen kinetic conditions only a limited number of products are formed. This means that with a limited number of elementary reactions, there are only a small number of adjustable parameters in the simulation. Rate constants were adjusted manually until the calculated composition matched the measured concentrations. The number of experimental measurements and the error margins were generally insufficient to warrant the use of statistical methods, such as least squares, in assessing the goodness of fit.

Modeling results

The central part of the mechanism for the CH_3/C_3H_6 system is presented in Fig. 2. The dominant reaction paths for the formation of the essential products are shown. This reaction scheme was

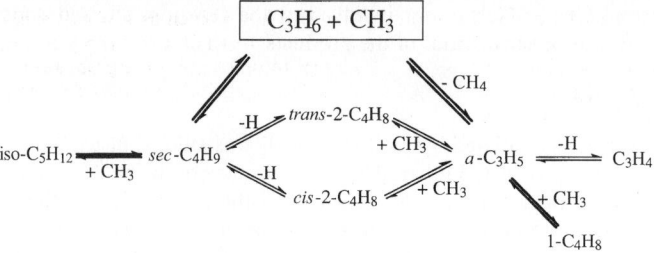

Fig. 2 Central part of the reaction mechanism.

obtained from a sensitivity analysis of the yields of the main products. Obviously iso-C_5H_{12} is the consecutive product of the intermediate sec-C_4H_9 radical produced from terminal addition of CH_3 radical to propene (C_3H_6). The non-terminal addition of CH_3 to C_3H_6 leads to the iso-C_4H_9 radical, contributing nothing measurable to the product formation. The products 1-C_4H_8 and C_3H_4 are the consecutive products of the allyl radicals (a-C_3H_5), which are formed by methyl radicals abstracting primary H-atoms from C_3H_6. Trans-2-C_4H_8 and cis-2-C_4H_8 can result from methyl addition to allyl and by decomposition of sec-C_4H_9 radicals. The competition between methyl radical addition to a-C_3H_5 and decomposition of the a-C_3H_5 radical under H atom production leads to 1-C_4H_8 and allene (a-C_3H_4), which can isomerize to propyne (p-C_3H_4), respectively. Similarly, there is a competition between methyl radical addition to sec-C_4H_9 and the decomposition of sec-C_4H_9 radicals with H atom production leading to iso-C_5H_{12}, trans-2-C_4H_8 and cis-2-C_4H_8, respectively.

The modified rate coefficients were obtained by adjusting the rate constants of those reactions to which the selectivities to the products were sensitive, so that the experimental product yields were finally matched. The quality of the fit was well within the experimental uncertainty of the analytical measurements. To model the trends of the relative product yields in our experiments, we changed the rate coefficients of some of the reactions listed in Table 3; in each case the original value from the mechanism or literature and the value used are shown.

The time dependence of the formation of the stable products 1-C_4H_8, iso-C_5H_{12}, trans-2-C_4H_8, cis-2-C_4H_8 and C_3H_4 at 800 K is depicted in Fig. 3. Clearly a stationary state is established for all products after ~2 ms. In Fig. 4 the fate of the main radicals, the initiator radical CH_3 and the intermediates sec-C_4H_9 and C_3H_5 is shown for 800 K. During the first few milliseconds their concentrations decrease rapidly and stationary states are established for each of them after ~10 ms.

Since the methyl precursor decomposes completely within the first few milliseconds of the laser heating pulse, the formation of stable products is independent of the applied reaction times of 50 ms. This result validates our experimental procedure whereby the products were accumulated after

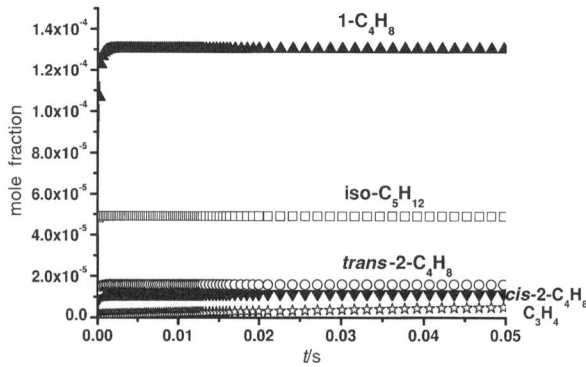

Fig. 3 Calculated temporal evolution of the mole fractions of the main products 1-C_4H_8, iso-C_5H_{12}, trans-2-C_4H_8, cis-2-C_4H_8 and C_3H_4 (as labelled). ($T = 800$ K, $[C_3H_6]_0/[DTBP]_0 = 9$.)

Table 3 Details of the most important rate coefficients written as $k = AT^b \exp(-E/RT)$. The units of A are a combination of cm^3, mol, s, and K depending on the reaction order and the temperature coefficient b. For methyl radical recombination the pressure dependence was calculated using the Troe formula[36] for the fall-off with F_{center}, also the low and high pressure limit of the rate coefficient is given

Reaction	A	b	E/kJ mol^{-1}	Reference
Methyl radical production				
$C_8H_{18}O_2 \rightarrow 2\ C_4H_9O$	2.04E+15	0.0	156.5	34
	3.98E+14	0.0	156.5	This work
$C_4H_9O \rightarrow CH_3 + CH_3COCH_3$	1.10E+14	0.0	61.2	35
$CH_3COCH_3 \rightarrow CH_3CO + CH_3$	7.94E+17	0.0	353.0	38
Reactions of methyl radicals with propene				
$C_3H_6 + CH_3 \rightarrow CH_4 + C_3H_5$	2.21	3.5	23.75	39
	3.3	3.5	23.75	This work
sec-$C_4H_9 \rightarrow C_3H_6 + CH_3$	2.30E+14	0.0	137.0	40
	4.00E+14	0.0	139.0	This work
iso-$C_4H_9 \rightarrow C_3H_6 + CH_3$	2.00E+13	0.0	125.34	41
tert-$C_4H_9 \rightarrow C_3H_6 + CH_3$	1.00E+16	0.0	193.0	40
Decomposition of propene				
$C_2H_4 + CH_2 \rightarrow C_3H_6$	7.24E+13	0.0	0.0	32
$C_3H_6 \rightarrow C_3H_5 + H$	5.00E+12	0.0	296.82	42
	5.00E+13	0.0	334.71	21
	1.00E+13	0.0	326.0	This work
$C_3H_6 \rightarrow C_2H_3 + CH_3$	1.10E+21	−1.2	408.8	25
Reactions of methyl radicals				
$CH_3 + CH_3(+M) \rightarrow C_2H_6(+M)$				
k_0	3.63E+41	−7.0	11.6	43
$F_{center} = (1 - 0.62)\exp(-T/73) + 0.62\exp(-T/1180)$				43
k_∞	3.61E+13	0.0	0.00	43
Reactions of sec-butyl radicals				
sec-$C_4H_9 + CH_3 \rightarrow$ iso-C_5H_{12}	2.00E+12	0.0	−4.88	This work
sec-$C_4H_9 \rightarrow$ trans-2-$C_4H_8 + H$	4.57E+12	0.0	142.2	44
	5.00E+13	0.0	158.7	32
	6.00E+12	0.0	158.7	This work
sec-$C_4H_9 \rightarrow$ cis-2-$C_4H_8 + H$	5.00E+13	0.0	158.7	32
	4.17E+12	0.0	145.5	44
	5.00E+12	0.0	158.7	This work
sec-$C_4H_9 \rightarrow$ 1-$C_4H_8 + H$	1.58E+13	0.0	165.7	45
	2.00E+13	0.0	169.2	This work
Reactions of allyl radicals				
1-$C_4H_8 \rightarrow C_3H_5 + CH_3$	1.00E+16	0.0	305.4	45
	3.00E+16	0.0	305.4	This work
trans-2-$C_4H_8 \rightarrow C_3H_5 + CH_3$	6.50E+14	0.0	298.3	This work
cis-2-$C_4H_8 \rightarrow C_3H_5 + CH_3$	1.00E+16	0.0	334.7	46
	1.254E+15	0.0	298.3	This work
$C_3H_5 + CH_3 \rightarrow C_3H_4 + CH_4$	2.10E+11	0.0	0.00	32
$C_3H_5 + C_3H_5 \rightarrow C_3H_6 + C_3H_4$	6.02E+10	0.0	−1.10	32
$C_3H_5 + H \rightarrow C_3H_4 + H_2$	1.80E+13	0.0	0.0	32
$C_3H_5 \rightarrow C_3H_4 + H$	0.398E+14	0.0	293.1	32
Isomerisations				
1-$C_4H_8 \rightarrow$ trans-2-C_4H_8	4.00E+11	0.0	251.0	32
1-$C_4H_8 \rightarrow$ cis-2-C_4H_8	4.00E+11	0.0	251.0	32
cis-2-$C_4H_8 \rightarrow$ trans-2-C_4H_8	3.98E+13	0.0	259.4	47
	1.00E+13	0.0	259.4	This work

periodic laser heating followed by convectional cooling. The resulting relative product yields have been compared with the experimental values. Reasonable agreement was found for the different initial ratios of $[C_3H_6]_0/[DTBP]_0$ equal to 9 and 100 as shown in Fig. 5 and 6. According to our modeling the concentration of the a-C_3H_5 radical in our temperature range of 750–1000 K is

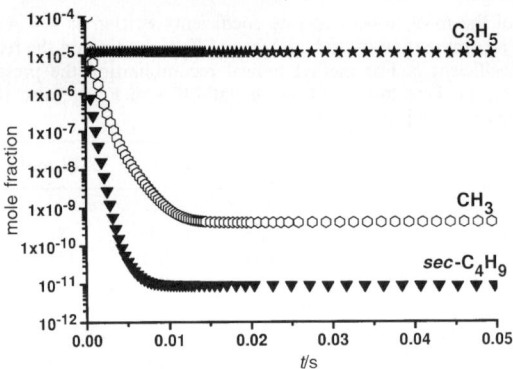

Fig. 4 Calculated temporal evolution of the mole fractions of the methyl radical (CH_3) and of the intermediates sec-C_4H_9 and C_3H_5 ($T = 800$ K, $[C_3H_6]_0/[DTBP]_0 = 9$).

governed by three routes: (1) by the production of the CH_3 radicals via the decomposition of the DTBP and additionally via the decomposition of $(CH_3)_2CO$ at higher temperatures, (2) by its formation via the abstraction route $CH_3 + C_3H_6$, and (3) by the consumption in subsequent reactions (see Fig. 2). The formation and destruction are not directly linked to the sec-C_4H_9 production. The formation of a-C_3H_4 is sensitive to the concentrations of the radicals a-C_3H_5 and CH_3, but independent of the sec-C_4H_9 reaction sub-mechanism.

The model suggests that in our temperature range the concentration of sec-butyl is sensitive to its own production rate and to the methyl production and the recombination of methyl radicals to ethane.

The formation of iso-C_5H_{12} is sensitive to the rate coefficients for the addition of methyl radicals to sec-C_4H_9 radicals and to the rate coefficients for the reaction of methyl radicals to ethane.

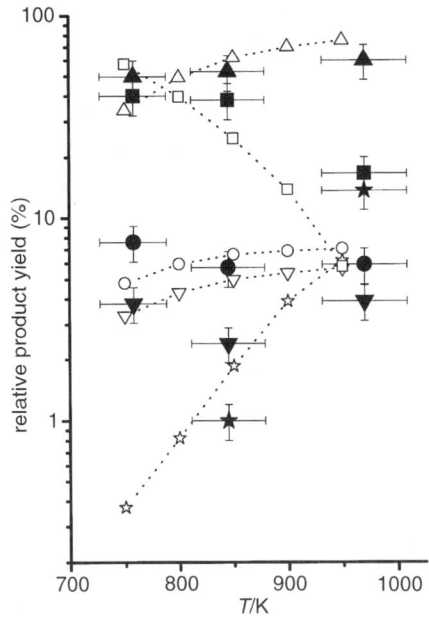

Fig. 5 Relative product yields as a function of temperature for $[C_3H_6]_0/[DTBP]_0 = 9$ (open symbols: calculation using proposed mechanism; closed symbols: experimental data with error bars; (△) 1-C_4H_8; (□) iso-C_5H_{12}; (○) trans-2-C_4H_8; (▽) cis-2-C_4H_8; (*) a-C_3H_4).

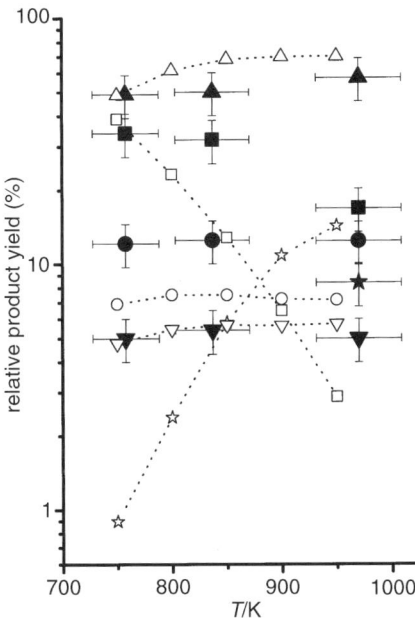

Fig. 6 Relative product yields as function of temperature for $[C_3H_6]_0/[DTBP]_0 = 100$ (open symbols: calculation using proposed mechanism; closed symbols: experimental data with error bars; (△) 1-C_4H_8; (□) iso-C_5H_{12}; (○) trans-2-C_4H_8; (▽) cis-2-C_4H_8; (*) a-C_3H_4).

The iso-C_5H_{12} concentration is only sensitive to the development of sec-C_4H_9 radicals at lower temperature.

Small amounts of C_2H_4, 1,3-C_4H_6 and C_5H_8 are formed and can be explained by C_3H_8 and iso-butane, both being present as negligible by-products and as impurities in the mixture:

$$iso\text{-}C_4H_{10} + CH_3 \rightarrow tert\text{-}C_4H_9 + CH_4$$

$$tert\text{-}C_4H_9 \rightarrow C_2H_4 + C_2H_5$$

$$C_3H_8 + CH_3 \rightarrow iso\text{-}C_3H_7 + CH_4$$

$$iso\text{-}C_3H_7 \rightarrow C_2H_4 + CH_3$$

$$C_2H_4 + CH_3 \rightarrow C_2H_3 + CH_4$$

$$C_2H_3 + C_2H_3 \rightarrow 1,3\text{-}C_4H_6$$

$$1,3\text{-}C_4H_6 + CH_3 \rightarrow C_4H_5 + CH_4$$

$$C_4H_5 + CH_3 \rightarrow C_5H_8$$

Conclusion

The application of laser-induced heat transfer by fast V–T energy transfer from SF_6 to argon, combined with the hydrocarbon reacting after initiation by methyl radicals was used to study reactions of propene in the medium temperature range 750–1000 K. This reaction system has a well-characterized temperature, determined by different procedures; also the production of chemical species was measured by GC-MS analysis. These techniques have been proved for this model system by the good agreement between experiment and modeling. It is interesting to note, that the main reaction scheme adopted stems from and has been validated by studies of the oxidation of both butane and pentane. Thus the observed agreement between our experimental findings and the modeling using this reaction mechanism supports the quality of the rate constants used. Furthermore, this established reaction mechanism, together with its rate coefficients, can be incorporated as a building block into the mechanisms for the formation of higher unsaturated hydrocarbons up to aromatics *via* the species C_3H_5, a-C_3H_4 and p-C_3H_4.

Acknowledgements

E.G. and H.H. gratefully acknowledge financial support by the Deutsche Forschungsgemeinschaft (SFB 551 "Kohlenstoff aus der Gasphase: Elementarreaktionen, Strukturen, Werkstoffe"). K.H. and B.J. are thankful for financial support by the Deutsche Forschungsgemeinschaft within the Sonderforschungsbereich 357 ("Molekulare Mechanismen unimolekularer Prozesse"). Both groups also acknowledge the financial support given by the Fonds der Chemischen Industrie.

Appendix

CH_3	methyl radical
CH_4	methane
CF_4	tetrafluoromethane
C_2H_2	ethine
C_2H_6	ethane
p-C_3H_4, CH_3C_2H	prop-1-yne
a-C_3H_4, CH_2CCH_2	allene, propa-1,2-diene
a-C_3H_5, CH_2CCH_3	allyl
p-C_3H_5, CH_3CHCH	propynyl
C_3H_6	propene
c-C_3H_6	cyclopropane
C_3H_8	propane
1-C_4H_8, $CH_2CHCH_2CH_3$	but-1-ene
2-C_4H_8, $CH_3CHCHCH_3$	but-2-ene
1-C_4H_9	n-butyl radical, 1-butyl radical
sec-C_4H_9, 2-C_4H_9	sec-butyl radical, 2-butyl radical
tert-C_4H_9, $(CH_3)_3CCH_3$	tert-butyl radical
n-C_4H_{10}	n-butane
iso-C_4H_{10}, $(CH_3)_2CHCH_3$	isobutane
iso-C_5H_{12}	isopentane, 2-methylbutane
CH_3COCH_3, $(CH_3)_2CO$	propanone, acetone
$(CH_3)_3COOC(CH_3)_3$, (tert-$C_4H_9O)_2$	DTBP, di-*tert*-butyl peroxide

References

1 C. K. Westbrook and F. L. Dryer, *Prog. Energy Combust. Sci.*, 1984, **10**, 1.
2 K. M. Leung and R. P. Lindstedt, *Combust. Flame*, 1995, **102**, 129.
3 S. G. Davis and C. K. Law, in *Twenty-Seventh Symposium (International) on Combustion*, The Combustion Institute, Pittsburgh, 1998, p. 521.
4 U. Alkemade and K. H. Homann, *Z. Phys. Chem. Neue Folge*, 1989, **161**, 19.
5 S. E. Stein, J. A. Walker, M. M. Suryan and A. Fahr, in *Twenty-Third Symposium (International) on Combustion*, The Combustion Institute, Pittsburgh, 1991, p. 85.
6 J. A. Miller and C. F. Melius, *Combust. Flame*, 1992, **91**, 21.
7 J. M. Roscoe, A. R. Bossard and M. H. Back, *Can. J. Chem.*, 2000, **78**, 16.
8 R. Günther and G. Janisch, *Chem. Ing. Tech.*, 1971, **43**, 975.
9 S. G. Davis, C. K. Law and H. Wang, *Combust. Flame*, 1999, **119**, 375.
10 S. D. Thomas, A. Bhargava, P. R. Westmoreland, R. P. Lindstedt and G. Skevis, *Bull. Soc. Chem. Belg.*, 1996, **105**, 501.
11 A. Burcat and K. Radhakrishnan, *Combust. Flame*, 1985, **60**, 157.
12 P. Dagaut, M. Cathonnet and J. C. Boettner, *J. Phys. Chem.*, 1988, **92**, 661.
13 P. Dagaut, M. Cathonnet and J. C. Boettner, *Combust. Sci. Technol.*, 1992, **83**, 167.

14 R. D. Wilk, N. P. Cernansky, W. J. Pitz and C. K. Westbrook, *Combust. Flame*, 1989, **77**, 145.
15 M. S. Stark and D. J. Waddington, *Int. J. Chem. Kinet.*, 1995, **27**, 123.
16 *Pyrolysis: Theory and Industrial Practice*, ed. L. F. Albright, B. L. Crynes and W. H. Corcoran, Academic Press, New York, 1983.
17 G. A. Chappell and H. Shaw, *J. Phys. Chem.*, 1968, **72**, 4672.
18 A. Burcat, *Fuel*, 1975, **54**, 87.
19 J. H. Kiefer, M. Z. Al-Alami and K. A. Budach, *J. Phys. Chem.*, 1982, **86**, 808.
20 V. S. Rao and G. B. Skinner, *J. Phys. Chem.*, 1989, **93**, 1869.
21 Y. Hidaka, T. Nakamura, H. Tanaka, A. Jinno, H. Kawano and T. Higashihara, *Int. J. Chem. Kinet.*, 1992, **24**, 761.
22 S. G. Davis, C. K. Law and H. Wang, in *Twenty-Seventh Symposium (International) on Combustion*, The Combustion Institute, Pittsburgh, 1998, p. 305.
23 S. G. Davis, C. K. Law and H. Wang, *J. Phys. Chem. A*, 1999, **103**, 5889.
24 A. C. Kinsman and J. M. Roscoe, *Int. J. Chem. Kinet.*, 1994, **24**, 191.
25 W. Tsang, *J. Phys. Chem. Ref. Data*, 1991, **20**, 221.
26 M. Miyoshi and R. K. Brinton, *J. Chem. Phys.*, 1962, **36**, 3019.
27 F. O. Rice, *J. Am. Chem. Soc.*, 1931, **53**, 1959.
28 D. R. Powers and W. H. Corcoran, *Ind. Eng. Chem. Fundam.*, 1974, **13**, 351.
29 W. H. Corcoran, in *Pyrolysis: Theory and Industrial Practice*, ed. L. F. Albright, B. L. Crynes and W. H. Corcoran, Academic Press, New York, 1983, p. 47.
30 E. Goos, H. Hippler, K. Hoyermann and B. Jürges, *Phys. Chem. Chem. Phys.*, 2000, **2**, 5127.
31 A. Cornu and R. Massot, *Compilation of Mass Spectral Data*, Heyden, London, 2nd edn., 1975.
32 C. Chevalier, M. Nehse and J. Warnatz, in *Twenty-Sixth Symposium (International) on Combustion*, The Combustion Institute, Pittsburgh, PA, 1996, p. 773.
33 M. J. Perona and D. M. Golden, *Int. J. Chem. Kinet.*, 1973, **5**, 55.
34 M. I. Sway, *J. Chem. Soc., Faraday Trans.*, 1991, **87**, 2157.
35 C. Fittschen, H. Hippler and B. Viskolcz, *Phys. Chem. Chem. Phys.*, 2000, **2**, 1677.
36 R. G. Gilbert, K. Luther and J. Troe, *Ber. Bunsen-Ges. Phys. Chem.*, 1983, **87**, 169.
37 R. J. Kee, F. M. Rupley, J. A. Miller, M. E. Coltrin, J. F. Grcar, E. Meeks, H. K. Moffat, A. E. Lutz, G. Dixon-Lewis, M. D. Smooke, J. Warnatz, G. H. Evans, R. S. Larson, R. E. Mitchell, L. R. Petzold, W. C. Reynolds, M. Caracotsios, W. E. Stewart and P. Glarborg, *Chemkin Collection III*, Reaction Design, Inc., San Diego, CA, 1999.
38 S. H. Mousavipour and P. D. Pacey, *J. Phys. Chem.*, 1996, **100**, 3573.
39 W. Tsang, *J. Phys. Chem. Ref. Data*, 1991, **20**, 221.
40 J. Warnatz, in *Combustion Chemistry*, ed. W. C. Gardiner, Jr., Springer, New York, 1984.
41 W. Tsang, *J. Phys. Chem. Ref. Data*, 1990, **19**, 1.
42 P. Barbe, R. Martin, D. Perrin and G. Scacchi, *Int. J. Chem. Kinet.*, 1996, **28**, 829.
43 D. L. Baulch, C. J. Cobos, R. A. Cox, P. Frank, G. Hayman, Th. Just, J. A. Kerr, T. Murrells, M. J. Pilling, J. Troe, R. W. Walker and J. Warnatz, *J. Phys. Chem. Ref. Data*, 1994, **23**, 847.
44 W. Tsang, *J. Am. Chem. Soc.*, 1985, **107**, 2872.
45 A. M. Dean, *J. Phys. Chem.*, 1985, **89**, 4600.
46 P. Jeffers and S. H. Bauer, *Int. J. Chem. Kinet.*, 1974, **6**, 763.
47 D. Masson, C. Richard and R. Martin, *Int. J. Chem. Kinet.*, 1976, **8**, 37.

General Discussion

Prof. Balint-Kurti and **Mr Cole** opened the discussion of Prof. Troe's paper: In his presentation Prof. Troe has stressed the importance of using J-conserving theories within the context of RRKM theory. These J-conserving theories give rise to centrifugal barriers which are absolutely essential in the theory. Prof. Troe's presentation concentrates on the treatment of the transitional modes and stresses their behaviour at large fragment separations. We have been examining this problem for two model systems (NH_2 + NO and CH_3 + OH) within the context of variational J-conserving RRKM theory. We have computed the vibrational frequencies of all the modes at each geometry along the reaction path as the fragments dissociate and have used standard J-conserving variational RRKM theory. As is well known this theory, in which all vibrational modes are treated within a harmonic approximation, breaks down at large fragment separations. In order to allow for this breakdown we have implemented a quantum phase space method in which the fragments are treated as free rotors, but their rotational angular momenta are coupled to their relative orbital angular momentum to give a definite total angular momentum for the systems. This approach also gives a sum-of-states at a particular fragment separation which can be compared with that obtained from the standard J-conserving RRKM theory for the same total angular momentum J, fragment separation R, and total energy, E. In the spirit of variational RRKM theory the lower of the two predictions of the sum-of-states for a given R, J and E should be the more correct of the two and our intention was to use this lower value. This approach would enable us to avoid the major effects of the break-down of the standard RRKM theory at large interfragment separations. The critical distance for a given E and J is that at which the sum-of-states is a minimum. Surprisingly we have found, for the two cases which we have examined, that the critical distances are always relatively small and fall within the range where the standard RRKM theory gives the lower sum-of-states value. Our conclusion is therefore, that the variation of the vibrational frequencies during the bond breaking process is of great importance and must be evaluated as a function of the bond stretching coordinate as the bond is broken.

Prof. Troe responded to Prof. Balint-Kurti and Mr Cole: Apparently you embrace sums-of-states along the reaction path by J-conserving variational harmonic oscillator RRKM and by phase space theory. A more accurate treatment is given by statistical adiabatic channel (SACM) calculations where rovibrational eigenvalues along the reaction coordinate are calculated explicitly (see, *e.g.*, our treatment of the OH + OH-system in ref. 1). In this case, neither the harmonic oscillator, nor the J-coupled free rotor limits, but the more correct intermediate case is obtained. Your observation that the channel pattern at the minimum of the sum-of-states corresponds more to an oscillator situation, not only for conserved modes but also for transitional modes, should be taken with caution. Our experience is that the pattern of adiabatic channel maxima, which leads to an activated complex pseudo-sum-of-states, for transitional modes and normal anisotropies of the potential, more corresponds to that of rotors with effectively increased rotational constants (in comparison to those of the corresponding free rotors, see, *e.g.*, ref. 2) than to oscillators. In any case, your final statement about the necessity, of evaluating vibrational frequencies of transitional modes as a function of the length of the breaking bond is an essential element of the SACM concept since its earliest formulation.[3]

1 A. I. Maergoiz, E. E. Nikitin and J. Troe, *J. Chem. Phys.*, 1991, **95**, 5117; A. I. Maergoiz, E. E. Nikitin and J. Troe, *J. Chem. Phys.*, 1995, **103**, 2083; A. I. Maergoiz, E. E. Nikitin and J. Troe, *Z. Phys. Chem.*, 1991, **172**, 129.
2 A. I. Maergoiz, E. E. Nikitin, J. Troe and V. G. Ushakov, *J. Chem. Phys.*, 1998, **108**, 5265; A. I. Maergoiz, E. E. Nikitin, J. Troe and V. G. Ushakov, *J. Chem. Phys.*, 1998, **108**, 9987.
3 M. Quack and J. Troe, *Ber. Bunsen-Ges. Phys. Chem.*, 1974, **78**, 240.

Prof. Golden asked: The details discussed by Prof. Troe are interesting. What is the sensitivity to this level of treatment in any combustion model? Given the uncertainty in other rate parameters and in the experimental data, does this level of detail make any difference to any quantity that a model would be used to calculate?

Prof. Troe replied: It has been proven useful to characterize falloff curves of dissociation and recombination reactions in terms of limiting low and high pressure rate constants and by broadening factors of the intermediate reduced falloff curves. The present work addresses the question of how broadening factors F_{cent}^{SC} at the center of the falloff curves, in the presence of two transitional modes and various numbers of conserved modes, can be estimated quickly. In this sense, it also allows for a check whether an experimental, empirically fitted, F_{cent}^{SC} is meaningful or not. In low temperature recombination reactions, where only transitional modes contribute to F_{cent}^{SC}, correct treatments of rotational effects become particularly important (see, *e.g.*, my treatment of the $HO + NO_2 \rightleftharpoons HONO_2$ system in ref. 1).

1 J. Troe, *Int. J. Chem. Kinet.*, 2001, submitted.

Dr Klippenstein said: I am unclear as to the reason for the concern about the treatment of orbital angular momentum barriers. Microcanonical *J*-conserving variational transition state theory, which is what we generally use in our single well master equation simulations, correctly treats the rotational dependence of the dissociation rates. With this approach it is not necessary to make separability assumptions or to explicitly consider the orbital barriers and so we do not. We could tabulate rotational barriers as a function of angular momentum, but again it is not necessary for our purposes and so we do not. Instead, such barriers are implicitly considered *via* appropriate determinations of the *E* and *J* dependence of the number of available states.

Prof. Troe responded: Microcanonical *J*-conserving variational transition state theory is not necessarily correct. First, dynamical, nonadiabatic, couplings (see, *e.g.*, ref. 1) generally escape this treatment. Second, it is not clear what sum-of-states should be varied along the reaction coordinate (see, *e.g.*, ref. 2). Finally, just giving the results of a variational transition state/master equation treatment will hide the origin of uncertainties, let it be in the collision model, or in the potential, or elsewhere. The treatment becomes much more transparent and useful for other researchers if the various contributions are separated, *i.e.* if strong and weak collision treatments are compared, if the used potential is given explicitly or, at least, if it is characterized by its centrifugal barriers $E_0(J)$, and if limiting low pressure and high pressure rate constants are given. Otherwise, a reader has just the choice to believe or not to believe the results; he has no chance to judge the quality of the treatment and to transfer parts of it to other applications, *e.g.*, transfer *T*-averaged to *E*- and *J*-resolved rate constants when an inverse Laplace transformation cannot be made.

1 L. B. Harding, J. Troe and V. G. Ushakov, *Phys. Chem. Chem. Phys.*, 2000, **2**, 631.
2 M. Quack and J. Troe, *Ber. Bunsen-Ges. Phys. Chem.*, 1874, **78**, 240.

Dr Klippenstein said: You have stated that it is not possible for variational transition state theory (VTST) to correct for the differences between your classical trajectory simulations and statistical adiabatic channel model calculations of capture rate constants. In fact, this is not true. The statistical adiabatic channel model implicitly assumes that the reaction coordinate is the separation between the centers-of-mass of the two reacting fragments. If one similarly restricts the reaction coordinate within VTST simulations, as in the early work of Wardlaw and Marcus,[1] then you are correct that one would observe similar differences from trajectory simulations. However, the implementation of more general reaction coordinates, as in our variable reaction coordinate VTST formalism, can correct for the non-adiabaticities that you use classical trajectory simulations to study. Thus, there is no inherent reason that your classical trajectory results should differ significantly from the results of optimized variable reaction coordinate VTST simulations, and I know of no evidence to suggest that they do.

1 D. M. Wardlaw and R. A. Marcus, *Adv. Chem. Phys.*, 1988, **70**, (Part I), 231.

Prof. Troe responded: It is certainly true that some problems of early variational transition state theory have been removed by optimized variable reaction coordinate (VRC)-VTST. However, for fundamental reasons, I doubt that all dynamical, non-adiabatic, effects can be accounted for in this way. This is not a problem related to the choice of the reaction coordinate. A comparison with our trajectory calculations for the capture of O_2 in collisions with T, D, H and Mu on an *ab initio* potential[1] would be indicative.

1 L. B. Harding, J. Troe and V. G. Ushakov, *Phys. Chem. Chem. Phys.*, 2000, **2**, 631.

Prof. Golden commented: I would point out that my colleagues and I have found that simple RRKM models based on the Gorin model of bond scission/radical combination reactions are often quite adequate to explain, codify and extrapolate the extant data for systems where little hope exists of finding accurate potential energy surfaces in the next few years. These models include treating the transitional modes as "restricted rotors" and using the Waage–Rabinovitch approximation to account for centrifugal effects, as well as simple pseudo strong collision for energy transfer. Anharmonicities can also be employed in the density of states calculations.

Prof. Troe replied: An RRKM treatment based on a restricted rotor-Gorin model will approach the present treatment if centrifugal barriers are calculated in the same way, *i.e*, not by using Lennard-Jones potential but *ab initio* potentials or at least Morse potentials, and if hindrance parameters are related to our rigidity factors derived from the anisotropy of the potential. On the given level of uncertainty, our approach is not more complicated to use than the Waage–Rabinovitch, restricted rotor, Gorin model; on the other hand, it employs more realistic potentials and provides more realistic predictions of the hindrance parameters and of the apparent moments of inertia of the activated complex.

Prof. Kohse-Höinghaus asked: Of how much value would detailed studies of non-reactive energy transfer (vibrational, rotational) under appropriate (combustion) conditions be in regard of your investigation?

Prof. Troe answered: Non-reactive, vibrational and rotational, energy transfer under combustion conditions can be treated by capture calculations such as used in high pressure unimolecular dissociation/recombination rate theory, if the collisional encounter leads to chemically bound intermediates. Such cases indeed are known, *e.g.* the energy transfer between $OH(v = 1)$ and CO involving HOCO complexes[1] where high pressure recombination and vibrational relaxation rates were shown to coincide.

1 D. Fulle, H. F. Hamann, H. Hippler and J. Troe, *J. Chem. Phys.*, 1996, **105**, 983.

Prof. Pilling† asked: For the HO_2 system cited what is the number and nature of the transitional modes? Eqn. (32) implies two, but atom–diatom systems are usually described using one transitional mode.

Prof. Troe answered: The treatment of this work is for atoms, adding to linear molecules of arbitrary numbers of oscillators and forming linear adducts, or for the reverse dissociations. In this case one has two transitional modes. The RRKM calculations of Fig. 12 (of our paper) were made for artificial linear HO_2, *i.e.* for two transitional modes, whereas the trajectory calculations used the true *ab initio* potential. It is indeed a problem to specify the number of transitional modes to be used in simplified models. Only knowledge of the geometry of the true minimum energy path on the accurate potential can help: atom–diatom systems can correspond to one or two transitional modes or to something in between.

---

† Also Dr S. H. Robertson (Accelrys, Cambridge, UK).

Prof. S. C. Smith said: (1) Two-dimensional (E, J) master equation (ME) solutions are available.[1] These may be useful in determining weak collisional broadening factors with inclusion of angular momentum effects.

(2) Densities of states. HO_2 densities are known exactly from direct quantum mechanical calculation. However, $\rho(E,J)$ models such as that which Prof. Troe has developed are very important for general applications. We are working on another model for densities of states of van der Waals dimer clusters which is based on an extension of the classical Monte Carlo phase space counting algorithms developed for loose (barrierless) transition states.

1 S. J. Jeffrey, K. E. Gates and S. C. Smith, *J. Phys. Chem.*, 1998, **100**, 7090.

Prof. Troe responded: (1) In my view, two-dimensional (E,J) master equation treatments (as your reference and also ref. 1) will only become practically relevant when truly reliable knowledge about rovibrational energy transfer becomes available. This is generally not the case at present.

(2) It is indeed very important to devote more effort to the calculations of accurate rovibrational energy levels on exact potential energy surfaces. Only a few systems have been investigated in sufficient detail such that more empirical approximate models have to be used to estimate anharmonicity contributions. These models, however, urgently need validation.

1 J. Troe, *J. Chem. Phys.*, 1977, **66**, 4745; J. Troe, *Z. Phys. Chem.*, 1987, **154**, 73.

Dr Miller addressed everyone: I agree with Prof. S. C. Smith that it is important to calculate accurate eigenvalues and eigenvectors at low temperature; these contain useful information. However, as long as one only wants the time history of the system, an extremely robust method of solving the problem is simply to integrate the master equation as a set of ODEs (ordinary differential equations) using a stiff ODE integrator such as VODE.[1] I have implemented this approach several times in the past,[2–5] and the method has been included as an option in VARIFLEX.[6] Also, it is common that the eigenpair that governs the reaction of interest is not one that is computed badly. For example, in the $^1CH_2 + C_2H_2$ reaction that Prof. Smith considers, it is the eigenvector $|g_4\rangle$ with the fourth largest eigenvalue λ_4 (*i.e.* the fourth smallest in absolute value) that governs the reaction at low temperature. Our calculations in ordinary precision arithmetic at $T = 300$ K give an accurate value of λ_4 at all pressures, but give good eigenvectors only at higher pressures (*e.g.*, $p = 1$ atm). The situation improves as the temperature increases.

In any event there are two advantages to calculating the eigenvalues and eigenvectors of the transition matrix of the master equation as Prof. Smith (and we) formulates it. The first one is practical. If we have these eigenpairs, we can construct the time evolution operator for the system as

$$\hat{T} = \sum_i e^{\lambda_i t} |g_i\rangle\langle g_i|, \qquad (1)$$

so that the problem is solved for any set of initial conditions corresponding to a given temperature and pressure, *i.e.*

$$|w(t)\rangle = \sum_i e^{\lambda_i t} |g_i\rangle\langle g_i|w(0)\rangle, \qquad (2)$$

where $|w(0)\rangle$ may correspond to $^1CH_2 + C_2H_2$ or any one of the stabilized C_3H_4 isomers. This is a computational advantage and allows us to study several chemical processes with one set of eigenpairs.

The second advantage is one of physical insight. The eigenvalues and eigenvectors individually have physical meaning. Normally one can associate a transition state with an eigenpair. Note that only four of the multitude of eigenpairs in eqn. (2) describe chemical change. In the present case λ_4 corresponds to the transition state separating $^1CH_2 + C_2H_2$ from cyclopropene, λ_3 to that between cyclopropene and propyne, λ_2 to that between cyclopropene and allene, and λ_1 to the transition state connecting propyne to $C_3H_3 + H$. At low temperature and sufficiently high pressure, the λs become the 'fundamental relaxation rates' for the reaction between the chemical components that are separated by the corresponding transition states. Under such conditions one can deduce both forward and reverse rate constants from knowledge of the λs and the corresponding equilibrium constants. Three different types of cases arise, as illustrated for the present reaction by λ_2 (λ_3 is the same), λ_4 and λ_1:

$$a - C_3H_4 \rightleftharpoons c - C_3H_4 \qquad {}^1CH_2 + C_2H_2 \rightleftharpoons c - C_3H_4$$

$$k_f = \frac{-\lambda_2 K_{eq}}{1 + K_{eq}} \qquad k_f = \frac{-\lambda_4 K_{eq}}{1 + n_{C_2H_2} K_{eq}}$$

$$k_r = \frac{-\lambda_2}{1 + K_{eq}} \qquad k_r = \frac{-\lambda_4}{1 + n_{C_2H_2} K_{eq}}$$

$$p - C_3H_4 \rightleftharpoons C_3H_3 + H$$

$$k_f = -\lambda_1$$

$$k_r = -\lambda_1/K_{eq}$$

The number density of acetylene appears above because the addition reaction is only pseudo first order, and the dissociation to $C_3H_3 + H$ is different from the others because the bimolecular products are assumed to be an infinite sink. At lower pressures multiple products appear but the basic idea is the same.

If one follows the time history of the system below 1500 K, as shown in Prof. Smith's movie, one can see four distinct time scales. All three stabilized products, as well as $C_3H_3 + H$, are formed from ${}^1CH_2 + C_2H_2$ on a time scale corresponding to $-1/\lambda_4$ (this is the reaction we are actually interested in). Cyclopropene and propyne equilibrate on a time scale $-1/\lambda_3$, and these two equilibrate with allene on a time scale $-1/\lambda_2$. Dissociation of the three equilibrated isomers takes place on a time scale of $-1/\lambda_1$.

At low temperature the rate constant for the ${}^1CH_2 + C_2H_2$ reaction is equal to $-\lambda_4/n_{C_2H_2}$. For a pressure of one atmosphere, at $T \approx 1500$ K the rate constant changes to $-\lambda_3/n_{C_2H_2}$, a change that is brought about by a shift in equilibrium of the reaction ${}^1CH_2 + C_2H_2 \rightleftharpoons c - C_3H_4$ in favor of ${}^1CH_2 + C_2H_2$, i.e. $c - C_3H_4$ begins to dissociate as fast as it is formed. This causes the reactant (${}^1CH_2$) consumption to be controlled by a different transition state, that corresponding to λ_3. At $T \approx 2500$ K, the $-\lambda_3/n_{C_2H_2}$ and $-\lambda_2/n_{C_2H_2}$ curves cross with no shift in the rate constant, indicating that the transition state between a-C_3H_4 and c-H_3H_4 never controls the rate. At still higher temperatures there is another jump of the rate constant from $-\lambda_3/n_{C_2H_2}$ to $-\lambda_1/n_{C_2H_2}$, caused by a shift in equilibrium analogous to that described above.

1 P. N. Brown, G. D. Byrne and A. C. Hindmarsh, *SIAM J. Sci. Stat. Comput.*, 1989, **10**, 1038.
2 J. A. Miller and D. W. Chandler, *J. Chem. Phys.*, 1986, **85**, 4502.
3 D. W. Chandler and J. A. Miller, *J. Chem. Phys.*, 1984, **81**, 4105.
4 J. A. Miller and S. J. Klippenstein, *J. Phys. Chem. A*, 2001, in press.
5 D. K. Hahn, S. J. Klippenstein and J. A. Miller, *Faraday Discuss.*, 2001, **119**, 79.
6 S. J. Klippenstein, A. F. Wagner, R. C. Dunbar, D. M. Wardlaw, S. H. Robertson and J. A. Miller, Variflex Version 1.08m, 2000.

Prof. S. C. Smith commented: Dr Miller's comments are a useful complement to the discussion in our paper. I note that the analysis of the quality or lack thereof of certain crucial eigenvalues and eigenvectors is not always trivial. Eigenvectors do not always converge accurately under the same conditions as the corresponding eigenvalue, such that one might have a good idea of the overall rate constant without necessarily having good population projections. Whether one can manage with regular double precision calculations, and which aspects of the calculations can be trusted and which cannot, is a matter which may often take careful study from one case to another as conditions are varied. This tedious task is avoided in our high precision algorithm. Our aim in this work has been to try to develop a robust and general algorithm which can be used with confidence under any conditions and can be easily generalized to two-dimensional master equation calculations. Direct time integration is certainly an alternative to high precision matrix diagonalization, but this effectively only gives one access to relatively short timescales, since a very large amount of numerical effort is involved to integrate out to long times.

Prof. Troe continued the discussion of Prof. S. C. Smith's paper: In your system several barrier crossing processes can be studied separately such as the thermal isomerisation of cyclopropene (work by R. Walsh). Have you included that information in your modelling?

Prof. S. C. Smith answered: We have not explicitly considered the results of Dr Walsh's study, but this certainly could be done without further modification of our master equation code.

Dr Kaiser asked: It seems that the schematic potential energy surface of the CH_2–C_2H_2 system is a little bit incomplete. Although addition of carbene to acetylene should lead predominantly to cyclopropene, a minor pathway might be the insertion of singlet carbene into a C–H bond of acetylene to form methylacetylene. Could you estimate the role of the addition *vs.* insertion pathways, and possibly the branching ratios? Further, allene is certainly no dead-end species as it can fragment without an exit barrier (except the C–H bond strength) to propargyl plus atomic hydrogen. How will your results change if you include this additional pathway?

Prof. S. C. Smith answered: The possibility of direct insertion of singlet carbene into the C–H bond of acetylene is not included in our present scheme, nor is the loss of H from allene. Our focus has been on solving the challenge of performing accurate calculations at low temperatures with the existing (incomplete) kinetic scheme. We thank Dr Kaiser for his suggestions and will explore the effect of these modifications to the kinetic scheme in future work.

Dr Klippenstein said: In collaboration with Harding we have recently studied the high pressure limit of the rate constant for addition of H atoms to propargyl radical *via* variable reaction coordinate transition state theory employing a multi-reference configuration interaction based potential energy surface.[1] This study suggests that the addition to the CH side to form allene occurs with a rate that is 2/3 of that for the addition to the CH_2 side to form propyne. Thus, in your model the allene potential well should also be directly connected to C_3H_3 + H.

1 S. J. Klippenstein and L. B. Harding, *Proc. Combust. Inst.*, 2000, **28**, 1503.

Prof. S. C. Smith responded: We thank Dr Klippenstein for his comments, which support those of Dr Kaiser in connection with the possibility of significant allene dissociation to propargyl plus H. As indicated previously, this will be incorporated into the kinetic scheme in future work.

Prof. Troe commented: When one thinks about where your matrix inversion code would be most useful, one might consider an application to ultrafast multiple well reactions in liquid phase where time scales of collisional stabilization are of similar order of magnitude as time scales of fast intramolecular arrangement processes.

Prof. Pilling said: There is an important aspect of the potential energy surface that requires further consideration. The reaction of $^1CH_2$ + C_2H_2 includes not only the chemical reactions involving the C_3H_4 isomers and H + C_3H_3, but also the collision induced intersystem crossing (CIISC) to give $^3CH_2$. CIISC occurs on collision of CH_2 with inert gases *via* 'doorway' states, of mixed singlet triplet character, coupled with collision induced rotational relaxation. The mechanism that applies in reactive collisions is not clear.

Wagner and coworkers have extensively examined the yield of triplet in reactions of $^1CH_2$, using LMR, and have shown that for reaction with C_2H_2 at 300 K, 20% of the total rate of removal of the singlet leads to triplet formation.[1] They have also shown that the energy barrier for reaction of $^3CH_2$ with C_2H_2 is 27 kJ mol^{-1},[2] which is less than the singlet triplet energy difference (37.7 kJ mol^{-1}). Thus there is the possibility of a transition from the attractive $^1CH_2$ + C_2H_2 potential energy surface to the triplet surface correlating with $^3CH_2$ + C_2H_2. Prof. Lin's paper refers to singlet to triplet surface crossing for CH_3 + OH/CH_2 + H_2O. Is there any clear procedure for introducing such a process into a master equation description of $^1CH_2$ + C_2H_2?

1 W. Hack, M. Koch, H. Gg. Wagner and A. Wilms, *Ber. Bunsen-Ges. Phys. Chem.*, 1988, **92**, 674.
2 T. Boehland, F. Temps and H. Gg. Wagner, *Proc. Twenty-first. Symp. (Int.) Combust.*, The Combustion Institute, Pittsburgh, 1986, p. 841.

Prof. S. C. Smith responded: It is relatively straightforward to extend the master equation formulation to incorporate a thermalised triplet methylene population, coupled to the thermalised singlet population *via* an estimated pseudo first order collision induced intersystem crossing (CIISC) rate, and to the propyne isomer with RRKM theory estimates for the association of triplet methylene with acetylene. The effective CIISC rate (as a function of temperature) would most likely have to be inferred from whatever experimental data is available.

Prof. Balint-Kurti commented: Prof. Pilling has asked if there are ways of treating spin forbidden intersystem crossings which may play a role in the reactions of systems such as CH_3 + OH. The answer is yes. There are now *ab initio* computer codes (*e.g.* MOLPRO) which permit the calculation of the spin–orbit coupling matrix elements which determine the transition probabilities between the singlet and triplet surfaces. Jeremy Harvey of the University of Bristol has developed interesting new RRKM based methods[1] for computing the rates of such spin forbidden transitions. These methods involve heating the lowest point on the singlet–triplet crossing seam and using this as a critical geometry for the RRKM calculations.

1 J. N. Harvey and M. Aschi, *Phys. Chem. Chem. Phys.*, 1999, **1**, 5555.

Dr Mebel said: In my opinion, the potential energy surface of C_3H_4 used for kinetic calculations of the $CH_2 + C_2H_2$ reaction is not complete.[1] First, allene can also dissociate to form the propargyl radical C_3H_3 + H without an exit barrier. Thus, both allene and propyne can give propargyl + H without an exit barrier in the sense that these decompositions are endothermic but the reverse reactions H + C_3H_3 with H addition both to the "head" and "tail" of propargyl have no barriers. The allene → C_3H_3 + H channel should certainly be included into kinetic calculations. Additionally, H_2 eliminations from allene and propyne can also play some role in the reaction.

1 A. M. Mebel, W. M. Jackson, A. H. H. Chang and S. H. Lin, *J. Am. Chem. Soc.*, 1998, **120**, 5751.

Prof. Kohse-Höinghaus opened the discussion of Prof. Golden's paper: Would you have any advice on how to treat higher hydrocarbons, such as those needed in mechanisms for PAH and soot formation, where there may not be experimental evidence? How would you ensure that databases would, while being consistent among themselves, not be off as a whole?

Prof. Golden responded: We have suggested using DFT methods as the only practical calculational procedure. We have offered some group additivity based correction where data exists to allow this. Possibly these can be checked with an occasional very expensive high level calculation. It would also be good if experiments were to be targeted on molecules that could supply the missing groups.

Prof. Lin commented: We have recently attempted to establish an efficient composite scheme for prediction of heats of formation based on B3LYP/aug-cc-pVTZ//B3LYP/cc-pVDZ energies with small 'higher level corrections (HLC)'.[1] The scheme works well for [C,H,O]-containing species, but it does poorly for N-containing species, such as NH_x, with large errors (± 3 kcal mol^{-1}).

1 Y. M. Choi and M. C. Lin, unpublished work.

Prof. Golden responded: We look forward to seeing the publication. In general we have not found that B3LYP calculations are very sensitive to basis sets. What does 'work well' mean? Were these values extracted from isodesmic reactions?

Prof. Troe asked: The described quantum methods still leave substantial uncertainties. Have they led to real improvements over the old fully empirical group additivity based tables?

Prof. Golden answered: Yes and no. The group additivity values come from data. The calculations agree with the data within a few kcal mol^{-1}. Thus we have a certain comfort when using calculated values in cases not measured. (*Caveat emptor!*)

Dr Hessler said: The bootstrap technique, see for example ref. 1, has been used to estimate the uncertainty of predictions by randomly discarding some of the data in a database, re-evaluating the parameters of the model, and then predicting the values of the discarded data. Have you applied these ideas to estimate the uncertainty of your calibration scheme?

1 B. Efron and R. J. Tibshirani, *An Introduction to the Bootstrap*, Chapman & Hall, New York, 1993.

Prof. Golden answered: No we have not.

Prof. I. W. M. Smith said: Prof. Golden provides extensive comparisons between results of his calculations and the experimental values for the heats of formation for many free radicals. I believe that there are two main sources for the experimental values; first the data obtained by himself and Benson over many years *via* the rates of bromination and iodination, and second a smaller, but significantly different set of values determined chiefly by Gutman and co-workers.[1] Is there anything in these comparisons to suggest that one set is in better agreement with his calculations than the other?

1 J. Berkowitz, G. B. Ellison and D. Gutman, *J. Phys. Chem.*, 1994, **98**, 2744.

Prof. Golden responded: We are not the first to see that calculations yield the higher BDEs for alkane C–H bonds consistent with the work of Gutman and others. We had hoped that the calculations presented here would shed light on this same question with respect to other BDEs. Unfortunately, the uncertainties in the calculations are too large.

Dr Mebel asked: In your paper, you calculated bond dissociation energies based on energies of a molecule and a radical formed after the bond cleavage. Would you expect higher accuracy of the results and their better agreement with experiment if you used isodesmic reactions for the calculations?

Prof. Golden answered: Of course, but these types of reactions can not always be formed with three known values for some of the species considered here.

Prof. Pilling asked: Prof. Golden has commented on inconsistency in thermodynamic databases. What is needed to ensure that such inconsistencies are eliminated?

Dr Hessler responded: Recently Branko Ruscic and his colleagues have generated a mathematical network that describes the enthalpy of formation of CH_2 and the bond dissociation of the methyl radical.[1] They first use this network to identify outliers, *i.e.* results that are not consistent with the other measurements, and then determine the 'best-fit' set of thermochemical parameters by a non-linear least squares analysis. Scientists have known about these networks for a long time. However, modern computers now provide an opportunity to put all of the thermodynamic measurements in a single database. As new information is added to the database all of the parameters that depend upon this information may be easily updated. Although it is more complex, this same approach may be extended to kinetic rate measurements.

1 B. Ruscic, M. Litorja and R. L. Asher, *J. Phys. Chem. A*, 1999, **103**, 8625.

Prof. Golden added: I for one have longed for the 'Great Spreadsheet in the Sky' that would enable self-consistent evaluation of data. If this is possible, I would love to see it done! Of course funding at a significant level would be necessary.

Prof. S. C. Smith said in conclusion: The microcanonical rate coefficients for dissociation from propyne to propargyl in our present kinetic scheme were obtained by inverse Laplace transform (ILT) of temperature dependent recombination rates for a closely related system, $C_3H_5 + H$.[1] Dr Klippenstein's earlier comments indicate unequivocally, however, that in the case of recombination of H plus propargyl there is significant branching to form both allene and propyne. Thus, the microcanonical rate coefficients inferred from the ILT method used previously will likely overestimate the rate constants for dissociation of propyne, while of course the dissociation of allene to propargyl is not represented in our present scheme.

1 M. A. Blitz, M. S. Beasley, M. J. Pilling and S. H. Robertson, *Phys. Chem. Chem. Phys.*, 2000, **2**, 805.

Prof. Golden said in conclusion: One, including me, might have thought that since the C–H bond dissociation energy (BDE) in ethylene is about 110 kcal mol^{-1} that the correct way to judge the C–H BDE in allene was to subtract the propargyl resonance energy from that number. Likewise, a way to describe the BDE in propyne to give the propargyl radical would be to subtract the propargyl resonance energy from the value of 101 kcal mol^{-1} that represents a primary C–H BDE. The latter is essentially correct, but allene is apparently not a molecule with two independent π bonds. The heat of hydrogenation of the first π bond to yield propene, is about 10 kcal mol^{-1} less exothermic than the heat of hydrogenation of the bond in propene itself. Thus the two π bonds repel each other by some 10 kcal mol^{-1} and despite what I said earlier, the C–H BDE in allene is less than that in ethylene.

Dr Miller opened the discussion of Prof. Lin's paper: I have three comments.

(1) In one part of your paper you indicate that you use a single exponential down model for $P(E,E')$ with $\langle \Delta E_d \rangle = 40$ cm^{-1}, whereas in another part of the paper you say that the grain size in your master equation calculations is 100 cm^{-1}. If this is the case, you do not even come close to resolving the energy transfer function, and it is unlikely that your calculations are very accurate.

(2) The inversion method in VARIFLEX was not intended for high temperatures, and its accuracy there is highly suspect. However your high-temperature calculations on the dissociation of methanol may have benefitted from a cancellation of errors. For methane dissociation (similar to CH$_3$OH), at $T = 3000$ K I determined using the eigenvector method that the lower limit of the computational domain must be 36 000 cm^{-1} below the dissociation limit (*i.e.* $E_{lower} = -36 000$ cm^{-1}) to calculate an accurate rate constant. If I use this lower limit on E and switch to the inversion method, I calculate a rate constant that is a factor of seven too small. However, if I raise E_{lower} to $-15 000$ cm^{-1}, which you use in your calculation, I make less than a 20% error.

(3) The evidence is overwhelming that the reaction (2) OH + CH$_3$ $\rightleftharpoons$ ^{1}CH$_2$ + H$_2$O is slightly endothermic (*i.e.* by less than 1 kcal mol^{-1}), and not exothermic by 1.6 kcal mol^{-1} as you indicate in your paper. At the last DOE-BES Combustion Contractors Meeting[1] Branko Ruscic claimed to have determined all the relevant heats of formation for this reaction to ± 0.1 kcal mol^{-1}, from which he deduced that reaction (2) is endothermic by 0.39 ± 0.2 kcal mol^{-1} at $T = 0$ K. One might argue with this precision, but an exothermic reaction (2) is completely inconsistent with the low-temperature experiments reported in ref. 46–48 of your paper, which include the observation that k_{-2} is almost 10^{14} cm^3 mol^{-1} s^{-1} at room temperature. It is worth noting that Ruscic's value of $\Delta H_2°$ (0 K) is almost identical to the value of 0.38 ± 0.48 kcal mol^{-1} that Pilling and coworkers deduced from their master equation modeling of several experiments. Making the appropriate correction in the thermochemistry should also allow you to predict correctly the product distribution in the dissociation of CH$_3$OH observed by Dombrowsky, *et al.* (ref. 9 of your paper), since such a correction would make CH$_3$ + OH more accessible energetically than ^{1}CH$_2$ + H$_2$O.

Reaction (2) is extremely important in methane and natural gas flames,[2] because it is the primary CH$_3$-consuming reaction from slightly lean conditions up to an equivalence ratio of roughly $\Phi = 1.3$. An increased value of k_2, as implied by your thermochemistry, would have significant effects on properties of these flames, increasing burning velocities[2] and yields of prompt NO. The ^{1}CH$_2$ formed produces greater yields of ^{3}CH$_2$ and CH. The reaction of ^{3}CH$_2$ and O$_2$ is chain branching, thus increasing burning velocities, and the increased CH results in more prompt NO. We should be careful not to confuse the issues concerning reaction (2) with bad thermochemistry.

1 B. Ruscic, *Photoionization Studies of Transient Metastable Species, Twenty-Second Annual Combustion Research Conference Book of Abstracts*, U.S. Department of Energy, Office of Basic Energy Sciences, pp. 264–267.
2 M. N. Bui-Pham and J. A. Miller, *Twenty-fifth Symp. (Int) Combust.*, The Combustion Institute, Pittsburgh, 1994, p. 1309.

Prof. Lin responded: (1) The grain size used in our calculations depends on the third-body involved. For He, we employed 30 cm^{-1} and for heavier masses such as N$_2$, Ar and SF$_6$, we used 100 cm^{-1}.

(2) There is an error in the text on the computational domain: $E_{lower} = -33\,000$ cm^{-1}, rather than $-15\,000$ cm^{-1}, was used. The calculation with the lower limit of $-36\,000$ cm^{-1} at 1600 and 2500 K showed that the differences between the former and latter limits amount to less than 1% at 1600 K and about 1% at 2500 K. When we switched the inversion method to the eigenvector method, with the $E_{lower} = -33\,000$ cm^{-1}, no difference was noted in the predicted rates with both lower limits.

(3) Both G2[1] and G2M[2] methods over-predict the atomization energy of CH_3OH (by 1.5 and 0.8 kcal mol^{-1}, respectively). On average, the highest-level G2M method predicts heats of formation for the 32 first-row G2 testing species with an absolute error of 0.88 kcal mol^{-1}.[2] Therefore, one can expect an error of ± 1 kcal mol^{-1} for predicted heats of reaction of small molecular systems such as CH_3OH. The relative energies shown in Table 1 and Fig. 2 of our paper are within the expected average error. The G2M method over-predicts $D_0(CH_3-OH)$ by 1.7 kcal mol^{-1}, comparing with the latest dissociation energy by Ruscic,[3] 90.16 ± 0.18 kcal mol^{-1}, although it predicts reasonably the heat of reaction for $CH_3OH \rightarrow {}^1CH_2 + H_2O$ (see Table 1 of our paper).

The result of our new calculation with the experimental value of $D_0(CH_3-OH)$, taking the endothermicity of 0.39 kcal mol^{-1} for $CH_3 + OH \rightarrow {}^1CH_2 + H_2O$, with and without multiple reflections above the $H_2C \cdots OH_2$ molecular complex, agrees with experimental rates within the scatter shown in Fig. 9a of our paper.

1   L. A. Curtiss, K. Raghavachari, G. W. Trucks and J. A. Pople, *J. Chem. Phys.*, 1991, **94**, 7221.
2   A. M. Mebel, K. Morokuma and M. C. Lin, *J. Chem. Phys.*, 1995, **103**, 7414.
3   B. Ruscic, private communication.

Prof. Pilling commented: I would like to comment further on the central importance of the difference in the energies of $CH_3 + OH$ and ${}^1CH_2 + H_2O$. In a recent experimental paper,[1] we used this energy difference, ΔE, as a variable parameter in fitting the magnitude and pressure dependence of the overall rate coefficient for $CH_3 + OH$, using a master equation model. We also used the rate coefficient for ${}^1CH_2 + H_2O$ and the yield of OH in that reaction.[2] We obtained $\Delta E = (1.6 \pm 2.0)$ kJ mol^{-1}. Using all of the fitted parameters, we then found that our model reproduced very satisfactorily the low pressure overall rate coefficients of Deters et al.[3] for $CH_3 + OH$, which shows a $\sim 40\%$ fall from the higher pressure limit in the ~ 1 Torr range. Over this range, the predominant channel changes from stabilization to formation of ${}^1CH_2 + H_2O$; this behaviour is, once again, very sensitive to the value of ΔE.

1   R. De Avillez Pereira, D. L. Baulch, M. J. Pilling, S. H. Robertson and G. Zeng, *J. Phys. Chem. A*, 1997, **101**, 9681.
2   W. Hack, H. Gg. Wagner and A. Wilms, *Ber. Bunsen-Ges. Phys. Chem.*, 1988, **92**, 620.
3   R. Deters, M. Oting, H. Gg. Wagner, F. Temps, B. Laszlo, S. Dobe and T. Berces, *Ber. Bunsen-Ges. Phys. Chem.*, 1998, **102**, 58.

Dr Smith said: In view of the apparent significance of the information presented at the recent DOE-BES Combustion Contractors Meeting in US, it would be helpful to learn just what was said there.

Dr Miller replied: As I mentioned in my initial comment, Branko Ruscic argued convincingly that he had determined the heats of formation for the species involved in the reaction, ${}^1CH_2 + H_2O \rightleftharpoons CH_3 + OH$, to an accuracy of ± 0.1 kcal mol^{-1}, from which he deduced that this reaction was exothermic by 0.39 ± 0.2 kcal mol^{-1}. This result is consistent with the low-temperature experiments cited in Prof. Lin's paper and with Mike Pilling's previous analysis. It is not consistent with Prof. Lin's electronic-structure calculations.

Prof. Troe said: The recent version of the heat of formation of OH[1] should have some influence on the kinetics of the methanol system, such as it brings rates of the reactions of $H + O_2$ and of $HO + O$ into much better internal consistency.[2]

1   B. Ruscic, D. Feller, D. A. Dixon, K. A. Peterson, L. B. Harding, R. L. Asher and A. F. Wagner, *J. Phys. Chem. A*, 2001, **105**, 1.

2 J. Troe and V. G. Ushakov, *J. Chem. Phys.*, 2001, **115**, 3621.

Prof. Troe asked: Your $\langle \Delta E \rangle_{down}$ values of 40 cm^{-1} which are much smaller than Pilling's value of 230 cm^{-1} for helium and which correspond to a very unusually small $\langle \Delta E \rangle_{total}$, might be due to an overestimation of other factors contributing to the low pressure recombination rate constants. Could you guess which factors that could be?

Prof. Pilling‡ said: The figure of 230 cm^{-1} for $\langle \Delta E \rangle_{down}$ cited by Pilling *et al.* for He is slightly higher than the usual value of ~ 200 cm^{-1}. The value of 40 cm^{-1} used by the authors seems small by comparison.

Prof. Lin responded to Prof. Miller, Prof. Pilling and Prof. Troe: The value of $\langle \Delta E \rangle_{down}$, 40 cm^{-1}, for the He-deactivation of the excited CH$_3$OH formed by CH$_3$ + OH appears to be low. In Fig. 1 presented here we show the results obtained by using higher values (120 and 150 cm^{-1}) for comparison. Although the effect of higher values is small at higher temperatures, it is noticeable at the lowest temperature studied, 290 K, by Pilling and co-workers.[1]

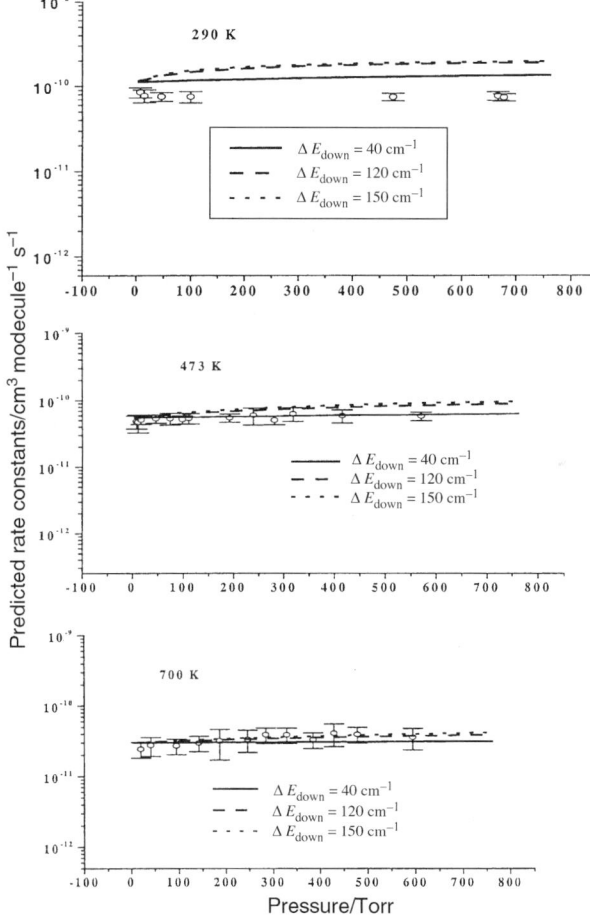

Fig. 1 Predicted rate constants for CH$_3$ + OH → products in comparison with the experimental data at different temperatures and different values of $\langle E \rangle_{down}$ employed.

‡ Also Dr S. H. Robertson (Accelrys, Cambridge, UK).

It should be mentioned, however, that our recent calculation[2] for $CH_3 + O_2 + He \rightarrow CH_3O_2 + He$ and that of Tardy[3] for $CH_3 + CH_3 + He \rightarrow C_2H_6 + He$ favour $\langle \Delta E \rangle_{down} = 70$ cm^{-1}. Interestingly, for $CH_3 + O_2$, Pilling and co-workers[4] concluded that a broad range of $\langle \Delta E \rangle_{down}$, 40–285 cm^{-1}, for Ar could reasonably fit experimental data, with 40 cm^{-1} giving the optimal fit.

1 R. De Avillez Pereira, D. L. Baulch, M. J. Pilling, S. H. Robertson and G. Zeng, *J. Phys. Chem. A*, 1997, **101**, 9681.
2 R. S. Zhu, C.-C. Hsu and M. C. Lin, *J. Chem. Phys.*, 2001, **115**, 195.
3 D. Tardy, private communication.
4 M. Keiffer, M. J. Pilling and M. J. C. Smith, *J. Phys. Chem.*, 1987, **91**, 6028.

Prof. Plane said: Rollason and I have recently completed a study of FeO recombining with H_2O, CO_2 and O_2 in the presence of either He or N_2.[1] Application of RRKM theory indicates that $\langle \Delta E \rangle_{down}$ for He is about 200 cm^{-1}, compared to ≈ 500 cm^{-1} for N_2. The He value is obviously much higher than the value of 40 cm^{-1} reported in Prof. Lin's paper. However, the value of $\langle \Delta E \rangle_{down}$ is rather sensitive to the parameters chosen to estimate the collision frequency with the third body, so this discrepancy may be less than it appears.

1 R. Rollason and J. M. C. Plane, *Phys. Chem. Chem. Phys.*, 2000, **2**, 2335.

Dr Miller said: An important point that we all tend to overlook (or ignore) is that the value of $\langle \Delta E_d \rangle$ or $\langle \Delta E \rangle$ that we deduce from any experiment may depend on the form of $P(E,E')$ that we assume. The differences between a single-exponential-down model and a double-exponential-down model are likely to be greatest for chemically activated problems or thermally activated multichannel reactions. A number of years ago David Chandler and I found significant differences from such models for photo-activated systems,[1,2] which are very similar to chemically activated ones.

1 J. A. Miller and D. W. Chandler, *J. Chem. Phys.*, 1986, **85**, 4502.
2 D. W. Chandler and J. A. Miller, *J. Chem. Phys.*, 1984, **81**, 445.

Prof. Balint-Kurti and **Mr Cole** commented: We have also been using *ab initio* methods to study the $NH_2 + NO$ reaction. We have used the B3LYP density functional method with a cc-pvdz basis set to compute all the stationary points on the surface, and also to compute the minimum energy reaction and break-up pathways. In addition we have computed the variation of the vibrational frequencies along these paths. At all the critical geometries, such as local minima, transition states and reactants and products, we have computed energies using the quadratic CI method, with a cc-pvqz basis. The energies of the reaction paths and stationary points were scaled so as to fit these quadratic CI values. With this *ab initio* input we have performed *J*-conserving variational RRKM calculations and have used a steady state model to compute overall rate constants. The lower curve in Fig. 2 shows our *ab initio* computed branching ratios for the production of OH as a function of temperature. The upper curve in the figure is an analytic fit to the experimental data taken from the work of Park and Lin.[1]

The *ab initio* calculations reproduce the increase of the branching ratio with temperature very well, they are however consistently lower than the experimental values throughout the whole temperature range. At low temperatures around 300 K the experimental branching ratios level off at around 10%. This is hard to reconcile with our theoretical results because we find that the energy of the products $N_2H + OH$ lies above that of the $NH_2 + NO$ reactants. This fact implies that the branching ratio must tend to zero as the temperature decreases to zero. A non-zero value of the branching ratio at 0 K would imply that the reaction to give $N_2H + OH$ is exothermic and does not possess any barrier whose energy is higher than the energy of the reactants.

1 J. Park and M. C. Lin, *J. Phys. Chem. A*, 1999, **103**, 8906.

Prof. Lin replied: Our recent calculations[1] show that there are low-lying excited states near the $HN_2 + OH$ dissociation limit. Their participation in the $NH_2 + NO$ reaction may lead to a non-statistical temperature dependence for the OH branching ratio observed experimentally.

1 D. Chakraborty and M. C. Lin, unpublished work.

Prof. Van Tiggelen opened the discussion of Dr Klippenstein's paper: I am pleased to see (Fig. 2) that the sigma shape of the curve of the branching ratio *vs.* temperature is predicted nicely from

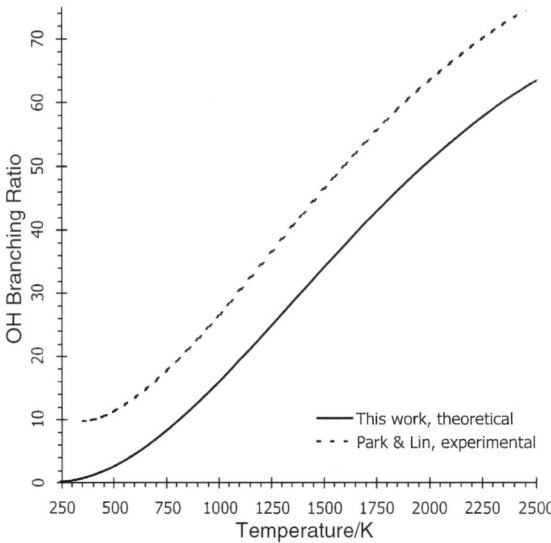

Fig. 2 *Ab initio* branching ratio for OH production from NH_2 + NO reaction.

your computation. Would it be possible to get a more steep increase of that ratio? Which thermochemical data are dominant in describing the steepness? What kind of adjustment in the energy barriers are required to obtain an absolute rate constant more flat in the high temperature range (1000–2000 K), but still keeping the rapid decrease in the 400–800 K range. Rather high values of the overall rate constant are necessary to propagate a flame in NH_3–NO mixtures, see for instance Brown and Smith's paper[1] well as ref. 13 of your paper.

1 M. J. Brown and D. B. Smith, *Twenty-fifth Symp.* (*Int.*) *Combust.*, The Combustion Institute, Pittsburgh, 1994, pp. 1011–1018.

Dr Klippenstein responded: As illustrated in the paper, the shape of the branching fraction curve in the intermediate temperature regime depends primarily on the combination of the HNN + OH endothermicity and the height of the ts3/4 barriers. The shape also depends on the temperature dependence of the HNN + OH association rate constant. However, it is difficult to obtain curves that are significantly steeper at intermediate temperatures (600 to 1500 K) without destroying the flatness at low temperatures. At higher temperatures (*e.g.*, above 1500 K) vibrational anharmonicities (beyond the hindered rotor corrections that were explicitly included in the analysis) and/or non-statistical deviations (since there are then significant contributions from energies for which the dissociation rates of the intermediate complexes exceed 10^{12} s^{-1}) might play a role in yielding somewhat larger branching fractions. There may be similar corrections for the total rate constant. However, it seems unlikely that any such corrections to the total rate constant could be large enough to provide agreement with the values from ref. 13 in our paper.

Prof. Golden said: A new study at Stanford[1] has extended the temperature range for the overall rate constant so that the range is now 1250–2500 K. The values for the branching ratio from 1350 to 17 700 K are in agreement with ref. 2.

1 S. Song, R. K. Hanson, C. T. Bowman and D. M. Golden, *Int. J. Chem. Kinet.*, 2001, **33**, in press.
2 J. A. Miller and S. J. Klippenstein, *J. Phys. Chem.*, 2000, **104**, 2061.

Prof. Plane asked: The experimental data plotted in Figs. 9–11 of your paper show that the branching fraction α at 300 K varies by just a few percent, whereas at higher temperatures the spread is much greater. Does this imply accidentally good agreement from a smaller set of measurements at 300 K? Should α be remeasured at this temperature, and perhaps down to 200 K?

Dr Klippenstein answered: It is indeed interesting that the spread of the measurements is smaller at 300 K. Unfortunately, I have no idea as to why that would be. From our perspective, it would indeed be valuable to have further measurements of the branching ratio near room temperature and lower as you suggest.

Dr Smith asked: If the problem is matching the measured branching ratio at low temperatures, is your comment a call for more experiments in that region? If so, over what temperature range?

Dr Klippenstein replied: Yes, we would like to see more experimental measurements of the branching ratio at low temperatures. Precise measurements at any temperature below about 400 K should help to delineate what the HNN + OH endothermicity really is, and thus what the correct low temperature behaviour is.

Prof. Lin said: The OH branching ratio at room temperature is well-established; it has been corroborated by several different experimental methods, including our mass spectrometric measurements for H_2O and CO_2 (formed by OH + CO) which gave the value, 0.10 ± 0.02.[1]

1   J. Park and M. C. Lin, *J. Phys. Chem.*, 1996, **100**, 3317.

Dr Klippenstein replied: While the room temperature branching ratio does appear to be well established, it is difficult for any theoretical model to reconcile the experimental values over the 300 to 500 K range with the also well-established rapid rise at temperatures from 800 to 1800 K. If one allows the room temperature value to be 0.05, which only deviates from your result by 2.5 times your stated error bars, then, as illustrated in Fig. 11 of our paper, one is able to satisfactorily reproduce the remainder of the temperature dependence up to 2000 K.

Dr Taatjes opened the discussion of Prof. Temps's paper: I have a question concerning the carbon balance in your experiments, specifically the ratio of CO production to ketene consumption. In the mechanism of Table 2 (in your paper) nearly every reaction which removes ketene produces CO. The only significant exception is reaction 15.2, $OH + CH_2CO \rightarrow CH_3 + CO_2$. However, this represents only 2/7 of the removal of ketene by OH, and OH production represents in turn only $\sim 10\%$ of the $^3CH_2 + NO$ reaction. Yet your $\Delta CO/-\Delta CH_2CO$ ratios are only ~ 0.7. What happens to CH_2CO which does not produce CO? Also, are there any contributions from photolysis of reaction products by the Hg lamp?

Prof. Temps responded: The mass balances, including the carbon balance, depend, of course, on the absolute calibrations of the FTIR sensitivities. In response to the question, we note first that due to the strong absorptions, but very small changes in the CH_2CO and NO pressures, the absolute depletions of these compounds could not be determined with high precision. Indeed, within the experimental scatter, the NO profiles were almost flat, which is the reason why we did not even consider them in the experiments at 100 mbar. We therefore specified all product yields with respect to the CO formed. The $\Delta CO/-\Delta CH_2CO$ ratio is about 0.7 (see Fig. 3 and Table 3 of the paper). As mentioned by Taatjes, only a small part of the extra loss of the CH_2CO is due to the reaction with OH. The main loss term for the CH_2CO is the reaction with HCNO. HCNO is a very unstable species. Its calibration was difficult as one had to work rapidly to avoid excessive decomposition of the prepared pure HCNO samples.

The 'dark' reactions of CH_2CO with HCNO (and possibly also with other acidic products, including HCN or HONO) were very slow, but for photolysis times of 1 h (corresponding to a duration of an experimental run of 2 h), they did play a role. We estimated a loss of HCNO of up to 20%, resulting in an additional loss of CH_2CO of about 18%. Together with the other CH_2CO loss processes, including the reaction with OH, this brings the $\Delta CO/-\Delta CH_2CO$ ratio to a corrected value close to 0.95, *i.e.* close to unity given the difficulty of determining ΔCH_2CO mentioned above.

We did not observe any evidence for a significant degree of photolysis of the reaction products. We would like to add that we have recently initiated time-resolved product measurements for $CH_2 + NO$. These experiments were performed using pulsed excimer laser photolysis in a tubular slow flow reactor in connection with mass spectrometric detection of the HCNO and HCN. The first results are in good agreement with the FTIR data presented here.[1]

1 G. Eshchenko, Th. Köcher and F. Temps, to be published.

Prof. Cheskis asked: Could you say something about the reaction of methylene in the single electronic state? What do your calculations predict for branching ratios of different channels of the reaction of singlet methylene with NO?

Prof. Temps answered: The experimental results obtained at a high pressure of M = Ar (570 mbar) as inert bath gas are for the reaction of triplet methylene ($^3CH_2$) with NO, as the singlet methylene ($^1CH_2$) produced by the ketene photolysis is rapidly deactivated to the ground state. However, the measurements conducted with M = He at a lower total pressure (100 mbar), where more of the $^1CH_2$ reacts with NO prior to deactivation, gave essentially the same ratio of HCNO to HCN. This indicates little dependence of the product yields on the CH_2 electronic state.

On theoretical grounds, the reactions of both $^3CH_2$ and $^1CH_2$ with NO are assumed to proceed on the same (doublet) potential energy hypersurface. In view of the calculated similar threshold energies of the channels to HCNO and HCN and the computed $k(E,J)$ curves given in the paper, we expect rather similar product yields for CH_2 in the two spin states. The extra excitation energy resulting from the electronic excitation of the $^1CH_2$ should not make a large difference. Although we did not perform full calculations with thermal averaging, these conclusions were confirmed by calculated product distributions of $^1CH_2$ for selected fixed values of E and J.

Note that the calculated steady state energy distribution functions for $T = 1200$ K (see, e.g., Fig. 7 of the paper) reach well beyond the singlet electronic excitation energy. Nevertheless, the predicted temperature dependence of the product branching ratio (Fig. 10 of the paper) was only weak.

Dr Parker communicated: In Fig. 4 of the paper, have you considered using a weighted linear least squares fit? It is obvious that the fitted line that you have underweighs the smaller-valued data points in H_2CCO and NO decays.

Prof. Temps communicated in response: The full lines in Fig. 4 of the paper are not fitted lines but the results of the numerical simulations of the reaction mechanism detailed in Table 2. Thus, there were no adjustable parameters used in the figure. Corresponding numerical simulations were carried out for the product profiles of all other experimental runs, giving similarly nice pictures.

The experimental results for the branching ratios for HCNO and HCN formation were derived from the experimental data for short reaction times, as explained in the paper. One should keep in mind that the data points at short times show some statistical scatter. However, they were taken for analysis because they are not affected to the same extent by consecutive reactions as the points for long photolysis times. At long times, the (presumably heterogeneous) reaction of HCNO with CH_2CO comes into play.

Prof. S. C. Smith commented: One of the difficulties of modelling chemical activation reactions which are clearly sensitive to angular momentum effects is that it is, to date, very difficult to carry out two-dimensional ME calculations to explore the possible effect of weak collisional relaxation on branching ratios. As a way of rationalising ourselves through this, there are two simple approaches (or scenarios).

(a) We argue that the lifetime of the intermediate is shorter than the mean collision free time, and conclude that collision relaxation should not be important such that microanonical modelling (master equation or steady state) followed by Maxwell–Boltzmann averaging suffices.

(b) We include strong collisional modelling (possibly with incorporation of a collision efficiency) and look to see if there is any sensitivity to the pressure. If we see no pressure dependence, we again conclude that collisional relaxation is unimportant. Strong collisional modelling can, however, be misleading because it gives qualitatively incorrect time evolution of the population. Branching ratios in reactions with competitive channels can be sensitive to the shape of the population distribution, and we should bear in mind that proper weak collisional simulations might show pressure effects on the competitive branching ratios which do not show up with the simpler rationalizations summarized above.

Dr Miller responded: For the $NH_2 + NO$ reaction (I suspect the same is true for $CH_2 + NO$), the mean time between collisions is larger than the RRKM lifetimes for any of the complexes, even for the lowest energy from which the complexes can be formed from $NH_2 + NO$.[1] However, I agree that the bimodal energy distribution implied by a pseudo strong-collider model can lead to errors, even very large ones for some reactions. However, here I would like to describe a simple phenomenon that we have observed in master equation calculations at various times, a phenomenon that cannot be described by a pseudo strong-collider model even by adjusting the collision efficiencies.

Fig. 3 is a potential energy profile for the $C_2H_5 + O_2$ reaction. At the collisionless limit the only products are $C_2H_4 + HO_2$. As the pressure is increased, a pseudo strong-collider model predicts that the rate coefficient for the reaction (1) $C_2H_5 + O_2 \rightarrow C_2H_4 + HO_2$ decreases, because collisions increasingly favor stabilization (2) $C_2H_5 + O_2 \rightarrow C_2H_5O_2$.

However, master equation calculations indicate that k_a actually *increases* slightly with pressure. This increase is a consequence of weak collisions that deactivate the $C_2H_5O_2^*$ complexes into the gray region of Fig. 3, from which they can no longer redissociate to $C_2H_5 + O_2$, and in which they do not live long enough to suffer more collisions and end up as stabilized $C_2H_5O_2$. Thus $C_2H_5O_2^*$ complexes that redissociate to $C_2H_5 + O_2$ in the collisionless regime end up forming $C_2H_4 + HO_2$ with a slight increase in pressure.

Fig. 4 shows this effect for $C_2H_5 + O_2 \rightarrow C_2H_4 + HO_2$ at 298 K. The circles are results from master equation calculations and the solid line is a representation of those calculations in the Troe format, through which F_{cent} was chosen to reproduce the master equation results in the center of the pressure range. The solid line is the same behavior one would get from a strong-collider model

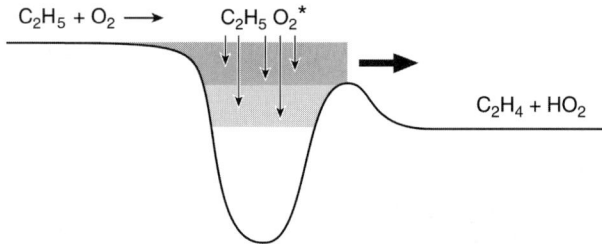

Fig. 3 Schematic diagram of the potential for $C_2H_5 + O_2 \rightarrow C_2H_4 + HO_2$. $C_2H_5O_2^*$ complexes deactivated into the dark gray region (and light gray region if tunneling is included) cannot return to $C_2H_5 + O_2$ and are not yet 'stabilized.'

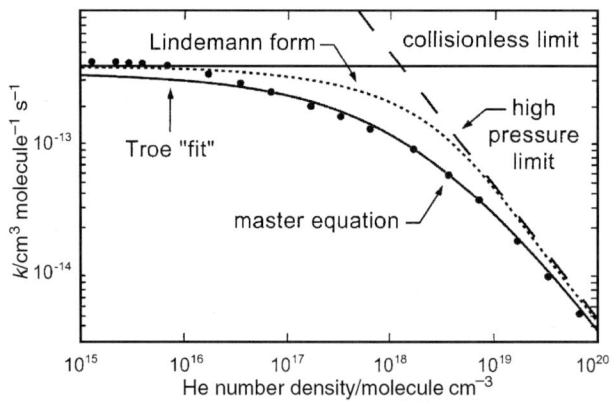

Fig. 4 Pressure dependence of rate constant for $C_2H_5 + O_2 \rightarrow C_2H_4 + HO_2$ at $T = 298$ K.

in which the collision efficiency was chosen to reproduce the master equation rate constants at intermediate pressures. In the present case, the maximum value of k_a is approximately 10% larger than that at the collisionless limit, but the points deviate from the line in Fig. 4 by about 30%. The latter is a reasonable assessment of the error one might incur in the present case from a pseudo strong-collider model and its associated bimodal energy distribution. However, there is no guarantee that the errors cannot be much larger in other cases.

It is interesting that the same types of collisions were observed by Ian Smith and his collaborators[2] in their experimental investigation of the $CH + H_2$ reaction, but with a different effect. At 744 K these investigators observed the total rate coefficient, a combination of the two reactions,

$$CH + H_2 \rightarrow CH_3 \qquad (3)$$

and

$$CH + H_2 \rightarrow CH_2 + H, \qquad (4)$$

to decrease with increasing pressure. They attributed this effect to deactivation of CH_3^* complexes with enough energy to go on to $CH_2 + H$ into an energy range where they could only go back to reactants (reaction (4) is slightly endothermic, analogous to the gray region of Fig. 3).

1   J. A. Miller, *Twenty-sixth Symp. (Int.) Combust.*, The Combustion Institute, Pittsburgh, 1996, p. 461.
2   R. A. Brownsword, A. Canosa, B. Rowe, I. W. M. Smith, D. W. A. Stewart, A. C. Symonds and D. Travers, *J. Chem. Phys.*, 1997, **106**, 7662.

Prof. Troe commented: The issue of collisional stabilization in multiple well systems and its treatment by a master equation was raised. On the foundations of the treatment by Seymour Rabinovitch of chemical activation systems, which correspond to the multiple well systems considered here, I have provided simplified solutions of step-ladder models and master equations.[1] They relate collision efficiencies γ_c (which are not identical with collision efficiencies β_c in thermal activation) through an equation $\gamma_c/(1 - \gamma_c^{0.7}) \approx -\langle \Delta E \rangle S/(\langle E_{ac}\rangle - E_o)$ with the average energy $\langle \Delta E \rangle$ transferred per collision, the excess excitation energy $\langle E_{ac} \rangle - E_o$ and the exponent S of the specific rate constants $k(E) \alpha (E - E_o)^{s-1}$ of the fastest process depopulating the stabilization well.

1   J. Troe, *J. Phys. Chem.*, 1983, **87**, 1800.

Prof. Golden said: The standard methods of dealing with chemical activation systems, pioneered by Rabinovitch involving psuedo strong collision assumptions, seems to be adequate to describe many systems. I think this means that once a system has experienced a single collision, it is deactivated sufficiently that the next most likely event is another collision, rather than product formation.

Prof. Temps responded to Prof. S. C. Smith's comment: As discussed in the paper, we are convinced that our experimental and computational results for the $CH_2 + NO$ reaction apply to the low pressure limit. Collisional effects should not play a role up to pressures below 10 bar (see section 3.2 of the paper).

However, in response to the comment, we would like to point out that the framework for treating chemical activation experiments (including radical–radical-reactions) with collisional deactivation has been well laid out since the work of Kohlmeier and Rabinovitch.[1] Based on master equation modelling, Troe has presented expressions which allow for weak collision effects.[2,3] It is true that the situation is much more complicated for multiple channel–multiple well reactions, as the $CH_2 + NO$ reaction. That is one of the fascinating aspects of radical–radical reactions and, indeed, a major motivation for our work. In our opinion, two-dimensional master equation models would be very desirable. However, they require allowance for weak collision vibrational and rotational energy transfer (VET and RET), respectively. While RET is in general much faster than VET, little is known experimentally about their combined effect, *i.e.* vibrational–rotational energy transfer (VRET) at the high vibrational energies of the molecules of interest.[4]

1   G. H. Kohlmeier and B. S. Rabinovitch, *J. Chem. Phys.*, 1963, **38**, 1709; G. H. Kohlmeier and B. S. Rabinovitch, *J. Chem. Phys.*, 1963, **39**, 490.

2   J. Troe, *J. Phys. Chem.*, 1983, **87**, 1800.
3   J. Troe, *Proc. Combust. Inst.*, 1998, **27**, 167.
4   F. Temps, S. Halle, P. H. Vaccaro, R. W. Field and J. L. Kinsey, *J. Chem. Soc., Faraday Trans 2*, 1988, **84**, 1457.

Prof. Kohse-Höinghaus opened the discussion of Ms Goos's paper: Would any of the adjusted rate coefficients in your paper directly influence the level of propargyl in propene (or other) flames?

In our recent investigations of propene flames[1] and modeling of this system,[2] there seem to be remaining deviations between measured and simulated propargyl radical concentrations, although benzene is quite well predicted. A similar pattern was recently noted by Lindstedt[3] when modeling our pentadiene flame investigated under similar conditions. In this regard, a comment would be much appreciated.

1   B. Atakan, A. T. Hartlieb, J. Brand and K. Kohse-Höinghaus, *Twenty-seventh Symp. (Int.) Combust.*, The Combustion Institute, Pittsburgh, 1998, p. 435.
2   H. Böhm, A. Lamprecht, B. Atakan and K. Kohse-Höinghaus, *Phys. Chem. Chem. Phys.*, 2000, **2**, 4956.
3   P. Lindstedt, personal communication (at this meeting).

Prof. Hoyermann responded: The experimental product distribution of the reaction system (propene + CH_3) is well described by the given mechanism (Fig. 2 and Table 3 of our paper) without inclusion of the C_3H_3 radical, thus no direct information on the C_3H_3 chemistry can be deduced. Our similar studies on the systems (1-propyne + CH_3) and (1,2-propadiene + CH_3), where the C_3H_3 is greatly involved with different precursors, gave equal amounts of formed benzene at an otherwise totally different product spectrum. This complements the observations of Kohse-Höinghaus and Lindstedt. In an ongoing evaluation of our experimental results (1-C_3H_4 + CH_3, 1,2-C_3H_4 + CH_3, C_2H_2 + CH_3, and their mixtures) we will consider the suggestions on alternative paths to C_3H_3 to benzene formation.

Prof. Plane asked: This technique clearly produced a steep temperature gradient between the optical axis of the CO_2 laser and the cold reactor walls. Does this affect the estimation of the average temperature at which the reactions are initiated, and perhaps constrain the precursors which can be used?

Dr Desgroux asked: Could you comment on the influence of the radial distribution of temperatures on your measurements? Particularly what kind of 'average' temperature do you use?

Prof. Hippler and **Prof. Hoyermann** replied to Prof. Plane and Dr Desgroux: The radial temperature distribution reflected in the formed C_3H_6 via the isomerization reaction c-$C_3H_6 \rightarrow C_3H_6$ is shown in Fig. 5. The average temperature is essentially that deduced by the chemical thermometer. For a detailed discussion see ref. 30 of our paper. The high activation energy of the decomposition of the precursor di-*tert*-butyl peroxide (156.5 kJ mol^{-1}) leads to the formation of CH_3 radicals mainly at the highest temperatures, switching off the reactions automatically in the cold parts.

Prof. Golden commented: I worry whenever I see that in the formulation of a mechanism, some rate constants are described with two parameters and others with three parameters. This is probably a result of the fact that many values in the literature measured at low to modest temperatures do not reflect Arrhenius curvature. However when these linear fits are extrapolated to higher temperatures they may lead to underestimation of rate constant values. When this is combined with assigning three parameters to other rate constants I wonder if we have created a situation where values cross on the inverse temperature plot, when they should be parallel or close to same.

Prof. Hippler and **Prof. Hoyermann** replied: This general warning has to be taken seriously by any kineticist and modeller. In our particular case we only state agreement between the absolute literature rate coefficients and our model within the limited temperature range of 750–1000 K.

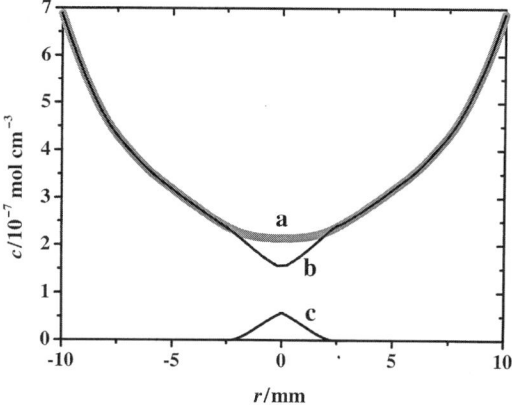

Fig. 5 Radial concentration profiles in the reactor c-C_3H_6 concentration in the temperature field without (a) and after the isomerization reaction c-$C_3H_6 \rightarrow$ propene (b). The concentration profile of propene is given by (c).

Dr Kaiser commented: The analysis of the CH_3–C_3H_6 system looks really interesting. It might be a good idea to investigate how the model predictions will change if you include the reaction of hydrogen poor transient species such as atomic carbon (C(^3P_j)) with ethylene. Recent crossed beam studies and theoretical investigations of this important reaction suggested the formation of four products: three C_4H_5 isomers and the propargyl radical (Fig. 6).[1,2] If you investigate the secondary reactions of these radicals, you might form substituted benzenes, cyclopentadienyl radicals, as well as seven and eight membered rings (Fig. 7). The story gets even more fascinating if you could include two C_5H_5 isomers (vinyl propargyl: Fig. 8) which are reaction products of atomic carbon with various C_4H_5 isomers into your model.[3–6] This can bring you to, for example,

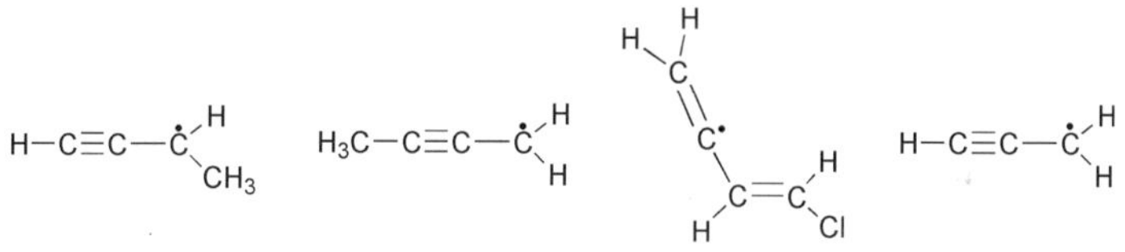

Fig. 6 Four products: three C_4H_5 isomers and the propargyl radical.

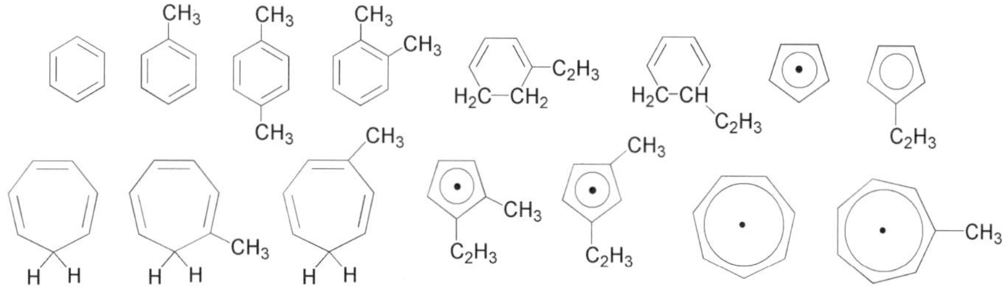

Fig. 7 Secondary reaction products: substituted benzenes, cyclopentadienyl radicals and seven and eight membered rings.

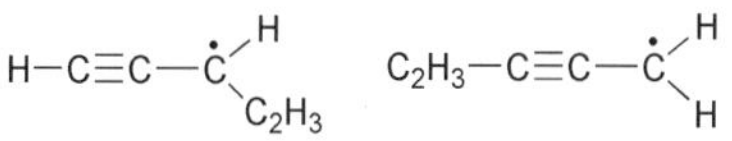

Fig. 8 Two vinyl propargyl (C_5H_5) radicals.

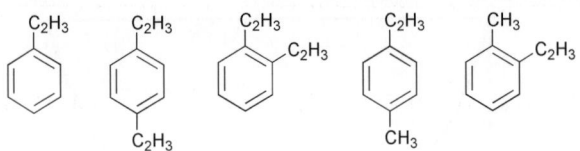

Fig. 9 Mono- and disubstituted benzene molecules.

disubstituted benzene molecules (Fig. 9). In particular, the *ortho*-divinylbenzene is very interesting as this species might undergo ring closure to form a $C_{10}H_{10}$ bicycle.

1. R. I. Kaiser, D. Stranges, H. Bevsek, Y. T. Lee and A. G. Suits, *J. Chem. Phys.*, 1997, **106**, 4945.
2. R. I. Kaiser, N. Nguyen, H. Le and A. M. Mebel, *Astrophys. J.*, 2001, in press.
3. R. I. Kaiser, H. Y. Lee, A. M. Mebel and Y. T. Lee, *Astrophys. J.*, 2001, **548**, 852.
4. I. Hahndorf, H. Y. Lee, A. M. Mebel, S. H. Lin, Y. T. Lee and R. I. Kaiser, *J. Chem. Phys.*, 2001, **113**, 9622.
5. L. C. L. Huang, H. Y. Lee, A. M. Mebel, S. H. Lin, Y. T. Lee and R. I. Kaiser, *J. Chem. Phys.*, 2001, **113**, 9637.

Energy transfer in combustion diagnostics: Experiment and modeling

Andreas Brockhinke* and Katharina Kohse-Höinghaus

Physikalische Chemie I, Fakultät für Chemie, Universität Bielefeld, Universitätsstr. 25, D-33615 Bielefield, Germany. E-mail: brockhinke@pc1.uni-bielefeld.de

Received 5th March 2001
First published as an Advance Article on the web 9th October 2001

Laser induced fluorescence (LIF) of OH (A $^2\Sigma^+$) is measured in several atmospheric-pressure flames using a short-pulse laser system (80 ps duration) in conjunction with an intensified streak camera. The two-dimensional signal-detection technique allows one to simultaneously monitor rotational and vibrational relaxation as well as electronic quenching. Rotationally-resolved LIF spectra affected by energy transfer are compared with the results of a rate-equation model and are found to be in reasonably good agreement. It is shown that a significant contribution of fluorescence detected by broad-band techniques is due to levels populated by vibrational energy transfer (VET). Implications for picosecond LIF techniques for the time-resolved, quench-free detection of OH are discussed. A detailed analysis is presented for fluorescence spectra originating from levels populated by VET after excitation of states in the OH (A $^2\Sigma^+$, $v' = 2$) level.

1 Introduction

Energy transfer is an ubiquitous process in chemistry. Considering the possibilities of motion of a molecule and its interactions with its environment, energy redistribution on the microscopic scale takes place, depending on the quantum states and degrees of freedom of the molecule and the nature of its surroundings. Reaction kinetics may need information on inter- and intramolecular energy transfer, often concerning electronic ground states. In contrast, energy transfer in and from the excited electronic state is considered in the present contribution, where the influence of collisional processes on the LIF detection of small molecules in high-temperature combustion is investigated. The OH radical is the most exhaustively studied diatomic radical in this respect, because of its importance in combustion and atmospheric chemistry, therefore, its spectroscopy is very well-known.[1–3] Many of the numerous combustion studies presented at the biannual International Combustion Symposia[4,5] have used LIF of OH in characterizing flame structures, in measuring temperatures or in providing experimental data for kinetic model validation. Typically, an excited-state population in a specific quantum state is prepared using a pulsed laser, and the OH-LIF signal is detected in the A–X band to measure OH concentrations or to monitor the flame temperature. Depending on the chemical composition, temperature and the pressure of the combustion environment, collisional processes lead to a redistribution of the energy and population during the lifetime of the excited electronic level.[6]

Quantitative OH-LIF measurements in flames at atmospheric pressure using nanosecond lasers typically require, at least, information on electronic quenching as a function of the excited state quantum numbers and collision partner mixture—a requirement not easily met in turbulent combustion studies. Even broadband detection schemes which are often used in two-dimensional imaging experiments will need information on the spectral structure of the signal. Here, quenching competes with vibrational (VET) and rotational energy transfer (RET), processes which also depend on the quantum states, temperature, pressure and chemical composition. In order to keep track of all individual state-to-state transfer steps, hundreds of respective rate coefficients may easily be involved. Often, experimental strategies are thus employed which are thought to minimize the influence of collisions, including saturated LIF and predissociative LIF.[7–11] However, both approaches may be affected strongly by energy transfer in the ground state.[11,12]

Recently, the use of picosecond lasers has been suggested for quench-free measurements[13–17] with the idea being to either excite and detect the fluorescence in a very short time interval with respect to typical collision times (of the order of 100 ps in atmospheric pressure flames) or to observe the complete temporal decay of LIF signals. Picosecond OH concentration measurements were performed in turbulent, non-premixed $CH_4/H_2/N_2$ flames.[18] Also, some energy transfer coefficients for OH have been measured by Nielsen et al.[19] and Beaud et al.[20] for the $A\,^2\Sigma^+$, $v' = 2$ and $v' = 1$ vibrational states, respectively, using picosecond LIF. In this context, Beaud et al.[20] have been the first to report polarization-dependent energy transfer coefficients in OH-LIF. Whereas an early paper[21] had already analyzed polarization effects in OH-LIF spectra observed in an atmospheric pressure flame, the influence of polarization has largely failed to attract further attention in LIF combustion measurements, with only a few exceptions.[7,22] An interesting question in picosecond OH-LIF studies using polarized excitation or detection is thus the collisionally-induced reorientation or depolarization process,[19,22,23] a potential additional complication. In a collision-free environment, polarization ratios for individual lines may be calculated following ref. 21 and fluorescence intensity distributions may be derived which are unaffected by collisions. However, it is intuitively clear that collisions will tend to destroy any preferential orientation, and again, this process may depend on the chemical nature of the collisional environment and on the quantum states involved.

In an attempt to analyze potential collisional influences for typical LIF experiments in combustion diagnostics, we have been using a combined approach of experiment and simulation.[22,24–31] A computer code (LASKIN) has been developed which uses kinetic equations to describe state-to-state energy transfer in the OH $A\,^2\Sigma^+$ state.[28,29,31] After experimental investigation of RET in $v' = 0$ and $v' = 1$[25,27] and VET[30] between these two levels for some combustion-relevant collision partners, we have recently been able to satisfactorily simulate the predominant trends observed in OH-LIF experiments under a wide range of conditions.[31] In spite of the comparatively large available data base existing for OH, remaining key questions were identified in this study. These include a lack of information on energy transfer in the higher vibrational levels, $v' = 2$ and $v' = 3$, from which branching ratios for VET with one or more vibrational quanta could be derived; also, the interdependence of vibrational and rotational relaxation, i.e. the "nascent" rotational distribution associated with a VET step, is not well known.

In view of these open questions and the unresolved question of collisional depolarization, we have recently developed a novel experimental approach using, simultaneously, time-, wavelength- and polarization-resolved picosecond OH-LIF.[22] With a further improved apparatus and sensitivity, the present study will report the first rotationally-resolved energy transfer measurements exciting the OH ($A\,^2\Sigma^+$, $v' = 2$) state in a well-defined combustion environment. In conjunction with the simulation by the LASKIN program, the influence of VET and RET in typical rotationally resolved and broadband LIF combustion diagnostics experiments will be discussed.

2 Experimental

The experimental set-up used for the experiments described in this paper basically consists of two parts: a picosecond laser and a detection system basing on a streak camera. Both are shown schematically in Fig. 1. The system consists of a tunable Ti : sapphire oscillator (Spectra Physics, "Tsunami") and a three-stage amplifier (Spectra Physics, model TSA-50). The oscillator is pumped by an argon ion laser and runs at a repetition rate of 80 MHz. Pulse durations of nominally 3 or 80 ps may be selected; the resulting line width is close to the Fourier transform limit. The amplifier system consists of one regenerative amplification stage and two double-pass travelling-wave amplifiers; two frequency-doubled Nd : YAG lasers running at a repetition rate of 10 Hz are used as pump sources. The amplified output, also at 10 Hz repetition rate, can then be frequency

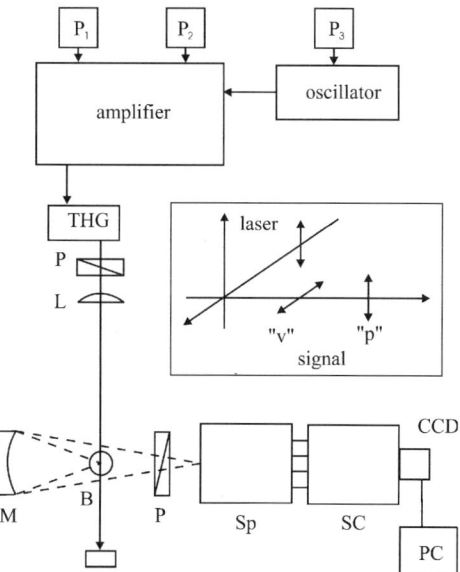

Fig. 1 Experimental set-up. The oscillator is a mode-locked Ti:sapphire laser. P_1: Ar$^+$ pump laser; P_2, P_3: Nd:YAG pump laser; THG: frequency tripling unit; P: polarizer; L: lens; B: burner; M: mirror; Sp: spectrograph; SC: streak camera; CCD: detector camera; PC: computer. The inset shows the geometry of laser and signal polarization directions.

doubled, tripled and quadrupled. In order to reach frequencies other than those obtainable by harmonic generation, a Raman shifter (10 bar H_2) is used. Thus, nearly the complete wavelength range from 200 to 900 nm can be covered. With the 80 ps pulse length option, pulse energies up to 5 mJ are achievable in the UV.

The UV laser beam is mildly focused into the flame using a 2 m focal length lens with the focal point located 20 cm downstream of the observation location to further reduce the power density and to avoid saturation. In some experiments, the polarization purity of the beam is enhanced by a Rochon-type polarizer (Laser Components; air-spaced UV quartz version) located before the lens. The LIF signal is collected by a large aperture (f number: $f/1$), spherical, concave mirror with Al:MgF$_2$ coating, in the usual 90 degrees excitation–detection geometry. The mirror is aligned to image the laser focal region onto the spectrograph entrance slit. A Glan–Thompson polarizer (Halle; selected calcite crystal, cut angle optimized for UV wavelengths) is mounted in a rotation stage directly in front of the entrance slit assembly.

The spectrograph is a 275 mm focal length, $f/4$ aperture model (Spectra Pro 275, Acton Research) equipped with a set of three gratings mounted on a turret. In this work, the gratings with 1800 and 3600 grooves mm^{-1} were used. The spectrograph is mounted with the entrance slit oriented parallel to the laser beam's propagation direction. The exit plane is imaged onto the entrance slit of a streak camera (Hamamatsu Model C2830) so that the slit is parallel to the wavelength axis. On the output side of the camera, a gated microchannel plate image intensifier is mounted. Streak ranges between 0.5 and 10 ns are selectable. The spectrally and temporally dispersed signal is eventually detected with a cooled, slow-scan CCD camera which provides a 12 bit dynamic range. The raw data consists of a two-dimensional image, where the dimensions correspond to the wavelength and timescales. The spectral system response of the detection system was calibrated using deuterium and halogen lamps.

For some experiments, the spectrometer was substituted with a set of colored glass filters (UG11 and WG305) which form a bandpass from 295 to 395 nm and the streak camera was rotated by 90° so that its entrance slit was oriented parallel to the laser beam's propagation direction. This geometry allows line-wise concentration measurements along a 5 mm line.

Two different burner types are used to generate OH radicals: A Hencken burner (square matrix, 25 mm × 25 mm) was used for H_2/air flames. It provides a relatively uniform flow field, has little heat transfer to the burner body and is thus easy to characterize. A commercial welding

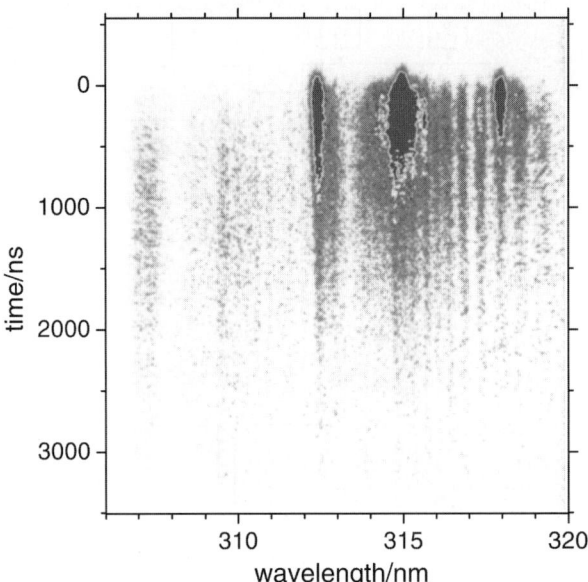

Fig. 2 Time- and wavelength-resolved raw image obtained in a H$_2$/O$_2$ flame after excitation of the A–X (1–0) R$_1$(7) transition (image already corrected for background and detection efficiency).

torch was used for hydrogen/oxygen flames. This offers the advantage of much higher temperature and a correspondingly larger OH number density, at the expense of less well-defined conditions.

Using an input pulse of nominally 80 ps duration, the temporal resolution of the detection system was tested by observing Rayleigh scattered light. Averaging over 1200 pulses results in a signal of about 90 ps FWHM, indicating a jitter of $\Delta t \leqslant 30$ ps. For the experiments performed with the 3600 g mm^{-1} grating, the temporal signal is broadened to about 140 ps due to different path lengths in the spectrometer; this effect is described in detail in ref. 22. Overall, the temporal resolution of the complete system is very well matched to the typical time between collisions in atmospheric pressure flames ($\tau_c \approx 100$ ps). The laser bandwidth was checked with excitation scans of different lines. Typical line widths are about 1 cm^{-1} in the UV, or 0.33 cm^{-1} with respect to the Ti : sapphire fundamental frequency scale. Doppler broadening (about 0.3 cm^{-1}) contributes to this linewidth. The present laser system is thus well suited to the resolution of individual lines in the OH spectrum.

For the measurements described in this paper, only a part of the laser energy (0.3 to 1 mJ) was used in order to avoid saturation. The entrance slit of the spectrograph is set at 150 μm width, the streak camera entrance slit is opened up to 200 μm width, and the streak range is set to 5 ns. With these settings, typical acquisition times are 5–15 min (3000–9000 laser pulses). For calibration, the Rayleigh signal scattered from room air is recorded to provide the "$t = 0$" point, and a measure of temporal resolution. Additionally, the dark signal of the camera is acquired at regular intervals.

Fig. 2 shows a typical time- and wavelength-resolved raw image obtained in a H$_2$/O$_2$ flame after excitation of the (1–0) R$_1$(7) line. The background is already subtracted, and the image has been corrected for the varying detection efficiency of the camera. Intensities are given in a linear grey scale. Fluorescence from the directly populated levels can be clearly discerned by the high intensity at the time of the excitation laser pulse. As time progresses, additional lines from levels populated by energy transfer (both RET and VET) appear.

3 Results and discussion

3.1 Rotationally resolved LIF spectra

For a more detailed study of energy transfer processes, Fig. 3 shows wavelength-resolved spectra extracted from the raw data in Fig. 2 for two different time intervals (top: $-120 < t/\text{ps} < 80$,

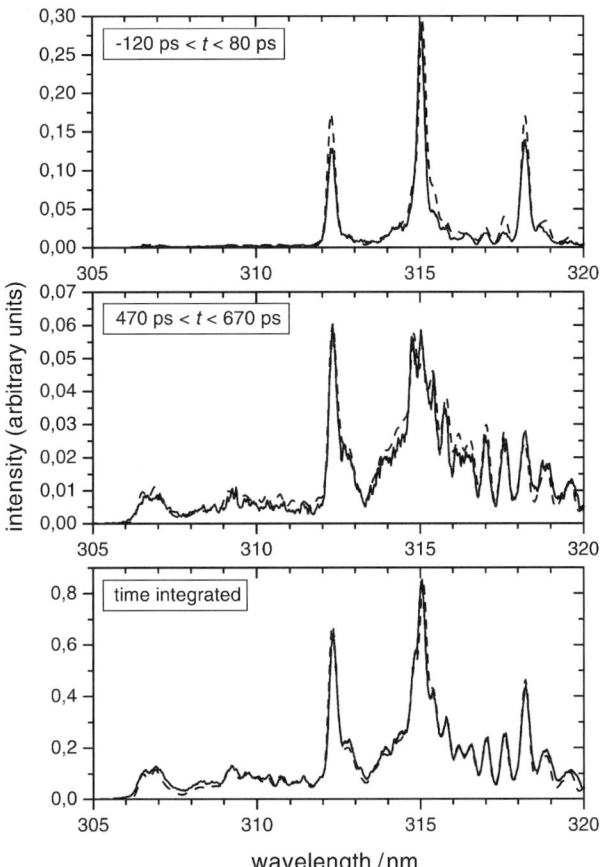

Fig. 3 Fluorescence spectra for different time intervals after excitation of the (1–0) $R_1(7)$ line (solid) and comparison with LASKIN simulations (broken lines). The experimental profiles are extracted from the raw data presented in Fig. 2.

middle: $470 < t/\text{ps} < 670$) and for the fully time-integrated case (bottom). Excitation was in the (1–0) $R_1(7)$ line. For the first time interval, the spectrum is clearly dominated by the major P-, Q- and R-lines originating from the directly populated level. The weak lines visible between 316 and 318 nm are due to the small amount of energy transfer present even this close to the exciting laser pulse. Roughly 600 ps later, the fluorescence spectrum has changed dramatically. Now, fluorescence stemming from several levels populated by energy transfer can be observed. The major features are now the band heads of the R_1 and Q_1 branches at 312 and 315 nm, respectively. Note especially that a significant amount of fluorescence is visible at wavelength shorter than 312 nm. This is due to vibrational energy transfer into the $v' = 0$ level. Analysis of the temporal evolution of the LIF signals (cf. Fig. 2) shows that the fluorescence from levels populated by VET reaches its maximum at times around 1 ns, which is significantly later than that from levels populated by RET. Since the population in the $v' = 0$ state decays more slowly than that of the initially populated level, the contribution of lines originating from $v' = 0$ to the time-integrated fluorescence is relatively high ($\sim 15\%$, cf. Fig. 3, bottom). Implications of this effect for broadband measurements will be discussed in the next section.

For a simulation of energy transfer processes, the composition of the surrounding gas has to be known. Taking into account the mixing with ambient air always present when using the small-diameter welding torch, the premixed fuel used in the experiments above consists of approximately 40% H_2, 40% N_2 and 20% O_2. Using the code of Gordon et al.,[32] the following species concentrations have been calculated assuming adiabatic conditions: 47.5% N_2, 40.8% H_2O, 5% H_2,

2.8% OH, 1.5% O_2, 1.3% H and 0.5% O. The adiabatic flame temperature is $T_{adiab}= 2708$ K. These values are used as input parameters for a simulation using the LASKIN package. Results of the calculation are plotted as broken lines in Fig. 3; intensities are normalized to the strongest line in each spectrum. For the fully time-integrated case, comparison of experimental and modeled spectra shows a very good agreement both for lines populated by RET and by VET. Some discrepancies still exist for fluorescence originating from the directly populated level. This can be seen most easily in the upper frame of Fig. 3, where the intensity of the P- and R-lines is over-predicted by about 30%. This is mainly due to polarization effects that occur since the laser is linearly polarized and since the spectrometer detection efficiency is also polarization dependent. For the collision-free case, these effects have already been discussed in detail by Doherty and Crosley.[21] In contrast, only a few experiments have been performed to date to study the temporal evolution of the anisotropy.[19,20,22] Work to include polarization effects in the simulation code is currently under way,[23] however, it can be seen that even in the current state of development, LASKIN is a reliable tool for predicting the shape and temporal evolution of spectra affected by energy transfer.

3.2 Influence of energy transfer on broadband LIF measurements

In most cases, broadband LIF measurements are used for minor species concentration determination. This means that, in contrast to the measurements described above, fluorescence is integrated over a wider spectral range (commonly by use of colored-glass or dielectric filters) before it is passed to the detector. The major advantage of this approach is a significantly better signal-to-noise ratio (SNR) that even allows single-pulse and/or spatially resolved measurements if a suitable detector is used. Recently, short-pulse lasers and detection systems that allow quench-free LIF measurements to be performed have become available. The basic idea of these techniques is always that, at times very close to the exciting laser pulse (that is, at $t = 0$), the signal is not affected by collisions. Several signal detection techniques have been suggested. The easiest is to use a detector with a gate interval of the order of several hundred picoseconds.[15] However, our results show that even for very short gate times, signals are affected by energy transfer and quenching (*cf.* Fig. 3, top). Thus, this technique is quenching-insensitive to some extent, but not truly quench-free. More sophisticated approaches analyze the temporal decay of the LIF signal, either by evaluating several gated intervals[16] or by observing the full temporal evolution of signals by means of a streak camera.[17] Extrapolation of the observed signal intensity backwards to $t = 0$ (that is, to the center of the exciting laser pulse) then allows one to determine completely quench-free results. This evaluation procedure is described in detail in ref. 17.

Our time- and wavelength-resolved measurements along with the numerical simulations allow us to study in detail how collision-induced processes, in particular VET, might influence even these quench-free measurements. For this, we used LASKIN to model a time-integrated fluorescence spectrum after excitation of the OH A–X (1–0) $Q_1(7)$ line in a H_2/air flame at $\Phi = 1.0$ and $T_{adiab} \approx 2400$ K. The results are shown in Fig. 4 (solid line). It can be seen, that besides fluorescence from the directly populated level, a multitude of lines populated by RET in the $v' = 1$ state is present. Also, a significant contribution of fluorescence from the $v' = 0$ state which is populated by VET is observed. Since the (0–0) and the (1–1) band partially overlap and transmission curves of filters are usually not steep enough, discrimination of these bands is generally not possible. In a typical experiment for minor species concentration, only the sum of these signals will be detected. To illustrate this, the transmission curve of a typical filter combination (UG11/WG305) used for broadband LIF measurements is plotted along with the spectrum in Fig. 4 (dashed line).

Fig. 5 shows the time-integrated, state-resolved population distribution for the spectrum plotted in Fig. 4. The major part of the population (62%) is found in levels populated by RET; only about 13% remains in the directly excited level $J' = 7$. Quite a notable fraction, *i.e.* 19% of the population, is observed in the $v' = 0$ state which is populated by VET. Again, this emphasizes the significant contribution of VET to the shape and intensity of LIF spectra. The results of this simulation compare favorably with the observed spectra shown in Fig. 3 (where the directly adjacent $J' = 8$ level has been populated). Since the Einstein A coefficient for spontaneous emission is about 50% higher in the $v' = 0$ level than in the initially populated $v' = 1$ level,[2] the contribution of VET to the total fluorescence is even higher (in this case, more than 25%).

Whereas the contributions of the different vibrational states are readily detected in our

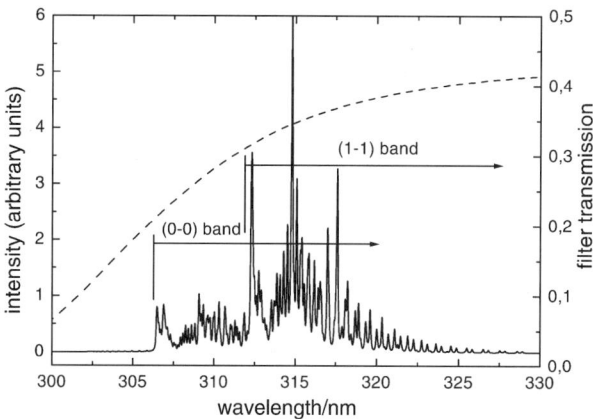

Fig. 4 LASKIN simulation of the fluorescence spectrum after excitation of the OH A–X (1–0) $Q_1(7)$ line in a H_2/air flame, $\Phi = 1.0$, $T_{adiab} \approx 2400$ K (solid line). The dashed line shows the transmission curve of a typical filter combination (UG11/WG305) used for broadband LIF measurements.

wavelength-resolved measurements, only the integrated fluorescence from these states is visible when broadband detection is employed. When performing time-resolved measurements, this has significant consequences: Since levels in $v' = 0$ are populated by a relatively slow process (VET) and their relative contribution to the total fluorescence is higher than that originating from $v' = 1$ (due to the difference in the A coefficient), the total fluorescence will no longer be a single exponential, but a more complex function of time.

To illustrate this deviation from the single-exponential decay, the following simplified model can be used: Consider a two-level system with equal rate coefficients for spontaneous emission and quenching (combined in k_q) in both levels, and allow for energy transfer $0 \leftarrow 1$ between these levels with a rate k_{VET}. Excitation is in the $v' = 1$ level. In this case, populations can be described by the following differential equations:

$$\frac{dN_1}{dt} = -(k_q + k_{VET})N_1$$

$$\frac{dN_0}{dt} = k_{VET} N_1 - k_q N_0$$

Solutions are:

$$N_1 = N_1^0 e^{-(k_q + k_{VET})t}$$
$$N_0 = N_1^0 e^{-k_q t} - N_1^0 e^{-(k_q + k_{VET})t}$$

The LIF intensity can be calculated according to:

$$I_{LIF}(t) = \eta_0 N_0(t) + \eta_1 N_1(t)$$

Here, the weighting factors η_i of the two fluorescence contributions are proportional to the Einstein A coefficients and/or the detection efficiency. By substituting the expressions for N_i, it can be seen that the solution is a bi-exponential decay. In Fig. 6, time-resolved plots of the LIF intensity are shown for three different cases. For $\eta_0 = \eta_1$, the expected single-exponential function (solid line) is observed. For $\eta_0 = 1.5\eta_1$, some deviations occur: the intensity is higher for all times, and the profile is nearly linear at short times (up to $t = 0.5$ ns). As mentioned above, this case closely resembles the excitation of OH in the A–X (1–0) band with broadband detection of fluorescence in the 0–0 and 1–1 band. The deviations from a single-exponential become even more pronounced for $\eta_0 = 3\eta_1$. Here, the maximum fluorescence occurs at $t = 0.3$ ns (and not at $t = 0$ as in the other cases).

It should be pointed out that the methods for quench-free measurements described above will no longer be reliable for spectra significantly affected by VET, since the extrapolation of the

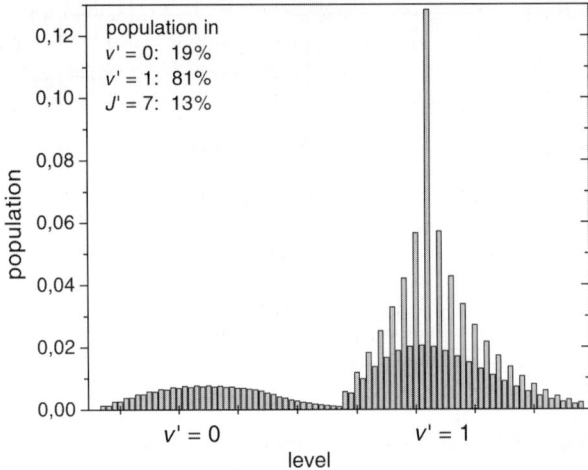

Fig. 5 Time-integrated, state-resolved population distribution (based on the simulation shown in Fig. 4). Rotational quantum number increases from left to right ($J' = 0$ to $J' = 20$ is shown); for each J', population in both fine-structure components is shown independently.

observed signal intensity backwards to $t = 0$ becomes more complicated. This is most severe for the detection technique using several gated intervals,[16,18] since, here, a single-exponential decay curve has been explicitly assumed in the data evaluation routine. If the full temporal evolution of signals is monitored by means of a streak camera,[17] deviations of the LIF intensity from the expected single-exponential curve can be detected and, in principle, be taken into account. However, reliability of the fit, and thus that of the concentration measurement will suffer if the experimental data have to be fitted to a curve with several additional parameters.

In addition to the mechanism described above, there may be further reasons for a deviation of the observed temporally-resolved LIF intensity from the expected single exponential. Generally, energy transfer to other levels will lead to a non- or multi-exponential decay if quenching rates or Einstein A coefficients in the populated levels are different, or if the filter efficiency is non-uniform. Additionally, polarization effects[22,23] or saturation in the excitation step might lead to deviations.

For OH detection there is, however, an easy way to compensate for the effects of the higher Einstein A coefficient in the $v' = 0$ level. If a filter with a lower transmission at lower wavelengths is used, fluorescence from $v' = 0$ will be attenuated and the effect of the different Einstein coeffi-

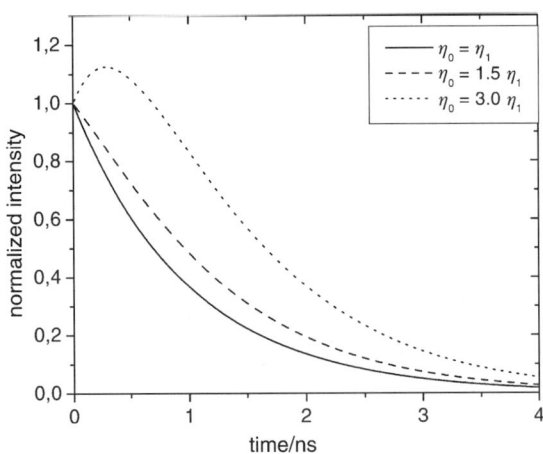

Fig. 6 Effect of different Einstein coefficients and/or detection efficiency of levels populated by energy transfer on wavelength-integrated, time-resolved measurements.

cients is (at least partially) compensated. In our work, we found that the combination of two Schott colored glass filters of types WG305 and UG11 is well suited for this purpose (compare the transmission curve in Fig. 4). In consequence, no deviation from single-exponential behavior was observed during our quantitative minor species concentration measurements.[17]

3.3 Detailed discussion of spectra populated by VET

For a more detailed analysis of the role of vibrational energy transfer for the temporal evolution of OH LIF spectra, we decided to probe the $A\,^2\Sigma\ v' = 2$ level. This has the advantage, that the transfer of more than one vibrational quantum can be observed. Fig. 7 shows part of the rotationally resolved, time-integrated fluorescence spectrum after excitation of A–X (2–0) $Q_1(8)$. Here, the strongest fluorescence line is the (2–2) $R_1(7)$ transition at 318.6 nm. Other lines originating from the directly populated level would appear at longer wavelengths and are thus not visible in Fig. 7. Next, a multitude of lines from levels populated by VET are seen: lines belonging to the (1–1) system appear at wavelengths $\lambda \geqslant 312.5$ nm, whereas lines from the (0–0) system appear at wavelengths $\lambda \geqslant 306.5$ nm. Note, that the spectra partially overlap. Other distinct features are the R_1 and R_2 band heads for the individual vibrational levels; their position is marked in Fig. 7.

In order to study the effects of the collisional environment on the rotational structure of spectra where several vibrational levels are populated by VET, a systematic series of measurements has been performed for different flame conditions and excitation lines. Fig. 8 shows a detailed view of the (1–1) part of the time-integrated fluorescence spectra at two flame stoichiometries ($\Phi = 0.5$ and $\Phi = 1$) in a H_2/air flame. Excitation was in the transition A–X (2–0) $P_1(5)$; to facilitate comparison, intensities are normalized to the R_1 band head. It is clearly seen, that the rotational structure in both spectra is very similar. Even for individual rotational lines, deviations are usually well below 10%. Equally, similar spectra are found for other flame stoichiometries, indicating that the influence of collision partners and temperature on the rotational structure is small.

Next, we compared spectra obtained with three different excitation lines, namely the $P_1(5)$, $Q_1(8)$ and $R_1(11)$ line. Even though slightly larger differences occur than in Fig. 8 (up to 20% for individual rotational lines), the overall structure changes only insignificantly. Especially, no preference for the directly populated rotational level was observed, indicating that the rotational population distribution is scrambled completely during a VET step. Comparison of spectra originating from the $v' = 0$ level (that is, after transfer of two vibrational quanta) show an even greater similarity. Considering the signal-to-noise ratio in our measurements, the spectra are virtually indistinguishable.

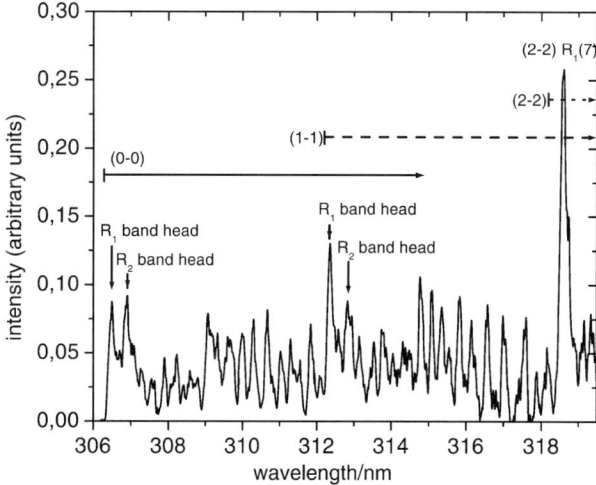

Fig. 7 Rotationally resolved, time-integrated fluorescence spectrum of the (0–0) and (1–1) band populated by VET after excitation of A–X (2–0) $Q_1(8)$.

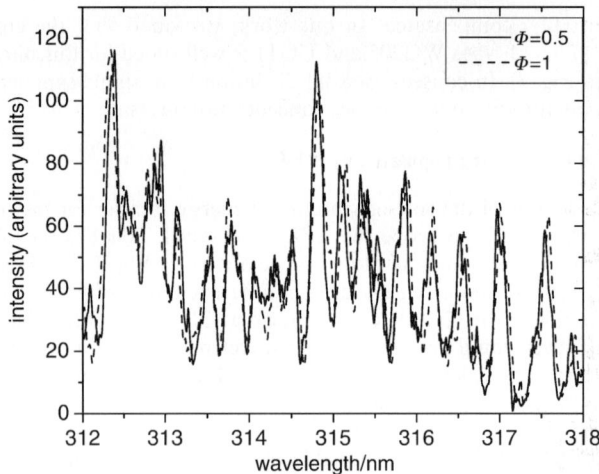

Fig. 8 Detailed view of the (1–1) part of the time-integrated fluorescence spectra at two flame stoichiometries in H_2/air flames. Excitation in the transition A–X (2–0) $P_1(5)$. The spectra are normalized to the R_1 band head.

For further analysis, we compared measured spectra with simulated ones using the LIFBASE code.[33] This is shown in Fig. 9 for an experimental spectrum after excitation of (2–0) $Q_1(8)$ (solid line) and a simulation for an apparent temperature of $T_{a0} = 7000$ K (broken line); covering the spectral interval $306 \leqslant \lambda/\text{nm} \leqslant 312$, which corresponds to the (0–0) fluorescence band. The overall agreement between both spectra is good. Note, that T_{a0} is not the true flame temperature (that would be $T_{\text{adiab}} = 2400$ K), but it reflects the rotational distribution in the "hot" band populated by VET. A similar analysis has been performed for the (1–1) band of the fluorescence spectrum. Here, because of the overlap between the vibrational bands originating from $v' = 1$ and $v' = 0$ (cf. Fig. 7), the spectrum fitted for the (0–0) portion has to be subtracted first. Comparison with spectra generated by LIFBASE showed that the (1–1) band can be described similarly well by the numerical code. However, the apparent temperature of the rotational distribution in $v' = 1$ is $T_{a1} = 4000$ K in this case. This value compares favorably with results by Crosley et al., who obtained $2250 \leqslant T_a/\text{K} \leqslant 3280$ in the (0–0) band populated by VET after excitation of A–X (1–0) in a lean methane/air flame.[34]

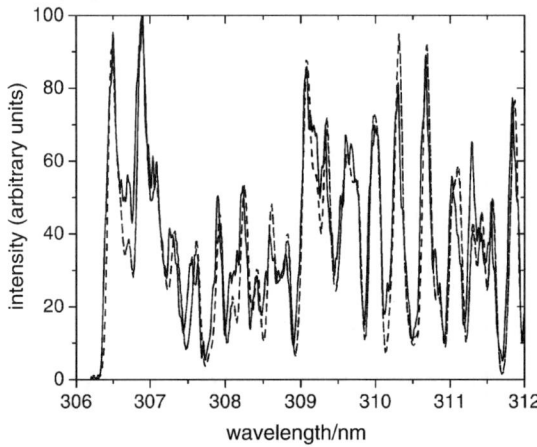

Fig. 9 Comparison between an experimental spectrum populated by VET after excitation of (2–0) $Q_1(8)$ (solid line) and a spectrum simulated by LIFBASE (broken line, $T_{a0} = 7000$ K). The spectra are normalized to the R_1 band head.

Evaluation of these fits allows the determination of the time-averaged population in the vibrational states $v' = 2$, 1 and 0, respectively. For a stoichiometric H_2/air flame, the ratio of the populations is 66 : 18 : 16; for $\Phi = 0.5$ this changes to 51 : 26 : 23. The reason for that is the higher abundance of H_2O as an efficient quencher in the former flame. The short fluorescence lifetime in this environment allows only a limited amount of energy transfer. In contrast, N_2 is the dominant collision partner in the lean flame; it is much more efficient for VET.[35–37] Both results are consistent with previous findings.[31] Note, that the ratio of the relative time-integrated population in both levels populated by VET remains constant.

Finally, we have analyzed the temporal evolution of spectra populated by VET. Whereas a small change in the spectrum with time is observed for $v' = 1$, the shape of the rotational structure in $v' = 0$ is nearly constant over time. This suggests, that a significant part of the $v' = 0$ level is populated by two consecutive VET steps rather than by one step alone with $\Delta v = -2$. In conjunction with previous results[31,38,39] it seems likely that, for OH, both single- and multiple-quantum vibrational energy transfer contributes to the population in $v' = 0$.

4 Summary

Two-dimensional, time- and wavelength-resolved LIF spectra of OH in atmospheric-pressure flames have been obtained using a short-pulse laser in conjunction with streak detection. This approach allows the simultaneous observation of time-resolved fluorescence from the directly populated level as well as from several levels populated by RET and VET and thus facilitates the investigation of energy transfer processes. Rotationally resolved spectra after excitation in the A–X(1–0) band are in good agreement with the results obtained by simulations using the rate equation model LASKIN. Small deviations, especially for fluorescence originating from the directly populated level, can be attributed to polarization effects, which are not yet included in the model. However, it can be seen that even in the current state of development, LASKIN is a reliable tool for prediction of shape and temporal evolution of spectra affected by energy transfer.

Measurements and simulations show, that the $v' = 0$ level is significantly populated by VET and that fluorescence from this level may contribute up to 30% of the total fluorescence detected by broad-band techniques. Since Einstein A coefficients are higher in this level, this may seriously affect picosecond-LIF techniques for the time-resolved, quench-free detection of OH. It is shown, that VET may lead to non- or multi-exponential decay in this case. However, these effects can be taken into account and compensated for, if suitable filters are used for the fluorescence detection.

Additionally, we report the first rotationally-resolved energy transfer measurements after excitation of OH (A, $v' = 2$) states in well-defined, atmospheric-pressure flames. Detailed analysis of LIF from levels populated by VET shows that for the H_2/air and H_2/O_2 flames investigated in this contribution, the overall shape of the fluorescence spectra is insensitive to the collision environment, the temperature and the excitation line. For our flames, it can be described by a LIFBASE simulation with a high apparent temperature (4000 and 7000 K for the $v' = 1$ and $v' = 0$ levels, respectively). Differences in the temporal evolution of fluorescence spectra originating from $v' = 1$ and $v' = 0$ suggests, that for OH, both single- and multiple-quantum vibrational energy transfer contribute to the population in $v' = 0$.

Acknowledgement

Contributions of U. Rahmann, U. Lenhard and A. Bülter to the work presented in this paper are gratefully acknowledged. Financial support was provided by the DFG under contract 1363/9-1.

References

1. G. H. Dieke and H. M. Crosswhite, *J. Quant. Spectrosc. Radiat. Transfer*, 1962, **2**, 97.
2. J. Coxon, *Can. J. Phys.*, 1980, **58**, 933.
3. J. Luque and D. R. Crosley, *J. Chem. Phys.*, 1998, **109**, 439.
4. *Proc. Combust. Inst.*, 1998, **27**.
5. *Proc. Combust. Inst.*, 2000, **28**, in press.
6. A. Schiffman and D. W. Chandler, *Int. Rev. Phys. Chem.*, 1995, **14**, 371.
7. P. Andresen, A. Bath, W. Gröger, H. W. Lülf, G. Meijer and J. J. ter Meulen, *Appl. Opt.*, 1988, **27**, 365.

8 K. Kohse-Höinghaus, *Prog. Energy Combust. Sci.*, 1994, **20**, 203.
9 A. C. Eckbreth, *Laser Diagnostics for Combustion Temperature and Species*, Gordon and Breach, Amsterdam, 1996.
10 E. W. Rothe, Y. Gu, A. Chryssostomou, P. Andresen and F. Bormann, *Appl. Phys. B*, 1998, **66**, 251.
11 J. W. Daily and E. W. Rothe, *Appl. Phys. B*, 1999, **68**, 131.
12 A. A. V. Kliner and R. L. Farrow, *J. Chem. Phys.*, 1999, **110**, 412.
13 A. Dreizler, R. Tadday, P. Monkhouse and J. Wolfrum, *Appl. Phys. B*, 1993, **57**, 85.
14 M. Köllner and P. Monkhouse, *Appl. Phys. B*, 1995, **61**, 499.
15 F. Bormann, T. Nielsen, M. Burrows and P. Andresen, *Appl. Phys. B*, 1996, **62**, 601.
16 M. W. Renfro, S. D. Pack, G. B. King and N. M. Laurendeau, *Appl. Phys. B*, 1999, **69**, 137.
17 A. Brockhinke, A. Bülter, J.-C. Rolon and K. Kohse-Höinghaus, *Appl. Phys. B*, 2001, **72**, 491.
18 M. W. Renfro, W. A. Guttenfelder, G. B. King and N. M. Laurendeau, *Combust. Flame*, 2000, **123**, 389.
19 T. Nielsen, F. Bormann, M. Burrows and P. Andresen, *Appl. Opt.*, 1997, **36**, 7960.
20 P. Beaud, P. P. Radi, D. Franzke, H.-M. Frey, B. Mischler, A.-P. Tzannis and T. Gerber, *Appl. Opt.*, 1998, **37**, 3354.
21 P. M. Doherty and D. R. Crosley, *Appl. Opt.*, 1984, **23**, 713.
22 A. Brockhinke, W. Kreutner, U. Rahmann, K. Kohse-Höinghaus, T. B. Settersten and M. A. Linne, *Appl. Phys. B*, 1999, **69**, 477.
23 A. Brockhinke, A. Bülter, U. Lenhard, U. Rahmann and K. Kohse-Höinghaus, *Appl. Phys. B*, submitted.
24 A. Jörg, U. Meier and K. Kohse-Höinghaus, *J. Chem. Phys.*, 1990, **93**, 6453.
25 A. Jörg, U. Meier, R. Kienle and K. Kohse-Höinghaus, *Appl. Phys. B*, 1992, **55**, 305.
26 R. Kienle, A. Jörg and K. Kohse-Höinghaus, *Appl. Phys. B*, 1993, **56**, 249.
27 M. P. Lee, R. Kienle and K. Kohse-Höinghaus, *Appl. Phys. B*, 1994, **58**, 447.
28 R. Kienle, M. P. Lee and K. Kohse-Höinghaus, *Appl. Phys. B*, 1996, **62**, 583.
29 R. Kienle, M. P. Lee and K. Kohse-Höinghaus, *Appl. Phys. B*, 1996, **63**, 403.
30 A. T. Hartlieb, D. Markus, W. Kreutner and K. Kohse-Höinghaus, *Appl. Phys. B*, 1997, **65**, 81.
31 U. Rahmann, W. Kreutner and K. Kohse-Höinghaus, *Appl. Phys. B*, 1999, **69**, 61.
32 S. Gordon and B. J. McBride, *NASA SP-273*, 1976.
33 J. Luque and D. R. Crosley, *LIFBASE*: Database and spectral simulation program (version 1.15), SRI International Report MP 99-009, 1999.
34 D. R. Crosley and G. P. Smith, *Appl. Opt.*, 1983, **22**, 1428.
35 R. A. Copeland, M. L. Wise and D. R. Crosley, *J. Phys. Chem.*, 1988, **92**, 5710.
36 P. H. Paul, *J. Quant. Spectrosc. Radiat. Transfer*, 1994, **51**, 511.
37 P. H. Paul, *J. Phys. Chem.*, 1995, **99**, 8472.
38 K. L. Steffens, J. B. Jeffries and D. R. Crosley, *Opt. Lett.*, 1993, **18**, 1355.
39 K. L. Steffens, PhD Thesis, University of Stanford, 1994.

Temperature fields during the development of autoignition in a rapid compression machine

John F. Griffiths,[a] John P. MacNamara,[†a] Caroline Mohamed,[‡a] Benjamin J. Whitaker,[a] Jinfeng Pan[§b] and Christopher G. W. Sheppard[b]

[a] *School of Chemistry, The University, Leeds, UK LS2 9JT.*
  E-mail: johng@chem.leeds.ac.uk
[b] *School of Mechanical Engineering, The University, Leeds, UK LS2 9JT*

Received 2nd March 2001
First published as an Advance Article on the web 26th September 2001

Temperature and concentration fields have been investigated in the cylindrical combustion chamber of a rapid compression machine (RCM) by schlieren photography, chemiluminescent imaging and planar laser induced fluorescence of acetone and of formaldehyde in a 2-dimensional sheet across the diameter. The timescale of particular interest was up to 10 ms after the piston has stopped. Experiments were performed in non-reactive and reactive conditions. Acetone was seeded in non-reactive mixtures. Combustion was studied first in a system containing di-*tert*-butyl peroxide vapour in the presence of oxygen. The decomposition of di-*tert*-butyl peroxide generates methyl radicals, which are then oxidised if oxygen is present. The overall reaction is exothermic and is characteristic of a conventional thermal ignition. In addition, chemiluminescence, resulting from CH_2O^*, accompanies the oxidation process. The combustion of n-pentane was then investigated at compressed gas temperatures that spanned the range in which there is a negative temperature dependence of the overall reaction rate, typically 750–850 K. The response to thermal feedback in this more complex thermokinetic system can be the opposite of the "thermal runaway" that accompanies di-*tert*-butyl peroxide combustion. The purpose of making comparisons between these two types of systems was to show how the temperature field generated in the RCM is modified in different ways by the interaction with the chemistry and to discuss the implications of this for the spatial development of spontaneous ignition. As the piston of the RCM moves it shears gas off the walls of the chamber. This probably creates a roll-up vortex, but more importantly it also collects gas from the walls and moves it across the cylinder head pushing it forward into a plug at the centre. Thus, soon after the end of compression there is an adiabatically heated gas which extends virtually to the wall, but this incorporates a plug of colder gas at its core. Diffusive transport will occur, but the timescale is relatively slow, and the effect hardly shows until at least 10 ms post-compression. The consequence of "thermal runaway" on a timescale that is compatible with the development of this temperature field is that the reaction rate in the adiabatically compressed toroidal region accelerates faster than in the core, and goes to completion first. A somewhat similar pattern emerges during

---

† Present address: Environmental Protection Agency, Pottery Road, Dun Laoghaire, Ireland.
‡ Present address: Combustion Systems—Engineering Research, Rolls-Royce Ltd, Derby, UK.
§ Present address: Department of Engineering, University of Sussex, UK.

DOI: 10.1039/b102002l

n-pentane combustion when the initial condition is set at the lower end of the negative temperature dependent range. By contrast, at adiabatically compressed gas temperatures close to the upper end of the negative temperature dependent region, the reaction rate in the cooler core develops faster than that in the surrounding zone, and the temperature difference is rapidly smoothed out. This does not lead to spatial homogeneity in all respects, however, because different rates and extents of reaction generate different concentrations of intermediates. This stratification has implications for the eventual spatial evolution of spontaneous ignition.

Introduction

This paper is concerned with investigations of the spatial structure associated with the development of spontaneous ignition, or autoignition, in a combustion chamber following the compression of gaseous reactants to high temperatures and pressures. Imaging techniques have been used (i) to reveal temperature inhomogeneities in the combustion chamber of a RCM during the post-compression period as a result of the motion generated as the piston moves forward and (ii) to show how an interaction of chemistry with these inhomogeneities can affect the spatial development of spontaneous ignition.

Hitherto, there has been a general belief that there is an adiabatically heated "core gas" following rapid compression, and that the spontaneous ignition of reactive mixtures will be determined by the conditions in the "core gas", both in terms of the ignition delay and the seat of its development.[1,2] Whether or not this perception is appropriate is fundamentally important to the application of results from RCM studies of ignition for validation of the predictions of kinetic models that exploit zero-dimensional numerical codes. Codes of this kind cannot allow for interactions between the developing chemistry and spatial temperature variations induced by gas motion following the compression stroke.

There is reason to be concerned because experimental and theoretical studies have demonstrated that a roll-up vortex, created at the interface of the piston crown and the cylinder wall, can create a flow field that is initiated in the relatively cold boundary layer and may penetrate the "core gas".[3,4] This possibility has been supported by computational fluid dynamic simulations of spontaneously reacting mixtures subjected to the gas motion induced by compression ahead of a piston in a cylinder.[5-8] Moreover, these latter studies indicated that the chain—thermal interactions associated with alkane oxidation could cause the onset of spontaneous ignition to be induced in the initially colder regions rather than the adiabatically heated gases. Subsequently, we obtained experimental evidence from schlieren imaging of the spatial temperature evolution in the combustion chamber of the RCM that there are complicated temperature profiles in reactive mixtures. However, the understanding of the results from this work has become clearer only through more recent studies in which other imaging techniques have been used. The combination of these experimental studies is reported here.

Goals of the present work

In order to illustrate the interaction of chemistry with the temperature field that is generated in the RCM and how it may be affected, we contrast the behaviour of a chemical system which exhibits an overall positive temperature dependence of the reaction rate throughout a wide temperature range, di-*tert*-butyl peroxide (2-*tert*-butylperoxy-2-methyl propane) decomposition in the presence of oxygen, with a system that exhibits an overall negative temperature dependence (or negative temperature coefficient (ntc)) throughout much of the temperature range accessible in the RCM, n-pentane + oxygen. The existence of the ntc in alkane combustion is encapsulated in the dependence on temperature of the ignition delay (or time to ignition) in closed systems,[9] as exemplified in Fig. 1. The reference temperature for RCM experiments is usually taken to be the temperature reached at the end of compression, which signifies an "initial condition". The timescale of principal interest is within 10 ms after the piston has stopped moving, during which the spatial temperature distribution is modified. For other closed vessel or flow reactor experiments the refer-

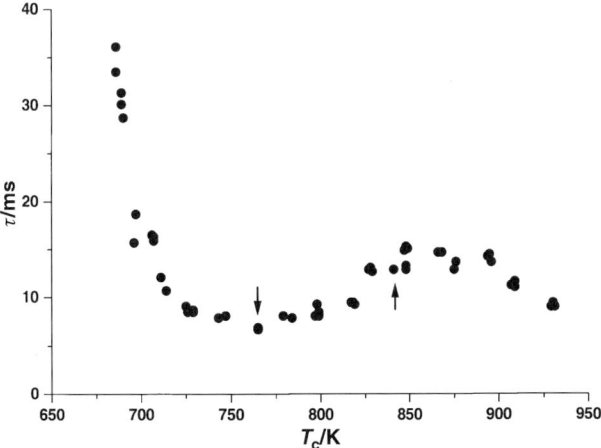

Fig. 1 Ignition delay time (τ) as a function of compressed gas pressure (T_c) for a stoichiometric mixture containing n-pentane.[10] The conditions that have been investigated in the present work are marked by arrows.

ence temperature would normally be the wall temperature, which represents a "boundary condition".

A number of experimental techniques were employed in the present work to examine the spatial development in temperature and chemical composition in a cross-sectional view of the cylindrical combustion chamber, as follows: (i) schlieren image photography, (ii) planar laser induced fluorescence (PLIF) of acetone (2-propanone), (iii) PLIF of formaldehyde and (iv) chemiluminescent emissions from CH_2O^* ($^1A''$). Both schlieren photography and PLIF imaging of acetone, included as a tracer in non-reactive gases, yield information about spatial variations in temperature within the cross-sectional view. However, acetone can be used only in inert gas mixtures to identify features of the temperature field because, once radical concentrations begin to increase in an active environment (and especially those of the most reactive species, such as OH), the acetone itself is susceptible to radical attack and the fluorescence intensity then becomes dependent on variations in both concentration and temperature. In addition, acetone is a product of di-*tert*-butyl peroxide decomposition in the presence of oxidation, so chemical information is derived from acetone fluorescence in this case, although the intensity cannot readily be calibrated to give a quantitative measurement because there are also accompanying temperature changes. The PLIF from formaldehyde, as a molecular product of both systems under investigation, can also yield information on the progress of reaction and spatial variations of it. In addition, from both reactions there is chemiluminescent emission from excited formaldehyde, CH_2O^* ($^1A''$), which originates in radical–radical interactions involving CH_3O.[11] The change of emission intensity is a monitor of a shift in the CH_3O concentration and is also representative of the spatial and temporal variations in free radical activity. Whereas the PLIF signals were obtained at a well-defined plane cutting across the chamber determined by the location and thickness of the laser beam, the chemiluminescent and schlieren images represented an integration of the signals obtained over a certain depth of focus of the ICCD camera lens within the combustion chamber, which was about 2 mm.

Chemical background

The essence of the decomposition of di-*tert*-butyl peroxide in the presence of oxygen is that *tert*-butoxyl radicals, formed in the initiation, readily decompose to methyl radicals and acetone and it is primarily the methyl radicals that are susceptible to oxidation. When oxygen is present in equimolar proportions with the peroxide, the reaction is represented by the simplified overall stoichiometry:[12]

$$(CH_3)_3COOC(CH_3)_3 + O_2 = 2(CH_3)_2CO + CH_2O + CH_3OH, \quad \Delta H^\circ_{298} = -394.5 \text{ kJ mol}^{-1}.$$

The peroxide itself is not susceptible to significant free radical attack, so there is no feedback mechanism that leads to chain branching at this initial composition. The reaction behaves as a "classical" thermal ignition,[12] although the combustion becomes rather more complicated when higher proportions of oxygen are present with di-*tert*-butyl peroxide.[8,12]

As is generally familiar, the low temperature oxidation of alkanes in the temperature regime 600–850 K is driven by exothermic chain branching processes. A generalised mechanism for alkanes having three or more carbon atoms may be outlined in simplified form as below. RH represents an alkane and R the alkyl radical. Propagating free radicals, such as OH and HO_2, are represented as X, and AO, BO and intermediate molecular products. QO_2H is a generalised representation of a hydroperoxyalkyl radical, from which diperoxy species (O_2QO_2H) and ketohydroperoxides ($OQ'O_2H$) may be formed.[13] Ground state formaldehyde may be formed as an intermediate product of QO_2H decomposition or by further oxidation following alkyl radical decomposition, as well as from methoxyl radicals. Thus its origins are rather more diverse than those which lead to the electronically excited CH_2O^* and chemiluminescence.

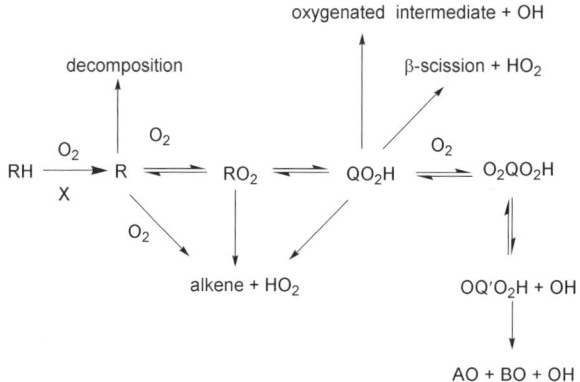

A very important consequence of the R/RO_2 equilibrium is that as the temperature increases through the range 750–850 K the overall reaction rate exhibits a ntc because there is a switch away from the vigorous branching, and OH propagated mechanism, to the essentially non-branching HO_2 propagated route. Moreover, in general the HO_2 propagation leading to H_2O_2 formation is not strongly exothermic and so there is not much thermal feedback once the temperature approaches about 850 K. At this temperature the rate of hydrogen peroxide decomposition to OH radicals is relatively slow and so further chain branching is not readily promoted until still higher temperatures are reached through supplementary heat releasing processes.

In a non-isothermal system the ntc confers a two-stage character on spontaneous ignition, the first stage being the transition between the branching and essentially non-branching mechanisms involving RO_2 as T increases. The combustion processes in the first stage are accompanied by weak, chemiluminescent emissions from CH_2O^*, predominantly formed in the radical–radical interactions of CH_3O and indicating the momentary occurrence of the OH/RO_2 radical activity. The subsequent reactions, driven by the decomposition of hydrogen peroxide and an acceleration in the rate of temperature change, constitute the second and final stage of ignition.

However, if the initial reactant temperature is set at 850 K or above, there is no scope for the low temperature reactions involving R/RO_2 to take place. Thus a single stage ignition must then develop. There are several important qualifications that follow. Given that hydrogen peroxide is regarded as an important precursor to the development of the final stage of ignition,[14] not only does the reactant temperature determine the ignition delay, but also the amount of hydrogen peroxide that is present is relevant. This has been demonstrated with respect to the development of multiple stage, oscillatory ignition of ethanal in flow reactors.[15,16] Thus, in the present context the duration of the ignition delay is affected by the qualitative nature of the chemistry that is able to take place. That is, low temperature reactions that are capable of generating H_2O_2, and also that raise the temperature as a result of heat release, are able to cause a shorter time to ignition than that which occurs in a system which is raised initially to a higher temperature in the RCM, but as a result of which has had no opportunity for any hydrogen peroxide to be formed. One may construe this to be a simplified basis for the relationship between an ntc encapsulated in the kinetic mechanism and the ntc that is reflected in the ignition delay. There is also a rather more subtle implication; If the temperature within the combustion volume is not initially uniform, the rate of hydrocarbon oxidation may be affected in a way that is counter intuitive. That is, reaction may be able to go faster in a colder region if the ntc prevails.

Previous studies of the spatial development of autoignition

The visualisation of light output during hydrocarbon autoignition has traditionally been conducted using photomultiplier tubes and more recently with optical fibre devices.[17] High speed natural light photography has also proved to be remarkably informative over many years.[18,19] In addition, interferometry,[20,21] shadowgraphy[18] and schlieren imaging[22,23] have been applied to the study of the development of ignition in combustion chambers and other devices. A seminal study of the development of autoignition during the combustion of several hydrocarbons in an RCM, using both natural light and schlieren photography, was that of Livengood and Leary.[24] More recently, single point temperature measurements have been made also in an RCM using Rayleigh scattering,[25,26] and we have developed the application of Rayleigh scattering to a 1-D line study.[27]

Non-intrusive laser based imaging techniques have been developed and widely used to examine flame structure and to measure temperature, molecular and radical concentrations, and also flow and turbulence in engines and other combustion systems.[28] Image intensifiers have often been employed as a means of detecting very low light intensities over short time intervals when coupled to a ICCD camera. Two-dimensional, double pulse, PLIF of ground-state formaldehyde was used by Bauerle et al.[29] to detect the location and development of hot spots in the end-gas following spark ignition in a single cylinder reciprocating engine.

Apparatus and experimental methods

The rapid compression machine

A full description of the RCM and its operation can be found elsewhere.[30] In summary, fuel vapour is premixed with oxygen as required and proportions of inert diluents (nitrogen, argon or carbon dioxide) are then chosen in order to achieve a specific value of the heat capacity ratio, $\gamma = C_p/C_v$, for the mixture. In conjunction with control of the cylinder wall temperature, a range of compressed gas temperatures ($T_c = 500-1000$ K) can be attained following rapid compression at a compression ratio of 10.50 (± 0.15). When the combustion of fuels in air is being investigated the total inert gases comprise the equivalent of the proportion of nitrogen in air (79%).

Reactants were admitted to the combustion chamber at an initial pressure of 33 kPa, then compressed by a stainless steel piston driven by high pressure air in a driving chamber. The final volume was about 30 cm^3, the combustion chamber being 4.5 cm diameter × 1.9 cm depth. Pressure–time data during the compression stroke (22 $\pm$ 1 ms) and throughout the post-compression period were measured by a pressure transducer (Kistler 601A, natural frequency 100 kHz) and recorded on a digital oscilloscope (Tektronix, TDS 220) together with a timing mark, where appropriate, for the laser firing sequence.

In order to allow a laser beam to traverse the diameter of the combustion chamber, two 1.0 cm diameter, optically flat, fused silica windows were located on opposite sides of the chamber. The end of the chamber was fitted with an optically flat, fused silica window (5.0 cm diameter × 3.0 cm thick) that allowed a full view of the chamber cross-section for the recording of natural light emission, schlieren photography or laser induced fluorescence. The piston crown was highly polished to an optically flat, mirror finish. This gave less interference from scattered light than a piston crown made from brushed, black anodised, aluminium. The mirror also served to generate a schlieren image by reflection of an incident beam *via* the quartz window. Since there was always some condensation of reaction products within the combustion chamber, prior to each experiment it was essential to clean all surfaces scrupulously with the windows removed. Where appropriate the laser beam was realigned after their replacement to ensure a path through the cell that was perpendicular to the windows.

Diagnostic techniques

Schlieren photography. Schlieren images were obtained using a low power, continuous wave, laser beam (He/Ne, 25 mW). An expanded incident beam was transmitted through the quartz window at the combustion cylinder head and reflected back from the mirrored piston crown to focus at an aperture and then onto a screen *via* an imaging lens. Refraction of the reflected beam, which resulted from density gradients in the test gas, revealed where temperature gradients existed in the cross-sectional image. The development was recorded by ciné photography of the screen image (Hitachi, 5000 frames s^{-1}).

Chemiluminescence imaging. Single shot, two-dimensional pictures were recorded by an image intensified charge coupled device camera (ICCD camera, Princeton Instruments, 576-G), across a full field of view of the cross-section of the chamber *via* the quartz window. The image intensifier lens was focused on the mid-plane of the combustion chamber, with a depth of focus of about 2 mm. In order to follow the development of spontaneous ignition, repeated experiments were performed at the same reaction conditions with increasing delays set on the imaging timing signal. In addition, a photomultiplier was used to observe the chamber through a small Perspex port located in the side wall so that a continuous record of the light output could be obtained simultaneously with the ICCD camera snapshot. The normal gating interval of the intensifier to study the CH_2O^* emission was 0.5 ms, in order to spatially resolve the particularly weak light output. The much higher intensity emission from the final stage of spontaneous ignition of pentane (and originating from different chemiluminescent sources) was not investigated in this work.

Laser induced fluorescence. The laser system (Nd : YAG, Quantel 680) was frequency doubled, tripled or quadrupled as required, by non-linear optical mixing. Single-mode operation and linewidth (~ 500 MHz) were monitored by means of a high finesse etalon placed in a reflected portion of the main beam. To minimise pulse-to-pulse fluctuations, the laser output was stabilised by continuous firing at a 10 Hz repetition rate, the laser flashlamp and Q-switch being triggered externally. In order to cope with the shot-to-shot temporal variation of the RCM resulting from the electro-mechanical operation ($+/- 1$ ms), an "enabling" signal, generated from the initial firing signal for the RCM, was used to couple the machine timing to the laser trigger sequence. Simultaneosly, the external clock was intercepted with a pre-set, variable delay to synchronise the next laser pulse to the required image point in the combustion process. The delayed pulse also triggered the ICCD gating electronics so that the camera was exposed only for a little longer than the duration of the laser pulse as it passed through the chamber. Once the camera had been triggered the external clock reverted to its 10 Hz mode.

After frequency mixing, the main beam was steered into the combustion chamber by three, high quality, dichroic mirrors to remove any residual IR radiation in the beam. For the purpose of the PLIF imaging, each single-shot pulse (6–7 ns) formed a laser sheet (width 8 mm × 100 µm depth) using a cylindrical beam expanding telescope in conjunction with a spherical lens to focus the beam in the horizontal plane with its vertical face to the end window. For the acetone fluorescence experiments the frequency doubled laser output, at 532 nm, was further doubled (266 nm) using an external frequency doubling crystal to obtain the fourth harmonic of the Nd : YAG fundamental. For formaldehyde PLIF, the laser sheet was generated by mixing the frequency doubled output with the fundamental frequency to obtain light at 355 nm. Typically the incident energy per laser pulse was 100 mJ at 532 nm, 80 mJ at 355 nm and 20 mJ for the 266 nm experiments.

The PLIF signal obtained from acetone is sensitive to both temperature and concentration change. At 266 nm the fluorescence quantum yield is reduced by 50% over the temperature range 600–800 K.[31,32] Thus spatial variations in temperature that arise in compressed inert gas seeded with acetone are shown as a diminished fluorescence signal in hotter regions. By contrast, reactive systems in which acetone is generated show an enhancement in the fluorescence intensity. In isothermal conditions this signal is proportional to the concentration. However, as noted above, there is a temperature increase associated with the formation of acetone as a result of the exothermic decomposition of di-*tert*-butyl peroxide. Consequently the LIF signal intensity has origins generated from opposing effects and so we do not have a quantitative calibration of acetone yield from it in these circumstances. Since the S_0–S_1 transition is not saturated by the laser as it traverses the combustion chamber in the present experiments then, in a uniform concentration field, the fluorescence signals show a Beer–Lambert absorption decay. Corrections can be made for the Beer–Lambert absorption in non-reactive conditions, and we have done so in the results given later, but not in circumstances where there are both temperature and concentration changes as a result of chemical reaction.

Results

The continuous pressure and chemiluminescence records are given first. Next we report results of the spatial imaging of chemiluminescence from di-*tert*-butyl peroxide combustion and from n-pentane combustion. The results obtained from schlieren photography during the development of spontaneous ignition of n-pentane are then given. Finally, the PLIF of acetone as tracer in a non-reactive gas and as a product from di-*tert*-butyl peroxide decomposition, and the PLIF studies of formaldehyde in both of the combustion systems are presented.

Pressure change following compression and chemiluminescence from CH_2O*

The pressure change and chemiluminescence associated with the combustion of di-*tert*-butyl peroxide (2 mol.%) in the overall composition $1.00\ C_4H_9OOC_4H_9 + 1.00\ O_2 + 15.80\ N_2 + 32.00\ CO_2$, at $T_c = 560$ K, are shown in Fig. 2. There are heat losses from the gas to the vessel walls, which cause the pressure eventually to decay to that of the cold, compressed gas. However, given that there is a relatively small overall change in mole numbers in the final, constant volume, combustion chamber, the pressure change displayed in this curve is related to temperature change within the reacting mixture. Thus the pressure record conforms to that from an exothermic reaction that exhibits an autoacceleration in rate as a result of thermal feedback. The maximum rate of pressure change corresponds to the maximum rate of reaction and coincident with this is the maximum intensity of the light output. However, as we shall see, the irregular decay of the chemiluminescent emission may be attributed to spatial variations in behaviour.

The pressure records, for two-stage ignition of n-pentane in the compositions $C_5H_{12} + 8.0\ O_2 + 29.7\ (N_2 + Ar)$ to give $T_c = 765$ and 840 K, are shown in Fig. 3. Also given is the total light output during the course of reaction at each of these temperatures. These compressed gas temperatures represent conditions that are close to the minimum and maximum respectively of the ignition delay associated with the ntc region (see Fig. 1). Both pressure records show a two-stage development of reaction after the piston has stopped, but the first stage activity, as shown in the rate and extent of the pressure rise, is much stronger at the lower compressed gas temperature. In fact, this activity progressively diminishes as T_c is increased such that, for pentane combustion at this composition and pressure in the RCM, the maximum duration of the ignition delay in the ntc region occurs at the compressed gas temperature for which the "first stage" chemistry is no longer

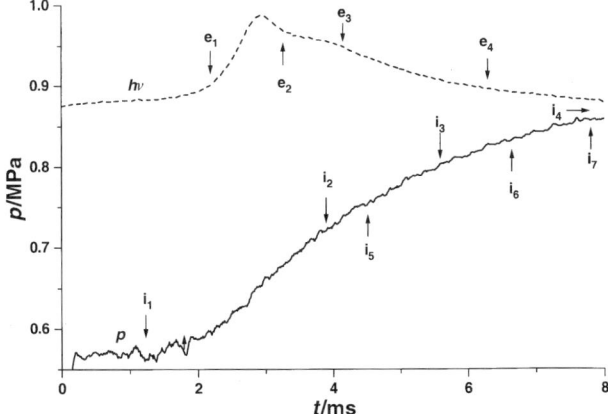

Fig. 2 Records of the pressure change (solid line, p) and chemiluminescence intensity (broken lines, $h\nu$) vs. time following the rapid compression of $1.00\ C_4H_9OOC_4H_9 + 1.00\ O_2 + 15.80\ N_2 + 32.00\ CO_2$ to 560 K. Zero on the timescale signifies the moment at which the piston stops. The intensity of emission was not calibrated. The arrows indicated on each curve refer to the times at which ICCD camera images of the chemiluminescence (e_1–e_4), PLIF of acetone (i_1–i_4) and PLIF of formaldehyde (i_5–i_7) were obtained.

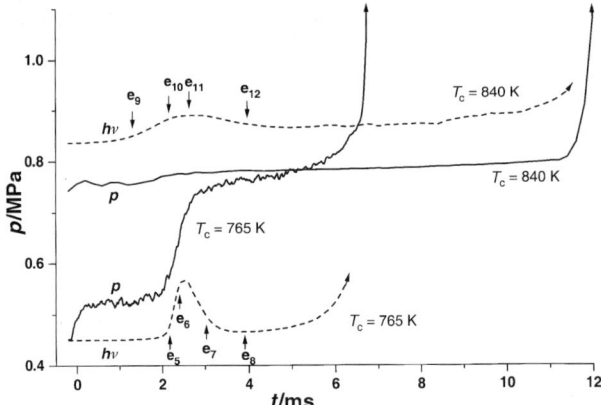

Fig. 3 Records of the pressure change (solid line, p) and chemiluminescence intensity (broken line, $h\nu$) vs. time following the rapid compression of two mixtures containing n-pentane in the proportions $C_5H_{12} + 8.0\ O_2 + 13.3\ N_2 + 16.8$ Ar and $C_5H_{12} + 8.0\ O_2 + 2.7\ N_2 + 27.4$ Ar to give the compressed gas temperatures 765 and 840 K when the cylinder temperature was set at 313 K. Zero on the timescale signifies the moment at which the piston stops. The intensity of emission was not calibrated. The numbers marked on each curve, (e_5–e_8) and (e_9–e_{12}), refer to the times at which ICCD camera images of the chemiluminescence were obtained.

able to develop.[30] Chemiluminescence from CH_2O^* is associated with the first stage reaction and is correspondingly weaker at the higher temperature.

As with di-*tert*-butyl peroxide, the change in pressure broadly represents a change in gas temperature. Since these reactions were performed at the same gas density, the crossover of the pressure records during the second stage development corresponds, on a spatially averaged basis, to the limiting temperature reached during combustion after the R/RO_2 equilibrium has shifted essentially to the non-branching mode.[30] A deceleratory increase in the reactant pressure signifies a falling rate of heat release as the reactant temperature increases. The relatively slow pressure change that follows the first stage is a consequence of reaction that takes place at an exceedingly low rate and with comparatively low radical concentrations associated with it. This may also account for the fall in chemiluminescent emission intensity after the first stage.

Spatial resolution of chemiluminescence

The counterparts of the continuous chemiluminescence displayed in Fig. 2 and 3 are successive images that show its spontaneous development within the cross-sectional field of view of the ICCD camera (Fig. 4). The images were obtained from a single shot in a series of individual experiments under the same conditions and not from multiple images obtained in a single experiment. During the combustion of di-*tert*-butyl peroxide + O_2 the chemiluminescence occupies most of the field of view at an early stage, but the highest intensity is in a concentric ring towards the edge (Fig. $4e_1$). Thereafter there is a contraction of the emission from the periphery, and the last appearance of chemiluminescence, coincident with the final stage of pressure change associated with reaction, occurs at the centre of the chamber (Fig. $4e_2$–e_4). The timing of these images is marked on Fig. 2.

Chemiluminescence from CH_2O^*, associated with n-pentane combustion from $T_c = 765$ K, is shown in Fig. 4 as the intensified images e_5–e_8, at times marked in Fig. 3. By contrast to the di-*tert*-butyl peroxide combustion, at no stage does there appear to be significant light output from the core gas. Although chemiluminescence from n-pentane combustion at $T_c = 840$ K is very weak, its spatial structure was identified with the image intensifier set at a high gain (Fig. $4e_9$–e_{12}). In this case the peak of the emission is confined mainly to the centre of the combustion chamber (e_9–e_{12}), which signifies that at these conditions the greatest chemical activity tends to be localised in this region during the first stage of combustion.

Schlieren images during n-pentane combustion

Sequences of frames from schlieren photography during n-pentane combustion, corresponding to the experimental pressure records in Fig. 3, are shown in Fig. 5 and 6. The first frame of each

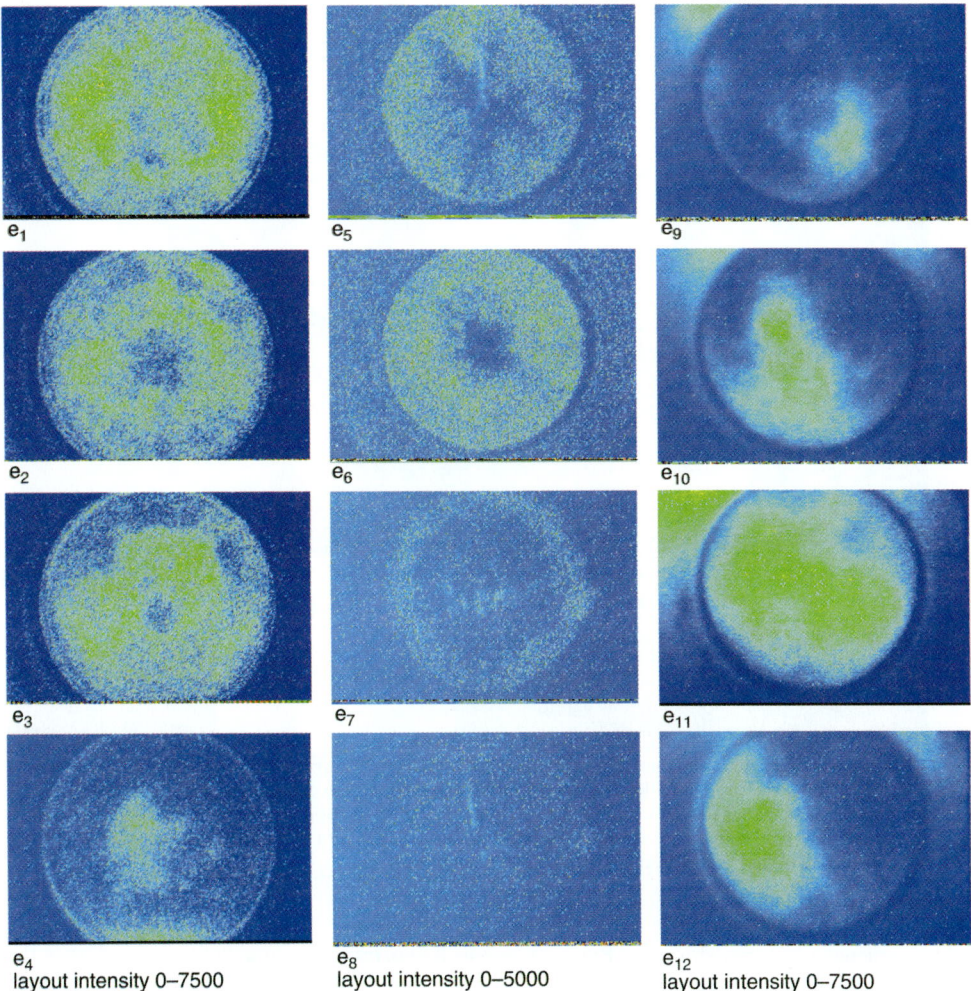

Fig. 4 ICCD images of the spatial variation in chemiluminescent emission at successive times as reaction proceeds following rapid compression. Each set is self-consistent, but to accommodate the different intensities of the emission, the images were obtained at different gain factors of the image intensifier and are displayed over different ranges for the digital layout. (e_1–e_4), di-*tert*-butyl peroxide combustion; (e_5–e_8), n-pentane combustion at 765 K; (e_9–e_{12}), n-pentane combustion at 840 K. The images shown in each series were obtained in separate experiments.

figure corresponds to the end of compression and the timing of successive frames is shown. Alternate frames were taken from the ciné film, so the images are displayed at 0.4 ms intervals. The final frame in each figure corresponds to an image taken during the development of the final stage of spontaneous ignition, at the time indicated.

At $T_c = 765$ K (Fig. 5), in the interval 2.4–2.8 ms there is a development of temperature gradients both in the centre of the cross-sectional view and at a concentric ring closer to the edge, shown by the lighter zones. The reaction is approaching the end of the first-stage activity at this time, and these features remain prominent throughout the development of the second stage of reaction. The final frame, at 6.4 ms into the post-compression period, shows that multiple, hot centres have developed from the edge of the combustion chamber, one of which was initiated slightly before the others and has propagated further across the chamber. There is still some evidence of the central feature from earlier in the sequence and it appears to have been displaced towards the edge by the propagating front.

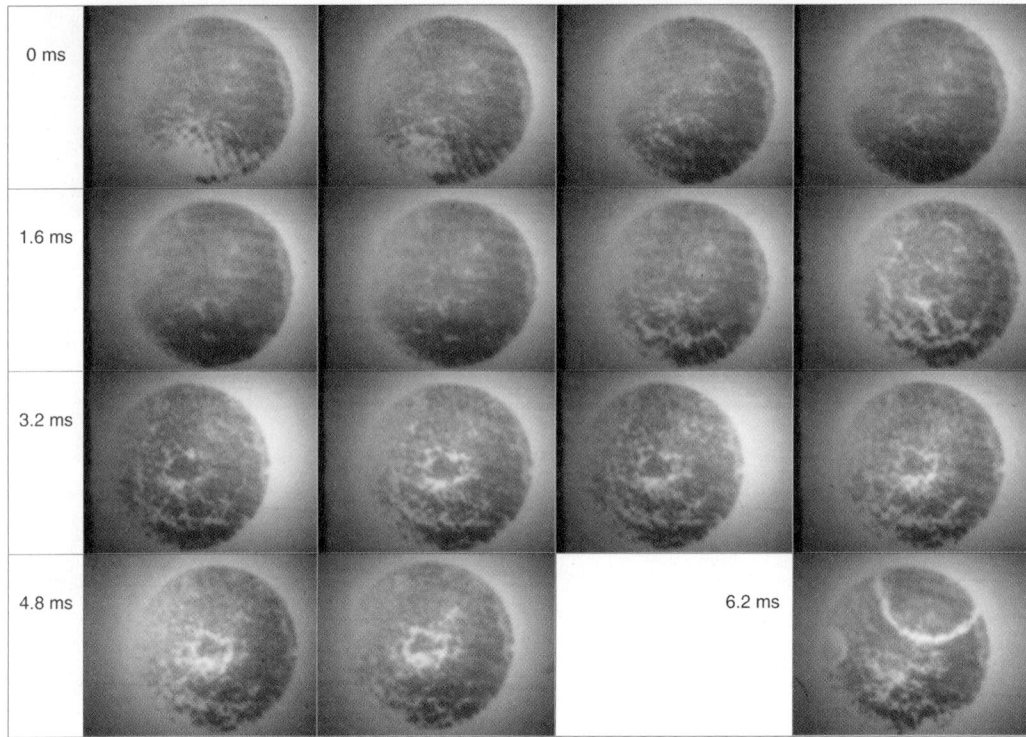

Fig. 5 A sequence of schlieren images taken from frames of a ciné film during the combustion of n-pentane at a compressed gas temperature of 765 K. The first frame corresponds to the end of compression, as indicated by the time, and the light patch in the lower left-hand area is believed to result from movement of the RCM as the piston stops. The final frame corresponds to a late stage of development of ignition. The growth of multiple ignition centres is identified in this frame.

The sequence of frames shown in Fig. 6 ($T_c = 840$ K) corresponds to a similar period after the piston has stopped as those given in Fig. 5. The final frame, which represents the thermal gradients associated with the development of the final stage of spontaneous ignition occurs at 11.8 ms post-compression. Although filmed on the same day, minor alterations of the optical set-up as a result of cleaning the system and re-setting the camera result in differences in the overall appearance of the images from different films. However, the most striking features of Fig. 6, relative to those of Fig. 5, are the lack of any marked structural change or the appearance of any significant temperature gradients throughout the interval to beyond 5 ms post-compression. Ignition centres become prominent in the final image, at 11.8 ms, but in this case they are located close to the centre of the field of view.

An inference from these observations is that the spatial variations in chemiluminescence, shown in Fig. 4, relate to temperature inhomogeneities in the combustion chamber following the end of compression. Moreover, the system evolves in qualitatively different ways when n-pentane oxidation is initiated at the two temperatures 765 and 840 K. Although a complex qualitative structure shows in the schlieren photographs, no quantitative interpretation can be put on the magnitude of the temperature gradients and it is not possible from this information alone to establish in which direction the gradients change.

PLIF of acetone as a tracer following compression in non-reactive gas

An alternative route to a spatial record that is purely thermal in origin arises from the intensity of fluorescence from acetone in a non-reactive gas.[31,32] In the present experiments, a mixture containing 1% acetone by volume (0.01 $C_4H_9OOC_4H_9 + 0.99\ N_2$) was compressed rapidly to yield

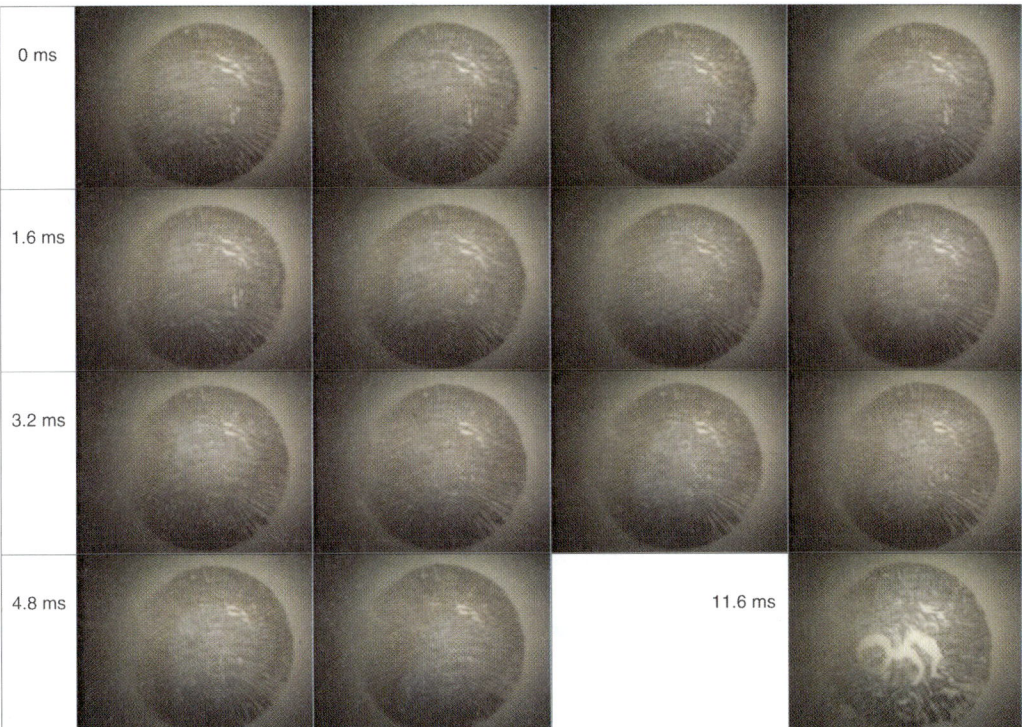

Fig. 6 A sequence of schlieren images taken from frames of a ciné film during the combustion of n-pentane at a compressed gas temperature of 840 K. The first frame corresponds to the end of compression, as indicated by the time. The final frame corresponds to a late stage of development of ignition. The growth of multiple ignition centres is identified in this frame.

an adiabatic temperature of 740 K. A PLIF image obtained at the moment the piston has stopped shows an exponential decay that is consistent with a Beer–Lambert absorption as the laser sheet traverses a uniform concentration field (Fig. 7a). The statistical record has been binned across 20 pixels of the laser sheet in the vertical direction and also corrected for the Beer–Lambert absorption.

There is a similar treatment of the image and data obtained 2 ms after the piston has stopped (Fig. 7b). In this case not only does the signal decay from left to right, which is attributable to a Beer–Lambert absorption of the incident beam, but also there is an enhanced intensity across the centre of the field of view. In order to emphasise this feature, the accompanying statistical analysis of the intensity as a function of pixel number has been corrected also for the Beer–Lambert absorption. The sharp transitions below pixel 75 and above pixel 540 signify the edges of the field of view of the laser beam by the camera, but this includes part of the recesses in which the entry and exit windows are located at the sidewall, in the pixel ranges 75–100 and 515–540 respectively. Thus the diameter of the combustion chamber falls within the range of pixels 100–515, which is equivalent to 9 pixels mm^{-1}. A slight asymmetry of the ICCD camera means that the centre of the chamber corresponds to pixel 305. From an interpolation of the data given by Thurber *et al.*,[31,32] the intensity of the corrected emission at the centre relative to that on either side (Fig. 7b) signifies that at 2 ms after the piston has stopped the temperature of the core gas on the central cross-section of the chamber is approximately 40 K below that of the surrounding gas. The temperature gradient adjacent to the core of this non-reactive mixture is between 20 and 40 K mm^{-1}.

From this fluorescence study the core structure that develops early in the schlieren images is identified as a temperature dip and that it takes several milliseconds after the piston has stopped for the relatively cool plug to the reach the central plane of the chamber. It seems very likely that the bulk of the gas is adiabatically heated under rapid compression but, during piston motion,

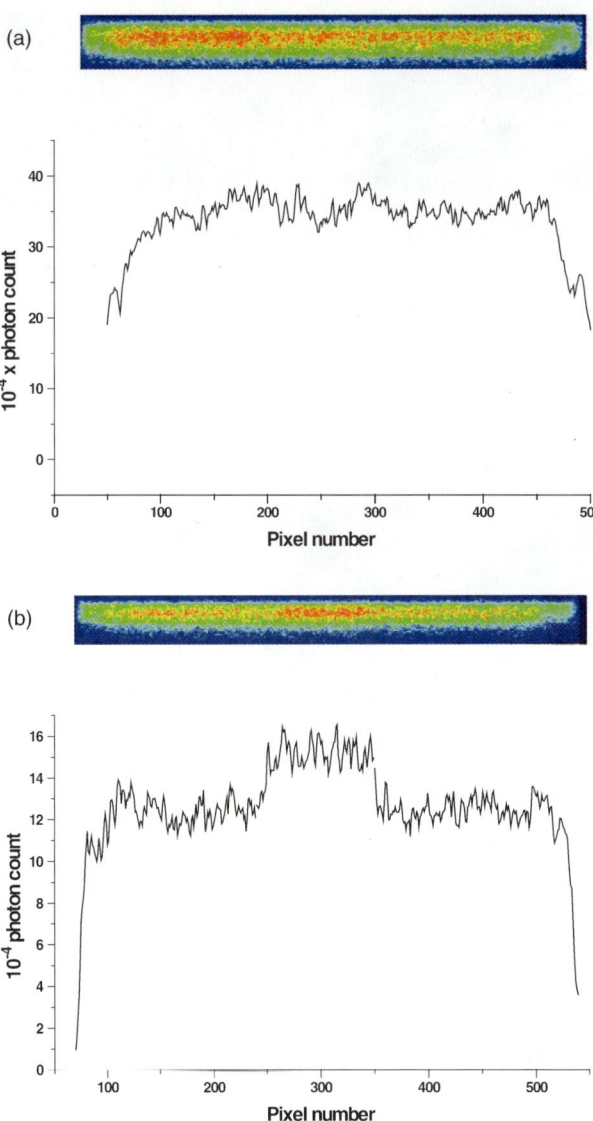

Fig. 7 The PLIF signals from an acetone tracer (1 mol.%) in N_2 imaged in a sheet across the central plane of the combustion chamber. An ICCD image and the photon count is shown at the moment the piston stops (a) and at 2 ms into the post-compression period (b). The ordinate represents the signal integrated across the width of the beam and is corrected for the Beer–Lambert absorption. The incident laser beam was not calibrated. The pixel number on the abscissa is related to the position in the field of view across the diameter of the chamber (see text).

flow is induced across the surface of the piston as a result of gas being sheared off the cylinder wall and this causes cold gas eventually to penetrate the core.[3]

PLIF of acetone during the combustion of di-*tert*-butyl peroxide

The PLIF of acetone without tracer addition is possible during di-*tert*-butyl peroxide decomposition because it is formed as a primary product. Thus the progress of reaction can be followed in successive images, although the information is not fully quantitative because the reaction is exothermic and so the fluorescence intensity is affected by both temperature and concentration change. A sequence of images obtained during the combustion of the mixture 1.00 $C_4H_9OOC_4H_9$ + 1.00 O_2 + 15.80 N_2 + 32.00 CO_2 is shown in Fig. 8 (see also Fig. $2i_1-i_4$). There is very little acetone formed immediately after the piston has stopped (*e.g.* Fig. 8, 1.2 ms), but by 4.0 ms there is a considerable concentration in a toroidal region surrounding the core. There is little evidence of acetone having been formed in the centre of the chamber. By 5.6 ms into the post-compression period the region through which acetone has been formed has increased, but it is only by 12 ms post-compression, when reaction is complete, that the concentration is uniformly distributed throughout the cross-section, although this appears as a decaying intensity as a result of a strong Beer–Lambert absorption of the incident laser beam. The overall record of events is consistent with the creation of the low temperature core prior to the onset of a significant extent of decomposition of the di-*tert*-butyl peroxide.

PLIF of ground state formaldehyde

Formaldehyde is formed both as the product of methyl radical oxidation following di-*tert*-butyl peroxide decomposition and also as an intermediate product of the low temperature combustion of n-pentane, so PLIF studies of it allow comparisons and contrasts to be made of the kinetic response to the stratified temperature field generated in the RCM. The sequences of images for product formation are shown in Fig. 9 from di-*tert*-butyl peroxide and in Fig. 10 from n-pentane. The exposure of the camera to capture fluorescence (150 ns) is too short for any accompanying chemiluminescence to interfere with the image.

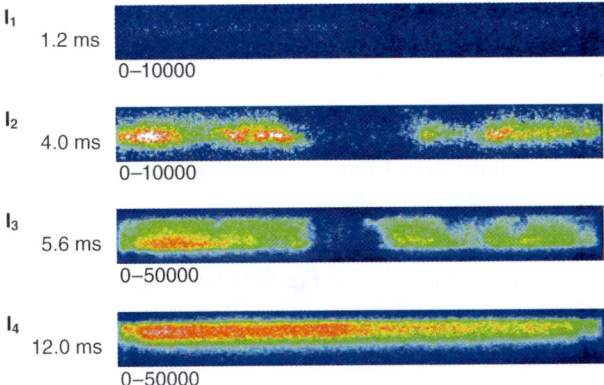

Fig. 8 PLIF images of acetone generated as a reaction product from the exothermic decomposition of di-*tert*-butyl peroxide in the presence of oxygen and compressed to 560 K at 1.2, 4.0, 5.6 and 12.0 ms into the post-compression period, as marked in Fig. 2. The layout is scaled to the intensity ranges shown.

Fig. 9 PLIF images of formaldehyde generated as a reaction product from the exothermic decomposition of di-*tert*-butyl peroxide in the presence of oxygen and compressed to 560 K at 4.4, 6.5 and 7.8 ms into the post-compression period, as marked in Fig. 2. The dark, vertical lines at the edge of each image represent the location of the piston ring. The layout is scaled to the same intensity range (0–30 000).

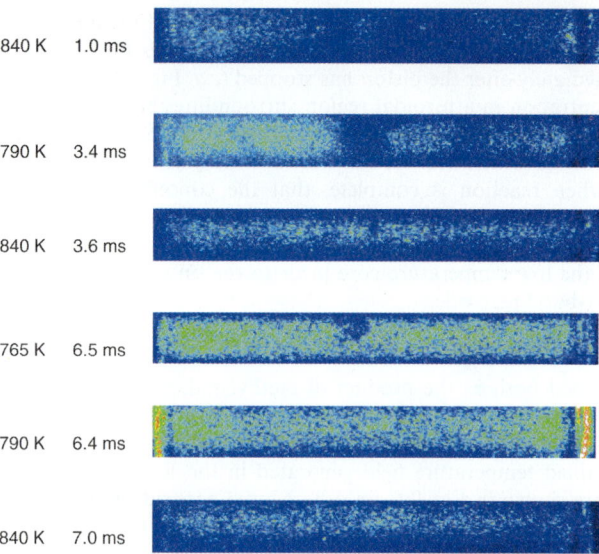

Fig. 10 PLIF images of formaldehyde generated as an intermediate product from the combustion of n-pentane. Results are shown for reaction at the compressed gas temperatures of 765, 790 and 840 K. The layout intensity is scaled to the range (0–7000).

As shown in Fig. 9 (at times noted in Fig. 2i_5–i_7) only very weak fluorescence from formaldehyde is detected early in the post-compression period during di-*tert*-butyl peroxide combustion, but by 4.4 ms the signal is well developed, and it shows that the most vigorous reaction is associated with a "toroidal" zone surrounding the core of the gas. This pattern continues until later stages of reaction, such that significant concentrations of formaldehyde gradually accumulate throughout the cross section of the chamber as shown in successive images at 6.5 and 7.8 ms. The development in the images of CH_2O, shown in Fig. 9, is consistent with the evolution of acetone as a reaction product (Fig. 8).

Only very weak PLIF images of formaldehyde production during n-pentane combustion were obtained in the early reaction stages at the lowest compressed gas temperature investigated. However, more information was accessible at $T_c = 840$ K and we have supplemented this from intermediate temperature studies at $T_c = 790$ K, as shown in Fig. 10. The interpretation of these data is more difficult than those from di-*tert*-butyl peroxide combustion because CH_2O has diverse kinetic origins, and different reactions leading to it may be more prominent at different temperatures. Nevertheless, contrasts in overall response of the chemistry can be distinguished. That is, at $T_c = 790$ K, by 3.4 ms there is limited evidence of the existence of the colder core and after a similar time following compression to 840 K it can hardly be distinguished. Moreover, even at 6.5 ms, there is a vestige of the cold core when n-pentane is compressed to 765 K, whereas reaction following compression to 790 and 840 K has caused it to disappear entirely by a similar stage. These patterns of behaviour are also in marked contrast to that which emerges from formaldehyde produced by methyl radical oxidation during di-*tert*-butyl peroxide combustion, whereby the cooler core remains in evidence to about 8 ms (Fig. 9).

Discussion

The accumulated information from the range of experimental techniques reported here shows that, following rapid compression of pre-mixed reactive gases in a cylinder, there is a development of the temperature field during the period after the piston has stopped which appears to be consistent with the predictions from earlier work.[3–8] There is little spatial variation in temperature across the central plane of the chamber at the moment the piston stops, but within several milliseconds a cold plug of gas is able to penetrate the core, and we may presume that it continues to progress across the full depth of the chamber, so that the adiabatic heated gas becomes restricted to a toroidal region surrounding the core.

Drawing on the earlier interpretation,[3] we may surmise that the gas motion leading to this behaviour originates in the boundary layer that is peeled away from the wall of the chamber is the piston progresses through the cylinder. The evolution is consistent with the machine design and operation: no bulk "tumble" or "swirl" is generated within the combustion chamber, but the piston speed, 10 m s^{-1} (or 1 cm ms^{-1}), is sufficiently high to create turbulence. However, a characteristic flow pattern is formed and the highest possible gas velocity, in a direction perpendicular to the crown, is determined by the piston speed. The gas motion must decay rapidly once the piston has stopped as a result of viscous drag, as has been demonstrated in computational fluid dynamic (CFD) simulations.[5] Given that cooler gas is already being swept across the piston surface during the compression stroke, its penetration to the centre of the combustion chamber within 2 ms after the piston has stopped seems entirely consistent. The temperature within the cooler region is estimated from the acetone tracer measurements to be about 40 K below that of the adiabatically compressed gas.

There are a number of issues to consider that are consequent upon this behaviour. First, there is a question regarding the extent to which heat transfer within the gas, by conduction and convection, continues to modify the spatial temperature distribution during the post-compression interval. Heat is dissipated to the chamber walls and eventually the adiabatically compressed gas, whether reactive or non-reactive, cools to the wall temperature. However, within the timescale of interest, the boundary layer is likely to be most significantly affected by heat loss, and one might envisage that the initial flow pattern permits the toroidal structure to be sustained for at least a few milliseconds.

Exothermic reaction which gives a heat release rate on a timescale compatible with the evolution of the temperature field must cause there to be an interaction and modification of the temperature field as a result of thermal feedback. Moreover, the response is affected by the overall nature of the chemistry, specifically with regard to its temperature dependence. The combustion of di-*tert*-butyl peroxide exhibits a positive temperature dependence which is governed by the rate of decomposition of the peroxide, for which $k_\infty = 2 \times 10^{15}\, e^{(-18280/T)}$ s^{-1}. The half-life at 560 K is 50 ms and this falls to 0.5 ms by about 650 K. Thus, reaction initiated by compression to 560 K must accelerate rapidly as a result of thermal feedback. However, the cooler gas that penetrates the core of the combustion chamber very soon after the piston has stopped cannot react on a similar timescale and, given that $t_{1/2} > 500$ ms at 520 K, the rate and extent of reaction is likely to be very limited indeed if the assessment of a 40 K temperature difference is reasonable. Thus, as shown in the PLIF images of acetone (Fig. 8) and of formaldehyde (Fig. 9), the colder core survives for at least 6 ms. There is a gradual degradation, however, and this cooler zone can no longer be distinguished by 12 ms into the reaction. The extent to which gas mixing has contributed to this cannot be interpreted from the present results, but the magnitude of the half-life at temperatures below the adiabatic compression temperature would suggest that mixing and heat transport are contributory to the demise of the cooler core. Whatever the cause, chemiluminescence shows that reaction takes place within the central region in the later stages (Fig. 4).

The evolution is somewhat different when the combustion is associated with reaction that exhibits a negative temperature dependence of rate. For n-pentane compressed with oxygen to a temperature not exceeding 765 K, the reaction is most active in the early stages within the adiabatically heated region (see Fig. 1). Thus, just as with di-*tert*-butyl peroxide, "thermal runaway" begins in the toroidal zone surrounding the core. This has the consequence of slowing the overall rate in that region because of the associated temperature increase and there is scope for the more slowly reacting gas in the core to "catch up". We may construe that the much reduced extent of the core at 6.5 ms determined by the formaldehyde PLIF (Fig. 10), relative to that during di-*tert*-butyl peroxide combustion, as supporting evidence. When n-pentane is compressed with oxygen to an adiabatic temperature approaching 850 K, the most reactive gas exists in regions that have been unable to attain that temperature. Thus, the core is the most active zone, which means that it is eroded more rapidly than at low T_c. The evidence for this is seen in the PLIF images at 790 K and 840 K, and the enhanced central activity is displayed in the chemiluminescence accompanying reaction (Fig. 4c).

The contrasting behaviour following combustion of n-pentane following compression to either low or high temperatures in the ntc region is also supported by the qualitatively different development of the density gradients in the schlieren images (Fig. 5 and 6). However, a comparison between the acetone tracer study and the schlieren photographs would suggest that, using the present apparatus, the schlieren image may not be sufficiently sensitive to be able to distinguish temperature a gradient as low as 20–40 K mm^{-1}. We make this inference because the density gradients become prominent at $T_c = 765$ K only late in the development of the first stage ($t > 2.4$ ms), at which the temperature of the toroidal zone has been raised significantly relative to that in the core. By contrast, at $T_c = 840$ K no temperature gradients surrounding the core can be clearly identified over a 5 ms post-compression interval. The evidence from formaldehyde PLIF would suggest that, by this stage, the cooler core structure will have been destroyed.

On the face of it numerical simulation of spontaneous combustion in a rapid compression machine using detailed chemical kinetics in zero models appears to be reasonably robust, provided that the heat loss rates to the chamber walls can be characterised satisfactorily. In a positive temperature dependent reaction regime the adiabatically compressed core gas dominates the rate development of reaction even when spatial inhomogeneities develop as a result of gas motion following rapid compression. In the negative temperature dependent reaction regime the temperature inhomogeneities appear to be smoothed out during the chemical development, causing the system to approach the ideal. However, the kinetics involved in the two-stage development of spontaneous ignition of alkanes and other organic substances may have adverse consequences when the effect of the ntc is predominant. That is, the faster progress of reaction in the initially cooler zone yields higher proportions of reactive intermediates than elsewhere. This means that the second stage development is then controlled by activity in this "sensitised" region, and this controls the duration of the ignition delay. The onset of hot ignition within the central region, shown in the final frame of Fig. 6, would seem to support this interpretation.

Acknowledgement

The authors wish to thank EPSRC for financial support (GR/K97189/01).

References

1. R. Minetti, M. Ribaucour, M. Carlier, C. Fittschen and L. R. Sochet, *Combust. Flame*, 1994, **96**, 201.
2. J. S. Cowart, J. C. Keck, J. B. Heywood, C. K. Westbrook and W. J. Pitz, *Proc. Combust. Inst.*, 1990, **23**, 1055.
3. R. J. Tabaczinski, D. P. Hoult and J. C. Keck, *J. Fluid Mech.*, 1970, **42**, 249.
4. D. Lee and S. Hochgreb, *Combust. Flame*, 1998, **114**, 531.
5. J. F. Griffiths, Q. Jaio, M. Schreiber, J. Meyer and K. F. Knoche, *Proc. Combust. Inst.*, 1992, **24**, 1809.
6. J. F. Griffiths, D. J. Rose, M. Schreiber, J. Meyer and K. F. Knoche, *I. Mech. E.*, 1992, **C448/030**, 29.
7. M. Schreiber, A. Sadat-Sakak, C. Poppe, J. F. Griffiths, P. Halford-Maw and D. J. Rose, *SAE Technical Paper*, 1993, 932758.
8. J. F. Griffiths, D. Rose, M. Schreiber, J. Meyer and K. F. Knoche, *Combust. Flame*, 1992, **91**, 209.
9. J. F. Griffiths and C. Mohamed, in *Comprehensive Chemical Kinetics*, ed. M. J. Pilling, Elsevier, Amsterdam, 1996, vol. 35, p. 545.
10. A. Cox, J. F. Griffiths, C. Mohamed, H. J. Curran, W. J. Pitz and C. K. Westbrook, *Proc. Combust. Inst.*, 1996, **26**, 2685.
11. R. S. Sheinson and F. W. Williams, *Combust. Flame*, 1973, **21**, 221.
12. J. F. Griffiths and C. H. Phillips, *Combust. Flame*, 1990, **81**, 304.
13. R. W. Walker and C. Morley, in *Comprehensive Chemical Kinetics*, ed. M. J. Pilling, Elsevier, Amsterdam, 1996, vol. 35, p. 1.
14. C. K. Westbrook, *Proc. Combust. Inst.*, 2000, **28**, 1563.
15. C. Gibson, P. Gray, J. F. Griffiths and S. M. Hasko, *Proc. Combust. Inst.*, 1984, **20**, 101.
16. J. F. Griffiths, *Adv. Chem. Phys.*, 1986, **64**, 203.
17. U. Spicher, U. H. P. Kolmeier, *SAE Technical Paper*, 1986, 861532.
18. T. Hyashi, M. Taki, S. Kojima and T. Kondo, *Automotive Eng.*, 1985, **93**, 56.
19. G. Konig, R. R. Maly, D. Bradley, A. C. Lau and C. G. W. Sheppard, *SAE Technical Paper*, 1990, 902136.
20. D. C. Bull, D. B. Pye and C. P. Quinn, *Combust. Flame*, 1976, **27**, 399.
21. M. Tanake, M. T. Bolik, C. Eigenbrod, H. J. Rath, J. Salo and M. Kono, *Proc. Combust. Inst.*, 1996, **26**, 1637.
22. T. Male, *Proc. Combust. Inst.*, 1949, **3**, 721.

23 R. Herweg and R. R. Maly, *SAE Technical Paper*, 1992, 922243.
24 J. C. Livengood and W. A. Leary, *Ind. Eng. Chem.*, 1951, **43**, 2797.
25 P. Desgroux, L. Gasnot and L. R. Sochet, *Appl. Phys. B*, 1995, **61**, 69.
26 P. Desgroux, R. Minetti and L. R. Sochet, *Combust. Sci. Technol.*, 1996, **113**, 93.
27 J. Clarkson, J. F. Griffiths, J. P. MacNamara and B. J. Whitaker, *Combust. Flame*, 2001, **125**, 1162.
28 A. C. Eckbreth, *Laser Diagnostics for Combustion, Temperature and Species*, Gordon and Breach, Amsterdam, The Netherlands, 1996.
29 B. Bauerle, J. Warnatz and F. Behrendt, *Proc. Combust. Inst.*, 1996, **26**, 2619.
30 J. F. Griffiths, P. Halford-Maw and D. J. Rose, *Combust. Flame*, 1993, **95**, 291.
31 M. C. Thurber, F. Grisch and R. K. Hanson, *Opt. Lett.*, 1997, **22**, 251.
32 M. C. Thurber, F. Grisch, B. J. Kirby, M. Votmeier and R. K. Hanson, *Appl. Opt.*, 1998, **37**, 4963.

NO reburning study based on species quantification obtained by coupling LIF and cavity ring-down spectroscopy

Xavier Mercier, Laure Pillier, Abderrahman El Bakali, Michel Carlier, Jean-François Pauwels and Pascale Desgroux*

Laboratoire de Cinétique et Chimie de la Combustion—UMR CNRS 8522, Centre d'Etudes et de Recherches Lasers et Applications, Université des Sciences et Technologies de Lille, 59655 Villeneuve d'Ascq Cedex—France. E-mail: pascale.desgroux@univ-lille1.fr

Received 6th March 2001
First published as an Advance Article on the web 15th October 2001

NO reburning is studied in a low pressure (15 hPa) premixed flame of CH_4–O_2 seeded with 1.8% of NO. Measurements were carried out by using cavity ring-down spectroscopy (CRDS) and laser induced fluorescence (LIF) techniques. The temperature profile was obtained by OH-LIF thermometry in the A–X (0–0) band. The OH profile was determined by LIF and calibrated by single pass absorption. The NO concentration profile was obtained by LIF in the A–X (0–0) band and corrected for Boltzmann fraction and quantum yield variations. The absolute concentration profile was determined in the burned gases by CRDS allowing a direct experimental determination of the NO reburning amount. Finally CH and CN mole fraction profiles were obtained by CRDS by exciting rotational transitions in the B–X (0–0) bands of CH and CN around 387 nm. We found a peak mole fraction of 29 ppm for CH and 3.3 ppm for CN. This last result is in contrast with a previous study of W. Juchmann, H. Latzel, D. L. Shin, G. Peiter, T. Dreier, H. R. Volpp, J. Wolfrum, R. P. Lindstedt and K. M. Leung, *XXVIIth Symposium (International) on Combustion*, The Combustion Institute, Pittsburgh, 1998, p. 469, performed in a similar flame, which reported much lower levels of CN. In that study the absolute concentration of CN was indirectly obtained by LIF calibrated by Rayleigh scattering. In a second part, experimental species profiles are compared with predictions of the GRI 3.0 mechanism. Comparison between experimental and predicted profiles shows a good agreement particularly for CN and NO species. A qualitative analysis of NO reburning is then performed.

Introduction

The understanding of the chemical mechanisms involved in NOx formation is a subject of high interest in the field of combustion, notably as regards the pollution problems generated by this species. Thanks to the recent implementations of laser diagnostic techniques, many studies devoted to NO measurements have been performed in the last few years that have allowed the development of models able to predict the formation of this compound in different kinds of flames. However several uncertainties remain, notably concerning the minor species involved in the NOx reburning processes. Quantitative measurements of species like CH or CN, which are of major importance in NOx chemistry, are now possible by laser diagnostics.[1–4] The implementation of the chemical mechanisms of NOx formation needs reliable experimental quantitative data on these radicals. It requires techniques of high sensitivity with an adequate spatial resolution. Previous studies[4,5] have shown that the coupling of the CRDS and LIF techniques is well suited for monitoring species concentration profiles in low pressure premixed flat flames. Indeed, CRDS offers the possibility of directly measuring the absolute concentration of minor species in flames with a very good accuracy and a spatial resolution of about several hundred μm. This technique can be regarded as an absorption method of very high sensitivity, allowing the determination of spatially integrated measurements of the concentration of species. Its application to premixed flat flames implies that these flames are nearly monodimensional.

The concentration of relatively abundant species is more conveniently determined by LIF calibrated by single-pass laser absorption. LIF is also useful for measuring temperature profiles. Although the coupling of CRDS and LIF techniques is a promising procedure, this coupling has not yet been used to carry out flame kinetic studies including several species.

In this work, we used this CRDS–LIF coupling as well as single pass absorption to study a premixed stoichiometric CH_4–O_2 flame seeded with NO (1.8%), with the aim of supplying accurate experimental data for the kinetic validation of models. Experimental mole fraction profiles of OH, CH, CN and NO are compared with the computed ones obtained by the premix code. The calculations were performed by using mechGRI 3.0[6] and the temperature profile was measured by OH-LIF thermometry. The choice of this flame was motivated by the conclusions of a recent paper of Juchmann et al.[7] In that work, quantitative measurements of species (OH, CH_3, CH, NO, CN) were obtained by coupling LIF, Rayleigh and multipath absorption techniques. The authors reported a non-significant CN concentration, in disagreement with the predictions of several mechanisms available in the literature. The results of our study are in good agreement with those of Juchmann et al.[7] with the exception of the CN profile. Moreover, some discrepancies also appear concerning the amount of NO reburning. An analysis of these discrepancies is tentatively performed in this paper.

Experimental

Experimental set-up

Experiments were carried out in a premixed CH_4/O_2 flame seeded with NO (1.8%) stabilized on a 6.8 cm diameter burner at low-pressure (15 hPa). The characteristics of the flame are displayed in Table 1. They are quite similar to those of the flame studied by Juchmann et al.,[7] with the exception of the pressure (15 hPa instead of 13 hPa) because of a vacuum limitation of our low-pressure cell equipment. The burner was translated vertically with an accuracy estimated to be around 30 μm. Laser measurements were performed perpendicularly to the vertical axis and provided local (by LIF) or spatially integrated measurements along the laser direction (by CRDS) of species concentration and temperature. Species profiles were measured by using three techniques schematically represented in Fig. 1. Most of the measurements (OH, CH, CN) were performed with a Quantel Nd-YAG laser pumping a dye laser supplying a 0.225 cm^{-1} bandwidth. NO was probed using a Continuum optical parametric oscillator (OPO) laser system providing a narrow bandwidth (0.2 cm^{-1}) between 220 and 250 nm.

LIF experiments were performed in the linear regime of excitation. Fluorescence was spectrally filtered using a 0.25 m monochromator and the LIF signal was recorded by a head-on photomultiplier (PMT: Philips XP2020Q), digitized, averaged and stored by an oscilloscope (Lecroy 9354A,

Table 1 Flame characteristics

Reactant gas	Flow rate/standard l min^{-1}	Concentration (mol.%)
CH_4	1.71	32.7
O_2	3.42	65.5
Dopant: NO	0.092	1.8

Burner diameter: 6.8 cm
Pressure: 15 hPa (11.5 Torr)

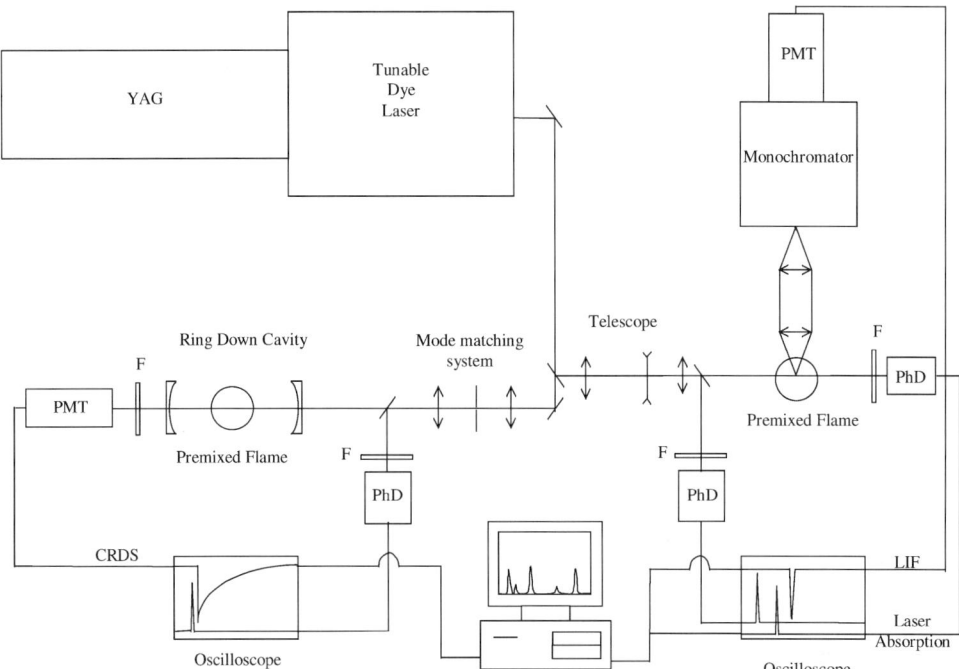

Fig. 1 Experimental arrangement: CRD and LIF–absorption set-up are separated for more convenient illustration. In practice, the CRD mirrors are removed when LIF or single-pass absorption measurements are performed. F: interferential filter, PhD: photodiode and PMT: photomultiplier.

8 bit, 500 Mhz bandwidth, 1 GS s^{-1} sampling rate) and later saved to a microcomputer. The entrance slit was parallel to the laser axis and the output slit was adjustable to provide a bandpass adapted to the fluorescence band according to the species studied. The recorded signal was normalized by the laser intensity, which was monitored by a photodiode (Hamamatsu S1722).

Concerning the CRD set-up, the laser beam (dye or OPO laser) was shaped with a system of a pinhole and two lenses to match approximately the TEM$_{00}$ transverse mode of the optical cavity, comprising two identical 25 mm diameter plano-concave mirrors separated by 40 cm (radius of curvature: 25 cm, coated at 387 nm for CN and CH with $R = 0.998$ and coated at 245 nm for NO with $R = 0.99$). The burner was placed at the center of the cavity. The waist diameter of the laser beam inside the cavity was estimated to be around 300 µm and was nearly constant along the flame length. The light transmitted by the cavity was recorded by the same photomultiplier used for LIF detection, placed just behind the second cavity mirror. A selected interferential filter was placed in front of the PMT to reject continuous background. The CRD signal was stored by the oscilloscope described above allowing the determination of the decay time of the laser pulse in the cavity. We then used specific software that provides, after a mathematical treatment, the absolute concentration of the measured species.

Single-pass laser absorption was only used to calibrate in absolute concentration the OH profile performed by LIF. The laser beam was collimated with a telescope and reduced to a diameter of 1 mm and then sent into the flame. The spatial resolution was good enough to make OH measurements in the burnt gases of the flame. The laser intensity fluctuations were monitored by a photodiode (Hamamatsu S1722) and then digitized and stored by the same apparatus as described for the CRDS measurements.

In this paper, we shall demonstrate the potential of the coupling of CRDS–LIF techniques for kinetic studies of minor species in a flame. All the techniques we used for temperature and species concentrations measurements as well as spectroscopic details are presented in Table 2.

Experimental method

The determination of concentration or mole fraction profiles results from several successive steps. First, the relative profile of a selected rotational population of the species is obtained by probing an adequate spectroscopic transition. This profile might be obtained either by CRD (integrated measurement) or by LIF (local measurement). Then this profile is converted into an absolute population profile by performing an additional measurement which consists in determining the absolute population at a given position in the flame (in the burnt gases for OH and NO and at the peak concentration location for intermediate species like CH and CN). This calibration results from an integrated absorptivity measurement of the excited transition performed by scanning the laser across the line. It can be obtained either by CRD or by single-pass laser absorption. Finally, the rotational population profile is converted into a mole fraction profile by taking into account the Boltzmann fraction variations and the temperature variations across the flame.

CH and CN species profiles were determined by CRDS, NO by LIF calibrated by CRDS and OH by LIF calibrated by single-pass laser absorption. The temperature profile was obtained by OH-LIF thermometry. The choice of the different methods is explained hereafter.

The spatial resolution and the sensitivity of the CRDS technique have been found to be well suited to providing reliable concentration profiles in premixed flames stabilized at low pressure.[8–10] This method is based on the measurement of the decay time of a laser pulse injected inside an optical cavity comprised of two highly reflective mirrors and undergoing hundreds or thousands of round-trips between them.[11] A detector placed just behind the second mirror, records the succession of pulses transmitted by the cavity after each round trip. Because the intensity of the pulse confined in the cavity undergoes attenuation at each round trip, the detector records the decay time waveform that can theoretically be fitted by an exponential. The measured ring-down time $\tau(\omega)$ of the cavity at frequency ω can be related to the total cavity losses, which are an algebraic sum of all the different losses according to the formula:

$$\tau(\omega) = \frac{t_r}{2[(1 - R(\omega)) + \alpha(\omega)l_s]}$$

where t_r is the round trip time of a light pulse in the cavity, $1 - R(\omega)$ is the reflection loss for the cavity mirrors and $2\alpha(\omega)$ is the round-trip absorbance for a sample present in the cavity with absorption coefficient $\alpha(\omega)$ and length l_s. The total cavity losses per round-trip are: $L(\omega) = 2[(1 - R(\omega)) + \alpha(\omega)l_s]$. The mirrors losses can be easily quantified from the decay time of the empty cavity. The loss per pass $L_s(\omega_o)$ due to the absorbing species and supposed to be homogeneous along the length l_s of the sample is: $L_s(\omega_o) = N_i \sigma_{ij} l_s$ where N_i is the number density of the species in its quantum level i, σ_{ij} is the peak cross section of the transition $j \leftarrow i$ and ω_o is the line centre frequency.

Table 2 Detection schemes

Species	Profile determination	Calibration method
OH	LIF Excitation: A–X (0–0) $R_2(7)$ (307 nm) Collection: A–X (0–0) band	Single pass absorption A–X (0–0) $R_2(7)$ (307 nm)
NO	LIF Excitation: A–X (0–0) $Q_1(26)$ (225.4 nm) Collection: A–X (0–2) band (247 nm)	CRDS A–X (0–2) $Q_1(33)$ (245 nm)
CH	CRDS B–X (0–0) $R_1(10)$ (387 nm)	CRDS B–X (0–0) $R_1(10)$ (387 nm)
CN	CRDS B–X (0–0) $R_1(28)$ (387 nm)	CRDS B–X (0–0) $R_1(28)$ (387 nm)
Temperature Measurement: LIF thermometry of OH Excitation: A–X (0–0) R branch Collection: A–X (0–0)		

Absolute concentration profiles of CH and CN species were obtained by CRDS. For both radicals the temporal decays were perfectly exponential and losses were easily deduced from the decay times of the pulse inside the cavity. This perfect exponential was not reached in the case of highly absorbing species like OH radical. Indeed, as shown in several papers,[12–15] CRD signals can exhibit multiexponential decays under highly absorbing conditions using a moderately narrowband pulsed dye laser. In that case, losses are no longer linear with the absorbance and thus, extraction of quantitative measurements requires some corrections.[15] Under our flame conditions, CRD losses due to OH species were affected by such non-linear effects and hydroxyl radical was preferably probed by LIF calibrated by single pass absorption. LIF was also preferred for determination of the relative NO concentration profile because of specific problems described in the next section and linked to the background absorption of "cold" NO surrounding the flame.

Although spatially integrated temperature measurement by CRDS could be obtained[15,16] in our flame, we have chosen the LIF thermometry technique to describe the flame temperature profile because it provides local measurement at the center of the flame.

Flame modelling

The calculation of the mole fraction profiles of the different species was realized by using the PREMIX and related flames codes. The mechanism used for the flames' modelling procedure is the last version of the Gas Research Institute mechanism (GRI 3.0).[6] It takes into account 53 species and 325 reversible reactions. This mechanism was tested over very wide conditions (laminar flames, laminar flame velocities, autoignition delays). In particular, it is well adapted to describe the methane oxidation chemistry as well as the processes of NOx formation and consumption. As reported in this paper, we have only tested the capabilities of the GRI 3.0 mechanism to provide $CH_4/O_2/NO$ flame chemistry for the different species measured in the study (i.e. OH, CH, CN and NO). For these calculations, the experimental temperature profile was considered. We present in a further section a comparison between our experimental measurements and the mole fraction profile calculated by the model.

Temperature profiles

Temperature measurements have been performed by OH-LIF thermometry by exciting the $R_1(4)$, $R_2(10)$, $R_2(9)$, $R_2(8)$, $R_2(12)$, $R_1(3)$, $R_2(7)$, $R_1(2)$, $R_2(6)$ transitions of the (A–X) (0–0) band. The broadband fluorescence signal $SF_{J''J'}$, following preliminary $J' \leftarrow J''$ excitation was collected over the entire (0–0) and (1–1) bands. Rotational populations $N_{J''}$ were obtained according to the following relation:

$$SF_{J''J'}(c) = G N_{J''} B_{J'J''} \frac{A}{A+Q} E(c)$$

where G is a constant depending on the collection efficiency, A and $B_{J''J'}$ are the Einstein coefficients for spontaneous emission and absorption respectively. $A = \Sigma_l A_{lJ'}$, where l denotes rotational levels of lower electronic state towards which allowed rotational transitions occur. Q is the total collisional quenching rate, $\Sigma A_{lJ'}/(\Sigma A_{lJ'} + Q)$ is the fluorescence quantum yield, and $E(c)$ is the laser intensity at the measurement point. The temperature is then obtained from the Boltzmann law. Fluorescence signals were corrected for rotational-dependent effects, linked to the absorption of the laser and the autoabsorption of the fluorescence, according to a procedure described elsewhere.[17] Corrections reached 150 K in the burnt gases. Measurements were sampled promptly after the laser pulse to minimize the effect of quantum yield variations with rotational levels.[18] Variations of the fluorescence decay rate were measured to be about 20% over the wide range of rotational energies investigated under our flame conditions. With such a variation, the temperature in the burnt gases, corrected for the variation of the fluorescence decay rate, was found to be 200 K lower than the temperature calculated by neglecting the influence of the quenching rate variation (but including spontaneous emission rates variations). By contrast, the first corrected temperature was found to be nearly identical to the temperature calculated by assuming a constant quantum yield. This indicates that the rotational-level dependence with quenching and

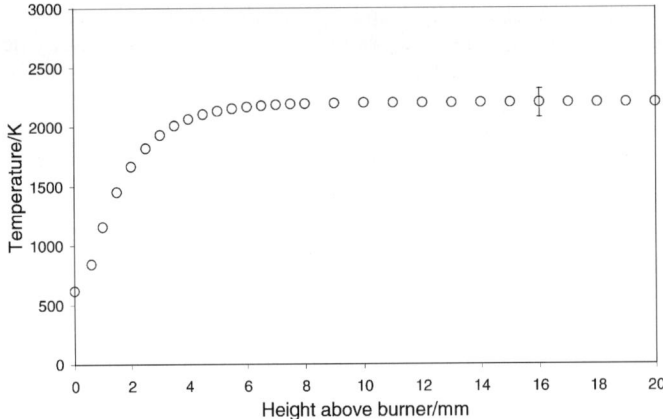

Fig. 2 Experimental temperature profiles provided by OH-LIF thermometry. The temperature at the burner surface was estimated to be 600 K.

spontaneous emission rates partially cancels. It is difficult to speculate about which measurement reflects the true temperature. Indeed, Kienle et al.[19] have shown that the true temperature would be obtained by measuring the LIF intensity at a given delay after the laser pulse; the value of this delay depending on the flame conditions (*i.e.* on the energy transfer processes). Their conclusions were based on a detailed rate equation model (the LASKIN program) which takes proper account of energy transfers during the LIF process. They particularly pointed out the need for effective Einstein coefficients after rotational energy transfers instead of the state-specific Einstein A coefficients usually considered. Their simulations indicate that temperatures not corrected for quantum yield variations were systematically underestimated. Although simulations still suffer from the lack of accurate spectroscopic and collisional cross section data, LIF thermometry assisted by a computational code is certainly the most accurate procedure to minimize the temperature error. In the absence of such a code at the laboratory, the selected temperature profile was obtained from the mean of the temperature profiles, corrected or not for fluorescence decay rate variations. The resulting profile is shown in Fig. 2. The final uncertainty was estimated to be ± 120 K ($\pm 5\%$) in the burnt gases. The temperature could be determined very close to the burner surface due to the significant OH concentration. The experimental temperatures serve as input data in the modelling calculation process. Several tests were performed to study the influence of the error of the temperature measurement on the modelled concentration profiles. Results indicate that the mole fractions may vary by a few percent, mainly due to a density effect. Nevertheless, the conclusions concerning the chemistry are unchanged; all the mole fractions of the investigated species varying in the same manner with the temperature.

Comparison between experimental and modelled mole fraction profiles

OH profile

The relative profile of OH was determined by LIF in the linear regime of excitation on the $R_2(7)$ line of the $A\,^2\Sigma^+$–$X\,^2\Pi$ (0–0) band around 307 nm. The fluorescence signal was collected in the (0–0) band. This line was chosen because of the weak variation of its Boltzmann fraction with temperature (about 7% in the temperature range (1200–2200 K) of the flame). These variations were taken into account in the final step leading to the mole fraction profile. The OH signal was averaged over 500 laser shots and the fluorescence signal was normalized by the laser intensity measured by a photodiode. We also measured the profile of the fluorescence decay rate of OH ($\Sigma A + Q$). Although the variations of this term were weak (less than 10%), we have taken them into account. The number density profile was then calibrated by single-pass laser absorption by probing the same rotational line. This measurement was made in the burnt gases at 13 mm above the porous burner by determining the integrated absorptivity of the $R_2(7)$ line. This region of the flame is characterized by very weak gradients of temperature and concentration and thus is well suited for spatially integrated absorption measurements. The absolute number density of OH was found to be $N_{J''} = 5.36 \times 10^{19}$ m^{-3}, which gives a mole fraction of 0.049 by using the temperature (2200 K) measured at 13 mm. The mole fraction profile of OH is shown in Fig. 3. One can notice a presence of OH very close to the porous burner, which indicates an important reactivity in this zone. As shown in Fig. 3, the agreement between our experimental and calculated profiles is very good. Furthermore, our experimental mole fraction (0.049 in the burnt gases) is in excellent agreement with that determined by Juchmann (0.047) (see Table 3).

CH profile

The CH profile determination was performed by CRDS in the B–X(0–0) vibrational band around 387 nm by probing the $R_1(10)$ line, which is characterized by a relatively constant Boltzmann fraction over the range of measured temperatures. Moreover, this line does not present any interference with CN transitions ((B–X) (0–0)) as can be seen in Fig. 4. Losses per pass were determined from the decay time of the CRDS signal averaged over 300 laser shots. The profile of the rotational population ($N'' = 10$) of CH was obtained by measuring the profile of the losses per pass at the peak of the $R_1(10)$ line (on-resonance losses) and by subtracting the off-resonance losses profile under the same flame conditions. This profile was calibrated into an absolute value by determining the integrated absorptivity of the $R_1(10)$ line at the peak of the CH profile (3.4 mm above the burner). The obtained peak number density $N_{J''} = 2.58 \times 10^{16}$ m^{-3} gives a mole fraction of 2.9×10^{-5} ($T = 2000$ K) in excellent agreement with that obtained by Juchmann et al.[7] (2.7×10^{-5}). A comparison of our CH profile with that calculated using the GRI 3.0 mechanism is shown in Fig. 5. Whereas the shape and the position of the experimental profile are perfectly

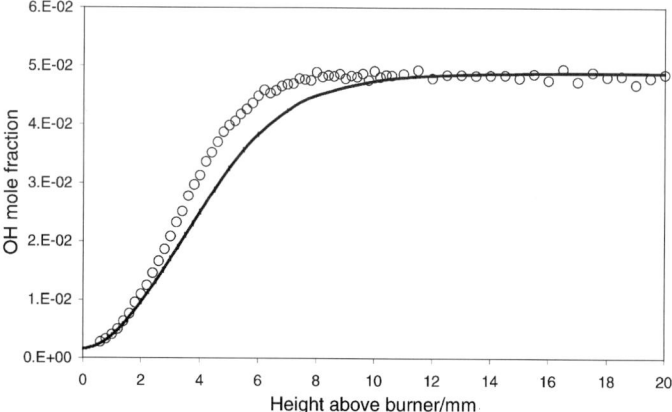

Fig. 3 Measured (○) and calculated (——) OH mole fraction profiles. The experimental profile has been calibrated in the burned gases by single pass absorption.

Table 3 Mole fractions

	Our work		Juchmann's work
Species	Experimental measurements	Calculated values	Experimental measurements
OH (final)	0.049	0.05	0.047
NO (cold gases)	0.018 (estimated)	0.0148	0.018 (estimated)
NO (final)	0.0165	0.0142	0.009
CH (peak)	2.9×10^{-5}	4×10^{-5}	2.7×10^{-5}
CN (peak)	3.3×10^{-6}	3.2×10^{-6}	8.3×10^{-7}

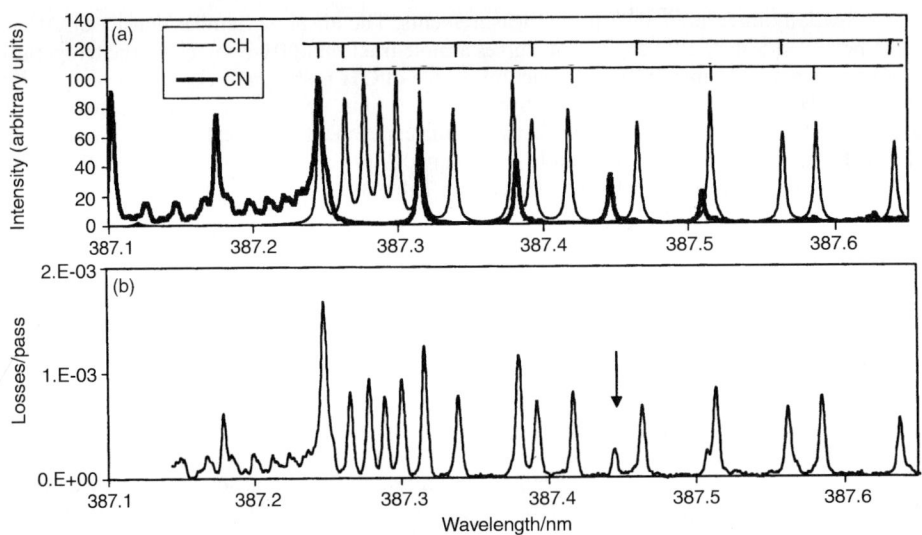

Fig. 4 (a): Superimposed CH and CN spectra independently calculated with the LIFBASE code.[20] (b): Experimental spectrum obtained in the CH_4–O_2–NO flame exhibiting R lines of the (B–X) (0–0) CH band and CN lines of the B–X (0–0) band.

reproduced, the predictions indicate a peak CH mole fraction 27% higher than the experimental value. As regards the excellent reproducibility of the experiments realized by CRDS, we estimate the uncertainty of the measurements to be 15% (taking into account the statistical error, the temperature error, the imperfect homogeneity of the flame, the error linked to the alignment accuracy of the beam into the cavity *etc.*) while the uncertainty of Juchmann's measurements (performed by coupling LIF and Rayleigh scattering techniques) were estimated by the authors to be 30%. Thus, the difference between the calculated value of the CH peak mole fraction (4×10^{-5}) and the experimental one (2.9×10^{-5}) is outside the experimental error range. The overestimation from the model calculation has already been observed in previous studies related to CH measurements in different kinds of flames[1,2,5] using the GRI 2.11 and GRI 3.0 mechanisms. With the new GRI 3.0 version, for which some rate constants implying the CH radical have been modified, Berg et al.[21] obtained a very satisfactory agreement for a CH_4/air flame ($\Phi = 1.07$) while the authors noticed an overestimation under rich flame conditions and an underestimate under poor flame conditions. This tendency is confirmed by the paper of Thoman and McIlroy[5] in

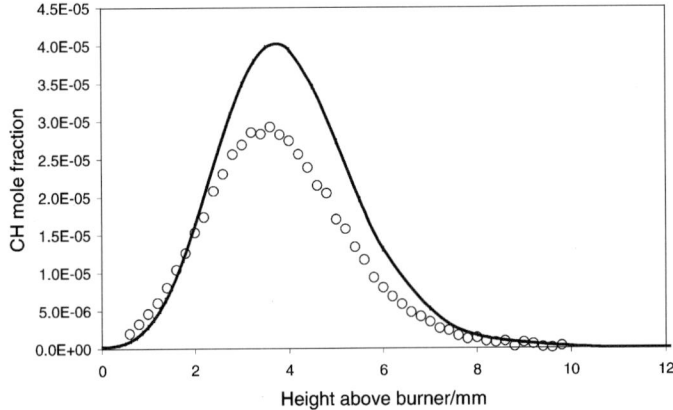

Fig. 5 Measured (○) and calculated (———) CH mole fraction profiles. The experimental profile has been calibrated by CRDS at the peak position.

which they present a comparison between experimental CH profiles (obtained by CRDS in $CH_4/O_2/Ar$ low-pressure flames) with different calculated profiles provided by several mechanisms (including the GRI2.11 and GRI3.0). They observed that both these mechanisms overpredict the formation of CH and that the overestimation is more pronounced for rich flame conditions. In conclusion, they suggest that some refinements of the methane oxidation chemistry used by the models is necessary and notably advise that the $CH + O_2$ reaction rate as a function of temperature should be carefully examined. Finally, the GRI3.0 mechanism has not been tested, to our knowledge, in non-diluted flames (for very high temperature conditions). Therefore, it seems that additional studies of $CH_4/O_2/NO$ flames, with different richness, might contribute to a model implementation.

CN profile

The measurement procedure for the CN profile is identical to that for the CH radical. Fig. 6 represents the experimental spectrum of CN issued from the B–X (0–0) band around 387 nm. We chose to determine the CN profile by recording the losses per pass of the $R_1(28)$ line (see Fig. 6). Although there is a partial overlap with the $R_2(28)$ line, the integrated absorptivity of the $R_1(28)$ line could be extracted after a very accurate deconvolution procedure (performed with the ORIGIN software). This transition presents a Boltzmann fraction that does not vary significantly within the temperature range of the flame. Calibration of the losses per pass profile was made by CRDS at 4.8 mm above the porous burner (*i.e.* at the peak of the CN profile) and led to a peak number density $N_{J''} = 1.91 \times 10^{15}$ m^{-3} which converts into a mole fraction of 3.3×10^{-6} by considering the experimental temperature (2100 K). This value is four times greater than that reported by Juchmann et al.[7] (8.3×10^{-7}) (Table 3). Our value seems to be confirmed by analysing the spectrum in Fig. 4. The line indicated by an arrow corresponds to the sum of the $R_1(2)$ and $R_2(2)$ lines of CN. A fast comparison of their intensity, corrected for their respective Einstein coefficients for absorption, with that of the $R_1(10)$ line of CH leads to a CN concentration 15 times lower than the CH one at 3.4 mm above the burner. This value is in good agreement with the previous value measured at 4.8 mm by taking into account the shape of the experimental CN profile. Moreover, the CRD technique provides direct absolute concentration measurements in contrast to the indirect method used by Juchmann et al.[7] As pointed out by the authors, saturation of the LIF signal is easily reached for CN species. This implies the use of very low laser

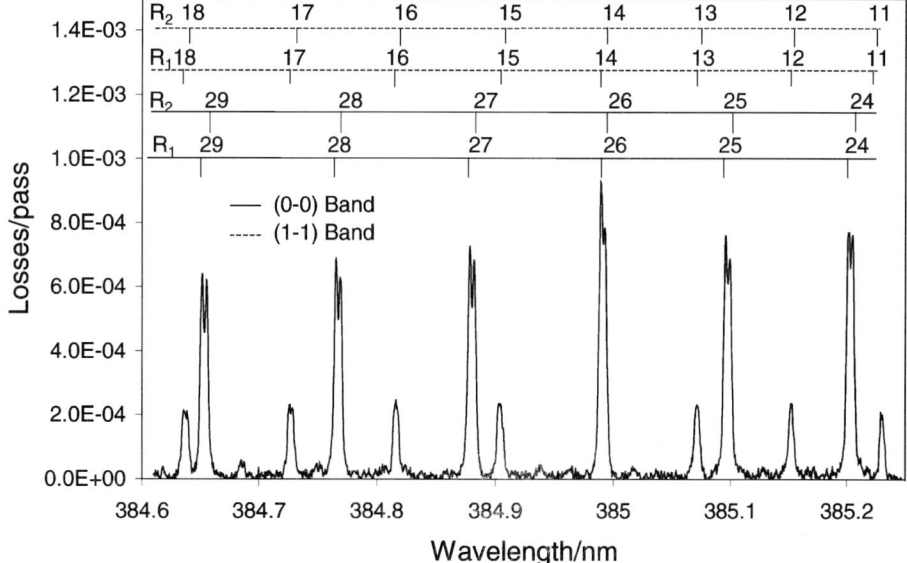

Fig. 6 Part of the spectrum of the (0–0) and (1–1) R bands of CN provided by CRDS in the CH_4–O_2 flame seeded with NO.

energies leading to measurements affected by an important background noise. Moreover, partial saturation of LIF signals would have led to an underestimate of the CN concentration. Such drawbacks do not affect the CRD measurements, which appear to be more reliable, especially for minor species detection.

Finally, our experimental mole fraction profile of CN has been compared to the profile calculated by using the GRI 3.0 mechanism (see Fig. 7). Shape, position and peak mole fraction of CN are perfectly reproduced by modelling in contrast to the work of Juchmann et al.[7] CN species essentially result from the reaction of HCN with H, O and OH. In the $CH_4/O_2/NO$ flame, HCN molecule is mainly formed by the reburning reactions between NO and CH and also with CH_2, CH_3 and HCCO. The excellent prediction of CN by using the GRI 3.0 mechanism might indicate that the reaction rates involved in the CN chemistry are well taken into account under our flame conditions.

NO profile

The relative profile of NO was determined by LIF by exciting the $Q_1(26)$ line of the $A\ ^2\Sigma^+ - X\ ^2\Pi$ (0–0) band at 225.4 nm. The fluorescence signal was collected over the entire A–X (0–2) vibrational band. We chose this line because of the weak variation of the Boltzmann fraction within our temperature range. The collection band was chosen to reduce the contribution of the continuous emission issued from the flame and to limit the possible interference with the Schumann–Runge bands of the molecular oxygen.[22] The relative concentration profile of NO was obtained by measuring the peak intensity value of the fluorescence signal averaged over 500 laser shots and normalized by the laser intensity.

Fig. 8 shows the resulting concentration profile of NO in the flame calibrated as an absolute value according to a procedure described next. The steep decrease just above the burner is mainly due to the gas density decrease correlated with the temperature increase. By taking into account the temperature profile, one obtains the mole fraction profile, not corrected for quantum yield variations. The shape of this mole fraction profile shows a slight decrease with the height above the burner. The conversion of concentration into mole fraction was particularly sensitive to the temperature gradient. Thus, values reported in the first 2 mm are affected by a large uncertainty. The mole fraction profile has been corrected for the quantum yield variations as shown in Fig. 8. An important effect of the quantum yield variations can be seen in the flame front region. The experimental profile of the fluorescence decay rate $\Sigma A + Q$ was determined by recording the fluorescence lifetime for different heights above the burner. As can be seen in Fig. 9, the experimental profile is astonishingly well reproduced by the calculation performed by taking into account the most important quenchers present in the flame (O_2, H_2O, CO, CO_2 and NO) and their respective collisional data available in the literature.[23] Individual quenching rate for each collider was calcu-

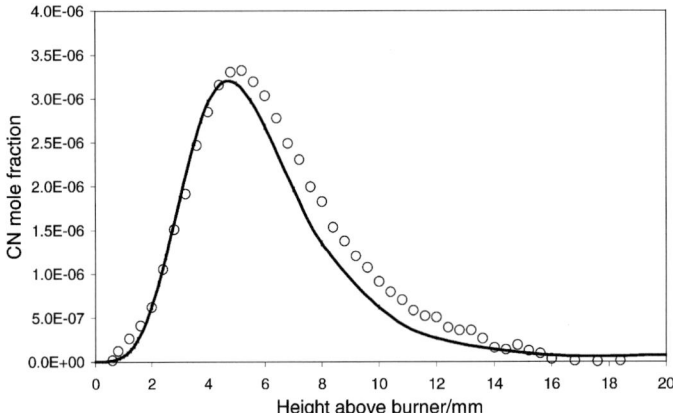

Fig. 7 Measured (○) and calculated (——) mole fraction CN profiles. The experimental profile has been calibrated by CRDS at the peak position.

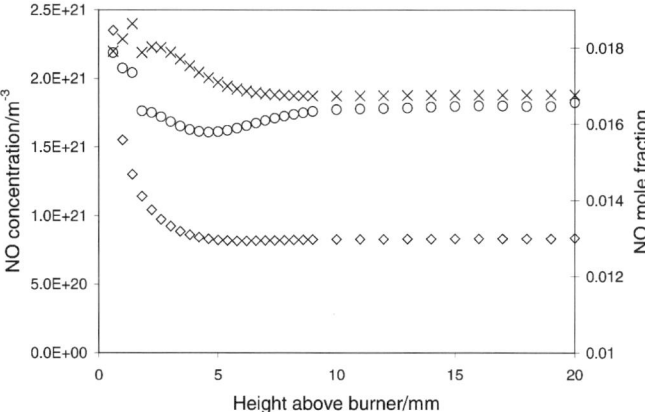

Fig. 8 Experimental profiles of NO: (◇) absolute concentration profile (m^{-3}), (×) mole fraction profile without quantum yield correction and (○) mole fraction profile corrected for quantum yield variations. The (×) profile has been slightly vertically shifted for more convenient illustration. The final NO mole fraction in the burnt gases is 0.0165.

lated according to the formula: $q_i = \Sigma \sigma_i v_i N_i$ where σ_i corresponds to the efficient collisional cross section of the species i, v_i is its average collision velocity and N_i its absolute concentration calculated by the GRI 3.0. Finally, the theoretical fluorescence decay rate ($\Sigma A + Q$) was obtained by summing the individual quenching rates and by considering the state-specific Einstein A coefficients subsequent to the $Q_1(26)$ excitation.

Calibration of the NO profile by absorption techniques is not an easy task because this stable species is present all around the flame in the low-pressure cell. Therefore, the line-of-sight absorption measurement must be corrected for the background absorption originating from the "cold" NO. Juchmann et al.[7] have reduced this effect by placing their multipath arrangement used for absorption as close as possible to the burner head, however, this procedure is very delicate. Another method consists of probing high rotational levels having a significant Boltzmann fraction within the flame temperature range. However, the radial temperatures measured by a thermocouple along the laser axis between the edge of the burner and the windows of the cell were found to be too high to omit the correction for "cold" NO absorption. Such a correction is described in detail elsewhere.[24] Under our flame conditions, which exhibit particularly hot temperatures, a correction of about 40% was calculated when probing the $Q_1(33)$ line of NO in the A–X (0–2) band. We failed to obtain accurate quantitative measurements in the (0–0) and (0–1) bands because of too strong background absorption. The integrated absorptivity of the $Q_1(33)$ line was

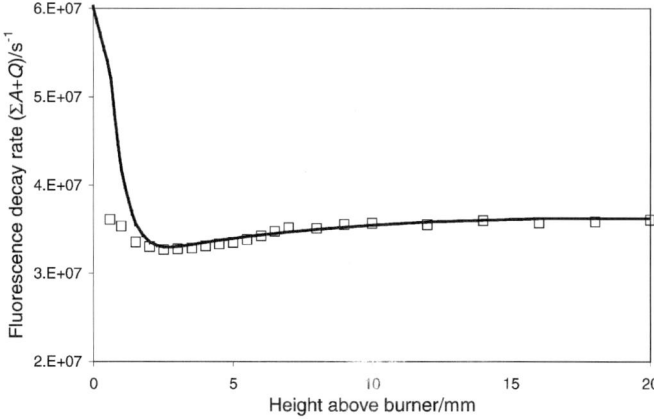

Fig. 9 Measured (□) and calculated (——) profiles of the NO fluorescence decay rate ($\Sigma A + Q$) in the CH_4–O_2–1.8% NO flame.

measured by CRDS in the burnt gases of the flame (15 mm). We found an absolute concentration of 8.32×10^{20} m^{-3} *i.e.* a mole fraction of 0.0165 for a temperature of 2200 K. Considering the global uncertainties of the NO calibration procedure (estimated to be $\pm 40\%$), the measured NO calibration mole fraction in the burnt gases appears to be very close to the one injected in the cold mixture (0.018). Such a weak consumption of NO was confirmed by the modelling (GRI 3.0 mechanism).

As shown in Fig. 10, the model predicts, at the burner surface, a mole fraction of 0.0148, which might be assigned to a very fast NO consumption (injected mole fraction = 0.018). In fact, the model predicts in our case (flame front very close to the burner) an important reactivity just above the burner, confirmed experimentally by a significant OH mole fraction (Fig. 3) near the porous burner. This prompt reactivity leads to an increase in the number of moles at the burner surface position and so to an artificial decrease in the NO mole fraction by a dilution effect. Indeed, we estimated from the model that only 2% of the injected NO has effectively been involved in chemical reactions at the porous burner. This value is derived from the ratio of the mole fraction of nitrogenous compounds other than NO (HCN, NO_2 ...) with the total mole fraction of nitrogenous compounds at $x = 0$ given by the model. Thus, the decrease in the NO mole fraction near the burner, as calculated by the model $X_{NO} = 0.0148$, is mainly due to a dilution effect whose importance had already been highlighted by Berg *et al.*[25] We have also analysed the dilution effect by replacing NO by Ar and by simulating the argon mole fraction profile in a $CH_4/O_2/1.8\%Ar$ flame with the same temperature profile as that used for the NO-seeded flame. We can make this assumption because of the poor amount of the substituted gas, which should not significantly modify the temperature profile. We chose to replace NO by Ar because of its negligible reactivity. Under these conditions, an argon mole fraction of 1.4% has been calculated by the model at the porous burner. This value is very close to that obtained for the NO species. This can be explained by an increase in the number of moles due to the high chemical reactivity of the flame. In our case, the increase in the number of moles due to the reactivity is particularly pronounced because the cold reactive mixture is not diluted in nitrogen or argon (in contrast with standard flames like methane–air *etc.*). We also found that the prompt reactivity was clearly linked to the high temperatures very close to the burner. Calculations that we made on the $CH_4/O_2/1.8\%Ar$ flame also showed that the dilution effect is more or less constant (only a few percent variation) over the whole height of the flame. In principle, we should consider this effect when analysing the experimental mole fraction profiles.

Detailed flame structures studies performed by gas chromatography–mass spectrometry and molecular beam mass spectrometry techniques lead to the knowledge of the most important species of the flame. The concentration profiles (N_i) obtained can then be converted into mole fraction profiles (X_i) along the whole flame by imposing the total mole fraction to be equal to one. In contrast, flame studies carried out by laser diagnostics lead to the knowledge of a few species only. Thus, the dilution effect can only be considered by means of a modelling calculation.

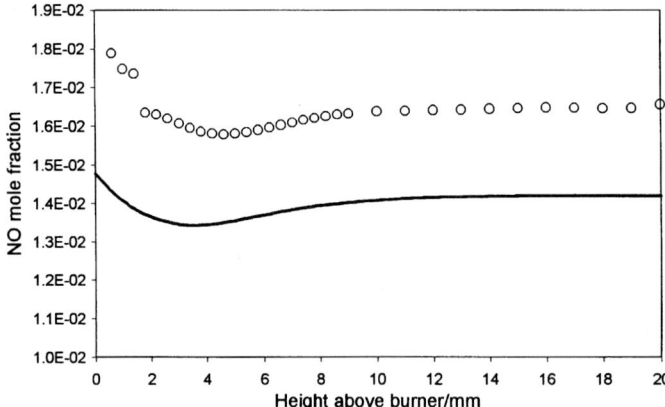

Fig. 10 Measured (○) and calculated (——) NO mole fraction profiles. The experimental profile has been calibrated in the burned gases by CRDS.

A more extensive study of these dilution effects is still in progress. First calculations have indicated that they were more or less constant along the whole flame. Thus, the shape of the NO profile is well representative of the NO chemistry and not of a dilution effect. These aspects are discussed in the following final section.

NO consumption discussion

In the present study, we were able to measure the relative NO profile from the burnt gases to 0.6 mm above the porous burner. A mole fraction variation of 10% has been observed in the flame front. This tendency is in excellent agreement with the measurements performed by Williams and Fleming[26] in a $CH_4/O_2/Ar$ flame seeded with NO. In $CH_4/O_2/N_2$ flames seeded with NO, Berg et al.[25] noticed a NO consumption of about 10% in a poor flame ($\Phi = 0.8$) and of 40% in a rich flame ($\Phi = 1.27$). However, the authors do not present any NO profiles in these flames. Juchmann et al.[7] do not report NO LIF measurements near the burner. Their NO mole fraction in the burnt gases was determined to be equal to 0.009 by using a multipath absorption technique in the (0–0) band. Thus, considering an initial mole fraction of 0.018, one obtains a NO reburning efficiency of about 50%. As pointed out in the previous section, NO calibration by absorption in the (0–0) band is highly sensitive to the apparent increase in the absorption path length arising from the persistence of NO in the low-pressure cell. Although the authors took care to reduce this pathlength by placing their optical device very close to the burner, accurate calibration of NO in the (0–0) band is affected by large uncertainties (not specified in their paper[7]) as pointed out by Williams and Fleming.[26]

In this study, we took care to use a systematic correction procedure for determination of the absolute concentration of NO but the mole fraction value of 0.0165 measured in the burnt gases by CRDS should be treated with caution because of uncertainties ($\pm 40\%$) related to the calibration procedure.

However, we are very confident about the shape of the relative NO profile, particularly in the flame front region. It is clear from Fig. 10 that the shape of the experimental NO profile is in very good agreement with the calculated one. In order to explain the inverted bell-shaped of the NO mole fraction profile, two zones of the flame have to be considered (denoted as zone 1 and zone 2 in Fig. 11). The first corresponds to a fast consumption of NO and the second to a re-formation of this compound. From the $x = 0$ position to $x = 4.6$ mm, a decrease in the mole fraction of NO is observed and can be explained qualitatively by considering the reactions occurring between NO and the CH_i species, namely corresponding to the prompt NO mechanisms. Indeed, from the experimental profiles of CH and NO reported in Fig. 11, it can be noticed that the maximum of the CH profile ($x = 3.4$ mm), is very close to the position of the minimum of the NO profile

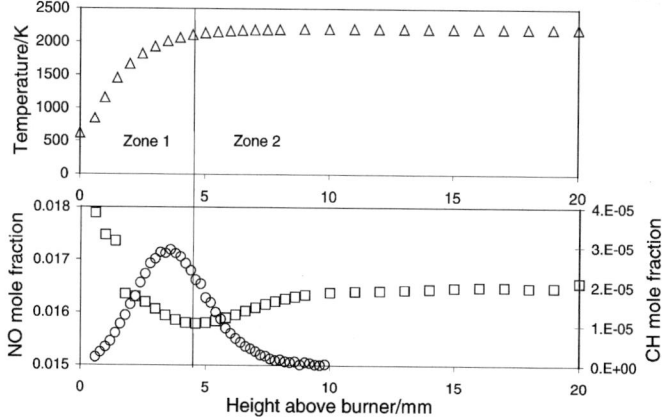

Fig. 11 Mole fraction profiles of CH (○), NO (□) and temperature profile (△) measured in the CH_4–O_2–1.8% NO flame.

($x = 4.6$ mm). It is well known that all the CH_i species are formed and interact around the flame front zone and that the CH radical is well representative of this zone. Therefore, according to our measurements, it appears that the prompt NO processes (CH_i + NO), leading to the destruction of NO at moderate temperature, are predominant in the first denoted zone. This consumption is partly attenuated by the production of thermal NO. These chemical interactions are confirmed by the GRI3.0 model, which accurately foresees the peak position and the shape of the profiles of CH and NO species (see Fig. 5 and 10). The second zone of the flame (Fig. 11) highlights an increase in the mole fraction of NO until equilibrium is reached. This shape is linked to the combination of two simultaneous phenomena explained hereafter. First, it can be seen in Fig. 11 that the formation of NO is intimately related to the decrease of CH (and CH_i species) that no longer contribute to the destruction of NO. Moreover, this effect is linked to the increase in the temperature of the flame that makes the production of thermal NO (mainly via the reaction N + OH = NO + H, N atoms being produced via the reaction Ch + NO = N + HCO) more important than in the first zone. The combination of both these factors leads to the production of NO in the burnt gases. As for the first zone, the experimental profile shape of NO is well reproduced by the GRI 3.0 model and the competition between prompt and thermal NO processes seems to be very well taken into account by the NO chemical mechanisms involved in this model.

Conclusion

This paper reports OH, CH, CN, NO mole fraction profiles measured in a stoichiometric CH_4–O_2 flame seeded with 1.8% NO and stabilized at 15 hPa. Absolute concentration profiles could be obtained by advantageously coupling LIF and CRDS techniques. A moderately narrow-band tunable laser and inexpensive mirrors (characterized by a reflectivity between 0.99 and 0.997) were well suited for the measurements performed in this study. The spatial resolution of the CRD technique was sufficient to describe accurately the concentration profiles of intermediate species like CH and CN radicals in the investigated low-pressure flame. The measurement accuracy of the maximum concentration is estimated to be $\pm 15\%$ for these species. An additional error ($\pm 5\%$) due to the temperature determination has to be considered for the peak mole fraction determination. The OH mole fraction profile was obtained by LIF calibrated by single-pass absorption. The mole fraction reached in the burnt gases was determined with about the same accuracy as for the CH and CN species. The LIF–absorption method was selected because CRD signals of the highly absorbing OH species were affected by non-exponential time decays. Finally, the NO mole fraction was measured by LIF calibrated by CRD in the burnt gases. Absorption measurements were found to be seriously disturbed by the presence of the stable NO species in the whole low-pressure cell. This drawback has been largely limited by probing a spectral transition in the $v'' = 2$ vibrational state of the A–X band of NO. Nevertheless, the global uncertainty of the absolute measurement of NO was estimated to be $\pm 40\%$. In contrast, the relative profile of NO, recorded by LIF corrected for quantum yield variations, was obtained with a relative error estimated to be $\pm 5\%$. The NO mole fraction in the burnt gases (0.0165) was found to be about 10% lower than the seeded mole fraction (0.018), indicating that reburning processes are apparently not very significant in this stoichiometric flame.

Our results were compared to those obtained by Juchmann et al.[7] in a quite similar flame using an indirect method of calibration for CH and CN radicals (LIF–Rayleigh) and the LIF–absorption coupling for OH and NO species. A very good agreement between both studies is found for OH and CH species but a more important NO consumption ($X_{NO} = 0.009$ in the burnt gases) is reported in the study of Juchmann et al.[7] However owing to the large uncertainties of both methods used for NO calibration, it is difficult to draw conclusions concerning this species. By contrast large discrepancies were observed for CN species which cannot be attributed to the experimental error. We believe that for such a minor species, the direct CRD technique provides more reliable quantitative measurements than the LIF–Rayleigh method. Indeed the CN absorption spectrum could be recorded by CRD with an excellent signal to noise ratio in the NO-seeded flame. By contrast, CN measurements performed in a methane–air flame[27,28] were found to be more subject to sensitivity problems.

Our experimental mole fraction profiles were very well reproduced by the predicted ones, calculated using the GRI 3.0 mechanism. In particular the CN chemistry appears to be well taken into account with excellent agreement between the experimental and calculated mole fraction profiles of this species. We notice however an overprediction of the CH concentration. Concerning the NO species, the experimental inverted bell-shaped profile was very well predicted. Both experimental and predicted NO profiles indicate a decrease of about 10% of the NO mole fraction between the burner surface and the burnt gases. In fact this weak consumption of NO results from two competitive effects between NO reburning processes in the flame front region and thermal NO production due to the high temperature of the flame. Thus, the GRI 3.0 mechanism appears well suited to predict NO reburning processes. As the thermal NO formation appears to be well reproduced, it might indicate that the temperature measurements performed by OH-LIF thermometry were also reliable. Further investigations of NO reburning in rich CH_4–O_2 flames are needed to validate the GRI 3.0 under non-diluted rich flame conditions.

References

1 J. Luque and D. R. Crosley, *Appl. Phys. B*, 1996, **63**, 91.
2 I. Derzy, V. A. Lozovsky and S. Cheskis, *Chem. Phys. Lett.*, 1999, **306**, 319.
3 X. Mercier, P. Jamette, J. F. Pauwels and P. Desgroux, *Chem. Phys. Lett.*, 1999, **305**, 334.
4 X. Mercier, E. Therssen, J. F. Pauwels and P. Desgroux, *Chem. Phys. Lett.*, 1999, **299**, 75.
5 J. W. Thoman and A. McIlroy, *J. Phys. Chem. A*, 2000, **104**, 4953.
6 G. P. Smith, D. M. Golden, M. Frenklach, N. W. Moriarty, B. Eiteneer, M. Goldenberg, C. T. Bowman, R. Hanson, S. Song, J. Gardiner, V. Lissianski and Z. Qin, *Meth. Comb. Kin. Mech, Version 3.0*, available at http://www.me.berkeley.edu/gri-mech/
7 W. Juchmann, H. Latzel, D. L. Shin, G. Peiter, T. Dreier, H. R. Volpp, J. Wolfrum, R. P. Lindstedt and K. M. Leung, *XXVIIth Symposium (International) on Combustion*, The Combustion Institute, Pittsburgh, 1998, p. 469.
8 A. McIlroy, *Chem. Phys. Lett.*, 1998, **296**, 151.
9 S. Cheskis, I. Derzy, V. A. Lozovsky, A. Kachanov and D. Romanini, *Appl. Phys. B*, 1998, **66**, 377.
10 J. J. Scherrer, K. W. Aniolek, N. P. Cernansky and D. J. Rakestraw, *J. Chem. Phys.*, 1997, **107**, 6196.
11 A. O'Keefe and D. A. G. Deacon, *Rev. Sci. Instrum.*, 1988, **59**, 2544.
12 P. Zalicki and R. N. Zare, *J. Chem. Phys.*, 1995, **102**, 2708.
13 R. T. Jongma, M. G. H. Boogarts, I. Holleman and G. Meijer, *Rev. Sci. Instrum.*, 1995, **66**, 2821.
14 I. Labazan, S. Rudic and S. Milosevic, *Chem. Phys. Lett.*, 2000, **320**, 613.
15 X. Mercier, E. Therssen, J. F. Pauwels and P. Desgroux, *Combust. Flame*, 2001, **125**, 656.
16 G. Meijer, G. H. Boogaarts, R. T. Jongma, D. H. Parker and A. M. Wodtke, *Chem. Phys. Lett.*, 1994, **217**, 112.
17 P. Desgroux, L. Gasnot, J. F. Pauwels and L. R. Sochet, *Appl. Phys. B*, 1995, **61**, 401.
18 K. J. Rensberger, J. B. Jeffries, R. A. Copeland, K. Kohse-Hoinghaus, M. L. Wise and D. R. Crosley, *Appl. Opt.*, 1989, **28**, 3556.
19 R. Kienle, M. P. Lee and K. Kohse-Höinghaus, *Appl. Phys. B*, 1996, **62**, 583.
20 J. Luque and D. R. Crosley, *LIFEBASE: Data and Spectral Simulation Program, Version 1.2*, SRI International Report MP 98-021, 1998.
21 P. A. Berg, D. A. Hill, A. R. Noble, G. P. Smith, J. B. Jeffries and D. R. Crosley, *Combust. Flame*, 2000, **121**, 223.
22 P. Desgroux, P. Devynck, L. Gasnot, J. F. Pauwels and L. R. Sochet, *Appl. Opt.*, 1998, **37**, 4951.
23 P. H. Paul, J. A. Gray, J. L. Durant, Jr. and J. W. Thoman, Jr., *Chem. Phys. Lett.*, 1996, **259**, 508.
24 L. Pillier, X. Mercier, J. F. Pauwels and P. Desgroux, in preparation.
25 P. A. Berg, G. P. Smith, J. B. Jeffries and D. R. Crosley, *XXVIIth Symposium (International) on Combustion, The Combustion Institute*, Pittsburgh, 1998, p. 1377.
26 B. A. Williams and J. W. Fleming, *Combust Flame*, 1994, **98**, 93.
27 X. Mercier, L. Pillier, J. F. Pauwels and P. Desgroux, *C. R. Acad. Sci. Paris*, t.2, Série IV, 2001, in press.
28 J. Luque, J. B. Jeffries, G. P. Smith and D. R. Crosley, *Combust. Flame*, in press.

Laser absorption spectroscopy diagnostics of nitrogen-containing radicals in low-pressure hydrocarbon flames doped with nitrogen oxides

V. A. Lozovsky,[a,b] I. Rahinov,[a] N. Ditzian[a] and S. Cheskis*[a]

[a] *School of Chemistry, Sackler Faculty of Exact Sciences, Tel Aviv University, Ramat Aviv, Tel Aviv 69978, Israel*
[b] *Semenov Institute of Chemical Physics, Russian Academy of Sciences, 4 Kosygin str., Moscow 117977, Russian Federation*

Received 2nd March 2001
First published as an Advance Article on the web 15th October 2001

Absolute concentration profiles of NH_2 and HNO have been measured in low-pressure methane/air flat flames doped with small amounts of NO and N_2O. Addition of a small amount of nitrogen oxides does not alter significantly the flame speeds, temperature profiles and other parameters of the relatively well-understood methane/air flames. Intracavity laser absorption spectroscopy (ICLAS) and cavity ring-down spectroscopy (CRDS) are high-sensitivity techniques used to measure absolute concentrations of minor species in flames. In this work ICLAS is used to monitor NH_2 and HNO, whereas CRDS is used for temperature measurements using OH spectra in the UV range. The (090)–(000) and (080)–(000) bands of the $\tilde{A}\,^2A_1$–$\tilde{X}\,^2B_1$ electronic transition of NH_2 and (100)–(000) and (011)–(000) bands of the $\tilde{A}\,^1A''$–$\tilde{X}\,^1A'$ transition of HNO are used. Methane flames of different equivalence ratios are used. NH_2 and HNO are observed in the flame as well as in the zone surrounding the flame, closer to the walls of the low-pressure chamber where the burner is located. An absorption originating from the species in this zone can affect substantially the results of line-of-sight experiments. A slow flow of nitrogen through the optical window holders was added in order to separate the spectra of HNO originating from the central flame zone. Calculations based on the commonly used GRI-Mech chemical mechanism predict two maxima in the HNO concentration profile in the NO doped flames. The first is located in the vicinity of the burner, and the second is closer to the luminescence flame zone. We were able to observe the first maximum, and its measured location agrees well with prediction. On the other hand, GRI-Mech strongly underpredicts the observed absolute concentration of HNO in this maximum. The measured absolute concentrations of NH_2 are in reasonable agreement with the GRI-Mech predictions.

Recently, there has been increased attention given to the application of sensitive absorption spectroscopy methods, especially CRDS, to combustion diagnostics.[1–9] One of the advantages of absorption spectroscopy is the possibility of using it to determine absolute concentration measurements of reactive intermediates in flames. These measurements provide important information for the creation and testing of chemical mechanisms of combustion. ICLAS and CRDS are high-sensitivity methods of absorption spectroscopy that sometimes provide much better sensitivity

DOI: 10.1039/b101981n

than *e.g.* laser induced fluorescence (LIF). In our group ICLAS and CRDS have been used to measure absolute concentration profiles in low-pressure premixed hydrocarbon flat flames. ICLAS provides a better sensitivity than CRDS[1,8] and allows a relatively broad spectral range to be recorded during one laser pulse. On the other hand with CRDS laser frequency conversion (for example frequency doubling) is possible, and therefore more radicals can be detected. Radicals absorbing light in the UV spectral range (OH, CH, NH), cannot be observed by ICLAS, at least at the moment.

In this work we studied the absorption of NH_2 and HNO in low-pressure hydrocarbon flames doped with a small amount of N_2O or NO. Both species are very imortant for NO_x combustion chemistry. Fast reaction of NH_2 with NO is the basis of the mechanism of the thermal De-NO_x process, wherein ammonia is added in the post-combustion zone to reduce the NO_x emission level in the exhaust gases.[10] This radical also plays an important role in the reburning process.[11] The sensitivity of LIF for NH_2 is lower than for many other species due to very effective quenching of the upper electronic state in collisions with other molecules. Even for rare gases the quenching rate constant is close to the gas kinetic value. NH_2 radical was detected by laser spectroscopy only in ammonia flames or in flames with nitrous oxide as an oxidizer.[12-15] Absorption spectroscopy has been used in several studies of NH_2 at high temperatures[12,16-18] and the absorption coefficient has been measured.

Nitrosyl hydride, HNO, is an important intermediate in the mechanism of formation of NO_x pollutants during combustion.[19] It also plays an important role in the thermal De-NO_x process[19,20] and in the mechanism of thermal reduction of NO by H_2.[21] The $\tilde{A}\,^1A''$–$\tilde{X}\,^1A'$ transition of HNO was first observed by Dalby[22] in absorption and has since been studied extensively by absorption,[23-26] emission[27-30] and LIF.[31,32] Despite the extensive investigation of HNO, there are few reported studies of its kinetics using spectroscopic detection. HNO has not been detected in flames in its ground state. A possible reason for this is that the fast quenching of the upper electronic level[31] decreases the sensitivity of LIF, which is usually the method of choice for such studies.

Recently we used CRDS and ICLAS for measurements of the concentration profiles of OH, HCO, $^1CH_2$, CH in low-pressure hydrocarbon flames[1,33-36] and of NH, NH_2 and HNO[37-39] radicals in those flames doped with nitrogen oxides. The experimental results were compared with the calculations based on the PREMIX code[40] and the GRI-Mech chemical mechanism.[41] For most species reasonable agreement between the experimental results and calculations was found, at least in the profile shapes and the positions of the concentration maxima. However, for NH_2 and HNO essential absorption was observed even at relatively high distances above the burner (up to 55 mm), strongly contradicting the calculations. We have proposed[38] that this absorption is caused by the presence of species studied in the zone surrounding the flame. In this work we modified our apparatus in order to decrease the influence of that zone. We injected a slow flow of inert gas close to the windows of the vacuum chamber flushing out the distant zone of the chamber. The HNO spectrum in the flames doped with N_2O was found to disappear as a result of this procedure, proving that the HNO spectrum in the N_2O-doped flame originated entirely from the zone surrounding the flame. In this work we have performed similar experiments with hydrocarbon flames doped with NO instead of N_2O and we were able to observe HNO spectra originating from the central, flame zone.

Experimental

The schematic diagram of the experimental apparatus used in this work is shown in Fig. 1. The apparatus is based on a McKenna flat flame stainless steel burner (diameter 6 cm) which is located in a pressure controlled vacuum chamber. Flames with different equivalence ratios, $\varphi = 0.8$, 1.0 were used in these experiments. Gas flows (CH_4, O_2, N_2, N_2O, NO and N_2 as shroud gas) were regulated by calibrated mass-flow controllers (1259 MFC by MKS instruments). A dry ice/acetone trap ($-63\,°C$) was used in experiments involving NO in order to remove traces of NO_2. The absence of NO_2 (the trace NO_2 concentration was less than 10^{-5} Torr) was verified spectroscopically by ICLAS. A feedback valve controller (model 252/253, MKS instruments) with an exhaust throttle valve regulates the gas pressure in the chamber. All the experiments described

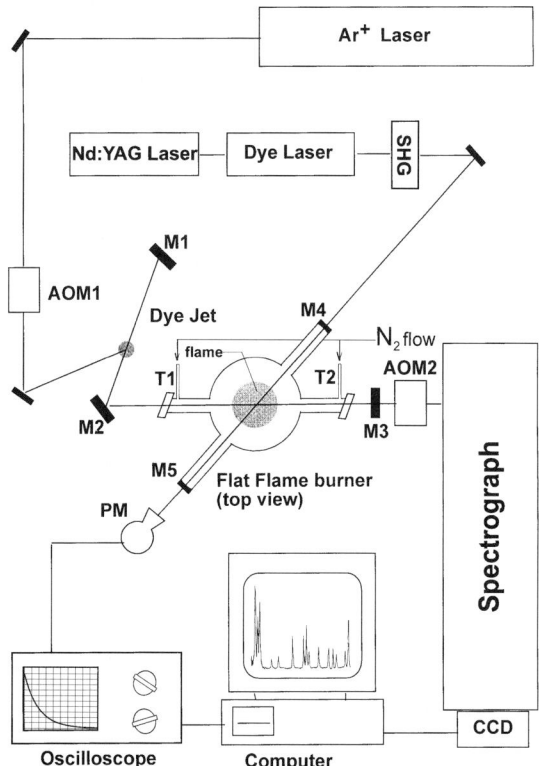

Fig. 1 Schematic diagram of the experimental set-up for combined ICLAS and CRDS measurements in a flat flame. The cavity of the ICLAS dye laser is formed by mirrors M1, M2 and M3. T1 and T2 are thin stainless steel tubes connected to optical window holders. AOM1 and AOM2 are acousto-optical modulators. CCD is a diode array. The CRDS cavity is formed by mirrors M4 and M5. SHG is a KDP crystal used for doubling. PM is a photomultiplier.

here are conducted at 30 Torr chamber pressure. The height of the burner is adjusted relative to the fixed optical axes using a linear motion feedthrough (Huntington Labs, VF-156-4) with an accuracy of 0.05 mm.

ICLAS and CRDS are combined in the same experimental set-up using different optical ports in the vacuum chamber. In ICLAS the burner system is located inside the cavity of a homemade dye laser and is isolated from the other parts of the cavity by two glass windows at the Brewster angle. The windows are 12.6 mm thick with 1° wedge to minimize interference fringes in the laser spectrum. Thin stainless steel tubes (T1 and T2) are connected to optical window holders. Nitrogen flow (10–300 sccm, regulated by calibrated needle valve) is supplied through T1 and T2 in order to remove intermediate combustion species from the ICLAS optical window holders. Hereafter we will refer to this flow as a "window flow".

We used ICLAS based on a quasi-cw dye laser. The dye laser has a three-mirror configuration. Mirrors M1 and M2 have 75 mm radii of curvature, mirror M3 is planar. The total length of the cavity, L, is 93 cm. A thin pellicle beamsplitter (PBS-2 Newport) is used as a broadband tuning element. However, in most of the experiments we were able to work without any tuning element in the cavity of the laser. We changed the type and concentration of the dye and the alignment of the laser in order to adjust the wavelength range. The dye laser is pumped by an argon-ion laser (Coherent model Innova 90-6) working with 300–400 mW at the 514.5 nm line (the excess pumping power over the threshold for dye laser operation is usually $\eta = 1.1$–1.3). The DCM and Kiton Red dyes are used. The diameter (FWHM) of the laser beam in ICLAS, which dictates the spatial resolution of the method, was measured using a 2D-CCD camera (SensiCam SVGA, PCO

CCD Imaging) aligned on the pathway of the laser beam emerging from the cavity through the M3 planar mirror. Using a known pixel size of the CCD we were able to estimate the diameter of the beam to be ⩽600 µm (a typical plot is shown in Fig. 2). It has been shown in several studies (see *e.g.* the review of Baev *et al.*[42] and references therein) that if the narrowband absorber is placed inside the cavity of a broadband multimode laser with homogeneous broadening, the time evolution of the laser intensity at a given wavenumber v and generation time, t_g, (which represents the time interval from the beginning of the laser generation to the moment of observation) is governed by the following expression:

$$I(v, t_g) = I_e(v, t_g)\exp[-n\sigma(v)L_{eq}(t_g)] \quad (1)$$

where $I_e(v, t_g)$, "the spectral envelope", is a relatively slowly varying function of the wavenumber; $\sigma(v)$ is the narrow-band absorption cross section; and n is the concentration of the absorbing species. The function $I_e(v, t_g)$ is equivalent to the incident intensity function in classical absorption spectroscopy. The equivalent optical path length, L_{eq}, is a linear function of the time of laser generation, t_g:

$$L_{eq}(t_g) = \frac{l}{L} ct_g \quad (2)$$

where l is the optical length of the absorbing compound, flame diameter in our case, L is the length of the laser cavity and c is the velocity of light.

The following scheme is employed to control the generation time, t_g: The pumped light is chopped by the acousto-optical modulator AOM1 (AFM-40 IntraAction Corp.). This modulator allows us to produce dye laser pulses of controlled duration. After a delay t_g from the beginning of the generation pulse the second modulator AOM2 (AFM-40 IntraAction Corp.) sends the laser beam to the entrance slit of the spectrograph during a sampling time of about 10 µs. A spectrum of the laser radiation is analyzed at high resolution with a 1 m monochromator (SPEX 1000 M)

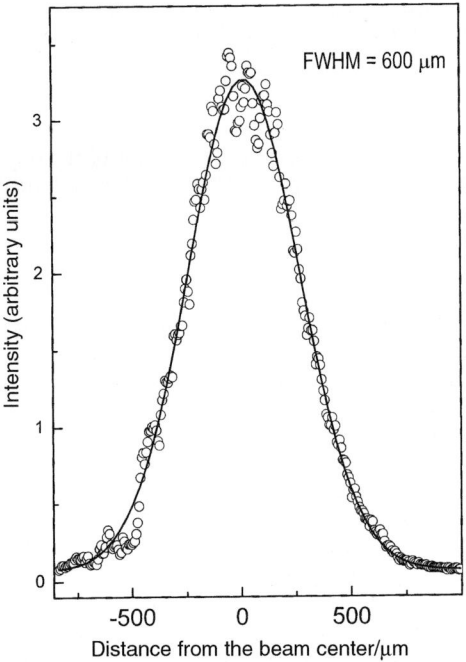

Fig. 2 Typical spatial profile of the dye laser beam. The solid line is the Gaussian best-fit to experimental points.

with a 100 grooves mm^{-1} echelle grating working at the 29th order and with a 2048 element photodiode charge coupled device (PCCD) (ALTON model LS-2000). The information from the PCCD is sent to a computer for processing and storing. The spectral resolution is about 0.003 nm as measured by observing the He–Ne laser line. The exposure time of the PCCD is 0.2 s which permits averaging of ~ 2000 spectra (the pulse repetition rate was 11230 Hz). The background signal of the PCCD (usually less than 1–2% of the total signal) was subtracted "on the fly" for each exposition time. An additional computer averaging of 50 PCCD signals spectra was used in the experiments.

Due to the high sensitivity of the method, ICLAS spectra contained besides radical spectral lines, some parasitic structure, including spectral features from atmospheric absorption (water overtone spectra and molecular oxygen $b\,^1\Sigma_g^+ - X\,^3\Sigma_g^-$ transitions), the spectral modulation of the laser losses, and interference on the flat window of the PCCD. In order to avoid the influence of this structure all experimental spectra were divided by a spectrum recorded at 50 mm above the burner (without the presence of NO and N_2O) which does not contain any radical spectra. This spectrum corresponds to the "blank" spectrum of conventional spectroscopy. Calibration of the spectrograph was done using the known atmospheric water absorption spectrum from the HITRAN database.[43]

CRDS is a highly sensitive absorption spectroscopic method based on the measurement of the intensity decay rate of a laser pulse injected into an optical cavity built with high-reflective mirrors. CRDS may be considered as a multipass absorption technique, with an equivalent absorption path length given by the number of passages inside the cavity before the injected intensity decays by a factor of e. This path length may be of the order of tens of km in the visible range, where the best mirrors are available. In contrast to standard multipass absorption methods, in CRDS the probe volume is confined within the fundamental cavity mode. This provides a spatial resolution of a fraction of a millimetre, sufficient for measurements in low-pressure flat flames.

The optical port of the CRDS spectrometer contains two long arms connected to the low-pressure chamber by stainless steel bellows. These arms are closed by two concave mirrors M4 and M5 (radii of curvature: 0.5 m) forming the ring-down cavity. A slow flow of nitrogen (190 sccm) was added through the thin tubes near the mirrors in order to reduce contamination of the mirrors by combustion products. CRDS is based on a Nd:YAG pumped dye laser (Continuum Corp. Nd-60). The laser beam of the second harmonic of the dye laser is injected into the cavity through the mirror M4. The laser linewidth is rather narrow, 0.08 cm^{-1} as measured by a Burleigh WA-4500 wavemeter. We observed the ring-down signal through mirror M5 using a photomultiplier (Hamamatsu R-166 or 1P28). The signal was digitized by a Tektronix TDS 220 oscilloscope and transferred to the PC computer for processing after averaging 128 consecutive laser pulses.

Calculations

Calculations of the species concentration profiles were performed using the PREMIX code[40] and the GRI-Mech methane combustion mechanism.[41] The PREMIX code models steady, isobaric, laminar, one-dimensional premixed flames. In order to take into consideration heat losses in the real burner system, the measured measured temperature profiles are used as an input for the model. In this work the OH CRDS spectra measured at different distances above the burner are used for this purpose. Most of the calculations have been done using version 2.11 of the GRI-Mech chemical mechanism. We also performed some calculations using a newer version (3.0) of the mechanism. This version includes a set of additional reactions related to propane and C_2 oxidation products. We did not find strong qualitative differences in the HNO and NH_2 concentration profiles calculated by the different versions and therefore we limit our discussion in this paper to the 2.11 version. The KINALC postprocessor[44] was used in order to estimate the relative contributions of different reactions at various locations in the flame. KINALC carries out three types of analysis: processing sensitivity analysis results, extracting information from reaction rates and stoichiometry, and providing kinetic information about the species. KINALC can carry out rate-of-production analysis and gives a summary of important reactions.[42]

Results and discussion

The NH_2 and HNO spectra

One of the advantages of ICLAS is the possibility of observing relatively broad spectral range involving lines of several species simultaneously. This is of particular interest in the study of non-stationary or turbulent flames. Fig. 3 shows the ICLAS spectra measured in the range close to 643 nm (between 15 520 and 15 570 cm^{-1}) in different conditions. This range includes lines of at least four different species: NH_2, HNO, $^1CH_2$, NO_2 and CN† and is hereinafter referred to as the 643 nm range. In this work the 643 nm range is used for measurements of the concentration of NH_2 and HNO. Let us consider the spectroscopy of these species in greater detail.

The NH_2 spectrum related to the $^2A_1-^2B_1$ transition covers the range from 390 to 830 nm.[45,46] The spectrum consists of a long progression of bands involving transition to the different levels of the bending vibration of NH_2. Probably the most studied transition is (090)–(000) which includes the $^pQ_{1,N}(7)$ line at 16 739.90 cm^{-1}. This line is rather intense and is not overlapped by lines with different rotational quantum numbers. For this line several measurements of the absorption coefficient have been made.[16,18] Fig. 4(a) shows the fragment of the NH_2 spectrum including this line

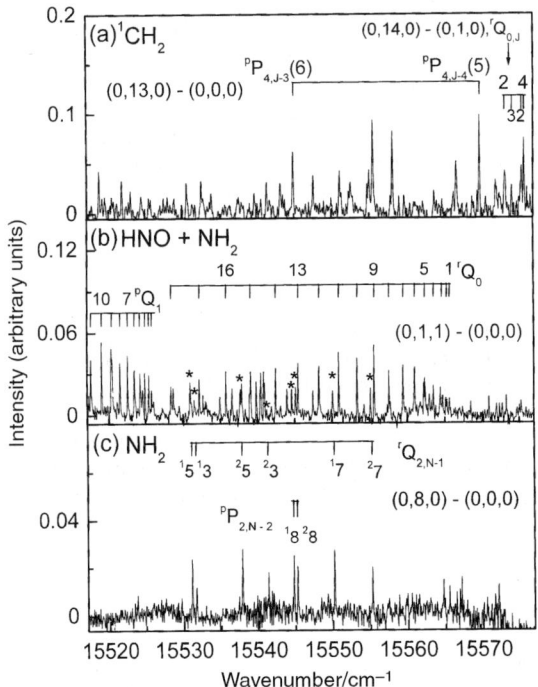

Fig. 3 ICLAS spectra of $^1CH_2$, HNO and NH_2 recorded in the 643 nm spectral range in the stoichiometric ($\varphi = 1$) $CH_4/O_2/N_2$ flames generation times is 75 μu. (a) The $^1CH_2$ spectrum in the maximum of the $^1CH_2$ concentration profile at 7.8 mm location above the burner: 100 sccm of nitrogen added through optical window holders. The line rotational assignments of the $^1CH_2$ are given as $^{\Delta K_a}\Delta J_{K_a K_c}$. (b) The HNO and NH_2 spectra at 52 mm above the burner in the flame doped with 0.17% NO. No nitrogen flow through the optical window holders was added. The lines marked with asterisks are the most intensive NH_2 lines with their assignments shown in (c). Only the assignments for the most intense rotational lines of the HNO are shown. (c) NH_2 spectrum recorded at the same conditions as in (b) along with 100 sccm of nitrogen added through the optical window holders. The notation of line rotational assignments of NH_2 is as recommended by Ross et al.[46]

---

† Recently we have observed CN lines belonging to the transition A–X (4–0 and 5–1) at some locations of the NO-doped flames. The flame spectroscopy of CN and its behavior in the doped hydrocarbon flames will be discussed elsewhere.

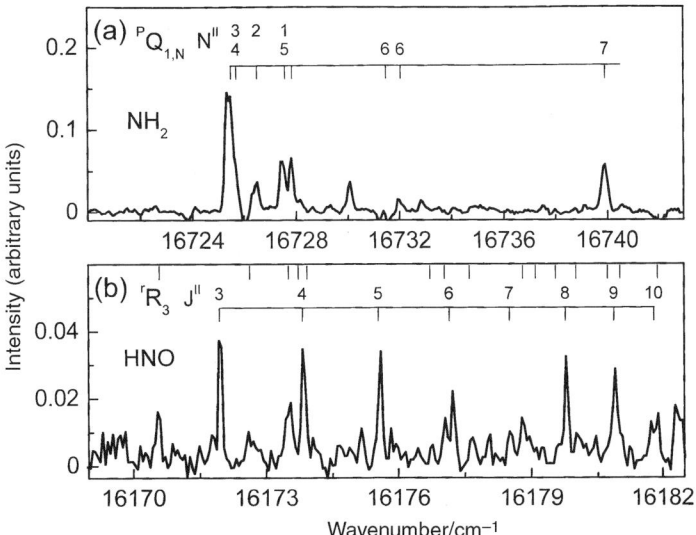

Fig. 4 The NH$_2$ and HNO spectra in the spectral ranges which include the lines with reported absorption coefficients: (a) The NH$_2$ ICLAS spectrum in the 598 nm spectral range recorded at 21 mm above the burner in the stoichiometric ($\varphi = 1.0$) CH$_4$/O$_2$/N$_2$ flame doped with 1.7% of N$_2$O. Rotational branch is denoted as $^{\Delta K_a}\Delta N_{K_a K_c}$.[45] (b) The ICLAS HNO spectrum near 618 nm in the R-branch region of the $K' = 4$–$K'' = 3$ subband of the ^{1}A″ (100)–^{1}A′ (000) transition. The rotational branches of the HNO are given as $^{\Delta K}\Delta J_{K''}$. Ticks in the uppermost part of the figure mark NH$_2$ line positions with reported rotational assignments that are not shown here. The spectrum was recorded at 48 mm above the burner surface in the near stoichiometric ($\varphi \approx 1.0$) CH$_4$/O$_2$/N$_2$ flame doped with 3.2% of N$_2$O. Laser generation time is 35 μs.

observed in the methane/air flame ($\varphi = 1$) doped with 1.7% of N$_2$O. The NH$_2$ spectrum is rather complex and the NH$_2$ lines can be found in numbers of spectral ranges. The 643 nm spectral range includes many NH$_2$ lines as well as lines of different species present in our flames, HNO, ^{1}CH$_2$ and, as we found recently, CN. Fig. 3(a) shows the ICLAS spectrum observed at the location closed to the luminescence zone ($h = 7.8$ mm) of the methane/air flame without nitrogen oxides being added. This spectrum contains only ^{1}CH$_2$ lines and does not contain lines of the nitrogen-containing species. The concentrations of those species formed as a result of the reaction with the molecular nitrogen in the flame are several orders of magnitude less than that observed in the flames doped with nitrogen oxides. Fig. 3(b) shows ICLAS spectra measured in the same flame ($\varphi = 1$) doped with 0.17% of NO at a high location above the burner ($h = 52$ mm). At this location no ^{1}CH$_2$ spectrum is observed, but many lines of HNO and NH$_2$ can be seen. The injection of the "window flow" of nitrogen suppresses the HNO spectrum, and only NH$_2$ lines can be seen in Fig. 3(c). The lines of the NO$_2$ electronic spectrum can also be seen in the 643 range under the appropriate conditions.[39] The comparison of the intensities of the NH$_2$ lines in this spectral range with that of the (090–000)Σ $^PQ_{1,N}$(7) line allows us to calculate absorption coefficients for the observed lines and use these lines for the concentration measurements. The most intense NH$_2$ lines in this range are about three times weaker than the (090–000)Σ $^PQ_{1,N}$(7) line. Nevertheless the sensitivity, which is reached using the NH$_2$ lines from this range, is sufficient for flame measurements.

The Ã ^{1}A″–X̃ ^{1}A′ band system of HNO is located in the red and near-infrared regions of the spectrum. The electronic origin is 13 154.38 cm^{-1}, and to date 12 vibrational levels of the Ã ^{1}A″ state have been observed in the region 550–770 nm via transition from the (000) level of the X̃ ^{1}A′ state.[22,23,25] The absorption cross section has been reported only for the rotational line R(6) of the $K' = 4$–$K'' = 3$ subband of the (100)–(000) band.[47] This cross section was estimated based on the HCO radical cross section, suggesting total conversion of HCO formed by flash photolysis of acetaldehyde to HNO via the reaction HCO + NO → HNO + CO.[47] The ICLAS spectrum observed in this range is shown in Fig. 4(b). The HNO lines in the 643 range are about 3 times more intense than in the (100)–(000) range. Comparison of different lines in both regions of the spectrum allowed us to determine the relative intensity of the lines and consequently to use more intense (011)–(000) spectra for concentration measurements.[39]

The spectra observed in the 643 nm region contain lines which are observed in flames without nitrogen oxides added. These lines seem to be the $^1CH_2$ lines. They display a similar intensity dependence on the distance from the burner and some of them can be unambiguously assigned to the known transitions in the $^1CH_2$ spectra. The other lines are not assigned and it is not surprising. The $\tilde{b}\,^1B_1$–$\tilde{a}\,^1A_1$ transition of the singlet methylene has many bands in the region between 11 000 and 22 000 cm^{-1}. This spectrum has been studied extensively at room temperature,[48–52] but because of the strong perturbation of the ã and $\tilde{b}$ states by the triplet ground state and by each other, most of the lines in the spectrum have not yet been assigned. For example, Petek et al.[52] observed about 10 000 transitions in the range of 15 000 to 19 000 cm^{-1}, but only 477 of them were assigned. At elevated flame temperatures the spectrum is more complicated since more lines are observable.

The presence of HNO in the zone surrounding the flame

According to the calculations based on the GRI-Mech the radical concentration profiles of all radicals except those of H, O and OH have sharp maxima close to the location of the flame front. The radical concentrations at relatively large distances above the burner (more than 30 mm) fall to negligible values. Calculations show that at 40 mm the NH_2 concentration is less than 0.1% of the maximum value, and that of HNO is 0.2% of its maximum value. Similar behavior was observed for HCO, $^1CH_2$ and CH radicals, in agreement with the mechanism predictions. In contrast with that, we observed a substantial absorption of NH_2 and HNO at 45 mm above the burner.[38,39] Fig. 5 shows the dependence of the NH_2 and HNO absorption on the distance above the burner in the N_2O- and NO-doped flames. In the N_2O-doped flame, HNO displays practically no depen-

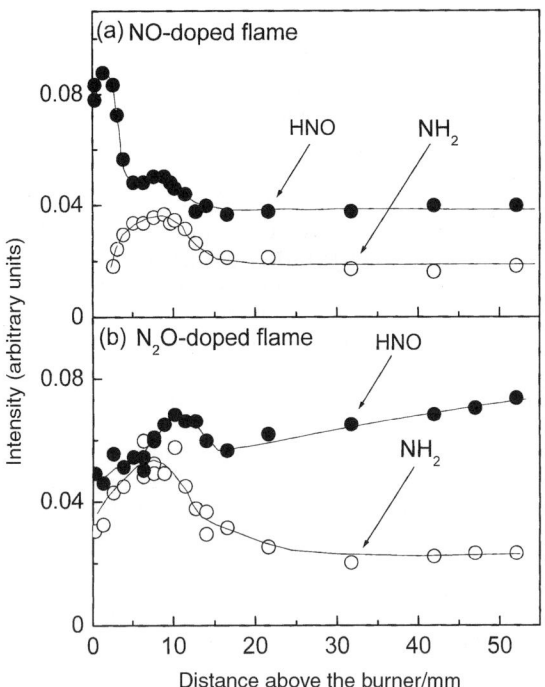

Fig. 5 The HNO (solid circles) and NH_2 (open circles) absorbance dependence on the distance above the burner in the stoichiometric ($\varphi = 1$) $CH_4/O_4/N_2$ flame doped with (a) 0.17% of NO, (b) 1.7% of N_2O. No nitrogen flow was added through the optical window holders. The $^rQ_{2,\,J-1}(^15)$ line of NH_2 and the $^PQ_1(7)$ line of HNO were used for absorbance measurements. Laser generation time is 75 µs.

dence on the distance above the burner. In the NO-doped flame, a strong peak of the HNO absorption profile is observed in the vicinity of the burner, lower than the luminescence zone. The NH_2 absorption profile has a peak close to the location of the luminescence zone in both N_2O- and NO-doped flames. As was mentioned above, in all flames substantial absorption is observed at large distances above the burner. We proposed[39] that this absorption is due to the presence of HNO and NH_2 in the zone surrounding the flame.

In order to prove this suggestion we added a flow of nitrogen through the optical window holders as was described in the Experimental section A, relatively slow flow decreases the HNO concentration dramatically, demonstrating that at least part of the HNO absorption originates from HNO located close to the optical windows. Fig. 6 shows the dependence of the HNO absorption at 55 mm above the burner on the nitrogen flow added. The flow of 25 sccm causes a twofold decrease in the HNO absorption, and the HNO absorption vanishes at 200 sccm. Note, that the experiments shown in Fig. 6 have been performed with 2.3% NO. This concentration is more than ten times higher than that used in most of the experiments. At 0.17% NO we cannot observe HNO absorption at high distances above the burner even with 50 sccm of the window flow. The total flow of gases through the burner is about 4400 sccm, therefore a 50 sccm flow does not disturb the flame. The evidences for this are shown in Fig. 7 which shows the $^1CH_2$ absorption profiles observed at different window flows. The flow does not affect either the maximum value or the location of the $^1CH_2$ profile.

The HNO absorption is not detectable after addition of the 100 sccm window flow in the N_2O-doped flame. Thus all HNO absorption, which is observed in the N_2O-doped flame before addition of the window flow, originates from HNO located in the zone close to the wall and windows of the vacuum chamber, as was proposed previously.[39]

In the NO-doped flame the window flow also eliminates the HNO absorption at high locations above the burner. Fig. 3 provides a good illustration of such behavior. The spectrum (c) is measured when the 100 sccm of window flow is added and it does not contain the HNO lines. However, in contrast with the N_2O-doped flame, the HNO absorption can be seen at locations

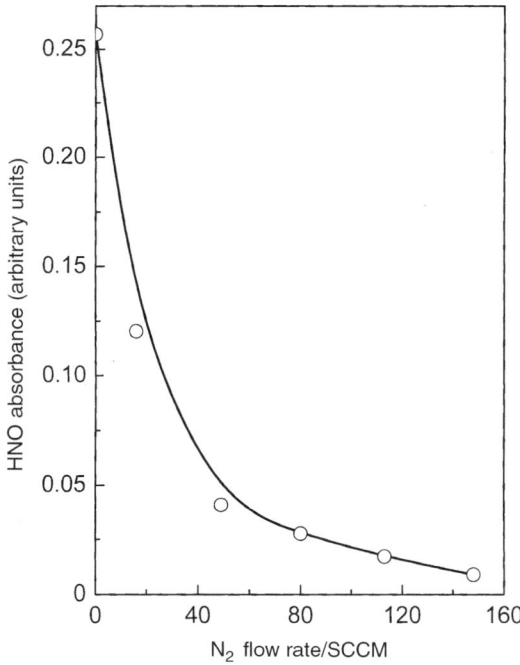

Fig. 6 The HNO absorbance measured 55 mm above the burner surface as a function of the nitrogen flow rate through the optical window holder. Experiments were performed in the stoichiometric ($\varphi = 1$) $CH_4/O_2/N_2$ flame doped with 2.3% NO. The $^pQ_1(10)$ line of HNO was used.

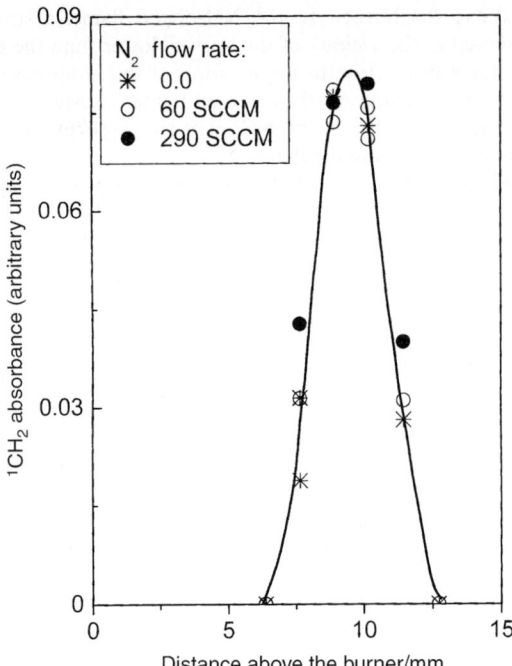

Fig. 7 The $^1CH_2$ absorbance profile measured in the stoichiometric ($\varphi = 1$) $CH_4/O_2/N_2$ flame doped with 1.7% of N_2O for three different nitrogen flow rates added through the optical window holder. The $^pP_{4, J-4}(5)$ line was used for absorbance measurements. Laser generation time is 75 µs.

close to the burner even with the window flow added. This shows that this absorption is caused by HNO from the central flame zone, and the window flow allows us to observe the HNO absorption of pure flame origin.

The HNO profile in the flame doped with NO

Fig. 8 and 9 show the comparison of the HNO profiles measured in NO-doped flames with different equivalence ratios ($\varphi = 0.8$ and 1.0) with the results of calculations using GRI-Mech. The HNO absorption at the locations close to the burner initially decreases when the window flow is added. When the window flow is increased, the HNO absorption is unaffected, until the flame becomes disturbed (when the flow exceeds ~1000 sccm). The HNO profile has two maxima according to the GRI-Mech predictions. The first is caused by diffusion of the H atoms from the flame front zone to the colder preflame zone where they reacts with NO forming HNO.

$$H + NO + M = HNO + M \tag{3}$$

This reaction provides 99% of the total HNO production rate at this location. (The KINALC[44] estimation.)

The second maximum is related to the hotter flame front zone where most of the radical concentrations also have their maxima. According to the GRI-Mech, the main reactions forming HNO in this location are:

$$O + NH_2 = H + HNO \tag{4}$$

(49.5% of total production rate), and

$$OH + NH = H + HNO \ (25\%) \tag{5}$$

Reaction (3) provides only ~4% of the total HNO production rate at this location.

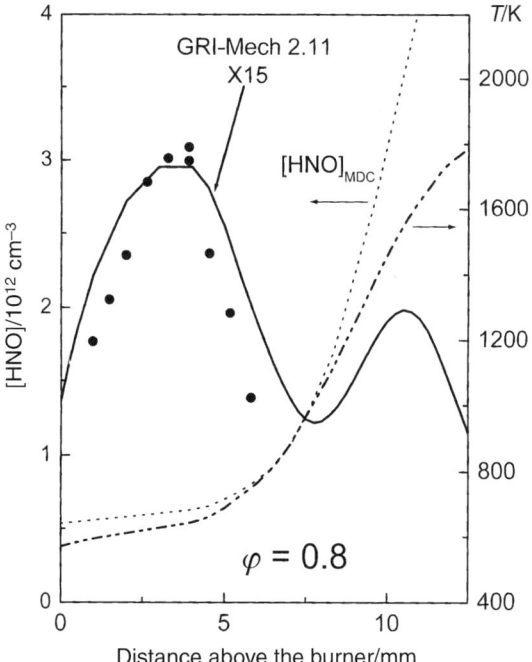

Fig. 8 The HNO concentration profile measured in the lean ($\varphi = 0.8$) $CH_4/O_2/N_2$ flame doped with 0.17% of NO. The solid line is the result of the GRI-Mech 2.11 calculations multiplied by a factor of 15. The "window flow" is 100 sccm. The dotted line is the estimated HNO minimum detectable concentration, $[HNO]_{MDC}$. Laser generation time is 75 μs. The dash-dot-dot line fits the experimental temperature profile. The $^pQ_1(7)$ line was used.

This flame front HNO maximum is predicted also for the N_2O-doped flame and is located at approximately the same place. The "first" maximum is not predicted for the N_2O-doped flames, since NO is absent in the preflame zone of the N_2O-doped flame. In accordance with those predictions we do observe the first maximum in the "cold" preflame zone of the NO-doped flames as can be seen in Fig. 8 and 9, and do not observe HNO in the N_2O flame. Unfortunately, we cannot observe the second maximum in either the NO- or N_2O-doped flames. The partition function of HNO increases rapidly with temperature, decreasing the individual rovibrational level populations and resulting in the appropriate degradation of the sensitivity. The dash-dot line in Fig. 8 and 9 shows the estimated minimum detectable concentration level which depends on the temperature, and consequently on the location. The temperature profiles for the flames studied have been measured using CRDS of the OH radicals as was described previously.[2] Assuming the flame diameter to be equal to the burner diameter we can relate measured absorption to the absolute concentration values. In order to do this the absorption coefficient of HNO based on the measurements of Cheskis *et al.*[47] is used.

The absolute HNO concentration obtained is about an order of magnitude higher than predicted by GRI-Mech. According to this mechanism, the concentration of HNO in the zone of the first, "preflame", maximum is controlled mainly by two reactions: reaction (3) and:

$$H + HNO = H_2 + NO \qquad (6)$$

In this case the HNO concentration does not depend on the H atom concentration and is a function of the NO concentration and the ratio of the rate constants of reactions (3) and (6). In order to fit the experimental value, the rate constant of reaction (6) must be reduced significantly in comparison with the value of $4.5 \times 10^{11} T^{0.72} \exp(-330/T)$ used in GRI-Mech 2.11. Note, that there are no reliable experimental measurements of this rate constant, and the value used in GRI-Mech 2.11 is obtained by calculations based on the variational transition state theory.[10]

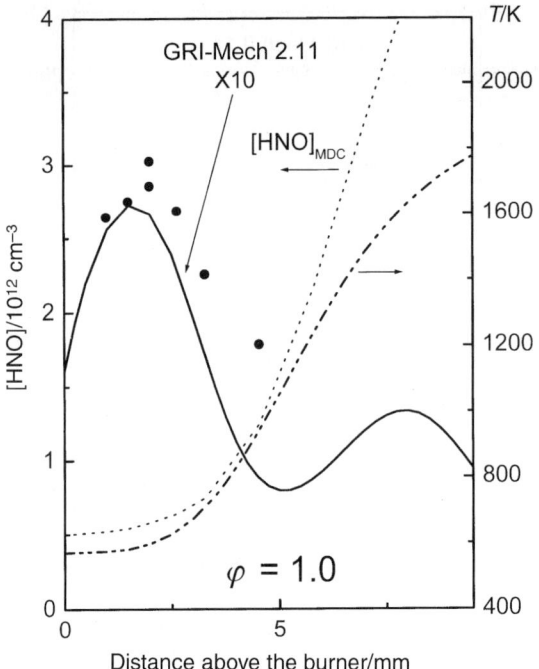

Fig. 9 The HNO concentration profile measured in the stoichiometric ($\varphi = 1$) $CH_4/O_2/N_2$ flame doped with 0.17% of NO. The solid line is the result of the GRI-Mech 2.11 calculations multiplied by factor of 10. The "window flow" is 100 sccm. The dotted line is the estimated HNO minimum detectable concentration. $[HNO]_{MDC}$. Laser generation time is 75 μs. The dash-dot-dot line fits the experimental temperature profile. The $^PQ_1(7)$ line was used.

Additional experiments are needed in order to verify the only measurement of the HNO absorption cross section. Useful information can be obtained if the HNO concentration is measured in the second, flame front maximum of the concentration profile. In order to reach this goal the sensitivity must be increased by at least an order of magnitude. The HNO absorption in the spectral range of the (000)–(000) transition, close to 750 nm, is expected to be significantly higher than that measured in the 643 nm range (corresponds to (011)–(000) transition). Efforts to expand our studies to this range are underway. Additional measurements of the reaction (6) rate constant would also be useful.

The NH_2 profile

In contrast to HNO, the NH_2 concentration depends much less on the window flow. Fig. 10 shows the dependence of the NH_2 and HNO absorption on window flow at the location high above the burner. Note, that flows shown in this figure are much higher than those in Fig. 6. Even at 1000 sccm flow the NH_2 spectrum is observable. That flow already disturbs the flame and higher flows usually just extinguish the flame. This observation shows that the radial location of NH_2 is different from that of HNO. NH_2 seems to be located much closer to the center of the flame and much further from the windows. The NH_2 absorption profiles obtained with a window flow of 100 sccm (the conditions of Fig. 8 and 9) are very similar to that without window flow (one of such profiles is shown in Fig. 5). We can suggest that at least part of the observed absorption at the locations close to the maximum of the profile is due to absorption from the central flame zone. In order to compare our observation with the GRI-Mech calculations we estimated the absolute NH_2 concentration in the maximum of the concentration profile based on the assumption that all absorption at this location is due to the absorption in the 6 cm diameter flame zone. The results are shown in Table 1 along with results of the calculations. Note, that the same NH_2 concentrations are observed for the NO concentration that is about 10 times lower than that of N_2O. This

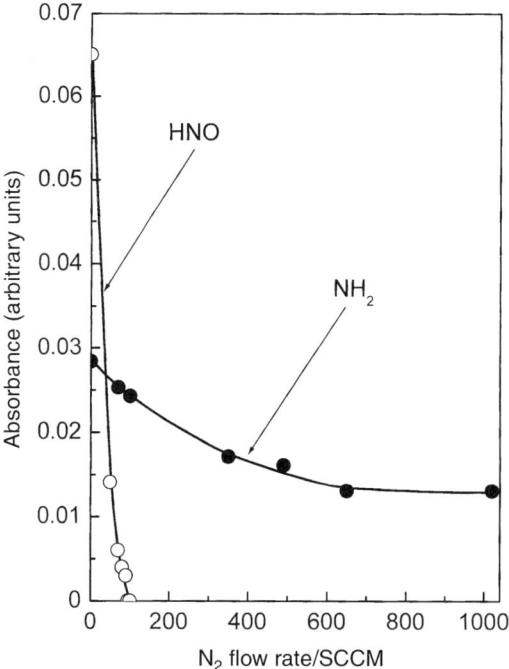

Fig. 10 The NH$_2$ (solid circles) and HNO (open circles) absorbances as a function of the window flow rate at 52 mm above the burner for the near stoichiometric ($\varphi = 1.04$) CH$_4$/O$_2$/N$_2$ flame doped with 1.7% of N$_2$O. The $^PQ_1(7)$ line of HNO and the $^rQ_{2,J-1}(^15)$ line of NH$_2$ were used.

agrees well with the prediction of the GRI-Mech. The absolute value of the NH$_2$ concentration is also in reasonable agreement with the predictions, taking into account that the observed absorption seems to include also the absorption of NH$_2$ which is located outside the central flame zone. Studies of the radial concentration dependence of NH$_2$ combined with 2-D model calculations are needed for a better understanding of the NH$_2$ flame chemistry. Obviously the observed presence of NH$_2$ far from the burner is of great interest for such applications as the De-NO$_x$ and reburning processes where NH$_2$ plays an important role.

Conclusions

Concentration profiles of HNO and NH$_2$ have been measured in methane/air flames doped with minor amounts of NO and N$_2$O. It was found that a substantial amount of HNO is located in the zone surrounding the flame, close to the optical windows of the vacuum chamber. The addition of a slow nitrogen flow into this area of the vacuum chamber removes HNO from this zone enabling observation of the HNO absorption of pure flame origin. In the NO-doped flame we were able to observe the first, low temperature maximum of the HNO profile predicted by GRI-Mech. The

Table 1 Measured and predicted NH$_2$ concentrations in 10^{11} molecule cm^{-3}

Doped gas	Mole fraction (%)	NH$_2$ concentration	
		Experiment	Calculations
NO	0.17	8.3	7.5
N$_2$O	1.7	12	6.3

GRI-Mech underpredicts the absolute concentration in this maximum evaluated from the experimental data using the absorption cross section measured by Cheskis *et al.*[47] The window flow only slightly affects the absorption of NH_2 indicating that NH_2 is present in the flame much closer to the center of the flame than HNO.

Acknowledgements

This research was supported by the Israel Science Foundation (Grant No. 574/00) and by the James Franck, German-Israeli Binational Program in Laser Matter Interaction.

References

1. S. Cheskis, *Prog. Energ. Combust. Sci.*, 1999, **25**, 233.
2. S. Cheskis, I. Derzy, V. A. Lozovsky, A. Kachanov and D. Romanini, *Appl. Phys. B*, 1998, **66**, 377.
3. G. Meijer, M. G. H. Boogaarts, R. T. Jongma, D. H. Parker and A. M. Wodtke, *Chem. Phys. Lett.*, 1994, **217**, 112.
4. M. D. Wheeler, S. M. Newman, A. J. Orr-Ewing and M. R. Ashfold, *J. Chem. Soc., Faraday Trans.*, 1998, **94**, 337.
5. R. Evertsen, R. L. Stolk and J. J. ter Meulen, *Combust. Sci. Technol.*, 2000, **157**, 341.
6. A. McIlroy, *Isr. J. Chem.*, 1999, **39**, 55.
7. X. Mercier, E. Therssen, J. F. Pauwels and P. Desgroux, *Chem. Phys. Lett.*, 1999, **299**, 75.
8. J. J. Scherer and D. J. Rakestraw, *Chem. Phys. Lett.*, 1997, **265**, 169.
9. J. J. Scherer, D. Voelkel and D. J. Rakestraw, *Appl. Phys. B*, 1997, **64**, 699.
10. A. M. Dean, J. W. Bozzelli, in *Gas-Phase Combustion Chemistry*, ed. W. C. Gardiner, Springer, Berlin, 1999, p. 125.
11. P. Glarborg, M. U. Alzueta, K. Dam-Johansen and J. A. Miller, *Combust. Flame*, 1998, **115**, 1.
12. M. S. Chou, A. M. Dean and D. Stern, *J. Chem. Phys.*, 1982, **76**, 5334.
13. R. A. Copeland, D. R. Crosley and G. P. Smith, *Proc. Combust. Inst.*, 1984, **20**, 1195.
14. R. M. Green and J. A. Miller, *J. Quant. Spectrosc. Radiat. Transfer*, 1981, **26**, 313.
15. K. N. Wong, W. R. Anderson, J. A. Vanderhoff and A. J. Kotlar, *J. Chem. Phys*, 1987, **86**, 93.
16. K. Kohse-Hoinghaus, D. F. Davidson, A. Y. Chang and R. K. Hanson, *J. Quant. Spectrosc. Radiat. Transfer*, 1989, **42**, 1.
17. M. Votsmeier, S. Song, D. F. Davidson and R. K. Hanson, *Int. J. Chem. Kinet.*, 1999, **31**, 445.
18. M. Votsmeier, S. Song, D. F. Davidson and R. K. Hanson, *Int. J. Chem. Kinet.*, 1999, **31**, 323.
19. J. A. Miller and C. T. Bowman, *Prog. Energ. Combust. Sci.*, 1989, **15**, 287.
20. J. A. Miller and P. Glarborg, *Int. J. Chem. Kinet.*, 1999, **31**, 757.
21. E. W. G. Diau, M. C. Lin, Y. He and C. F. Melius, *Prog. Energ. Combust. Sci.*, 1995, **21**, 1.
22. F. W. Dalby, *Can. J. Phys.*, 1958, **36**, 1336.
23. J. L. Bancroft, J. M. Hollas and D. A. Ramsay, *Can. J. Phys.*, 1962, **40**, 322.
24. J. Pearson, A. J. Orr-Ewing, M. N. R. Ashfold and R. N. Dixon, *J. Chem. Soc., Faraday Trans.*, 1996, **92**, 1283.
25. J. Pearson, A. J. Orr-Ewing, M. N. R. Ashfold and R. N. Dixon, *J. Chem. Phys.*, 1997, **106**, 5850.
26. K. Boraas, C. Douketis and J. P. Reilly, *Spectrochim. Acta Part A*, 1992, **48**, 1529.
27. N. I. Butkovskaya and D. W. Setser, *J. Phys. Chem. A*, 1998, **102**, 9715.
28. T. Ishiwata, I. Tanaka and H. Akimoto, *J. Phys. Chem*, 1978, **82**, 1336.
29. I. Jano, I. H. Elhag and A. Townshend, *Anal. Chim. Acta*, 1990, **230**, 151.
30. D. A. Ramsay and Q. S. Zhu, *J. Chem. Soc., Faraday Trans.*, 1995, **91**, 2975.
31. R. N. Dixon, M. Noble, C. A. Taylor and M. Delhoume, *Faraday Discuss. Chem. Soc.*, 1981, **71**, 125.
32. S. Mayama, K. Egashira and K. Obi, *Res. Chem. Intermed.*, 1989, **12**, 285.
33. S. Cheskis, *J. Chem. Phys.*, 1995, **102**, 1851.
34. S. Cheskis, I. Derzy, V. A. Lozovsky, A. Kachanov and F. Stoeckel, *Chem. Phys. Lett.*, 1997, **277**, 423.
35. I. Derzy, V. A. Lozovsky and S. Cheskis, *Chem. Phys. Lett.*, 1999, **306**, 319.
36. V. A. Lozovsky, S. Cheskis, A. Kachanov and F. Stoeckel, *J. Chem. Phys*, 1997, **106**, 8384.
37. I. Derzy, V. A. Lozovsky and S. Cheskis, *Isr. J. Chem.*, 1999, **39**, 49.
38. I. Derzy, V. A. Lozovsky, N. Ditzian, I. Rahinov and S. Cheskis, *Proc. Combust. Inst.*, 2000, **28**, 1741–1748.
39. V. A. Lozovsky and S. Cheskis, *Chem. Phys. Lett.*, 2000, **332**, 508.
40. R. J. Kee, J. F. Grcar, M. D. Smooke and J. A. Miller, *A Fortran Program for Modeling Steady Laminar One-Dimensional Premixed Flames*, SAND85-8240, Livermore, CA, 1991.
41. C. T. Bowman, R. K. Hanson, D. F. Davidson, W. C. Gardiner, V. Lissianski, G. P. Smith, D. M. Golden, M. Frenklach, H. Wang and M. Goldenberg, http://www.me.berkeley.edu/gri_mech
42. V. M. Baev, T. Latz and P. E. Toschek, *Appl. Phys. B*, 1999, **69**, 171.

43 L. S. Rothman, C. P. Rinsland, A. Goldman, S. T. Massie, D. P. Edwards, J. M. Flaud, A. Perrin, C. Camy-Peyret, V. Dana, J. Y. Mandin, J. Schroeder, A. McCann, R. R. Gamache, R. B. Wattson, K. Yoshino, K. V. Chance, K. W. Jucks, L. R. Brown, V. Nemtchinov and P. Varanasi, *J. Quant. Spectrosc. Radiat. Transfer*, 1998, **60**, 665.
44 T. Turanyi, http://www.chem.leeds.ac.uk/Combustion/kinalc.htm
45 K. Dressler and D. A. Ramsay, *Philos. Trans. R. Soc. London, Ser. A*, 1959, **251**, 553.
46 S. C. Ross, F. W. Birss, M. Vervloet and D. A. Ramsay, *J. Mol. Spectrosc.*, 1988, **129**, 436.
47 S. G. Cheskis, V. A. Nadtochenko and O. M. Sarkisov, *Int. J. Chem. Kinet.*, 1981, **13**, 1041.
48 M. N. R. Ashfold, M. A. Fullstone, G. Hancock and G. W. Ketley, *Chem. Phys.*, 1981, **55**, 245.
49 I. Garcia-Moreno and C. B. Moore, *J. Chem. Phys.*, 1993, **99**, 6429.
50 W. H. Green, Jr., I.-C. Chen, H. Bitto, D. R. Guyer and C. B. Moore, *J. Mol. Spectrosc.*, 1989, **138**, 614.
51 G. Herzberg and J. W. C. Johns, *Proc. R. Soc. London Ser. A*, 1966, **295**, 107.
52 H. Petek, D. J. Nesbit, D. C. Darwin and C. B. Moore, *J. Chem. Phys.*, 1987, **86**, 1172.

Experimental and modelling study of sulfur and nitrogen doped premixed methane flames at low pressure

Kevin J. Hughes,† Alison S. Tomlin, Valerie A. Dupont and Mohammed Pourkashanian

Department of Fuel and Energy, University of Leeds, Leeds, UK LS2 9JT

Received 5th March 2001
First published as an Advance Article on the web 21st September 2001

Laser-induced fluorescence (LIF) has been used to observe NS and NO in methane/oxygen/argon laminar flames at low pressure doped with ammonia and sulfur dioxide. NS profiles as a function of height above the burner have been measured for rich flames. The effect of adding various amounts of sulfur dioxide on the observed NO in the burnt gas region has been investigated for a variety of stoichiometries. The experimental measurements have been compared with PREMIX simulations using a detailed elementary reaction mechanism for nitrogen- and sulfur-containing species in a methane flame. Sensitivity analysis has been employed to highlight the important reactions for NS, NO and SO_2. The results demonstrate significant uncertainties in currently best available rate data for important reactions involving sulfur-containing species.

1. Introduction

Understanding the mechanisms for the formation of nitrogen oxides and the interaction of sulfur compounds is of fundamental importance in the design and development of low NO_x combustion systems for sulfur-containing fuels. Many previous studies have established that the presence of sulfur in fuels can have a significant influence on the formation and destruction of NO with both depletion and enhancement of NO occurring under different experimental conditions.[1-10] In several of these experimental studies, attempts to model the observations have been made using detailed chemical mechanisms. These mechanisms however, have been constructed often with speculation as to the identity and rate coefficients of possible reactions of nitrogen- and sulfur-containing species that may be required in order to explain experimental observations. There is currently no detailed chemical mechanism that can predict even qualitatively the observed non-linear effects which occur over the range of different experimental conditions previously investigated. In order to try to understand the mechanistic basis of these effects, an improved chemical scheme is required that is capable of describing the interaction of sulfur-containing species with hydrocarbon oxidation and NO_x chemistry. The validation of such a scheme relies on experimental data in well-defined simple systems without the complications of turbulent flow conditions. Such data are at present sparse for sulfur-containing flames with two major consequences for the development of successful chemical schemes. The first is that few data are available for the evaluation of current schemes over a wide range of stoichiometries, especially for intermediate species. The second problem stems from the fact that much of the sulfur chemistry is poorly characterised from the point of view of evaluated elementary rate data. For many of the reactions, estimates

† Current address: School of Chemistry, University of Leeds, Leeds, UK LS2 9JT.

DOI: 10.1039/b102061g

have been used in previous studies. The optimisation of such rate data is often a useful way of improving such a mechanism in the short term where evaluated data are unavailable. Optimising sulfur schemes such as the one used in this work would however be very difficult given the large number of uncertain reactions when compared with available target data sets for simple flame scenarios. This investigation attempts to improve current understanding of sulfur–nitrogen interactions in flames by providing additional experimental data obtained from an investigation of a low pressure laminar methane flame doped with ammonia and sulfur dioxide. A modelling study of the system has also been conducted using an updated sulfur mechanism[11–13] combined with detailed chemical mechanisms describing methane oxidation[14,15] and NO_x chemistry.[16,17]

2. Experimental

A water-cooled flat flame sintered burner was constructed, consisting of a 6 cm diameter sintered burner with a 0.5 cm diameter outer ring sinter through which a shroud gas flow of argon could flow. The burner was housed in a stainless steel chamber of 20 cm internal diameter and 40 cm height. The top half of the chamber was jacketed to enable water-cooling. Access for the laser was by means of two brewster-angled fused silica windows. An iris was used to reduce the diameter of the laser beam down to approximately 0.2 mm as it passed over the burner. Viewing access was by means of two 10 cm diameter Spectrosil-B grade fused silica windows at 90° to the path of the laser. Radiation loss corrected temperature profiles were obtained from a type R Pt/Pt–13%Rh thermocouple constructed from 50 μm diameter wire with a bead diameter of approximately 125 μm, estimated to be approximately 200 μm after coating with silica to eliminate the possibility of catalytic reactions on the surface. The laser probe radiation was generated from an excimer (Questek series 2000 at 308 nm) pumped dye laser (Lambda Physik FL3002). The output from the dye laser was doubled to produce the light used to probe NO and NS by laser induced fluorescence (LIF). NS has been observed previously in the reaction zone of an atmospheric pressure methane flame doped with ammonia and hydrogen sulfide,[18] and LIF spectroscopy of various electronic states of NS has been investigated in a low pressure discharge flow system.[19] From this work it was decided to probe NS on the $C\,^2\Sigma^+ \leftarrow X\,^2\Pi$ system on the Q1 bandhead at approximately 230.3 nm, due to the lack of interference from other species present in the flame, and the convenience of being able to use the same laser dye for NO detection. A typical NS LIF excitation spectrum over the wavelength range 229.5–231.6 nm is shown in Fig. 1. Only a limited range of experimental conditions was found to be suitable for the detection of NS, and results are presented

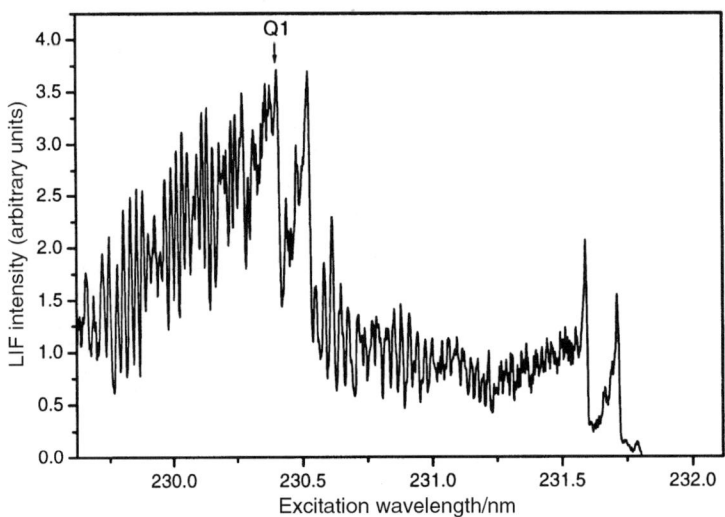

Fig. 1 NS LIF excitation spectrum showing the position of the Q1 bandhead.

Table 1 Summary of experimental conditions of the flames in which relative [NS] profiles are measured

Flame	ϕ	P/Torr	Mass flow rate/g cm^{-2} s^{-1}	Composition (%)				
				Ar	CH$_4$	O$_2$	NH$_3$	SO$_2$
1	1.4	40	0.002 144	33.18	26.75	38.23	1.32	0.52
2	1.6	40	0.002 121	33.18	28.88	36.1	1.32	0.52
3	1.6	100	0.002 121	33.18	28.88	36.1	1.32	0.52

for three sets of conditions as given in Table 1. NO was probed using the P$_1$(26) transition of the A $^2\Sigma \leftarrow$ X $^2\Pi$ system at 225.88 nm at a fixed height of 25 mm above the burner surface unless stated otherwise. Table 2 lists the range of experimental conditions at which NO was measured. The laser induced fluorescence signal was collected from the path of the laser over the burner from a region of approximately 3 cm in length centered on the middle of the burner, *via* a fused silica lens and slit through a 220 nm interference filter onto a photomultiplier. The signal from the photomultiplier was integrated by a boxcar averager(SRS250), digitized and stored on a personal computer. Profiles as a function of height above the burner were recorded by adjusting the height of the burner within the chamber. Typically 100 laser shots were averaged for both the LIF and background signals at each height, in order to produce the actual LIF signal for the species being investigated. In all of the LIF profiles presented, error bars given are of one standard deviation calculated from this averaging and subtraction. Gas flows into the burner were controlled by the use of mass flow controllers (MKS 1179A) for each gas to give total flows in the range of 2600–4600 sccm, consisting of argon, methane, oxygen, ammonia and sulfur dioxide. The pressure inside the chamber was monitored by a 0–100 Torr range Baratron pressure transducer (MKS Baratron type 626), and controlled by adjustment of a bellows valve between the chamber and a rotary vacuum pump (Edwards E2M 80) to maintain a pressure between 40–100 Torr.

3. Modelling

The CHEMKIN[20] program was used to simulate the experimental conditions detailed in Tables 1 and 2 by using the PREMIX[21] module, designed to simulate one-dimensional laminar premixed flames, with the constraint that it uses the measured temperature profiles. Relative profiles of [NS] as a function of height above the burner were simulated, along with the effects of different sulfur dioxide dopant levels on the relative NO concentration at 25 mm above the burner surface. The detailed chemical mechanism consists of the Leeds methane oxidation mechanism[14,15] and extensions to it that describe both NO$_x$[16,17] and SO$_x$[13] chemistry in combustion systems, to produce a combined mechanism of 897 irreversible reactions and 77 species. The Leeds methane oxidation mechanism[14,15] is almost exclusively based on evaluated rate coefficients of elementary chemical reactions. The NO$_x$ extension[16,17] is also based on evaluated rate coefficient data wherever possible, though to a lesser extent than the methane mechanism. Both mechanisms have been tested

Table 2 Summary of experimental conditions of the flames in which SO$_2$ effects on relative [NO] are measured

Flame	ϕ	P/Torr	Mass flow rate/g cm^{-2} s^{-1}	Composition (%)				
				Ar	CH$_4$	O$_2$	NH$_3$	SO$_2$
4	0.7	40	0.004 301	65.66	8.75	24.88	0.3	0–0.3
5	1.0	40	0.004 252	65.68	11.23	22.49	0.3	0–0.3
6	1.4	40	0.004 199	65.52	13.93	19.93	0.3	0–0.3
7	0.7	40	0.002 247	33.17	16.80	48.18	1.32	0–0.52
8	1.0	40	0.002 191	33.18	21.65	43.32	1.32	0–0.52
9	1.3	40	0.002 147	33.18	25.59	39.39	1.32	0–0.52
10	1.6	40	0.002 11	33.18	28.88	36.1	1.32	0–0.52

against a wide variety of data including that from the study of laminar flames, and found to perform satisfactorily in most circumstances. The SO_x extension is based on the mechanism compiled by Glarborg et al.[11,12] from their study on moist carbon monoxide oxidation in the presence of sulfur dioxide. This has been augmented by the inclusion of additional reactions of sulfur-containing species appropriate for a methane oxidation environment, along with all reactions describing sulfur–nitrogen interactions that have previously appeared in the literature. The SO_x reaction mechanism of 226 irreversible reactions and 22 sulfur-containing species is much more uncertain than the methane or NO_x components, with approximately 70% of the reactions being estimates or based on low temperature measurements only with extrapolation to high temperatures. In many cases, those SO_x reactions that have been measured at high temperatures are restricted to the results from a single study, with no independent verification and no evaluation. Normalised first order sensitivity coefficients, defined as $w_{j,i} = (A_i/Z_j)\partial Z_j/\partial A_i$ were calculated using PREMIX[21] where Z represents temperature, flame velocity or species mass fraction and A_i the pre-exponential factor for reaction i. This sensitivity information was processed by use of the program KINALC[22] in order to highlight those parts of the reaction mechanism that are important with respect to the experimental measurements.

4. Results and discussion

4.1. NS LIF Profiles

Relative NS LIF profiles as a function of height above the burner were obtained for the three sets of conditions as given in Table 1, and are shown in Fig. 2–4, along with the simulated relative profile of [NS] in each case and the measured temperature profile corrected for radiation losses. By adjusting the conditions between flames 1 and 2 in the same experiment, and observing the change in the NS LIF signal at a height of 25 mm above the burner surface, the relative quantity of NS in flames 1 and 2 at this position could be estimated. Here the assumption is made that the LIF signal between the two experiments can be directly compared by neglecting any effects caused by differences in quenching of the excited NS between the two different sets of conditions. This is a reasonable assumption as there will be a small difference in the composition between the two flame stoichiometries. The assumption may be less valid throughout the individual flame profiles, as the gas composition varies from unburned reactants plus argon diluent, to combustion products plus argon diluent. These changes in composition may lead to relative differences between the actual NS concentration profile and that inferred from the measured LIF signal. The results of the comparison of flames 1 and 2 are shown in Fig. 5, where the measured NS LIF signal for each flame is normalised so that the relative concentration for flame 2 at 25 mm is 1, and the ratio

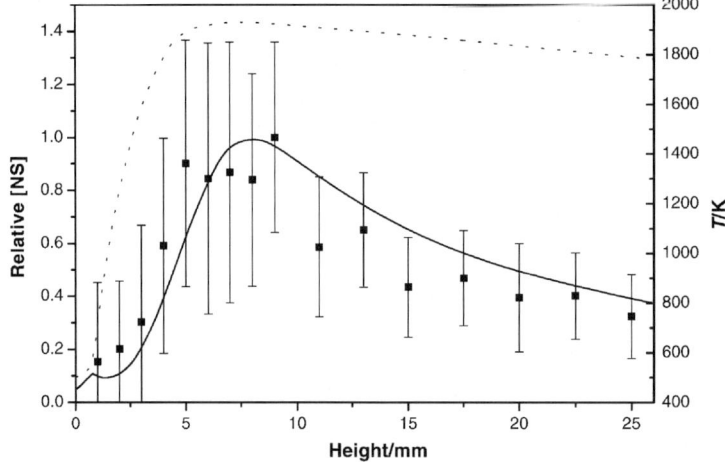

Fig. 2 Measured NS LIF and temperature profiles, and simulated [NS] profile for flame 1. (■) LIF measurements, (- - -) temperature measurement, (——) [NS] simulation.

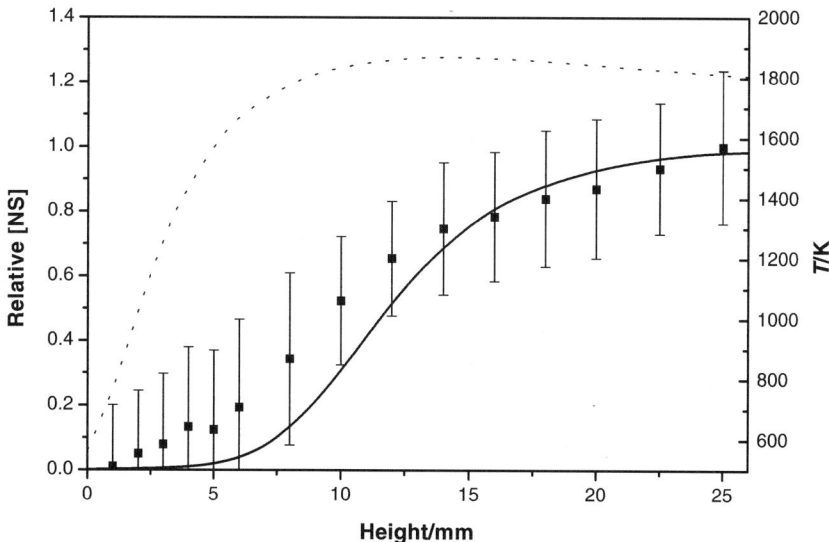

Fig. 3 Measured NS LIF and temperature profiles, and simulated [NS] profile for flame 2. (■) LIF measurements, (— —) temperature measurement, (——) [NS] simulation.

between flames 1 and 2 at 25 mm is equal to that observed experimentally. The simulated profiles are then both scaled by the same factor in order to give a simulated value of flame 2 at 25 mm of 1.

4.2. Sulfur dioxide effects on NO

For the flames detailed in Table 2, the effect of different quantities of sulfur dioxide dopant on the observed NO LIF signal at 25 mm above the burner surface were measured. Assumptions made in relating LIF signals to species concentrations will be valid in these experiments, as the addition of a small fraction of sulfur dioxide dopant is not expected to have any significant effect on the

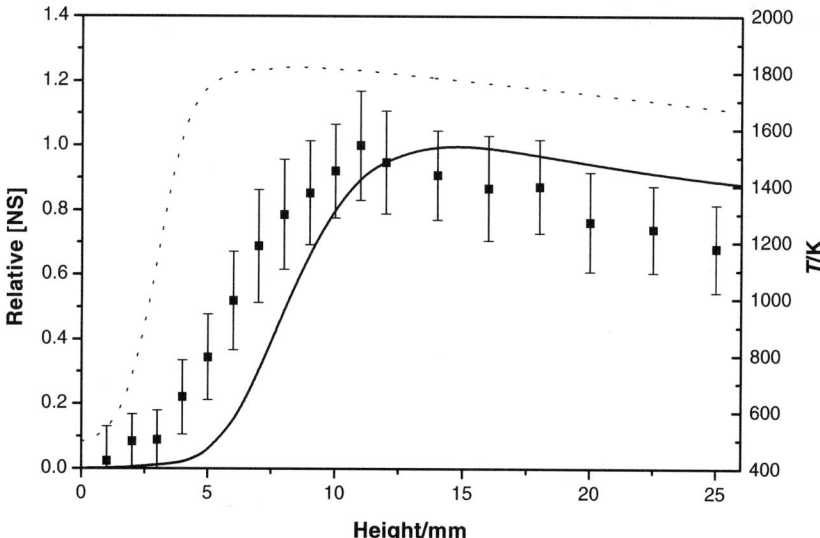

Fig. 4 Measured NS LIF and temperature profiles, and simulated [NS] profile for flame 3. (■) LIF measurements, (— —) temperature measurement, (——) [NS] simulation.

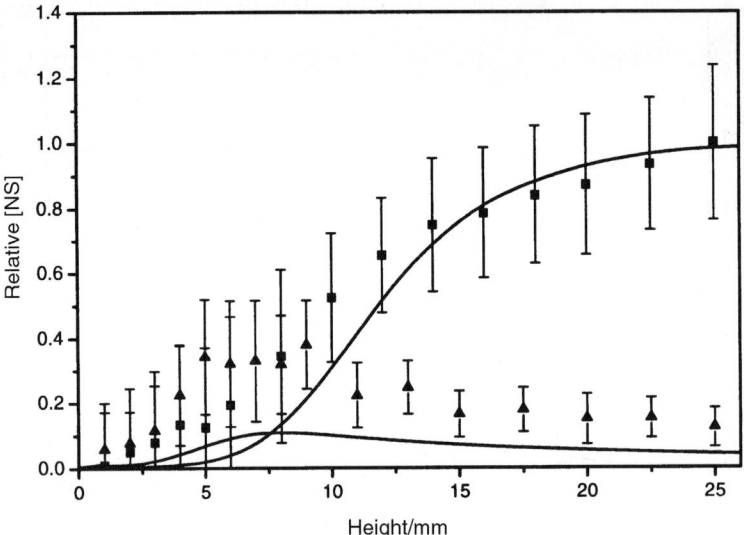

Fig. 5 Comparison of measured and simulated [NS] profiles in flames 1 and 2. (■) $\phi = 1.6$, (▲) $\phi = 1.4$, (———) model.

quenching rate of excited NO. Fig. 6 summarises the results obtained for flames 4–6 which had a high percentage of argon diluent. Fig. 7 and 8 summarise the results for flames 7–10 containing a low percentage of argon diluent. For the flames containing a high percentage of argon diluent, the effect of adding sulfur dioxide is to cause a significant reduction in NO for all of the stoichiometries investigated. For the flames containing a low percentage of argon diluent, a variety of behaviour is observed, from reductions in NO at lean and stoichiometric conditions, to an increase in NO in rich flames. Effects of different diluent gases and total mass flow rates have also been demonstrated in previous studies by Wendt et al.[23] and Corley and Wendt[24] and may possibly be due to the influence on the resulting temperature profiles in the flame. For the higher diluent gas

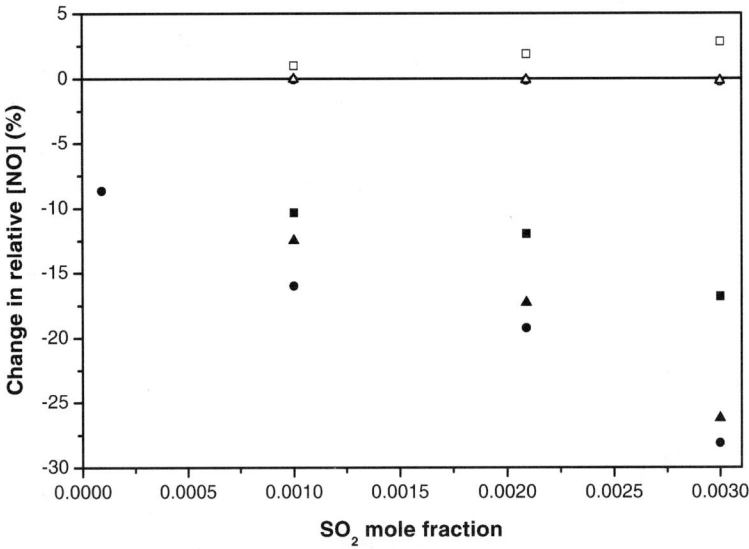

Fig. 6 Measured and modelled effect of added sulfur dioxide on NO for flames 4, 5 and 6. (■) $\phi = 1.4$, measurement; (□) $\phi = 1.4$, simulation; (▲) $\phi = 1.0$, measurement; (△) $\phi = 1.0$, simulation; (●) $\phi = 0.7$, measurement; (○) $\phi = 0.7$, simulation.

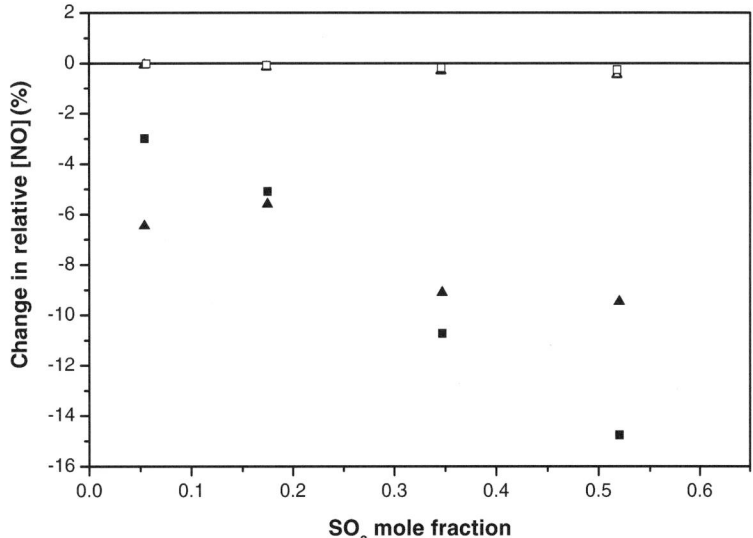

Fig. 7 Measured and modelled effect of added sulfur dioxide on NO for flames 7 and 8. (■) $\phi = 0.7$, measurement; (□) $\phi = 0.7$, simulation; (▲) $\phi = 1.0$, measurement; (△) $\phi = 1.0$, simulation.

concentration in these experiments, the temperature close to the burner is higher, but with a shallower temperature rise and a shift in the maximum further away from the burner as illustrated in Fig. 9 for flames 4 and 7. Such variations in the temperature profiles for the two cases will influence the production and consumption rates of NO in the flame and therefore the impact of adding SO_2.

4.3. Sulfur dioxide LIF profile

A weak fluorescence signal was observed between 225 and 230 nm, peaking at approximately 228 nm, that was ascribed to sulfur dioxide.[25] A sulfur dioxide LIF profile for flame 3 as a function of

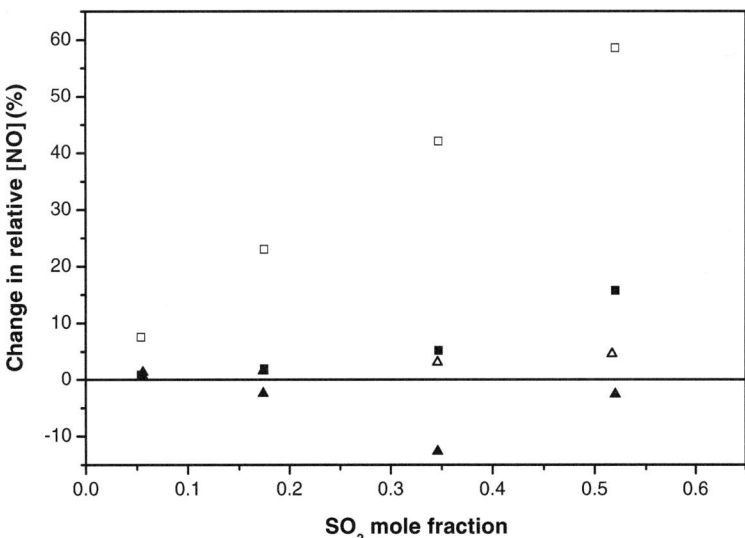

Fig. 8 Measured and modelled effect of added sulfur dioxide on NO for flames 9 and 10. (▲) $\phi = 1.3$, measurement; (△) $\phi = 1.3$, simulation; (■) $\phi = 1.6$, measurement; (□) $\phi = 1.6$, simulation.

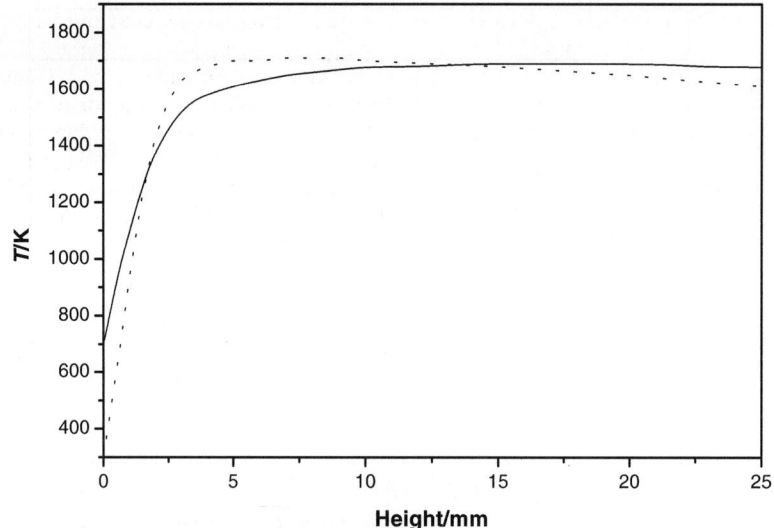

Fig. 9 Measured temperature profiles, (——) flame 4, (– – –) flame 7.

height above the burner was measured as shown in Fig. 10. Due to the low intensity of the fluorescence signal, the iris constricting the laser beam had to be opened to approximately 0.5 mm diameter in order to obtain a measurable signal, and large error bars resulted for each measurement. However the behaviour observed was as expected, and is similar to a previous experimental measurement of a sulfur dioxide profile in a laminar hydrogen/oxygen low pressure flame in that it shows a steady drop of sulfur dioxide which then levels out in the burnt gas region.[26]

4.4. Simulated NS profiles

Fig. 2–5 show both the experimental normalised NS LIF profiles, and the normalised NS concentrations as calculated with PREMIX. The experimental results show a fall-off of [NS] in flames 1

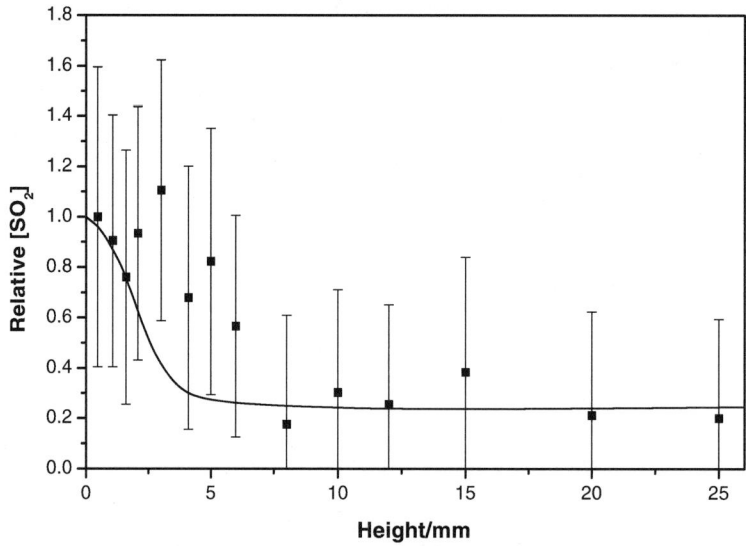

Fig. 10 Measured sulfur dioxide LIF profile, and simulated sulfur dioxide concentration profile for flame 3. (■) LIF measurements, (——) simulation.

and 3 that is not seen in flame 2. This behaviour is qualitatively reproduced by the model. The height above the burner of maximum [NS] is also well predicted, especially in flame 1, although in flame 3 there is an offset of 3 mm to a higher distance, and in both flames 2 and 3 some underprediction close to the burner. An area in which the model is deficient is in predicting the relative NS concentration between stoichiometries of 1.4 and 1.6 at 40 Torr, as shown in Fig. 5. This could be attributed to assumptions made regarding the interpretation of the LIF signals as discussed in Section 4.1 but is more likely due to deficiencies in the chemical mechanism. Fig. 11 and 12 show the sensitivities of the most important reactions with respect to [NS] of sulfur-containing species

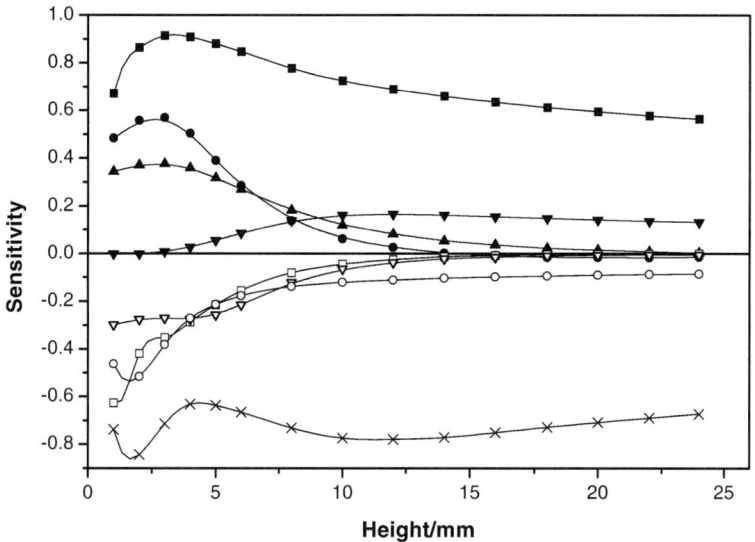

Fig. 11 Sensitivities of reactions of sulfur-containing species with respect to NS in flame 1. (■) $SH + NH \rightarrow NS + H_2$, (●) $NH + SO \rightarrow NO + SH$, (▲) $SO_2 + H \rightleftharpoons SO + OH$, (▼) $SO + N \rightleftharpoons NS + O$, (▽) $S + O_2 \rightarrow SO + O$, (○) $NS + O \rightarrow S + NO$, (□) $SO + O_2 \rightarrow SO_2 + O$, (×) $NS + OH \rightarrow SH + NO$.

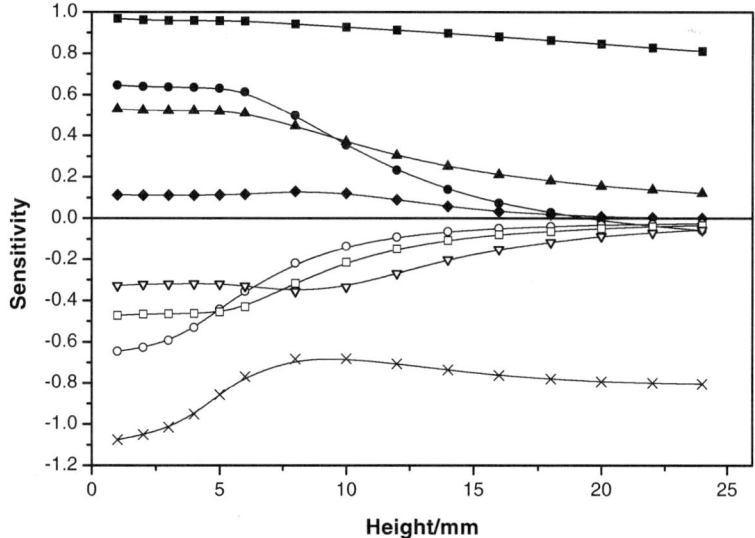

Fig. 12 Sensitivities of reactions of sulfur-containing species with respect to NS in flame 2. (■) $SH + NH \rightarrow NS + H_2$, (●) $NH + SO \rightarrow NO + SH$, (▲) $SO_2 + H \rightleftharpoons SO + OH$, (◆) $H + SO \rightleftharpoons S + OH$, (▽) $S + O_2 \rightarrow SO + O$, (○) $NS + O \rightarrow S + NO$, (□) $SO + O_2 \rightarrow SO_2 + O$, (×) $NS + OH \rightleftharpoons SH + NO$.

for the relative [NS] profiles given in Fig. 5. The sensitivity trends observed in each case are similar, with the same reactions important to both stoichiometries, with significant differences only appearing for the less sensitive reactions. This allows little scope for adjusting the simulated predictions for one stoichiometry by the optimisation of rate coefficients for sulfur-containing reactions currently contained in the mechanism without adversely affecting the other. Any change made to a particular rate coefficient to improve the simulation in one case, will have a similar effect in the other case leading to a worsening of its prediction. Improvements to these relative predictions may therefore require either a reconsideration of the NO_x scheme or the consideration of additional sulfur-containing reactions for which no current rate data are available. Previous studies of the NO_x scheme[16,17] have demonstrated that it reproduces reasonably well species profiles for low pressure methane flames over a range of stoichiometries. Given the known uncertainties in the sulfur chemistry, its improvement is perhaps more likely to lead to a corresponding improvement in predictions of the profiles presented here.

4.5. Simulated sulfur dioxide effects on NO

The results given in Fig. 6–8 show the experimental results and simulations of the effect of sulfur dioxide on the NO LIF signal. The simulations qualitatively reproduce most of the observed trends as a function of stoichiometry, but the quantitative agreement is poor. The model fails to predict any significant effect under lean and stoichiometric conditions, whereas for rich conditions it over-predicts the increase in NO observed in flame 10. For flames 6 and 9 the model is predicting an increase in NO where experimentally a decrease is observed. Fig. 13 and 14 give the sensitivities of the most important reactions of sulfur-containing species with respect to NO as a function of height above the burner for flames 4 and 6, which represent the extremes of stoichiometry in the high argon diluent flames containing 0.3% of sulfur dioxide. Fig. 15 and 16 show the same with respect to the low argon diluent flames 7 and 10 in Table 2. Important reactions of sulfur-containing species relating to these simulations will be discussed below.

4.6. Simulated sulfur dioxide profile

Fig. 10 shows the experimental sulfur dioxide LIF profile and the normalised simulated concentration profile for flame 3. Overall the agreement is good, with the extent of the depletion of sulfur dioxide well predicted by the model. Fig. 17 and 18 give the sensitivities of the most important

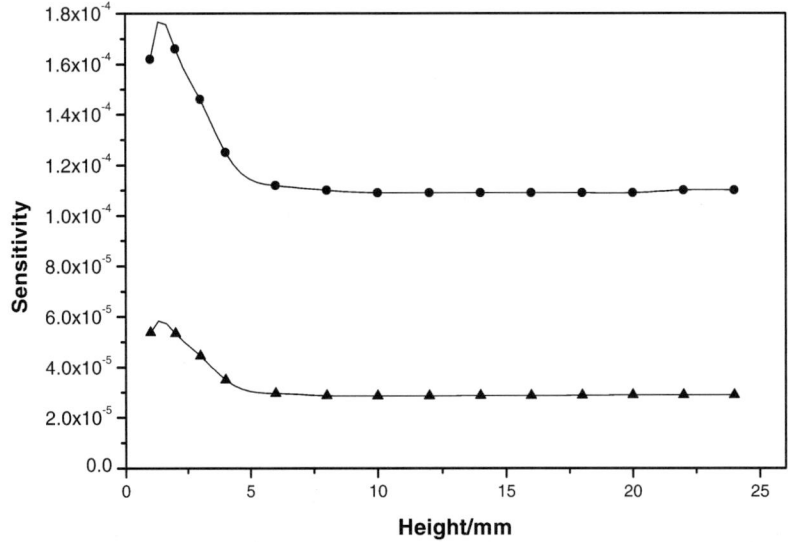

Fig. 13 Sensitivities of reactions of sulfur-containing species with respect to NO in flame 4. (▲) $SO_2 + H \rightleftharpoons SO + OH$, (●) $NH + SO \rightarrow NO + SH$.

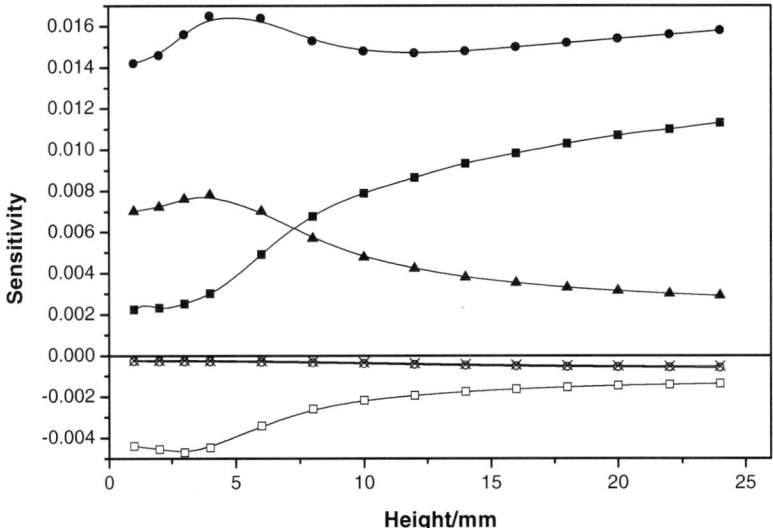

Fig. 14 Sensitivities of reactions of sulfur-containing species with respect to NO in flame 6. (●) NH + SO → NO + SH, (■) N + SO → NO + S, (▲) $SO_2 + H \rightleftharpoons SO + OH$, (□) $SO + O_2 \rightleftharpoons SO_2 + O$, (○) SO + OH + M → HOSO + M, (×) SO + H + M → HSO + M.

sulfur reactions with respect to sulfur dioxide, and also the major reactions producing and consuming sulfur dioxide.

4.7. Important reactions of sulfur-containing species

The sensitive reactions highlighted in Fig. 11–17, show a relatively small number of reactions of sulfur-containing species being of significance with respect to NS, NO and sulfur dioxide. The most important of these, and the origin of their rate coefficients are discussed below.

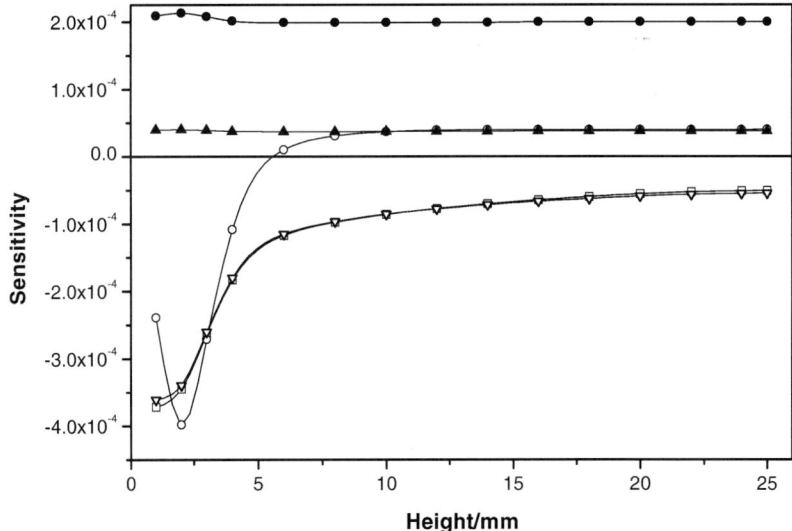

Fig. 15 Sensitivities of reactions of sulfur-containing species with respect to NO in flame 7. (●) NH + SO → NO + SH, (▲) $SO_2 + H \rightleftharpoons SO + OH$, (□) $SO_2 + O + M \rightarrow SO_3 + M$, (▽) $H + SO_2 + M \rightarrow HOSO + M$, (○) $SO_2 + OH + M \rightarrow HOSO_2 + M$.

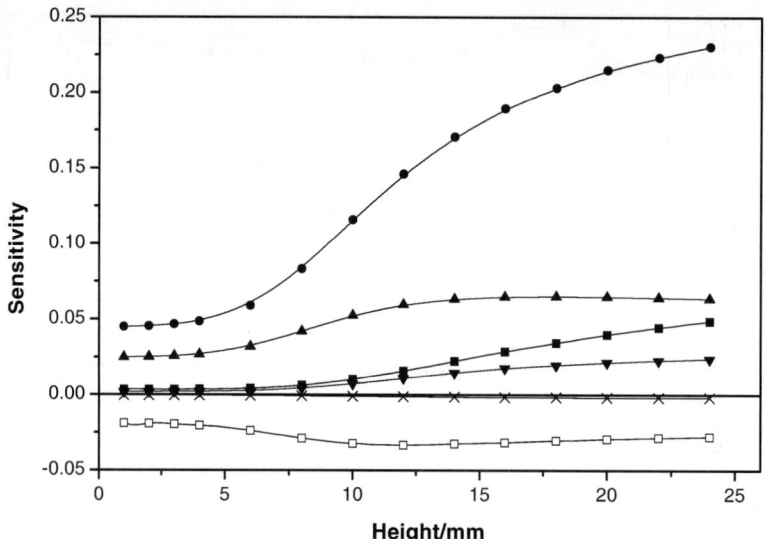

Fig. 16 Sensitivities of reactions of sulfur-containing species with respect to NO in flame 10. (●) NH + SO → NO + SH, (▲) SO_2 + H ⇌ SO + OH, (■) SH + NH → NS + H_2, (▼) N + SO → NO + S, (□) SO + O_2 → SO_2 + O, (×) S + H_2 ⇌ SH + H.

4.7.1. SO_2 + H ⇌ SO + OH. This is the dominant reaction for removal of sulfur dioxide in the flame, and is important with respect to all of the observations. Its rate coefficient is determined from the relationship between the forward and reverse reactions and the equilibrium constant, given by $K = k_f/k_r$. Given a measurement of the reverse rate coefficient k_r, and the equilibrium constant K calculated from the heats of formation and entropies of the reactants and products, k_f can be determined. Previously the rate coefficient of SO + OH had been assumed to be temperature independent and based on the two measured values obtained at room temperature[27,28]

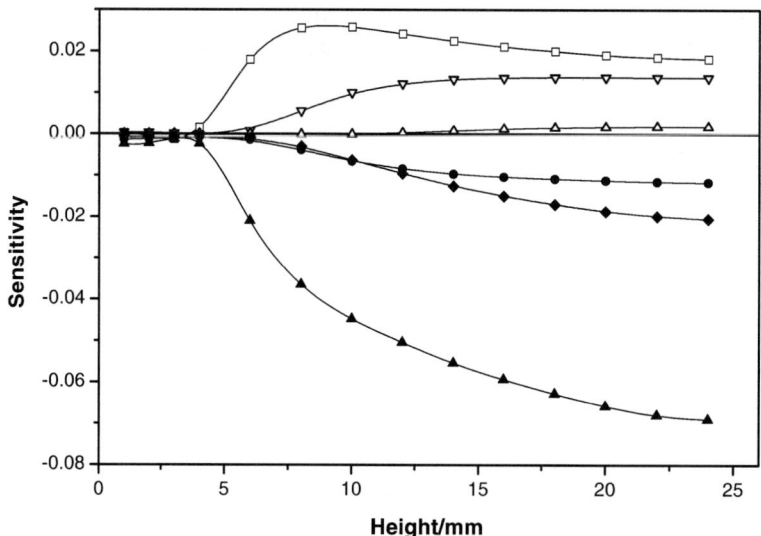

Fig. 17 Sensitivities of reactions of sulfur-containing species with respect to SO_2 in flame 3. (□) SO + O_2 → SO_2 + O, (▽) S + O_2 → SO + O, (△) 2SO ⇌ SO_2 + S, (●) NH + SO → NO + SH, (◆) H + SO ⇌ S + OH, (▲) SO_2 + H ⇌ SO + OH.

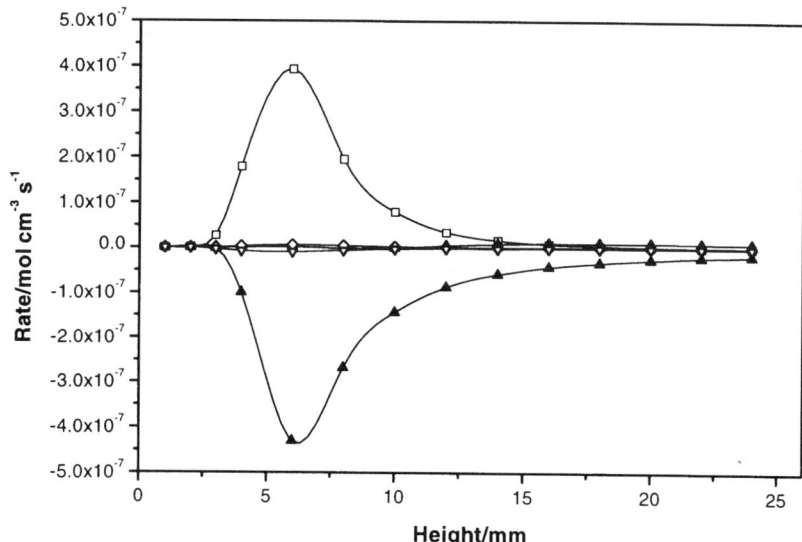

Fig. 18 Rates of production of sulfur dioxide in flame 3. (▲) $SO_2 + H \rightleftharpoons SO + OH$, (□) $SO + O_2 \rightleftharpoons SO_2 + O$, (△) $2SO \rightleftharpoons SO_2 + S$, (◇) $HOSO + H \rightarrow SO_2 + H_2$, (▽) $H + SO_2 + M \rightarrow HOSO + M$.

of 7×10^{13} and 5×10^{13} cm^3 mol^{-1} s^{-1}. A recent experimental investigation over the temperature range 295–703 K showed the reaction had negative temperature dependence, with a room temperature value of 5×10^{13} cm^3 mol^{-1} s^{-1} dropping by a factor of 3 up to 703 K.[29] This temperature dependent rate coefficient was used in the current reaction mechanism, leading to a reduction in the rate coefficient at high temperatures for the forward reaction. Due to the high sensitivity of all of the model simulations to this reaction it therefore merits further study and a more detailed theoretical investigation based on a master equation analysis[30] of this system is currently being undertaken. The reaction proceeds *via* two intermediate structures, HSO$_2$ and HOSO[31,32] linked as shown in the schematic potential energy diagram in Fig. 19. It follows that estimates based on its reverse reaction and thermodynamics may not be adequate.

4.7.2. $SO + O_2 \rightleftharpoons SO_2 + O$. This is the dominant reaction that converts SO back to sulfur dioxide. The rate coefficient expression used in this work is based on a fit to the two recent high temperature direct determinations of Tsuchiya *et al.*[33] and Woicki and Roth.[34] The expression is also consistent with the low temperature recommendation of Atkinson *et al.*[35] This reaction is

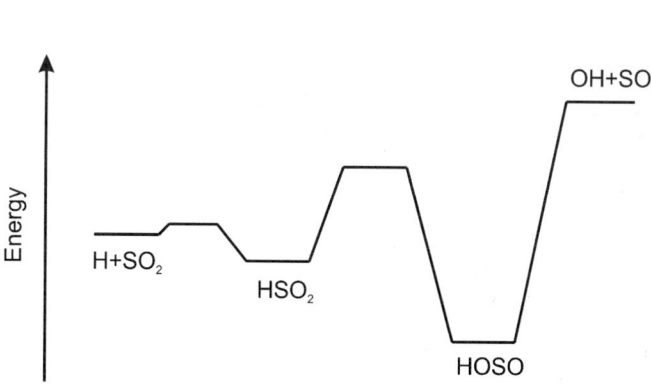

Fig. 19 Schematic potential energy diagram for the $H + SO_2$ system.

better characterised than many of the reactions involving sulfur-containing species, and therefore there is reduced scope for improving model predictions by adjusting or improving this rate coefficient expression.

4.7.3. NH + SO → NO + SH. This reaction is important in terms of both [NS] and [NO] sensitivities and was initially tentatively suggested by Pfefferle and Churchill[6] with an estimated rate coefficient of 1×10^{13} cm^3 mol^{-1} s^{-1}. No published data yet exist concerning this reaction, however recent unpublished work[36] based on flash photolysis with LIF detection of NH over the temperature range 298–703 K gives a temperature independent rate coefficient of 3×10^{13} cm^3 mol^{-1} s^{-1} for reaction of NH with SO. The latter rate coefficient was therefore used in the reaction mechanism, although the exact nature of the products is unknown.

4.7.4. SH + NH → NS + H$_2$. This reaction was initially proposed by Chagger et al.[37] with an estimate for both its product channel and rate coefficient of 1×10^{13} cm^3 mol^{-1} s^{-1}. No literature data exist on this reaction, and it is only of importance with respect to the simulated [NS] profile. Its sensitivity as a function of height above the burner shows little variation, therefore its main effect will be to influence the overall NS concentration, with little effect on the shape of the [NS] profile. Without accurate absolute NS concentration profiles with which to compare against the model, the real significance of this reaction is difficult to assess.

4.7.5. N + SO → products. This reaction is important in terms of [NS] and [NO] sensitivities but no experimental data exist in the literature for this reaction or its product channels. Both of the product channels and rate coefficients included in our mechanism, producing NO + S and NS + O, originate from the work of Wendt et al.[38] These reactions were included by Wendt et al. in order to improve the modelling of the effect of sulfur dioxide on NO and nitrogen in their study of a rich, moist, carbon monoxide/argon/oxygen flame. Since the rates were estimated there is scope for possible improvement of the simulations presented here by better characterised rate expressions for each product channel of this reaction.

4.7.6. H + SO ⇌ S + OH. This reaction is important in terms of [SO$_2$] sensitivities. The rate coefficient expression used for this reaction is derived from the equilibrium constant calculated from the heats of formation and entropies of the reactants and products along with the only published experimental measurement of the reverse reaction. The reverse reaction was measured at room temperature in a discharge flow system with detection by electron spin resonance spectroscopy.[28] The assumption of temperature independence may not be true for this reaction.

4.7.7. NS + OH → SH + NO. This is the dominant reaction responsible for the removal of NS. No experimental data exist in the literature for its rate coefficient or product channels. The rate coefficient in this mechanism is derived from the reverse reaction rate coefficient and the equilibrium constant. The reverse reaction was initially proposed by Chagger et al.[37] with an estimated rate coefficient. Calculation of the enthalpy change for this reaction has led the authors to modify the activation energy from that initially proposed by Chagger et al.[37] to one which is compatible with the forward reaction having no barrier, although this assumption may not be valid. This reaction requires much better characterisation.

4.7.8. S + O$_2$ → SO + O. This reaction is significant in contributing to the removal of NS, and inhibiting the removal of sulfur dioxide. The rate coefficient used in this mechanism is derived from a fit to the high temperature direct measurement of Tsuchiya et al.[33] and the low temperature recommendation of Atkinson et al.[39] As it is based on a direct measurement at high temperatures, this reaction is better characterised than many of the sulfur-containing species reactions, and therefore offers reduced scope to improve model predictions through its adjustment.

5. Conclusions

Relative profiles of [NS] and sulfur dioxide have been determined using LIF for rich methane flames doped with ammonia and sulfur dioxide providing additional data of use in the evaluation of chemical mechanisms under development to investigate the interactions of sulfur- and nitrogen-containing species in flames. Simulations of both the relative [NS] and [SO_2] profiles using PREMIX coupled with a methane/NO_x/SO_x mechanism derived from currently available rate data, show qualitatively good agreement with the experimental measurements. The qualitative effect of changing pressure and stoichiometry on the fall-off of [NS] in the burnt gas region is reproduced by the model, as is the depletion of sulfur dioxide in the burnt gas region.

The influence of adding various amounts of sulfur dioxide on [NO] in the burnt gas region has also been determined for a range of stoichiometries using LIF. A variety of effects were observed, ranging from large percentage reductions for lean flames, to large percentage increases for rich flames containing low amounts of argon diluent. The factors influencing the effect of added SO_2 were (i) the amount of SO_2 added, (ii) the overall flame stoichiometry and (iii) the level of diluent gas affecting the overall flow rates and concentrations of the reactive gases and the resulting temperature profiles. Simulations of these experimental studies highlight significant deficiencies within the sub-scheme describing reactions of sulfur-containing species. Apart from the rich flames, the model predicts very low percentage changes in [NO] in the burnt gas region for all levels of sulfur dioxide added, and in some cases it also predicts the wrong qualitative trends. Some scope exists for the improvement of predictions through either improved measurements or theoretical modelling of certain elementary reactions within the sub-scheme. Sensitivities have been used to highlight those reactions of the highest importance for [NS], [NO] and sulfur dioxide profiles within the flame. Such sensitivities however are based on concentrations and rates as predicted by the current scheme. They do not give information on possible important elementary reactions and product channels that have not been included in the mechanism because of the lack of rate data available in the literature. Substantial further work is required to identify possible additional reactions influencing the interaction of sulfur- and nitrogen-containing species and to establish their rates either by measurement or theoretical modelling.

Acknowledgements

We wish to acknowledge M. J. Pilling and M. A. Blitz for contributions to the experimental measurements, T. Turanyi and P. Glarborg for contributions to the modelling work and EPSRC for financial support under grant GR/M19925.

References

1 J. A. Miller, M. C. Branch, W. J. Mclean, D. W. Chandler, M. D. Smooke and R. J. Kee, *Proc. Combust. Inst.*, 1984, **20**, 673.
2 C. F. Cullis, R. M. Henson and D. L. Trimm, *Proc. R. Soc. London, Ser. A*, 1966, **295**, 72.
3 D. J. Halstead and D. R. Jenkins, *Trans. Faraday Soc.*, 1969, **65**, 3013.
4 J. O. L. Wendt and J. M. Ekmann, *Combust. Flame*, 1975, **25**, 355.
5 L. D. Pfefferle and S. W. Churchill, *Combust. Sci. Technol.*, 1986, **49**, 235.
6 L. D. Pfefferle and S. W. Churchill, *Ind. Eng. Chem. Res.*, 1989, **28**, 1004.
7 S. I. Tseregounis and O. I. Smith, *Combust. Sci. Technol.*, 1983, **30**, 231.
8 S. I. Tseregounis and O. I. Smith, *Proc. Combust. Inst.*, 1984, **20**, 761.
9 E. Hampartsoumian and W. Nimmo, *Combust. Sci. Technol.*, 1995, **110**, 111, 487.
10 W. Nimmo, E. Hampartsoumian, K. J. Hughes and A. S. Tomlin, *Proc. Combust. Inst.*, 1998, **27**, 1419.
11 P. Glarborg, D. Kubel, K. Dam-Johansen, H. M. Chiang and J. W. Bozzelli, *Int. J. Chem. Kinet.*, 1996, **28**, 773.
12 M. U. Alzueta, R. Bilbao and P. Glarborg, *Combust. Flame*, submitted.
13 K. J. Hughes and A. S. Tomlin, http://www.chem.leeds.ac.uk/Combustion/sox.htm.
14 K. J. Hughes, T. Turányi, A. Clague and M. J. Pilling, *Int. J. Chem. Kinet.*, in press.
15 K. J. Hughes, T. Turányi and M. J. Pilling, http://www.chem.leeds.ac.uk/Combustion/methane.htm.
16 K. J. Hughes, A. S. Tomlin, E. Hampartsoumian, W. Nimmo, I. G. Zsély, M. Ujvári, T. Turányi, A. Clague and M. J. Pilling, *Combust. Flame*, 2001, **124**, 573.
17 K. J. Hughes, T. Turányi, A. S. Tomlin and M. J. Pilling, http://www.chem.leeds.ac.uk/Combustion/nox.htm.
18 J. B. Jeffries and D. R. Crosley, *Combust. Flame*, 1986, **64**, 55.
19 J. B. Jeffries, D. R. Crosley and G. P. Smith, *J. Phys. Chem.*, 1989, **93**, 1082.
20 R. J. Kee, F. M. Rupley and J. A. Miller, *A Fortran Chemical Kinetics Package for the Analysis of Gas Phase Chemical Kinetics, SAND89-8009B*, Sandia National Laboratories, 1991.

21 R. J. Kee, J. F. Grcar, M. D. Smooke and J. A. Miller, *A Fortran Program for Modeling Steady Laminar One-Dimensional Premixed Flames, SAND85-8240*, Sandia National Laboratories, 1985.
22 T. Turányi, http://www.chem.leeds.ac.uk/Combustion/kinalc.htm.
23 J. O. L. Wendt, J. T. Morcomb and T. L. Corley, *Proc. Combust. Inst.*, 1979, **17**, 671.
24 T. L. Corly and J. O. L. Wendt, *Combust. Flame*, 1984, **58**, 141.
25 H. Okabe, *Photochemistry of Small Molecules*, Wiley-Interscience, New York, 1978.
26 M. R. Zacharia and O. I. Smith, *Combust. Flame*, 1987, **69**, 125.
27 R. W. Fair and B. A. Thrush, *Trans. Faraday Soc.*, 1969, **65**, 1557.
28 J. L. Jourdain, G. LeBras and J. Coimbourieu, *Int. J. Chem. Kinet.*, 1979, **11**, 569.
29 M. A. Blitz, K. W. McKee and M. J. Pilling, *Proc. Combust. Inst.*, 2000, **28**, 2491.
30 M. J. Pilling, S. H. Robertson and N. J. B. Green, in *Advances in Chemical Kinetics and Dynamics*, ed. J. R. Barker, JAI Press, Connecticut, 1995, vol. 2A, p. 209.
31 A. J. Frank, M. Sadilek, J. G. Ferrier and F. Turecek, *J. Am. Chem. Soc.*, 1997, **119**, 12343.
32 A. Goumri, J. D. R. Rocha, D. Laakso, C. E. Smith and P. Marshall, *J. Phys. Chem. A*, 1999, **103**, 11328.
33 K. Tsuchiya, K. Kamiya and H. Matsui, *Int. J. Chem. Kinet.*, 1997, **29**, 57.
34 D. Woiki and P. Roth, *Int. J. Chem. Kinet.*, 1995, **27**, 57.
35 R. Atkinson, D. L. Baulch, R. A. Cox, R. F. Hampson, J. A. Kerr and J. Troe, *J. Phys. Chem. Ref. Data*, 1992, **21**, 1125.
36 M. A. Blitz and M. J. Pilling, personal communication.
37 H. K. Chagger, P. R. Goddard, P. Murdoch and A. Williams, *Fuel*, 1991, **70**, 1137.
38 J. O. L. Wendt, E. C. Wootan and T. L. Corley, *Combust. Flame*, 1983, **49**, 261.
39 R. Atkinson, D. L. Baulch, R. A. Cox, R. F. Hampson, J. A. Kerr, M. J. Rossi and J. Troe, *J. Phys. Chem. Ref. Data*, 1997, **26**, 1329.

General Discussion

Dr Whitaker opened the discussion of Prof. Kohse-Höinghaus's paper: In your paper you describe experiments in which you excite to $v' = 2$ and look at the 0–0 band fluorescence and observe a very high apparent rotational temperature. Similarly for the 1–1 band. Now the $v' = 2$ state exhibits fairly rapid predissociation so I am curious to know if this is what could be responsible for the high apparent rotational temperatures observed—because I don't understand how VET on its own could account for this effect.

Prof. Kohse-Höinghaus responded: Our understanding of this process is that VET and RET are closely coupled, and that an individual VET step does not necessarily end in the same rotational level in the lower vibrational state. High apparent temperatures are observed since high rotational levels are populated by combined VET/RET steps. Additionally, for OH, RET in higher rotational levels is slower than in the lower ones.

Apparent high temperatures are also observed in $v' = 0$ after excitation of $v' = 1$,[1,2] which is not predissociative. Predissociation is just another loss term, as quenching, in this experiment; it does not influence the population in the lower vibrational levels.

1 A. T. Hartlieb, D. Markus, W. Kreutner and K. Kohse-Höinghaus, *Appl. Phys. B*, 1997, **65**, 81.
2 D. R. Crosley and G. P. Smith, *Appl. Opt.*, 1983, **22**, 1428.

Prof. Stuhl asked: In your publication you say that you try 'to avoid saturation'. Further, you expect that polarization might affect some results. My question is: why don't you work under saturation conditions? In this case, polarization effects are expected to be negligible.[1]

1 R. Altkorn and R. N. Zare, *Ann. Rev. Phys. Chem.*, 1984, **35**, 265.

Prof. Kohse-Höinghaus replied: In our contribution, we wish to *study* the effects of polarisation, since not much is known about the temporal decay of these effects; picosecond laser experiments now offer the time resolution to investigate the influence of collisions on the prepared population under defined polarisation conditions in combustion experiments. We expect these effects to contribute significantly to the LIF signals observed in typical experiments for concentration or temperature measurement, where *linear* LIF is most often employed. Saturated LIF experiments are much less suitable for quantitative measurements, since saturation may affect the population distribution in the ground state and lead to population re-circulation, which would complicate the interpretation. Additionally, it is very hard to maintain constant saturation over the complete excitation volume.

Prof. Cheskis asked: Did you check the pressure dependence? How does your model work at different pressures? Does your model take into account the slower rotational relaxation of rotational levels with higher J numbers?

Prof. Kohse-Höinghaus answered: We applied the model at different pressures, ranging mostly from reduced pressure (near 1 mbar) to atmospheric. The model seems to represent most conditions tested so far quite well. Additional examples can be found in ref. 1 and 2. Some tentative modelling experiments with LASKIN have been performed at higher pressures, but reliable experimental data for comparison are still scarce in that regime.

The change in rotational energy transfer rate coefficient with rotational quantum number has been considered in the model, see ref. 1 and 3 for more detail.

1   U. Rahmann, W. Kreutner and K. Kohse-Höinghaus, *Appl. Phys. B*, 1999, **69**, 61.
2   A. Brockhinke, W. Kreutner, U. Rahmann, K. Kohse-Höinghaus, T. B. Settersten and M. A. Linne, *Appl. Phys. B*, 1999, **69**, 477.
3   R. Kienle, M. P. Lee and K. Kohse-Höinghaus, *Appl. Phys. B*, 1996, **63**, 403.

Prof. Greenhalgh asked: (1) How does your system compare with Roger Farrow's distributed feedback dye laser (DFDL) system? (2) Please can you explain what you mean by quench-free measurements, since you seem to indicate that this is a goal rather than a reality? (3) Please recommend the best transition for probing in a practical measurement.

Prof. Kohse-Höinghaus answered: (1) At the time we purchased our regeneratively amplified mode-locked laser (in 1995), it was the only commercially available picosecond laser system which met our specifications (that is, which produced Fourier-limited pulses with a high-enough power to allow two-dimensional wavelength–time imaging even when exciting a multi-photon transition). Recent advances in the DFDL development showed that they can meet these specifications as well. However, they are still more unstable and less flexible than our all-solid-state system and to our knowledge, they still can't be bought off the shelf.

(2) In the paper itself, we explain that rotational and vibrational redistribution may, even on very short time scales, lead to the erroneous assumption that collisions would not influence the result. Thus, experiments using the conventional 'gated-detection' approach are 'quench-insensitive' at most. Only experiments where the complete temporal evolution of the LIF signal is monitored can be truly 'quench-free'. One approach is described in our paper, another can be found in ref. 1.

(3) In our opinion, exciting the OH A–X 1–0 transition and collecting the fluorescence in the 1–1 and 0–0 transitions has several advantages for many practical applications, which have been often exploited and described in previous work. The 0–0 transition is more susceptible to optical density/re-absorption problems. Transitions with $v' \geqslant 2$ are attractive mainly because they can be pumped with cheap excimer lasers and are predissociated (which alleviates part of the quenching problem). However, due to problems with collisional redistribution in the ground state, these transitions are not recommended for quantitative measurements.

1   M. W. Renfro, S. D. Pack, G. B. King and N. M. Laurendeau, *Appl. Phys. B*, 1999, **69**, 137.

Prof. Hippler asked: (1) What are the important colliders for energy transfer (ET) and quenching of OH(A) in a flame? (2) Are there estimates for radicals and atoms doing ET or quenching? (3) For what kind of flames does the LASKIN model work well?

Prof. Kohse-Höinghaus replied: (1) In many flames of hydrocarbons burning in air, water and nitrogen are the two most important colliders. This may not hold for very fuel-rich conditions.

(2) In low-pressure flames with high H-atom or OH radical concentrations, as investigated in ref. 1, H-atom quenching has been estimated to contribute with a rate constant of 8×10^{-10} cm^3 s^{-1}.

(3) Typically, the flames described above (hydrocarbon–air) as well as hydrogen–air flames, can be modelled with surprising success. However, LASKIN can also predict the qualitative energy transfer behaviour under quite different conditions, including flow reactor experiments. LASKIN has been tested in the pressure range from a few mbar to several bar.

1   J. B. Jeffries, K. Kohse-Höinghaus, G. P. Smith, R. A. Copeland and D. R. Crosley, *Chem. Phys. Lett.*, 1988, **152**, 160.

Prof. Wolfrum opened the discussion of Prof. Griffiths's paper: Are the PLIF images of formaldehyde shown in Fig. 9 and Fig. 10 of your paper corrected for the temperature dependence of the ground state population?

Dr Whitaker responded: The PLIF images of H_2CO shown in the paper are primary data. Of course, you are absolutely right to point out that the formaldehyde fluorescence signal will be a function of the temperature field as well as the concentration field. Klein-Douwel et al.[1] have shown that the effect of the partition function is minimal for 370 nm excitation of the $\tilde{A}\,^1A_2$–$\tilde{X}\,^1A_1$ 4^1_0 vibronic band. However, in our case, *i.e.* excitation at 355 nm, there is a strong dependence of the fluorescence signal from the 4^1_0 band on temperature.[2] Just as for the acetone PLIF images presented in our paper, at higher temperatures the partition function effects reduce the observed LIF signals. This is simply a reflection of the fact that at higher temperatures a greater number of rotational states are energetically accessible. Collisional quenching, which to a first approximation will scale as $1/\sqrt{T}$, has the opposite effect of enhancing the signal at elevated temperatures. It is known that the relatively strong Raman signal from any methane that might be present can be a source of interference in LIF measurements of formaldehyde (shift $\sim$ 50 nm). Another known problem associated with the determination of formaldehyde concentration fields by LIF is interference from polycyclic aromatic hydrocarbons (PAHs). The combined effects of these phenomena are difficult to quantify (although the latter two are unlikely to be of any significance in our case). Of course, if the temperature field was known *a priori* it would be simple enough to calculate the partition function correction.

The implication of your question with respect to the interpretation of the images presented for n-pentane combustion in Fig. 10 of our paper is interesting. The bottom three images, for T_c = 765, 790 and 840 K respectively, are obtained at a time delay where the pressure record indicates that the reacting mixture has attained approximately the same average temperature in all three cases (see Fig. 3 in the paper). The images show approximately uniform fluorescence yield across the chamber in the three cases but that the intensity becomes weaker through the sequence from T_c = 765 to 840 K. Our interpretation is that this reflects the decrease in the extent of reaction resulting from the increased ignition delay, and not a decrease of fluorescence quantum yield as a result of increased nominal adiabatic temperature of the compressed charge. The results obtained at earlier times for T_c = 840 K, where the heat release is quite slow, are also consistent with this interpretation. The result for T_c = 790 K at 3.4 ms is, however, harder to interpret. Under these conditions we expect that the core gas in the centre of the chamber 1–2 ms after the end of the compression stroke is about 750 K, close to the bottom of the ntc curve (Fig. 1 of our paper), and consequently that the core gas should react (slightly) faster than the periphery. Yet we observe a weaker fluorescence signal in the centre of the chamber. However, as we mention in the paper, the interpretation is clouded by the diverse kinetic origins of formaldehyde. As Fig. 4 e_7 of the paper shows there is a considerable chemiluminescent signal from excited state formaldehyde some 3 ms after the end of the compression at the edge of the chamber, albeit at the compressed temperature of 765 K.

1 R. J. H. Klein-Douwel, J. Luque, J. B. Jeffries, G. P. Smith and D. R. Crosley, *Appl. Opt.*, 2000, **39**, 3712.
2 J. E. Harrington and K. C. Smyth, *Chem. Phys. Lett.*, 1993, **202**, 196.

Dr Morley said: Rapid compression machines were originally designed to provide an ideal environment and many sets of data, particularly ignition delay as a function of temperature, and have been presented assuming ideality. You have demonstrated, and to some extent quantified the temperature inhomogeneity. Is it possible to develop a methodology based on this understanding to correct these earlier results? I'm thinking of autoignition delays which occur in the colder regions in systems with negative temperature coefficients being erroneously assigned to the peak temperature.

Prof. Griffiths responded: Of course, the experimental results are what they are. The normal procedure is to use the adiabatically compressed gas temperature as the reference temperature, regardless of the thermal development that follows as a result of heat release and heat loss. The important implication of your comment is that modelling, in which a spatially uniform temperature is assumed, cannot be compared directly with these results when seeking validation of the model. Our best estimate is that, within several milliseconds after the piston has stopped, the core gas is about 40 K lower than the adiabatically heated gas surrounding it.

The simplest qualitative interpretation from our observations might be that the longest measured delay in the negative temperature dependent range, and/or the compressed gas temperature range over which the negative temperature dependence of ignition delay exists, is shorter than would be the case for an experiment in ideal conditions. Simulations of ignition delay from a two zone model, with a core at 40 K lower than the toroidal region surrounding it, might be compared with the predictions from a spatially uniform model as a first step to understanding the quantitative implications.

As you have mentioned [see below], the role up vortex can be suppressed by careful design of the piston crown. Also the gas motion that is created must be dependent on piston speed. So different machines will cause different types and extents of departure from ideality. For example, the long compression time (60 ms) at proportionately lower speed, that is used by workers at Lille, will cause a different spatial temperature distribution after the end of compression from that in our machine. Their temperature measurements by Rayleigh scattering and thermocouples indicate that the adiabatic region probably resides in the core, but I would suggest that there may be variations in temperature at the periphery, resulting from heat transfer to the wall. So whatever methodology is developed, it would have to be specific to a particular rapid compression machine.

Prof. Greenhalgh asked: How quiescent is the gas in the rapid compression machine when the piston stops? Also the image at 2 ms later shows some structure. Would stirring help and what role might shear stresses play in the autoignition?

Prof. Griffiths answered: There must certainly be significant gas motion at the moment the piston stops since the piston speed during compression is about 10 m s^{-1}, but it tends to die out within the first 10 ms. The penetration of a colder plug into the core from off the piston face, very soon after the piston has stopped, is consistent with the generation of a roll-up vortex.[1] Our evidence is that this colder gas reaches the central plane of the combustion chamber only after about 2 ms, as you have noted.

Undoubtedly stirring does have an effect on the combustion process. We have shown previously[2,3] that the development of autoignition is affected by enhancing the gas motion during the post compression period, but the consequence seems to be connected mainly with modification of the heat loss rate to the chamber walls. Those studies predated our ability to investigate the temperature field, so we do not know how the spatial development was affected. It is easiest to enhance or modify the gas motion on the relevant timescale by controlling the gas that is being pushed ahead of the piston, such as by squeezing it through a mesh. I am inclined to think that shear stress might have a part to play in autoignition particularly in turbulent non-premixed gases.

1 R. J. Tabaczinski, D. P. Hoult and J. C. Keck, *J. Fluid Mech.*, 1970, **42**, 249.
2 J. F. Griffiths and W. Nimmo, *Nature*, 1986, **322**, 46.
3 J. Franck, J. F. Griffiths and W. Nimmo, *Twenty-first Symp. (Int.) Combust.*, The Combustion Institute, Pittsburgh, 1986, p. 447.

Dr Morley said: Rapid compression machines can be made more ideal by suppressing the roll up vortex. Lee and Hochgreb (ref. 4 of the paper by Griffiths *et al.*) have shown that by incorporating a crevice in the piston that swallows the cold boundary layer, the cold gas can be prevented from mixing into the adiabatic core gas in the chamber. They verified the effectiveness of this approach by CFD modelling and showed that, with the modifications, the pressure could be calculated using a thermodynamic model, which was inadequate in the unmodified system.

Prof. Griffiths responded: This is an important experimental development. We have not explored its potential but it would certainly be interesting to demonstrate the response using our non-invasive methods to study the temperature distribution. Part of our interest lies in the behaviour close to the walls of the chamber, because there is evidence that autoignition may originate there is certain circumstances. Consequently, to date we have maintained a flat piston crown extending to the circumference of the cylinder so that optical aberrations in that important region are minimised.

Dr Taatjes said: I notice that your mechanism for autoignition includes a branching step involving O_2 addition to a QOOH species. In this connection I would like to point out that the calculations in our paper (presented here), as shown in the figure for isobutyl + O_2, confirm that the isomerization from RO_2 to QOOH *via* a six-membered ring proceeds over a relatively small

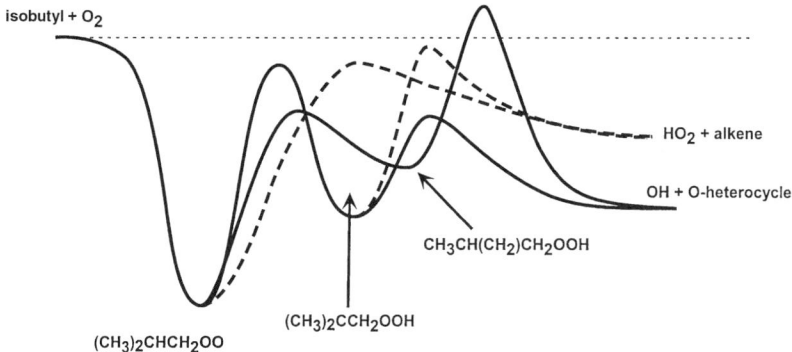

Fig. 1 Reaction paths for the oxidation of isobutane.

barrier. However, the well for QOOH is very shallow, and the barrier in the exit channel to form OH plus an oxetane is above the energy of the R + O_2 reactants (see Fig. 1 here). Therefore, the fate of this QOOH species will be to rapidly return to the more stable RO_2 form, and our calculations indicate that the equilibrium will greatly favor RO_2. I would like to ask what level of stability of QOOH is necessary in order for present models to accurately predict low temperature oxidation and autoignition phenomena? I would also invite general comments on the consequences of the reversibility of the $RO_2 \rightleftharpoons$ QOOH isomerization.

Prof. Griffiths responded: It is believed that low temperature autoignition of alkanes evolves through chain branching *via* a small fraction of QOOH forming the O_2QOOH species, coupled to self-heating. Thus the kinetic issue that you have addressed in the context of the 'stability of QOOH' is highly relevant. Your analysis presented here of the elementary reactions of R and RO_2 also includes other interactions that determine the existence of the overall negative temperature dependence of the reaction rate during alkane oxidation, which means that your fundamental studies constitute a very important contribution to the correct interpretation of the global behaviour.

The fraction of QOOH leading to branching (and hence the relative reactivity of different alkanes), must be affected by the ease of the RO_2/QOOH isomerization, the stability of RO_2 preceding it and (in a similar way) the stability of O_2QOOH following it. It would be necessary to turn to numerical studies based on comprehensive kinetic schemes, such as those by Westbrook and co-workers,[1] to quantify the fraction of QOOH involved in chain branching. Reversibility of $RO_2 \rightleftharpoons$ QOOH is assumed for all species in the computations, and I believe that the consensus would support that interpretation.

The database for RO_2/QOOH isomerizations obtained from experimental work is still quite limited, being principally that derived by Pilling *et al.*[2] The activation energies for $RO_2 \rightleftharpoons$ QOOH range from 62 to 176 kJ mol^{-1} (15 to 42 kcal mol^{-1}).

For the *sec*-C_4H_9 in the 1,4p and *tert*-C_4H_9 in the 1,4t transitions, the activation energies for the forward processes are (in your units) 36.6 and 28.2 kcal mol^{-1}, in excellent agreement with and reflect a similar spread for transfer of H from different sites as those predicted from your calculations (37.0 and 29.7 kcal mol^{-1}, respectively). This is very encouraging. For the 1,5s transition in n-C_4H_9 the data from Pilling *et al.*[2] are not quite as close (26.3 *vs.* 22 kcal mol^{-1}).

The *ab initio* calculations should certainly be exploited for more accurate quantitative interpretation of these complex processes within the overall alkane mechanisms and the origins of discrepancies or uncertainties resolved through them. The overall consequence of different values for the activation energies should also be put to numerical test. We could investigate the latter through numerical experiments on n-C_4H_{10} and i-C_4H_{10} combustion using a unified approach to kinetic modelling of alkane ignition in the rapid compression machine.[3] Although the reduced form of the kinetic scheme means that there are empirical elements in a number of rate coefficients, there is sufficient detail for us to test variations of parameters, with specific reference to your calculated values and recommendations for the butyl radical reactions.

To be consistent we would have to change the qualitative structure of the scheme to include the independent elimination channel for alkene + HO_2, which is not incorporated in the scheme as it is set up at present. This must have a significant effect, requiring correction of other parameters to give quantitative accord with experiment. Perhaps you can comment on the implications that this alteration might make.

1   H. J. Curran, P. Gaffuri, W. J. Pitz and C. K. Westbrook, *Combust. Flame*, 1998, **114**, 149.
2   K. J. Hughes, P. A. Halford-Maw, M. J. Pilling and T. Turanyi, *Twenty-fourth Symposium (Int.) on Combust.*, The Combustion Institute, Pittsburgh, 1992, p. 645.
3   J. F. Griffiths, K. J. Hughes, M. Schreiber and C. Poppe, *Combust. Flame*, 1994, **99**, 533.

Dr Taatjes replied: If the alkene + HO_2 channel is direct elimination from RO_2 (as appears to be the case), its activation energy no longer need be tied to that of the isomerization step. This would have a couple of advantages. First, the calculated activation energy for elimination seems to be independent of the type of hydrogen eventually taken to form HO_2, as can be seen in Tables 5 and 6 of our paper. This is because the reaction coordinate for the elimination is mostly heavy-atom motion, so the bond energy of the 'abstracted' H atom does not contribute. The isomerization transition state energy does, in fact, depend on the type of C–H bond attacked as you note. So separate activation energies for the QOOH and HO_2 production steps seems like a good idea. Also, then the alkene production comes from the more stable isomer in the $RO_2 \rightleftharpoons QOOH$ pair, and is somewhat decoupled from the QOOH stability question.

Our calculations suggest that the QOOH may be less stable than some of the models appear to assume. For example, while the calculated energy for the 1,5s transition state is calculated to be ~ 22 kcal mol^{-1} above the n-$C_4H_9O_2$ well, not significantly different from the activation energy in your model, this transition state is only 8.5 kcal mol^{-1} above the QOOH well, a little more than half the QOOH $\rightarrow RO_2$ activation energy in your model. That may still provide sufficient stability to permit the chain-branching step to occur as modelled. The numerical experiments you propose would be most helpful in determining how sensitive the models are to that reverse activation energy.

Prof. Pilling[†] commented: A few years ago we studied the reaction between the neopentyl radical and O_2, by following, using LIF, the time dependence of OH formed from dissociation of $C_5H_{10}OOH$ (QOOH).[1] The advantage of this system, as a vehicle for studying the kinetics of the $RO_2 \rightarrow QOOH$ reaction, arises from the absence of a competing HO_2 channel.

The reaction mechanism is shown in Fig. 2 here. There are three channels for reaction of QOOH, one of which depends on reaction with O_2, while the other two involve dissociation.

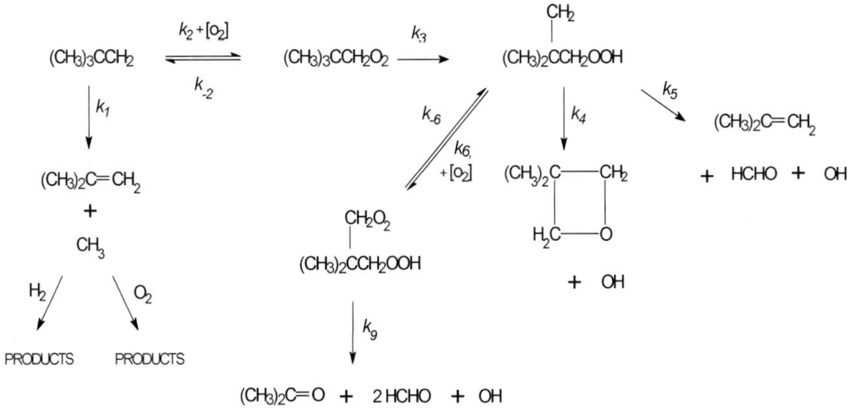

Fig. 2 Mechanism for the oxidation of neopentyl radicals.

---

† Also Dr A. R. Clague (University of Leeds), Dr S. J. Griffiths (University of Leeds), Dr K. H. Hughes (University of Leeds) and Dr D. W. Stocker (University of Leeds).

While the rate coefficient for the reverse reaction, QOOH → RO_2, k_{-3}, is substantially greater than the rate coefficient for the forward reaction, k_3, the evidence at the time suggested that the removal of QOOH to form the forward products, via reactions 4, 5 and 6, is sufficiently rapid that QOOH is removed irreversibly and the reverse isomerisation via reaction (−3) can be neglected. Since then we have re-evaluated the equilibrium constant, K_3, and found that the ratio, R, of the rates of the forward and reverse reactions of QOOH, lies in the range 3–8, over the range of $[O_2]$ used experimentally and at a temperature of 700 K. R is defined as $(k_4[O_2] + k_5 + k_6)/k_{-3}$. The rate coefficients k_5 and k_6 were obtained from measurements of Walker and coworkers,[2] k_4 was equated to the rate coefficient for neopentyl + O_2 and k_{-3} was equated to k_3/K_3, with K_3 determined by group additivity[3] and k_3 obtained by fitting the experimental data.

The OH LIF profile is biexponential (Fig. 3) with the reciprocal time constant for the rise ($-\lambda_+$) related to the rate coefficients for reactions (1)–(5) and that for the decay to the removal of OH by diffusion and reaction with the neopentyl radical precursor. The profiles were studied over a range of $[O_2]$, and Fig. 4 shows a plot of $|\lambda_+|$ vs. $[O_2]$ at 700 K. Two fits are shown, one assuming that reaction (3) is irreversible (i.e. $R \gg 1$) and the other using the best estimates for k_{-3}–k_5. The two

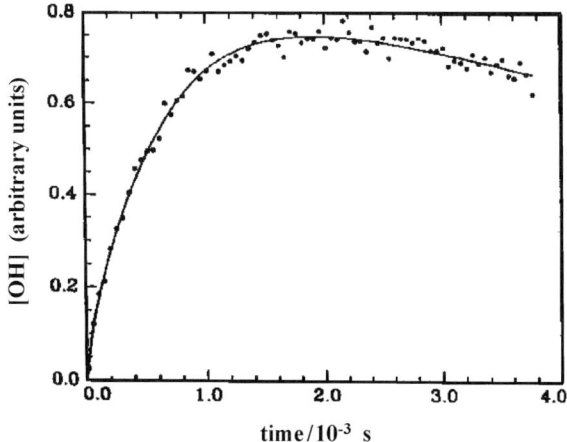

Fig. 3 Time profile of OH fluorescence.

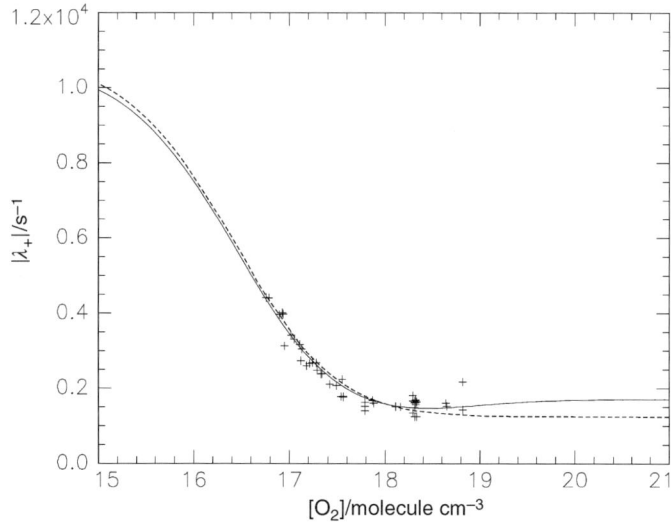

Fig. 4 Comparison of the fits to the experimental data at 700 K with an irreversible model (dashed curve) and a reversible model (solid curve).

Table 1 Comparison of fitted k_3 values obtained from assuming reversible and irreversible kinetics for reaction (3)

T/K	$10^{-3} \, k_3/\text{s}^{-1}$ irreversible model	$10^{-3} \, k_3/\text{s}^{-1}$ reversible model
660	0.30 ± 0.03	0.28 ± 0.03
670	0.33 ± 0.04	0.37 ± 0.04
690	0.85 ± 0.17	1.24 ± 0.24
700	1.23 ± 0.09	1.72 ± 0.14
715	1.64 ± 0.25	2.14 ± 0.20
730	2.05 ± 0.33	3.01 ± 0.50
750	4.12 ± 0.81	3.76 ± 0.76

approaches show a comparable quality of fit and only deviate significantly from one another at higher [O_2]—on the basis of the experimental data alone it is not feasible to establish the value of R.

Table 1 compares the values of k_3 obtained from analysis of the data (i) assuming that reaction (3) is irreversible (*i.e.* $R \gg 1$) and (ii) using the best estimates for k_{-3}–k_5. The latter values are slightly larger than the former.

1 K. J. Hughes, P. A. Halford-Maw, M. J. Pilling and T. Turanyi, *Twenty-fourth Symp. (Int.) Combust.*, The Combustion Institute, Pittsburgh, 1992, p, 645.
2 R. R. Baldwin, M. W. Hisham and R. W. Walker, *J. Chem. Soc., Faraday Trans.*, 1982, **78**, 1615.
3 D. W. Stocker and M. J. Pilling, in preparation.

Dr DeSain commented: You stated that you have previously observed only OH formation from the reaction of neopentyl (C_5H_{11}) + O_2.[1] As you know the C_5H_{11} + O_2 mechanism is different from that of most R + O_2 reactions in that the direct elimination to form the conjugate alkene and HO_2 is impossible. However, the formation of HO_2 in Cl-initiated neopentane oxidation has recently been experimentally observed by Dr Taatjes and myself. The HO_2 time profile in neopentane oxidation is somewhat different from those in other R + O_2 studies[2] in that no prompt HO_2 formation is observed (see Fig. 5 presented here). At 673 K the HO_2 yield is 9.5% (referenced to a 100% signal from the $CH_2OH + O_2$ reaction). This yield is also somewhat smaller than the HO_2

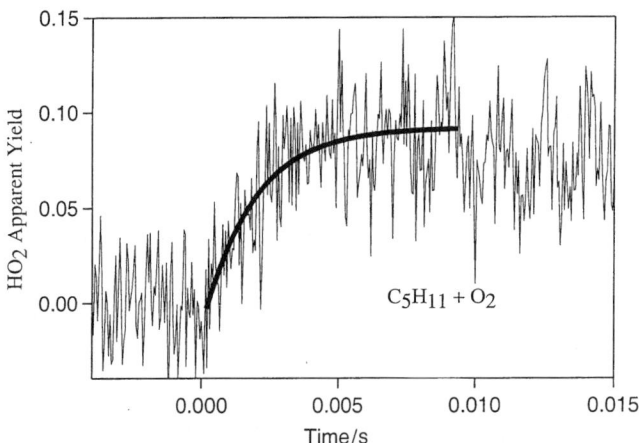

Fig. 5 The time resolved infrared FM signals for HO_2 from neopentane oxidation taken at 673 K and 59.3 Torr. The trace is obtained by an experimental method similar to that in ref. 2. The amplitude has been scaled by the amplitude of the HO_2 signal from the reference reaction $CH_2OH + O_2$ to obtain an apparent yield. The trace has been corrected for $HO_2 + HO_2$ reaction as described in ref. 2. The line through the $C_5H_{11} + O_2$ trace is an exponential fit to the HO_2 formation which corresponds to a rate constant of formation of $540 \pm 100 \, \text{s}^{-1}$.

yields observed in other alkane oxidation studies at this temperature. The observation of HO_2 formation should be of interest for validating neopentane oxidation models at these temperatures.

1 K. J. Hughes, P. D. Lightfoot and M. J. Pilling, *Chem. Phys. Lett.*, 1992, **191**(6), 581.
2 J. D. DeSain, C. A. Taatjes, J. A. Miller and S. J. Klippenstein, *Faraday Discuss.*, 2001, **119**, 101.

Dr Taatjes added: I would add that one interpretation of the HO_2 production in this system is that it results from the oxidation of formaldehyde produced in reactions of the QOOH species, *e.g.*,

$$(CH_3)_2C(CH_2)CH_2OOH + O_2 \rightarrow (CH_3)_2CO + 2CH_2O + OH$$

Reactions of OH or other radicals with formaldehyde will produce HCO, which will rapidly react with excess O_2 to make HO_2. If this mechanism is correct, modelling of the HO_2 formation in neopentane oxidation could provide data on the $RO_2 \rightleftharpoons QOOH$ isomerization complementary to the direct measurements of OH production in neopentyl + O_2 from Pilling and co-workers. This analysis is presently underway in our laboratories.

Prof. Greenhalgh addressed Prof. Griffiths: Would filtered Rayleigh scattering assist the interpretation of the acetone images since this would give some temperature information?

Dr Whitaker and **Prof. Griffiths** responded: We have recently reported elsewhere some results obtained by filtered Rayleigh scattering under both non-reactive and reactive conditions in our RCM and compared these to acetone LIF measurements.[1] We found that even in non-reactive mixtures Rayleigh scattering was not sufficiently sensitive to detect the slightly colder core gas, the roll-up vortex, that is present just after the end of the compression stroke. However, when acetone was seeded at 1% into the non-reactive mixture the existence of the colder core was clearly visible in the PLIF images. By comparison with data from Hanson and co-workers[2,3] we were able to estimate a temperature difference of approximately 40–50 K between the core gas and the surrounding gas from the LIF images when the mixture was compressed to a notional adiabatic temperature of 717 K. In other experiments on the exothermic decomposition of di-*tert*-butyl peroxide vapour, Rayleigh measurements were able to clearly observe the effects of the presence of the roll-up vortex. In this case, the gas mixture towards the periphery of the chamber reacts faster than the gas in the core and, since the reaction is exothermic, locally heats it, and we were able to observe the Rayleigh signal reducing as a function of time in the surrounding gas as a consequence. From these measurements we concluded that, although a useful technique, filtered Rayleigh scattering was rather less sensitive for thermal imaging than acetone tracer LIF measurements.

There are two additional comments that we would like to add. Firstly, our filtered Rayleigh measurements were made using an I_2 filter at ~ 532 nm by tuning the output of an injection seeded frequency doubled Nd:YAG laser into coincidence with a strong molecular absorption. Iodine vapour is easy to handle and 532 nm is a convenient wavelength, but it would obviously be better to perform Rayleigh scattering measurements at a shorter wavelength because the signal scales as λ^{-4}. Unfortunately, there seems to have been little work to date in identifying suitably narrow atomic or molecular transitions at the higher harmonics of the Nd:YAG laser that could be used as filters[4] (although, recently, Hg vapour has been used to filter the output of a frequency tripled Ti:sapphire laser[5]). Secondly, the interpretation of Rayleigh scattering measurements is complicated in reactive mixtures because the scattering cross-section depends on the (changing) molecular polarisability as well as the temperature.

1 J. Clarkson, J. F. Griffiths, J. P. MacNamara and B. J. Whitaker, *Combust. Flame*, 2001, **125**, 1162.
2 M. C. Thurber, F. Grisch and R. K. Hanson, *Opt. Lett.*, 1997, **22**, 251.
3 M. C. Thurber, F. Grisch, B. J. Kirby, M. Votmeier and R. K. Hanson, *Appl. Opt.*, 1998, **37**, 4963.
4 R. B. Miles, W. R. Lempert and J. N. Forkey, *Meas. Sci. Technol.*, 2001, **12**, R33.
5 A. P. Yalin and R. B. Miles, *J. Thermophys. Heat Transfer*, 2000, **14**, 210.

Prof. Lindstedt commented: There are two issues that come to the fore. (i) In the flame under consideration the CH + O_2 channel is of key importance and is responsible for 30–60% of the oxidation of CH. The good agreement shown between the current experimental study and that of

Juchmann et al.[1] for this species coupled with the apparent success of the high temperature determinations by Rohrig et al.[2] and Markus et al.[3] in the context of flame modelling raises an interesting inconsistency with room temperature determinations for this channel.[4] (ii) It may be pointed out that the measured peak CN mole fraction of 3.2 ppm also agrees comparatively well with the predictions of 4.7 ppm (GRI Mech. 2.11) 2.8 ppm (Lindstedt) and 10 ppm (Warnatz) presented by Juchmann et al.[1]

1 W. Juchmann, H. Latzel, D. L. Shin, G. Peiter, T. Dreier, H. R. Volpp, J. Wolfrum, R. P. Lindstedt and K. M. Leung, *Twenty-seventh Symp. (Int.) Combust.*, The Combustion Institute, Pittsburgh, 1998, p. 469.
2 M. Rohrig, E. L. Petersen, D. F. Davidson, R. K. Hanson and C. T. Bowman, *Int. J. Chem. Kinet.*, 1997, **29**, 781.
3 M. W. Markus, P. Roth and T. Just, *Int. J. Chem. Kinet.*, 1996, **28**, 171.
4 A. Bergeat, T. Calvo, F. Caralp, J.-H. Fillion, G. Dorthe and J. C. Loison, *Faraday Discuss.*, 2001, **119**, 67.

Prof. Kohse-Höinghaus commented: I would like to acquaint the audience with recent results of Burak Atakan and Tobias Hartlieb[1] with respect to NO reburning in fuel-rich propene flames. In these flames at three different stoichiometries, the NO measurements agree quite well with model simulations. The reaction flow analysis reveals that under these conditions, CH + NO reaction plays only a minor role while both NO and HCN are more sensitive to HCCO + NO reactions. Among radicals to be monitored, it might thus be useful to consider HCCO.

1 B. Atakan and T. Hartlieb, *Appl. Phys. B*, 2000, **71**, 697.

Prof. Kohse-Höinghaus opened the discussion of Dr Desgroux's paper: (1) Your flame is almost stoichiometric (actually even slightly leaner because of the admixed NO). For study of (fuel-rich) reburning chemistry, would you not rather investigate richer flames? Have you done so?

(2) With respect to your CH measurement and the disagreement with the model, you attribute this potentially to the temperature dependence of the CH + O_2 reaction at high temperatures. With your temperature uncertainty of ± 120 K, would this affect the CH profile enough so that model and measurement would be in better agreement?

Dr Desgroux responded: (1) The validation of reburning chemistry is also an important task in stoichiometric flames. But you are right that more extensive studies are required in richer flames. We believe that the CRDS-LIF coupling will also be well suited in such flames.

(2) Tests have been performed by imposing different temperature profiles within the uncertainty range (± 120 K in the burnt gases). An important effect on the CH mole fraction profile (peak value and position) is effectively observed. But the difference with the experimental mole fraction (also affected by the temperature change due to the conversion of the concentration into a mole fraction) is still outside the experimental error. The smallest difference (28% with respect to the experimental value) is obtained for the coldest temperature profile. But in that case the peak position of the CH mole fraction (4.05 mm) is 0.65 mm downstream of the experimental peak.

Prof. Kohse-Höinghaus asked: Your temperature profile needs high accuracy in the gradient where you observe the NO 'dip'. Could you comment on the superiority of using OH LIF rather than LIF of your seeded NO for this purpose?

Dr Desgroux replied: At the location of the NO dip (around 4 mm) the OH concentration is sufficient to make accurate OH LIF temperature measurements. Near the burner the temperature measurements are less accurate due to the cumulative effects of a weak OH concentration and of an important temperature gradient. NO-LIF measurements would not have been affected by sensitivity problems in the vicinity of the burner surface. However we believe that NO excitation spectra would have been perturbed by self-absorption from the cold NO surrounding the burner, which was not isolated by a coflow. Strong absorption from the cold NO was shown in our burner enclosure using the CRDS technique.

Dr Whitaker asked: I would like to follow up on this problem of the temperature gradient. In CRDS in an open cavity containing a flame and therefore refractive index gradients, I imagine there will be an effect on the cavity hold time due to beam walk. Can you characterise these effects and more importantly correct them?

Dr Desgroux answered: The presence of refractive index gradients might effectively modify the beam walk. As long as the perturbation is weak, such an effect acts like supplementary off-resonance losses and the time decay of the CRD signal is still exponential. Under these conditions the net losses due to the absorbing species are obtained after subtraction of the off-resonance losses including the thermal effects but also broadband absorption (or scattering) from the flame and losses due to the mirrors.

Prof. Cheskis asked: Why did you not use CRDS for the OH calibration? It is possible if the weaker lines are used.

Dr Desgroux replied: CRDS on OH weak lines (S_{21} lines for example) would have been possible. But with our laser bandwidth, the exponential behavior would have been obtained only for high rotational levels (*i.e.* corresponding to a weak Boltzmann fraction and thus to a weak absorbance) leading to a temperature-sensitive measurement.[1] We have found that it is more accurate to work with the single-pass absorption technique performed on a weak-temperature dependent transition in the specific case of the abundant OH species.

1 X. Mercier, E. Therssen, J. F. Pauwels and P. Desgroux, *Combust. Flame*, 2001, **125**, 656.

Mr Zsély opened the discussion of Prof. Cheskis's paper: You used GRI-Mech version 2.11. Why didn't you use version 3.0 instead of 2.11?

Prof. Cheskis responded: Both versions of the GRI-Mech demonstrate similar qualitative behaviour for HNO and NH_2. Both versions predict two maxima in the HNO profile and do not predict NH_2 at high locations above the burner. Fig. 6 shows the concentration profiles of HNO and NH_2 calculated using both versions of the mechanism. Absolute concentration of HNO in the

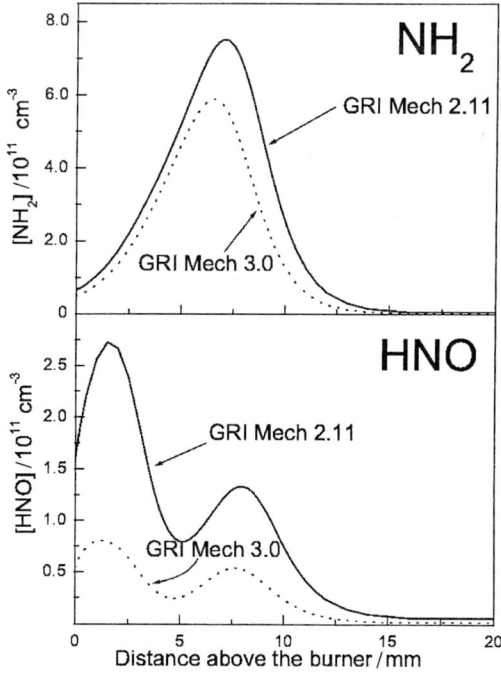

Fig. 6 The HNO and NH_2 concentration profiles calculated using versions 2.11 and 3.0 of GRI-Mech. The calculations were performed for stoichiometric ($\varphi = 1.0$) doped with 7.2 cm^3 of NO.

first maximum is lower in version 3.0 because of two times higher value for H + HNO reaction and 50% lower value for H + NO + M reaction used in this version. Thus, the discrepancy with our experimental results is higher for version 3.0.

Prof. Kohse-Höinghaus commented: (1) I would like to express some concern about the build up of radicals in the burner housing, *i.e.* in the zone surrounding the flame. The assumption of a homogenous 60 mm absorption length and 'top-hat' temperature profile will not be appropriate in regions as high as 50–60 mm above the burner. Outer zones with 'cold' radicals will potentially distort spectra and affect the concentration measurement, mixing of the hot combustion gases with the N_2 shroud flow may add to these uncertainties. Did you change the shroud flow too? A one-dimensional model will probably not represent your measurement condition.

(2) Also, I would like to ask about the uncertainty of the HNO absorption cross section which was based on an HCO calibration using the reaction of HCO + NO → HNO + CO. As a total impression, it seems premature from these measurements to state that models may or may not be in agreement, unless the effects of absorption path length and temperature as well as absorption coefficient were properly quantified.

Prof. Cheskis replied: (1) I agree with your comment that a one-dimensional model does not represent the experiment correctly. For HNO we were able to improve this situation using the nitrogen flow through the window holders preventing accumulation of HNO in the vicinity of the windows. On the other hand, the NH_2 concentration is not affected by this flow substantially. It demonstrates that NH_2 is located closer to the center of our vacuum chamber. We changed the shroud flow over a wide range and did not observe any influence of the shroud flow on the HNO and NH_2 concentrations.

(2) The cross section of HNO was measured in only one study and that was more than 20 years ago. In this work the concentration of HNO was measured on the basis of the HCO concentration, assuming that all HCO formed as a result of the acetaldehyde photolysis, converts to HNO *via* the reaction HCO + NO → HNO + CO. I think that the possible error is less than 100%, taking into account that the rate constant of the reaction of HCO + HCO, which is directly related to the HCO concentration, was obtained in this work in reasonable agreement with literature data. The uncertainties in the optical length and in the temperature increase the total error in the measured HNO concentration. On the other hand the discrepancy found, which is of the order of magnitude, probably could not be explained by these errors alone, and some corrections in the mechanism will be needed. I agree that in order to correct the mechanism we will need a more accurate value of the HNO cross section as well as three dimensional measurements in flames. I believe that the direct measurement of the H + HNO rate constant will also be very useful.

Prof. Lindstedt said: The rate for the reaction H + HNO ⇌ NO + H_2 is uncertain and the determination by Soto and Page[1] ($4.5 \times 10^{11} T^{0.72} e^{330/T}$) used in GRI-Mech 2.11 may be somewhat fast. However, it has been shown[2] that for a range of flames the early estimate by Bulewicz and Sugden[3] (5×10^{12}) results in persistent over-predictions of HNO in cases where the channel matters. There are obvious multiple uncertainties and it would be good to see the sensitivity to the (arguably) lower limit rate.[3]

1 M. R. Soto and M. J. Page, *Chem. Phys.*, 1992, **97**, 7287.
2 R. P. Lindstedt, F. C. Lockwood and M. A. Selim, *Combust. Sci. Technol.*, 1994, **99**, 253.
3 E. M. Bulewicz and T. M. Sugden, *Proc. R. Soc. London, Ser. A*, 1962, **277**, 143.

Prof. Cheskis responded: Unfortunately there is no reliable measurement of the rate constant for the H + HNO reaction. The sensitivity of the calculated profile of HNO to the value of that rate constant is illustrated in Fig. 7, (presented here) where calculations were carried out for the same system with the only difference being the H + HNO rate constant. The over-prediction of HNO which was mentioned by Lindstedt *et al.*[1] varied from 40% to three times that for flames where HNO concentrations have been measured and can be caused by different reasons. It should be noted that experimental values of the HNO concentration were obtained by mass spectroscopy.[2,3] In order to obtain absolute concentrations the authors calibrated the signal using the

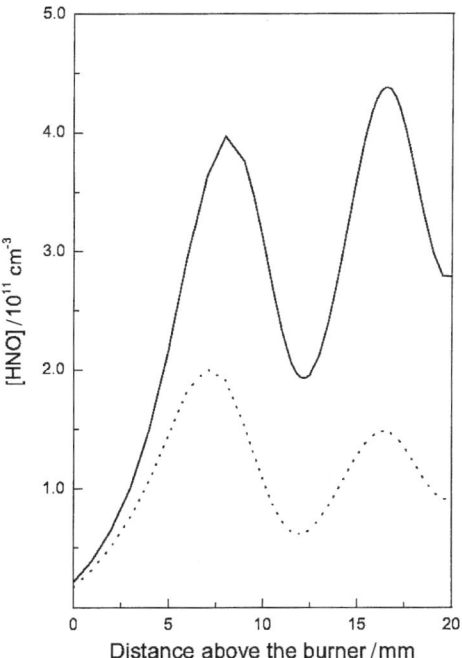

Fig. 7 The HNO concentration profile calculated for two different expressions for the rate constant of the reaction of H + HNO. Solid line is the result of calculation with $k = 5 \times 10^{12}$. The dotted line is for $k = 4.5 \times 10^{11} \, T^{0.72} \exp(330/T)$ as adopted in GRI-Mech 2.11. The calculations were performed for methane–air flame with equivalence ratio $\varphi = 1.2$ doped with 7.2 cm^3 of NO.

ionization cross section of stable compounds having a molecular weight close to the investigated species. The accuracy of that procedure is questionable and this uncertainty can be also involved in the over-prediction reported by Linstedt et al.[1]

1   R. P. Lindstedt, F. C. Lockwood and M. A. Selim, *Combust. Sci. Technol.*, 1994, **99**, 253.
2   J. Bian, J. Vandooren and P. J. Van Tiggelen, *Twenty-first Symp. (Int.) Combust.*, The Combustion Institute, Pittsburgh, 1986, p. 953.
3   J. Bian, J. Vandooren and P. J. Van Tiggelen, *Twenty-third Symp. (Int.) Combust.*, The Combustion Institute, 1990, p. 379.

Dr Miller asked: (1) What is the source of the rate constant for H + NO + M → HNO + M in GRI-Mech? Peter Glarborg et al. measured the rate constant a couple of years ago[1] for $1000 < T/K < 1170$, and my impression at the time was that the rate constant from 300 K up to these temperatures was not uncertain by more than a factor of two. Such an uncertainty would not be consistent with this rate constant being the cause of the discrepancy between the model predictions and your measurements.

(2) How do you calculate transport properties in the flame model? Do you include thermal diffusion? One might expect the HNO predictions early in the flame to be quite sensitive to the transport model.

1   P. Glarborg, M. Østberg, M. U. Alzueta, K. Dam-Johansen and J. A. Miller, *Twenty-seventh Symp. (Int.) Combust.*, The Combustion Institute, Pittsburgh. 1998, p.219.

Prof. Cheskis replied: (1) The GRI-Mech 2.11 used $k = 9 \times 10^{19} \, T^{-1.32} \exp(-370/T)$ from ref. 1. This value was halved in version 3.0. This change, along with a twofold increase in the rate constant of the reaction of H + HNO → H$_2$ + NO, results in the lower HNO concentration predicted by version 3.0. For lower temperatures (~ 300 K) the recommendation of Glaborg et al.[2] is close to that used in GRI-Mech 2.11, at higher temperatures this work gives a value that is three times lower than that adopted in the GRI-Mech 2.11. The uncertainty of the rate constant of the reaction of H + HNO is higher than that of H + NO + M and the discrepancy between the measurements and predictions depends on the ratio of these constants which uncertainty is obviously higher than those of both constants.

(2) We used the PREMIX code, which allows one to take into account thermal diffusion. Thermal diffusion increases substantially the concentration of H atoms in the preflame zone, but the HNO concentration increases only by several percent. The reason is the concurrent acceleration of the reactions of H + NO and H + HNO which form and consume HNO respectively.

1   W. Tsang and R. Herron, *J. Phys. Chem. Ref. Data*, 1991, **20**, 623.
2   P. Glarborg, M. Østberg, M. U. Alzueta, K. Dam-Johansen and J. A. Miller, *Twenty-seventh Symp. (Int.) Combust.*, The Combustion Institute, Pittsburgh, 1998, p. 219.

Prof. Lin said: The rate constants for both H + HNO and OH + HNO have been calculated by Page and co-workers.[1,2] In addition to the commonly assumed exchange processes, other facile indirect product channels should perhaps be included in kinetic modeling of HNO concentration profiles. For example:

$$H + HNO + M \rightarrow H_2NO + M$$

$$\rightarrow HNOH + M$$

$$OH + HNO \rightarrow HN(O)OH^* \rightarrow HONO + H$$

$$\rightarrow H_2O + NO.$$

1   M. R. Soto, M. Page and M. L. McKee, *Chem. Phys.*, 1991, **153**, 415.
2   M. Page and M. R. Soto, *J. Chem. Phys.*, 1993, **99**, 7709.

Prof. Cheskis replied: I agree with you that more detailed kinetic modelling including different channels of the HNO reactions is needed.

Prof. Wolfrum asked: Could you perform some tomography experiments with your set up to measure deviations from the 1D-flame behaviour? In your broadband absorption you should also be able to discriminate groups of absorption lines originating from cooler and hotter parts of your absorption path to correct for contribution outside the flame.

Prof. Cheskis replied: We are planning such tomographic experiments now. In our opinion it is the best way to overcome the disadvantages of the line of sight techniques. Simultaneous measurements of the absorption originated from the levels with different energy might help to solve this task.

Prof. Kohse-Höinghaus said: Certainly it would be advisable in your measurements to specify your actual absorption path. In following the suggestion of using tomographical reconstruction, you know that it is also done to monitor laser-induced fluorescence at the same time for an independent observation of the absorption path. It is certainly difficult for some radicals, but you could maybe comment for your own system.

Also, would the use of two-photon absorption techniques in CRD assist in placing the actual zone of measurement more into the center of the flame where your (one-dimensional) model calculates the concentrations?

Prof. Cheskis replied: It was demonstrated that the combination of CRDS and LIF provides very important information and offers in several cases an alternative to the tomographic measurements. Unfortunately, the sensitivity of LIF is not sufficient in order to use this technique for CH_2, NH_2, *etc.* On the other hand, the space profiles can vary for different species, as was demonstrated in this work for HNO and NH_2. Thus, the measurements of radial profiles of one radical cannot be used for another radical directly. Your idea of using two-photon absorption is very interesting. However, one must keep in mind that the use of two beams in order to localize the monitoring zone will decrease the optical length, and therefore the sensitivity of the line of sight method.

Dr Miller asked: At what distance above the burner does the HNO mole fraction peak?

Prof. Cheskis responded: The first maximum of HNO is located at ~ 3 mm above the burner.

Prof. Wolfrum asked: The first HNO-maximum should be located before the flame front due to the counter diffusion of H-atoms towards the cold NO. Did you observe this?

Prof. Cheskis responded: Yes, we do observe it. And its location is in good agreement with predictions of the GRI-Mech.

Dr Miller said: It is known that jetting of the gases through the porous-plug burner can cause the flow near the burner not to be one-dimensional. Could this be the cause of the discrepancy between the model predictions and experiments for HNO? Also, the boundary condition at the burner surface in the PREMIX code is a continuity of mass *fluxes*, which implies a discontinuity of the *mole fractions* at the surface. Obviously, this is an idealization and, for *highly stabilized flames*, could produce errors near the surface. However, this problem is easily alleviated by turning up the flow rates in the experiment and pushing the flame zone away from the burner as far as possible.

Prof. Cheskis responded: The position of the flame zone was different for flames with different equivalence ratios. However, in all flames studied we observed good agreement in the location of the first HNO peak, and the same disagreement in absolute value. In our opinion the main two reasons for this disagreement could be the uncertainties in the HNO absorption cross section and in the HNO rate constants.

Prof. Pilling opened the discussion of Dr Hughes's paper: Hughes et al. have emphasized the importance of $H + SO_2$ in removing SO_2 under flame conditions. We have examined this reaction using a master equation model,[1] with the potential energy profile shown as Fig. 19 in Hughes et al. and the transition state parameters calculated by Marshall et al.[2] We also used our direct measurements of $OH + SO$, over the temperature range 300–700 K,[3] to characterize the loose transition state between $OH + SO$ and $HOSO$.

At low temperatures (~ 300 K), the reaction is limited to the formation of HSO_2, with a pressure dependent rate coefficient of $\sim 10^{-13}$ cm^3 molecule^{-1} s^{-1} at 1 atm. At ~ 400 K, HSO_2 dissociates to regenerate the reactants and the overall rate coefficient falls.

At higher temperatures (~ 1000 K), formation of HOSO becomes the major channel, with, at low pressures, some direct formation of $OH + SO$ *via* energized HOSO*, without stabilization. As the temperature is increased further, HOSO becomes unstable, again regenerating reactants. Finally, and at the limiting temperatures established in the flames reported by Hughes et al., the reaction forms $OH + SO$ directly, *via* the final loose transition state, with a rate constant that is independent of pressure and that is related to the rate coefficient for $OH + SO$, *via* the equilibrium constant. The intermediates HSO_2 and HOSO play no role in the reaction under these conditions.

We are currently determining rate coefficients for the individual steps in the mechanism, from the results of the master equation calculations.[1]

1 M. A. Blitz, K. J. Hughes and M. J. Pilling, in preparation.
2 A. Goumri, J. D. R. Rocha, D. Laakso, C. E. Smith and P. Marshall, *J. Phys. Chem. A*, 1999, **103**, 11328.
3 M. A. Blitz, K. W. McKee and M. J. Pilling, *Proc. Combust. Inst.*, 2000, **28**, 2491.

Dr Tomlin responded: It is possible from your calculations to establish the relative contributions from the reactions $SO_2 + H \rightarrow$ products and the third body reaction $H + SO_2 + M \rightarrow HOSO + M$? This may be important since our sensitivity studies show that for a lean methane flame $SO_2 + H \rightleftharpoons SO + OH$ has a positive sensitivity with respect to NO whereas $H + SO_2 + M \rightarrow HOSO + M$ has a negative one which also varies significantly with distance from the burner.

Prof. Pilling said: As the temperature increases, over the first 5 mm above the burner, the formation of HOSO will be important and the rate coefficient will be pressure dependent. At the highest temperatures, where OH + SO are formed directly, the rate coefficient is independent of pressure. The results of the master equation calculations can be analysed to produce temperature and pressure dependent rate coefficients for the individual steps which will allow a detailed assessment of the relative rates to be made.

Dr Desgroux asked: Could you tell me the precise initial (relative) values of the NO concentrations that you measured in the different flames before the injection of SO_2?

Could you also comment on how accurate the prediction of NO was compared to your measurements without SO_2 in the different flames?

What is the accuracy of your NO (relative) measurements?

Dr Hughes replied: We have measured the relative LIF signals in the different flames at a height of 25 mm above the burnerTable 2 presented here. However these may not directly provide the relative NO concentrations, as surface, the results being summarised in corrections for the different flame temperatures on the rotational population distribution of NO need to be made, and different quenching rates of excited NO in the different flames may have to be accounted for. These measurements were made by adjusting the detection system to give a large signal at $\phi = 0.7$, then in a single run, in order to minimise the effect of any drift in laser output power, data were collected for the different flame stoichiometries, both with and without NH_3 doping of the flame. The NO LIF signal was averaged over typically 100 laser shots, and corrected for the baseline signal observed in the absence of NH_3, which was collected in a subsequent run. The quoted errors represent one standard deviation obtained from the averaging of the NO signal and the background signal. Without further experiments, it is important to note that the numbers from one table cannot be compared with the other table.

For comparison, predicted relative NO values for each flame at 25 mm above the burner surface have been calculated from the simulations and are shown in the final column of Table 2. For flames 4, 5 and 6 containing 66% of argon, the agreement is very good, but less so for flames 7–10 containing 33% of argon. Currently the origin of this discrepancy is unknown. One possible contributory factor is that flames 7–10 exhibit a greater temperature difference, and having less argon will have a greater contribution from different quenching rates for excited NO, therefore larger corrections to convert the LIF signals to relative concentrations for flames 7–10 will be required.

Prof. Plane said: The LIF intensity is assumed to be proportional to the radical concentration. Did you establish that the species were thermally equilibrated in the rotational levels that you probed? How reliable is the temperature measurement using the thermocouple, and were you able to compare this with OH LIF thermometry, for example?

Dr Hughes replied: We were not able to establish whether the species were thermally equilibrated in the rotational levels that we probed. In the case of NS and SO_2 we did not have a sufficiently detailed assignment and analysis of the spectra to enable us to do so. For NO it was

Table 2 Relative LIF signals in different flames

Flame	ϕ	Relative NO LIF signal	Predicted relative [NO]
66% Argon			
4	0.7	1 ± 0.134	1
5	1.0	0.84 ± 0.122	0.864
6	1.4	0.58 ± 0.12	0.53
33% Argon			
7	0.7	1 ± 0.2	1
8	1.0	0.853 ± 0.189	0.736
9	1.3	0.621 ± 0.155	0.337
10	1.6	0.214 ± 0.124	0.071

not necessary, as we were not trying to measure a concentration profile in the flame, but only the effect on NO at a fixed point in the flame of the addition of small quantities of SO_2.

The temperature measurement using the thermocouple has some uncertainty due to the fact that the thermocouple needs to be coated with silica to minimise catalytic surface reactions, and temperature corrections for radiation loss made. These corrections depend on assumptions regarding the diameter of the coated thermocouple and its emissivity. Additionally under some of the flame conditions the coating was not totally stable, and needed to be periodically replaced. Ideally we would have liked to compare it with OH LIF thermometry, but time constraints prevented us from performing these measurements.

Prof. Wolfrum asked: Did you perform fluorescence lifetime measurements for the sulfur containing radicals?

Dr Hughes answered: No explicit measurement of the fluorescence lifetime of NS was made, the only observation being that there was no obvious change as a function of position in the flame.

Dr Seakins said: You mentioned in your talk that you looked for SO but did not observe any characteristic fluorescence. Does this allow you to put any limiting values on the concentration of SO. If so, is this limiting value consistent with the model calculations? In the paper, a number of the reactions that you identify as being important involve SO, would future SO observations be a good test of your model?

Dr Hughes replied: Theoretically, it may be possible to infer a limiting value of SO, but this would require a detailed investigation of the detector efficiency, laser power and linewidth, knowledge of the SO spectrum and quenching rates of excited SO in the flame conditions. We were not in a position to perform all of this, and even so, we would still have the problem of overlapping fluorescence observed from SO_2 when interpreting the maximum SO level. Future SO observations would be a useful test of the model, and may be possible at a different wavelength.

Dr Kaiser said: Although not many C_2 molecules exist in hydrogen rich, sulfur containing fuel flames, it could be important to investigate the role of dicarbon reactions with H_2S in hydrogen and oxygen poor combustion flames. Our recent crossed beam study on the C_2–H_2S system found three reaction pathways, *i.e.* formation of HCCS, HSCC and CCS.[1] In particular, the HCCS radical could be of potential importance to form—upon multi-step reactions with acetylene and vinyl—sulfur heteroaromatic molecules (see Fig. 8 presented here).

1 R. I. Kaiser, Y. T. Lee and Y. Osamura, *J. Phys. Chem. A*, submitted.

Dr Hughes responded: Given the current state of knowledge of the kinetics and product channels of elementary reactions involving sulfur species, it is difficult to comment on the importance of species such as HCCS and their reactions. A search of the NIST chemical kinetic database[1] shows no entries for HCCS. It is unlikely however that C_2 molecules exist in large quantities in methane flames as opposed to higher hydrocarbon flames and therefore we do not believe that such reactions are the major source of error in our simulations.

1 *Standard Reference Database 17*, Version 7.0 (Web Version), http://kinetics.nist.gov/index.php.

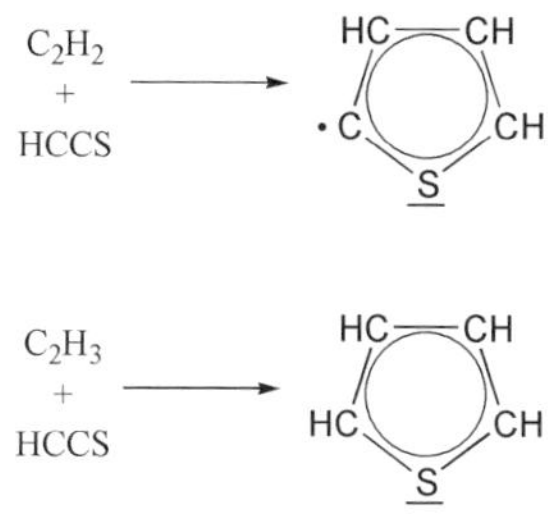

Fig. 8 Possible reactions of the HCCS radical in sulfur-rich hydrocarbon flames.

Prof. Troe asked: Do you find any evidence for the formation of S_2? Would S_2 be a kinetically relevant species?

Prof. Cheskis also asked: Did you observe the presence of S_2 molecules in your flame, in particular did you observe light emission of electronically excited sulfur molecules? The presence of sulfur in the flame usually produces a blue–violet light emission.

Dr Hughes responded: A fluorescence signal consistent with S_2 was observed by LIF in experiments that were designed to detect SH although detailed experiments were not carried out. A violet light emission from the flame was observed when it was doped with sulfur. In the current model and for the conditions simulated however, reactions of S_2 did not show high sensitivities and so are not likely to be of importance.

Prof. Wolfrum asked: Could you observe the formation of solid products during the experiment in your burner?

Dr Hughes replied: When the flame was doped with SO_2, there was a slow build up of a solid deposit on the surface of the burner.

Detailed surface reaction mechanism in a three-way catalyst

Daniel Chatterjee, Olaf Deutschmann and Jürgen Warnatz

Interdisciplinary Center of Scientific Computing (IWR), Heidelberg University, Im Neuenheimer Feld 368, D-69120 Heidelberg, Germany

Received 1st March 2001
First published as an Advance Article on the web 18th October 2001

Monolithic three-way catalysts are applied to reduce the emission of combustion engines. The design of such a catalytic converter is a complex process involving the optimization of different physical and chemical parameters (in the simplest case, *e.g.*, length, cell densities or metal coverage of the catalyst). Numerical simulation can be used as an effective tool for the investigation of the catalytic properties of a catalytic converter and for the prediction of the performance of the catalyst. To attain this goal, a two-dimensional flow-field description is coupled with a detailed surface reaction model (gas-phase reactions can be neglected in three-way catalysts). This surface reaction mechanism (with C_3H_6 taken as representative of unburnt hydrocarbons) was developed using sub-mechanisms recently developed for hydrogen, carbon monoxide and methane oxidation, literature values for C_3H_6 oxidation, and estimates for the remaining unknown reactions. Results of the simulation of a monolithic single channel are used to validate the surface reaction mechanism. The performance of the catalyst was simulated under lean, nearly stoichiometric and rich conditions. For these characteristic conditions, the oxidation of propene and carbon monoxide and the reduction of NO on a typical Pt/Rh coated three-way catalyst were simulated as a function of temperature. The numerically predicted conversion data are compared with experimentally measured data. The simulation further reveals the coupling between chemical reactions and transport processes within the monolithic channel.

Introduction

Today three-way catalysts are used extensively to reduce the emissions of combustion engines. The majority of automotive catalytic converters have a monolithic structure, which is coated with an alumina washcoat that supports the noble metal such as platinum, palladium and rhodium. To achieve a large catalytic surface area, the substrates consist of numerous parallel channels with a diameter of approximately 1 mm.

The experimental characterization of the catalytic performance of the converter is time-consuming and requires a large experimental set-up. Numerical simulation offers an interesting alternate method for the investigation of the catalytic activity of a converter. This method is also efficient in analyzing the transient flow and thermal phenomena in the catalytic converter and may help to understand the complex interactions between the flow field and the catalytic surface chemistry.

In recent years, several proposals were made for the numerical simulation of catalytic converters.[1–6] In these studies, global models for the chemistry were used, neglecting the various single reactions occurring on the surface. Thus, they are descriptive only in the range of conditions

DOI: 10.1039/b101968f

used for fitting of the global rate coefficients, but not predictive at extrapolated experimental conditions (see *e.g.*, ref. 7).

An alternate approach is the description of the chemical reaction by a set of "quasi-elementary" steps describing the reactions on a molecular level ("quasi" because of the restriction by the mean-field assumption; see *e.g.*, ref. 8). The main advantage of these detailed reaction mechanisms is their potential to predict the behavior of the chemical system at external conditions not accessible to experiments. The disadvantage of using elementary chemical reactions is the large number of reaction equations which demands a large computational capacity. Furthermore, the rate coefficients of all the elementary steps have to be known.

The model presented here considers the different adsorption positions (platinum or rhodium) on the metallic catalyst surface. However, on rhodium only surface reactions between NO, CO and O_2 are considered.[10] The kinetic data of the mechanism was taken either from the literature or from fits to experiments. Parts of the surface reaction mechanism have already been used for numerical modeling of catalytic ignition,[11] simulation of total and partial oxidation of light hydrocarbons on platinum,[9] and modeling the CO–O_2 and NO–CO reactions on rhodium.[10,12]

Experimental

The catalyst used in this study is a commercially available three-way catalyst. It contains 50 g ft^3 of metal (Pt/Rh = 5 : 1) impregnated on a ceria stabilized γ-alumina washcoat. The washcoat is supported by a cordierit monolith with a cell density of 62 cells cm^{-2} and a wall thickness of 0.165 mm.

The experimental determination of the conversion rate and therefore the performance of the three-way catalyst was carried out in a laboratory-scale tube reactor (details in ref. 13). The oxygen concentration was varied in order to simulate a stoichiometric, a rich and a lean exhaust gas mixture (see Table 1), where C_3H_6 is chosen as representative of the hydrocarbon in exhaust gas. The λ_{Ox}-values are defined as the inverse of the equivalence ratio.[14] The volume flow of the artificial exhaust gas into the sample was 15 l min^{-1} at standard conditions (25 °C). A uniform axial inlet velocity of 1.35 m s^{-1} corresponds to this volume flow. The CO_2 concentration in the inlet exhaust gas mixture was zero due to experimental reasons. Tests have shown that there was no difference in the conversion rate of CO, HC and NO between measurements with and without CO_2 in the inlet gas.

After the experimental studies of the temperature-dependent conversion rate the sample was investigated with H_2 chemisorption in order to obtain the properties of the active metal phase. The active metal surface of the catalyst was determined to 28 m^2 g^{-1}, the dispersion of the conditioned catalyst was 33%. Calculation of the ratio of active metal surface area and geometrical surface area of the catalyst leads to a value of 70. The ratio of the platinum to rhodium surface was taken from the literature as 3 : 1 for a catalyst with a noble metal composition of Pt/Rh with 5 : 1 which corresponds to the catalyst we used for our investigations.[15]

Table 1 Composition of the simulated exhaust gas and concentration of the gas species used for the simulations at nearly stoichiometric mixture ($\lambda_{Ox} = 0.9$), rich mixture ($\lambda_{Ox} = 0.5$) and lean mixture ($\lambda_{Ox} = 1.8$)

	Concentration (vol.%)		
Species	$\lambda_{Ox} = 0.9$	$\lambda_{Ox} = 0.5$	$\lambda_{Ox} = 1.8$
CO	1.4200	1.4200	1.4200
O_2	0.7700	0.4000	1.6000
C_3H_6	0.0450	0.0450	0.0450
NO	0.1000	0.1000	0.1000
N_2	balance	balance	balance

Mathematical model and numerical simulation

Flow-field simulation

In this study, not only a more accurate chemical reaction model is used, but also a detailed description of the flow field. One single channel of the monolith, assumed to be a tube reactor, is modelled by the two-dimensional Navier–Stokes equations. The flow-field simulation is based on the commercially available CFD-code FLUENT.[16] This code is well established and can easily be used to set up fluid-flow problems and to solve them. However, modeling of detailed chemistry in current versions is practically impossible because of the limited number of reactions allowed and because of the difficulties with handling complex chemistry, leading to a extremely stiff differential equation set. Furthermore, FLUENT's surface reaction model does not take the surface coverage into account which is essential for this work.

Thus, FLUENT was coupled with the chemistry module DETCHEM,[17] modeling chemical reactions in the gas phase and on the surface using elementary step reaction mechanisms. This computational tool has already been used successfully to model catalytic partial oxidation processes in monolithic reactors.[18]

Conservation equations

The basic conservation equations for laminar flow fields are, due to the axial symmetry of the problem, 2-dimensional formulations of conservation of total mass, axial momentum, radial momentum, the species conservation equation, and thermal energy. The density is computed *via* the ideal gas law. The mixture viscosity and thermal conductivity as well as the diffusion coefficient $D_{i,M}$ of species i in the mixture depend on the local composition and on the temperature; they are calculated using the kinetic gas theory.[14] The specific heat $c_{p,i}$ at constant pressure of species i is modeled as a polynomial function of temperature.[14] Details on the flow-field calculations are given in ref. 13.

The flow enters the computational domain with a known velocity, gas composition and temperature. A flat profile of the axial velocity u and a vanishing radial velocity v are assumed in the simulation at the inlet boundary. At the reactor exit, an outlet boundary is applied at which values for all variables are extrapolated from the interior cells adjacent to the outlet.

Surface chemistry model and its coupling with the flow model

A detailed multi-step reaction mechanism is used to model the catalytic reactions and to calculate the surface mass fluxes. The surface coverage of the species on the catalyst is also calculated as a function of the position in the channel. The mechanism includes only surface chemistry, gas phase chemistry can be neglected because of the low pressure and temperature as well as the short residence time of the species inside the catalyst.

Chemical reactions on the catalytic reactor wall lead to the following boundary conditions:

$$\eta F \dot{s}_i M_i = (j_i + \rho w_i v_{St}); \qquad i = 1, \ldots, N_g, \qquad (1)$$

where F is the ratio between catalytic active surface area and geometric surface (see Table 2), $\dot{s}_i$ the creation or depletion rate of species i by adsorption and desorption processes, M_i is the molar mass of species i, j_i the diffusive flux, ρ the mixture mass density, w_i the mass fraction of species i in the gas phase adjacent to the surface, v_{St} the Stefan velocity, and N_g the number of gas-phase species.

The active catalytic surface was experimentally determined. η is the effectiveness factor to account for pore diffusion within the washcoat. In this work, the effectiveness factor is related to the diffusion of the species CO (lean and stoichiometric mixtures) or O_2 (rich mixtures) within the pores of the washcoat, because the CO and O_2 reactions, respectively, are the most crucial ones in the surface kinetics. The effectiveness factor depends on the porosity, the gas-phase species concentration at the washcoat, the surface reaction rates, and diffusion coefficients using the Thiele module approach.[19]

The state of the catalytic surface is described by its temperature and the coverages of adsorbed species which vary in the flow direction. External subroutines calculate the surface coverages Θ_i

Table 2 Input data used for the simulation studies

Noble metal composition Pt/Rh	5 : 1
Noble metal loading	50 g ft^3
Active metal surface	28 m^2 g^{-1}
Surface ratio Pts/Rhs	3 : 1
Metal dispersion	33%
Ratio active metal surface/geometrical surface	70
Surface site density Γ	2.72 × 10^{-9} mol cm^{-2}
Channel diameter	1.0 mm
Channel length	29 mm
Temperature range	100–600 °C
Velocity (temperature-dependent)	1.35 ms s^{-1} at 25 °C

(the fraction of surface sites covered by species i) at each computational cell at the tube wall. The same applies for the calculation of the surface mass fluxes (left-hand term of eqn. (1)).

The chemical source terms for gas-phase species due to adsorption/desorption and for the surface species (that are actually the adsorbed species) are given by the expression

$$\dot{s}_i = \sum_{k=1}^{K_s} v_{ik} k_k^{(s)} \prod_{j=1}^{N_g+N_s} c_{s,j}^{v'_{ik}}; \quad i = 1, \ldots, N_g + N_s, \quad (2)$$

where K_s is the number of elementary surface reactions (including adsorption and desorption), v_{ik} (right side minus left side) and v'_{ik} (left side of reaction equation) are the stoichiometric coefficients, k_f is a rate coefficient and N_s is the number of surface species. The surface concentration c_{sj} of an adsorbed species is given, *e.g.*, in mol m^{-2} and is the product of surface coverage Θ_i and the surface site density Γ (see Table 2). The temperature dependence of the rate coefficients is described by a modified Arrhenius expression:

$$k_k^{(s)} = A_k T^{\beta_k} \exp\left(-\frac{E_{a,k}}{RT}\right) \prod_{i=1}^{N_s} \Theta_i^{\mu_{ik}} \exp\left(\frac{\varepsilon_{ik} \Theta_i}{RT}\right). \quad (3)$$

This expression takes an additional coverage dependence into account, using the parameters μ_{ik} and ε_{ik}.

The rate coefficient for adsorption processes is calculated from the initial sticking coefficient S^0, which is the sticking probability at vanishing coverage,

$$k_k^{(s)} = S_i^{(0)} \frac{1}{\Gamma^\tau} \sqrt{\frac{RT}{2\pi M_i}}, \quad (4)$$

where τ is the number of occupied adsorption sites of species i, R the gas constant and T the temperature.

The steady state solution is the point of interest here. Hence, the time variation of the surface coverage Θ_i is zero:

$$\frac{\partial \Theta_i}{\partial t} = \frac{\dot{s}_i}{\Gamma} = 0; \quad i = N_g + 1, \ldots, N_g + N_s. \quad (5)$$

In DETCHEM, this equation system is solved to obtain surface coverages and surface mass fluxes. Here, FLUENT provides the concentration of the gas-phase species and the temperature at each computational cell with a catalytic wall as boundary.

In the DETCHEM module, coverages and surface mass fluxes are calculated for each "global" iteration, keeping the local species concentrations and temperature constant. The algebraic equation system (eqn. (5)) is solved by a time integration of the corresponding ODE system until a steady state is reached. An implicit method based on LIMEX[20] is used for the time integration. The coverage data of the former iteration are used as initial conditions for the next step.

Under-relaxation of the variation of the surface mass fluxes may be necessary, *e.g.*, if a species largely produced on the surface has a high sticking coefficient, such as CO here.

Surface reaction mechanism

The conversion reactions of the harmful exhaust gases into harmless components inside a monolithic three-way catalyst can be written globally as

$$CO + 1/2 O_2 \rightarrow CO_2$$

$$C_nH_m + (n + m/4)O_2 \rightarrow nCO_2 + m/2 H_2O$$

$$CO + NO \rightarrow CO_2 + 1/2 N_2.$$

This kind of global chemistry has been used in the studies described in the literature. The surface reaction mechanism, however, consists of numerous elementary reaction steps. Thus, a more detailed approach to modelling the surface chemistry is applied here. The reaction scheme consists of 61 elementary (or as "elementary" as possible) reaction steps and 31 chemical species, *e.g.*, dissociative oxygen adsorption, non-dissociative adsorption of C_3H_6, CO and NO, the steps for formation of carbon dioxide (CO_2), water (H_2O) and nitrogen (N_2) and desorption of all species. Some activation energies (*e.g.* oxygen desorption) are coverage-dependent due to interactions between adsorbed species.

It is assumed that all species are adsorbed competitively. The model also considers the different adsorption places (platinum or rhodium) on the metallic catalyst surface. However, on rhodium only surface reactions between NO, CO and O_2 are considered.[10] The kinetic data of the mechanism were taken either from the literature or from fits to experiments.

Parts of the surface reaction mechanism have already been used for numerical modeling of catalytic ignition,[1] simulation of total and partial oxidation of light hydrocarbons on platinum[9] and modeling the CO–O_2 and NO–CO reactions on rhodium.[10,12]

The surface reaction mechanism for a Pt/Rh three-way catalyst consists of three parts: a C_3H_6 oxidation mechanism on Pt/Al_2O_3 (Table 3(a)), a mechanism for NO reduction on platinum (Table 3(b)) and a reaction mechanism for NO reduction and CO oxidation on rhodium (Table 3(c)). The mechanisms of C_3H_6 oxidation and NO reduction on Pt were developed using experiments on lean DeNOx in an integral reactor,[21–23] additional experiments without NO were used especially to extract the kinetic data for the $C_3H_5(s)$ formation. However, it has been modified with respect to the following: OH adsorption and desorption are disregarded; the NO reaction system is simplified; the desorption energy of CO(s) is decreased by 10 kJ mol^{-1} and thermodynamic data for reactions (32), (47) and (49) are modified; N_2O and NO_2 formation, adsorption and desorption are disregarded; the pre-exponential for the NO(s) dissociation is decreased and NO reduction and CO oxidation on rhodium are added.

Results and discussion

In this section, the results of the simulations are discussed. First, the predicted conversion rates of the pollutants C_3H_6, CO and NO as function of temperature for three different gas mixtures (see Table 1) will be shown and compared with experimentally measured data. Then, the interaction of transport and chemistry within a single channel is described as revealed by the two-dimensional simulation.

Temperature dependence of conversion rates

In the simulation, the conditions used experimentally and described above are applied. Table 3 summarizes the simulation input data. The gas flows at a uniform inlet velocity into the cylindrical tube. Due to the isothermal sample temperature, the channel wall is assumed to be isothermal. In Fig. 1, the conversion rates of CO, C_3H_6 and NO are shown as a function of temperature for a lean inlet gas mixture. The conversion starts at 300 °C and increases up to 100% for CO and C_3H_6 at 500 °C and 400 °C, respectively. The NO conversion shows a maximum at 360 °C and then decreases.

Table 3 Surface reaction mechanisms of C_3H_6 oxidation on Pt, NO reduction on Pt and NO reduction and CO oxidation on Rh. The units are A [mol cm s], E_a [kJ mol^{-1}]. Θ_i describes the variation of the activation energy or reaction order, respectively, with the coverage of species i (see eqn. (3)). $CC_2H_5(s)$ is a propylidyne ($Pt{\equiv}C\text{-}C_2H_5$), $C_3H_5(s)$ is a propenyl ($Pt\text{-}CH_2\text{-}C_2H_3$) and $C_3H_6(s)$ is bound to the Pt as $CH_3\text{-}CH\text{-}CH_2$

No.	Reaction	A/S^0	β/μ	E_a/ε	Ref.
(a) C_3H_6 oxidation on Pt					
A. Adsorption					
1	$O_2 + Pt(s) + Pt(s) \rightarrow O(s) + O(s)$	$S^0 = 7.00 \times 10^{-02}$	0.0	0.0	9, 24[a]
2	$C_3H_6 + Pt(s) + Pt(s) \rightarrow C_3H_6(s)$	$S^0 = 9.80 \times 10^{-01}$	0.0	0.0	[b]
3	$C_3H_6 + O(s) + Pt(s) \rightarrow C_3H_5(s) + OH(s)$	$S^0 = 5.00 \times 10^{-02}$	0.0	0.0	[c]
	$\Theta_{Pt(s)}$		−0.9		
4	$H_2 + Pt(s) + Pt(s) \rightarrow H(s) + H(s)$	$S^0 = 4.60 \times 10^{-02}$	0.0	0.0	9, 24
	$\Theta_{Pt(s)}$		−1.0		
5	$H_2O + Pt(s) \rightarrow H_2O(s)$	$S^0 = 7.50 \times 10^{-01}$	0.0	0.0	9, 24
6	$CO_2 + Pt(s) \rightarrow CO_2(s)$	$S^0 = 5.00 \times 10^{-03}$	0.0	0.0	9
7	$CO + Pt(s) \rightarrow CO(s)$	$S^0 = 8.40 \times 10^{-01}$	0.0	0.0	9, 25
B. Desorption					
8	$O(s) + O(s) \rightarrow O_2 + Pt(s) + Pt(s)$	$3.70 \times 10^{+21}$	0.0	232.2	[b]
	$\Theta_{O(s)}$			90.0	
9	$C_3H_6(s) \rightarrow C_3H_6 + Pt(s) + Pt(s)$	$1.00 \times 10^{+13}$	0.0	72.7	26
10	$C_3H_5(s) + OH(s) \rightarrow C_3H_6 + O(s) + Pt(s)$	$3.70 \times 10^{+21}$	0.0	31.0	[c]
11	$H(s) + H(s) \rightarrow H_2 + Pt(s) + Pt(s)$	$3.70 \times 10^{+21}$	0.0	67.4	11
	$\Theta_{H(s)}$			6.0	
12	$H_2O(s) \rightarrow Pt(s) + H_2O$	$1.00 \times 10^{+13}$	0.0	40.3	11
13	$CO(s) \rightarrow CO + Pt(s)$	$1.00 \times 10^{+13}$	0.0	136.4	27[l,m]
	$\Theta_{CO(s)}$			33.0	
14	$CO_2(s) \rightarrow CO_2 + Pt(s)$	$1.00 \times 10^{+13}$	0.0	27.1	9
C. Surface reaction					
C.1 $C_3H_5(s)$ Oxidation [global reaction]					
15	$C_3H_5(s) + 5O(s) \rightarrow 5OH(s) + 3C(s)$	$3.70 \times 10^{+21}$	0.0	95.0	[d,e]
C.2 $C_3H_6(s)$ Consumption					
16	$C_3H_6(s) \rightarrow H(s) + CC_2H_5(s)$	$1.00 \times 10^{+13}$	0.0	75.4	26
17	$CC_2H_5(s) + H(s) \rightarrow C_3H_6(s)$	$3.70 \times 10^{+21}$	0.0	48.8	[f]
18	$CC_2H_5(s) + Pt(s) \rightarrow C_2H_3(s) + CH_2(s)$	$3.70 \times 10^{+21}$	0.0	108.2	26
19	$C_2H_3(s) + CH_2(s) \rightarrow Pt(s) + CC_2H_5(s)$	$3.70 \times 10^{+21}$	0.0	3.2	[f]
20	$C_2H_3(s) + Pt(s) \rightarrow CH_3(s) + C(s)$	$3.70 \times 10^{+21}$	0.0	46.0	9[g]
21	$CH_3(s) + C(s) \rightarrow C_2H_3(s) + Pt(s)$	$3.70 \times 10^{+21}$	0.0	46.9	9
C.3 CH_x Consumption					
22	$CH_3(s) + Pt(s) \rightarrow CH_2(s) + H(s)$	$1.26 \times 10^{+22}$	0.0	70.4	9, 28
23	$CH_2(s) + H(s) \rightarrow CH_3(s) + Pt(s)$	$3.09 \times 10^{+22}$	0.0	0.0	28
24	$CH_2(s) + Pt(s) \rightarrow CH(s) + H(s)$	$7.00 \times 10^{+22}$	0.0	59.2	9, 28[h]
25	$CH(s) + H(s) \rightarrow CH_2(s) + Pt(s)$	$3.09 \times 10^{+22}$	0.0	0.0	28
26	$CH(s) + Pt(s) \rightarrow C(s) + H(s)$	$3.09 \times 10^{+22}$	0.0	0.0	28
27	$C(s) + H(s) \rightarrow CH(s) + Pt(s)$	$1.25 \times 10^{+22}$	0.0	138.0	9, 28
C.4 C_2H_x Oxidation					
28	$C_2H_3(s) + O(s) \rightarrow Pt(s) + CH_3CO(s)$	$3.70 \times 10^{+19}$	0.0	62.3	[f]
29	$CH_3CO(s) + Pt(s) \rightarrow C_2H_3(s) + O(s)$	$3.70 \times 10^{+21}$	0.0	196.7	29
	$\Theta_{O(s)}$			−45.0	
30	$CH_3(s) + CO(s) \rightarrow Pt(s) + CH_3CO(s)$	$3.70 \times 10^{+21}$	0.0	82.9	[f]
31	$CH_3CO(s) + Pt(s) \rightarrow CH_3(s) + CO(s)$	$3.70 \times 10^{+21}$	0.0	0.0	[f]
C.5 CH_x Oxidation					
32	$CH_3(s) + O(s) \rightarrow CH_2(s) + OH(s)$	$3.70 \times 10^{+21}$	0.0	36.6	30[i]
33	$CH_2(s) + OH(s) \rightarrow CH_3(s) + O(s)$	$3.70 \times 10^{+21}$	0.0	25.1	30
34	$CH_2(s) + O(s) \rightarrow CH(s) + OH(s)$	$3.70 \times 10^{+21}$	0.0	25.1	30
35	$CH(s) + OH(s) \rightarrow CH_2(s) + O(s)$	$3.70 \times 10^{+21}$	0.0	25.2	30[i]
36	$CH(s) + O(s) \rightarrow C(s) + OH(s)$	$3.70 \times 10^{+21}$	0.0	25.1	30
37	$C(s) + OH(s) \rightarrow CH(s) + O(s)$	$3.70 \times 10^{+21}$	0.0	224.8	30[i]
C.6 H, OH, H_2O Surface Reactions					
38	$O(s) + H(s) \rightarrow OH(s) + Pt(s)$	$3.70 \times 10^{+21}$	0.0	11.5	11
39	$OH(s) + Pt(s) \rightarrow O(s) + H(s)$	$5.77 \times 10^{+22}$	0.0	74.9	11[j]

Table 3 Continued

No.	Reaction		A/S^0	β/μ	E_a/ε	Ref.
40	$H(s) + OH(s) \to H_2O(s) + Pt(s)$		$3.70 \times 10^{+21}$	0.0	17.4	11
41	$H_2O(s) + Pt(s) \to H(s) + OH(s)$		$3.66 \times 10^{+21}$	0.0	73.6	11[j]
42	$OH(s) + OH(s) \to H_2O(s) + O(s)$		$3.70 \times 10^{+21}$	0.0	48.2	11
43	$H_2O(s) + O(s) \to OH(s) + OH(s)$		$2.35 \times 10^{+20}$	0.0	41.0	11[j]
C.7 CO oxidation						
44	$CO(s) + O(s) \to CO_2(s) + Pt(s)$		$3.70 \times 10^{+20}$	0.0	108.0	9, 11[k,n]
		$\Theta_{CO(s)}$			33.0	
		$\Theta_{NO(s)}$			−90.0	
45	$CO_2(s) + Pt(s) \to CO(s) + O(s)$		$3.70 \times 10^{+21}$	0.0	165.1	f
		$\Theta_{O(s)}$			−45.0	
46	$C(s) + O(s) \to CO(s) + Pt(s)$		$3.70 \times 10^{+21}$	0.0	0.0	30
		$\Theta_{CO(s)}$			−33.0	
47	$CO(s) + Pt(s) \to C(s) + O(s)$		$3.70 \times 10^{+21}$	0.0	218.5	30[i]
		$\Theta_{O(s)}$			−45.0	
(b) NO reduction on Pt						
A. Adsorption						
48	$NO + Pt(s) \to NO(s)$		$S^0 = 8.50 \times 10^{-01}$	0.0	0.0	31–34[b]
B. Desorption						
49	$NO(s) \to NO + Pt(s)$		$1.00 \times 10^{+16}$	0.0	140.0	35–37[b]
50	$N(s) + N(s) \to N_2 + Pt(s) + Pt(s)$		$3.70 \times 10^{+21}$	0.0	113.9	38[o]
		$\Theta_{CO(s)}$			75.0	
C. NO surface reactions						
51	$NO(s) + Pt(s) \to N(s) + O(s)$		$5.00 \times 10^{+20}$	0.0	107.8	32[p,n]
		$\Theta_{CO(s)}$			−3.0	
52	$N(s) + O(s) \to NO(s) + Pt(s)$		$3.70 \times 10^{+21}$	0.0	128.1	f
		$\Theta_{O(s)}$			45.0	
(c) NO reduction and CO oxidation on Rh						
A. Adsorption						
53	$O_2 + Rh(s) + Rh(s) \to O(Rh) + O(Rh)$		$S^0 = 1.00 \times 10^{-02}$	0.0	0.0	39
		$\Theta_{Rh(s)}$		−1.0		
54	$CO + Rh(s) \to CO(Rh)$		$S^0 = 5.00 \times 10^{-01}$	0.0	0.0	39
55	$NO + Rh(s) \to NO(Rh)$		$S^0 = 5.00 \times 10^{-01}$	0.0	0.0	39
B. Desorption						
56	$O(Rh) + O(Rh) \to O_2 + Rh(s) + Rh(s)$		$3.00 \times 10^{+21}$	0.0	293.3	40
57	$CO(Rh) \to CO + Rh(s)$		$1.00 \times 10^{+14}$	0.0	132.3	40
		$\Theta_{N(Rh)}$			41.9	
		$\Theta_{CO(Rh)}$			18.8	
58	$NO(Rh) \to NO + Rh(s)$		$5.00 \times 10^{+13}$	0.0	108.9	39
59	$N(Rh) + N(Rh) \to N_2 + Rh(s) + Rh(s)$		$1.11 \times 10^{+19}$	0.0	136.9	12[q]
		$\Theta_{N(Rh)}$			16.7	
C. NO/CO surface reactions						
60	$CO(Rh) + O(Rh) \to CO_2 + Rh(s) + Rh(s)$		$3.70 \times 10^{+20}$	0.0	59.9	39
61	$NO(Rh) + Rh(s) \to N(Rh) + O(Rh)$		$2.22 \times 10^{+22}$	0.0	79.5	39

[a] Literature value without temperature dependence of S^0. [b] Mean value of literature data. [c] Estimate taken from the comparison of simulation and experiment (C_3H_6-oxidation on Pt/Al_2O_3). [d] Derived from the activation energy of C–H fission in $CH_2CH_3(s)$.[29] [e] Complex fast reaction; takes place, if sufficient O(s) is available for reaction (3). [f] Activation energy derived from enthalpies of formation. [g] Literature value without increase of activation energy with $\Theta_{C(s)}$. [h] Literature value increased by 0.3 kJ mol^{-1} to get thermodynamic consistency. [i] Activation energy fitted to enthalpies of formation of OH(s) and O(s) used here. [j] Equilibrium reactions used in ref. 11. [k] Mean of the literature values (3.70×10^{21}, 105.0 kJ mol^{-1}; 3.70×10^{19}, 117.6 kJ mol^{-1}) combined with a $\Theta_{NO(s)}$ dependence of the activation energy. [l] Activation energy larger than that from TPD (130.5 kJ mol^{-1}). [m] Activation energy decreased by 10 kJ mol^{-1} in comparison to surface mechanism on Pt/Al_2O_3. [n] Pre-exponential factor derived in this work. [o] Variation of the activation energy with $\Theta_{CO(s)}$ considered. [p] Decreased activation energy and pre-exponential in comparison to literature values (7.40×10^{23}, 119 kJ mol^{-1}); coverage dependence introduced. [q] Activation energy decreased in comparison to the literature value (153.6 kJ mol^{-1}).

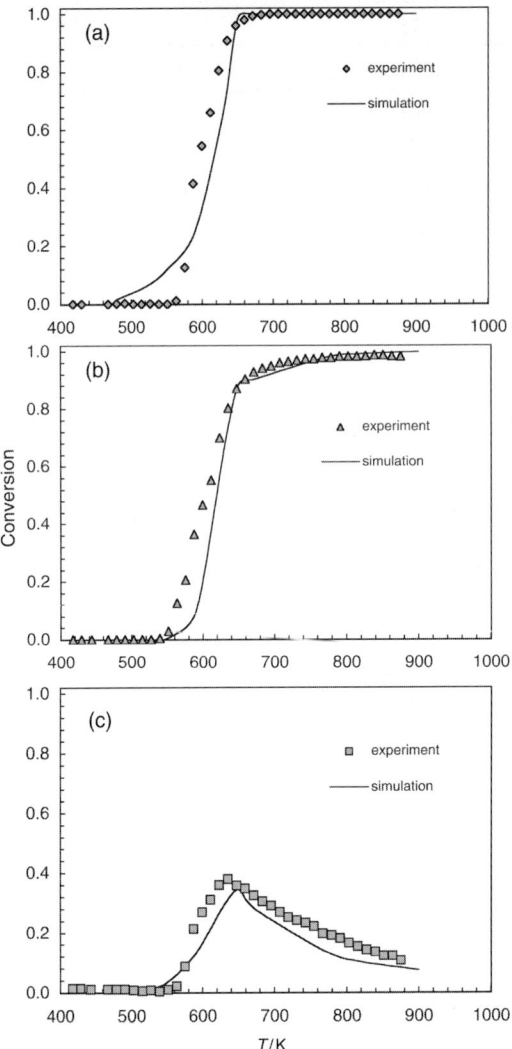

Fig. 1 Conversion of (a) CO, (b) C_3H_6 and (c) NO at lean conditions with increasing temperature; comparison of experimental and calculated data.

The predicted conversion rates of all three species agree well with the experimentally measured data. Especially the temperature behavior of the NO conversion and the slow increase of the C_3H_6 conversion in the temperature range above 360 °C. The simulation shows some CO conversion at temperatures lower than 300 °C probably because, in the reaction mechanism, reactions for hydrocarbons on rhodium are not included. Therefore, C_3H_6 is not able to block the CO oxidation on Rh in contrast to the CO oxidation on Pt.

The conversion rates as a function of the temperature for the nearly stoichiometric mixture are shown in Fig. 2. Again, the conversion of CO, C_3H_6 and NO starts at 300 °C, but increases more slowly with temperature compared to the lean mixture. Because of the lack of O_2 in the mixture, CO conversion is not complete. The conversion of C_3H_6 is complete at 500 °C, which indicates that C_3H_6 can compete with CO for O_2. Also, for temperatures higher than 450 °C, NO reduction is complete. Concerning the conversion rates of CO and NO and the competition between CO and C_3H_6 for O_2, the simulation results agree well with the experimental data. Only the C_3H_6

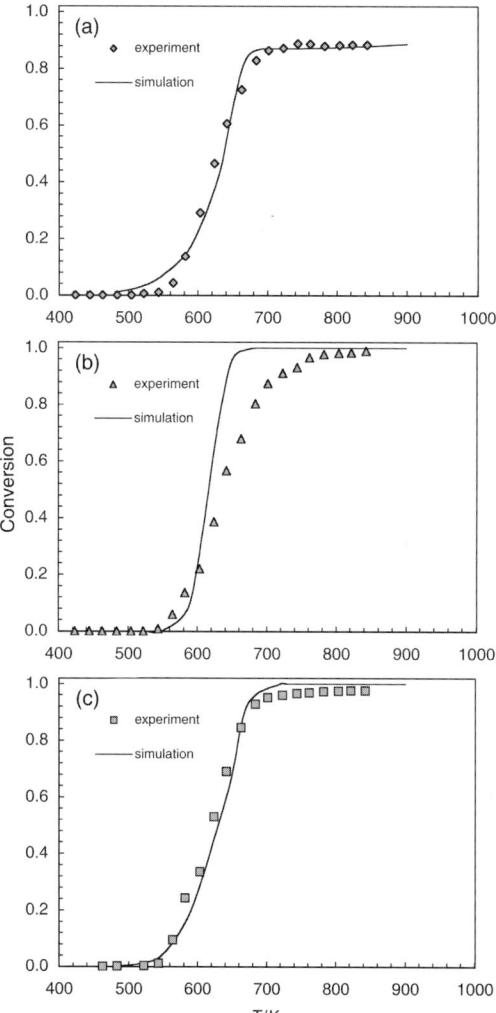

Fig. 2 Conversion of (a) CO, (b) C_3H_6 and (c) NO at nearly stoichometric conditions with increasing temperature; comparison of experimental and calculated data.

conversion rate shows deviations between 340 and 460 °C. Compared to the experimental data the predicted conversion rate is too high.

The rich mixture, shown in Fig. 3, contains 0.4 vol.% O_2, which leads to only a maximum of CO conversion of *ca.* 33%. The conversion of NO reaches 100% for temperatures higher than 570 °C. In comparison with the results of the two other mixtures, the increase in the conversions of CO, C_3H_6 and NO with temperature is slower, in particular the C_3H_6 conversion rate.

Regarding the CO and NO conversion, the simulations agree with the experimental values. However, large deviations exist between the predicted C_3H_6 conversion and the experimental data. These deviations can be explained by the fact that in the rich regime a wider variety of surface species, *e.g.*, partial oxidation products of C_3H_6, resides on the catalytic surface, which can reduce the oxygen coverage and lead to different reaction paths. In the reaction mechanism used here, only a limited number of possible surface species are included. Therefore further reactions and surface species have to be included in order to improve the prediction of the C_3H_6 conversion at richer mixtures.

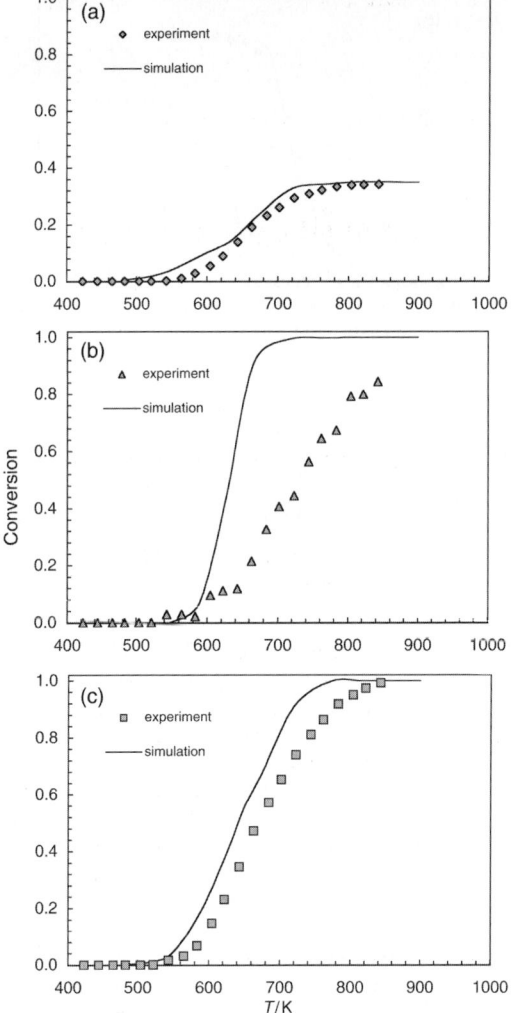

Fig. 3 Conversion of (a) CO, (b) C_3H_6 and (c) NO at rich conditions with increasing temperature; comparison of experimental and calculated data.

Transport and chemical reaction within a single channel

In Fig. 4, the mass fractions of C_3H_6, CO, CO_2 and NO for the lean mixture (see Table 1) at 407 °C within the channel are shown. The mass fraction profiles show that most of the C_3H_6 is converted within the first centimetre. In this axial range CO is nearly completely converted. The NO conversion is limited to the first centimetre and then ceases.

This behavior can be explained by means of the surface coverages. The calculated coverages of the most relevant surface species on platinum and rhodium are shown as a function of the axial position along the channel in Fig. 5 and 6, respectively. The coverages are defined in respect of the whole catalytic area, consisting of rhodium and platinum.

In Fig. 5 it is revealed that at the axial distance of 1.4–1.6 cm the surface coverage state varies strongly because of the decreasing CO concentration in the gas phase, from a mainly CO(s)-covered state to an O(s)-covered state. This variation is initiated by the decreasing CO/O_2 ratio in

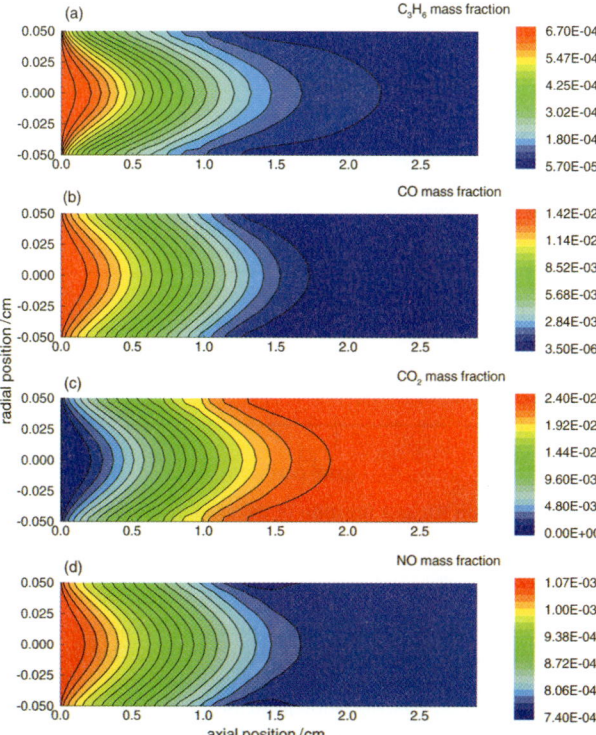

Fig. 4 Flow fields of the mass fraction of the species (a) C_3H_6, (b) CO, (c) CO_2 and (d) NO at 400 °C. Different scales are used in axial and radial direction for visual clarity; lean mixture.

the gas phase. During this transition the number of free platinum sites Pt(s) increases, which allows more NO to be adsorbed and dissociated. When the surface reaches the O(s)-covered state, the number of free platinum sites is decreased and the equilibrium of NO dissociation is shifted to NO(s), resulting in a vanishing NO conversion.

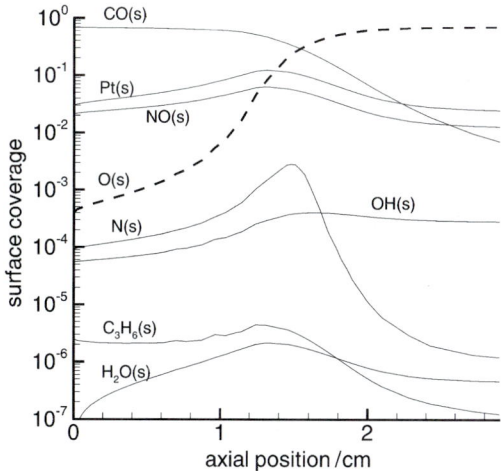

Fig. 5 Surface coverage at 400 °C on Pt as a function of the axial position; lean mixture; (s) denotes surface species.

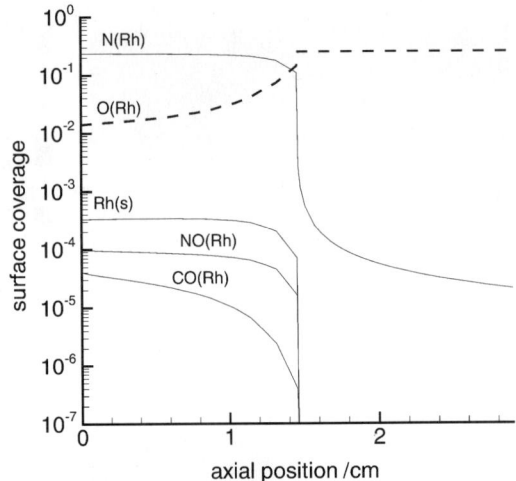

Fig. 6 Surface coverage at 400 °C on Rh as a function of the axial position; lean mixture; (s) denotes surface species.

The rhodium surface goes from a N(s)-covered state to an O(s)-covered state. This transition phenomenon can be seen in Fig. 6. The O(s)-covered surface state on rhodium also prevents NO conversion. Below 1.4 cm, the surface is mainly covered with N(s) and active for NO conversion.

With increasing reaction temperature the transition point moves toward the channel entrance, which reduces the area that is active for NO conversion. In Fig. 3 the resulting decrease in NO conversion with increasing temperature can be seen.

λ-Window

A typical feature of the three-way converter is the "λ-window" behaviour, that means, C_3H_6, CO and NO are simultaneously converted with high efficiency in a narrow range around the stoichiometric air/fuel ratio. Fig. 7 presents the predicted conversions for the catalytic converter at 500 °C for different gas mixtures. This temperature is a typical catalyst inlet temperature for partial load.[41] The experimental data have already been presented in Fig. 1–3. The simulation well predicts the "λ-window" in the lean region. For the rich regime, as already shown in Fig. 3, the CO

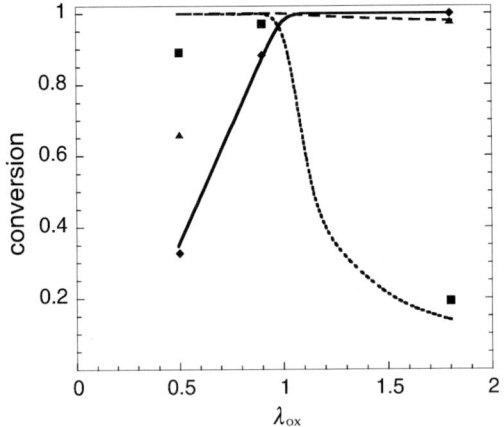

Fig. 7 λ-window at 500 °C; comparison of experimental ((---) C_3H_6, (···) NO, (——) CO) and calculated ((▲) C_3H_6, (■) NO, (◆) CO) data.

conversion is also well predicted, while there are major deficiencies for C_3H_6 conversion, as discussed above.

Conclusions

A two-dimensional flow-field description, including a detailed reaction mechanism for the conversion of CO, C_3H_6, O_2 and NO has been used to simulate the exhaust gas treatment in a platinum/rhodium-coated single channel of a typical three-way catalytic converter. The simulation is based on the CFD code FLUENT and the chemistry module DETCHEM, which were coupled for the simulations performed.

The computational tool was used to predict conversion rates at lean, nearly stoichiometric and rich conditions as a function of temperature. The calculated data are compared with experimentally measured data and a good agreement was achieved. Only for rich conditions is the prediction of C_3H_6 conversion too high. Furthermore, the interaction of transport and kinetics within a single channel were revealed by the simulation.

Numerical simulation of the emission reduction inside a monolithic three-way catalyst offers an efficient method to investigate converter performance. As a first step on the way to a complete model, a reliable multi-step reaction mechanism has to be developed. A surface reaction mechanism consisting of 61 reactions among 31 species was used and validated by a comparison with experimental data. This mechanism will finally allow simulations to predict emissions over a broad range of conditions.

Outlook

The simulation presented in this paper is carried out for a single channel in a monolithic catalytic converter. In the ongoing research, several extensions will be made for a more accurate model for the real catalytic converter. The next step in using numerical simulation is to expand the model from a single channel model to the total monolith. A simulation of the total monolith using FLUENT requires too much computer time and memory. Therefore, a boundary-layer model will be used as an alternate method to stimulate the behavior of the single channel. This future work will also take the thermal conductivity of the catalyst into account for the calculation of the heat distribution of the whole monolithic converter.

The final goal is the simulation of the behavior of an automotive catalytic converter during a whole test cycle using models that are based on the real physical and chemical processes. The test cycle simulation will include the light-off behavior of the converter.

In order to improve the prediction quality for rich mixtures, further reactions have to be included in the reaction mechanism.

Acknowledgements

The authors thanks Sven Kureti and Oliver Görke, Karlsruhe University, for their experimental measurements of the converter performance. Financial support of this work has been provided by the DFG (SFB 359), the European Union (Brite/Euram Programme), and the Fonds der Chemischen Industrie. J. Braun, T. Hauber, H. Többen and P. Zacke (J. Eberspächer GmbH & Co.) are acknowledged for their valuable advice.

References

1   G. C. Koltsakis, P. A. Konstantinidis and A. M. Stamatelos, *Appl. Catal. B*, 1997, **12**, 161.
2   T. Kirchner and G. Eigenberger, *Catal. Today*, 1997, **28**, 3.
3   G. P. Ansell, P. S. Bennett, J. P. Cox, J. C. Frost, P. G. Gray, A.-M. Jones, R. R. Rajaram, A. P. Walker, M. Litorell and G. Smedler, *Appl. Catal. B*, 1996, **10**, 183.
4   A. L. Boehman, *Numerical Modeling of NO Reduction over Cu-ZSM-5 under Lean Conditions*, SAE 1997-07-52. SAE, Warrendale, PA, USA.
5   C. N. Montreuil, S. C. Williams and A. A. Adamczyk, *Modeling Current Generation Catalytic Converters: Laboratory, Experiments and Kinetic Parameter Optimization—Steady State Kinetics*, SAE 920096 (1992). SAE, Warendale, PA, USA.

6 S.-J. Jeong and W.-S. Kim, *A Numerical Approach to Investigate Transient Thermal and Conversion Characteristics of Automotive Catalytic Converters*, SAE 1998-08-81. SAE, Warrendale, PA, USA.
7 J. Warnatz, *Proc. Combust. Inst.*, 1992, **24**, 553.
8 R. Kissel-Osterrieder, F. Behrendt and J. Warnatz, *Proc. Combust. Inst.*, 1998, **27**, 2267.
9 D. K. Zerkle, M. D. Allendorf, M. Wolf and O. Deutschmann, *J. Catal.*, 2000, **196**, 18.
10 S. H. Oh, G. B. Fisher, J. E. Carpenter and W. Goodmann, *J. Catal. C*, 1986, **10**, 360.
11 O. Deutschmann, R. Schmidt, F. Behrendt and J. Warnatz, *Proc. Combust. Inst.*, 1996, **26**, 1747.
12 E. I. Altman and R. J. Gorte, *J. Catal.*, 1988, **113**, 185.
13 J. Braun, T. Hauber, H. Többen, P. Zacke, D. Chatterjee, O. Deutschmann and J. Warnatz, *Influence of Physical and Chemical Parameters on the Conversion Rate of a Catalytic Converter: A Numerical Simulation Study*, SAE 2000-01-0211. SAE, Warrendale, PA, USA.
14 J. Warnatz, U. Mass and R. W. Dibble, *Combustion*, Springer, Heidelberg, 1996; 2nd edn., 1999; 3rd edn., 2001.
15 E. Rogermond, N. Essayem, R. Frety, V. Perrichon, M. Primet, M. Chevrier, C. Gouthier and F. Mathis, *J. Catal.*, 1999, **186**, 414.
16 FLUENT 4.4, Fluent Inc., Lebanon, 1997.
17 DETCHEM, Version 1.2, O. Deutschmann, IWR, University of Heidelberg, Germany, 1998.
18 O. Deutschmann and L. D. Schmidt, *AIChE J.*, 1998, **44**, 2465.
19 M. Baerns, H. Hofmann and A. Renken, *Chemische Reaktionstechnik*, Georg-Thieme-Verlag, Stuttgart, 1992, vol. 1.
20 P. Deuflhard, E. Hairer and J. Zugk, *Num. Math.*, 1987, **51**, 501.
21 D. Chatterjee, E. Frank, J. Warnatz and W. Weiswelier, *Modellierung der NOx-Reduktion in sauerstoffreichen Abgasen an ausgewählten Katalysator-Systemen*, FVV Information Meeting on Engines, Forschungsvereinigung Verbrennungskraftmaschinen, Frankfurt, 2000, vol. R508.
22 D. Chatterjee, E. Frank, J. Warnatz and W. Weiweiler, *DeNOx-Modell: Modellierung der selectiven katalysierten NOx-Reduktion*, Forschungsvercinigung Verbrennungskraftmaschinen, Frankfurt, 2001, vol. R703.
23 D. Chatterjee, Dissertation, Ruprecht-Karls-Universität, Heidelberg, 2001.
24 M. Rinnemo, O. Deutschmann, F. Behrendt and B. Kasemo, *Combust. Flame*, 1997, **111**, 312.
25 C. T. Campbell, G. Ertl, H. Kuipers and J. Segner, *Surf. Sci.*, 1981, **107**, 207.
26 Y. L. Tsai, C. Xu and B. E. Koel, *Surf. Sci.*, 1997, **385**, 37.
27 E. I. Altman and R. J. Gorte, *J. Catal.*, 1988, **110**, 191.
28 M. Wolf, O. Deutschmann, F. Behrendt and J. Warnatz, *Catal. Lett.*, 1999, **61**, 15.
29 A. T. Bell, in *Reaction Energetics: Theory and Application to Heterogeneous Catalysis, Chemisorption and Surface Diffusion*, ed. E. Shustorovich, VCH-Verlagsgesellschaft, Weinheim, 1991.
30 M. Wolf, Dissertation, Ruprecht-Karls-Universität, Heidelberg, 2000.
31 E. G. Seebauer, A. C. F. Kong and L. D. Schmidt, *Surf. Sci.*, 1986, **176**, 134.
32 M. Gruyters, A. T. Pasteur and D. A. King, *J. Chem. Soc., Faraday Trans.*, 1996, **92**, 2941.
33 M. Brandt, G. Mueller, H. Zagatta, O. Wehmeyer, N. Boewering and U. Heinzmann, *Surf. Sci.*, 1995, **331–333**, 30.
34 C. E. Wartnaby, A. Struck, Y. Y. Yeo and D. A. King, *J. Phys. Chem.*, 1996, **100**, 12483.
35 D. T. Wickham, B. A. Banse and B. E. Koel, *Surf. Sci.*, 1989, **223**, 82.
36 H. Miki, T. Nagase, T. Kioka, S. Sugai and K. Kawasaki, *Surf. Sci.*, 1990, **225**, 1.
37 R. J. Gorte, L. D. Schmidt and J. L. Gland, *Surf. Sci.*, 1981, **109**, 367.
38 E. Shustorovich and A. T. Bell, *Surf. Sci.*, 1993, **289**, 127.
39 S. H. Oh, G. B. Fisher, J. E. Carpenter and D. W. Goodeman, *J. Catal. C*, 1986, **10**, 360.
40 O. Deutschmann and L. D. Schmidt, *AIChE J.*, 1998, **44**, 2465.
41 E. S. J. Lox and B. H. Engler, in *Handbook of Heterogeneous Catalysis*, ed. G. Ertl, H. Knötzinger and J. Weitkamp, VCH-Verlagsgesellschaft, Weinheim, 1997, vol. 4, p. 1586.

A study of the reaction of oxygen with graphite: Model chemistry

Raymond Backreedy, Jenny M. Jones, Mohammad Pourkashanian and Alan Williams*

Department of Fuel and Energy, Leeds University, Leeds, UK LS2 9JT

Received 5th March 2001
First published as an Advance Article on the web 7th September 2001

A considerable amount of research has been directed towards the mechanism of oxidation of graphite as a model reaction system and because of its industrial importance. A number of recent studies have been concerned with *ab initio* molecular orbital calculations on graphite including model chemistry and the reactions with molecular oxygen. This study is concerned with oxidation steps involving the attachment of molecular oxygen to the graphene, the formation of carbon monoxide and, in particular, the subsequent oxidation reactions.

Introduction

The combustion of carbon is a reaction of immense importance not only because of its industrial significance but because it is a model reaction for an important generic heterogeneous reaction. The combustion of coal is responsible for nearly 40% of the worlds' electricity production and char combustion accounts for about half of that amount. Clearly an understanding of the combustion mechanism of carbon is of great importance.

As combustion of coal char proceeds the largely amorphous carbon graphitises and becomes less reactive, resulting in some unburned carbon remaining in the ash; the losses of unburned carbon in ash are about 1% of all coal burned.[1] In order to support ongoing experimental research directed towards the graphite oxidation process, theoretical studies on the molecular chemistry of the representative elements from graphite have been initiated.

The reactivity of graphite to oxygen is also of significance to industry generally where graphite is used in many high temperature applications including nuclear reactor construction.

The combustion of carbon is complex and whilst many experimental and theoretical studies have been made it has been difficult to unravel the detailed chemistry. The availability of computer programs for *ab initio* molecular orbital calculations facilitates this type of research and we have used the technique here. Whilst these techniques have previously been applied largely to small molecules, methods are now available that allow the application to larger structures.

The kinetics of graphite oxidation

There have been numerous studies of the oxidation of carbons generally and graphite specifically and these have been the subject of numerous reviews, for example ref. 2–4. Graphite has a π-bonded inert planar surface and reactive edge surfaces which can have zigzag or armchair formats and these have different oxidative reactivities. Functional groups such as oxygen atoms bound to these edges as semiquinones or carbonyls can have a significant influence, as can heteroatoms in the graphite structure. Most chars produced by combustion processes are turbostratic[1] as indeed are most industrial graphites and the graphite sheets can have different bonding energies as a consequence.

The overall oxidation reaction of carbon involves the steps:

$$C + 0.5O_2 = CO \qquad (1)$$

and

$$CO + O_2 = CO_2 \qquad (2)$$

In studies of these reactions it has been observed that: 1. The oxidation of graphite can be described by a single overall activation energy that is constant over a wide temperature range.[3–6] 2. The reaction order with respect to oxygen varies and is zero at low temperatures (400 K) and tends to unity at high temperatures (say 2500 K). At about 1000 K it is 0.5.[4,7–10] 3. At high temperatures the major initial combustion product is CO, but at low temperatures both CO and CO_2 are formed.[7,9,10] The CO_2/CO ratio has two maxima between 850 and 1650 K[9,10] although in most cases the CO is later converted to CO_2 either in the boundary layer of the carbon particle or in the ambient gases. In order to explain these experimental observations, models based on the adsorption of oxygen on active sites have been developed over a number of years and only some pertinent references are given here.[2,3,7,11–13]

Two types of interaction of oxygen with the surface carbon have been identified, see for example ref. 11 and 12. Type A results in the adsorption of oxygen without the gasification of carbon, that is, the reaction can be represented by $-C + O_2 = -C(O_2)$, and Type B, where there is gasification, namely $-C + O_2 = -(CO) + CO$. Here $-C$ is a surface carbon. The activation energy of the former is 10–80 kJ mol^{-1} and for the latter 40–140 kJ mol^{-1}.[12] The Type B mechanism dominates at a temperature of about 900 K.

The main features of carbon oxidation are usually set out as below:

$$2C_f + O_2 = 2C(O) \qquad (3)$$

$$C(O) = CO + C_f \qquad (4)$$

$$C(O) + C(O) = CO_2 + C_f \qquad (5)$$

The surface carbons where gasification takes place are identified as active sites, C_f, but they are not specifically identified in the literature as individual structures and are usually considered to involve defects in the structure. These active sites have a range of reactivities as pointed out above. The mechanism is effectively a pseudo-catalytic mechanism where the active sites are regenerated by a turnover mechanism.

It has recently been pointed out[7] that the mechanism is not consistent with all the observed features and has been adapted to involve a modified reaction (3) that invokes the concept of tightly and weakly bound oxygen atoms

$$2C_f + O_2 = 2(C-O)_t \qquad (6)$$

and followed by the steps:

$$(C-O)_t = (C-O)_w \qquad (7)$$

$$(C-O)_w = CO \qquad (8)$$

$$2(C-O)_w = 2C_f + O_2 \qquad (9)$$

Here we have used the terminology of ref. 7. It is possible to conclude[7] that, at higher temperatures, $E_1 = (E_8 + E_6/2 - E_9/2)$ and values were assigned as follows: $E_6 = 60$,[11] $E_8 = 300$[7] and $E_9 = 225$–425 kJ mol^{-1}.[12]

It has also been shown that the rate of production of CO, R, resulting from oxidation of the carbon is given by:

$$R = k\Theta_w = k_8(k_6/k_9)^{0.5}[O_2]^{0.5}/\{1 + (k_6/k_9)^{0.5}[O_2]^{0.5}(1 + k_7^{-1})\} \qquad (i)$$

where Θ_w is the fractional coverage of the weakly bound oxygen atoms, and this would imply an oxygen dependence of 0.5 in this temperature range. Deductions were also made of the values of the rate constants and the heats of combustion and a consistent picture emerged.

The activation energies for carbon oxidation

There have been numerous studies of carbon and graphite oxidation at low temperatures of about 800 K (*e.g.* ref. 10–15) but few studies have been made at higher temperatures.[16] Smith[4] presented a compilation of reaction data for a wide range of carbons, which is shown in Fig. 1. He concluded that the intrinsic activation energy for all carbons is best written as 178.9 kJ mol^{-1} with an error of $\pm 5\%$ kJ mol^{-1}. Hargrave *et al.*[5] have obtained a similar value of 162 $\pm$ 6 for coal chars and it has been more recently concluded[6] that the best value for coal chars is $E = 155$ kJ mol^{-1}.

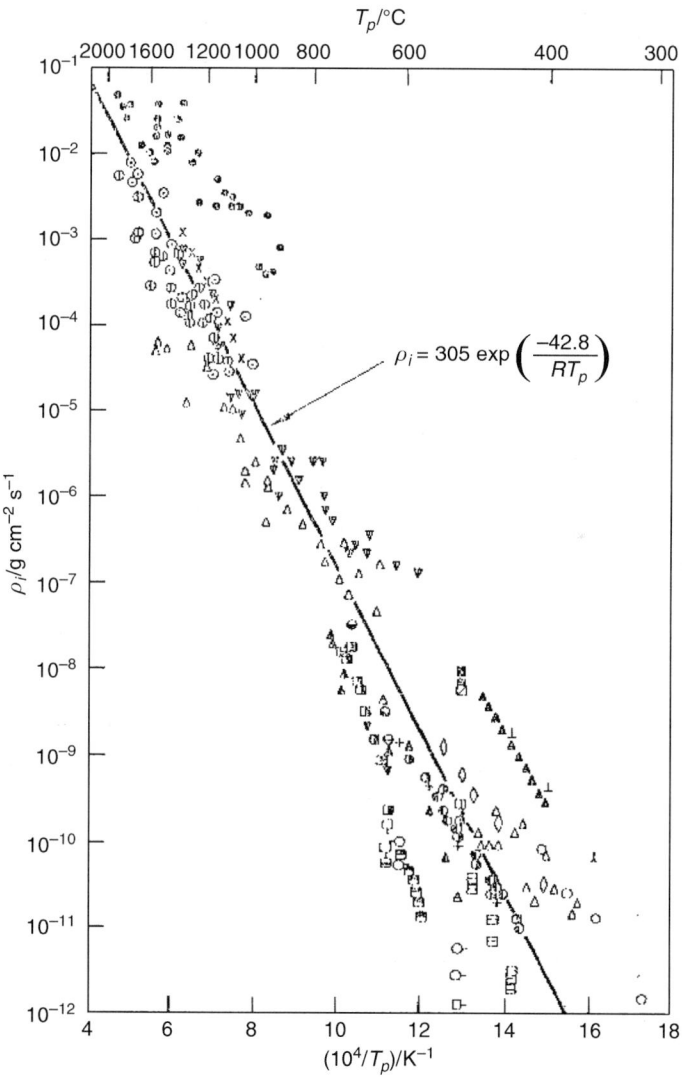

Fig. 1 Intrinsic reactivity of various carbons (oxygen pressure, 1 bar) compiled by Smith,[4] symbols as in ref. 4. (Reprinted with permission from ref. 4 © 1982, Combustion Institute.)

Smith[4] draws attention to the fact that the reactivities of carbons containing inorganic impurities lie above the curve and that pure carbons such as nuclear and spectroscopic carbons lie below but with the same activation energy. A carbon that was highly purified and annealed at very high temperature showed the lowest reactivity and this particular carbon had an activation energy of 250.8 kJ mol^{-1}.

Molecular modelling

By means of empirical and *ab initio* modelling it is possible to calculate[17–20] the geometric parameters for the structures set out in Fig. 2 for both the initial graphite model and the subsequent oxygen intermediates using GAUSSIAN 98 revision A8.[21] The molecular system used here is the graphene model $C_{25}H_9$ that is based on studies[17,18] that show that it replicates the behaviour of graphite. It behaves as a finite model section of the infinite solid and the use of H-atom termination enables it to replicate correctly the boundaries of the molecule.[17] It can be used to represent zigzag or armchair molecules but only the more reactive former case is studied here.

The calculation procedure involves checking the stability of the wavefunction. A stable wavefunction for the structure indicates that there is a certain minimum on the potential surface that

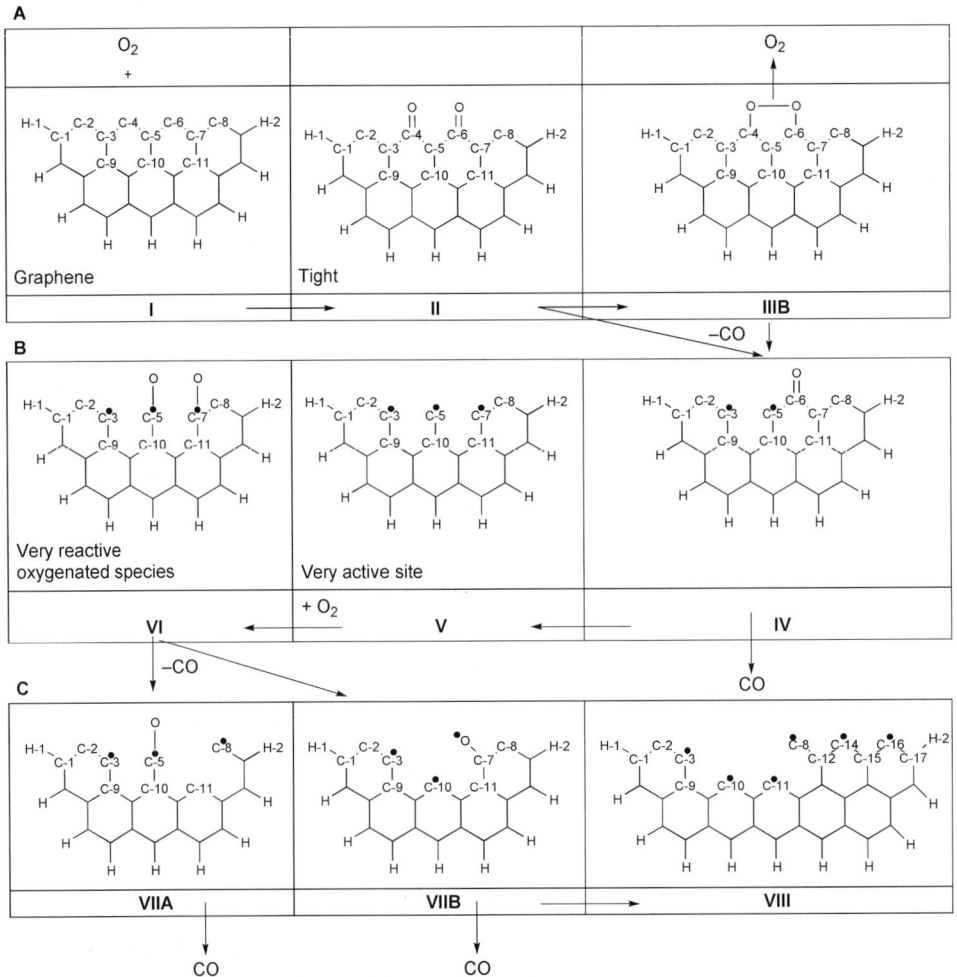

Fig. 2 Proposed mechanism of graphite oxidation.

corresponds to the expected structure for each of our models, *i.e.* for geometry optimisation or for frequency and thermochemical calculations.

We employed the unrestricted Hartree–Fock (UHF) method with a basis set of 3-21G-(d) for geometry optimisation followed by density functional method using the B3LYP with the basis set of 6-31 G(d) for single point energy calculation. The geometry and multiplicity of each structure, including those with oxygen were determined by optimising the spin for a given structure. All the structures in Fig. 2 were calculated using B3LYP/6 31G(d). The preliminary studies were carried out using STO–3G, a minimal basis set.[21] The geometrical structures obtained in this way were consistent with those obtained at a higher level.

The bond energies of specific critical bonds in the structures can in principle be obtained using the single point total SCF energies of the reacting and product structures. In the work undertaken here we have found that there can be considerable errors using this method and have only employed the technique to demonstrate trends. Chen and Yang came to the same conclusion[18] and used a different technique which we have not employed here.

The chemical model

Various attempts have been made to translate turnover models of the type given above into specific chemical mechanisms. Most recently Chen and Yang[18] proposed a low temperature model which involves both zigzag and armchair edge attack by molecular oxygen and also the formation of epoxy off-plane oxygen complexes.

We have restricted this study to only the high temperature mechanism where CO is the dominant product. In these circumstances, which involve the elimination of some of the more complex low temperature steps, it is useful to translate the above observations into a detailed chemical reaction mechanism.

The first stage involves the adsorption and desorption of oxygen, the formation of surface oxides and the formation of CO and CO_2 during their thermal desorption; these processes and their interpretation have been well described. We have focussed this work entirely on the formation of the chemically adsorbed species, and the formation of only CO by the high temperature mechanism.

Previous studies of the detailed modelling of the oxidation of carbon have been based on a 7-ring graphene molecule with four free electrons and the other carbon atoms terminated by hydrogen atoms and consisting of a single sheet of carbon.[17–20] This molecule is set out in Fig. 2 as structure **I**. Oxygen is adsorbed onto the active site and then forms an oxygenated species, the semiquinone molecule, structure **II**. Alternatively it can form a loosely bonded oxygen compound labelled **IIIB** or its precursor **IIIA**. Structure **IIIA** not shown in Fig. 2 is an intermediate between **II** and **III** in which the oxygen atoms are not bonded together and thus would have carbene-oxy groups in place of semiquinone. We propose that structures **II** and **III** can be identified with the active sites tight $(C–O)_t$ and loose $(C–O)_w$ bonds respectively. Thus structure **I → II** would be the equivalent of reaction (6) in the equation scheme set out above and by analogy its activation energy would be 60 kJ mol^{-1}. There is evidence to support this value theoretically.[20] Our estimates are not inconsistent with this value.

Chen and Yang[18] have suggested that the semiquinone (structure **II**) decomposes to give CO by the fission of two C–C bonds and this energy has been estimated to be about 320 kJ mol^{-1} per bond. Our studies, based on the kinetic analysis of Bews *et al.*,[7] suggest that there is an additional reaction route, involving the formation and subsequent decomposition of **IIIB** to give **IV** and CO by one route (*via* reaction (8) above) or alternatively to form molecular oxygen (*via* reaction (9)).

Computation of the bond lengths of graphene and its oxidation products (structures **II** and **IIIB**) using geometry optimisation results in the values set out in Table 1. The bond lengths for C1–C2, C2–C3 and C3–C4 are similar to the values of 142 pm quoted by Chen and Yang.[17] From the bond lengths it is suggested that the decomposition of **II** to yield CO needs slightly more energy than **III**. Therefore the decomposition of **III** is the rate-determining step, at least at low temperatures, but at high temperatures both will contribute to the formation of **IV**.

It is seen from the bond length data given in Table 1 that the C–C bond length (C3–C4 and C6–C7) external to the oxygen complex **II** increases compared to the graphene whilst the internal carbon bond (C4–C5 and C5–C6) decreases in length. This would imply that the extra-bond

Table 1 Bond lengths (pm) in graphene (**I**) and oxygen complexes **II** and **IIIB**[a]

Bond	Graphene	Semiquinone (**II**)	O2-Complex (**IIIB**)
C1–C2	142.9	139.0	138.3
C2–C3	142.9	138.1	146.8
C3–C4	142.2	148.0	137.9
C4–O	—	125.9	142.6
C4–C5	142.6	140.6	144.1
O–O	—	271.3	142.4
C5–C6	142.6	140.2	135.8
C6–C7	140.4	147.8	144.7
C7–C8	140.9	138.2	139.8

[a] See Fig. 2 for the atom numbering scheme.

undergoes fission preferentially. The ease of decomposition of **III** is easily seen; indeed **III** is an unstable molecule and as represented here is slightly non-planar. Both structure **II** and **III** exhibit considerable strain. **II** has some distortion from the plane but **III** is considerably more distorted and from the evidence it seems to be a very unstable structure.

The subsequent reaction mechanism

The next stage involves the formation of the structure **IV** that can decompose and lose a CO molecule and form structure **V**. It can be seen from Table 2 that in structure **IV** the bond C5–C6 is somewhat elongated resulting in easy fission. It should be noted that the carbonyl bond length (C6=O, 120.8 pm) is slightly shorter than that in semiquinone. This results in the formation of structure **V** which involves a dangling carbon. This dangling carbon can be identified with a super-reactive site. Evidence for this comes from Lahaye and coworkers[2] and from the ensuing discussion on the paper by Professor Bonnetain. It has been observed that carbon becomes very reactive at 950 °C and chemisorbs oxygen rapidly, and that the new carbon atoms become superficial and less labile than the previous ones. This would be in accord with the sequence set out in Fig. 2.

We propose that structure **V** can react with oxygen to give **VI**, which contains both a semiquinone and a carbonyl. Data on the bond lengths are given in Table 2. It is apparent that **VI** can decompose to give **VIIA** or **VIIB** with the bond length data suggesting the major route lies via **VIIA**. It is an unreactive site and in turn this is transformed to the reactive structure **VIII**. It should be noted that structure **VIII** is drawn in such a way that the aromatic rings extend beyond the original graphene model although the calculation was undertaken on the smaller molecule.

Table 2 Bond lengths (pm) in active sites and oxygenated products

Bond	IV	V	VI	VIII
C1–C2	142.0	140.6	141.6	139.7
C2–C3	142.0	138.3	137.4	125.3
C3–C9	142.0	141.2	142.4	139.5
C5–C6	150.9	—	—	—
C5–C10	136.7	150.2	133.2	—
C5–O	—	—	120.0	—
C7–O	—	—	123.1	—
C6–O	120.8	—	—	—
C6–C7	120.5	—	—	—
C7–C8	142.0	138.3	147.1	—
C8–C12	—	—	142.2	—

Consequently, at any time there is a distribution of reactive and less reactive sites, but the total active surface area is self-preserving except at the very beginning of the reaction and near the end when the graphite structure disintegrates. This is in accord with experimental observation.[2]

It is seen that the first seven steps result in the oxidation of the first row of the graphene leaving a dangling carbon in a reactive site. The next stage must involve the reaction of an oxygen molecule with a dangling carbon and a carbon with a free electron either side of it, that is, along the first layer of aromatic rings or it will start to form a cavity. The evidence from steric factors and from the electronic charges suggests that it reacts along the first layer of carbons by C–C fission and this then leaves the second row of four carbons with free electrons. That is, it is exactly the same situation as when the oxidation started but one layer of aromatic rings have been oxidised away. This is shown as structure **VIII** in Fig. 2.

Net atomic charge distribution calculations

Partial charges can provide insight into the relative reactivity of sites in a molecule as well as the overall molecular reactivity of the molecule. Partial charges can also be used to derive the electrostatic interaction energy component of force fields used in molecular mechanics calculations.

The partitioning of electronic charge can be determined using empirical methods or derived from the quantum mechanical wavefunction.[21] The methods available are the Mulliken population analysis and the atoms in molecules (AIM) method. It is thought that the latter calculations using extended basis sets and diffuse functions may be a more accurate approach but for computational reasons we used the Mulliken method here.

Fig. 3 and 4 show the distribution of net atomic charges for a number of the intermediates in the reaction scheme obtained in this way. It is interesting to note that the reactive sites for graphene tend to be more electronegative and Chen and Yang[18] found that the active site carbons in graphene have a negative charge. We have obtained slightly different numerical values to them, as shown in Fig. 3, and we attribute this to the fact that they have used the more accurate AIM method.

These edge sites with a negative charge are more liable to chemisorb oxygen atoms or molecules from the gas phase and form oxygen structures. The active site in structure **V**, the super-active site, has a high negative charge which is consistent with our earlier conclusions. There are a number of edge effects that can arise from adjacent semiquinone groups and heteroatoms in the carbon structure which can influence the way the charge is distributed in the molecule.

Structure **VIII** which we have assigned the role of the continuation active site does not have any special attributes in terms of charge. The structure is attempting to form a five-membered ring and

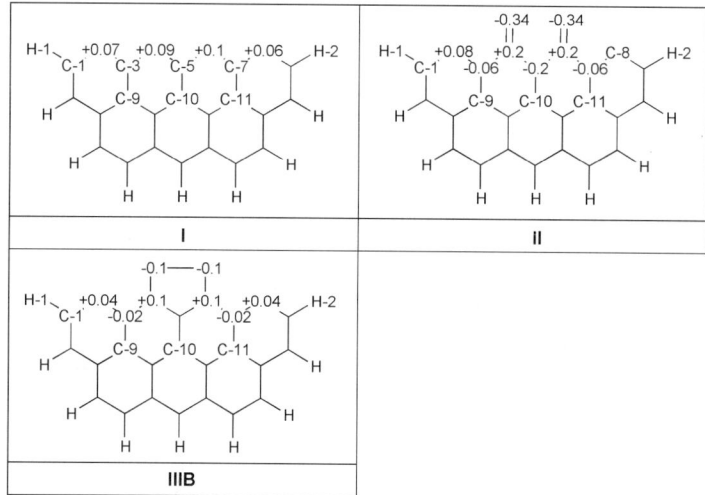

Fig. 3 Net charges for structures **I**, **II** and **IIIB**.

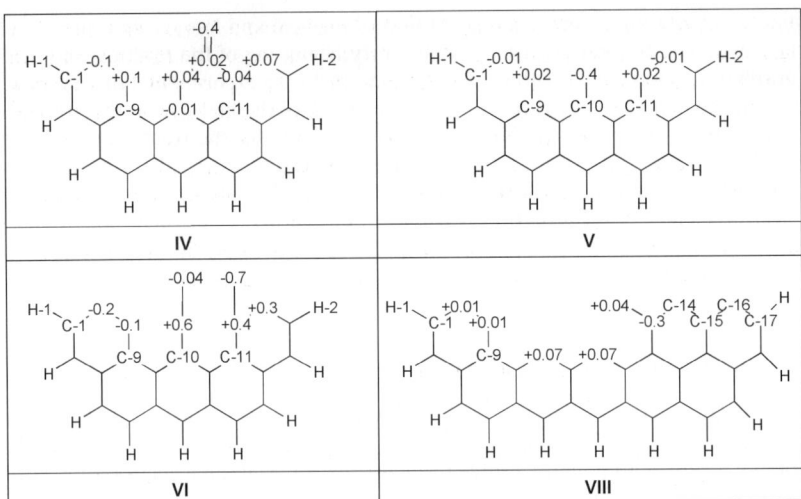

Fig. 4 Net charges for subsequent structures **IV, V, VI** and **VIII**.

consequently the carbon atom (C8) does not have a negative charge associated with it (it is slightly positive).

There are no experimental methods available to determine the charge distribution in the molecule and the values we have obtained are very much a function of the models used. We have attempted to use a number of different theory and basis sets and we tend to get some variations for the charges. Therefore the data shown in Fig. 3 and 4 indicate the general situation rather than provide for accurate information. Clearly further studies will have to be undertaken to obtain a deeper insight into the role of atomic charge distribution in active sites but the general pattern seems to have been identified.

Discussion

The chemical mechanism

The detailed mechanism of graphite oxidation is complex, especially in the lower temperature regime[15–19] where most thermogravimetric (TG) studies have been undertaken. As a consequence of this complexity we have studied only a simplified high temperature mechanism leading only to the formation of CO and which leaves out any CO_2 forming routes. We have not considered the inclusion of initial zigzag structures which would increase the randomness of the types of active sites produced during the progress of the reaction.

The basis of our understanding of the chemical processes is improved by assigning chemical structures to the turnover mechanism on the basis of evidence from chemisorption and flame studies. This involves the assumption that an active site consists of a free edge carbon site with enhanced negative charge and the formation of two types of primary oxygen complexes with differing levels of stability. The formation of these will result in a carbon surface with a distribution of active sites with different energies. This stochastic nature will be enhanced by the involvement of functional groups from armchair edges and the formation of semiquinone and basal epoxy groups as the reaction progresses. This would influence only the probability of reaction, that is the rate, but not the activation energy and this would be consistent with previous studies of annealing of carbon and coal chars.[1,23–26]

Thermochemical and kinetic factors

The implication for graphite oxidation is that if the analysis proposed by Bews et al.[7] is correct then we can set $E_1 = E_8 + E_6/2 - E_9/2$, that is $E_1 = 300 + 30 - (225 - 425) = 167 \pm 50$ kJ mol^{-1}. Therefore there is not a single value of E_1 and the values follow a probability distribution function with a mean value of 167 kJ mol^{-1}. If the conclusions of Chen and Yang[18] about an enhanced reactivity at low temperatures, *i.e.* the value of E_6 becomes smaller, the activation energy curves at low temperature will have a slightly different (lower) activation energy and consequently the plot of E_1 over a wide temperature range will not be linear. However, the data given in Fig. 1 suggest that the activation energy for the purified graphite becomes large at low temperatures and presumably the values of E_8 and E_9 change in some way that is not clear at present.

The above two conclusions have some significance to the combustion of carbons. First, the activation energy for the basic graphite or carbon material follows a distribution similar to a probability density function (pdf) and this will be greatly enhanced by the imperfections in the char structures. Secondly, there is danger in extrapolating the data to high temperatures.

The char morphology during combustion

There is experimental evidence[15,26] from studies on the morphology of char oxidation to support the arguments put forward for the mechanism set out in Fig. 2. At low oxidation temperatures, for example 1000 K, oxygen reacts in such a way that defects are enlarged during reaction and show pronounced anisotropy parallel to the (001) surface. This growth can come about *via* the mechanism proposed. It has been observed that at 1000 K oxidation does not lead to a significant increase in the disorder of the graphene layer. The number of defective graphene layers increases only slightly although the surface becomes disordered on a macroscopic scale. In studies that we have made we have injected 100 μm graphite particles into a flame and examined the surfaces of the product by high resolution scanning electron microscopy (SEM). We come to the same conclusion, that there is reaction by edge recession. This enables a repeat of the sequence leaving terraces in the graphite crystal. A similar course of events will take place at higher temperatures.

Conclusions

The following conclusions can be deduced:

1. Graphite can be modelled using a graphene molecule $C_{25}H_9$. An *ab initio* molecular orbital computational technique has been used and it has been proved to be a satisfactory way of examining complex molecular structure.

2. A reaction scheme has been proposed that replicates the general features that describe the reactivity of graphite and in particular the turnover model.

3. It has been possible to model active sites, the formation of carbon monoxide and the formation of new active sites as the reaction proceeds. One type of active site involving the dangling carbon atom has been designated a super-reactive site.

4. The mechanism proposed gives a plausible background to the combustion of coal char and industrial graphite and the work indicates that the single activation energy assigned to the oxidation is insufficient and should have a distribution of activation energies.

Acknowledgements

We wish to acknowledge support for this project from the DTI UK Cleaner Coal Programme. We would like to acknowledge assistance from Mr D Hainsworth and helpful discussions with Professors A. Hayhurst and B. Rand.

References

1. R. H. Hurt, *Proc. Combust. Inst.*, 1998, **27**, 2887.
2. J. Lahaye, J. Dentzer, P. Soulard and P. Ehrburger, in *Fundamental Issues in Control of Carbon Gasification Reactivity*, ed. J. Lahaye and P. Ehrburger, Kluwer, Dordrecht, 1991, p. 143.
3. R. H. Essenhigh, in *Chemistry of Coal Utilisation*, ed. M. A. Elliott, Wiley, New York, 1981, 2nd Suppl. vol., ch. 19, p. 1153.
4. I. W. Smith, *Proc. Combust. Inst.*, 1982, **19**, 1045.
5. G. Hargrave, M Pourkashanian and A.Williams, *Proc. Combust. Inst.*, 1986, **21**, 221.
6. A. Williams, M. Pourkashanian and J. M. J. Jones, *Proc. Combust. Inst.*, 2001, **28**, 2141.
7. I. M. Bews, A. N. Hayhurst, S. M. Richardson and S. G. Taylor, *Combust. Flame*, 2001, **124**, 231.

8   E. M. Suuberg, in *Fundamental Issues in Control of Carbon Gasification*, ed. J. Lahaye and P. Ehrburger, Kluwer, Dordrecht, 1991, 269.
9   S. Kulasekaran, T. M. Linjewile, P. Agarwal and M. J. Biggs, *Fuel*, 2000, **77**, 1549.
10 M. Takahashi, M. Kotaka and H. J. Sekimoto, *J. Nucl. Sci.*, 1994, **31**, 1275.
11 T. C. Brown, A. E. Lear and B. S. Haynes, *Proc. Combust. Inst.*, 1992, **24**, 1199.
12 M. C. Ma and B. S. Haynes, *Proc. Combust. Inst.*, 1996, **26**, 3119.
13 P. Salatino, F. Zimbardi and M. Paulicelli, *Proc. Combust. Inst.*, 1994, **25**, 527.
14 G. Tremblay, F. J. Vastola and P. L. Walker, *Carbon*, 1978, **16**, 35.
15 J. M. Thomas, in *Chemistry and Physics of Carbon*, ed. P. Walker, Marcel Dekker, New York, 1965, vol. 1, p. 121.
16 G. Dixon-Lewis, D. Bradley and S. E. Habik, *Combust. Flame*, 1991, **86**, 12.
17 N. Chen and R. T. Yang, *Carbon*, 1998, **36**, 1061.
18 N. Chen and R. T. Yang, *J. Phys. Chem. A*, 1998, **102**, 6348.
19 L. R. Radovic and K. A. Skakova, *American Chemical Society, Div. Fuel Preprints, 219th ACS National Meeting, San Francisco*, American Chemical Society, Washington, DC, 2000, vol. 45, p. 225.
20 T. Kyotani and A. Tomita, *American Chemical Society, Div. Fuel Preprints 219th ACS National Meeting, San Francisco*, American Chemical Society, Washignton, DC, 2000, vol. 45, p. 221.
21 GAUSSIAN 98, Revision A.8, M. J. Frisch, G. W. Trucks, H. B. Schlegel, G. E. Scuseria, M. A. Robb, J. R. Cheeseman, V. G. Zakrzewski, J. A. Montgomery, Jr., R. E. Stratmann, J. C. Burant, S. Dapprich, J. M. Millam, A. D. Daniels, K. N. Kudin, M. C. Strain, O. Farkas, J. Tomasi, V. Barone, M. Cossi, R. Cammi, B. Mennucci, C. Pomelli, C. Adamo, S. Clifford, J. Ochterski, G. A. Petersson, P. Y. Ayala, Q. Cui, K. Morokuma, D. K. Malick, A. D. Rabuck, K. Raghavachari, J. B. Foresman, J. Cioslowski, J. V. Ortiz, A. G. Baboul, B. B. Stefanov, G. Liu, A. Liashenko, P. Piskorz, I. Komaromi, R. Gomperts, R. L. Martin, D. J. Fox, T. Keith, M. A. Al-Laham, C. Y. Peng, A. Nanayakkara, M. Challacombe, P. M. W. Gill, B. Johnson, W. Chen, M. W. Wong, J. L. Andres, C. Gonzalez, M. Head-Gordon, E. S. Replogle and J. A. Pople, Gaussian, Inc., Pittsburgh PA, 1998.
22 G. Miessen, F. Behrent, O. Deutschmann and J. Warnatz, Joint British, French and German Combustion Institute Symposium, 'Numerical Studies of the Heterogeneous Combustion of Char Particles using Detailed Chemistry', Lyons, France, September, 1999.
23 K. H. van Heek and H. J. Muhlen, *Fuel*, 1985, **64**, 1405.
24 K. A. Davis, R. H. Hurt, N. Y. C. Yang and T. J. Headley, *Combust. Flame*, 1995, **100**, 31.
25 R. H. Hurt, J. K. Sun and M. Lunden, *Combust. Flame*, 1998, **113**, 181.
26 N. V. Russel, J. R. Gibbins and J. Williamson, *Fuel*, 1999, **78**, 803.
27 F. Atamny, J. Blocker, B. Henschke, R. Schogl, Th. Schedel-Niedrig, M. Keil and A. M. Bradshaw, *J. Phys. Chem.*, 1992, **96**, 4522.

Small-angle X-ray studies of soot inception and growth

Jan P. Hessler,[a] Soenke Seifert,[a] Randall E. Winans[a] and Thomas H. Fletcher[b]

[a] *Chemistry Division, Argonne National Laboratory, 9700 South Cass Avenue, Argonne, Illinois, 60439-4831 USA*
[b] *Chemical Engineering Department, Brigham Young University, Provo, Utah, 84602 USA*

Received 27th March 2001
First published as an Advance Article on the web 1st November 2001

The high spectral intensity of X-rays produced by the undulator at the Basic Energy Sciences Synchrotron Radiation Center of Argonne's Advanced Photon Source has allowed us to perform small-angle X-ray scattering (SAXS) studies of the initial distribution of soot particles formed by various fuels. SAXS provides an *in situ* probe of the morphology of soot in the region between 1 and 100 nm and complements the *ex situ* technique of electron microscopy. The basic aspects of SAXS and its potential are illustrated with measurement on a laminar flame of acetylene in air. The more complex fuel toluene has been studied in a flat-flame burner that supports a CH_4/H_2/air or CO/H_2/air diffusion flame stabilized by N_2 co-flow. This burner produces a nearly constant temperature region above the flame where the pyrolysis and combustion of the heavier fuels occurs. Kinetic information is obtained by performing measurements of the scattered intensity profile as a function of the height above the burner. These profiles have been reduced to give the mean radius and dispersion of a distribution of spherical particles. Mean radii between 0.8 and 18 nm have been observed. The smallest of these is a factor of ten smaller than previously detected with Lorentz–Mie scattering. Near 1550 K, the soot distribution found in toluene shows a distinct step behavior that is consistent with model calculations.

1. Introduction

The incomplete combustion of hydrocarbon fuels leads to the production of soot. A comprehensive theory or model that is capable of predicting the inception and growth of soot over a wide range of chemical and physical conditions is just beginning to emerge. Such a model is needed to help reduce the health hazards associated with soot production, improve the radiative transfer of energy in industrial applications, and devise efficient production processes that use soot in various applications. Research in soot inception and growth has been documented in various workshops[1] and symposia.[2]

Recent transmission electron microscopy studies[3] suggest that the initial stages of soot formation involve the growth of large aromatic hydrocarbons which produce nuclei (~0.5–2 nm). These nuclei grow to produce elementary particles (~4 nm) that have a well-defined size and are relatively transparent. The elementary particles produce sub-primary particles (6–9 nm) that then cluster together to form small chains. These chains merge to produce the well-known primary particles (20–50 nm) that coagulate to form the chain-like and fractal soot aggregates. Therefore, soot grows both by chemical reactions with species in the gas phase (a gas–solid interaction) and by the clustering and agglomeration of particles (a solid–solid interaction).

DOI: 10.1039/b102822g

Both *ex situ* and *in situ* techniques have been used to study soot formation. Optical techniques such as absorption,[4] induced luminescence[5] and wide-angle elastic scattering[6] are examples of *in situ* techniques. More precisely,[7] elastic Lorenz–Mie scattering provides information on the morphology of particles that range in size from 100 to 5000 nm while lower-resolution quasi-elastic scattering, which cannot provide information on morphology, can be used for particles from 1 nm to 10 µm. Laser-induced incandescence has been used to detect mean particle radii between 1.8 and 10 nm.[8] Cavity ring-down spectrometry has been used to detect soot precursors such as the phenyl radical.[9] The formation of larger precursor species has been studied in low-pressure flames by photoionization mass spectrometry[10] and gas-chromatography/mass spectrometry.[11] From these results we note that it is difficult to use *in situ* techniques and extract information on the structure of particles in the 1 to 100 nm. Electron microscopy has bridged this gap, but it is an *ex situ* technique and kinetic information is difficult to obtain. SAXS is ideally suited to provide *in situ* measurements of the morphology of soot formation in this important region. In 1986 England presented the first *in situ* SAXS studies of soot morphology.[12] Although he used a synchrotron radiation source, exposure times of minutes were needed to obtain a single scattering profile. We present here, for the first time, SAXS measurements of soot morphology that were obtained from a third-generation synchrotron source. It uses an undulator to enhance the intensity of the X-rays and, thereby, allows us to obtain scattering profiles in a few seconds. This improvement provides the opportunity to perform kinetic measurements of incipient soot formation. Such measurements will contribute to the development of a comprehensive model of soot inception and growth.

Below we give a brief overview, primarily to define notation and vocabulary, of the basic concepts of small-angle X-ray scattering. We then describe our experimental apparatus and present results for acetylene and toluene. These demonstrate that our *in situ* measurements agree with previous *ex situ* measurements and provide new information that may be associated with elementary particles (∼4 nm).

2. General discussion of small-angle X-ray scattering

Small-angle X-ray scattering is over sixty years old and dates from the classical works of Guinier, which were published in 1938 and summarized in the book by Guinier and Fournet.[13] Two other texts have been written[14,15] and the technique has been discussed at NATO workshops.[16]

2.1. Scattering intensity and contrast

All scattering experiments may be described in terms of the differential scattering cross section, $d\sigma_S/d\Omega$, which is given by

$$\frac{d\sigma_S}{d\Omega} = \frac{S(\theta, \phi)}{I_0} = I(q) \qquad (2.1)$$

where $S(\theta, \phi)$ is the average energy scattered per second per unit solid angle at the angle θ with respect to the incident beam and the azimuthal angle ϕ with respect to a vector property of the incident radiation. I_0 is the intensity (energy time^{-1} length^{-2}) of the incident radiation. Generally, in small-angle X-ray scattering studies the differential cross section is cylindrically symmetric about the direction of the incident radiation and the azimuthal angle ϕ is dropped. Also, the differential scattering cross section is referred to as the scattering intensity and is denoted by $I(q)$ where q is the scattering vector, $q = |\mathbf{q}| = (4\pi/\lambda)\sin(\theta/2)$. Therefore, the dimension of $I(q)$ is length squared even though it is referred to as an intensity. We also note that several different symbols are used in the literature to denote the scattering vector; these include $\mathbf{q}$, $\mathbf{h}$, $\mathbf{Q}$ and $\mathbf{s}$. The scattering intensity is given by

$$I(\mathbf{q}) = V \iiint P(\mathbf{r}) \exp(-i\mathbf{q} \cdot \mathbf{r}) d\mathbf{r} \qquad (2.2)$$

where

$$P(\mathbf{r}) = \frac{1}{V} \iiint \rho(\mathbf{r}_0)\rho(\mathbf{r}_0 + \mathbf{r}) d\mathbf{r}_0 = \langle \rho(\mathbf{r}_0)\rho(\mathbf{r}_0 + \mathbf{r}) \rangle. \qquad (2.3)$$

The function $P(r)$ is the autocorrelation function (Patterson's function of crystallography) and expresses the spatial dependence of the self-correlation of the scattering density, $\rho(r)$; averaged over the total irradiated volume V. Thus far, we have assumed that the scattering particles are in a vacuum. Unfortunately, soot is produced in a complex environment and only the difference in electron density between the soot particles, ρ_2, and the medium that contains the soot particles, ρ_1, can be detected. Therefore, the scattered intensity will be proportional to $(\rho_2 - \rho_1)^2 = \Delta\rho^2$. The quantity $\Delta\rho$ is referred to as the "contrast" of the particles with respect to the medium. SAXS studies in combustion environments present new problems because the soot particles of interest interact with the surrounding medium and change with both position and time.

In 1952 Porod showed that the integral

$$Q \equiv \frac{1}{V} \int_0^\infty q^2 I(q) \mathrm{d}q = 2\pi^2 \phi_1 \phi_2 \Delta\rho^2 \tag{2.4}$$

is independent of the geometry of the particles if both the scattering densities and volume fractions (ϕ_1 and ϕ_2) of the particles and medium are constant. Porod also showed that the intensity at large scattering angles ($q \gg 1/D$), where D is some characteristic length of the particles, reflects the structure of the interface. For an infinitely sharp interface the surface area per unit volume is given by

$$\frac{S}{V} = \frac{1}{2\pi(\Delta\rho)^2 V} \lim_{q \to \infty} q^4 I(q) = \frac{1}{2\pi(\Delta\rho)^2 V} K \tag{2.5}$$

where K is referred to as the Porod limit. Therefore, one of the first steps in the analysis of SAXS data is to generate a log–log plot of $q^4 I(q)$ vs. q, a "Porod plot". If such a plot produces a horizontal line a Porod region may be identified. Another important limiting case is the small-angle or low-q limit. In 1938 Guinier showed that for $qR_G \ll 1$

$$I(q) = G \exp\left(-\frac{q^2 R_G^2}{3}\right) \tag{2.6}$$

where R_G is the radius of gyration of the particle and the exponential prefactor $G = I_e N_p n^2$ where the number of particles in the scattering volume is N_p, the number of electrons in a particle is n, and the scattering factor is I_e which has a dimension of length squared. For a polydispersed system of particles R_G is the radius of gyration of the largest particle. Therefore, a plot of $\ln[I(q)]$ vs. q^2, which is referred to as a "Guinier plot", should be linear if Guinier's law applies.

2.2. Structural information

Information about the structure of the scattering particles is obtained from the behavior of the scattering intensity in the central region between the Guinier and Porod regions. For example, the scattering intensity for a homogeneous sphere of radius R is

$$I(q) = I_0 \left[3 \frac{\sin(qR) - qR \cos(qR)}{(qR)^3} \right]^2. \tag{2.7}$$

We note that the above expression has a series of zeros at $qR = 4.493, 7.725$, etc. A convenient expression for merging the Guinier and Porod regions has recently been proposed by Beaucage[17] and Beaucage and Schaefer.[18] For the simplest case they write

$$I(q) \approx G \exp\left(-\frac{q^2 R_G^2}{3}\right) + B\left[\frac{\{\mathrm{erf}(qR_G/\sqrt{6})\}^3}{q}\right]^P \tag{2.8}$$

where B is a prefactor specific to the region in which the exponent P falls and the error function is (erf). In the Porod region $P = 4$ and $B = 2\pi N_p \rho_e^2 S_p I_e$ where $\rho_e = n/V_p$ (the electron density of the particle) and S_p is its surface area. The prefactor G has the same form as above, $G = I_e N_p n^2$. The above equation may be extended to describe a system of particles that contains subparticles with a radius of gyration of R_s which agglomerate to form a large-scale structure of size R_g.

$$I(q) \approx G \exp\left(-\frac{q^2 R_G^2}{3}\right) + B \exp\left(-\frac{q^2 R_{sub}^2}{3}\right)\left[\frac{\{\text{erf}(qR_g/\sqrt{6})\}^3}{q}\right]^P$$

$$+ G_S \exp\left(-\frac{q^2 R_S^2}{3}\right) + B_S\left[\frac{\{\text{erf}(qR_S/\sqrt{6})\}^3}{q}\right]^{P_S}. \tag{2.9}$$

The third and fourth terms describe scattering from the sub-particles. The second term describes the mass-fractal, surface-fractal, or diffuse interfacial power-law regimes where the sub-particles agglomerate to form the large-scale structure, which is described by the first term. Generally, in the high-q cut-off for the intermediate power law region R_{sub} is identical to R_S. Of course, the above equation can be generalized to describe an arbitrary number of interrelated structural features.

Sorensen et al. have provided a more fundamental understanding of the scattering profile for aggregated assemblies with multiple length-scales.[19] They present results where the scattering profile has at least four distinct regions where $I(q) \propto q^{-2}$, q^{-4} and q^{-D_m} where D_m describes regions in recriprocal space where mass fractal aggregates dominate. Therefore, a log–log plot of $I(q)$ vs. q may contain several linear regions where the slope in a region will give information about a dominant structure. We have found it convenient to plot $d\{\ln[I(q)]\}/d\{\ln q\}$ as a function of q as an aid to identification of these regions. In many cases, significant regions in recriprocal space cannot be described by a single power law dependence on the scattering length. Often, these regions indicate systems that are polydispersed. For polydispersed systems the size distribution can be determined from scattering profiles if we assume the particles are a certain shape. In particular, we note the useful paper by Sheu that contains expressions for polydispersed systems of various shapes.[20]

3. Experimental details

Our small-angle X-ray scattering apparatus has been describe previously.[21] Briefly, it consists of a high-brilliance undulator, a double-crystal monochromator, and various components needed to transport, focus, define, and monitor the X-ray beam. Typically, 10^{12} X-rays per second are incident on the sample, which is locate 58 m from the undulator. A vacuum scattering chamber is used to transport the scattered X-rays to a mosaic nine-segment, two-dimensional CCD detector with an active area of 15 cm × 15 cm.[22] This detector has 3072 × 3072 pixels that are binned and digitized to produce an output file that contains 1536 × 1536 numbers with a precision of 1 part in 2^{16}. This image is then corrected for the dark background, pixel efficiency, and spatial inhomogeneities.[23] The location of the center of the X-ray beam on the detector is determined and then an azimuthal average is performed to produce $I(q)$ in increments of $\Delta q/q \approx 0.02$.

Fuels are injected into the center of a 25.4 × 25.4 mm laminar-flow flat-flame burner that is surrounded with a 6.3 mm wide co-flow region.[24] To obtain scattering profiles from liquid fuels such as toluene the burner must provide a bath gas that is warm enough to sustain combustion of these relatively low vapor liquids. A constant temperature region is maintained above the burner by using different mixtures of H_2 and CH_2 or H_2 and CO as primary fuels. For the results presented here the flow rates for H_2, CH_4, dry air, and N_2 were 11.8, 18.8, 148 and 10.5 cm^3 s^{-1}, respectively. This gives a combined flow velocity of 294 mm s^{-1} under standard conditions. A co-flow of N_2 at a velocity of 304 mm s^{-1} is used to stabilize the system. Under these conditions, the blue fluorescence from CH radicals near 408 nm is approximately 1 mm above the top of the burner and much less than a mm in thickness. Toluene or other liquid fuels are injected into the high temperature region above this flame through a 10 μm diameter quartz capillary at a rate of 0.15 ml min^{-1} or 2.5 μl s^{-1}. This flow rate is critical. When it is too high, soot will begin to grow on the tip of the capillary and when it is too low the system will oscillate. The location of the tip of this capillary with respect to the blue flame front, which is indicated by the CH florescence, is also critical. If this tip is too low the fuel does not have sufficient time to vaporize and small droplets may be seen running down the side of the capillary. If the tip is too high the structure of the flame just above the capillary become more complex. In all of the measurements presented here this height was adjusted so that the plume of the flame was regular and smooth over as large a distance above the tip as possible. Gaseous fuels, such as acetylene, are injected through a 50 μm i.d. tube along the same center line, however, only air and nitrogen are passed though the central region of the burner. The flow rates of these gases are adjusted to give the desired amount of soot. This burner is mounted on a base whose position is controlled by two steppermotors. One of these controls the vertical height of the burner with respect to the X-ray beam and the other its horizontal position.

One of the special problems with studying combustion systems is the need to obtain scattering information not only from the species of interest, but also from the background. In all of the experimental results presented here five different horizontal positions were sampled for each vertical height. The positions of the center of the burner with respect to the X-ray beam are -1, 0, $+2$, $+4$ and $+6$ mm. The measurements at $+6$ mm are sufficiently far from the soot forming plume that they provide scattering from the background only. An example of the scattering profiles obtained from an experiment with acetylene is shown in Fig. 1. The solid dots represent the profile from the center of the flame and the open circles the profile 6 mm from the flame center. The scattering intensity is obtained by subtracting the profile recorded at 6 mm from the other profiles obtained closer to the center line of the flame. We have compared the reproducibility of this scheme to other more traditional schemes and found that it provides more reproducible results. Therefore, all of the scattering profiles reported here are actually the difference between two profiles; the profile of interest minus the profile 6 mm from the flame.

Another problem that is inherent in our measurements stems from the fact that the density and composition of the soot particles in the flame depend upon both the distance along the direction of flow and the radial distance from this axis of symmetry. In mathematical terms, we are measuring the integral of a two-dimensional scattering function along the line of the X-ray beam. We can reconstruct this two-dimensional scattering function by recording several measurements at various positions and then performing a transformation on all of our results. This is analogous to, but much simpler than, medical tomography. This transformation is straightforward because of the cylindrical symmetry of the flame. However, we have not yet implemented the transform. Therefore, we must keep in mind that the results presented below are integrals of a two-dimensional scattering function that depends critically on the distance from the center of the flame.

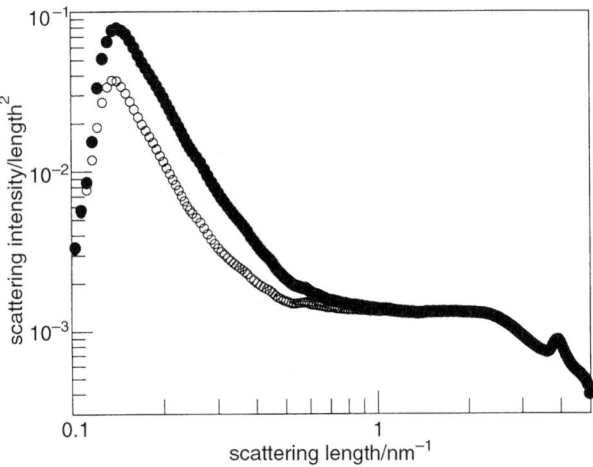

Fig. 1 Example of uncorrected scatterings profiles from an acetylene flame. The solid dots represent the scattering intensity obtained at the center of the flame and the open circles the intensity 6 mm from the center. Profiles at a given height above the burner are corrected for background scattering by subtracting the profile at 6 mm from all profiles measured closer to the center of the flame.

4. Results and discussion

Under fuel-rich conditions acetylene is known to produce copious amounts of soot and has been studied by many authors. The scattering profiles for several different vertical distances along the center line of the flame are shown in Fig. 2. This figure was generated by combining profiles obtained on different days for different regions of reciprocal space. The data at the lower scattering length were obtained with an X-ray energy of 9.997 keV and a camera length of 3043 mm while the higher scattering length data used 8.012 keV and a camera length of 843 mm. Therefore, the different values of the scattering length are due predominantly to changes in the distance from the flame to the detector not changes in the wavelength of the incident X-rays. The data taken on different days were combined by simply multiplying the profiles in the lower region by a constant so that the profiles are equal at 0.4 nm^{-1}. Clearly, the profiles at the lower scattering lengths, $q \leqslant 0.7$ nm^{-1}, behave similarly while at higher scattering lengths they differ significantly. The structural information contained in these profiles becomes evident when we examine a "Porod plot", Fig. 3. The region $0.3 \leqslant q/\text{nm}^{-1} \leqslant 0.7$ clearly follows Porod's law. This behavior is consistent with the observations of Sorensen et al.[19] who observe similar behavior in the region $0.1 \leqslant q/\text{nm}^{-1} \leqslant 1$. They show that this region contains approximately spherical monomers that are generally independent of the source of the soot. For $q \geqslant 0.7$ nm^{-1} the three upper curves ($t = 95$, 155 and 195 ms) begin to rise with different slopes and may be characterized by $I(q) \propto q^{-D_m}$ over a small region. These results may indicate the structure of species that are evolving into the larger particles that are observed in the Porod region just below 0.7 nm^{-1}. We note that they show no indication of approaching the q^{-4} behavior that must occur at higher scattering lengths. The most

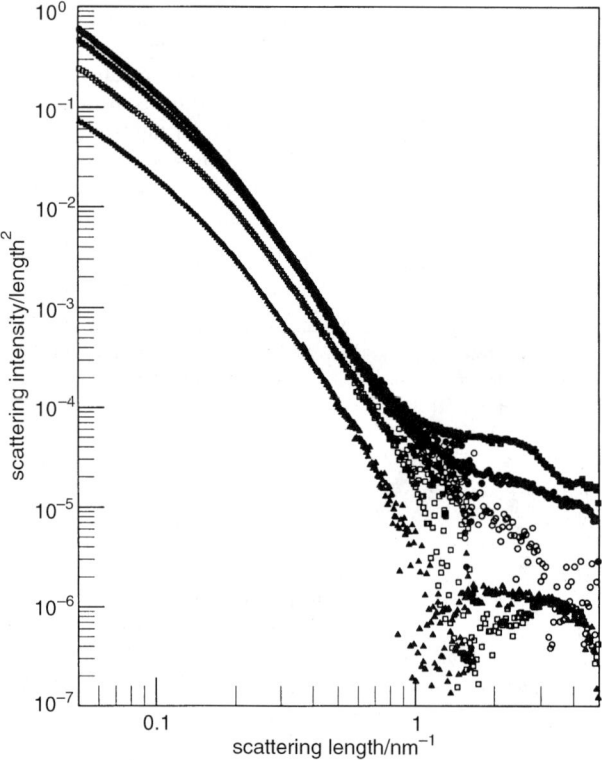

Fig. 2 Combined scattering profiles from an acetylene flame as a function of distance above the tip of the acetylene inlet tube. The solid triangle were taken 3.5 mm above the tip of the tube used to introduce the acetylene, the open squares 5.5 mm, the solid squares 9.5 mm, the open circles 15.5 mm, and the solid circles 19.5 mm. The flow rate of the oxygen was 3.9 SLPM giving a velocity of 100 mm s^{-1} at the surface of the burner, hence these positions correspond to reaction times of 35, 55, 95, 155 and 195 ms, respectively.

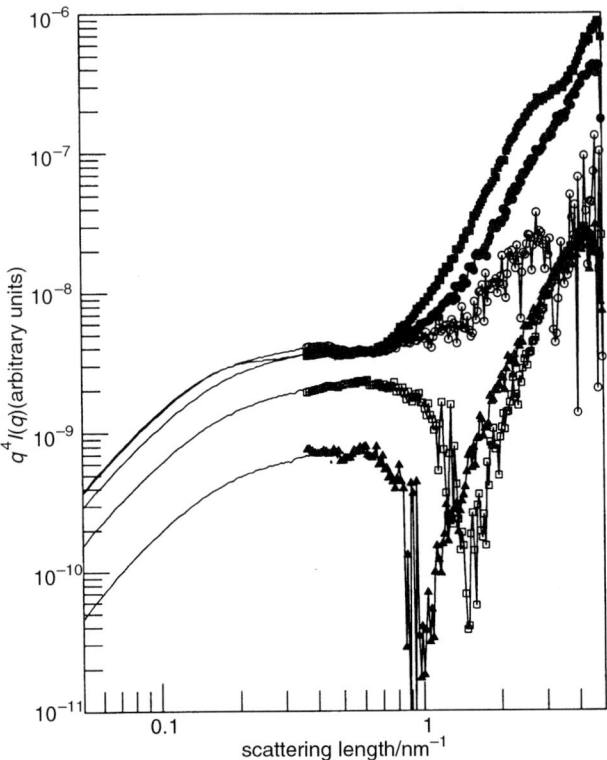

Fig. 3 Porod plots, $q^4 I(q)$ vs. q, of the scattering profiles shown in Fig. 2. The symbols have the same meaning as in Fig. 2. For clarity the symbols have not been plotted for the profiles recorded at the lower scattering lengths.

striking feature of the other two profiles is the fact that the lowest profile ($t = 35$ ms) appears to approach zero near 1 nm^{-1} while the second profile ($t = 55$ ms) approaches zero near 1.8 nm^{-1}. A third ($t = 95$ ms) may have a slight dip at 3 nm^{-1}. If we assume these dips are due to nearly monodispersed systems of particles and use the relationship $qR = 4.493$ we calculate that $R \approx 4.5$, 2.5 and 1.5 nm, respectively. To our knowledge, this behavior has not been observed before in SAXS studies of soot. However, almost all previous studies have been performed on soot that has been processed in some way and must be classified as *ex situ* measurements. This work, along with the pioneering work of England,[12] are the only *in situ* SAXS studies of soot formation. We note that these particles may be the "elementary particles" that de Stasio has observed in transmission electron micrographs.[3] He observed very regularly sized transparent shells (~4 nm) made of amorphous carbon and an inner graphitic nuclei (0.5 to 2 nm). Unlike the behavior at larger scattering lengths, where the scattering profiles differ significantly, their shapes in the region $q \leqslant 0.4$ nm^{-1} are strikingly similar. These similarities becomes even more apparent when we plot $d\{\ln[I(q)]\}/d\{\ln[q]\}$ as a function of the scattering length for the region $q \leqslant 0.4$ nm^{-1}, Fig. 4. Here, we see that the slopes change monotonically and smoothly, as we move to lower scattering lengths and they are independent of the height above the burner. Furthermore, all of the curves are approaching a value of -1.8, which is expected for fractal aggregates.

We now present profiles as a function of the distance from the center of the flame at a fixed height above the burner. Fig. 5 shows scattering profiles at high scattering lengths, where the behavior is interesting, taken at a height of 15.5 mm above the tip of the acetylene source ($t = 155$ ms) for various distances from the center of the burner. Curve (A) was taken at the center of the flame, and is identical to the data shown in Fig. 2, curves (B), (C) and (D) were taken 1, 2 and 4 mm from the center, respectively. We note a systematic decrease in the scattering intensity at

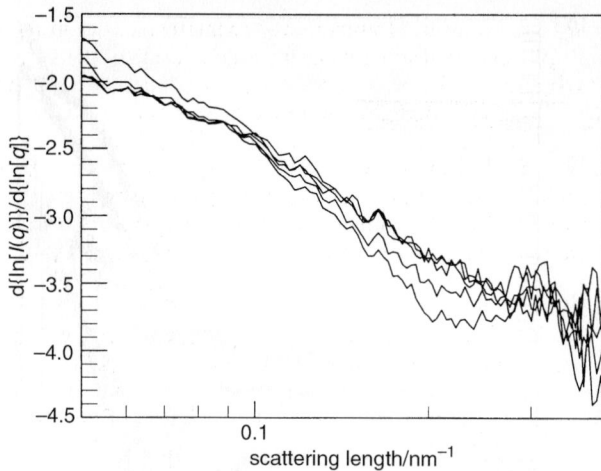

Fig. 4 Test for a power law region $d\{\ln[I(q)]\}/d\{\ln[q]\}$ as a function of scattering length for the five curves shown in Fig. 2 and 3. These results were obtained by numerical calculations in a spread sheet.

lower scatter lengths as we move away from the center of the flame and a systematic increase at higher values. Furthermore, a dip in curve (B) occurs near 0.95 nm^{-1} while for curve (C) it occurs at 0.6 nm^{-1}. If these dips correspond to nearly monodispersed spheres, their radii are 4.7 and 7.5 nm, respectively. This may indicate that the soot particles are increasing in size as they move from

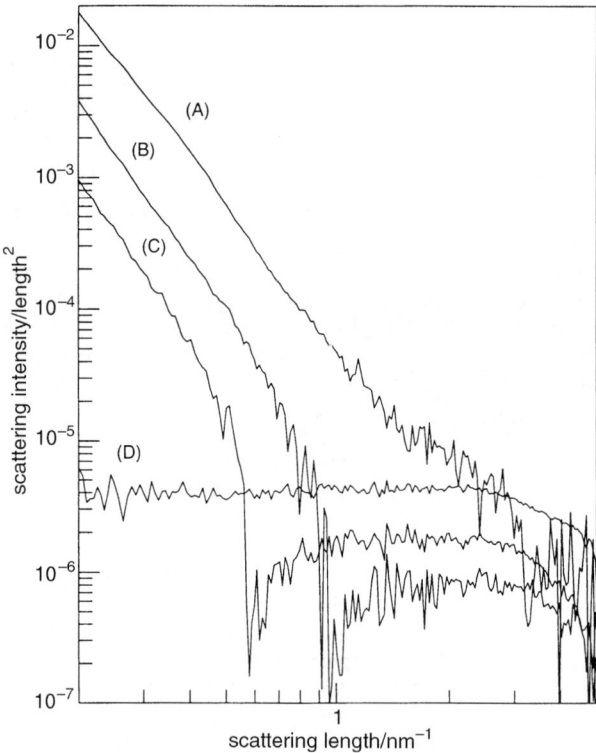

Fig. 5 Scattering profiles from an acetylene flame for different distances from the center of the flame and at a fixed height of 15.5 mm above the tip of the inlet tube. Curve A was taken on the center line, curves B, C and D were taken 1, 2 and 4 mm off center, respectively.

the center of the flame to the outside. However, the systematic increase in the scattering intensity at higher scattering lengths may indicate growth from the outside of the flame. Of course, no one would be surprised to find that different growth mechanisms dominate at different locations in the flame. The Porod plots calculated from the scattering profiles taken at different heights above the burner at 2 mm from the center of the flame are shown in Fig. 6. Whereas a Porod region near 0.5 nm^{-1} could be identified at all heights (reaction times) for profiles at the center of the flame (see Fig. 3) we note that 2 mm from the center line, Fig. 5, a Porod region near 0.5 nm^{-1} is not evident until 95 ms. The absence of a Porod region near 0.5 nm^{-1} and a nearly constant scattering profile become even more evident as we move to 4 mm from the center of the flame, Fig. 7. First we note that the scattering intensities are much weaker at low scattering lengths and relatively flat up to a scattering length of 2 nm^{-1}. The profiles at 55 and 95 ms both show Porod regions above 3.5 and 2.5 nm^{-1}, respectively. The profile at 155 ms is not steep enough at higher scattering lengths to support a Porod region. This indicates a different dominant growth mechanism at this location. Clearly, scattering information at higher scattering lengths and the transformation of these data into a radial function are needed before a consistent description of the inception and growth of small soot particles from acetylene can be given.

Toluene is also known as a fuel that efficiently produces soot and is one of the first fuels we studied. In fact, we have studied several other liquid fuels: methylcyclohexane, hexane and biphenyl in hexane. Their general behaviors are quite similar and indicate a polydispersed system for scattering lengths below 0.6 nm^{-1}. Fig. 8 shows a composite of four different scattering profiles from toluene near 1550 K taken at two different heights above the burner. The results were combined by simply forcing the profiles to agree at 0.5 nm^{-1}. Porod plots of these profiles are

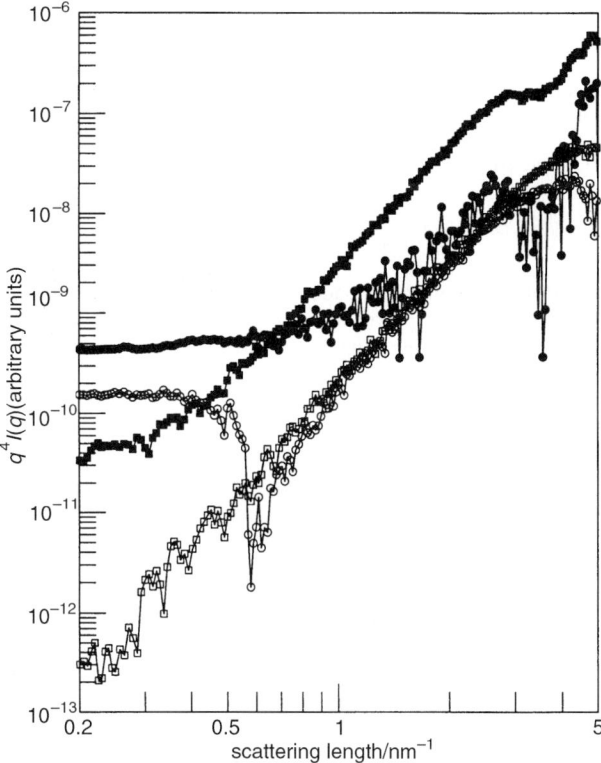

Fig. 6 Porod plots, $q^4 I(q)$ vs. q, for the scattering profiles at different heights above the inlet of the acetylene flame taken 2 mm from its center. The symbols that denote distances from the inlet tube are the same as in Fig. 2. The profile at the shortest distance, 3.5 mm, has been omitted because its intensity and signal-to-noise ratio are low.

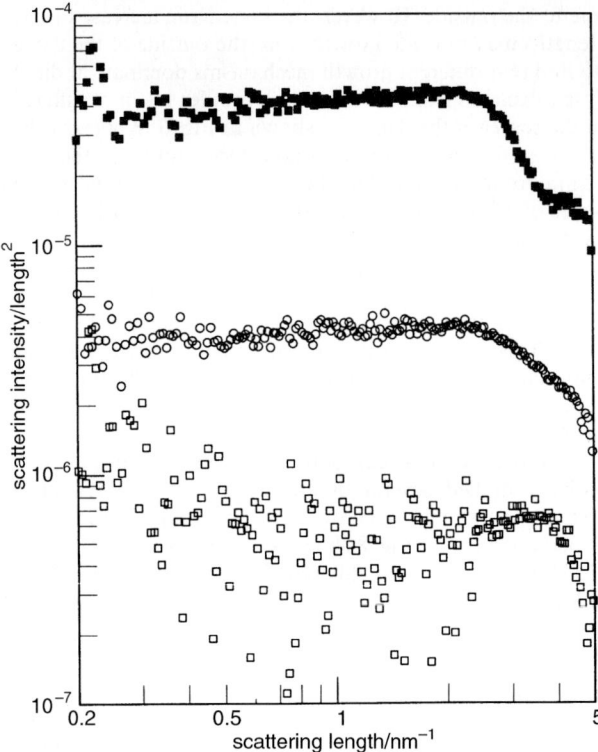

Fig. 7 Scattering profiles from an acetylene flame at various heights above the inlet tube at a fixed distance of 4 mm from the center of the flame. The symbols that denote distances from the inlet tube are the same as in Fig. 2. Profiles taken at 3.5 and 19.5 mm are not shown because their intensities and signal-to-noise ratios are low. The sharp decrease in scattering intensity for the curve at 9.5 mm, solid squares, is in a Porod region from $3 \lesssim q/\text{nm}^{-1} \lesssim 3.8$.

shown in Fig. 9. Again, Porod regions can be identified around 0.5 nm^{-1} and a sharp dip is found at 0.65 nm^{-1}, which may indicate particles with an R_g of 6.4 nm. The scattering intensity approaches a constant as the scattering length increases and may indicate scattering from predominantly smaller particles. The scattering intensity at lower scattering lengths, $q < 0.4$ nm^{-1}, resembles the behavior shown in Fig. 4, i.e. a region of a single power law in q could not be identified. Therefore, the scattering profiles for $0.075 \leqslant q/\text{nm}^{-1} \leqslant 0.4$ were reduced with the expression for a Schultz distribution of polydispersed spheres given by Sheu.[20] The Schultz distribution function is an approximation introduced by Zimm to describe light scattering from polymer solutions.[25] This distribution function is

$$f(r) = \frac{y^{z+1}}{z!} r^z \exp(-yr) \qquad (4.1)$$

where

$$z = \frac{\langle r \rangle^2}{\langle \Delta r^2 \rangle} - 1 \qquad (4.2)$$

and

$$y = \frac{\langle r \rangle}{\langle \Delta r \rangle^2} = \frac{z+1}{\langle r \rangle} \qquad (4.3)$$

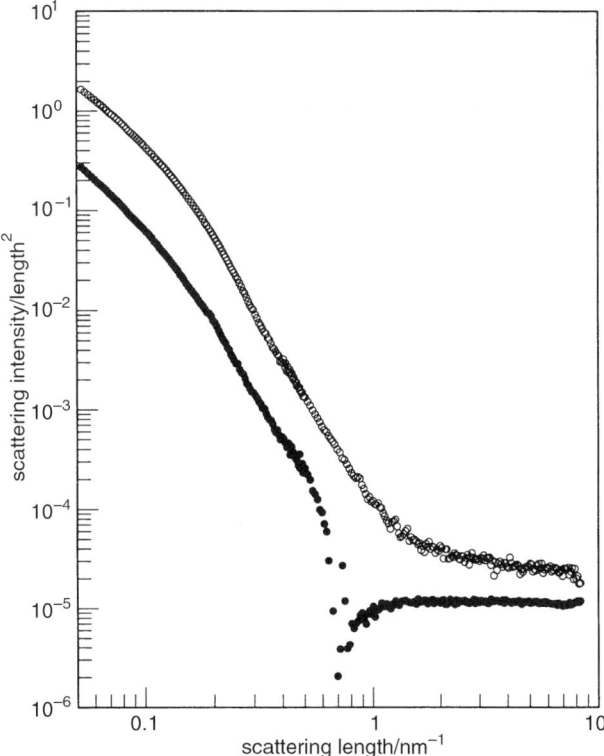

Fig. 8 Combined scattering profiles from toluene at two different heights above the burner.

Therefore, for spherical particles the distribution is characterized by a mean radius, $\langle r \rangle$, and its dispersion, $\langle \Delta r^2 \rangle$. At the time these analyses were performed, we simply assumed a constant scattering intensity at higher scattering lengths. The results of this reduction are summarized in Fig. 10. The solid line represents our simulation of soot formation that uses the mechanism and model developed by Frenklach and Wang.[26] To perform this simulation we removed all of the fluid dynamics from the problem by assuming a constant temperature and pressure environment with no diffusion. The early part of toluene dissociation was modeled by the reaction mechanism for the pyrolysis and oxidation of toluene given by Alexiou and Williams.[27] The remainder of the reaction mechanism and the application code used to used to calculate the moments of the soot distribution were taken directly from the web site at Berkeley.[28] The only point we would like to make with this comparison is that SAXS provides information in the early regions where the model is most sensitive to slight changes of parameters. Therefore, SAXS can provide data that is needed to validate models.

5 Acknowledgements

The work presented here would not have been possible without the cooperation of the staff of the Basic Energy Sciences Synchrotron Radiation Center at Argonne's Advanced Photons Source. Mark A. Beno, Guy Jennings and Jennifer A. Linton were particularly helpful. Stimulating conversations with Albert F. Wagner and a critical reading of the manuscript are gratefully acknowledged. This work was supported by the U.S. Department of Energy, Office of Basic Energy Sciences, Division of Chemical Sciences, under contract W-31-109-ENG-38.

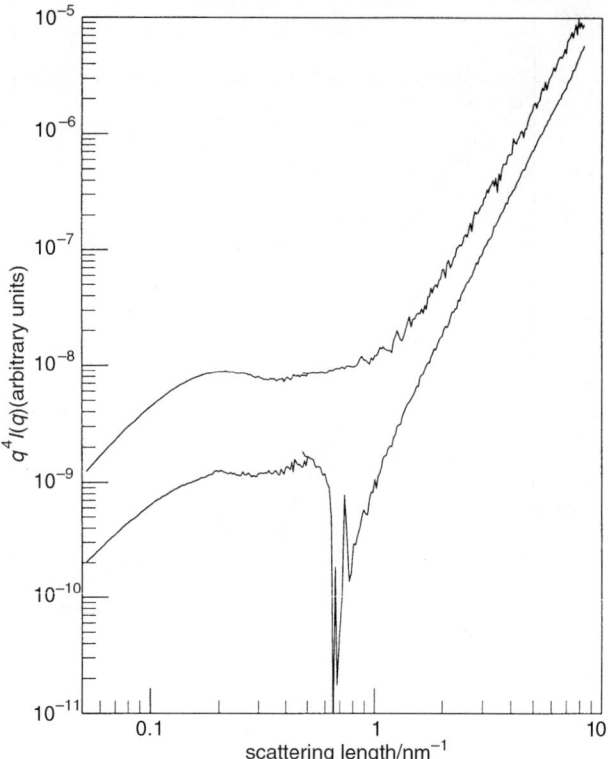

Fig. 9 Porod plots of the scattering intensities shown in Fig. 8.

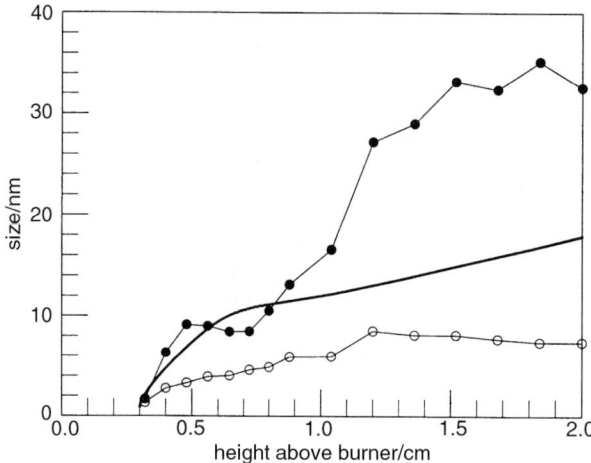

Fig. 10 Results of a reduction of the scattering intensities from a toluene flame. The solid circles represent the mean diameter of a Schultz distribution and the open circles the square root of the second moment or dispersion of the distribution. The thick solid line represents a greatly simplified simulation of the formation of soot by toluene. See the text for details.

References

1. *Soot Formation in Combustion, Mechanisms and Models*, ed. H. Bockhorn, Springer-Verlag, Berlin, 1994 and earlier workshops referenced therein.
2. *Proc. Combust. Inst.*, 1998, **27**, 1461–1632 as well as previous and subsequent symposia.
3. S. di Stasio, *Carbon*, 2001, **39**, 109.
4. JI. Zhu and M. Y. Choi, *Int. J. Heat Transfer*, 2000, **43**, 3299.
5. M. Y. Choi and K. A. Jensen, *Combust. Flame*, 1998, **112**, 485.
6. S. di Stasio, P. Massoli and M. Lazzaro, *J. Aersol. Sci.*, 1996, **27**, 897.
7. O. Glatter, in *Modern Aspects of Small-Angle Scattering*, ed. H. Brumberger, Kluwer, Dordrecht, 1995, pp. 107–180.
8. D. Woiki, A. Giesen and P. Roth, *Proc. Combust. Inst.*, 2000, **28**, 2531.
9. Y. M. Choi, W. S. Xia, J. Park and M. C. Lin, *J. Phys. Chem. A*, 2000, **104**, 7030.
10. A. Bhargava and P. R. Westmoreland, *Combust. Flame*, 1998, **113**, 333.
11. H. Richter, W. J. Grieco and J. B. Howard, *Combust. Flame*, 1999, **119**, 1.
12. W. A. England, *Combust. Sci. Technol.*, 1986, **46**, 83.
13. A. Guinier and G. Fournet, *Small-Angle Scattering of X-rays*, (translated into English by C. B. Walker), Wiley, New York and Chapman & Hall, London, 1955.
14. *Small-Angle X-Ray Scattering*, ed. O. Glatter and O. Kratky, Academic Press, London, 1982.
15. L. A. Feigin and D. I. Svergun, *Structure Analysis by Small-Angle X-ray and Neutron Scattering*, Plenum Press, New York, 1987.
16. *Modern Aspects of Small-Angle Scattering*, ed. H. Brumberger, Kluwer, Dordrecht, 1995.
17. G. Beaucage, *J. Appl. Crystallogr.*, 1995, **28**, 717.
18. G. Beaucage and D. W. Schaefer, *J. Non-Cryst. Solids*, 1994, **172–174**, 797.
19. C. M. Sornsen, C. Oh, P. W. Schmidt and T. P. Rieker, *Phys. Rev. E*, 1998, **58**, 4666.
20. E. Y. Sheu, *Phys. Rev. A*, 1992, **45**, 2428.
21. S. Seifert, R. E. Winans, D. M. Tiede and P. Thiyagarajan, *J. Appl. Crystallogr.*, 2000, **33**, 782.
22. I. Naday, S. Ross, E. M. Westbrook and G. Zentai, *Opt. Eng.*, 1998, **37**, 1235.
23. M. Stanton, W. C. Phillips and K. Kalata, *J. Appl. Crystallogr.*, 1992, **25**, 549.
24. J. Ma, T. H. Fletcher and B. W. Webb, *Energy Fuels*, 1995, **9**, 802.
25. B. H. Zimm, *J. Chem. Phys.*, 1948, **16**, 1093; B. H. Zimm, *J. Chem. Phys.*, 1948, **16**, 1099.
26. M. Frenklach and H. Wang, in *Soot Formation in Combustion, Mechanisms and Models*, ed. H. Bockhorn, Springer-Verlag, Berlin, 1994, pp. 162–192.
27. A. Alexious and A. Williams, *Combust. Flame*, 1996, **104**, 51.
28. K. L. Revzan, N. J. Brown and M. Frenklach, http://www.me.berkeley.edu/soot/.

Thermodynamic and kinetic issues in the formation and oxidation of aromatic species

Peter Lindstedt,*[a] Lourdes Maurice[b] and Michael Meyer[a]

[a] *Department of Mechanical Engineering, Imperial College of Science, Technology and Medicine, Exhibition Road, London, UK SW7 2BX. E-mail: p.lindstedt@ic.ac.uk*
[b] *SAF/AQRT, 1060 Air Force Pentagon, Washington, DC 20330-1060, USA*

Received 8th May 2001
First published as an Advance Article on the web 1st November 2001

The chemistry of aromatic species is discussed in the context of detailed kinetic modelling of benzene and butadiene flames and stirred reactors featuring ethylene and mixed aromatic/ethylene/hydrogen fuels. The development of reliable detailed mechanisms depends on the accuracy of the underlying hydrocarbon chemistry and the present paper highlights some current issues in the formation and oxidation of aromatics. In particular, uncertainties pertaining to the rates and product distributions of a range of possible naphthalene and indene formation sequences are discussed from the basis of improved predictions of key intermediates. The naphthalene formation paths considered include initiation via $C_5H_5 + C_5H_5$, $C_6H_5 + C_4H_4$ and $C_7H_7 + C_3H_3$ reactions and results are assessed in the context of a number of tentative detailed and simplified sequences. It is shown that a number of possible formation channels are plausible and that their relative importance is strongly dependent upon oxidation conditions. Particular emphasis is placed on the investigation of formation paths leading to isomeric C_9H_8 structures. The latter are typically ignored despite measured concentrations similar to those of naphthalene. The rates of formation of C_9H_8 compounds are consistent with sequences initiated by $C_6H_5 + C_3H_3$ and $C_6H_5 + C_3H_4$ leading to indene through repeated isomerisation reactions. The current work also shows that reactions of the type $C_9H_7 + CH_3$ and $C_9H_7 + {}^3CH_2$ provide a mass growth source that link five and six member ring structures.

Introduction

The formation and destruction of aromatic species in flames remains of practical significance through links with environmental and health issues. It is also well established that the development of detailed mechanisms for arbitrarily large poly-aromatic hydrocarbon (PAH) and mono-substituted aromatic (MSA) molecules constitutes a formidable task due to uncertainties in kinetic and thermochemical data (*e.g.* Richter and Howard[1]). The formulation of detailed kinetic mechanisms must typically be based on recurring generic patterns for PAH growth. The latter may be inferred from the chemistry of smaller aromatic species, such as benzene, naphthalene and indene. The chemistry of the first aromatic ring has received particular attention and the significance of the propargyl radical for benzene formation in flames of aliphatic fuels has been established (*e.g.* Miller and Melius,[2] Leung and Lindstedt[3] and Dagaut and Cathonnet[4]). Perhaps the predomi-

DOI: 10.1039/b104056c

nant difficulty in the high temperature chemistry of benzene (*e.g.* Lindstedt and Skevis[5]) can be found in the branching ratio between phenyl radical linearization/decomposition and the attack of molecular oxygen on the phenyl radical leading to the phenol/phenoxy system. Some issues pertaining to the latter have recently been reviewed by DiNaro *et al.*[6] Phenoxy leads directly to the cyclopentadienyl radical and studies aimed at reducing the uncertainties in the related C_6H_5/C_6H_5O and C_5H_5/C_5H_6 systems include work on the pyrolysis of cyclopentadiene (Roy *et al.*[7]) and phenol (Horn *et al.*[8]).

The potential importance of cyclopentadienyl recombination as a possible path to naphthalene has been highlighted in several studies (*e.g.* Dean[9] and Melius *et al.*[10]). The reaction C_5H_5 + CH_3, leading to benzene *via* fulvene as proposed by Lin and co-workers (*cf.* Moskaleva *et al.*[11] and Hahndorf *et al.*[12]), is another interesting pathway for the formation of C_6 ring structures. The channel has recently been investigated experimentally by Ikeda *et al.*[13] and Laskin and Lifshitz[14] have presented further possible indirect experimental support of the role of this path in mass growth. The evaluation of such reaction sequences—leading to a mass growth linkage between indene and naphthalene (*via* the analogous C_9H_7 + CH_3 path) in flames—requires a measure of confidence in formation pathways leading to the participating species. For the formation of naphthalene, and higher aromatics, the H-abstraction/C_2H_2-addition mechanism proposed by Frenklach and co-workers[15] has long been established as a possible major contributor and remains of direct relevance in diffusion flames (McEnally and Pfefferle[16]). Recent studies, however, have in addition outlined the potential importance of cyclopentadienyl radical recombination in premixed flames (*e.g.* Marinov *et al.*[17] and Richter *et al.*[18]). The addition of vinyl acetylene to the phenyl radical (Bittner,[19] Maurice[20] and Moriarty and Frenklach[21]) has also received attention and related paths (*e.g.* C_6H_5 + $CH_2CHCHCH$ and C_6H_5 + $HCCCHCH$) are considered in the present study.

The PAH formation and oxidation paths discussed above are assessed in the context of flames and jet stirred reactors featuring aliphatic and aromatic fuels. The present study includes the low pressure laminar premixed $C_6H_6/O_2/Ar$ (ϕ = 1.80) and 1,3-$C_4H_6/O_2/Ar$ (ϕ = 2.40) flames of Bittner and Howard[22] and Cole *et al.*[23] The stirred reactor studies feature the C_2H_4, $C_6H_6/C_2H_4/H_2$ and $C_7H_8/C_2H_4/H_2$ fuels investigated by Vaughn and co-workers.[24–26] The latter data sets provide the basis for a sensitivity based analysis of MSA and PAH formation and destruction paths in mixtures with increasing amounts of aromatics.

Modelling approach

The computational procedures used in the present study have been outlined elsewhere[27] and are not repeated here. The starting mechanism used in the present work consists of 166 species and 827 reversible reactions and is based on the C_1–C_{10} mechanism of Lindstedt and co-workers.[20,27,28] Lindstedt and Skevis[5,29] have, in earlier studies, extensively discussed features of the flames considered (*e.g.* the dynamics of benzene formation/destruction) and it has been shown that the starting mechanism does reproduce detailed flame structures comparatively well also for other flames.[30] There are, however, some significant updates applied in the modelling of the flame structures reported here. These include the $C_6H_4O_2$ oxidation sequences of Alzueta *et al.*[31] and the C_6H_5OOH chemistry of DiNaro *et al.*[6] Furthermore, the pyrolysis steps for the cyclopentadienyl radical (C_5H_5) by Roy *et al.*[7] replace the earlier proposal by Emdee *et al.*[32] A further key issue concerns the branching between oxidative attack on and the linearization of the phenyl radical with the latter leading to the C_2/C_4 chains through the subsequent decomposition. Roy *et al.*[7] adopted a global step for the thermal decomposition which, within the context of the current mechanism, results in a significant under-prediction of key species. Lindstedt and Skevis[5] preferred a two-step sequence featuring linearization of the phenyl radical with the rate of Rao and Skinner[33] and with the subsequent decomposition proceeding with the rate proposed by Dean.[9] Laskin and Lifshitz[14] also considered a two-step sequence in their more recent studies of the thermal decomposition of benzene and indene and their proposed rates have been adopted in the present work. An extensive sensitivity analysis is performed to assess the potential influence of (i) the rates of individual elementary reactions and (ii) thermodynamic properties on mass growth in the context of sequences such as those considered by Richter and Howard.[1] The basis is a standard "brute force" method and key results are highlighted. Logarithmic response sensitivities are

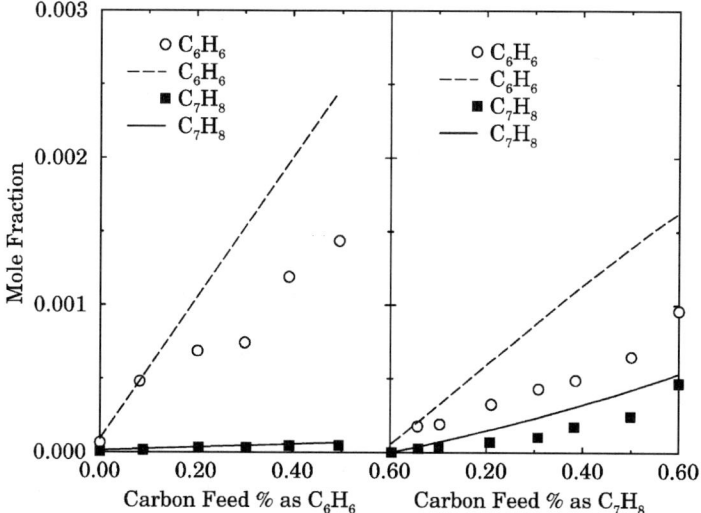

Fig. 1 Major aromatic species in aromatic/ethylene/hydrogen jet-stirred reactors ($T = 1630$ K, $\phi = 2.2$, $\tau = 6$ ms). Symbols indicate experimental data[24–26] and lines indicate computations.

defined as $\chi_k = \ln(Y_k/Y_{k_0})/\ln(5)$ where Y_{k_0} and Y_k are the baseline and perturbed concentrations and 5 is the perturbation factor. In the second part of the analysis, the heats of formations of each of the species of interest were adjusted by ±10%. Logarithmic response sensitivities were computed as $\chi = \ln(Y_k/Y_{k_0})$ where Y_{k_0} and Y_k are the concentrations computed with the baseline and perturbed heats of formation. The linear sensitivity of naphthalene and indene concentrations to alternative proposals for kinetic rate parameters is also evaluated for selected sequences. Species concentrations computed with the starting mechanism are in reasonable agreement with the experimental stirred reactor data[24–26] (Fig. 1–3). The key reaction channels are discussed below.

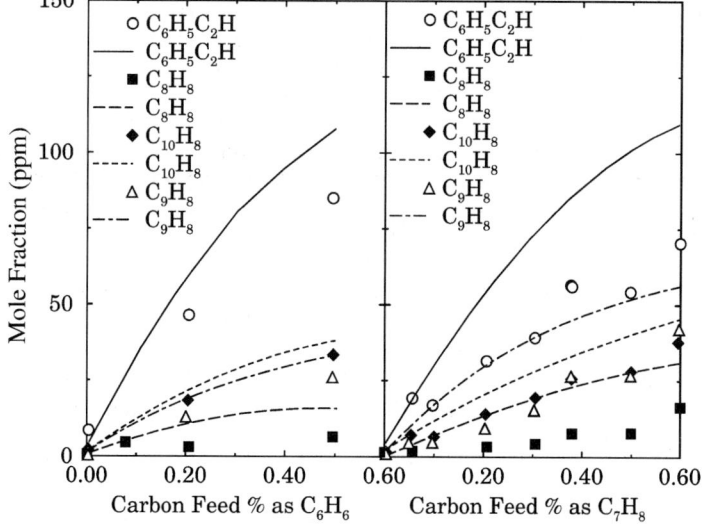

Fig. 2 Minor aromatic species in aromatic/ethylene/hydrogen jet-stirred reactors ($T = 1630$ K, $\phi = 2.2$, $\tau = 6$ ms). Symbols indicate experimental data[24–26] and lines indicate computations.

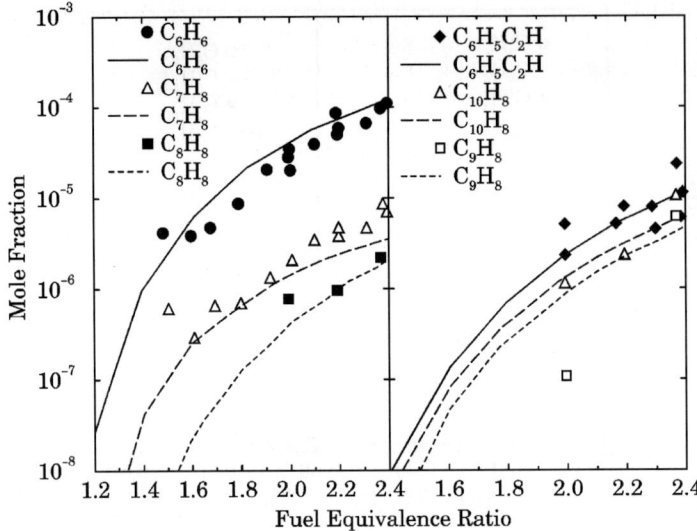

Fig. 3 Benzene, MSA and PAH species in ethylene jet-stirred reactors. ($T = 1630$ K, $\phi = 2.2$, $\tau = 6$ ms). Symbols indicate experimental data[24–26] and lines indicate computations.

The formation of mono-substituted aromatics

Mono-substituted aromatics (MSA) such as toluene, phenylacetylene and styrene form a basis for subsequent mass growth sequences. Primary toluene formation paths include the recombination of phenyl and methyl radicals and the addition of singlet and triplet methylene radicals to benzene. The rates for the individual channels follow the recommendations by Pamidimukkala et al.[34] (-612) and Kraus et al.[35] (606) and (607).

$$C_7H_8 = C_6H_5 + CH_3 \tag{612}$$

$$C_6H_6 + {}^1CH_2 = C_7H_8 \tag{606}$$

$$C_6H_6 + {}^3CH_2 = C_7H_8 \tag{607}$$

Toluene formation through Diels–Alder reactions, as suggested by Kern et al.,[36] is comparatively unimportant and destruction paths are in good agreement with earlier observations.[27] Consumption proceeds through H radical attack yielding the benzyl radical (613) and benzene (614) with a 6:1 branching ratio. The rates recommended by the CEC working group (Baulch et al.[37]) and Robaugh and Tsang[38] are assigned to the two channels. Aryl H atom abstraction (615), yielding the methylphenyl radical (p-C_7H_7), is important for subsequent PAH formation sequences.

$$C_7H_8 + H = C_7H_7 + H_2 \tag{613}$$

$$C_7H_8 + H = C_6H_6 + CH_3 \tag{614}$$

$$C_7H_8 + H = p\text{-}C_7H_7 + H_2 \tag{615}$$

The benzyl radical readily (85–90%) recombines with an H atom to yield toluene (-611), whereas isomerization to bicyclo[2.2.1]-hepta-2,5-dien-7-yl (i-C_7H_7) contributes 10–15%. The recommendation by Lindstedt and Maurice[27] is used for this channel and the rates for reactions (634) and (635) are adopted from Braun-Unkhoff et al.[39]

$$C_7H_7 + H = C_7H_8 \tag{-611}$$

$$C_7H_7 = i\text{-}C_7H_7 \tag{634}$$

$$i\text{-}C_7H_7 = C_7H_6 + H \tag{635}$$

The rate measured by Herzler and Frank[40] has been assigned to the primary phenylacetylene formation path (−686). In addition, thermal decomposition of the styryl radical (707), formed through the styrene channels discussed below, constitutes a minor phenylacetylene formation path. The rate is based on an analogy with the thermal decomposition of the vinyl radical.

$$C_6H_5 + C_2H_2 = C_8H_6 + H \tag{686}$$

$$C_8H_7 + M = C_8H_6 + H + M \tag{707}$$

Oxidation of phenylacetylene is based on O and OH radical attacks on the acetylenic bond through exothermic reactions ($-\Delta_rH \cong 75$, 78 and 111 for (689), (691) and (692) respectively). The rates are based on similarity with acetylene oxidation.[28]

$$C_8H_6 + O = C_6H_5 + HCCO \tag{689}$$

$$C_8H_6 + OH = C_6H_5 + CH_2CO \tag{691}$$

$$C_8H_6 + OH = C_6H_6 + HCCO \tag{692}$$

Phenylacetylene also undergoes aryl or side-chain H atom abstraction to form the ethynylphenylene (694) and phenylacetyl (695) radicals. Thermodynamically, aryl abstraction is favoured by around 68 kJ mol^{-1}. However, both channels are important in subsequent naphthalene growth sequences. The rate for (694) is adopted from Frenklach and Wang[15] whereas that for (695) is estimated based on analogy with acetylene oxidation.

$$C_8H_6 + OH = C_8H_5 + H_2O \tag{694}$$

$$C_8H_6 + OH = s\text{-}C_8H_5 + H_2O \tag{695}$$

Styrene is formed through thermal decomposition of the C_8H_9 radical (738), in turn formed through the thermal decomposition of ethylbenzene (731). The latter is formed predominantly *via* recombination of benzyl and methyl radicals (730). The rates for (731) and (738) are adopted from the work of Müller-Markgraf and Troe.[41]

$$C_8H_9 = C_8H_8 + H \tag{738}$$

$$C_8H_{10} = C_8H_9 + H \tag{731}$$

$$C_7H_7 + CH_3 = C_8H_{10} \tag{730}$$

Alternative styrene formation channels pass *via* (i) vinyl radical addition to benzene (702) and (ii) ethylene addition to the phenyl radical (703). The rates assigned to both channels are as recommended by Fahr and Stein.[42]

$$C_6H_6 + C_2H_3 = C_8H_8 + H \tag{702}$$

$$C_6H_5 + C_2H_4 = C_8H_8 + H \tag{703}$$

As observed for phenylacetylene, styrene levels were initially over-predicted and additional exothermic ($-\Delta_rH \cong 80$ kJ mol^{-1}) oxidation paths, *e.g. via* O radical attack, through reactions analogous to ethylene oxidation were adopted.

$$C_8H_8 + O = C_6H_5 + C_2H_3O \tag{720}$$

Side chain (717) and aryl (718) H atom abstractions to form the styryl and ethylphenyl radicals, again plausible building blocks for PAH formation, are also considered. The rate assigned to (717) is estimated based on analogy to ethylene while that adopted for (718) is based on analogy to aryl abstraction from phenylacetylene.[40]

$$C_8H_8 + H = C_8H_7 + H_2 \tag{717}$$

$$C_8H_8 + H = p\text{-}C_8H_7 + H_2 \tag{718}$$

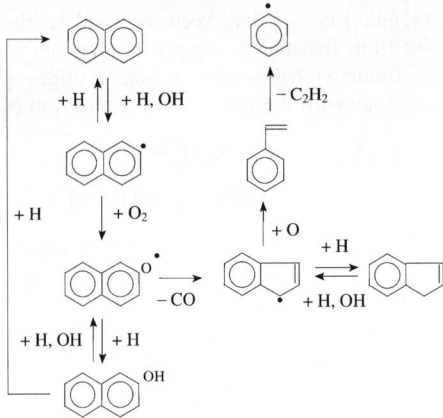

Fig. 4 Naphthalene and indene oxidation mechanisms.

The oxidation of naphthalene and indene

The oxidation of naphthalene and indene (Fig. 4) is based on that of benzene and cyclopentadiene with frequency factors adjusted to account for molecular size and reactive site differences. Computations show the dominant naphthalene destruction paths to be through H atom abstraction reactions (760) and (763). The addition of an O atom to the ring to form the naphthalenoxy radical (761) constitutes a secondary destruction path.

$$C_{10}H_8 + H = C_{10}H_7 + H_2 \tag{760}$$

$$C_{10}H_8 + OH = C_{10}H_7 + H_2O \tag{763}$$

$$C_{10}H_8 + O = C_{10}H_7O + H \tag{761}$$

Results of the sensitivity analysis for naphthalene destruction (Fig. 5) suggest that the response is nearly independent of fuel type, and that naphthalene concentrations are sensitive to molecular oxygen attack on the naphthyl radical.

$$C_{10}H_7 + O_2 = C_{10}H_7O + O \tag{754}$$

The rate assigned to (754) is based on the suggestion by Lin and Lin[43] who studied molecular oxygen attack on the phenyl radical in a shock tube in the temperature range 1030–1670 K and in the pressure range 0.32–0.54 bar. Herzler and Frank[44] studied the same reaction in the temperature range 900–1800 K and at pressures from 1.5 to 2.5 bar. A much faster rate (>10 times) was required to match the initial H and O atom generation in their experiment. The latter rate has

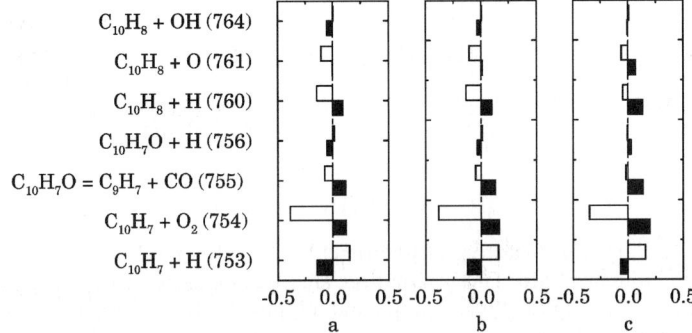

Fig. 5 Naphthalene concentrations logarithmic response sensitivities to naphthalene consumption paths in (a) ethylene (b) benzene/ethylene/hydrogen and (c) toluene/ethylene/hydrogen jet-stirred reactors ($T = 1630$ K, $\phi = 2.2$, $\tau = 6$ ms). The light bars represent the sensitivities for the rate constants increased by a factor of 5. The dark bars represent the sensitivities for the rate constants decreased by a factor of 5.

benefits for the modelling of benzene flames. However, in the context of naphthalene oxidation, significant (factor of 3) under-predictions of stirred reactor data results and a direct determination of (754) is evidently required. Predictions are also sensitive to the rates of H atom recombination with the naphthyl radical (753), H abstraction (760) and thermal decomposition of the naphthalenoxy radical (755).

$$C_{10}H_7 + H = C_{10}H_8 \tag{753}$$

$$C_{10}H_8 + H = C_{10}H_7 + H_2 \tag{760}$$

$$C_{10}H_7O = C_9H_7 + CO \tag{755}$$

Reaction (755) competes with H addition (756) to form naphthalenol, which in turn is recycled to naphthalene (−764). Reactions (756) and (764) are partially equilibrated and naphthalene concentrations remain essentially unaffected by changes to kinetic rate parameters.

$$C_{10}H_7O + H = C_{10}H_7OH \tag{756}$$

$$C_{10}H_7OH + H = C_{10}H_8 + OH \tag{−764}$$

The sensitivity of predicted naphthalene concentrations to the heat of formation of the naphthalenoxy radical was also determined and results are shown in Fig. 6a–c. Perturbing the heat of formation by ±10% results in only a modest (±5%) response in computed naphthalene concentrations.

Indene is consumed through H abstraction (766) and (768) and the indenyl radical is either recycled back to indene (765) or undergoes O addition and subsequent expulsion of CO (769) to form the styryl radical.

$$C_9H_8 + H = C_9H_7 + H_2 \tag{766}$$

$$C_9H_8 + OH = C_9H_7 + H_2O \tag{768}$$

$$C_9H_7 + H = C_9H_8 \tag{765}$$

$$C_9H_7 + O = C_8H_7 + CO \tag{769}$$

Predicted indene concentrations are sensitive (Fig. 7) to the rates of O attack on the indenyl radical (769), H abstraction (766) and H addition to the indenyl radical (765). The sensitivity of

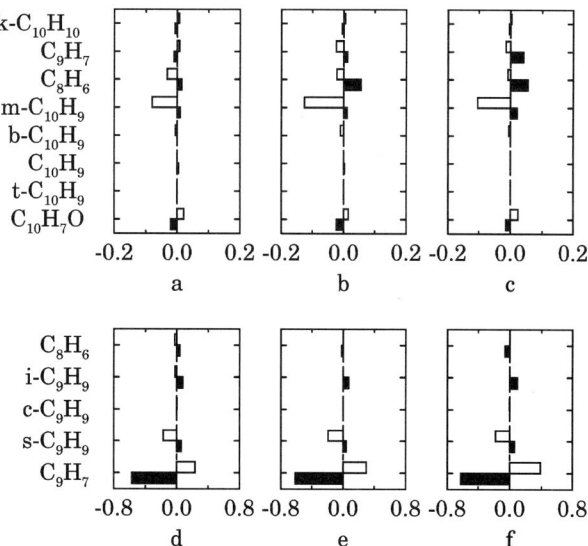

Fig. 6 Logarithmic response sensitivities for naphthalene concentrations (a–c) and indene concentrations (d–f) in (a,d) ethylene (b,e) benzene/ethylene/hydrogen and (c,f) toluene/ethylene/hydrogen ($T = 1630$ K, $\phi = 2.2$, $\tau = 6$ ms). The light bars represent the sensitivities for the heats of formation increased by 10%. The dark bars represent the sensitivities for the heats of formation decreased by 10%.

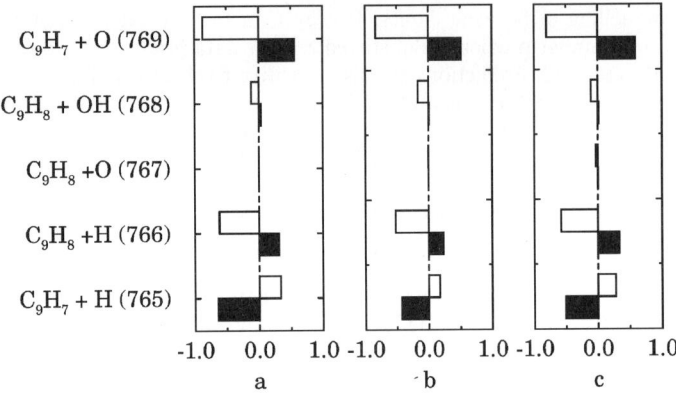

Fig. 7 Indene concentrations logarithmic response sensitivities to indene consumption paths in (a) ethylene (b) benzene/ethylene/hydrogen and (c) toluene/ethylene/hydrogen jet-stirred reactors ($T = 1630$ K, $\phi = 2.2$, $\tau = 6$ ms). The light bars represent the sensitivities for the rate constants increased by a factor of 5. The dark bars represent the sensitivities for the rate constants decreased by a factor of 5.

predicted indene concentrations to the heat of formation of the indenyl radical was also determined and results are shown in Fig. 6d–f and it is evident that predicted indene concentrations are highly sensitive to the thermodynamic data. The thermodynamic properties of the indenyl radical could not be determined through group additivity (e.g. Stein[45]) and are based on analogy with the cyclopentadienyl radical and cyclopentadiene. The estimated[20] heat of formation of indenyl radical (295 kJ mol^{-1}) is in fair agreement with the 286 kJ mol^{-1} suggested by Melius[46] on the basis of BAC-MP4 calculations. However, both values differ from the 259 kJ mol^{-1} proposed by Marinov et al.[47] and the topic requires further clarification.

The formation of naphthalene

A number of tentative naphthalene formation pathways are considered (e.g. Moskaleva et al.,[11] Maurice,[20] Moriarty and Frenklach,[21] Bittner and Howard[22] and Marinov et al.[47]). These include HACA type sequences, such as acetylene addition to the phenylacetylene and styryl radicals, leading to branched aromatics and two-ringed structures through subsequent cyclization reactions. Phenyl radical addition to the vinylacetylene triple bond leading to the phenyl-1,3-butadien-2-yl radical is considered along with alternative pathways for this reaction.[21] In the present work a multi-step sequence with phenylbutatriene and phenyl-butyn-3-ene as initial products and passing through phenyl-1,3-butadien-2-yl and on to naphthalene via a sequence of isomerisation reactions is investigated. Related sequences featuring initiation via C_6H_5 + $CH_2CHCHCH$ and C_6H_5 + $HCCCHCH$ are also considered along with reactions of the type C_5H_5 + CH_3 and C_9H_7 + CH_3 which provide a mass growth source that link five- and six-ringed structures.[11,13] The cyclopentadienyl recombination sequence leading to naphthalene is considered in the context of global reactions and two-step sequences exploring stabilisation of alternative intermediate structures.[10] Radical sequences featuring C_7H_7 + C_3H_3 and p-C_7H_7 + C_3H_3 are also considered. The contributions made to the formation of two-ringed structures by a number of the above paths are initially assessed in the context of the stirred reactor data of Vaughn and co-workers.[24–26] The potential key sequences are then explored further in the context of the low-pressure laminar premixed $C_6H_6/O_2/Ar$ ($\phi = 1.80$) and 1,3-$C_4H_6/O_2/Ar$ ($\phi = 2.40$) flames of Bittner and Howard[22] and Cole et al.[23]

The hydrogen abstraction–acetylene addition sequences

Computations of the stirred reactor configurations of Vaughn and co-workers[24–26] suggest that HACA type sequences (N_1 and N_2 in Fig. 8), proceeding via reactions (−686), (687), (742), (743), (753) and (−686), (688), (750), (751), (753), contribute 30% of the naphthalene formation in ethylene mixtures. However, in broad agreement with previous observations[47–49] these paths contrib-

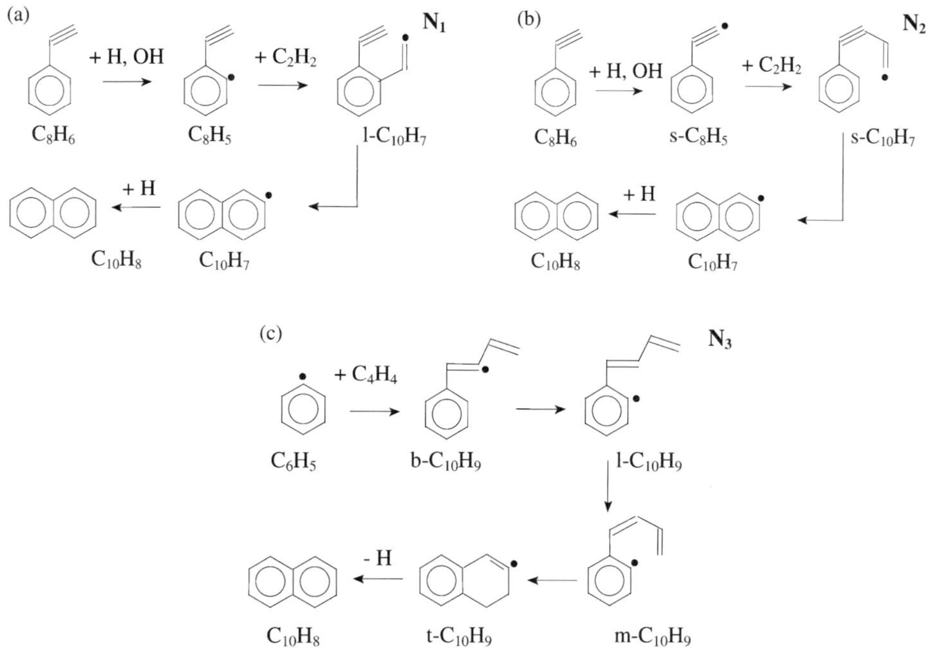

Fig. 8 Naphthalene formation sequences[19,20,22] N_1 (a), N_2 (b) and N_3 (c).

ute only about 10% in aromatic fuel blends. The individual reactions for both sequences exhibit some degree of partial equilibration with the exception of H atom recombination with the naphthyl radical (753). Phenylacetylene aryl abstraction is thermodynamically favoured over side-chain abstraction (N_2) and the kinetic parameters adopted here reflect the differences in endothermicity. However, computations show that the aryl abstraction reactions are more partially equilibrated and the net production of the s-C_8H_5 radical is competitive with that of the C_8H_5 radical. Furthermore, acetylene addition to s-C_8H_5 is more exothermic (-240 kJ mol^{-1}) than addition to C_8H_5 (-169 kJ mol^{-1}) and consequently both sequences contribute to naphthalene formation *via* the naphthyl radical. Computed naphthalene concentrations are not sensitive to the rates adopted for the removal of the C_8H_5 isomers *via* molecular oxygen attack (697) and (698) and destruction of phenylacetylene *via* O (689)–(690) and OH (691)–(693) attack. The heat of formation assigned to phenylacetylene (315 kJ mol^{-1}) in order to ensure consistency with adopted phenylacetylene formation (-686) and destruction (687) paths, differs from the value of 327 kJ mol^{-1}, recommended by Burcat and McBride,[50] and the value of 309 kJ mol^{-1} used by Wang and Frenklach.[51] The sensitivity of predicted naphthalene and indene concentrations to the heat of formation of phenylacetylene show (Fig. 6a–c) that if the heat of formation of phenylacetylene is increased by 10% then a decrease in naphthalene concentrations in ethylene mixtures of 7% is observed. As may be expected aromatic/ethylene/hydrogen mixtures remain largely unaffected. Decreasing the heat of formation of phenylacetylene by 10% results in a modest 3% increase in naphthalene concentrations in ethylene mixtures. However, a near ten-fold increase in phenylacetylene concentrations in mixtures with aromatic fuels is observed.

The vinylacetylene addition sequence

Vinylacetylene addition to the phenyl radical (Sequence N_3 in Fig. 8) may constitute a significant naphthalene formation path[22,52,53] because of the high endothermicity of the decomposition of the initial adduct (-781). The sequence (781), (782), (784), (785), (-778) contains two rate limiting steps (781) and (784). If it is assumed, as explored by Leung *et al.*[52] and Appel *et al.*,[53] that the *trans–cis* isomerization step (784) proceeds rapidly at flame temperatures then the initial adduct formation (781) is rate limiting (Fig. 9a–c).

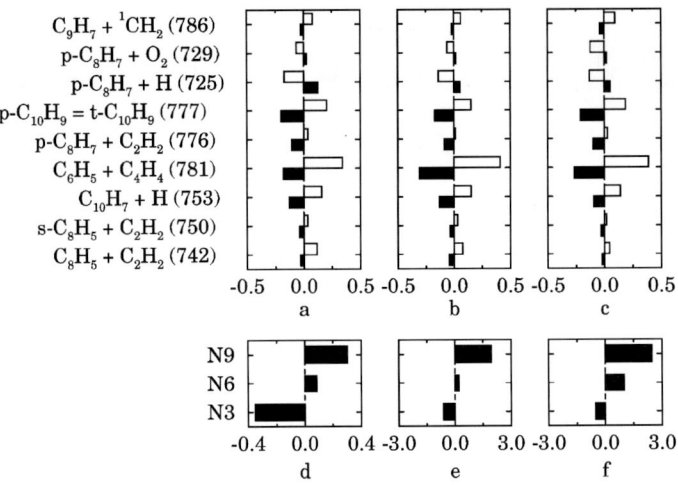

Fig. 9 Naphthalene concentrations logarithmic response sensitivities to naphthalene formation paths in (a) ethylene (b) benzene/ethylene/hydrogen and (c) toluene/ethylene/hydrogen jet-stirred reactors ($T = 1630$ K, $\phi = 2.2$, $\tau = 6$ ms). The light bars represent the sensitivities for the rate constants increased by a factor of 5. The dark bars represent the sensitivities for the rate constants decreased by a factor of 5 (a–c) or for alternative proposals for kinetic parameters (d–f).

$$C_6H_5 + C_4H_4 = b\text{-}C_{10}H_9 \qquad (781)$$

$$1\text{-}C_{10}H_9 = m\text{-}C_{10}H_9 \qquad (784)$$

However, if the upper limit[19–21] of 188 kJ mol^{-1} for the energy barrier of the isomerization reaction (784) is used ($k_{784} \sim 1.0 \times 10^{13} \exp(-22\,612/T)$ m^3 kmol^{-1} s^{-1}), then the contribution of this sequence to naphthalene formation is reduced to $\sim 5\%$ of the total. In the present work the barrier is retained. Naphthalene concentrations also show sensitivity to an increase in the heat of formation of m-$C_{10}H_9$, but remain unaffected by perturbations to the heat of formation assigned to t-$C_{10}H_9$ (Fig. 6). Based on the potential of this path Moriarty and Frenklach[21] considered a number of alternative channels. A two-step sequence passing *via* phenyl-butyn-3-ene with the latter species assumed to be in the steady state was proposed. In the current work three-step sequences passing *via* (838) phenylbutatriene (g-$C_{10}H_8$) and (839) phenyl-butyn-3-ene (j-$C_{10}H_8$) are used. Naphthalene is formed *via* H addition (840) and (841) leading to 1-hydro-naphthalen-2-yl (t-$C_{10}H_9$) *via* decomposition of the initial adduct (phenyl-1,3-butadien-2-yl).

$$C_6H_5 + C_4H_4 = g\text{-}C_{10}H_8 + H \qquad (838)$$

$$g\text{-}C_{10}H_8 + H = t\text{-}C_{10}H_9 \qquad (840)$$

$$t\text{-}C_{10}H_9 = C_{10}H_8 + H \qquad (-778)$$

Computations with the latter sequences suggest that the initiation steps (838) and (839) have a strong tendency to partially equilibrate and that the subsequent H addition exerts a significant overall influence. The rates adopted here arguably constitute the upper limit. Related sequences initiated *via* $C_6H_5 + CH_2CHCHCH$ (853), assumed for simplicity to lead directly to Z-1,3-butadienyl-benzene-2-yl (m-$C_{10}H_9$), and $C_6H_5 + HCCCHCH$ (852), leading to phenyl-butyn-3-ene, are also considered. The thermodynamic data for phenyl-butatriene ($\Delta_f H \simeq 390$ kJ mol^{-1}) and phenyl-butyn-3-ene ($\Delta_r H \simeq 385$ kJ mol^{-1}) follows the study of Moriarty and Frenklach.[21]

The styryl radical and ethylphenyl radical sequences

Acetylene addition to the styryl radical (Sequence N_4 in Fig. 9) is a minor naphthalene formation path and thermal decomposition of the styryl radical (-706) and (707) is typically favoured over acetylene addition (771). Acetylene addition to the ethylphenyl radical (Sequence N_5 in Fig. 9) is relevant in mixtures with aromatic fuels where styrene concentrations are up to an order of magnitude higher. The initial addition step (776) is assigned a rate of $4.0 \times 10^{10} \exp(-5088/T)$, by analogy to acetylene addition to the phenyl radical, and is typically strongly (>80%) partially equilibrated.

$$p\text{-}C_8H_7 + C_2H_2 = p\text{-}C_{10}H_9 \qquad (776)$$

$$p\text{-}C_{10}H_9 = t\text{-}C_{10}H_9 \qquad (777)$$

The cyclization step (777) is a possible though, due to its highly exothermic ($\Delta_r H \simeq -187$ kJ mol^{-1}) nature, unlikely rate limiting step (Fig. 9a–c). The relative contribution of this sequence is also sensitive to removal of the ethylphenyl radical (725) and (729).

$$p\text{-}C_8H_7 + H = C_8H_8 \qquad (725)$$

$$p\text{-}C_8H_7 + O_2 = C_6H_5O + CH_2CO \qquad (729)$$

The rates for these reactions are based on analogy with the phenyl radical and further refinement remains desirable. The contributions of Sequences N_4 and N_5 are broadly unaffected by thermodynamic parameters.

Propargyl radical addition to benzyl and methylphenyl radicals

Propargyl addition to the benzyl (Sequence N_6 in Fig. 10) and methylphenyl radicals (Sequence N_7) are interesting potential naphthalene formation paths. Colket et al.[54] propose a global step for naphthalene formation via recombination of the propargyl and benzyl radicals and suggest a frequency factor of 6.0×10^8 and no activation energy in order to match naphthalene concentrations detected in the pyrolysis of toluene in a shock tube.

$$C_7H_7 + C_3H_3 = C_{10}H_8 + 2H$$

The heats of formation for the benzyl and propargyl radicals suggest an energy barrier of ~ 25 kJ mol^{-1} for the global step leading to a 6-fold decrease in the corresponding rate at 1630 K. Formation of the initial adduct (787) and (794) is potentially rate limiting for both sequences and the rates assigned are based on analogy with the measurements by Alkemade and Homann[55] for propargyl radical recombination.

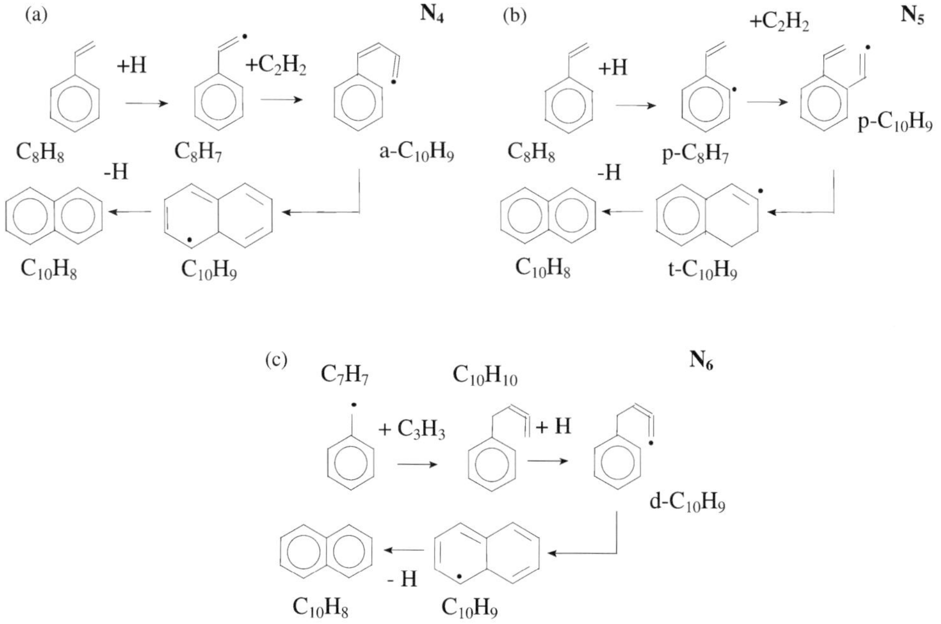

Fig. 10 Naphthalene formation sequences[19,20,22] N_4 (a), N_5 (b) and N_6 (c).

$$C_7H_7 + C_3H_3 = C_{10}H_{10} \quad (787)$$

$$p\text{-}C_7H_7 + C_3H_3 = f\text{-}C_{10}H_{10} \quad (794)$$

Adopting the global step for Sequence N_6 (Fig. 6d–e) results in a modest increase in computed naphthalene concentrations in ethylene and benzene/ethylene/hydrogen mixtures. However, naphthalene concentrations in toluene/ethylene/hydrogen mixtures increase more than a factor of 2 (Fig. 9f), which is not corroborated by experimental observations.[24–26]

The cyclopentadienyl radical sequence

Cyclopentadienyl radical recombination (Sequence N_9) leading to the formation of the bicyclic C_5H_5–C_5H_4 radical and to naphthalene through subsequent rearrangements is a possible route[9,10,17] for which there is qualitative support.[18,56] There is, however, significant doubt concerning the rate. The proposal by Dean[9] is 2 to 3 orders of magnitude slower that the initial global reaction ($k = 2.0 \times 10^{10} \exp(-2013/T)$ m^3 kmol^{-1} s^{-1}) proposed by Marinov et al.[47]

$$C_5H_5 + C_5H_5 = C_{10}H_8 + 2H$$

Computations with the above step appear prone to produce results contrary to measured trends in fuel mixtures featuring aromatic fuel components. Thus while the above global step results in a 30% increase in computed naphthalene concentrations in ethylene mixtures, naphthalene levels in aromatic/ethylene/hydrogen mixtures increase, as shown in Fig. 9d–f, in a manner which is not consistent with the experimental observations of Vaughn.[24–26] Richter et al.[18] also found it necessary to reduce the rate by around an order of magnitude to bring agreement with measured naphthalene levels in the benzene flame of Bittner and Howard.[22] Naturally, such adjustments are related to uncertainties in predicted C_5H_5 levels, as discussed below. However, the subsequent revision[17] of the barrier to 34 kJ mol^{-1}, in line with the recommendation by Melius et al.,[10] constitutes a sensible measure. The cyclopentadienyl route is further evaluated here in the context of alternative two-step mechanisms. The first considers the effects of stabilisation of the C_5H_5–C_5H_4 radical leading to a two step sequence.

$$C_5H_5 + C_5H_5 = C_5H_5\text{–}C_5H_4 + H \quad (802a)$$

$$C_5H_5\text{–}C_5H_4 = C_{10}H_8 + H \quad (803a)$$

Reaction (802a) is weakly endothermic (34 kJ mol^{-1}) and this barrier is used in the evaluation along with the proposed frequency factor for the global step. A competing path leads to the relatively stable fulvalene (C_5H_4=C_5H_4) and it has been suggested[20] that this route does not contribute significantly to naphthalene formation. Melius et al.[10] deduced a barrier for the conversion of C_5H_5–C_5H_4 to naphthalene of around 197 kJ mol^{-1} and it may be noted that with the second step rate limiting, an estimated rate constant $k_{803a} = 3.0 \times 10^{13} \exp(-23\,695/T)$ yields a value of 1.1×10^7 m^3 kmol^{-1} s^{-1} at 1600 K. The latter compares with 2.0×10^7 m^3 kmol^{-1} s^{-1} at the same temperature for the QRRK based estimate for the global step.[9] The common assumption is, however, that the initial step is rate limiting[1,17,56] and an alternative, more favourable, two-step sequence is also investigated.

$$C_5H_5 + C_5H_5 = C_{10}H_9 + H \quad (802)$$

$$C_{10}H_9 = C_{10}H_8 + H \quad (-773)$$

Reaction (773) is endothermic (~ 80 kJ mol^{-1}) and in the present work the rate assigned is based on similarity with H addition to the benzene ring[3] ($C_6H_6 + H = C_6H_7$). The alternative path (802) and (-773) yields a limiting rate constant of around 2.0×10^9 m^3 kmol^{-1} s^{-1} at 1600 K.

The formation of indene

Formation of indene is considered using several paths that include allene and propargyl addition to the phenyl radical and acetylene addition to the benzyl and methylphenyl radicals. Similarly to the naphthalene formation sequences, the above channels are labelled I_1 to I_5, where I_5 implies formation through naphthalene oxidation. Examples are shown in Fig. 11.

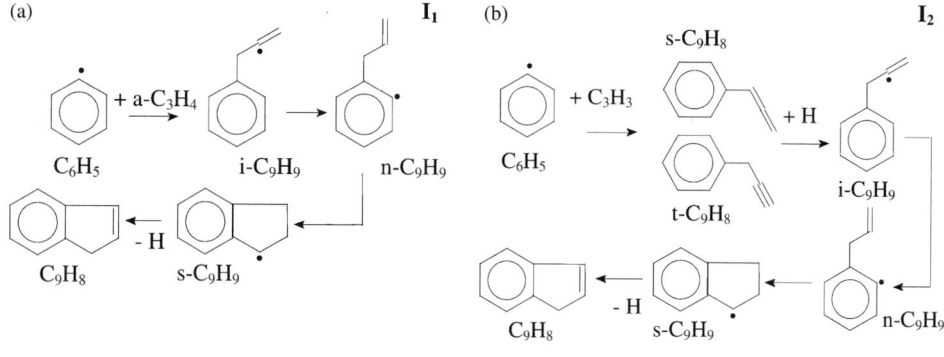

Fig. 11 Indene formation sequences[19,20,22] I_1 (a), I_2 (b) and I_3 (c).

The allene and propargyl radical sequences

Allene addition to the phenyl radical (Sequence I_1 in Fig. 11) is an important indene formation path. There are two rate-limiting steps in the sequence (810), (829), (830), (831). The first is the formation of the adduct (810) and the second is the H atom elimination from the s-C_9H_9 isomer (831) to stabilise indene. A frequency factor of 1.0×10^{10} is assigned to the exothermic adduct formation step. A frequency factor of 1.0×10^{13} and an energy barrier commensurate with the heat of reaction are assigned to the decomposition step (831).

$$C_6H_5 + a\text{-}C_3H_4 = i\text{-}C_9H_9 \tag{810}$$

$$s\text{-}C_9H_9 = C_9H_8 + H \tag{831}$$

The isomerization (829) and cyclization (830) steps are partially equilibrated and do not affect indene concentrations. Bittner[19] also suggests that H elimination from the i-C_9H_9 radical to form stable species with non-cyclic side chains (811) and (812) should be competitive with the indene stabilisation step (831).

$$i\text{-}C_9H_9 = t\text{-}C_9H_8 + H \tag{811}$$

$$i\text{-}C_9H_9 = s\text{-}C_9H_8 + H \tag{812}$$

These steps are not sensitive to thermodynamic properties and are linked directly to propargyl radical addition to the phenyl radical through (813) and (814). The reactions forming the (stable) C_9H_8 isomers are plausible and highly exothermic ($-\Delta_r H \simeq 384$ and 375 kJ mol^{-1}) reactions.

$$C_6H_5 + C_3H_3 = s\text{-}C_9H_8 \tag{813}$$

$$C_6H_5 + C_3H_3 = t\text{-}C_9H_8 \tag{814}$$

The side-chain substituted C_9H_8 isomers are also removed *via* side-chain oxidation reactions analogous to those of propyne and allene. The propargyl radical addition steps are assigned frequency factors of 1.0×10^{10} based on analogy with propargyl radical recombination.[55] Indene concentrations only show a modest sensitivity to increases in these rates (Fig. 12a–c). As shown in Fig. 6d–f,

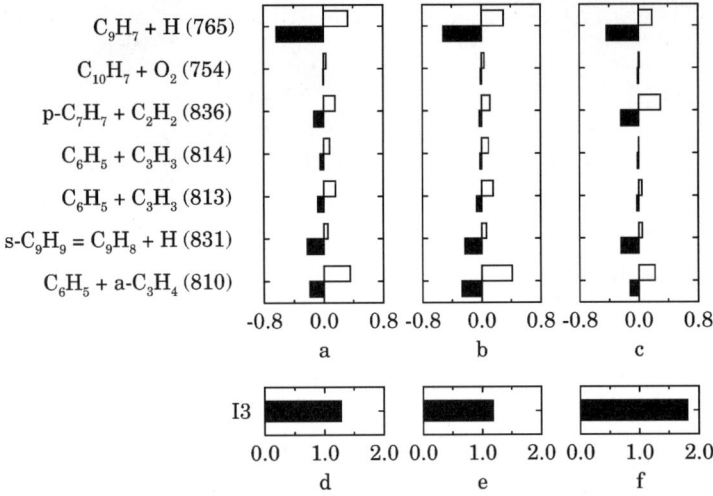

Fig. 12 Indene concentrations logarithmic (a–c) and linear (d–f) response sensitivities to indene formation paths in (a,d) ethylene (b,e) benzene/ethylene/hydrogen and (c,f) toluene/ethylene/hydrogen jet-stirred reactors ($T = 1630$ K, $\phi = 2.2$, $\tau = 6$ ms). The light bars represent the sensitivities for the rate constants increased by a factor of 5. The dark bars represent the sensitivities for the rate constants decreased by a factor of 5 (a–c) or for alternative proposals for kinetic parameters (d–f).

the relative contributions of Sequences I_1 and I_2 are decreased by decreasing the stability of s-C_9H_9 and increased by increasing the stability of i-C_9H_9.

The benzyl radical and methylphenyl radical sequences

Acetylene addition to the benzyl radical (Sequence I_3 in Fig. 11) is a potentially dominant indene formation path in toluene/ethylene/hydrogen mixtures. Colket et al.[53] proposed a global step to explain experimental observations of toluene pyrolysis products in a shock tube.

$$C_7H_7 + C_2H_2 = C_9H_8 + H$$

Adopting the above suggestion for Sequence I_3 leads to only a modest increase in computed indene concentrations in ethylene and benzene/ethylene/hydrogen mixtures (Fig. 12d–e). However, indene concentrations in toluene/ethylene/hydrogen mixtures increase by about a factor of 2 (Fig. 12f), which is not corroborated by the experimental observations.[24–26] It would therefore appear that the proposed reaction bypasses the energy barriers inherent in the intermediate steps and as a result over-predicts the contribution of the channel. If the rate limiting cyclization step (−831) is based on analogy with the corresponding cyclopentadienyl reaction then predictions of indene remain consistent with experimental observations.

$$\text{s-}C_9H_9 = C_9H_8 + H \tag{831}$$

Acetylene addition to the methylphenyl radical (Sequence I_4) is a principal indene formation path in toluene mixtures and a competitive path for the other fuels. The formation of the initial adduct (836) and the stabilisation of the s-C_9H_9 radical to indene (831) are rate limiting.

$$\text{p-}C_7H_7 + C_2H_2 = \text{f-}C_9H_9 \tag{836}$$

The cyclization step (837) is partially equilibrated and only the thermodynamic properties of the s-C_9H_9 radical affect indene concentrations (Fig. 6d–f).

Mass growth linkage of indene and naphthalene

A number of C_1 related steps have been proposed for the linkage of aromatic C_5 and C_6 ring structures and several are evaluated in the context of the current work. The sequences include methyl addition to the cyclopentadienyl radical, proposed as a formation route leading to benzene

Table 1 Selected rate constants. Rates in the form $k = AT^n\exp(-E_a/RT)$. Units are kmol, m^3, s, K and kJ mol^{-1}

No	Reaction	A	n	E	Ref.
606	$C_6H_6 + {}^1CH_2 = C_7H_8$	4.000E+10	0.00	36.33	35
607	$C_6H_6 + {}^3CH_2 = C_7H_8$	1.700E+10	0.00	36.33	35
611	$C_7H_8 = C_7H_7 + H$	5.600E+15	0.00	381.13	27
612	$C_7H_8 = C_6H_5 + CH_3$	8.910E+12	0.00	303.76	34
613	$C_7H_8 + H = C_7H_7 + H_2$	3.980E−01	3.44	13.05	37
614	$C_7H_8 + H = C_6H_6 + CH_3$	1.200E+10	0.00	21.43	38
615	$C_7H_8 + H = p\text{-}C_7H_7 + H_2$	2.500E+11	0.00	66.94	20
634	$C_7H_7 = i\text{-}C_7H_7$	3.160E+15	0.00	356.67	39
635	$i\text{-}C_7H_7 = C_7H_6 + H$	5.000E+15	0.00	165.00	39
686	$C_8H_6 + H = C_6H_5 + C_2H_2$	2.000E+11	0.00	40.11	40
687	$C_8H_6 + H = C_8H_5 + H_2$	2.700E+10	0.00	40.59	20
688	$C_8H_6 + H = s\text{-}C_8H_5 + H_2$	3.000E+10	0.00	116.40	20
689	$C_8H_6 + O = C_6H_5 + HCCO$	6.510E+03	2.09	6.54	20
691	$C_8H_6 + OH = C_6H_5 + CH_2CO$	3.000E+00	3.02	46.34	20
692	$C_8H_6 + OH = C_6H_6 + HCCO$	2.440E+00	3.02	46.34	20
694	$C_8H_6 + OH = C_8H_5 + H_2O$	2.100E+10	0.00	19.10	15
695	$C_8H_6 + OH = s\text{-}C_8H_5 + H_2O$	1.685E+04	2.00	58.58	20
702	$C_6H_6 + C_2H_3 = C_8H_8 + H$	7.940E+08	0.00	26.78	42
703	$C_6H_5 + C_2H_4 = C_8H_8 + H$	2.510E+09	0.00	25.93	42
706	$C_6H_5 + C_2H_2 = C_8H_7$	4.000E+10	0.00	42.30	20
707	$C_8H_7 + M = C_8H_6 + H + M$	2.000E+14	0.00	166.29	20
717	$C_8H_8 + H = C_8H_7 + H_2$	6.625E+02	2.53	51.21	est.
718	$C_8H_8 + H = p\text{-}C_8H_7 + H_2$	2.700E+10	0.00	40.59	est.
720	$C_8H_8 + O = C_6H_5 + C_2H_3O$	3.500E+10	0.00	11.85	est.
721	$C_8H_8 + O = C_8H_7 + OH$	7.550E+03	1.91	15.63	est.
725	$p\text{-}C_8H_7 + H = C_8H_8$	7.830E+10	0.00	0.00	est.
729	$p\text{-}C_8H_7 + O_2 = C_6H_5O + CH_2CO$	1.880E+09	0.00	31.25	est.
730	$C_7H_7 + CH_3 = C_8H_{10}$	1.000E+10	0.00	0.00	est.
731	$C_8H_{10} = C_8H_9 + H$	1.880E+09	0.00	31.25	41
738	$C_8H_9 = C_8H_8 + H$	3.160E+13	0.00	212.00	41
742	$C_8H_5 + C_2H_2 = l\text{-}C_{10}H_7$	4.000E+10	0.00	42.30	20
743	$l\text{-}C_{10}H_7 = C_{10}H_7$	1.000E+10	0.00	0.00	est.
750	$s\text{-}C_8H_5 + C_2H_2 = s\text{-}C_{10}H_7$	4.000E+10	0.00	42.30	est.
751	$s\text{-}C_{10}H_7 = C_{10}H_7$	1.000E+10	0.00	0.00	est.
753	$C_{10}H_7 + H = C_{10}H_8$	7.830E+10	0.00	0.00	20
754	$C_{10}H_7 + O_2 = C_{10}H_7O + O$	1.600E+09	0.00	31.25	est.
755	$C_{10}H_7O = C_9H_7 + CO$	3.600E+11	0.00	183.68	20
756	$C_{10}H_7O + H = C_{10}H_7OH$	2.000E+11	0.00	0.00	20
760	$C_{10}H_8 + H = C_{10}H_7 + H_2$	2.600E+11	0.00	66.94	20
761	$C_{10}H_8 + O = C_{10}H_7O + H$	2.500E+10	0.00	19.54	20
763	$C_{10}H_8 + OH = C_{10}H_7 + H_2O$	1.700E+05	1.42	6.07	20
764	$C_{10}H_8 + OH = C_{10}H_7OH + H$	1.400E+10	0.00	44.31	20
765	$C_9H_7 + H = C_9H_8$	5.000E+10	0.00	0.00	est.
766	$C_9H_8 + H = C_9H_7 + H_2$	5.000E+10	0.00	0.00	est.
767	$C_9H_8 + O = C_9H_7 + OH$	5.000E+10	0.00	0.00	est.
768	$C_9H_8 + OH = C_9H_7 + H_2O$	5.000E+10	0.00	0.00	est.
769	$C_9H_7 + O = C_8H_7 + CO$	1.000E+11	0.00	0.00	20
771	$C_8H_7 + C_2H_2 = a\text{-}C_{10}H_9$	4.000E+10	0.00	42.30	est.
773	$C_{10}H_8 + H = C_{10}H_9$	4.040E+10	0.00	18.04	20
776	$p\text{-}C_8H_7 + C_2H_2 = p\text{-}C_{10}H_9$	4.000E+10	0.00	42.30	est.
777	$p\text{-}C_{10}H_9 = t\text{-}C_{10}H_9$	1.000E+10	0.00	0.00	est.
778	$C_{10}H_8 + H = t\text{-}C_{10}H_9$	4.040E+10	0.00	18.04	20
781	$C_6H_5 + C_4H_4 = b\text{-}C_{10}H_9$	2.800E+10	0.00	42.30	est.
782	$b\text{-}C_{10}H_9 = l\text{-}C_{10}H_9$	1.000E+13	0.00	36.00	est.
784	$l\text{-}C_{10}H_9 = m\text{-}C_{10}H_9$	1.000E+13	0.00	188.00	text

Table 1 Continued

No	Reaction	A	n	E	Ref.
785	m-$C_{10}H_9$ = t-$C_{10}H_9$	1.000E+10	0.00	0.00	est.
786	C_9H_7 + $^1CH_2$ = t = $C_{10}H_9$	1.000E+11	0.00	0.00	20
787	C_7H_7 + C_3H_3 = $C_{10}H_{10}$	3.000E+10	0.00	0.00	20
794	p-C_7H_7 + C_3H_3 = f-$C_{10}H_{10}$	3.000E+10	0.00	0.00	20
802	C_5H_5 + C_5H_5 = $C_{10}H_9$ + H	2.000E+10	0.00	33.47	est.
810	C_6H_5 + a-C_3H_4 = i-C_9H_9	1.000E+10	0.00	0.00	est.
811	i-C_9H_9 = t-C_9H_8 + H	1.000E+13	0.00	150.00	est.
812	i-C_9H_9 = s-C_9H_8 + H	1.000E+13	0.00	138.00	est.
813	C_6H_5 + C_3H_3 = s-C_9H_8	1.000E+10	0.00	0.00	20
814	C_6H_5 + C_3H_3 = t-C_9H_8	1.000E+10	0.00	0.00	20
829	i-C_9H_9 = n-C_9H_9	1.000E+13	0.00	17.00	est.
830	n-C_9H_9 = s-C_9H_9	1.000E+10	0.00	0.00	est.
831	s-C_9H_9 = C_9H_8 + H	1.000E+13	0.00	137.00	est.
832	C_7H_7 + C_2H_2 = p-C_9H_9	1.000E+09	0.00	20.92	20
833	c-C_9H_9 = p-C_9H_9	1.000E+14	0.00	334.00	20
834	p-C_9H_9 = t-C_9H_8 + H	1.000E+13	0.00	137.00	est.
835	c-C_9H_9 = C_9H_8 + H	1.000E+13	0.00	90.00	est.
836	p-C_7H_7 + C_2H_2 = f-C_9H_9	1.000E+09	0.00	20.92	20
837	f-C_9H_9 = s-C_9H_9	1.000E+09	0.00	0.00	est.
838	C_6H_5 + C_4H_4 = g-$C_{10}H_8$ + H	2.800E+10	0.00	42.30	est.
839	C_6H_5 + C_4H_4 = j-$C_{10}H_8$ + H	2.800E+10	0.00	42.30	est.
840	g-$C_{10}H_8$ + H = t-$C_{10}H_9$	1.000E+11	0.00	0.00	est.
841	j-$C_{10}H_8$ + H = t-$C_{10}H_9$	1.000E+11	0.00	0.00	est.
842	C_9H_7 + CH_3 = C_9H_7-CH_3	1.000E+10	0.00	0.00	14
843	$C_9H_7CH_3$ = $C_9H_6CH_3$ + H	5.000E+15	0.00	314.00	14
844	$C_9H_6CH_3$ = $C_9H_6CH_2$ + H	5.000E+14	0.00	213.50	14
845	$C_9H_6CH_2$ = $C_{10}H_8$	8.000E+13	0.00	305.60	14
846	C_9H_7 + $^1CH_2$ = $C_9H_6CH_2$ + H	1.200E+11	0.00	0.00	est.
847	C_9H_7 + CH_2 = $C_9H_6CH_2$ + H	3.000E+10	0.00	0.00	est.
848	C_5H_5 + C_3H_3 = f-C_8H_7 + H	2.000E+10	0.00	26.00	est.
849	f-C_8H_7 = C_8H_6 + H	3.000E+13	0.00	242.20	est.
850	C_5H_5 + a-C_3H_5 = f-C_8H_9 + H	2.000E+10	0.00	26.00	est.
851	f-C_8H_9 = C_8H_8 + H	3.000E+13	0.00	172.40	est.
852	C_6H_5 + n-C_4H_3 = j-$C_{10}H_8$	1.000E+10	0.00	0.00	est.
853	C_6H_5 + t-C_4H_5 = m-$C_{10}H_9$ + H	1.000E+10	0.00	0.00	est.
854	C_6H_5 + a-C_3H_5 = i-C_9H_9 + H	1.000E+10	0.00	0.00	est.

via fulvene.[10,11] A directly analogous path (842)–(845), passing via methyl + indenyl to naphthalene, has subsequently been proposed by Laskin and Lifshitz[14] and the suggested rates and thermodynamic data (C_9H_7–CH_3, C_9H_6–CH_3, C_9H_6=CH_2) have been adopted.

$$C_9H_7 + CH_3 = C_9H_7\text{–}CH_3 \qquad (842)$$

$$C_9H_7\text{–}CH_3 = C_9H_6\text{–}CH_3 + H \qquad (843)$$

$$C_9H_6\text{–}CH_3 = C_9H_6\text{=}CH_2 + H \qquad (844)$$

$$C_9H_6\text{=}CH_2 = C_{10}H_8 \qquad (845)$$

Maurice[20] considered a step via $^1CH_2$ addition to indene and found this to be of some importance for all conditions studied and the current study therefore also considers the related steps proceeding via $^1CH_2$ (846) and $^3CH_2$ (847) addition.

$$C_9H_7 + {}^1CH_2 = C_9H_6\text{=}CH_2 + H \qquad (846)$$

$$C_9H_7 + {}^3CH_2 = C_9H_6\text{=}CH_2 + H \qquad (847)$$

The frequency factors for the above steps are based on analogy with methylene reactions with benzene (Böhland et al.[57]). An interesting issue does arise in the context of the possible generic

Table 2 Thermodynamic data for MSA and PAH species

Species	Structure	$\Delta_f H$/kJ mol^{-1}	S^0/J mol^{-1}
C_7H_7		215.5[20]	323.2[20]
p-C_7H_7		299.7[20]	339.0[20]
C_7H_8		50.0[20]	320.2[20]
C_8H_5		558.9[45]	356.4[45]
s-C_8H_5		627.0[20]	356.4[20]
C_8H_6		315.2[20]	333.2[20]
C_8H_7		389.0[45]	345.2[45]
p-C_8H_7		397.6[45]	368.4[45]
C_8H_8		148.4[20]	344.8[20]
C_9H_7		294.7	343.2
C_9H_8		163.2[45]	337.7[45]
s-C_9H_8		290.5[45]	370.2[45]
t-C_9H_8		299.4[45,est]	372.6[45,est]
i-C_9H_9		370.0[45,est]	394.9[45,est]
n-C_9H_9		386.5[45,est]	418.2[45,est]
s-C_9H_9		244.0[20]	337.7[20]

nature of sequences of the above type. The reaction $C_5H_5 + C_3H_3$ may, for example, lead to phenylacetylene and two such sequences are tentatively explored in the current work.

$$C_5H_5 + C_3H_3 = C_5H_4\text{=}C_3H_3 + H \tag{848}$$

$$C_5H_4\text{=}C_3H_3 = C_8H_6 + H \tag{849}$$

The barrier for (848) (26 kJ mol^{-1}) is consistent with H expulsion and the frequency factor 2×10^{10} m^3 kmol^{-1} s^{-1} is based on similarity with paths discussed above. The barrier for the

Table 2 Continued

Species	Structure	$\Delta_f H$/kJ mol^{-1}	S^0/J mol^{-1}
p-C$_9$H$_9$		377.9[45,est]	394.9[45,est]
c-C$_9$H$_9$		290.9[20]	394.9[20]
f-C$_9$H$_9$		358.0[45,est]	381.8[45,est]
C$_{10}$H$_7$		400.8[45,est]	361.7[45,est]
l-C$_{10}$H$_7$		617.3[45,est]	393.5[45,est]
s-C$_{10}$H$_7$		614.4[45,est]	405.8[45,est]
C$_{10}$H$_8$		150.6[45,est]	332.6[45,est]
C$_{10}$H$_7$O		115.2[20]	382.2[20]
C$_{10}$H$_7$OH		−25.7[45,est]	375.6[45,est]
C$_{10}$H$_9$		287.0[20]	338.0[20]
b-C$_{10}$H$_9$		425.0[20]	398.5[20]
l-C$_{10}$H$_9$		454.3[45,est]	421.8[45,est]
m-C$_{10}$H$_9$		451.0[20]	421.8[20]
p-C$_{10}$H$_9$		455.9[45,est]	399.7[45,est]
t-C$_{10}$H$_9$		269.0[est]	338.0[est]
C$_{10}$H$_{10}$		279.4[45,est]	420.0[45,est]
f-C$_{10}$H$_{10}$		259.5[45,est]	406.9[45,est]

second step is set to the heat of reaction. A directly analogous path leading to styrene via C_5H_5 + a-C_3H_5 has also been formulated (850) and (851). Table 1 and 2.

Computational results and discussion

Examples of predictions of intermediate species featuring in molecular growth sequences for the C_6H_6/O_2/Ar flame are given in Fig. 13 and 14. It should be noted that all species profiles in the benzene flame have been translated by 1.2 mm to account for probe effects. Concentrations of the propargyl radical, propyne, allene, ethylene and acetylene levels are reproduced with reasonable accuracy as shown in Fig. 13. Furthermore, the concentrations of key C_4 species are arguably within experimental uncertainties. Computed levels of C_4H_4 and C_4H_6, C_4H_5 and C_4H_3 isomers

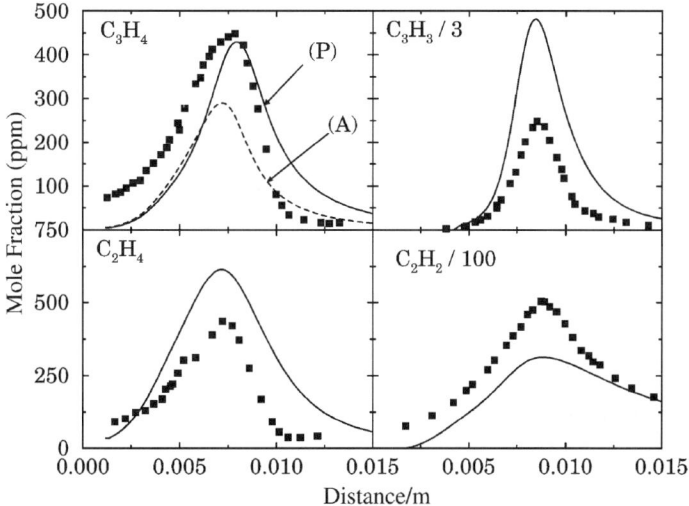

Fig. 13 Comparison between computed (lines) and experimental[19,22] (symbols) C_3H_4, C_3H_3, C_2H_4 and C_2H_2 in a rich ($\phi = 1.80$) C_6H_6/O_2/Ar flame. Isomeric forms are (P) = $HCCCH_3$ and (A) = H_2CCCH_2.

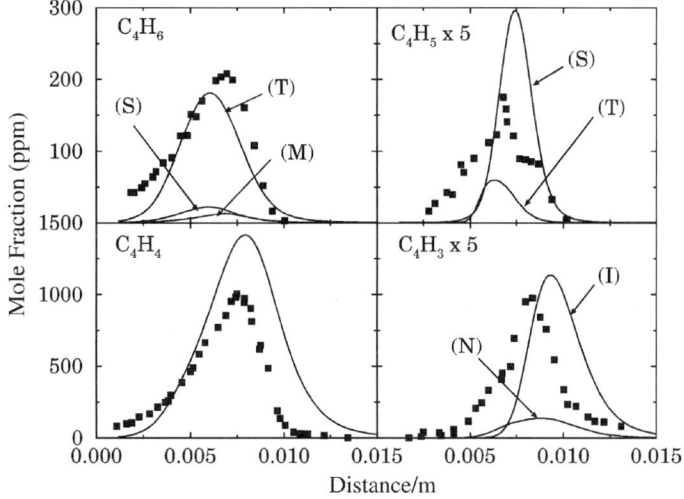

Fig. 14 Comparison between computed (lines) and experimental[19,22] (symbols) C_4H_6, C_4H_5, C_4H_4 and C_4H_3 in a rich ($\phi = 1.80$) C_6H_6/O_2/Ar flame. Isomeric forms are (T) = $CH_2CHCHCH$ and $CH_2CHCHCH_2$, (S) = CH_3CHCCH and $CHCHCCH_2$, (M) = $CHCCH_2CH_3$, (N) = $HCCCHCH$ and (I) = $HCCCCH_2$.

are compared with measurements in Fig. 14. The dominant isomers are 1,3-C_4H_6 (T), 1,2-C_4H_6 (S), 1,3-C_4H_5 (T), 1,2-C_4H_5 (S) and internally stabilised radicals such as i-C_4H_3. The thermodynamic data and lower hydrocarbon sub-mechanisms are described elsewhere.[28] Levels of MSA are reproduced in a predominantly satisfactory manner as is shown in Fig. 15. However, toluene appears to be a notable exception with concentrations over-predicted by a factor of 3. Poor predictions of the $CH_4/CH_3/^1CH_2/^3CH_2$ system could in principle be to blame given the dominant reaction channels discussed above. However, the experimentally observed peak concentration (e.g. $CH_4 \sim 1500$ ppm and $CH_3 \sim 1400$ ppm) compare reasonably well with the current computations ($CH_4 \sim 900$ ppm and $CH_3 \sim 1100$ ppm). The matter is thus discussed further in the context of the 1,3-C_4H_6/O_2/Ar flame results given below. It is also interesting to note that the tentatively proposed $C_5H_5 + C_3H_3$ and $C_5H_5 +$ a-C_3H_5 channels are competitive in this flame and contribute >30% to styrene and >10% to phenylacetylene formation. The channels appear worthy of further investigation.

The computed levels of the phenyl and cyclopentadienyl radicals show a notable improvement as compared to earlier studies[5,18] (Fig. 16). For example, the more recent study by Richter et al.[18] indicates errors close to 300% and 900% for C_6H_5 and C_5H_5 respectively. Comparisons between computed and experimental naphthalene and indene profiles are also shown for the benzene flame in Fig. 16. The agreement is very satisfactory. The dominant naphthalene formation channels are (802) and (773) cyclopentadienyl radical recombination (50%), HACA (N_1,N_2) sequences (25%) and (N_7) propargyl addition to the methylphenyl radical (5%). Sequences passing via $C_9H_6=CH_2$ (842)–(844), (846), (847) contribute around 5% of the total. It thus appears that there is no major inconsistency between the cyclopentadienyl radical recombination route (802) and (773) and the current benzene flame. However, the uncertainties remain significant and use of the alternative sequence (802a), (803a) effectively shuts down the path. Predictions of mass 116 are also satisfactory and phenyl radical addition to allene leading directly to the i-C_9H_9 radical contributes approximately 10% to the total indene formation rate. However, indene is predominantly (50%) formed by propargyl radical addition to the phenyl radical leading to s-C_9H_8 and t-C_9H_8 formation. The latter constitute the dominant C_9H_8 isomers in both flames and subsequently undergo H radical addition leading to i-C_9H_9 formation and isomerisation to indene. Naphthalene oxidation contributes around 15% of the total. In view of the importance of the C_3 based sequences, a new step $C_6H_5 +$ a-$C_3H_5 = $ i-C_9H_9 + H (854) was considered and found not to be competitive with a contribution <5%. Finally, acetylene addition to the methylphenyl radical also contributes

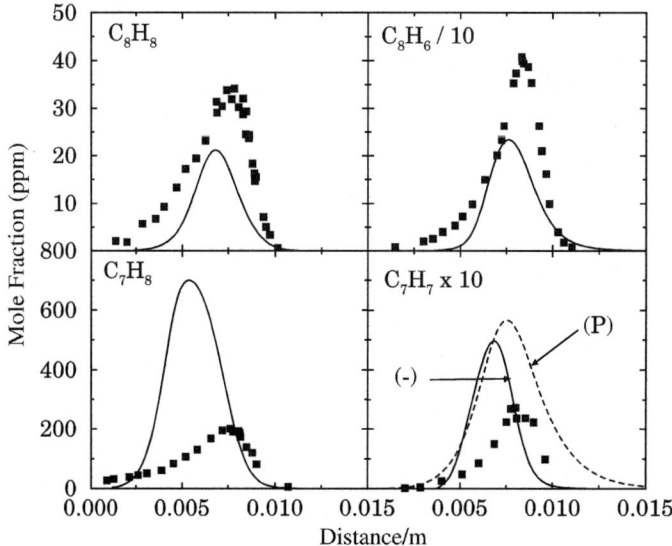

Fig. 15 Comparison between computed (lines) and experimental[19,22] (symbols) C_8H_8, C_8H_6, C_7H_8 and C_7H_7 in a rich ($\phi = 1.80$) C_6H_6/O_2/Ar flame. Isomeric forms are (—) benzyl radical and (P) methylphenyl radical.

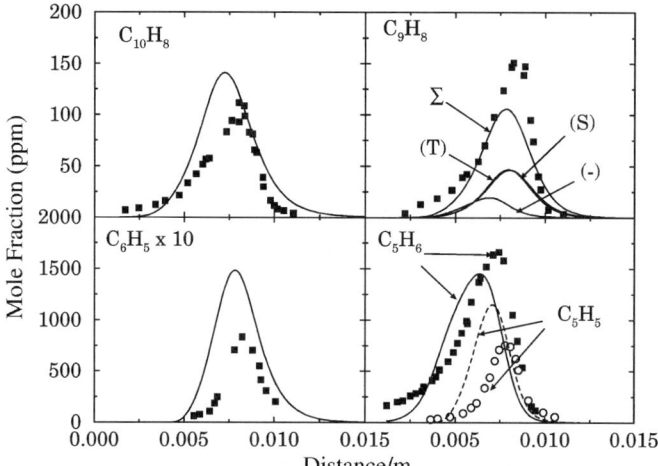

Fig. 16 Comparison between computed (lines) and experimental[19,22] (symbols) $C_{10}H_8$, C_9H_8, C_6H_5, C_5H_5 and C_5H_6 in a rich ($\phi = 1.80$) $C_6H_6/O_2/Ar$ flame. Isomeric forms are (—) indene, (S) = ϕ-CHCCH$_2$ and (T) = ϕ-CH$_2$CCH.

around 10%. It may be noted that the concentrations of the s-C_9H_8 and t-C_9H_8 isomers exceed those of indene by a factor of 2–3. This is perhaps initially somewhat surprising, though there is experimental evidence to suggest that the ratio computed may not be unreasonable.[58]

For the 1,3-$C_4H_6/O_2/Ar$ flame it may be noted that concentrations of the propargyl radical, propyne, allene, ethylene and acetylene levels are reasonably well reproduced as shown in Fig. 17. The major discrepancy can be found in the peak concentration of acetylene. However, past studies[29,56] have reported similar difficulties and it has been shown[59] that C_2H_2 levels are sensitive to the branching of the $C_2H_3 + O_2$ reaction. However, the latter topic has recently been discussed at some length[60] and the changes required in branching ratios in order to improve agreement with measurements are not justified. Levels of MSA are reproduced in a reasonable manner as is shown in Fig. 18. Somewhat surprisingly, given the similarity of the formation and destruction channels with the benzene flame discussed above, toluene levels are under-predicted.

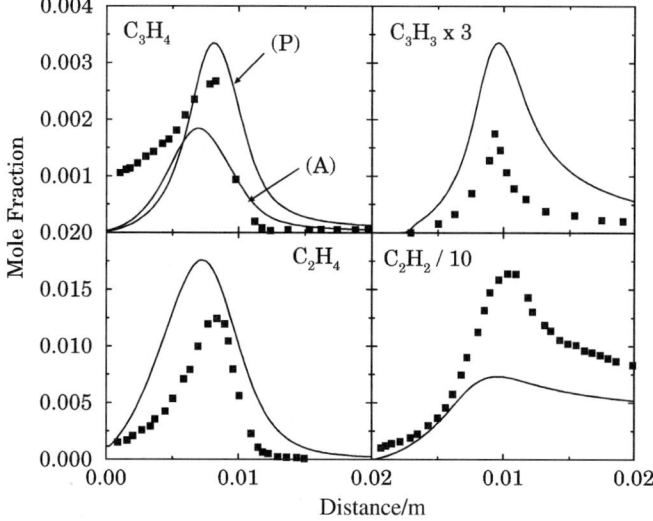

Fig. 17 Comparison between computed (lines) and experimental[23,57] (symbols) C_3H_4, C_3H_3, C_2H_4 and C_2H_2 in a rich ($\phi = 2.40$) 1,3-$C_4H_6/O_2/Ar$ flame. Isomeric forms are (P) = HCCCH$_3$ and (A) = H$_2$CCCH$_2$.

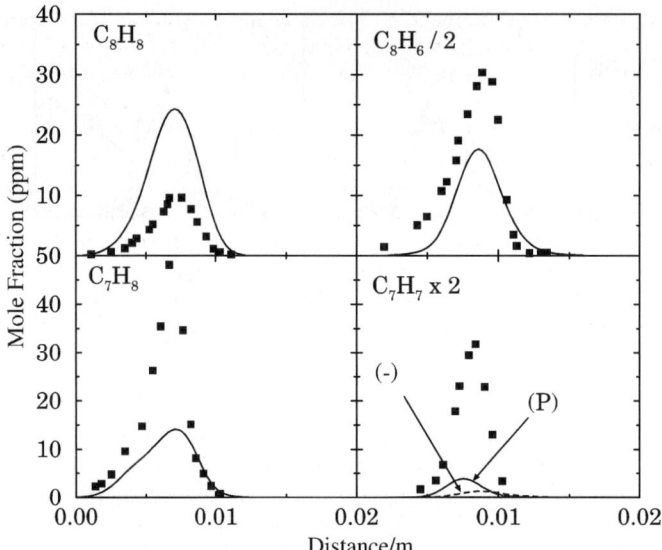

Fig. 18 Comparison between computed (lines) and experimental[23,57] (symbols) C_8H_8, C_8H_6, C_7H_8 and C_7H_7 in a rich ($\phi = 2.40$) 1,3-C_4H_6/O_2/Ar flame. Isomeric forms are (—) benzyl radical and (P) methylphenyl radical.

Further progress on this issue appears difficult at this point and it may also be noted that the MSA sub-mechanism provides reasonable predictions for other flames.[30] The computed levels of benzene, fulvene, vinylacetylene and the dominant 1,2-C_4H_5 isomer are shown in Fig. 19.

In the butadiene flame, $C_{10}H_8$ levels are somewhat under-predicted, while total C_9H_8 levels are again satisfactorily reproduced. Indene formation again proceeds predominantly *via* the C_6H_5 + C_3H_x routes (90%). The dominant naphthalene formation sequences are (N_5) acetylene addition to the ethylphenyl radical (30%), (N_4) vinylacetylene addition to the phenyl radical (25%) and cyclopentadienyl radical recombination (25%). Despite the 188 kJ mol^{-1} barrier for the *trans–cis*

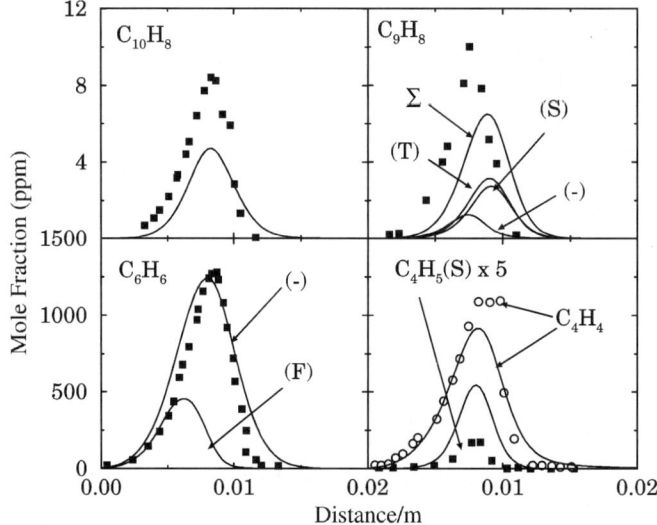

Fig. 19 Comparison between computed (lines) and experimental[23,57] (symbols) $C_{10}H_8$, C_9H_8, C_6H_6 and C_4H_5 in a rich ($\phi = 2.40$) 1,3-C_4H_6/O_2/Ar flame. Isomeric forms are C_9H_8: (—) indene, (S) = ϕ-CHCCH$_2$ and (T) = ϕ-CH$_2$CCH. C_6H_6: (—) = benzene and (F) = fulvene. C_4H_5: (S) = CH$_3$CHCCH.

isomerisation (784), the contribution made by this sequence is significant. The under-prediction of naphthalene in the butadiene flame is interesting and suggests that additional formation paths are probably required. Indeed, it may be readily shown that vinyl acetylene addition at rates close to those used by Maurice[20] and Appel et al.[53] do lead to a significant improvement. The additional $C_6H_5 + CH_2CHCHCH$ and $C_6H_5 + HCCCHCH$ sequences do, however, not contribute significantly. Furthermore, the three-step sequence passing via phenylbutatriene or phenyl-butyn-3-ene and leading on to naphthalene via phenyl-1,3-butadien-2-yl decomposition to 1-hydronaphthalen-2-yl is not significant. Further work is evidently required and, as shown by McEnally and Pfefferle[16] in their work on ^{13}C-labelled aromatic fuel dopants, uncertainties pertaining to the relative contributions of phenyl and cyclopentadienyl based sequences are perhaps even greater in diffusion flames. One implication of their work is that the cyclopentadienyl sequence does not dominate mass growth and the present work has shown the potential of a number of alternative sequences.

Conclusions

A multiple channel reaction mechanism featuring detailed kinetic paths for second-ring formation is proposed and evaluated through the study of a wide range of fuel mixtures containing aromatic components. Computations have also been performed in low-pressure laminar premixed benzene and buta-1,3-diene flames in order to assess the relative importance of proposed naphthalene and indene formation pathways. The study has shown that several competitive mechanisms for the formation of naphthalene and indene are plausible and that multiple second-ring formation paths are necessary to explain observed levels. It has also been shown that major and key intermediate species can be computed reasonably well and that the growth of benzene and MSA molecules are predicted with an accuracy similar to that achieved for precursor species. Furthermore, the latter applies for a wide range of fuel blends and conditions. Finally, computed levels of naphthalene and indene are not unreasonable. It is recognised that the second-ring formation models evaluated here are speculative. In particular, the thermodynamic and kinetic rate data used for five-membered ring species is conjectural. However, key reactions and thermodynamic parameters affecting the proposed second-ring formation sequences have been identified through an extensive sensitivity analysis. While more accurate determinations of rate constants and thermodynamic data remain necessary the present work forms a good basis for further developments.

Acknowledgement

The authors are grateful to Dr. G. Skevis for discussions.

References

1 H. Richter and J. B. Howard, *Prog. Energy Combust. Sci.*, 2000, **26**, 565.
2 J. A. Miller and C. F. Melius, *Combust. Flame*, 1992, **91**, 21.
3 K. M. Leung and R. P. Lindstedt, *Combust. Flame*, 1995, **102**, 129.
4 P. Dagaut and M. Cathonnet, *Combust. Flame*, 1998, **113**, 620.
5 R. P. Lindstedt and G. Skevis, *Combust. Flame*, 1994, **99**, 551.
6 J. L. DiNaro, J. B. Howard, W. H. Green, J. W. Tester and J. W. Bozzelli, *J. Phys. Chem. A*, 2000, **104**, 10576.
7 K. Roy, C. Horn, P. Frank, V. G. Slutsky and T. Just, *Twenty-Seventh Symposium (International) on Combustion*, The Combustion Institute, 1998, p. 329.
8 C. Horn, K. Roy, P. Frank and T. Just, *Twenty-Seventh Symposium (International) on Combustion*, The Combustion Institute, 1998, p. 321.
9 A. M. Dean, *J. Phys. Chem.*, 1990, **94**, 1432.
10 C. F. Melius, M. E. Colvin, N. M. Marinov, W. J. Pitz and S. M. Senkan, *Twenty-Sixth Symposium (International) on Combustion*, The Combustion Institute, 1996, p. 685.
11 L. V. Moskaleva, A. M. Mebel and M. C. Lin, *Twenty-Sixth Symposium (International) on Combustion*, The Combustion Institute, 1996, p. 521.
12 I. Hahndorf, H. Y. Lee, A. M. Mebel, S. H. Lin, Y. T. Lee and R. I. Kaiser, *J. Chem. Phys.*, 2000, **113**, 9622.
13 E. Ikeda, R. S. Tranter, J. H. Kiefer, R. D. Kern, H. J. Singh and Q. Zhang, *Proc. Combust. Inst.*, 2000, **28**, 1725.

14 A. Laskin and A. Lifshitz, *Twenty-Seventh Symposium (Int.) on Combustion*, The Combustion Institute, 1998, p. 313.
15 M. Frenklach and H. Wang, in *Mechanisms and Models of Soot Formation*, ed. H. Bockhorn, Springer-Verlag, Berlin, 1994, p. 165.
16 C. S. McEnally and L. D. Pfefferle, *Proc. Combust. Inst.*, 2000, **28**, 2569.
17 N. M. Marinov, W. J. Pitz, C. K. Westbrook, A. M. Vincitore, M. J. Castaldi and S. M. Senkan, *Combust. Flame*, 1998, **114**, 192.
18 H. Richter, W. J. Grieco and J. B. Howard, *Combust. Flame*, 1999, **119**, 1.
19 J. D. Bittner, Sc.D. Thesis, Massachusetts Institute of Technology, 1981.
20 L. Q. Maurice, PhD Thesis, University of London, 1996.
21 N. W. Moriarty and M. Frenklach, *Proc. Combust. Inst.*, 2000, **28**, 2563.
22 J. D. Bittner and J. B. Howard, *Eighteenth Symposium (International) on Combustion*, The Combustion Institute, 1980, p. 1105.
23 J. A. Cole, J. D. Bittner, J. P. Longwell and J. B. Howard, *Combust. Flame*, 1984, **56**, 51.
24 C. B. Vaughn, PhD Thesis, Massachusetts Institute of Technology, 1988.
25 C. B. Vaughn, J. B. Howard and J. P. Longwell, *Combust. Flame*, 1991, **87**, 278.
26 C. B. Vaughn, W. H. Sun, J. B. Howard and J. P. Longwell, *Combust. Flame*, 1991, **84**, 38.
27 R. P. Lindstedt and L. Q. Maurice, *Combust. Sci. Technol.*, 1996, **120**, 119.
28 R. P. Lindstedt and G. Skevis, *Combust. Sci. Technol.*, 1997, **125**, 73.
29 R. P. Lindstedt and G. Skevis, *Twenty-Sixth Symposium (International) on Combustion*, The Combustion Institute, 1996, p. 703.
30 R. P. Lindstedt, *Twenty-Seventh Symposium (International) on Combustion*, The Combustion Institute, 1998, p. 269.
31 M. U. Alzueta, M. Oliva and P. Glarborg, *Int. J. Chem. Kinet.*, 1998, **30**, 683.
32 J. L. Emdee, K. Brezinsky and I. J. Glassman, *J. Phys. Chem.*, 1992, **96**, 2151.
33 V. S. Rao and G. B. Skinner, *J. Phys. Chem.*, 1988, **92**, 2442.
34 K. M. Pamidimukkala, R. D. Kern, M. R. Patel, H. C. Wei and J. H. Kiefer, *J. Phys. Chem.*, 1987, **91**, 2148.
35 H. Kraus, C. Oehlers, F. Temps and H. Gg. Wagner, *J. Phys. Chem.*, 1993, **97**, 10989.
36 R. D. Kern, H. J. Singh and C. H. Wu, *Int. J. Chem. Kinet.*, 1988, **20**, 731.
37 D. L. Baulch, C. J. Cobos, R. A. Cox, C. Esser, P. Frank, J. A. Th. Just Kerr, M. J. Pilling, J. Troe, R. W. Walker and J. Warnatz, *J. Phys. Chem.*, 1992, **21**, 411.
38 D. Robaugh and W. Tsang, *J. Phys. Chem.*, 1986, **90**, 4159.
39 M. Braun-Unkhoff, P. Frank and Th. Just, *Ber. Bunsen-Ges. Phys. Chem.*, 1990, **94**, 1417.
40 J. Herzler and P. Frank, *Proceedings of the Anglo-German Combustion Symposium*, ISBN 0 952035006, 1993, p. 164.
41 W. Müller-Markgraf and J. Troe, *Twenty-First Symposium (International) on Combustion*, The Combustion Institute, 1986, p. 815.
42 S. Fahr and S. E. Stein, *Twenty-Second Symposium (International) on Combustion*, The Combustion Institute, 1988, p. 1023.
43 C.-Y. Lin and M. C. Lin, *J. Phys. Chem.*, 1986, **90**, 425.
44 J. Herzler and P. Frank, in *Soot Formation in Combustion: Mechanisms and Models*, ed. H. Bockhorn, Springer-Verlag, Berlin, 1994, p. 11.
45 S. E. Stein, *NIST Standard Reference Database 25, Version 2.0*, 1994.
46 C. F. Melius, *BAC-MP4 Thermochemical Database*, SANDIA National Laboratories, 1996.
47 N. M. Marinov, W. J. Pitz, C. K. Westbrook, M. J. Castaldi and S. M. Senkan, *Combust. Sci. Technol.*, 1996, **116**, 211.
48 M. B. Colket and D. J. Seery, *Twenty-Fifth Symposium (International) on Combustion*, The Combustion Institute, 1994, p. 883.
49 R. P. Lindstedt and G. Skevis, *207th ACS National Meeting*, 1994, American Chemical Society, Washington, DC, vol. 39, no. 1, p. 147.
50 A. Burcat and B. J. McBride, *Technion-Israel Institute of Technology Report No. TAE697*, 1994.
51 H. Wang and M. Frenklach, *Combust. Flame*, 1997, **110**, 173.
52 K. M. Leung, R. P. Lindstedt and L. Q. Maurice, *216th ACS National Meeting*, American Chemical Society, Washington, DC, 1998.
53 J. Appel, H. Bockhorn and M. Frenklach, *Combust. Flame*, 2000, **121**, 122.
54 M. B. Colket, R. J. Hall and M. D. Smooke, *Mechanistic Models of Soot Formation*, Final Report on AFOSR Contract F49620-91-C-0056, 1994.
55 U. Alkemade and K. H. Homann, *Z. Phys. Chem. (Neu Folge)*, 1989, **161**, 19.
56 A. Goldaniga, T. Faravelli and E. Ranzi, *Combust. Flame*, 2000, **122**, 350.
57 T. Böhland, K. Heberger, F. Temps and H. Gg. Wagner, *Ber. Bunsen-Ges. Phys. Chem.*, 1989, **93**, 80.
58 J. A. Cole, PhD Thesis, Massachusetts Institute of Technology, 1984.
59 G. Skevis, PhD Thesis, University of London, 1996.
60 R. P. Lindstedt and G. Skevis, *Proc. Combust. Inst.*, 2000, **28**, 1801.

The influence of fuel additives on the behaviour of gaseous alkali-metal compounds during pulverised coal combustion

H. Schürmann,[a] S. Unterberger,[a] K. R. G. Hein,[a] P. B. Monkhouse[b] and U. Gottwald[b]

[a] *Institute of Process Engineering and Power Plant Technology, University of Stuttgart, Pfaffenwaldring 23, 70569 Stuttgart, Germany.*
   E-mail: schuermann@ivd.uni-stuttgart.de
[b] *Institute of Physical Chemistry, University of Heidelberg, Im Neuenheimer Feld 253, 69120, Heidelberg, Germany. E-mail: d63@ix.urz.uni-heidelberg.de*

Received 6th March 2001
First published as an Advance Article on the web 18th October 2001

The alkali-metal vapour release during pulverised hard (bituminous) coal combustion was investigated in a semi-technical drop flow reactor in the temperature range 1100–1400 °C. Absolute concentrations of total gas-phase sodium and potassium species were determined using the *in situ*/on-line excimer laser induced fragmentation fluorescence technique (ELIF). Alkali-metal concentrations measured for the untreated coals were found to be in the range 0.1 to 4.7 ppm, depending on the temperature. As well as observing the temperature dependence, the effect of co-feeding defined amounts of silica and clay minerals was studied. In addition, to assist interpretation of ELIF measurements, ash samples were taken and analysed by scanning electron microscopy (SEM) and energy dispersive X-ray analysis (EDX). The additives lead to a pronounced binding of the alkali-metal species and suppression of the sharp temperature dependence observed without co-feeding. Therefore, the use of such getter materials can be confirmed as an effective way to remove corrosive alkali-metal species from the flue gas in pulverised coal combustion.

Introduction

Alkali-metal compounds constitute one of the main reasons for fouling and slagging in coal-fired boilers and down-stream components such as turbines. Condensed alkalis form sticky surfaces, onto which ash particles adhere, building up contaminating layers.[1] At higher temperatures, *e.g.* in the region of superheated surfaces, these layers can form slag and grow into stable deposits, which hinder optimal heat transfer. In power stations based on fluidised bed technology, agglomeration of the bed material (sand and ash) is a serious problem, since it prevents fluidisation. A further consequence of alkali-metal concentrations in the flue gas is high temperature corrosion of metallic construction materials. This can occur *e.g.* following sulfation of the alkali-metal species and deposition on superheated or convective surfaces.[1,3] Finally, sodium- or potassium-bound chlorine in contact with metallic plant materials (heat exchanger) can react with iron to form iron chloride, which has a high vapour pressure in this temperature region and leads to strong erosion of the wall material.[4]

If the flue gases are to be used hot, *e.g.* to run a gas turbine, knowledge of the concentrations and behaviour of gas-phase alkali-metal compounds under plant conditions will become a particularly important aspect of economically and environmentally sustainable energy production. Indeed, some turbine manufacturers fear damage even at ppb levels of alkali-metals in the flue gas concentration. Conventional measurements on alkali-metal species are mostly performed by wet chemical methods, whereby the alkali-metal-containing flue gas is led through washing bottles for 1–2 h and then analysed for its sodium and potassium content *e.g.* by flame spectrometry. This method is laborious, prone to errors and lacks temporal resolution. This means that a possible large, sudden increase in alkali-metal concentration will be averaged out. In the present work, a laser spectroscopic technique is used which determines the actual alkali-metal concentrations on-line and *in situ* (*i.e.* without sampling).[5-13] Sensitivities down to 0.1 ppb and time resolutions down to 12 s can be achieved. Furthermore, the described method is sensitive essentially only to gas-phase species of sodium and potassium.[7]

The objective of the present study was to investigate the effect of fuel composition and temperature on alkali-metal release under pulverised combustion conditions, whereby a German hard coal from the Ensdorf mine, Saarland was combusted in a laboratory-scale drop-flow reactor. Measurements were made *in situ* using the excimer laser induced fragmentation (ELIF) method and the results were related to the conditions and the actual fuel composition.

Alkali-metal species in coal

Coal contains alkali metals (sodium and potassium) in various chemical and physical forms. The alkali minerals present in coal can be in the form of organometallic salts, chlorides, sulfates, carbonates and silicates. The distribution of sodium and potassium between the organic and mineral coal substance and different alkali-metal species changes with coal rank. Bituminous coal contains predominantly mineral-bound alkali metals, mainly silicates, whereas alkali-metal species in the lower rank coals are predominantly organic-bound components and water soluble inorganic salts (*e.g.* chlorides). Alkali-metal distribution between the mineral and organic phases of coal directly influences the release of alkali-metal during combustion. Organic-bound alkali-metal tends to vaporise more easily while the mineral-bound alkali-metal may be retained within the coal aluminosilicate co-ordination environments.[1,2] This is illustrated by the following examples.

Wen *et al.*[2] studied alkali-metal volatilisation from an American hard coal in a micropyrolysis furnace with atomic absorption spectroscopy. Volatilisation of sodium and potassium is shown to occur in two phases, the first starting at 1150–1200 °C and extending up to 1540 °C, the second commencing at about 1870 °C. The first phase of release is deemed to be due to alkali-metal of organic, weakly bound origin, the second phase corresponded to alkali-metal of strongly bound, mineral origin. The organically bound part of the alkali-metal could also be largely removed by washing the coal in hot water. A further finding is that in the first phase, the release of sodium is much higher than that of potassium, showing that the latter is bound predominantly in the mineral phase. The influence of the binding form in the coal is also demonstrated in combustion experiments on pulverised coal by Neville and Sarofim[14] who obtained substantially higher gas-phase concentrations of sodium from a brown coal than from three different hard coals, in which sodium is mainly present in mineral form. The authors assumed that sodium is released by a simple evaporation process. They also found that coals with a stoichiometric excess of chlorine emitted more sodium into the gas-phase. Sodium is released up to 1630 °C, after which the ash melts. Then, an increased rebinding into the solid phase could be observed, a process favoured for fuels with a high silica content.

Behaviour of alkali metals during combustion

The behaviour of alkali-metal species in combustion has been discussed by various authors *e.g.* Osborne,[1] Thambimuthu[15] and Wibberley and Wall.[16] Based on model simulations linked to databases containing thermodynamic data, predictions of the chemical species most likely to be formed during combustion may be made as a function of temperature. The condensed-phase alkali-metal species are largely silicates and sulfates, in the gas-phase mainly NaCl and NaOH are found. Under pulverised combustion conditions and for coal chlorine contents above approximately 0.5 wt.% (as in this work), NaCl is the dominant gas-phase species up to about 1450 °C. Only above this temperature does NaOH become more dominant in the flue gas. On the other hand, for low chlorine contents, NaOH becomes the dominant species at much lower temperatures, *e.g.* for 0.05 wt.% Cl, the proportion of NaOH exceeds that of NaCl from about 1200 °C.[16]

Although predictions based on thermodynamic equilibrium calculations alone usually provide useful information, it has to be remembered that they do not necessarily produce an accurate picture of the proportions of trace elements in the respective phases. In recent years, more sophisticated models have emerged, see *e.g.* ref. 17. Factors such as coal-specific concentrations, element-specific volatilities and partitioning among different particle size classes must also be included. In order to verify models and to gain a complete picture of the partitioning process, a broad base of experimental data in all phases is needed. Since the ELIF method discriminates towards gas-phase species, the data obtained by this technique should be used together with information gained from complementary methods, *e.g.* wet chemical, surface ionisation and plasma assisted spectroscopy.[7]

Although a number of previous studies[5–13,18–23] describe experimental efforts to characterise alkali-metal release in solid-fuel conversion systems, there are few direct, on-line measurements and particularly few systematic studies down to 100 ppb and below, which is regarded as a critical level for many types of boiler and turbine materials. Clearly, the availability of such on-line data will be a requirement for assessing potential filter and getter materials as well as characterising different fuels and fuel blends during combustion with special regard to possible operational problems that result from high alkali-metal concentrations in the flue gas.

Experimental

Excimer laser induced fragmentation fluorescence (ELIF)

The ELIF method uses excimer laser light at 193 nm to photodissociate alkali-metal compounds and, simultaneously, to excite electronically the alkali-metal atoms formed. Fluorescence from the excited $Na(3^2P)$ or $K(4^2P)$ atoms is readily detected in the visible region. The ELIF method has been applied, as a truly *in situ* optical technique to analyse alkali-metal species in the flue gas, to several small-scale coal combustion and gasification systems.[5–13]

Experimental set-up. The complete ELIF system consists of the following parts: An optical access for introduction of the excitation light into the flue gas region and for collection of fluorescence emission. A pulsed UV light source, together with a suitable detector for continuous monitoring of the emitted energy; in the applications to date, an ArF-excimer laser at 193 nm was used. A detection system, using two channels, one for sodium and one for potassium consisting of optical filter, photomultiplier and electronics. A data control, acquisition and evaluation system.

Optical access. The set up for ELIF used for these measurements at the drop tube reactor at IVD, University of Stuttgart, is shown in Fig. 1. The optical access consisted of four ports holding quartz windows. Suprasil quartz is essential for the laser access because of the short laser wavelength, but is also preferred for thermal stability, therefore the detection windows were also made of this material. Windows were mounted in flanges of thermally/mechanically stable materials, to ensure that they withstood the actual operating conditions and to minimise heat loss by the flanges. The windows were also flushed continually with nitrogen, to keep them free of dust.

The geometry shown in Fig. 1 is the most usual one for fluorescence experiments, but other arrangements are feasible. Indeed, a single-window system was recently designed and demonstrated,[6,7] which enables the application of the laser beam near the wall of the flue gas pipe. In fact, it is only necessary for the beam to illuminate a small section (*ca.* 1 cm^3) of the flue gas to generate an ELIF signal. The single-window system is suitable for large flue gas ducts and conditions of high temperature and pressure.

Laser excitation; fluorescence detection. Alkali-metal compounds in the flue gas are generally photolysed using laser energy densities of several mJ cm^{-2} and repetition rates of 6.4 Hz. Then, if ELIF signals are averaged *e.g.* over 50 shots, a time resolution of 12 s is obtained. The laser energy entering and leaving the optical access is monitored constantly and provides a measure of the effective beam transmission. Fluorescence from excited sodium and potassium atoms is detected by two separate photomultipliers coupled to boxcar integrators, using suitable time gates. To suppress undesired radiation due *e.g.* to incandescence, atomic line filters (0.2 nm width/central

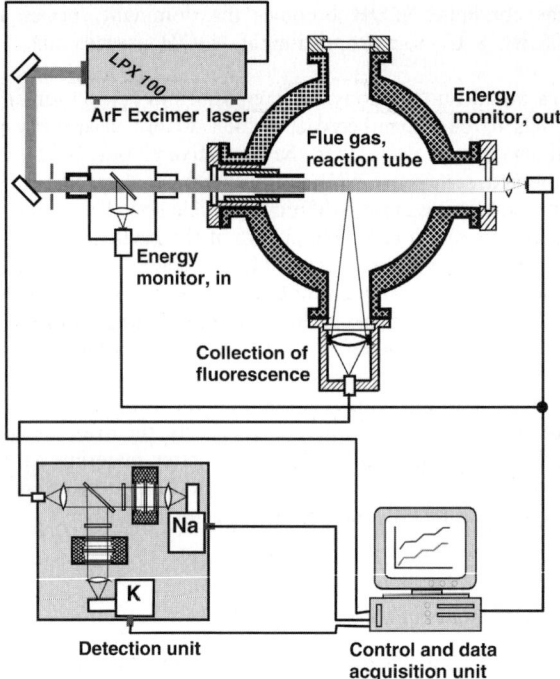

Fig. 1 Block diagram of the ELIF set-up.

wavelength 589 nm for sodium and 1 nm/768 nm for potassium) are placed in front of the detectors. Alternatively, a monochromator may be used to filter fluorescence light. Exchangeable neutral density filters avoid detector saturation at high alkali-metal concentration levels and reduce the remaining background radiation. In this set-up an optical fibre cable was used to transmit the fluorescence light from the optical access to the detection system.

Calibration. For calibration measurements, a quartz reference cell with four optical access ports was set up. The cell was encased in a tubular oven, heatable to 1200 °C. NaCl or KCl was put into the attached reservoir, which was heated separately, so that the vapour pressure of the salt could be varied independently.[24,25] To prevent salt condensation, the oven was constructed with effective thermal insulation material and the cell was maintained at least 50 °C hotter than the reservoir. Evacuated quartz tubes at the ports served both for thermal insulation and to minimize absorption of the laser beam in the hot oven atmosphere.

The gas-phase alkali-metal compounds MX are photodissociated by UV-laser light (193 nm), whereby simultaneously excited Na(3^2P) and K(4^2P) atoms, represented here by M* are formed:

$$MX + hv \rightarrow M^* + X$$

Fluorescence from M* is detected at 589 nm (sodium) and 768 nm (potassium), respectively. A calibration factor C_{det} can be determined using the alkali-metal number densities in the calibration cell obtained from thermodynamic data and the following relation for the fluorescence signal S_{fl}:

$$S^{fl} = C_{det} E_{laser} n_{MX} a_{abs} q$$

where E_{laser} is the effective laser energy in the cell, n_{MX} the number density of the alkali-metal salt MX, a_{abs} is the UV absorption cross section for production of M*, and q describes the collisional quenching effect:

$$q = A/(A + Q)^{-1} \quad \text{with} \quad Q = \sum n_i v_i \sigma_i$$

with A the inverse natural lifetime of the excited alkali-metal atom and Q the collisional quenching rate. The value of Q was determined using measured quenching cross sections σ_i,[25] the number densities n_i of the quenching partners i and the average relative velocities v_i of i and M*. The main quenching partners in the flue gas were N_2, CO_2, O_2 and H_2O with typical mole fractions of 0.73, 0.16, 0.035 and 0.075 respectively. The average laser energy (E_{laser}) at the measurement point was obtained from the effective transmission, which in the present set-up could be determined directly using energy detectors before and after the flue gas pipe. However, laser beam absorption by flue gases such as O_2 and CO_2 have been studied in detail[26] and those data can be used to calculate the net beam absorption in the flue gas.

It will be seen that the maximum concentrations evaluated in this work are around 4.7 ppm. Since the degree of photodissociation of the alkali-metal compounds of interest is less than 1%[12,24] (leading to very low atom concentrations) a correction for radiation trapping will not be required. For alkali-metal compound concentrations above approx. 20 ppm, corrections can be made using experimental calibration data together with Monte-Carlo modelling.[12]

Combustion test facility

The combustion process and the formation of pollutants in pulverized coal firings are the subject of extensive investigations carried out on experimental facilities at IVD. The experimental results discussed in this article were obtained from experiments with an atmospheric pressure drop flow reactor.

The furnace is illustrated in Fig. 2. The electrically heated ceramic tube is 2500 mm long and has a diameter of 200 mm. The electrical heating is divided into five different zones in order to adjust a constant wall temperature as well as a temperature profile along the furnace. The wall temperature can be set at up to 1500 °C. The burner, where the coal is injected into the reaction chamber, is located at the top of the furnace. The feeding system consists of a gravimetric conveyor and a screw feeder and the pulverized coal is fed to the burner by carrier air. The residence time of the flue gas and particles in the ceramic tube depends on the coal mass flow and the combustion air. The coal mass flow corresponds to approximately 0.5 kg h^{-1} for the different fuel mixtures investigated. The combustion air is injected through annular clearances and is divided into primary and secondary air. The facility provides a good environment for carrying out parametric investigations on the combustion process of pulverized fuels. The application of laser spectroscopic measurement systems such as ELIF is facilitated by different optical access ports at the test facility.

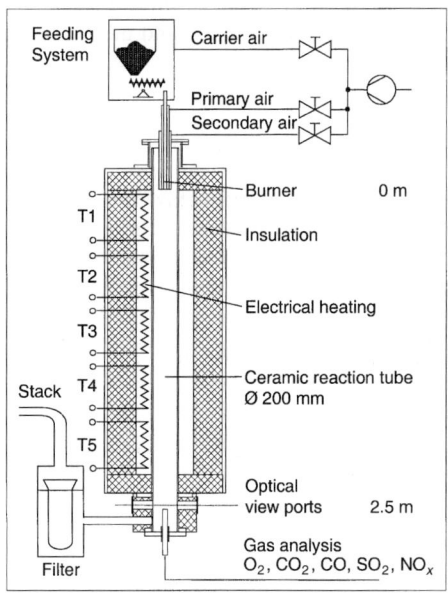

Fig. 2 Scheme of the test facility, drop flow reactor.

Table 1 Coal and additive analysis

	Ensdorf, raw coal (wt.%)	Clay (wt.%)	Aerosil (wt.%)
Moisture	3.46	9.53	
Volatile	7.03	89.57	
Ash	33.85		
Fixed C	55.66		
C	80.12	0.34	
N	1.7	0.04	
S	0.9	0.52	
O	1.41		
H	4.91		
Cl	0.47		
Moisture	3.46		
Ash	7.03		
Ash:			
SiO_2	38.7	58.9	>99.8
Al_2O_3	23.8	23.22	<0.05
Fe_2O_3	12.3	2.14	<0.01
TiO_2	0.9	1.47	<0.03
MgO	2.2	0.66	
CaO	8.9	0.07	
Na_2O	0.5	0.08	
K_2O	2.7	2.82	
MnO	0.15		
P_2O_5	0.1		
Others	9.75	0.21	

Fuel and fuel mixtures

A hard coal from the Ensdorf revier/Saarland, Germany was used as reference fuel for the experiments. The pulverised coal has an average particle size (d_{50}) of 61 µm. Various additives were added to the hard coal in order to investigate the effect of fuel composition on the release and binding behaviour of gas-phase alkali-metal species. For the experiments clay ($d_{50} = 5$ µm) from the brown coal open-cut mine Hambach (Rheinland), and industrially produced, pure SiO_2 (R972 Aerosil, Degussa Hüls; $d_{50} = 16$ nm) were used. The two additives were doped into the coal in a series of different proportions. The various fuel mixtures and fuel analyses are given in Tables 1 and 2.

The release and capture mechanisms of gaseous alkali metals are dependent on the residence time of the particles in the reactor. To ensure comparability of the measured results, the residence time was adjusted to 10 s by varying the fuel mass flow and combustion air. For all experiment settings the oxygen content of the flue gas was kept constant at 3 vol.%.

After the combustion of raw coal, doped coal ash samples were collected and analysed by a scanning electron microscope from Leitz-ISI, DS-130.

Table 2 Investigated fuel mixtures

	Ensdorf, raw coal (wt.%)	Clay (wt.%)	Aerosil (wt.%)
1	100	—	—
2	99	—	1
3	98.5	—	1.5
4	98	—	2
5	98	2	—
6	95	5	—

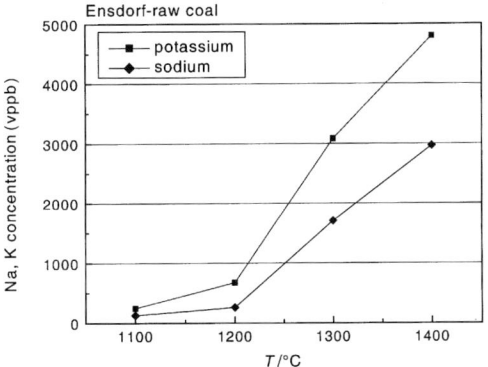

Fig. 3 Sodium and potassium concentration *vs.* temperature for Ensdorf raw coal.

Results

In general, a different furnace temperature was established and held constant each day. At each temperature, a series of measurements was performed in the presence and absence of the respective additives. In this way the dependence of flue gas alkali-metal concentration on temperature alone and on amount of additive was obtained. The results are summarised in Fig. 3–5. The values shown represent the concentrations determined when stable temperature conditions within the test facility had been reached. Factors contributing to the total uncertainty of the measurements, which amounts to 25–30%, are laser beam absorption by the flue gas components as well as collisional quenching effects, signal statistics, and the quality of the calibration measurements.

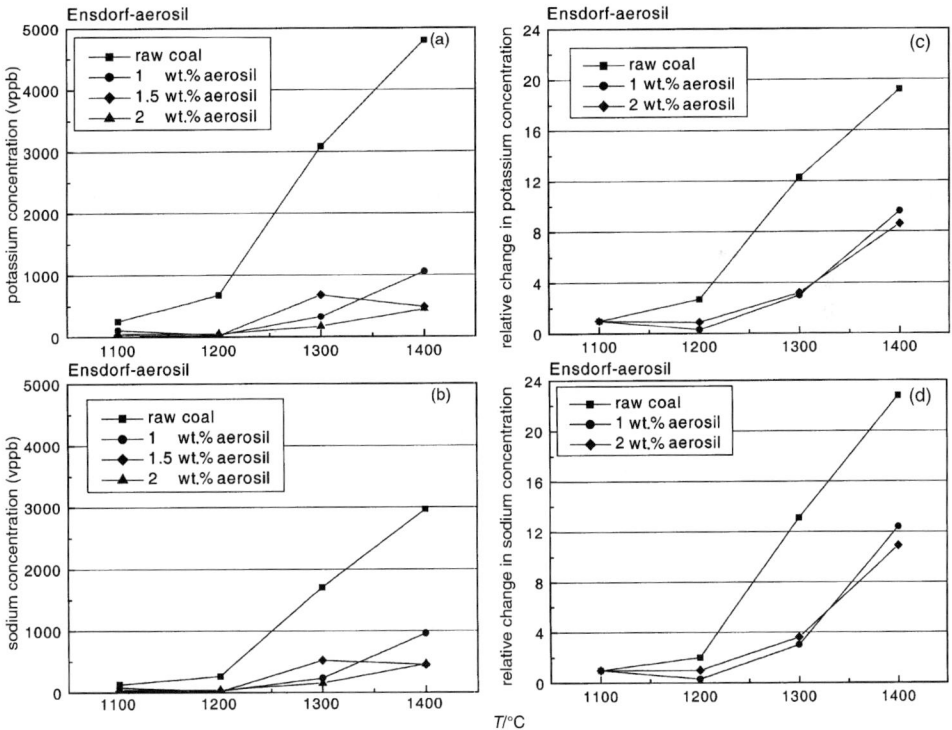

Fig. 4 (a) Potassium concentration *vs.* temperature for different fuel mixtures of raw coal and aerosil. (b) Sodium concentration *vs.* temperature for different fuel mixtures of raw coal and aerosil. (c) Relative change in potassium concentration *vs.* temperature for different fuel mixtures of raw coal and aerosil. (d) Relative change in sodium concentration *vs.* temperature for different fuel mixtures of raw coal and aerosil.

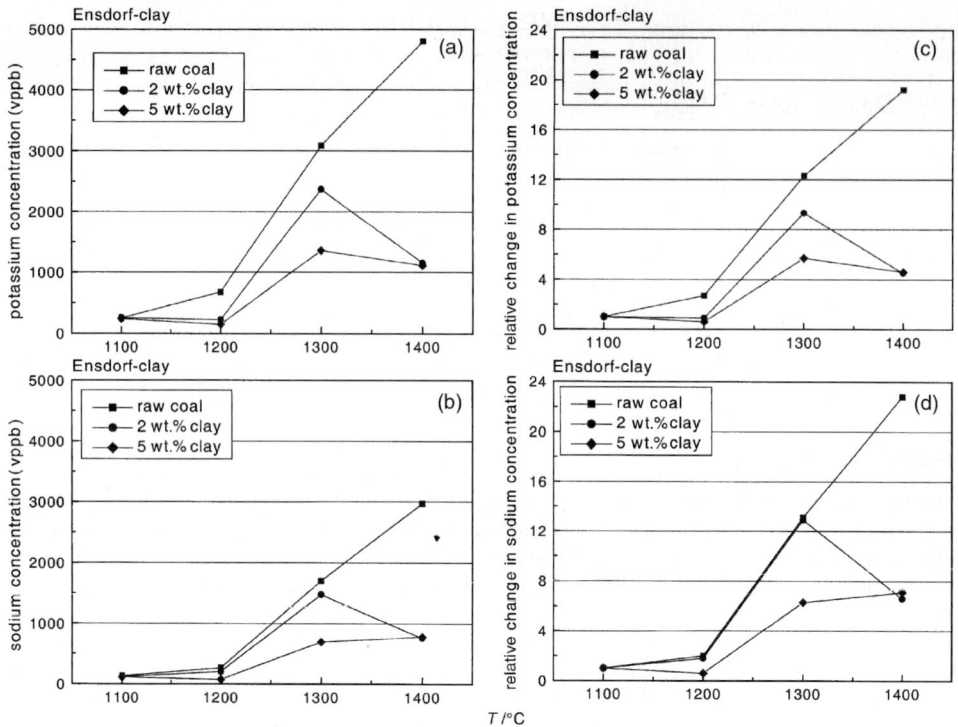

Fig. 5 (a) Potassium concentration *vs.* temperature for different fuel mixtures of raw coal and clay. (b) Sodium concentration *vs.* temperature for different fuel mixtures of raw coal and clay. (c) Relative change in potassium concentration *vs.* temperature for different fuel mixtures of raw coal and clay. (d) Relative change in sodium concentration *vs.* temperature for different fuel mixtures of raw coal and clay.

Raw coal

The average levels of alkali-metal release at the lowest temperature studied, 1100 °C, are 245 ppb for potassium and 120 ppb for sodium (Fig. 3). The higher value for potassium reflects the much higher amount 2.7 wt.% of this element in the coal compared to 0.5 wt.% of sodium. It is seen that the alkali-metal concentrations rise steadily with increasing temperature up to 4.7 ppm for potassium and 2.9 ppm for sodium at 1400 °C. The sharper increase in alkali-metal concentration between 1200 and 1300 °C than between 1100 and 1200 °C may be associated with the softening temperature of the ash, which for this coal is 1270 °C. The temperature behaviour shown in Fig. 3 is expected and is qualitatively similar to that obtained in previous studies.[8] However, the *absolute* values of the alkali-metal concentrations should not be compared since the coal used in ref. 8 was obtained from a different charge and had a somewhat different composition, as shown by its elemental and ash analysis.

Coal with additives

The addition of aerosil (silica) except at 1100 °C, leads to a significant decrease in measured alkali-metal concentration, due to binding of sodium and potassium with the silica (Fig. 4(a) and (b)). The largest effect was observed for a furnace temperature of 1400 °C, whereas the effect at 1100 °C was insignificant. For a temperature of 1400 °C the emission of gaseous potassium decreased from 4802 vppb (raw coal) to 447 vppb (2 wt.% aerosil). The data are also shown normalised to the concentrations at 1100 °C (Fig. 4(c) and (d)); in this way it can be seen that the addition of aerosil leads to a much less severe rise in alkali-metal emission with increasing temperature. Increasing the amount of aerosil from 1 to 2 wt.% in the fuel mixture had only a marginal effect.

Fig. 5(a) and (b) show the effect of adding clay mineral on the combustion process. Again, at the high temperatures, the binding effect is strong, whereas for lower temperatures, smaller or insignificant reductions were observed. A decrease from 4802 to 1107 vppb can be observed using fuel with 5 wt.% clay at high temperature. Fig. 5(c) and (d) show the corresponding effect on the temperature dependence normalised to the concentrations at 1100 °C.

Discussion

In theory, the ELIF signals measured at 589 and 768 nm would represent the sums of different compounds of sodium and potassium, respectively. However, as explained in the Introduction on the alkali-metal behaviour during combustion, the level of chlorine in the present coal (0.5 wt.%) is high enough for the dominant species in the flue gas to be the chlorides.[16] Hydroxides will only be important for very low chlorine concentrations and temperatures above about 1200 °C. It should be noted in this context that the release of alkali-metal is very sensitive to the amount of chlorine in the coal; this was demonstrated experimentally in several previous studies.[5,13,23] If on the other hand, conditions are such that the hydroxides are of importance, there is a possibility of discriminating them from the chlorides using an additional detection wavelength. Since the hydroxides have slightly smaller bond strengths, irradiation of NaOH and KOH with laser light of 193 nm leads to population of some higher atomic levels of sodium and potassium. In this way, several additional wavelengths can be detected for ELIF on the hydroxides. For the chlorides, the spectral energy is only sufficient to reach the $Na(3^2P)$ and $K(4^2P)$ levels. This method has been demonstrated experimentally for the sodium species under gasification conditions,[11] but requires some further development for use in a variety of practical applications.

There is reasonable evidence that provided relatively low laser energies are used, the ELIF method can provide a measure of the gas-phase species present in the flue gas. First, experiments by Helble et al.[13] directly on solid-phase alkali-metal species showed that no ELIF signal could be obtained. More recently, ELIF measurements were performed in the flue gas of a PFBC reactor simultaneously with measurements using surface ionisation (SI) and plasma spectroscopy (PEARLS) methods.[7] These two techniques measure alkali metals in both gas and particle form. In general, the concentrations measured by SI and PEARLS were higher than those determined by ELIF and the ratio between the respective data varied systematically with flue gas temperature.

In general, the initial alkali-metal release is mainly determined by the amount of alkali-metal bound in an easily volatilisable form, e.g. bound to the organic coal matrix, and the temperature of the combustion process. Binding or uptake reactions, on the other hand, can occur over the whole of the residence time where the flue gas and solid (or liquid) particles are in contact and depend on the chemical and physical properties of the getter material present, including the available getter surface, as well as the alkali-metal gas concentration and the process temperature. It should be noted that the measured alkali-metal concentrations represent the sum of these different processes. In practice, however, it can be said that qualitatively, an increase in measured concentration reflects mainly a release-controlled process, whereas a decrease can be ascribed predominately to an uptake-controlled reaction.

Gas-phase alkali-metal species released in the combustion process can be captured by components of the mineral substance in the fuel itself or in additives doped into the fuel. The alkali-metal species in the gas phase are strongly basic, therefore acidic or amphoteric compounds, which can undergo acid–base reactions are appropriate as getter materials/additives. The acid character of getters or fuel mineral substances derives mainly from SiO_2 and Al_2O_3. On the other hand, Steffin et al.[18] showed that the addition of CaO to the ash, i.e. increasing its basic character, drastically reduces the sorption property of the getter.

Reaction with silica additives

The reactions of sodium species with silica have been investigated previously e.g. ref. 19–22, 27 and 28 and were reviewed by Wall.[29] For sodium chloride, these reactions include

$$2NaCl + 1/2 O_2 + SiO_2 \rightarrow Na_2O \cdot SiO_2 + Cl_2 \tag{1}$$

$$2NaCl + H_2O + 2SiO_2 \rightarrow Na_2O \cdot 2SiO_2 + 2HCl. \tag{2}$$

Now, the effectiveness of silica as a binding agent depends on several factors. In general, the rate of NaCl capture is found to increase with temperature.[19,20,22,27–29] However, the extent of silicate formation is limited by progressive agglomeration and coalescence of the ash at higher temperatures, which can be seen from electron micrographs.[27,28] In addition, the surface area of the silica particles is probably important; in fact, improved retention of alkali-metal for additives with smaller particle size has been observed.[21,28] In the present work, the aerosil particles had an average diameter of 16 nm, and therefore were much finer than in previous investigations. Thus a strong binding effect can be expected.

To assist interpretation of the measurements with aerosil additive, scanning electron micrographs were recorded for the conditions of interest. As shown in Fig. 6(a), the addition of 2 wt.% aerosil and a reaction tube temperature of 1100 °C results in mainly unshaped and unmelted particles. Increasing the reaction tube temperature to 1400 °C, see Fig. 6(b), leads mainly to spherical and melted particles with a small diameter. The melting phase and the formation of finer particles support binding of gaseous alkali metals. A similar particle behaviour can also be

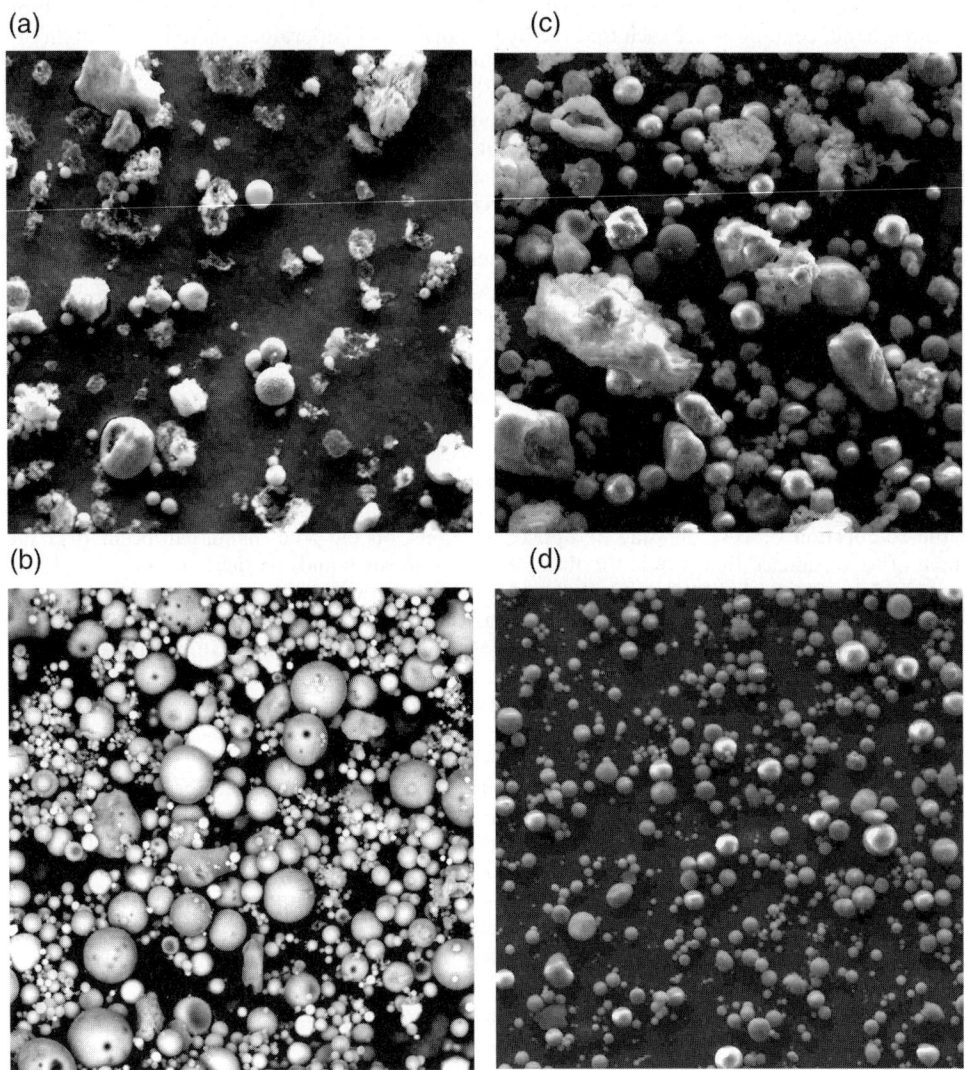

Fig. 6 (a) SEM photograph of ash sample from fuel mixture of raw coal with 2 wt.% aerosil at 1100 °C. (b) SEM photograph of ash sample from fuel mixture of raw coal with 2 wt.% aerosil at 1400 °C. (c) SEM photograph of ash sample from raw coal at 1100 °C. (d) SEM photograph of ash sample from raw coal at 1400 °C.

observed during the combustion of raw coal at the same temperatures, see Fig. 6(c) and (d). The binding of gaseous alkali metals in the ash during combustion of raw coal was not so effective compared with the raw coal/aerosil fuel mixture. Consequently the process of alkali-metal capture at high temperatures is mainly determined by the chemical composition of the mineral phase of the coal or coal/additive mixture. As a result of the EDX analysis it can be stated that the ash particles from fuel mixture raw coal/aerosil generally contain silica and aluminium, known as suitable elements for getter materials.[18,21,22] The ash particles of the raw coal additionally contain high amounts of Ca and Fe, which are both basic components of ash and which repress the getter effect.[18]

Reaction with clay additives

A number of investigations demonstrate the effectiveness of clay minerals containing both alumina and silica in binding alkali metals.[4,7,8,16–20] Furthermore, experiments by Steffin[18] on model substances in synthetic gas atmospheres showed that the uptake proceeds more efficiently under oxidising conditions. In the presence of O_2 (or H_2O), clays generally operate by incorporating Na (or K) into the aluminosilicate structure with the formation of *e.g.* nepheline or albite, whereby chlorine (or HCl, respectively) is released:

$$2NaCl + 1/2O_2 + Al_2O_3 \cdot 2SiO_2 \rightarrow Na_2O \cdot Al_2O_3 \cdot 2SiO_2 + Cl_2 \tag{3}$$

$$2NaCl + H_2O + Al_2O_3 \cdot 2SiO_2 \rightarrow Na_2O \cdot Al_2O_3 \cdot SiO_2 + 2HCl. \tag{4}$$

Under inert (N_2, CO_2) or reducing conditions (CO in N_2), gaseous alkali metals are bound reversibly into the mineral (metakaolin in the work of Steffin). Generally speaking, no chemical reaction occurs with the mineral material and the NaCl is released on heating.

Kyi and Chadwick[21] tested a number of mineral additives for their effectiveness in binding alkali metals released from lignite coals. They showed, by ashing coal samples in air and subsequent thermochemical analysis, that minerals with high contents of both alumina and silica were most effective at retaining NaCl, whereas minerals with only high silica or alumina amounts were often poor retainers. Various Al- and Si-containing getter materials were also studied by McLaughlin[30] and Uberoi *et al.*,[22] who found that minerals such as bauxite, montmorillonite, kaolin, emathlite, show very good uptake properties.

As well as the fundamental studies mentioned above, the effects of additives were confirmed in several combustion investigations. In an addition to the present work, ref. 8 describes further measurements in a drop-flow furnace, where a clay of high aluminosilicate content, which was added to the combustion process, led to effective alkali-metal gettering. Finally a study, performed on a bituminous coal under PFBC conditions and using the ELIF measurement method[5] employed kaolin, one of the getters recommended by Kyi and Chadwick, to suppress alkali-metal release. Again, a strong binding effect was observed.

Conclusions

The results presented in this work show a strong dependence of alkali-metal release on combustion temperature. With increasing reaction tube temperature increasing concentrations of gaseous alkali metals, sodium and potassium, are observed in the flue gas. The increase is particularly strong after the softening temperature of the ash has been reached. To suppress this release, co-feeding of coal with additives such as silica and clay minerals can be recommended. In the case of potassium, the emission of gaseous alkali metals at high temperatures decreases by a factor of 10 on addition of aerosil and a factor of 4.5 on addition of clay. Furthermore, SEM and EDX analysis showed that the binding effect of getter materials at high temperatures is generally influenced more by the chemical composition of the mineral phase in the coal and less by the particle behaviour and structure of the mineral phase.

Further work will involve screening of different types of coals/coal blends and additives as well as comparing ELIF data with measurements using other methods such as wet chemical, surface ionisation *etc.* in order to obtain information about the partitioning of alkali-metal into different phases under various conditions. In addition, with the drive towards renewable energy sources, similar studies in biomass combustion will be of interest, since there, large amounts of alkali-metal species are released due to the lack of mineral matter in the fuels.

Acknowledgements

Technical assistance by C. Klaiber is gratefully acknowledged.

References

1. G. A. Osborn, *Fuel*, 1992, **71**, 131.
2. C. S. Wen, L. H. Cowell, F. J. Smit, J. D. Boyd and R. T. LeCren, *Fuel*, 1992, **71**, 219.
3. S. H. D. Lee, K. Natesan and W. M. Swift, *Report ANL/FE-95-01, Argonne National Laboratory*, 1995.
4. M. Born and P. Seifert, "*Thermodynamische Berechnung zur chlorinduzierten Korrosion an Heizflächen von Feuerungsanlagen*", VGB Forschungsprojekte 145 and 153, 1997, ISSN 0937-0188.
5. U. A. Gottwald, P. B. Monkhouse and B. Bonn, *Fuel*, 2001, in press.
6. U. A. Gottwald and P. B. Monkhouse, *Appl. Phys. B*, 1999, **69**, 151.
7. V. Häyrinen, R. Hernberg, R. Oikari, U. Gottwald, P. Monkhouse, K. Davidsson, B. Lönn, K. Engvall, J. B. C. Pettersson, R. Kuivalainen and P. Lehtonen, *Proceedings, 6th International Conference on Circulating Fluidised Beds, Würzburg*, 1999, p. 873.
8. T. Reichelt, H. Spliethoff and K. R. G. Hein, *4th International Symposium and Exhibition on Gas Cleaning at High Temperatures, Karlsruhe*, 1999.
9. F. Greger, K. T. Hartinger, P. B. Monkhouse, J. Wolfrum, H. Baumann and B. Bonn, *Proc. Combust. Inst.*, 1996, **26**, 3301.
10. P. G. Griffin, R. J. S. Morrison, A. Campisi and B. L. Chadwick, *Rev. Sci. Instrum.*, 1998, **69**, 3674.
11. B. L. Chadwick, G. Domazetis and R. J. S. Morrison, *Anal. Chem.*, 1995, **67**, 710.
12. B. L. Chadwick and R. J. S. Morrison, *J. Chem. Soc., Faraday Trans.*, 1995, **91**, 1931.
13. J. J. Helble, S. Srinivasachar, D. O. Ham, A. A. Boni, O. Charon and A. Modestino, *Combust. Sci. Technol.*, 1992, **81**, 193.
14. M. Neville and A. F. Sarofim, *Fuel*, 1985, **64**, 384.
15. K. V. Thambimuthu, *IEA/53 Report*, IEA Coal Research, London, 1993.
16. L. J. Wibberley and T. F. Wall, *Fuel*, 1982, **61**, 87.
17. J. Helble, *Fuel Process. Technol.*, 2000, **63**, 125.
18. C. Steffin, W. Wanzl and K. H. van Heek, *Proceedings, 9th International Conference on Coal Science*, DGMK Tagungsbericht 9703, Hamburg, 1997, p. 1055; C. Steffin, Ph.D. Thesis, Essen, 1998.
19. P. O. Mwabe and J. O. L. Wendt, *Proc. Combust. Inst.*, 1996, **26**, 2447.
20. N. B. Gallagher, T. W. Peterson and J. O. L. Wendt, *Proc. Combust. Inst.*, 1996, **26**, 3197.
21. S. Kyi and B. L. Chadwick, *Fuel*, 1998, **78**, 845.
22. M. Uberoi, W. A. Punkjak and F. Shadman, *Prog. Energy Combust. Sci.*, 1990, **16**, 205.
23. S. H. D. Lee, F. G. Teats, W. M. Swift and D. D. Banerjee, *Combust. Sci. Technol.*, 1992, **86**, 327.
24. R. C. Oldenburg and S. L. Baughcum, *Anal. Chem.*, 1986, **58**, 1430.
25. K. T. Hartinger, S. Nord and P. B. Monkhouse, *Appl. Phys. B*, 1997, **64**, 363.
26. K. T. Hartinger, S. Nord and P. B. Monkhouse, *Appl. Phys. B*, 2000, **70**, 133.
27. E. R. Lindner and T. F. Wall, *Proc. Combust. Inst.*, 1990, **23**, 1313.
28. L. J. Wibberley and T. F. Wall, *Fuel*, 1982, **61**, 93.
29. T. F. Wall, *Proc. Combust. Inst.*, 1992, **24**, 1119.
30. J. McLaughlin, Ph.D. Thesis, University of Surrey, Guildford, 1990.

General Discussion

Prof. Cheskis opened the discussion of Prof. Warnatz's paper: You considered only surface reactions. Does it mean that gas phase reactions are not important? At what temperature and pressure must they be taken into account?

Prof. Warnatz responded: Gas phase reactions can be processed by the code as well, and the mechanisms are available.[1] However, it would need greater temperature and/or pressure[2] to have considerable contribution of the gas phase (*e.g.*, >1000 K at atmospheric pressure).

1   J. Warnatz, U. Maas and R. W. Dibble, *Combustion*, Springer, Heidelberg, 1996, 2nd edn. 1999, 3rd edn. 2001.
2   O. Deutschmann, L. D. Schmidt and J. Warnatz, in *Scientific Computing in Chemical Engineering II*, ed. F. Keil, W. Mackens, H. Voss and J. Werther, Springer, Berlin, 1999, pp. 368–375.

Prof. Golden asked: Is there a chemical role for H_2O in this process?

Prof. Warnatz answered: Water does not play an important role in surface chemistry, because its surface bond energy is small, leading to fast desorption (reaction (12) in Table 3).

Prof. Troe asked: Could you explain in more detail why the hydrocarbon profiles are not well modelled?

Prof. Warnatz replied: There is a severe lack of information on the detailed surface decomposition/oxidation of C_3H_6. We are sure that there are additional reaction paths with more blockage of the surface not listed in the reaction mechanism. This would retard chemical reaction and, thus, lead to a better agreement with the measurements on the fuel-rich side (see Fig. 3 and 7 of our paper).

Prof. Golden said: You stressed the need for data on clean surfaces. Isn't the surface inherently dirty?

Prof. Warnatz replied: Of course, the catalyst has to be cleaned from evident poisons (silicon, sulfur, *etc.*). But it is not adequate to speak of a 'dirty' surface; the surface is just prepared by the reaction, and this status normally is reproducible. The problem, however, is to get information on this surface structure during reaction. Prominent cases to show this behavior are reproducible reconstructions of the surface structure (see, *e.g.*, ref. 1 and 2).

1   R. Ducros and R. P. Merrill, *Surf. Sci.*, 1976, **55**, 227.
2   R. Kissel-Osterrieder, F. Behrendt and J. Warnatz, *Proc. Combust. Inst.*, 2001, **28**, 1323.

Prof. Lindstedt said: The apparent success of the 'mean field' approach under the current conditions is very interesting. One possible interpretation would be that structural effects on the surface might be less important than perceived. Could the authors possibly comment?

Prof. Warnatz answered: We use a 'mean-field' approximation, averaging the specific impact of different surface–catalyst structures, because of our limited knowledge of the structural effects. Nevertheless, this approach has been shown to provide a working tool for modeling heterogeneous catalysis at ambient conditions for several systems such as oxidation of light hydrocarbons. After preconditioning the catalyst, a relatively stable composition of surface structures is formed, and the reaction rates at atmospheric pressure and reasonably high temperatures seem to be determined rather by the overall availability of adsorbed species than by structural effects. The 'mean-field' approximation can even be applied if the effect of surface structures becomes significant and the associated kinetic data are known; here different mean-field based mechanisms can be set up for different surfaces (*e.g.*, crystal) structures.[1]

1 R. Kissel-Osterrieder, F. Behrendt, J. Warnatz, U. Metka, H.-R. Volpp and J. Wolfrum, *Proc. Combust. Inst.*, 2001, **28**, 1341.

Prof. Pilling asked: Prof. Warnatz has pointed out that the adsorption and desorption processes form the rate determining steps for many of the sub-mechanisms. Many of the available rate data derive from studies of idealized surfaces. How realistic are these data for the kinetic applications involved and how adequately can competitive adsorption be modelled?

Prof. Warnatz responded: As mentioned in the presentation, we use a 'mean-field' approach, averaging the specific influence of different sites. This is done because of lack of data for the different potential sites. The influence of different surface structures can be included, if sufficient data exist. This is, *e.g.*, the case for CO oxidation, where—as shown in the presentation—different mechanisms can be written down for step and terrace sites.[1] However, data on specific low-indexed surface planes are valuable and usable for polycrystalline surfaces as well (especially in a concept using step and terrace sites), because polycrystalline surface show a tendency to have low-indexed surfaces for energetic reasons.

1 R. Kissel-Osterrieder, F. Behrendt, J. Warnatz, U. Metka, H.-R. Volpp and J. Wolfrum, *Proc. Combust. Inst.*, 2001, **28**, 1341.

Dr Hessler said: If I recall correctly, many reactions on surfaces occur at imperfections such as glide planes or grain boundaries. Have you included these in your mechanism and are they important?

Prof. Warnatz replied: No, we have not included different sites on the surface. As mentioned in the presentation, we use a 'mean-field' approach, averaging the specific influence of different sites. This is done because of lack of data for the different potential sites. The influence of different surface structures can be included, if sufficient data exist. This is the case, *e.g.*, for CO oxidation, where—as shown in the presentation—different mechanisms can be written down for step and terrace sites (see ref. 1) or different time-dependent surface structures are present (see, *e.g.*, ref. 2).

1 R. Kissel-Osterrieder, F. Behrendt, J. Warnatz, U. Metka, H.-R. Volpp and J. Wolfrum, *Proc. Combust. Inst.*, 2001, **28**, 1341.
2 R. Kissel-Osterrieder, F. Behrendt and J. Warnatz, *Proc. Combust. Inst.*, 2001, **28**, 1323.

Prof. Hayhurst said: As you know, diffusion of reactants towards a surface (and also of products away from the catalyst) can dominate the overall kinetics at high temperatures or with a surface of low curvature. How did you ensure that diffusion between the gas-phase and the catalyst was not rate-controlling?

Prof. Warnatz replied: Gas-phase diffusion to and from the surface is considered in any case, because it is included in the Navier–Stokes equations solved. Transport limitations can easily be detected considering the temperature and concentration profiles in the boundary layer.[1]

1 J. Warnatz, U. Maas and R. W. Dibble, *Combustion*, Springer, Heidelberg, 1996, 2nd edn. 1999, 3rd edn. 2001.

Prof. Lin said: In the 1980s, we studied the dynamics of OH radical desorption from Pt surfaces, including wires, foils and (111) single crystals in the catalytic oxidation of H_2, H_2O and NH_3 by O_2, N_2O and NO_2.[1] We observed a strong dependence of desorption energies on the [fuel]/[oxidant] ratio. I wonder if you could account for the dependence with your kinetic model?

1 See for example, M. C. Lin and G. Ertl, *Ann. Rev. Phys. Chem.*, 1986, **37**, 587.

Prof. Warnatz replied: The strong dependence of OH desorption energy on the [fuel]/[oxidant] ratio could be explained using two different reaction mechanisms on two different sites, where one site (*e.g.*, a step) has a large OH desorption energy and one (*e.g.*, a terrace) has a low OH desorption energy.

Dr Hughes† said: In Table 3 of your paper, around twenty reactions describe the propene surface reaction mechanism, most of which seem to have estimated rate constants with fast kinetics. Would it not be more appropriate to replace these by three or four pseudo-global reactions whose rates have a better chance of being elucidated?

Prof. Warnatz responded: The use of pseudo-global reactions, of course, would give the same simulation results. However, the use of elementary reactions has the advantage, *e.g.*, to be able to identify rate-limiting steps by sensitivity analysis, which gives important information to experimentalists on urgently needed studies.

Dr Tomlin asked: Is it not true that the sensitivities themselves depend on the value of the rate constant used?

Prof. Warnatz replied: Yes, in principal sensitivities depend on rate coefficients, which can be quantitatively expressed by second-order sensitivities. In practice, no quantitative knowledge is used for sensitivities, just the information, whether they are large or negligible. Thus, second-order sensitivity is not important in practice; to be uncautious: I do not know any case, where they have proved to be important.

Dr Hessler said: At the CHEMKIN Workshop that proceeded the Combustion Symposium in Edinburgh last year I presented a short comment on how Monte Carlo techniques may be used to estimate the uncertainties of a simulation that result from the uncertainties of the rate coefficients and thermodynamic parameters in the reaction mechanism. Have you used this or another approach to estimate the uncertainties of your predictions?

Prof. Warnatz answered: We are glad to have a working model. It is too early to quantify potential uncertainties.

Prof. Golden said: As a point of information, the subject of the importance of considering the product of sensitivity and uncertainty has been addressed for atmospheric purposes by Smith and co-workers.[1]

1 G. P. Smith, M. K. Dubey, D. E. Kinnison and P. S. Connell, *J. Phys. Chem.*, 2001, **105**, 1449.

Prof. Warnatz replied: I have used similar pictures, plotting the product of sensitivity and uncertainty.[1]

1 J. Warnatz, *Twenty-fourth Symp. (Int.) Combust.*, The Combustion Institute, Pittsburgh, 1992, p. 553.

Prof. Lindstedt said: Two interesting issues not discussed in the paper relate to (i) the sensitivity of predictions and measurements to uncertainties in (surface) temperature distributions and thus heat transfer in the monolith and (ii) to the spatial resolution requirements. Clarification would be helpful.

Prof. Warnatz responded: (i) In the experiment, we tried to achieve isothermal conditions over the whole catalytic monolith. However, small spatial temperature variations, mainly caused by reaction heat release, could not be prevented. Based on heat balance studies, axial temperature measurements, and observed variations of the experimental data, we estimated an overall temperature uncertainty of approximately 30 K. In Fig. 1–3 of the paper, a vertical error bar of 30 K for the experimental data would illustrate this uncertainty.

---

† Also Dr V. A. Dupont (University of Leeds).

(ii) The predicted spatial profiles, as shown in Fig. 4 and 5 of the paper can not be determined experimentally for the time being. However, collaborating groups currently study the impact of flow velocity and catalyst length on conversion in the system used. Those results will also help us to evaluate the spatial (axial) profiles taking the residence time as variable.

Dr Hughes‡ said: At the time of the discussion, the experimental section led the reader to interpret that the experimental conversions shown in Fig. 1–3 were obtained at the exit of a tubular channel of monolith as described in the modelling section. If that was the case, when the conversion reaches a saturation value, the reader cannot know where in the tube this happens, whereas the model contains this information. This is of course an effect of residence time. I understand that sampling in such channels is impossible at this time, and that if the reactant flow were increased, it would become turbulent. However, further validation of the model with the measurement of reactants conversions at the end of the tube could be carried out by simply physically shortening the reactor tube (thereby lowering the residence time), until the saturation conditions are left. This does not involve any increased complexity of the experiments and would greatly enhance the confidence in the model.

Prof. Warnatz replied: Indeed, the experiment determines conversion at the exit of the monolith, hence no information about the profiles within the catalytic channel is provided. For further model evaluation, collaborating groups currently carry out experiments, in which the reactor length and the flow velocity (still ensuring laminar flow conditions) are varied, as this comment proposes.

Dr Morley asked: Practical catalysts are often operated with a modulation of the stoichiometry. This implies that storage (probably of oxidised species) on the catalyst will play a part. Would transient experiments be able to give additional information on the adsorbed species?

Prof. Warnatz replied: Yes, if the time scale of the transient experiment is corresponding to the time scale of the process determining the storage.

Dr Whitehead said: I would like to return to the question of the role of sublayers in surface reactions. We have heard in the discussion of the work of Prof. Warnatz about sublayers being sinks for oxygen but not otherwise playing a role in the surface chemistry. However, there is a difference between the situation in the three-way catalysts where the surface remains intact throughout the reaction and that occurring in the oxidation of graphite described by Prof. Williams where the surface is being continually removed by oxidation. Would Prof. Warnatz and Prof. Williams care to comment on the role played by sub-layer absorption and the porosity of the surface in the reactive processes that they have described and how their models can treat such effects?

Prof. Warnatz replied: We have included the role of pores in a one-dimensional reaction–diffusion model, as described in the paper. Reactions and/or transport in sublayers can be included in the treatment of surface processes (see SURFACE CHEMKIN computer package).[1]

An application not far from the problem considered here would be the simulation of NO storage catalysts at lean conditions.

1 M. E. Coltrin, R. J. Kee and F. M. Rupley, *Int. J. Chem. Kinet.*, 1993, **23**, 1111.

Prof. Williams responded to Dr Whitehead: We have not yet tried to solve the three dimensional case but we are assessing the situation because we believe that this is now a soluble problem.

Dr Whitaker said: Following on from Dr Whitehead's question on the porosity of chars. I would like to ask Dr Hessler if there is any possibility that an X-ray scattering experiment could provide us with any information on the internal structure of soot particles. The Perod plots you present in your paper give information about the size distribution of the 'monomeric' building

---

‡ Also Dr V. A. Dupont (University of Leeds).

blocks and the larger aggregates, but is there more information hidden in the data that might be used to deduce something about the structure of the aggregates and do you have any comment as to how this might be achieved?

Dr Hessler replied: In principle we should be able to say something about the internal structure of the soot particles. Recall, the scattered intensity as a function of the transferred momentum is directly related to the electron correlation function,

$$I(q) = \int P(r) \exp(+i2\pi qr) \, dr$$

where the Patterson function, $P(r)$, is the self-convolution of the electron density, $\mathcal{S}(r)$, of the particle,

$$P(r) = \mathcal{S}(r) * \mathcal{S}(-r) = \int \mathcal{S}(u)\mathcal{S}(u+r) \, du.$$

Therefore, to the extent that the electron correlation functions are different we should be able to measure this difference. However, in practice, this may be very difficult. I do not have any information that directly address your question. However, we have been thinking about this and similar questions. In particular, the analysis of scattering data is an inversion problem. We wish to invert the data to learn something about the system that is responsible for the scattering. As with all inversion problems we encounter uniqueness issues, *i.e.* have we found a unique description of the scattering system. Of course, one can never use experimental data, with its ever present uncertainties, to prove uniqueness. The approach we, and in particular Albert F. Wagner and others at Argonne, are taking is to use Monte Carlo techniques to generate structures and then ask if this generated structure is consistent with the observed data. With this approach we hope, at least, to put some bounds on the possible structures that are consistent with the experimental observations.

Prof. Hayhurst opened the discussion of Dr Williams's paper: You were kind enough to refer to some of my published work (ref. 7 in your paper) on the oxidation of graphite. My poster at this meeting reported that the activation energies for the oxidation of coal chars vary markedly with the rank of the parent coal. You have been identifying different sites in graphite; do you have any information on the active and inactive sites in coal chars?

Prof. Williams responded: The reactivities of different coal chars will depend on their method of preparation as well as their chemical composition and this influences both the pre-exponential factor and the activation factor. Differences in the method of preparation (pyrolysis conditions) determine the number of edge sites and hence the active sites which influences the pre-exponential factor and the overall reactivity.

The different activation energies for different rank coal chars probably arises from the fact that chars from lower rank coals have an increasing amount of O-atoms and metals present in the char matrix. The influence of oxygen and other heteroatoms within the carbon is twofold. Firstly, it creates a more disordered structure and hence increases the number of active sites. Secondly, it creates new types of active sites *i.e.* the C–O–C bond has a higher inherent reactivity than the $C_{aromatic}$–$C_{aromatic}$ bond. The first effect will increase the reactivity of the chars but may not result in a significantly different overall activation energy rather the pre-exponential factor would increase. However, it would result in a different distribution of activation energies.

The catalytic effects of certain metals (particularly calcium) in lignites on oxidation rate is well known and involves an oxygen transfer mechanism between the metal oxide and the carbon matrix which can be modeled along the lines described in our paper. The oxygen transfer mechanism has a much lower activation energy than the uncatalysed edge recession mechanism. If catalytic metals are present this would have a dominating factor in the activation energy and this is probably the situation in the case cited for the lignite char.

Prof. Wolfrum asked: What elementary steps are expected to take place during the CO_2 formation pathway during graphite oxidation?

Prof. Williams replied: There is some debate about this step and consequently we studied the higher temperature regime where the influence of this reaction is minimized.

A number of research groups have suggested the route;

$$-C(O) + -C(O) \rightleftharpoons CO_2 + -C \qquad (1)$$

But as can be seen from our Fig. 2 of our paper this involves a complex transition state in order to form the CO_2 molecule. Recently Hurt and Calo[1] have suggested the reaction;

$$O_2 + C(O) \rightleftharpoons CO_2 + C \qquad (2)$$

based on evidence from the interchange of oxygen isotopes, but this reaction also involves a complex transition state. A seemingly attractive alternative is the reaction of a dangling carbon atom directly with an oxygen molecule. However this route can only result in one third of the carbon atoms being converted to CO_2 and could not explain isotope exchange. It therefore seems most likely that the reaction involves the reaction between a carbonyl group on an edge site with an epoxide on the basal plane. In a coal char the other major route would involve a metal atom, such as calcium, which is known to enhance CO_2 formation and this could occur *via* the reaction of the metal oxide with a carbonyl group.

1 R. H. Hurt and J. M. Calo, *Combust. Flame*, 2001, **125**, 1139.

Prof. Hayhurst said: One complication which deserves mentioning is that Haynes[1] has studied the absorbed oxygenated species on carbons: there is a complex variety of such species. In this context Haynes also includes reactions of gas-phase oxygen with absorbed oxygen atoms in

$$O_2 + (C-O)_s \rightarrow CO + CO_2$$

giving gaseous products. Such a reaction gives the order of production of carbon oxides not to be of simple half order in O_2.

1 B. S. Haynes, *Combust. Flame*, 2001, **126**, 1421.

Prof. Lin commented: We have recently studied a series of reactions on the reconstructed Si(100)-2 × 1 surface using a hybrid density functional (B3LYP) method with full geometry optimization employing up to 15 Si atoms.[1–3] The results have been very encouraging. Similar calculations can be readily made for elucidation of the mechanism of graphite oxidation by O, OH and O_2.

1 F. Bacalzo-Gladden, PhD Dissertation, Emory University, May 2000.
2 F. Bacalzo-Gladden, X. Lu and M. C. Lin, *J. Phys. Chem. A*, 2001, **105**, 4368.
3 X. Lu and M. C. Lin, *Phys. Chem. Chem. Phys.*, 2000, **2**, 4213.

Dr Hessler said: Recently, Richter and Howard[1] have shown that molecules with three-dimensional compact structures such as coranulene and C_{60} play an important role in soot formation. Have you considered the reactivity on the edges or at dangling bonds of these three-dimensional structures?

1 H. Richter and J. B. Howard, *Prog. Comb. Sci.*, 2000, **26**, 565.

Prof. Williams replied: It seems that C_{60} reacts in a different way to graphite because the ratio of the surface to edge sites is different. Measurements of the temperature programmed oxidation of graphite and C_{60} show that the rates of reaction and the CO/CO_2 ratios are significantly different.[1] During low temperature combustion the CO/CO_2 ratio is much higher for C_{60} compared to that of graphite. In order to explain this an understanding of both the mechanism of C_{60} combustion and CO_2 formation are required. Two scenarios can be proposed. In the first, there is oxidation at one site in the fullerene creating a defect which then undergoes further oxidation by a mechanism analogous to edge recession in graphite, *i.e.* the hole in the football gets bigger and bigger. In this case the influence of the 5-membered ring and the curved structure on C(O) desorption would be important, and the experimental data suggests that desorption of surface oxides to form CO is faster in this case than it is in graphite. The second scenario is that the aromatic structure of C_{60} is extensively oxidized *via* epoxide formation, prior to the opening of the truncated icosahedron *i.e.* the football disintegrates. However, such a route is expected to result in high levels of CO_2 formation.

We know of no evidence relating to coranulene. We suggest that it is oxidized in a way similar to a PAH compound rather than to graphene even though it contains a five membered ring and is curved. This is because it is entirely terminated by hydrogen atoms and the tendency is to lose one hydrogen at a time by radical attack.

1   K. M. Thomas, J. M. Lewis, S. H. Bottrell and J. Foulkes, *Carbon*, 1994, **32**, 991.

Prof. S. C. Smith said: We have worked on related computational studies of coal char gasification in Brisbane. One of the issues which has arisen is that in coal char crystallites the basal planes are not clean structures—they have holes and irregularities which mean that one can have reactivity on the plane. The difficulty with this from the point of view of *ab initio* modelling is that we cannot yet accurately represent multiple graphene planes in order to be able to simulate such irregularities. Thus, I fully concur with Prof. Williams' comments that this important area of modelling remains wide open with many unsolved challenges.

Prof. Williams responded: We acknowledge this problem and have also examined by SEM the complex structures of graphite particles that have passed through flames. There are major difficulties in applying the basic model to such large structures using multiple graphene planes which will take considerable computer effort to solve.

Prof. Golden asked: Can the computational tools that have been developed with respect to the growth of solid state electronic devices be applied to this problem?

Prof. Williams answered: This may be a possibility and may help to solve the difficulties described above. The development of *ab initio* methods to large molecular models have recently been promoted by the pharmaceutical and solid state industries. So the possibility of the crossover of solid-state computational models to the combustion research area is, we feel, probably inevitable and indeed desirable.

Mr Senosiain asked: Have you tried using periodic boundary condition calculations for investigating the long-range and inter-plane effects?

Prof. Williams answered: We have not yet tried to solve long-range or inter-plane effects other than to consider their influence on the chemical steps that we have proposed in a general sense.

Prof. Pilling said: Prof. Williams has commented on the stochastic nature of the oxidation of graphite and of coal. How can stochastic aspects be introduced into models in which the chemical kinetics is described through deterministic rate equations?

Prof. Williams responded: There is a problem in applying stochastic models into large reaction mechanisms but although expensive in computer time it can be undertaken. This situation is frequently found in the pyrolysis of large organic molecules such as coal or in the annealing kinetics of carbons and chars.[1] Because of the complexity of such systems the reaction rates are often handled by averaged kinetics.

1   E. M. Suuberg, in *Fundamental Issues in Control of Carbon Gasification Reactivity*, ed. J. Lahaye and P. Ehrburger, NATO ASI Series E: Applied Sciences, Kluwer Academic Publishers, Dordrecht, 1990, vol. 192, p. 269.

Prof. Lindstedt said: The uncertainties in the activation barriers are very large ($225 < E_a/\text{kJ mol}^{-1} < 425$) and it would be interesting to know what the sensitivity to predictions would be in practice. How representative would graphene be of the type of coal chars discussed in Fig. 1.

Prof. Williams replied: Many of the current uncertainties in rate parameters result from the randomness induced in the preparation of the carbon samples. In the case of coal chars, these effects are complicated by the non-homogeneity of coal itself and hence of the char samples. Graphene thus can only play the role of being a starting sample model carbon and randomness would have to be imposed on a larger model structure to represent terraces and defects.

Prof. Kohse-Höinghaus asked: How representative would your graphene model be of a solid?

Prof. Williams answered: One important step in using *ab initio* models for solid systems is the correct choice and representation of a model for that solid system. Work by Chen and Yang[1] has addressed the selection and use of the most optimum molecular system for graphite models. Their work has shown that the molecular system, $C_{25}H_9$, is an adequate abstraction from the graphite solid based on good comparisons between predicted and experimental Raman vibrational frequencies and bond lengths and angles.

1 N. Chen and R. T. Yang, *Carbon*, 1998, **36**, 1061.

Mr Frankcombe said: In the paper you report only geometries of stationary points. Have you looked at any reaction paths out of these minima, such as C extraction or O migration?
These systems are very loose and very difficult to optimise. We cannot find geometries for the plain graphene at the HF + MP2 levels (as opposed to B3LYP) with no imaginary frequencies.

Prof. Williams answered: There are clearly difficulties in calculating reaction paths even with the simpler reaction leading to CO formation; the formation of CO_2 is much more complex. We attempted to calculate the energies of some of the intermediate species but were not able to get an accurate or stable solution, and we had difficulty in calculating the zero point energy solutions. The values that we did obtain are consistent with the mechanism and bond lengths that we put the paper.

Prof. Kohse-Höinghaus opened the discussion of Dr Hessler's paper: (1) In comparison to the different radii measured by competing techniques (volume-equivalent spheres, hydrodynamic radius *etc.*), which radius is determined by your technique? (2) Is there any sensitivity to a (changing) refractive index in the different regimes? (3) Is there any influence of the flow dynamics (as in dynamic light scattering)?

Dr Hessler responded: (1) In general, SAXS measures the radius of gyration, R_g, of the electron density distribution or correlation function of a particle. This is especially true in the Guinier region, $qR_g < 1$. One must keep in mind that the radius of gyration is nothing more than the first moment of the correlation function. For polydispersed systems we generally discuss the system in terms of its mean radius of gyration.

(2) No. The complex index of refraction is given by

$$n = 1 - \frac{r_e}{2\pi} \lambda^2 \sum_i n_i f_i(0),$$

where the classical radius of the electron is r_e, the wavelength is λ, the number of atoms per unit volume of type i is n_i, and the complex scattering factors for the forward direction are given by $f(0) = f_1 + if_2$. For hydrogen, carbon, nitrogen and oxygen f_1 is independent of energy above 1 keV and equal to 1, 6, 7 and 8, respectively. Therefore, in our experiments we will not observe any deflection of the X-ray beam as we would expect in experiments at optical wavelengths. The imaginary part of the index of refraction is also well behaved.

$$f_2 = \frac{\sigma_a}{2r_e \lambda}$$

where the atomic photoabsorption cross section is σ_a. Resonances for C, N and O occur near 300, 400 and 550 eV, respectively. Since our experiments are performed well above these energies, we will observe changes in the absorption as we change the energy, but nothing in addition to the change that is proportional to λ.

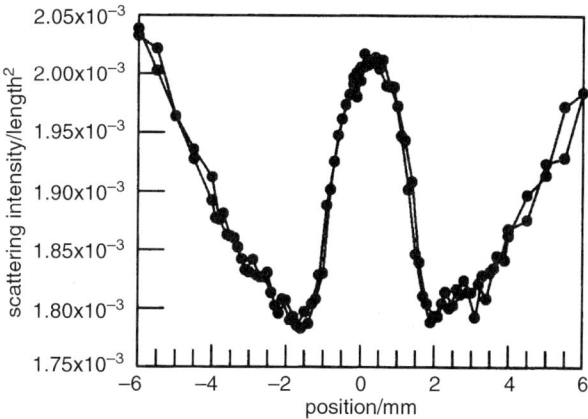

Fig. 1 Scattering intensity for soot produced by an acetylene diffusion flame.

(3) Yes, this was covered in the poster I presented on Monday evening. First, recall that all SAXS measurements depend upon contrast between the background medium and the scattering particles of interest. In our experiments, background scattering is produced by the air, fuel and any other chemical species in the scattering region. In most of the experiments we have performed, the ratio of the scattering intensity of the soot to the scattering intensity of the background varies from 3 to a low of about 10^{-6}. Therefore, if anything causes the scattering of the background to change by 1 part in 10^6, we must account for this change. The best example I know of was discovered in our experiments performed last March. At the present time, we assume the change in the background scattering is due to a change in the density of the background gas, which is easily observed in diffusion flames because there is a strong temperature gradient, and therefore, a density gradient in these systems. An example of this is shown in Fig. 1 where we show the raw scattering data at a single value of q for soot produced by an acetylene diffusion flame. We assume that the background scattering is parabolic in the region close to the flame.

Prof. Wolfrum said: Did you compare your data with the results of laser spectroscopic methods, like laser induced incandescence (LII)? Time and spectral resolved LII-data can also give information on soot particle size distributions.

Dr Hessler replied: At the present time we have not made such a comparison. However, we have always planned to make such comparisons. Unfortunately, the burner that we used was designed for a different experiment. Therefore, comparisons to systems that have been thoroughly investigated with optical techniques are not straightforward. In the future, we plan to perform SAXS experiments with burners that are similar to those that have been used by others to study soot formation. In addition, we plan to perform independent optical experiments with our new burner so we may make more complete comparisons to previous work.

Prof. Pilling said: Dr Hessler prefaced his presentation with his intention to use SAXS in shock tube studies of soot formation. What other probes does he plan to use to augment the X-ray measurements and what is the time resolution of the SAXS system at the Advanced Photon Source in the proposed shock tube applications?

Dr Hessler replied: We plan to characterize our burners with optical scattering with light from a Nd : YAG laser, 1064, 532 and 266 nm. Laser induced incandescence, and optical absorption at 632 nm. Unfortunately, most of these experiments can not be done simultaneously with the X-ray scattering experiments, but we can perform them with the same burner under as identical a set of operating conditions as can be achieved.

Our motivation for doing these studies came from the wide-angle scattering experiments on soot in shock tubes by di Stasio *et al.*[1] They studied soot from ethylene between 1800 and 2000 K and could not obtain any information below 1 ms or for particles sizes below 15 nm. We would like to perform these experiments with X-rays. However, before we can do this we must have a detector that has a time resolution in the microsecond range. We have designed such a detector, built a prototype, and tested it. Our preliminary results are encouraging. We have designed the time resolution to be about 3.68 µs, which is the time needed for an electron bunch to make one revolution around Argonne's Advanced Photon Source. This time resolution should be sufficient for most kinetic studies yet it will not require any special operating conditions of the X-ray source.

1   S. di Stasio, P. Massoli and M. Lazzaro, *J. Aerosol Sci.*, 1996, **27**, 897.

Prof. Lindstedt said: The paper is very interesting and the technique opens up the possibility of providing information relating to the importance of aromatic and non-aromatic contributions to mass growth for very incipient particles. When will the data be processed?

Dr Hessler responded: The data presented here was analysed in enough detail to design the next set of experiments, which were performed in March of this year. An important aspect of any scattering experiment is the fact that it is a line of sight measurement. When scattering measurements are performed at many different positions in a plane perpendicular to an axis of cylindrical symmetry, Abel inversion techniques may be used to calculate the scattering intensity as a function of the radial distance. This last March we performed SAXS measurements at 104 different positions with respect to the centre of a flame and at several planes along its height. These results were used to locate the centre of the flame, determine the functional dependence of the background scattering, as discussed in the reply to Prof. Kohse-Höinghaus's third question earlier, and inverted to give 31 sets of scattering intensity, $I(q)$, at different radial positions from the centre of the flame. The experiments on acetylene and toluene were discussed in the poster session on Monday evening. In addition, we performed measurements on a propylene diffusion flame. As discussed in the poster, the negative scattering intensity observed in diffusion flames was approximated by a parabolic function. Two examples of the scattering intensity as a function of the transferred momentum, q, are shown in Fig. 2(a). The solid circles represent inverted data at 1.3 mm from the flame centre and the open circles at 0.9 mm. A striking feature of this data is the insensitivity of the scattering intensity with respect to the radial position for $q < 0.8$ nm^{-1} while for $q > 1$ nm^{-1} it varies significantly in amplitude yet retains its general shape. We assume the data for $q > 1$ nm^{-1} represents scattering from small soot particles that are in the Guinier region, i.e. $I(q) = G \exp(-q^2 R_g^2/3)$. At each radial position we extract G and R_g, the radius of gyration. The radius of gyration is independent of the distance from the centre of the flame and equal to 0.19 nm. The prefactor, G, varies significantly with the distance from the flame centre and in certain regions goes to zero. The number density of the different particles may be estimated by performing the Porod integral, $\int q^2 I(q) dq$, over the range of the data. In Fig. 2(b), we show the radial dependence of the Porod integral for the small particles, which is calculated from the parameters G and R_g, and the integral assigned to the larger primary particles. The absence of small particles at different radial distances is found at other heights in the flame.

Before we can make detailed statements about the shape of the larger primary particles additional measurements must be performed at low q values. These data were taken early in August. Preliminary observations indicate that the primary particles also show radial oscillations in the number density.

Prof. Lindstedt asked: The oxidation/mass growth of soot has long been subject to uncertainties related to 'surface' as opposed to 'volume' effects with the latter term used here to imply a role of the internal structure of particles. Is there any prospect of providing information in relation to this issue through the use of techniques such as SAXS?

Prof. Williams answered: We have tried to avoid the complication of determining the role of the internal structure in our work by taking the most basic model using the intrinsic reaction rates which eliminate these effects. We know of no evidence that suggests that the internal surface of graphite is any different to the external graphene layers. Therefore the chemical mechanism should apply to combustion in both regimes. This may not be the case for soot where oxidation and

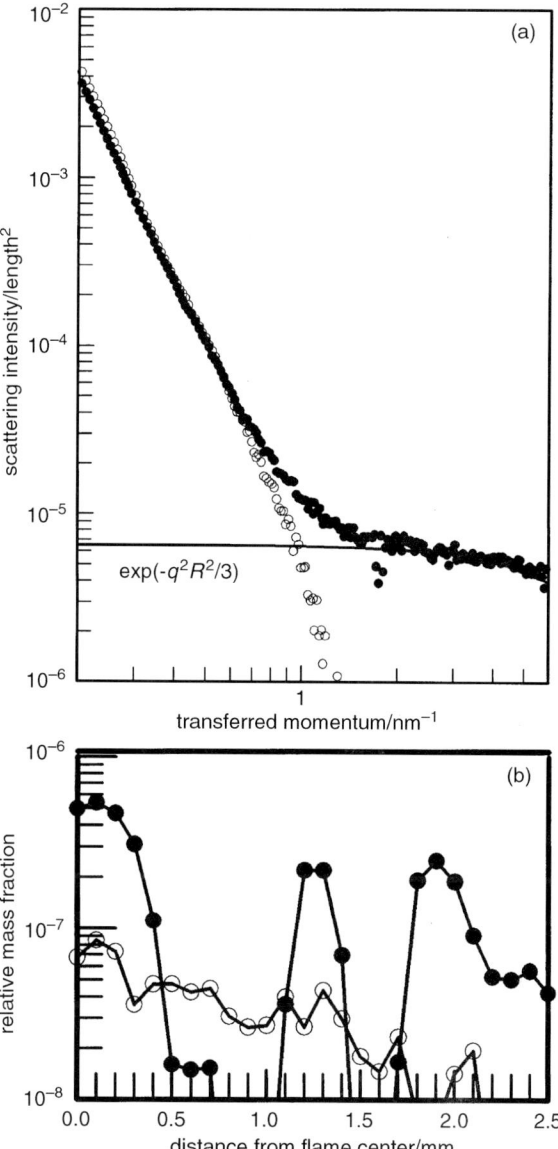

Fig. 2 (a) Two examples of scattering intensity as a function of transferred momentum; see text for details. (b) Radial dependence of the Porod integral for small particles, solid circles, and larger primary particles, open circles.

growth can occur simultaneously. There may be ways of achieving more detailed studies of the internal structure using SAXS which may shed more light on this problem.

Prof. Troe opened the discussion of Prof. Lindstedt's paper: The database for aromatic species is still very limited. Many more measurements of individual reaction rates are needed. At the same time more critical evaluations should be made. Apparently the CEC/IUPAC evaluations (ref. 37 of your paper and the update from 1994) are the only ones carried out with a sufficiently broad scope. It should be mentioned that a new update with better and more recommendations for aromatic species is in preparation.[1]

1   D. L. Baulch *et al.*, in preparation.

Prof. Lindstedt responded: Significant uncertainties prevail in measured and computed rate constants for several key reactions and the updated database is very welcome.

Prof. Troe commented: (1) The competing C–H and C–C bond splits in toluene dissociation as well as reactions of H atoms with toluene now are very well characterised such that the rate data in the table should be updated. It appears worth mentioning that the branching ratio for C–H and C–C bond splits in toluene are temperature and also pressure dependent.

(2) We found that rate constants of H atom combinations with many aromatic radicals are all close to 2×10^{11} l mol^{-1} s^{-1} such that this should be a better estimate than given in the table. The reverse C–H bond splits then should be estimated using the corresponding equilibrium constants.

Prof. Lindstedt replied: The branching of the C–H and C–C bond splits in the dissociation of toluene has long been problematic with significant discrepancies between different determinations (ref. 27 of our paper). The branching ratio applied in the current study does reproduce a range of more basic (*e.g.* shock tube) data. Uncertainties do, however, prevail and the new determination is welcome. The rates of H addition to aromatic radicals used in the current work are based on the recommendation (0.783×10^{11} m^3 kmol^{-1} s^{-1}) by the CEC Evaluation Group (ref. 37) for the model reaction $C_6H_5 + H \rightleftharpoons C_6H_6$. In some cases adjustments have been made within the uncertainty band and the rates used are in the range 0.500×10^{11} to 1.5×10^{11} m^3 kmol^{-1} s^{-1}. We will explore the impact of the new suggestion.

Prof. Troe commented: We have recently identified and measured ring opening rates in shock tube studies of biphenylene decomposition.[1] *Via* the equilibrium constants this leads to typical rate constants for ring closure forming polycyclic compounds. This may help in estimating ring closure rates in general.

1 A. E. Croce, K. Henning, K. Luther and J. Troe, *Phys. Chem. Chem. Phys.*, 1999, **1**, 5345.

Prof. Lindstedt replied: The quantification of reaction channels for complex aromatic systems leading to stable products entails many difficulties and the measured ring opening rates may indeed be helpful.

Prof. Kohse-Höinghaus said: I would like to comment on the importance of C_5-species pathways towards aromatic formation as observed recently by Burak Atakan and others in our group. In a family of fuel-rich non-sooting low-pressure flames burning C_2-, C_3- and C_5-fuels, C_5H_5 and C_5H_6 reactions were found to be of similar importance as the propargyl recombination (ref. 1). Detailed analysis of the flame structure with laser spectroscopy and molecular beam mass spectrometry has been performed to measure more than 30 species for each fuel/stoichiometry condition, and this database would be available for comparison with flame models (ref. 2–4).

1 A. Lamprecht, B. Atakan and K. Kohse-Höinghaus, *Proc. Combust. Inst.*, 2000, **28**, 1817.
2 B. Atakan, A. T. Hartlieb, J. Brand and K. Kohse-Höinghaus, *Proc. Combust. Inst.*, 1998, **27**, 435.
3 A. Lamprecht, B. Atakan and K. Kohse-Höinghaus, *Combust. Flame*, 2000, **122**, 483.
4 B. Atakan, available *via* Atakan@uni-bielefeld.de, unpublished work.

Prof. Klippenstein commented: Our recent master equation simulations for reactions such as $C_3H_3 + C_3H_3$ and $C_3H_3 + O_2$ provide some general results that can be useful in examining the feasibility of specific steps in a postulated mechanism. To illustrate some of these ideas I would like to briefly consider the benzyl ($C_6H_5CH_2$) + propargyl (CH_2CCH) reaction. Both the benzyl radical and the propargyl radical are resonantly stabilized. In contrast, the initial complex formed by addition of the propargyl radical to the CH_2 radical in benzyl has only the resonance stabilization of the benzene ring. As a result, this initial complex should be fairly weakly bound (*e.g.*, by about 60 kcal mol^{-1}) and the equilibrium of the initial formation reaction should be shifted towards the reactants at combustion temperatures. Also, the simple loss of an H atom from this initial addition complex to form a $C_6H_5C_4H_4$ compound should be significantly endothermic relative to the benzyl + propargyl reactants, perhaps by 20 to 30 kcal mol^{-1}.

If feasible, the isomerization of the addition complex to form a second C_6 ring with a hydrated napthalenic like $C_{10}H_{10}$ structure, would yield a somewhat more stable compound. However, the subsequent loss of an H atom from such a bicyclic structure would yield a resonantly stabilized radical that now lies lower in energy than the starting reactants. Furthermore, the increased resonance stabilization of such a radical (relative to the $C_{10}H_{10}$ species) should make the bond dissociation energy of the bicyclic $C_{10}H_{10}$ species quite small. As a result, the bicyclic $C_{10}H_{10}$ species is likely to be beyond its stabilization limit at quite low temperatures, certainly before reaching combustion temperatures. Correspondingly, the proper elementary reaction step should be bimolecular in nature, involving the formation of $C_{10}H_9 + H$ or, *via* equivalent arguments, $C_{10}H_8 + H_2$.

Similar considerations are applicable to the $^1CH_2 + C_6H_6$ reaction. In this case, a CH dissociation of the initially formed C_7H_8 product yields the resonantly stabilized benzyl radical (C_7H_7). The corresponding CH dissociation energy is about 88 kcal mol^{-1}, which is small enough that the C_7H_8 complex may be beyond its stabilization limit at combustion temperatures. The proper elementary reaction step would then be bimolecular in nature (*i.e.*, $C_6H_6 + {}^1CH_2 \rightarrow C_7H_7 + H$).

Prof. Lindstedt replied: There is no doubt that the master equation simulations will be helpful. For example, if it can be shown that isomerisation can lead to a direct ring closure route for the benzyl + propargyl channel then this may well have a significant impact in several systems. The $^1CH_2 + C_6H_6$ reaction is another well-chosen example for which the product distribution is not established at this point in time.

Dr Miller commented: I would like to say to Prof. Lindstedt that I really enjoyed reading his paper, and I believe he has done us all a valuable service in outlining a number of possible paths to indene and naphthalene formation in flames. It is important now for us to investigate these paths in detail, both theoretically and experimentally, and to put fundamentally sound rate constants and product distributions with the mechanisms proposed.

Prof. Lindstedt responded: A better characterisation of oxidation and mass growth sequences is essential. The practical importance of the chemistry of higher aromatics is now such that tentative detailed mechanisms have to be used to provide input to practical design calculations.

Prof. Hayhurst said: Looking at reactions (613) and (−611) and the analgous pair (753) and (760) I am reminded that Sugden[1] compared the reactions:

$$A + HZ \rightleftharpoons AZ + H \qquad (I)$$

$$A + Z + M \rightleftharpoons AZ + M \qquad (II)$$

where A is a metallic additive in a flame, Z represents an OH radical or an atom of hydrogen or a halogen and M is any molecule acting as a third body. He concluded that if the dissociation energy of AZ exceeds ~ 335 kJ mol^{-1}, the time constants are such that the reactions in (I) are unimportant under flame conditions. Also, if the dissociation energy of AZ is less than ~ 335 kJ mol^{-1}, then only the forward and reverse steps of (I) need to be considered. Does such a simple rule operate to simplify the kinetic schemes for sooting flames?

1 Sugden, *Trans. Faraday Soc.*, 1956, **52**, 1465.

Prof. Lindstedt answered: It would appear difficult to simplify the current system in a directly analogous manner. However, generic rate constants for individual reaction classes are useful.

Prof. Golden asked: How do you take pressure dependence into account? I would point out that often two-parameter representations of rate constants come from values outside the appropriate temperature range.

Prof. Lindstedt replied: In cases where falloff has been determined such effects are naturally included. For other reactions simple estimates have had to be used and more accurate determinations will be introduced as feasible. Many reactions are represented reasonably well by a standard Arrhenius expression in the relevant temperature range. However, the absence of data does present significant difficulties.

Prof. Lin commented: Our recent study on the reaction of H with C_6H_5[1] indicates that benzyne (C_6H_4) can be formed by both direct and indirect (association/elimination) metathetical mechanisms. A similar but more efficient reaction may occur between OH and C_6H_5. The large yield of C_6H_4 you detected may be attributed to these bimolecular reactions and the unimolecular decomposition of C_6H_5.[2]

1   A. M. Mebel, M. C. Lin, D. Chakraborty, J. Park, S. H. Lin and Y. T. Lee, *J. Chem. Phys.*, 2001, **114**, 8421.
2   L. K. Madden, L. V. Moskaleva, S. Kristyan and M. C. Lin, *J. Phys. Chem.*, 1997, **101**, 6790.

Prof. Lindstedt responded: The comprehensive studies by Lin and co-workers address parts of the C_6 system that require consolidation. It may in this context also be suggested that the $H + C_6H_6 \rightleftharpoons C_6H_5 + H_2$ channel is in need of further clarification. Our current mechanism suggests that $\sim 20\%$ of phenyl radical destruction passes *via* benzyne and that the dominant channel under fuel rich flame conditions is *via* $C_6H_5 + H$. The rate constant used for this step in the current mechanism (1.5×10^{11} m^3 kmol^{-1} s^{-1}) is an estimate by Leung and Lindstedt (ref. 3 of our paper) and the effects of the new determinations by Mebel *et al.* (2001) will be explored with interest.

Prof. Hayhurst communicated: I am intrigued that reactions of O_2 (*e.g.* reaction (754)) with large hydrocarbon radicals are important in these very fuel-rich flames. For how does O_2 persist in such a system?

Prof. Lindstedt communicated in response: There is a strong molecular oxygen gradient through reaction zones of fuel rich flames and the maximum rates of oxidation *via* related channels occur comparatively early on. For example, in the benzene flame the maximum rate for the $C_{10}H_7 + O_2$ reaction occurs at a temperature of ~ 1570 K. Large quantities of molecular oxygen prevail at this point.

Prof. Hayhurst communicated: There is a large variety of growth species in these flames, including C_2, C_3, C_4 and C_5 entities. Are there rapid equilibria which couple their concentrations?

Prof. Lindstedt communicated in response: In some cases a strong partial equilibration occurs for individual reactions as the post flame zone is approached. It is, however, in practice difficult to use this approximation in order to simplify the overall system.

Prof. Hayhurst opened the discussion of Mr Schürmann's paper: (1) Is it possible that $(NaCl)_n$ or $(NaOH)_m$ exist in your furnace; here n and m are small integers. Would your technique detect such species?
(2) Your technique is evidently effective at removing alkalis from combustion gases. However, it does leave the chlorine, which is corrosive and can give dioxins under some conditions. What would you recommend as a getter for chlorine?

Mr Schurmann responded: Thermodynamic equilibrium calculations (*e.g.* ref. 28 of our paper) show that in typical reactor flue gases, hardly any dimers should be present. In addition, ref. 24 shows that in the concentration and pressure range of interest, dimers make a negligible contribution to the ELIF signal. Ref. 24 also finds that the ease of photofragmentation for monomers is greater than that for dimers by about a factor of two for sodium and a factor of eight to nine for potassium.
   Regarding the fate of chlorine, in normal coal conversion, amounts of chlorine are small enough that species such as dioxins do not constitute a particular problem. In this case the amount of additional chlorine in the flue gas is in the range of some ppm. Most of the chlorine is transformed into HCl, which in modern power plants is removed by wet flue gas purification.

On the other hand, if waste, particularly plastic materials, is combusted or co-combusted with other fuels, much greater levels of chlorine will be present. If dioxins are formed, so far, the only ways to remove them are thermal degradation (long enough residence time at temperatures well above the dioxin formation region of 250–450 °C) or a flue gas adsorption process with active char, which is done when large amounts are present (*e.g.* in waste incineration of plastic materials like PVC). In fact, EU-directive 2000/76/EG of 4.12.2000, concerning waste combustion, requires that hazardous waste materials with a chlorine content bound in halogenated organic substances of more than 1% be burned at 1100 °C for at least two seconds.

Prof. Wolfrum said: The absolute amount of chlorine is relatively small in the coal types investigated by ELIF-methods. A quantitative comparison of the alkali content determined by three different methods (ELIF, plasma excitation and atomic line absorption, surface ionization) allowed a clear distinction of the alkali components in the gas phase and alkali components transported on the surface of particles.

Dr Morley asked: To what extent is equilibrium between the sodium species and silica attained in your experiments, or is kinetics important?

Mr Schurmann answered: In general, kinetics of reactions between gaseous alkali species and various minerals are of interest and were studied by several investigators, some of which are cited in the paper. Probably the most relevant of these is by Wibberley and Wall (ref. 16 of our paper) who investigated processes forming silicates in flue gas containing sodium, sulfur, chlorine, water vapour and silica particles, in which the reactions occurred on the time scale of seconds. The capture of alkali by silica is sensitive to temperature and increases with the combustion temperature. However, silicate formation may be reduced during post-combustion when ash fusion and particle coalescence take place. The extent of sodium silicate formation is also influenced by the contact between coal ash minerals, which in turn is dependent on the mode of sodium occurrence and the coal (and additive) particle size as well as by the concentration of silicate minerals present. Wibberley and Wall obtained reaction rates that were higher than theoretical rates calculated for the formation of certain thicknesses of silicate layer and thus concluded that the thickness of the layer formed in a PCC boiler with residence times around 2 s would probably be determined by thermodynamics rather than kinetic constraints. On the other hand, these authors used silica particle sizes of 35 and 68 μm respectively and our particles were ~16 nm. So far, we are not aware of any comparable kinetic study for such small particle sizes.

Prof. Hayhurst said: The residence time in your system looks to be ~5 s. Is this sufficiently long to produce molten particles and also for the gases to react in reaction (1)–(4) with these particles? Alternatively, from your measurements can one deduce any information about the kinetics of reactions like (1)–(4).

Mr Schurmann replied: In these experiments the residence time was adjusted to 10 s. In flames derived from pulverised coal the heating rate of a coal particle is approximately 10^4–10^5 K s^{-1}. In this way, formation of molten particles and reactions with alkali metals are possible.

It may be possible to gain some information about the total reaction rates if a systematic series of measurements were made in the appropriate concentration range of the additive. In the present work, too few data points are available to be able to extract such values.

Concluding Remarks

James A. Miller

Combustion Research Facility, Sandia National Laboratories, Livermore, CA 94551-0969, USA

Received 11th September 2001
First published as an Advance Article on the web 6th November 2001

Introduction

When Mike Pilling asked me last summer in Edinburgh to give the "Concluding Remarks" at this Discussion, I was so flattered that I did the unthinkable—I agreed to do it! It was only weeks later that I came to realize that I could not do justice to the task, a realization that occurred too late for me to escape graciously. Combustion Chemistry is a much broader topic than those normally chosen for Faraday Discussions, and I will not pretend that I am capable of commenting intelligently on all aspects of it. However, I can share some thoughts that I had as I read the papers and listened to the (frequently lively) discussion.

If I were to list topics that are currently active research areas in combustion chemistry, that list would look something like this:

1. Gas phase combustion: Elementary reactions. Aromatics, PAH, and soot. "Low-temperature" oxidation. NO_x (and SO_x) formation and control.
2. Coal combustion
3. Laser diagnostics
4. Heterogeneous processes
5. Theory and modeling

All these topics were covered by this Discussion. I have never before been to a meeting where molecular beam scattering and pulverized coal combustion were both discussed by the same group of people. Our Discussion clearly lived up to its billing, "Combustion chemistry: Elementary reactions to macroscopic processes."

Perhaps the most important development in combustion chemistry over the last 20 years or so is the increased role played by theory and modeling. The overwhelming majority of papers presented here were either totally theoretical in content or relied heavily on theory or a chemical kinetic model to interpret experimental results. In fact, it appears to have become accepted now that the goal of combustion chemistry is to develop chemical kinetic models that can be used to predict the results of macroscopic experiments, and ultimately the performance and emissions of practical combustion devices. Moreover, a large part of the information that goes into those chemical kinetic models comes from theory. The thermochemistry is based heavily on electronic structure calculations, and a large number of rate constants come from theoretical analyses that are used to extrapolate (at best) a small number of experimental data points. Of course, this must necessarily be the case in a field like combustion that encompasses such a wide range of temperatures and pressures, and that involves such large numbers of species and elementary reactions. However, such a situation places a premium on well chosen, reliable experiments and on accurate, physically realistic, theoretical analyses. Both experiments that provide input to the models and those that provide tests of reliability are important.

The remainder of my remarks address specific issues. They are not intended to be exhaustive, *i.e.* I make no attempt to summarize every point that was made at the meeting. The papers and their accompanying discussion speak for themselves. Rather, I hope to add to the discussion and place some of the research presented here in context. Others may have a different perspective. However, they are not the one whom Mike Pilling put on the spot.

Theoretical methods in general

We are approaching a state of nirvana for theorists. At least for some elementary reactions of importance, it is possible to start from scratch, calculate the important features of the potential energy surface (PES) from electronic structure theory, calculate microcanonical (or canonical) rate coefficients by statistical methods from that PES, and utilize those results (if necessary) in a master-equation analysis to predict rate constants and product distributions as a function of temperature and pressure. The results of such an analysis then, in principle, can be used in phenomenological descriptions of the macroscopic kinetics, *i.e.* CHEMKIN-based codes, to understand important issues in combustion chemistry, *e.g.* soot or NO_x formation. With a few experimental check points, such a prescription can be very successful. However, improvements in the methodology are still possible.

Senosiain et al.[1] discussed 3 different electronic-structure methods of calculating hydrocarbon bond dissociation energies (DFT-B3LYP, their own density functional method KMLYP, and CBS-Q), with the goal of using such methods as a basis for predicting such BDE's for large hydrocarbon molecules. Presumably the popular G2-like methods would be similar in performance to CBS-Q. None of the 3 methods is completely satisfactory. B3LYP is not sufficiently accurate, and CBS-Q is too computationally intensive to use on very large molecules such as those that one might encounter in modeling the growth of polycyclic aromatic hydrocarbons (PAH) in flames. However, the accuracy of the CBS-Q method appears to be comparable to the experimental uncertainty in the BDE's that were used for comparison, an encouraging result. The KMLYP method is intermediate in performance between the other two, more accurate than B3LYP and less computationally demanding than CBS-Q. The authors propose correcting the results from DFT calculations using group additivity. Such a method shows promise. It is probably the natural analog of Carl Melius's BAC-MP4 method[2] (bond additivity corrected Moller–Plessett fourth-order perturbation theory) for very large molecules.

The methods mentioned in the last paragraph are all single-reference methods. However, at least in principle, many problems require multiple-reference methods. Examples of the latter are resonance structures, transition states (*i.e.* saddlepoints), and the potential governing the breaking and forming of bonds. However, single reference methods are routinely used for all of these problems except for breaking and forming bonds. The paper by Fang et al.[3] is a good example of the use of multi-reference methods in conjunction with a sophisticated form of variational transition-state theory to make predictions about an important combustion reaction (NH_2 + NO). Carpenter[4] has recently cautioned against the routine use of single-reference methods for transition-state structures. He concluded that, because of its multi-reference character, the high point on the reaction path connecting vinylperoxy to HCO + CH_2O in the C_2H_3 + O_2 reaction is most likely predicted to be too high by about 7.6 kcal mol^{-1} in the analysis of Mebel et al.[5] The immediate implication of this result is that current chemical kinetic models probably underestimate the branching of C_2H_3 + O_2 to CH_2O + HCO at high temperature and overestimate that to vinoxy + O. The longer term implications are that care must be taken in interpreting transition state energies obtained using single-reference methods; perhaps heeding more closely the warnings of large spin contamination is a good idea.

The master equation (ME) is becoming an increasingly important tool in combustion chemistry. The papers by Frankcombe and Smith,[6] Hahn et al.,[7] Xia et al.,[8] DeSain et al.[9] and the posters by Jones et al. and Blitz et al. utilize it in analyzing elementary combustion reactions. The papers by Fang et al.[3] and Fikri et al.[10] make use of its "collisionless" form. Except for special cases, such as single-well dissociations and multiple-well problems in the collisionless limit, one is generally limited to solving the *one-dimensional* ME with energy as the independent variable. However, some progress is being made on two-dimensional (E,J) solutions.[11–14] Many reactions beg for a two-dimensional analysis, particularly those where "rotational channel switching" may be an issue.[15–17]

I would be remiss if I did not point out the limitations of our own two-dimensional model[12] (discussed in ref. 7) and that of Smith and Gilbert.[11] First, reduction of the 2-d problem to an equivalent 1-d problem is restricted to particular forms of the E, J transfer function, $P(E, E'; J, J')$. Most importantly, $P(E, E'; J, J')$ cannot depend on the initial angular momentum quantum number, *i.e.*

$$P(E, E'; J, J') = P(E, E')\varphi(E, J). \tag{1}$$

Such an approximation precludes taking into account certain effects, most notably the dependence of $\langle \Delta E_d \rangle$ on J', where $\langle \Delta E_d \rangle$ is the average energy transferred in a deactivating collision. Such a dependence has been identified several times in classical trajectory calculations.[18–20] It may be at least partially responsible for the observation made repeatedly that $\langle \Delta E_d \rangle$ increases with temperature in thermal dissociation/recombination reactions. That is, based on a 1-d ME analysis, $\langle \Delta E_d \rangle$ appears to increase with T, because higher temperatures excite higher angular momentum distributions, and the higher J's actually cause the larger values of $\langle \Delta E_d \rangle$. The second restriction of the "simple" 2-d models (*i.e.* those that reduce the problem to an equivalent 1-d ME) is that they are restricted to single-well dissociations—multiple product channels are okay, but all the dissociations must take place from the same well. Recombination rate constants can be obtained after-the-fact from the appropriate equilibrium constants, but the method fails in its present form if a reaction must be allowed to take place in both directions in a single ME calculation. Such a situation is normally the case even when one considers a single-well, chemically activated problem with both bimolecular and stabilized product channels. At the expense of more computational complexity, the methods can and should be extended to these more complicated problems. Perhaps a good problem to start with is the $CH_3 + OH$ reaction, investigated by Xia *et al.*[8] at this Discussion. This reaction has only one well (a useful simplification), but it has several exit transition states of similar energies, both loose and tight, an indication that angular momentum conservation could be important in determining the product distribution.

One cannot leave a discussion of the master equation without mentioning the most poorly characterized quantity in it, the energy transfer function $P(E, E')$. To start with, it must be recognized that $P(E, E')$ and Z, the collision rate, are inextricably tied together, because they always appear as the product $ZP(E, E')$ in the master equation. As a consequence, what one deduces from experiment for moments of $P(E, E')$, *e.g.* $\langle \Delta E_d \rangle$ or $\langle \Delta E \rangle$, depends on what one takes for Z. It has become common practice to assume that $Z = Z_{LJ}$, where Z_{LJ} is the Lennard-Jones collision rate, and that $P(E, E')$ can be approximated by a "single-exponential down" model. Neither of these approximations is particularly well justified. Probably a better set of assumptions would have $Z > Z_{LJ}$ and $P(E, E')$ modeled as a double exponential.[18] The redeeming point is that most master-equation calculations for thermal reactions are not overly sensitive to such details. These points and others are discussed at length in the reviews by Lendvay and Schatz[21] and Nordholm and Schranz.[20]

From a combustion chemistry perspective, the most important application of the master equation is to problems involving multiple wells. This topic is discussed in the next section.

The time-dependent, multiple-well master equation

Many reactions important in combustion, particularly those involved in the formation of aromatics, polycyclic aromatics, and soot, are complicated processes that take place over multiple, interconnected potential wells. Some are "collisionless", *i.e.* the intermediate complexes are so short-lived that they do not suffer any collisions under conditions that are normally of interest. Such reactions usually involve a small number of atoms and relatively shallow potential wells, resulting in small densities of states. The classic example of such a reaction is $NH_2 + NO$, the subject of the paper by Fang *et al.*[3] at this Discussion. However, it is more common that the intermediate complexes live long enough to suffer numerous collisions. In such cases, any of a number of stabilized or bimolecular products can result. For such reactions, it is necessary to solve the time-dependent, multiple-well master equation in all its complexity in order to understand and predict rate constants and product distributions.

Stephen Klippenstein and I have spent considerable effort solving multiple-well master equations and trying to understand how to determine rate constants from their solutions.[22–25] I want to summarize a few points here. It is usually desirable to start by formulating the problem in the general form,

$$\frac{d|w\rangle}{dt} = G|w\rangle, \tag{2}$$

where $|w\rangle$ is a vector containing the populations of all the relevant states, and G is a real, symmetric matrix, the transition matrix of the master equation. The solution to eqn. (2) then can be written as

$$|w\rangle = \sum_i e^{\lambda_i t} |g_i\rangle\langle g_i | w(0)\rangle, \qquad (3)$$

where $|w(0)\rangle$ contains the initial condition; and $|g_i\rangle$ and λ_i are eigenvectors and eigenvalues of G, i.e. $G|g_i\rangle = \lambda_i |g_i\rangle$, $i = 1, \cdots N$. The number of unknown populations (and number of eigenpairs), N, is typically a very large number. However, only a small number of these eigenpairs describe "chemistry". These are the ones with algebraically the largest eigenvalues. The rest are normal modes of relaxation of the internal energy. But just how many eigenpairs are chemically significant in any particular problem, and what is their chemical significance?

In our work to date we have emphasized the relationship between eigenpairs and transition states.[22–25] Such relationships are important, but the most general statement we can make about the number of chemically significant eigenpairs, N_{chem}, is that

$$N_{chem} = S - 1, \qquad (4)$$

where S is the number of chemically distinct configurations in the problem, i.e. the number of wells plus the number of sets of separated fragments. One can see that this is correct by realizing that each of the eigenpairs describes the establishment of a different "chemical equilibrium", at least at low temperature where the eigenvalues are separated in magnitude. The "first" of these normal modes to relax for any initial condition typically describes the establishment of equilibrium between 2 chemical configurations (although more than one product may be formed in the process); the second normal mode describes the equilibration of these 2 configurations with a third; the third mode to relax describes the equilibration of these 3 configurations with a fourth, and so on. A good example of this behavior is the $^1CH_2 + C_2H_2$ reaction discussed by Frankcombe and Smith in this Discussion.[6] In their model there are 5 chemically distinct configurations (allene, propyne, cyclopropene, $^1CH_2 + C_2H_2$, and $C_3H_3 + H$), and thus $N_{chem} = 4$, as I noted in my comment following their paper. The relationship of eqn. (4) to the simple cases of single-channel dissociation and isomerization reactions should be clear. In these cases $S = 2$, so $N_{chem} = 1$, and both the forward and reverse rate constants can be obtained from the single chemically significant eigenvalue, $-\lambda_1$, and the equilibrium constant.

How are the eigenpairs related to the transition states? Suppose we have a reaction coordinate diagram such as those shown as Fig. 1 in the papers by Frankcombe and Smith and Hahn et al.[7] Starting with any configuration, let us successively add configurations. If, for each added configuration, we add only one transition state to connect it to what we already have, there will be one transition state for each chemically significant eigenpair (CSE). However, if we add a transition state between 2 already existing configurations, there is no change in the number of chemically significant eigenpairs. In such cases there are more transition states than CSE's, and a given CSE can be correlated with more than one transition state. However, our experience thus far would indicate that it is unusual for the opposite to happen, i.e. we have not found a case where a transition state correlates strongly with more than one CSE.

How can eqn. (4) be violated? There are at least 2 ways:

1. Sometimes, at sufficiently high temperature, reaction takes place so rapidly that one or more of the chemically significant eigenvalues can be comparable in magnitude to the energy relaxation eigenvalues. In such cases some reaction may occur through these energy relaxation eigenpairs.

2. If one models more than one configuration as an "infinite sink", N_{chem} will be reduced correspondingly. If there are no infinite sinks, $\lambda_0 = 0$ and $|g_0\rangle$ contain populations corresponding to complete thermal and chemical equilibrium. If there is one infinite sink, its population as a function of time can be obtained from global conservation of mass, even though all the population ultimately ends up in this configuration. The case for more than one infinite sink (an approximation that one would probably want to make if the problem contains more than one set of bimolecular products) is more complicated, but it is still soluble in a straightforward way.

As Mike Pilling pointed out in his comment following our paper on $C_3H_3 + O_2$, there is no simple relationship between CSE's and rate constants. It should be noted that in any master-equation analysis, there are

$$N_k = \sum_{n=1}^{S-1} n = \frac{S(S-1)}{2} \tag{5}$$

forward rate constants (and an equal number of reverse rate constants or equilibrium constants) embedded in the eigenvalue spectrum. The consideration of the conservation property,

$$\left(\sum_{i=1}^{S} \Delta x_i\right)_j = 0 \quad j = 1, \ldots, \tag{6}$$

where Δx_{ij} is the change in population of the ith chemical configuration as a consequence of the time evolution of the jth eigenpair, is useful in sorting out the rate constants from the eigenvalues. It is quite common that $\Delta x_i \approx -1$ for the reactant in eqn. (6) for one eigenvector. In such cases it is straightforward to extract the rate constant and product distribution from that eigenpair. Other cases are more complicated, but probably still tractable. In these more complicated situations one may need to solve sets of nonlinear algebraic equations for the rate constants with the λ's, the equilibrium constants, and the various terms in eqn. (6) as input. In some cases it may be easier to avoid this latter complication by solving the same master equation with several sets of initial conditions.

Pressure dependence of multiple-channel, multiple-well reactions

When Professor Troe discussed fall-off effects in unimolecular reactions, it reminded me of some complicated pressure dependences that we have encountered recently, behavior that is not easily incorporated into CHEMKIN-based phenomenological models. I shall point out only two effects here, but they are ones that are likely to be very common in multiple-channel, multiple-well unimolecular reactions.

Fig. 1 shows the pressure dependence at various temperatures of the rate constant for the reaction,

$$C_2H_5O_2 \rightarrow C_2H_4 + HO_2, \tag{R1}$$

where $C_2H_5O_2$ is the ethylperoxyl radical.[24] Reaction (R1) has an intrinsic potential energy barrier (*i.e.* in the exothermic direction), and thus its transition state may be characterized as "tight". Ethylperoxyl has another dissociation channel that competes with (R1),

$$C_2H_5O_2 \rightarrow C_2H_5 + O_2, \tag{R2}$$

which is barrierless in the association direction. Its transition state may be characterized as loose. Reaction (R1) is favored energetically by 3 kcal mol^{-1}, and (R2) is favored entropically. Fig. 1 has two curves for each temperature: the solid curve is $k_1(T, p)$ including the effect of the competing reaction (R2) (this is the one that occurs in reality), and the dashed curves are the same rate constants ignoring the existence of the other channel. Both sets of curves were calculated from an appropriate master equation. At the low-pressure limit the two curves of the same temperature coincide, because the lower energy channel is the only one that matters under such conditions. Similarly, at the high pressure limit the dashed and solid curves coalesce again, because complete thermal equilibrium is attained (*i.e.* the energy distributions of the molecule are completely determined by collisions), and the two product channels do not interfere with each other. However, at intermediate pressures the $C_2H_5 + O_2$ channel, with its loose transition state, robs large amounts of flux from the $C_2H_4 + HO_2$ channel, leading to the odd shape displayed by the solid curves in Fig. 1; the maximum effect is almost 2 orders of magnitude at 1500 K. The options available in CHEMKIN for representing pressure dependence of rate constants anticipated only the behavior of the dashed curves in Fig. 1, not the solid curves. It is quite difficult to represent such behavior satisfactorily over wide ranges of temperature and pressure.

Another type of "complex" pressure dependence occurs in chemically activated reactions with multiple stabilization products. A good example is the $C_2H_3 + C_2H_2$ reaction, whose potential energy surface is displayed diagrammatically in Fig. 2.[22] Transition state 4 (TS-4) is sufficiently high in energy that it effectively blocks access to the i-C_4H_5 well, but complexes can be stabilized

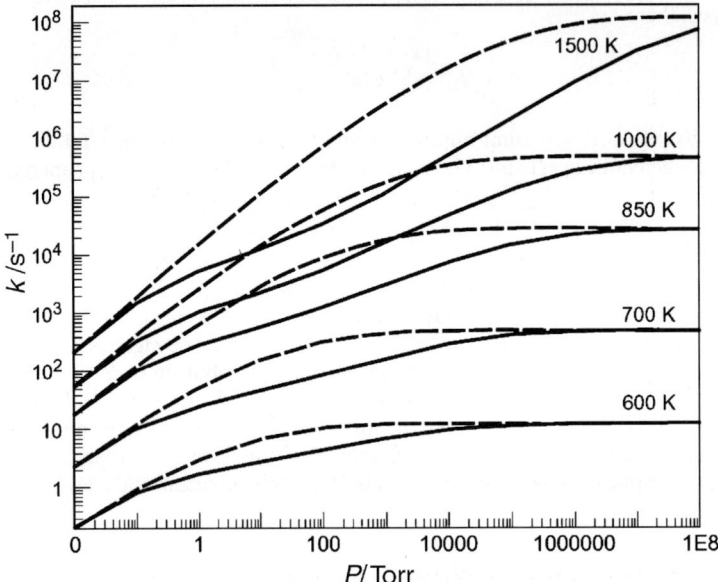

Fig. 1 Pressure dependence of the rate constant for the reaction, $C_2H_5O_2 \rightarrow C_2H_4 + HO_2$.[24] The solid curves include the effects of the competing reaction, $C_2H_5O_2 \rightarrow C_2H_5 + O_2$; the dashed curves do not.

as either n-C_4H_5 or c-C_4H_5, or they can go on to form vinylacetylene (C_4H_4) + H. Fig. 3 shows the pressure dependence of α_n and α_c, the branching fractions for formation of n-C_4H_5 and c-C_4H_5, respectively, at 300 and 400 K. The individual-channel rate constants are qualitatively the same. At low pressures, both α_c and α_n increase with pressure. However, α_c begins to decrease with pressure beyond 10 Torr at 300 K and beyond 100 Torr at 400 K, whereas α_n continues to increase. Such behavior occurs when stabilization of the more accessible product (n-C_4H_5), *i.e.* the one closer to the reactants on the potential energy surface, begins to limit the entry of complexes into the less accessible well. The pressure dependence of $C_2H_3 + C_2H_2 \rightarrow$ c-C_4H_5 is not easily

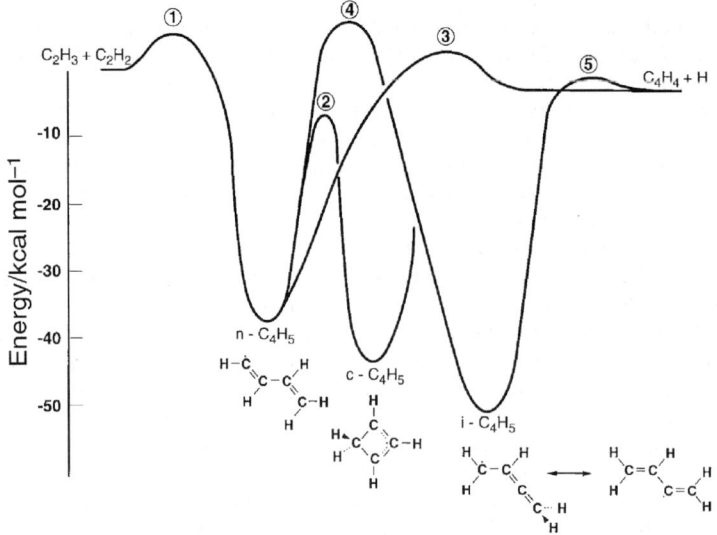

Fig. 2 Potential energy diagram for the $C_2H_3 + C_2H_2 \rightarrow$ products reaction.[22]

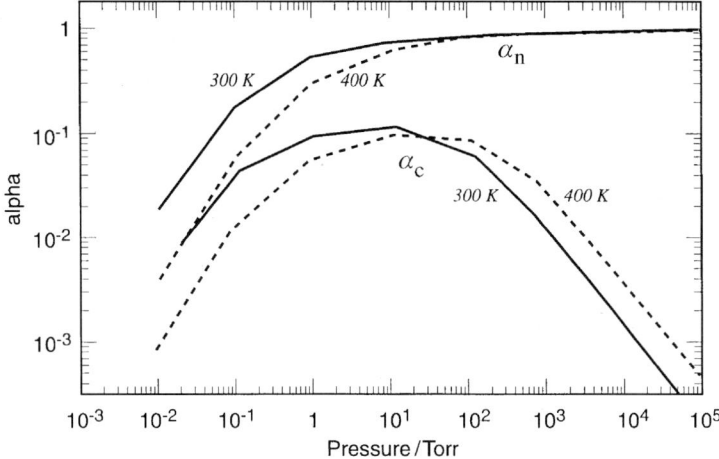

Fig. 3 Branching fractions for the $C_2H_3 + C_2H_2$ reaction at $T = 300$ and 400 K.[22]

modeled in the CHEMKIN format. At the very least, its rate coefficient would need to be represented as the sum of two rate coefficients, one that increases and one that decreases with pressure, the latter having negative values. Clearly, it would be better to avoid such artificial representations.

The examples discussed above are consequences of a reaction having multiple, competitive, dissociation or stabilization channels. Such reactions are actually the norm, not the exception. Master equation analyses of multiple-well systems will become routine in the next five years. It would be desirable to have a systematic, uniformly valid method of representing their results in phenomenological models. At least on the surface, methods discussed by Kazakov et al.[26] appear to show promise in this regard. However, the issue is still an open one.

Elementary reactions

Elementary reactions are the cornerstone of our field. Consequently, a large number of papers addressed this topic from one perspective or another. However, surprisingly few were of the traditional (experimental) type in which a rate constant and/or product distribution is measured as a function of temperature or pressure (or both). One should probably put the molecular beam papers in the "elementary reactions" category even though they do not provide rate constants directly. Such experiments are best viewed as sensitive probes of potential energy surfaces and are extremely powerful when used in conjunction with electronic structure calculations, as was shown by the 2 papers presented here. The posters included a number of excellent experimental investigations of elementary reactions (compensating for the lack thereof in the papers), including the Skinner Prize winning poster of Barnes et al.

The paper by Bergeat et al.[27] addressed the issue of the product distribution in the $CH + O_2$ reaction experimentally, although its most important contribution may be the accompanying electronic structure calculations, which show that the reaction can occur on both the doublet and quartet surfaces with no intrinsic potential energy barrier. The importance of the $CH + O_2$ reaction in combustion lies in its role in the nitrogen chemistry (prompt NO and reburning) as a "sink" for CH; the product distribution is only of secondary importance. The occurrence of reaction on both surfaces (on the doublet surface at low T, both at high T) could explain the unexpected temperature dependence of the rate constant, as I suggested in my comment. Stephen Klippenstein's direct dynamics calculations lend credence to this suggestion in that they show that HCO + O (a major low temperature product) can be formed *non-statistically* on the doublet surface at room temperature, *i.e.* the HCO + O products do not necessarily imply that reaction is taking place on the quartet surface.

I was impressed by my colleague Stephen Klippenstein's use of direct dynamics calculations as a complement to statistical methods in the analysis of elementary reactions. Such methods, even with relatively low-level quantum chemistry, are particularly useful for 3 purposes:
1. Discerning the plausibility of non-statistical products.
2. Determining product distributions in extremely exothermic reactions where a species such as HCO may actually be formed as H + CO.
3. Treating reactions such as $CH_3 + O \rightarrow CO + H_2 + H$, which may be statistical but are difficult to analyze with ordinary transition-state methods.

Heterogeneous chemistry

I cannot claim to be an expert on this topic, but both the papers presented at this Discussion in the general area of gas–surface interactions caught my attention. The paper by Backreedy et al.[28] discussed the oxidation of graphite and demonstrated that even such a complicated process is amenable to theoretical analysis. Research in this area has shown that graphite can be modeled satisfactorily as graphene, a molecule consisting of 7 six-membered rings with 4 free valence sites and the other carbons terminated by hydrogen atoms. Sequential attacks of molecular oxygen on free valence sites, followed by CO elimination and the accompanying surface degradation, can explain many features of "elementary" reactions associated with graphite oxidation, at least qualitatively. Quantitative predictions appear to be well within reach.

Equally impressive is the modeling of the surface chemistry in a three-way catalyst by Prof. Warnatz and his collaborators.[29] This type of modeling began in the chemical vapor deposition field, but Prof. Warnatz and his group have been instrumental in extending it into areas of combustion related to catalysis. The model presented here is noteworthy in that it was able to predict the general attributes of a three-way catalyst with very limited knowledge of the surface chemistry, an extremely encouraging result. He correctly characterized the situation as being analogous to that of gas phase combustion in the 1960's, when we had some idea of what steps were involved (*i.e.* the structure of the reaction mechanism), but we had relatively little knowledge of the rate constants.

NO_x and SO_x formation and control

The chemistry underlying NO_x formation and control in combustion is relatively well understood,[30–32] at least compared with that for soot formation. However, strict government regulation of NO_x emissions from combustion sources in many industrialized countries has prompted the search for improved control technologies.[31] The latter in turn places severe demands on our knowledge of the underlying chemistry and our ability to construct predictive kinetic models. Thus the chemistry of nitrogen in combustion continues to be an active research area.

Methods of NO_x control fall into two basic categories: combustion modification and aftertreatment.[30,31] Both topic areas were represented at this Discussion. The papers by Fikri et al.,[10] Mercier et al.[33] and Lozovsky et al.[34] address issues related to reburning,[30–32] a combustion modification technique that relies on combustion-generated, hydrocarbon free radicals to remove NO. The paper by Fang et al.[3] and the poster by Cole and Balint-Kurti provide theoretical analyses of the NH_2 + NO reaction, the pivotal step in the Thermal De-NO_x process,[38] which is the most widespread of the non-catalytic aftertreatment schemes.

Professor Temps presented an excellent paper on the CH_2 + NO reaction,[10] one of the reactions important in removing NO during reburning. The reaction has two plausible channels,

$$CH_2 + NO \rightarrow HCNO + H \qquad (R3)$$

and

$$CH_2 + NO \rightarrow HCN + OH. \qquad (R4)$$

Measurements at room temperature and theoretical analysis from room temperature to 2000 K show conclusively that the dominant products are HCNO + H (virtually 100%). In previous experimental work, Professor Temps[39] and others[40,41] have shown that the most important reburn reaction,

$$HCCO + NO \rightarrow HCNO + CO \quad (R5)$$

$$\rightarrow HCN + CO_2 \quad (R6)$$

is also dominated by HCNO formation, a result that has been confirmed by the recent theoretical analysis of Vereecken et al.[42] These results have important consequences in our reburn model.[35–37]

Fig. 4 is a reaction path diagram taken from my own work[35] on modeling reburn chemistry under plug-flow conditions at atmospheric pressure and temperatures in the range, $800 < T/K < 1500$. The diagram shows the paths to ketenyl (HCCO), overwhelmingly the dominant NO removal agent under these conditions regardless of the reburn fuel (C_2H_2, C_2H_4, C_2H_6, CH_4 or natural gas). Under more radical-rich conditions (e.g. higher temperatures, low-pressure flames or stirred reactors), reactions of CH_2, CH and C with NO play a greater role, but HCCO + NO is always a major contributor to NO removal. If HCNO + CO are the dominant products of HCCO + NO, the question that arises is "What happens to the HCNO?" Our analysis indicated that the reactions,

$$O + HCNO \rightarrow HCO + NO \quad (R7)$$

and

$$OH + HCNO \rightarrow HCOH + NO, \quad (R8)$$

are very fast and recycle the NO removed through (R5) back to NO, thus largely eliminating (R5) as an effective reburn reaction. This result, taken together with the experimental observation that NO is in fact converted to HCN and N_2 under the conditions of interest, and our statistical-theoretical analysis of the HCCO + NO reaction (based on the PES of Nguyen et al.[43]) caused us to construct a reaction mechanism with (R6) the dominant product channel, in conflict with subsequent and existing experiments.

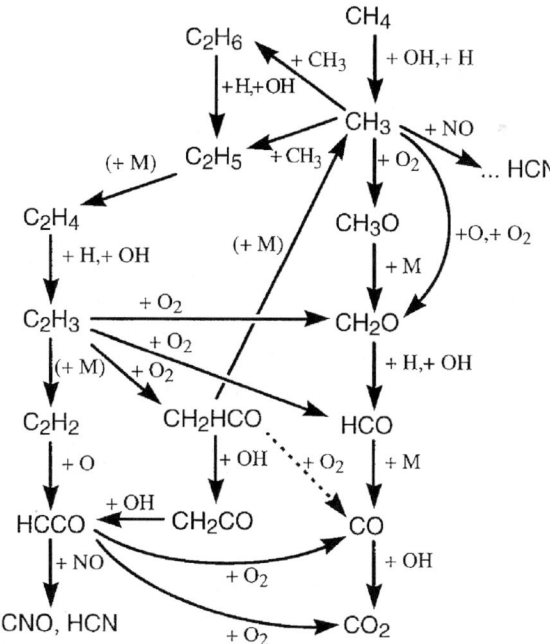

Fig. 4 Reaction path diagram showing paths to HCCO from C_2H_2, C_2H_4, C_2H_6 and CH_4 in plug-flow reactor studies of reburning.[35]

The dilemma described in the previous paragraph is probably resolvable by having a much larger rate constant in the model for the reaction,

$$HCNO + H \rightarrow HCN + OH, \tag{R9}$$

one that is large enough that (R9) can compete with (R7) and (R8) for HCNO. Our value for k_9 was chosen many years ago and was based on that for the analogous N_2O reaction (CH is isoelectronic with N),

$$N_2O + H \rightarrow N_2 + OH. \tag{R10}$$

Reaction (R9) takes place on the same potential as the $CH_2 + NO$ reaction studied by Temps and co-workers. By analogy with (R10), reaction (R9) should occur both through attack of the H on the carbon atom, followed by rearrangement, and by attack on the oxygen atom in HCNO. The latter is dominant at high T in the $N_2O + H$ reaction. Temps's surface apparently contains all the information required to calculate k_9 except the properties of the transition state for H attack on the oxygen. If it is any indicator at all, the barrier for H attack on the C in HCNO (barrierless) is significantly smaller than the barrier for its attack on the analogous N in N_2O,[44,45] perhaps indicating that k_9 is indeed much larger than k_{10}. In any event, we need a good rate constant for (R9).

The reaction between NH_2 and NO has fascinated me for years.[15,30,46-49] At low temperature the reaction occurs by breaking all three chemical bonds in the reactants and forming three completely new bonds in the products ($H_2O + N_2$), all in a single collision.[15] At high temperatures, the importance of the reaction stems from its ability to form free radicals, including the exotic NNH. A very large number of people have contributed to our understanding of this reaction, both experimentally and theoretically, as evidenced by the large number of investigations represented in Figs. 8, 9, 10 and 11 of our paper[3] at this Discussion and by the references in the paper. As a result, we might characterize this reaction and the Thermal De-NO_x process as relatively well understood.[48] In my opinion, work on Thermal De-NO_x and the $NH_2 + NO$ reaction stands as the best example we have in combustion of how theory, modeling and experiment can be used in concert to understand a practical combustion problem.

It has been known for a long time that sulfur in the fuel, during combustion, can effect the formation and destruction of NO_x.[50,51] However, there have been very few serious attempts to understand how this occurs. This problem is particularly important with the more stringent NO_x emissions regulations now in place and the continued use of high-sulfur fuels (mainly coal) in practical combustion devices. The paper by Hughes et al.[52] is a promising start to understanding this phenomenon and developing a reliable mechanism for quantifying its effects.

Aromatic hydrocarbons, polycyclic aromatic hydrocarbons and soot

This is probably the most active area of research in combustion chemistry right now, largely motivated by health concerns about emissions of small particles, carcinogenic and mutagenic PAH and "air toxics".[15,53-55] Species such as benzene and 1,3-butadiene, which are intimately tied to mechanisms of hydrocarbon growth in flames, were classified as air toxics by the New Clean Air Act of 1990 in the United States, and consequently their emissions are now regulated explicitly. At the same time, the underlying chemical mechanisms in this area are extremely complex, requiring knowledge of numerous species (including their various isomeric forms) and elementary reactions. Moreover, the critical elementary reactions are of the most complicated type, involving multiple interconnected isomers as products and/or intermediates. At this Discussion, the papers by Kaiser et al.,[56] Hahn et al.,[7] Smith and Frankcombe,[6] Goos et al.,[57] Hessler et al.,[58] Lindstedt et al.[59] and the poster by Isemer et al. explicitly addressed issues related to this general topic.

Chemical kinetic modeling of the formation of aromatics, PAH growth and soot formation from aliphatic fuels began with the work of Frenklach and co-workers in 1984.[60] These investigators introduced many important ideas, some of which are still prominent today. However, only more recently has a unifying theme emerged in this area, *i.e.* the importance of resonantly stabilized free radicals.[15,54,55,61-64] The simplest and probably most important resonantly stabilized free radical

(RSFR) is propargyl (C_3H_3). Research to date has largely focused on the formation of the first ring, *i.e.* the cyclization step, and most investigators agree that the recombination of propargyl radicals plays a major role (if not the dominant one) in this process,

$$C_3H_3 + C_3H_3 \rightarrow \text{benzene} \quad \text{(R11a)}$$
$$\rightarrow \text{fulvene} \quad \text{(R11b)}$$
$$\rightarrow \text{phenyl} + H \quad \text{(R11c)}$$

Our understanding of the $C_3H_3 + C_3H_3$ reaction is based primarily on the BAC-MP4 calculations of Miller and Melius,[61,65] shown diagrammatically in Fig. 5. However, only recently have Stephen Klippenstein and I tried to calculate rate constants and product distributions from this

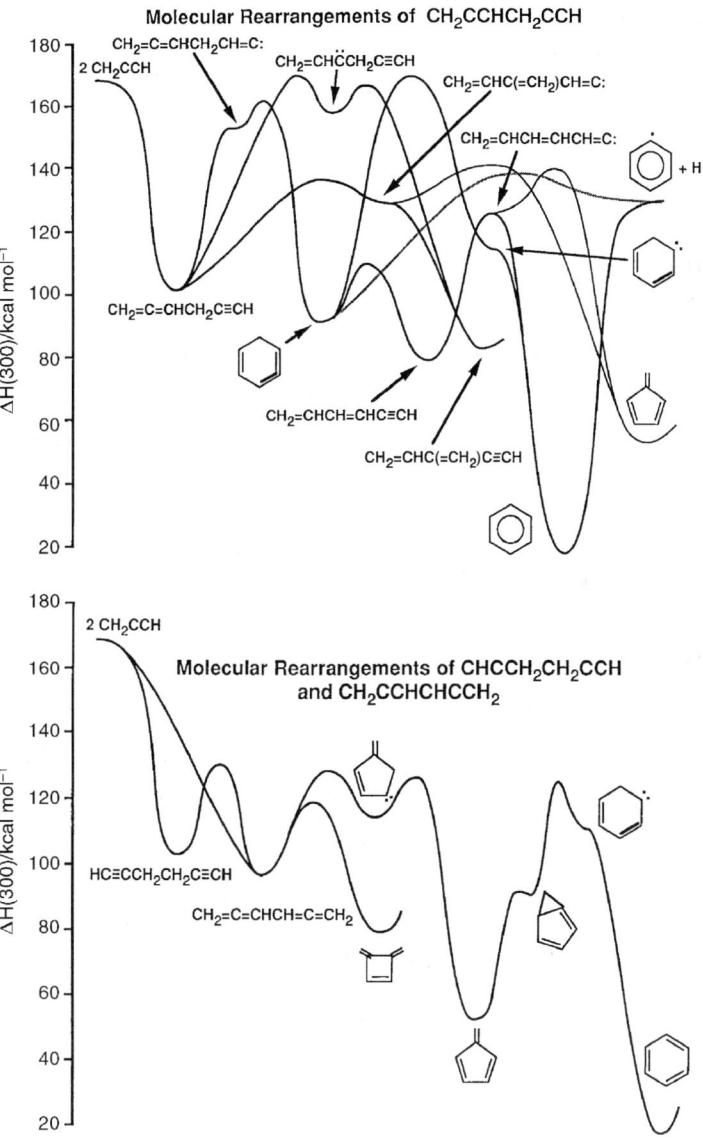

Fig. 5 Reaction coordinate diagram for propargyl radical recombination.[61]

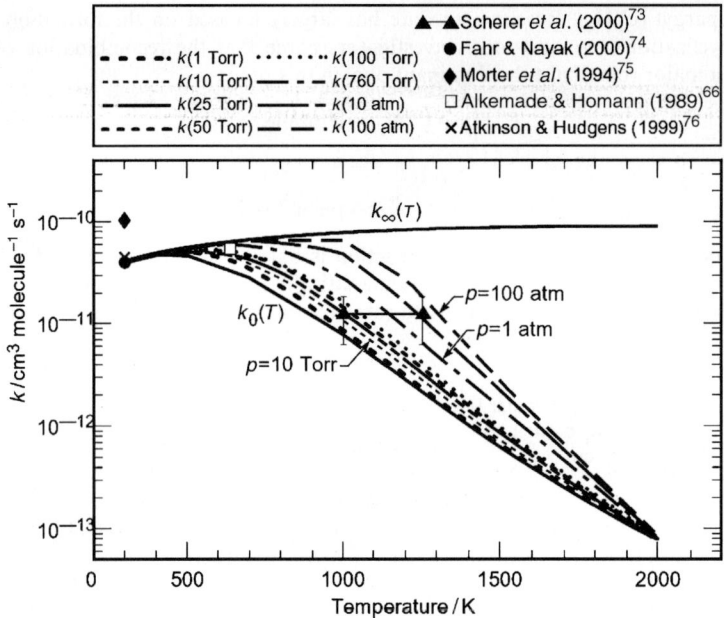

Fig. 6 Rate constant for $C_3H_3 + C_3H_3 \rightarrow$ products.[25]

potential using the time-dependent, multiple-well master equation.[25] Our results for the rate constant are shown in Fig. 6. The most important point to notice is that the rate constant drops off rapidly at high temperature. At atmospheric pressure and 1500 K, a temperature typical of the reaction zone in a laminar premixed flame, $k_{11} \approx 8 \times 10^{11}$ cm^3 mol^{-1} s^{-1}, which is significantly smaller than our initial proposal of $k_{11} = 1 \times 10^{13}$ cm^3 mol^{-1} s^{-1}. It is also smaller than the low-temperature measurements of the rate constant (2–7 $\times$ 10^{13}), and it is even smaller than that used in recent modeling[62] (3 $\times$ 10^{12}). If such a small rate constant proved to be correct, it would limit the importance of reaction (R11) as a cyclization step.

Fortunately, we have recently found a transition state that was overlooked by Miller and Melius. It is a 1,5 hydrogen transfer connecting 1,2,4,5-hexatetraene (CH$_2$=C=CHCH=C=CH$_2$ in Fig. 5(a)) to 1,3-hexadiene-5-yne (CH$_2$=CHCH=CHC≡CH in Fig. 5(b)). This transition state lies only 1 to 2 kcal mol^{-1} higher in energy than the transition states connecting 1,2,4,5-hexatetraene to fulvene in Fig. 5(a). Preliminary calculations including this transition state indicate that the rate constant at 1500 K increases by about a factor of 3 to 5, making it comparable in magnitude to that used in some of the most recent models. Moveover, the revised PES has some other desirable features, which I shall only mention here in passing:

1. It allows us to predict more accurately the product distributions of Alkemade and Homann,[66] which include 1, 3-hexadiene-5-yne, and those of Stein et al.,[67] the latter for the isomerization of 1,5-hexadiyne (HC≡CCH$_2$CH$_2$C≡CH in Fig. 5(a)).

2. It provides a path from $C_3H_3 + C_3H_3$ to benzene that does not go through fulvene. This is a desirable feature, because none of the chemically activated $C_3H_3 + C_3H_3$ experiments have detected fulvene as a product, although it is a major product in the thermally activated isomerization of 1,5-hexadiyne. With the revised PES, the different types of experiments may explain this difference in product distributions.

Our experience with $C_3H_3 + C_3H_3$ has made me leery of the idea that we know everything there is to know about the formation of the first ring. Other reactions worthy of consideration as cyclization steps are

$$C_3H_3 + C_2H_2 \rightarrow \text{c-}C_5H_5, \tag{R12}$$

$$\text{i-}C_5H_3 + CH_3 \rightarrow \text{benzene}, \tag{R13}$$

and

$$\text{i-}C_4H_5 + C_2H_2 \rightarrow \text{fulvene} + H, \tag{R14}$$

where c-C_5H_5 is cyclopentadienyl and i-C_5H_3 is the resonantly stabilized CH_2CCCCH radical.

Moriarty et al.[68] and Moskaleva and Lin[69] have proposed (R12) as an important cyclization step. Presumably, it would be followed by

$$\text{c-}C_5H_5 + CH_3 \rightarrow \ldots \text{fulvene} \tag{R15}$$

and

$$\text{fulvene} + H \rightarrow \text{benzene} + H, \tag{R16}$$

as described by Melius et al.,[70] leading to benzene and then to higher PAH's. My own quick-and-dirty equilibrium calculations indicate that the equilibrium of (R12) begins to shift to the left at $T \approx 1500$ K for a pressure of 1 atm. Such a shift would likely limit severely the potential of (R12) as an important cyclization step. However, the thermochemistry on which the equilibrium calculations are based is not sufficiently well established to be certain of the result.

The i-C_5H_3 + CH_3 reaction has been suggested by Pope and Miller[71] as a viable cyclization step in flames on the basis of the electronic-structure calculations of Mebel et al.[72] which show that benzene is accessible from i-C_5H_3 + CH_3 through a path with no intrinsic energy barrier. However, the highest energy point along the reaction path is only 7 kcal mol^{-1} below the reactants. Experience with C_3H_3 + C_3H_3 would suggest that, with such a small energy "window", k_{13} would not be sufficiently large at flame temperatures to be an effective cyclization step. Nevertheless, a more energetically favorable path may be discovered. I should note that i-C_5H_3 is relatively easily formed in flames from diacetylene,

$$^3CH_2 + C_4H_2 \rightarrow \text{i-}C_5H_3 + H. \tag{R17}$$

The last reaction I want to mention is (R14), i-C_4H_5 + C_2H_2 → fulvene + H, which Carl Melius and I discussed briefly in our 1992 paper.[61] Reaction (R14) cannot be very important in most acetylene flames, because C_4H_5 of any isomeric form is difficult to produce. However, that is not necessarily the case for other fuels. The appeal of (R14) is that cyclization can occur in the complex without any H transfers or any concerted, multi-centered processes. It could be particularly important in 1,3-butadiene flames, where i-C_4H_5 can be formed from 1,3-butadiene by direct abstraction.

In his paper at this Discussion,[59] Prof. Lindstedt has outlined several possible paths to the two-ring aromatics, indene and naphthalene, as well as possible competing oxidative paths. To summarize, the reactions proposed for formation of indene and naphthalene are the following:

indene:

$$\text{phenyl} + C_3H_3 \rightarrow \ldots \text{indene}$$

$$\text{phenyl} + \text{allene} \rightarrow \text{indene} + H$$

$$\text{benzyl} + C_2H_2 \rightarrow \text{indene} + H$$

naphthalene:

$$\text{c-}C_5H_5 + \text{c-}C_5H_5 \rightarrow C_5H_5C_5H_4 + H$$

$$C_5H_5C_5H_4 \rightarrow \text{naphthalene} + H$$

$$\text{benzyl} + C_3H_3 \rightarrow \ldots \text{naphthalene}$$

$$\text{phenyl} + C_4H_4 \rightarrow \ldots \text{naphthalene}$$

Except possibly for c-C_5H_5 + c-C_5H_5, none of these reactions have rate constants or mechanisms that are well established. Melius et al.[70] studied the cyclopentadienyl recombination theoretically, but they did not calculate rate constants from their potential. In addition to c-C_5H_5 + c-C_5H_5, the phenyl + C_3H_3 and benzyl + C_3H_3 look the most attractive to me, because they involve 2 resonantly stabilized free radicals. Such reactions have 3 desirable features:

1. The resonantly stabilized radicals might be expected to be present in flames in relatively large quantities, at least compared with ordinary free radicals.
2. Because the reactants are free radicals, there is no barrier to the initial association.
3. Again, because the reactants are radicals, the bond initially formed is relatively strong compared with that for a radical + molecule reaction. The stronger bond helps to support rearrangement. That is, the rearrangement barriers for such reactions are likely to lie lower on the PES, relative to the reactants, than for a radical + molecule reaction.

In any event, Professor Lindstedt has provided us with much food for thought, not to mention topics for research. I hope that in the next 5 to 10 years we will have a good quantitative description of the formation of the one- and two-ring aromatics. Perhaps the formation of larger PAH's will follow a simple generic mechanism, and life will become easier, even if less interesting.

Acknowledgements

I would like to thank my colleague Stephen Klippenstein for numerous productive conversations during the course of my writing these comments. My time was supported by the United States Department of Energy, Office of Basic Energy Sciences, Division of Chemical Sciences, Geosciences, and Biosciences.

References

1. J.-P. Senosiaian, J. H. Han, C. B. Musgrave and D. M. Golden, *Faraday Discuss.*, 2001, **119**, 173.
2. C. F. Melius, in *Chemistry and Physics of Energetic Materials*, NATO ASI 309, ed. S. Bulusu, 1990, p. 21; P. Ho and C. F. Melius, *J. Phys. Chem.*, 1990, **94**, 5120.
3. D.-C. Fang, L. B. Harding, S. J. Klippenstein and J. A. Miller, *Faraday Discuss.*, 2001, **119**, 207.
4. B. K. Carpenter, *J. Phys. Chem. A*, 2001, **105**, 4585.
5. A. M. Mebel, E. W. G. Diau, M. C. Lin and K. Morokuma, *J. Am. Chem. Soc.*, 1996, **118**, 9759.
6. T. J. Frankcombe and S. C. Smith, *Faraday Discuss.*, 2001, **119**, 159.
7. D. K. Hahn, S. J. Klippenstein and J. A. Miller, *Faraday Discuss.*, 2001, **119**, 79.
8. W. S. Xia, R. S. Zhu, M. C. Lin and A. M. Mebel, *Faraday Discuss.*, 2001, **119**, 191.
9. J. D. DeSain, C. A. Taatjes, J. A. Miller, S. J. Klippenstein and D. K. Hahn, *Faraday Discuss.*, 2001, **119**, 101.
10. M. Fikri, S. Meyer, J. Roggenbuck and F. Temps, *Faraday Discuss.*, 2001, **119**, 223.
11. S. C. Smith and R. G. Gilbert, *Int. J. Chem. Kinet.*, 1988, **20**, 329.
12. C. Raffy, S. J. Klippenstein and J. A. Miller, in preparation.
13. P. K. Venkatesh, A. M. Dean, M. H. Cohen and R. W. Carr, *J. Chem. Phys.*, 1999, **111**, 8313.
14. S. J. Jeffrey, K. E. Gates and S. C. Smith, *J. Phys. Chem.*, 1996, **100**, 7090.
15. J. A. Miller, *Proc. Combust. Inst.*, 1996, **20**, 461.
16. J. Troe, *J. Chem. Soc., Faraday Trans.*, 1994, **90**, 2303.
17. Th. Just, *Proc. Combust. Inst.*, 1994, **25**, 687.
18. N. J. Brown and J. A. Miller, *J. Chem. Phys.*, 1984, **80**, 5568.
19. A. Gelb, *J. Phys. Chem.*, 1985, **89**, 4189.
20. S. Nordholm and H. W. Schranz, in *Advances in Chemical Kinetics and Dynamics*, ed. J. R. Barker, JAI Press, Greenwich, CT, 1995, vol. 2, p. 245.
21. G. Lendvay and G. C. Schatz, in *Advances in Chemical Kinetics and Dynamics*, ed. J. R. Barker, JAI Press, Greenwich, CT, 1995, vol. 2, p. 481.
22. J. A. Miller, S. J. Klippenstein and S. H. Robertson, *J. Phys. Chem. A*, 2000, **104**, 7525; also *J. Phys. Chem. A*, 2000, **104**, 9806 (correction).
23. J. A. Miller, S. J. Klippenstein and S. H. Robertson, *Proc. Combust. Inst.*, 2000, **28**, 1479.
24. J. A. Miller and S. J. Klippenstein, *Int. J. Chem. Kinet.*, 2001, **33**, 654.
25. J. A. Miller and S. J. Klippenstein, *J. Phys. Chem. A*, 2001, **105**, 7254.
26. A. Kazakov, H. Wang and M. Frenklach, *J. Phys. Chem.*, 1994, **98**, 10598.
27. A. Bergeat, T. Calvo, F. Caralp, J.-H. Fillion, G. Dorthe and J.-C. Loison, *Faraday Discuss.*, 2001, **119**, 67.
28. R. Backreedy, J. M. Jones, M. Pourkashanian and A. Williams, *Faraday Discuss.*, 2001, **119**, 385.
29. D. Chatterjee, O. Dewschmann and J. Warnatz, *Faraday Discuss.*, 2001, **119**, 371.
30. J. A. Miller and C. T. Bowman, *Prog. Energy Combust. Sci.*, 1989, **15**, 287.
31. C. T. Bowman, *Proc. Combust. Inst.*, 1992, **24**, 859.
32. A. M. Dean and J. W. Bozzelli, in *Gas-Phase Combustion Chemistry*, ed. W. C. Gardiner, Jr., Springer-Verlag, 1996, pp. 125–341.
33. X. Mercier, L. Pillier, A. El Bakali, M. Carlier, J.-F. Pauwels and P. Desgroux, *Faraday Discuss.*, 2001, **119**, 305.

34 V. A. Lozovsky, I. Rahinov, N. Ditzian and S. Cheskis, *Faraday Discuss.*, 2001, **119**, 321.
35 J. A. Miller, J. L. Durant and P. Glarborg, *Proc. Combust. Inst.*, 1998, **27**, 235.
36 P. Glarborg, M. U. Alzueta, K. Dam-Johansen and J. A. Miller, *Combust. Flame*, 1998, **115**, 1.
37 L. Prada and J. A. Miller, *Combust. Sci. Tech.*, 1998, **132**, 225.
38 R. K. Lyon, *U.S. Pat.*, 1975, 3 900 554.
39 U. Eickhoff and F. Temps, *Phys. Chem. Chem. Phys.*, 1999, **1**, 243.
40 K. T. Rim and J. F. Hershberger, *J. Phys. Chem. A*, 2000, **104**, 293.
41 W. Boullart, M. T. Nguyen and J. Peeters, *J. Phys. Chem.*, 1994, **98**, 8036.
42 L. Vereecken, R. Sumathy and J. Peeters, work in progress, Poster 4-B10, presented at the Twenty-Eighth International Combustion Symposium, Edinburgh, August 2000.
43 M. T. Nguyen, W. Boullart and J. Peeters, *J. Phys. Chem.*, 1994, **98**, 8030.
44 J. A. Miller and C. F. Melius, *Proc. Combust. Inst.*, 1992, **24**, 719.
45 P. Marshall, A. Fontijn and C. F. Melius, *J. Chem. Phys.*, 1987, **86**, 5540.
46 J. A. Miller, M. C. Branch and R. J. Kee, *Combust. Flame*, 1981, **43**, 81.
47 J. A. Miller and P. Glarborg, in *Gas Phase Chemical Reaction Systems: Experiments and Models 100 years after Max Bodenstein*, ed. J. Wolfrum, H.-R. Volpp, R. Rannacher and J. Warnatz, Springer, Berlin, 1996, p. 318.
48 J. A. Miller and P. Glarborg, *Int. J. Chem. Kinet.*, 1999, **31**, 757.
49 J. A. Miller and S. J. Klippenstein, *J. Phys. Chem. A*, 2000, **104**, 2061.
50 S. I. Tseregounis and O. I. Smith, *Combust. Sci. Technol.*, 1983, **30**, 231.
51 S. I. Tseregounis and O. I. Smith, *Proc. Combust. Inst.*, 1984, **20**, 761.
52 K. J. Hughes, A. S. Tomlin, V. A. Dupont and M. Pourkashanian, *Faraday Discuss.*, 2001, **119**, 337.
53 H. Richter and J. B. Howard, *Prog. Energy Combust. Sci.*, 2000, **26**, 565.
54 M. J. Castaldi, N. M. Marinov, C. F. Melius, J. Hwang, S. M. Senkan, W. J. Pitz and C. K. Westbrook, *Proc. Combust. Inst.*, 1996, **26**, 693.
55 N. M. Marinov, W. J. Pitz, C. K. Westbrook, M. J. Castaldi and S. M. Senken, *Proc. Combust. Inst.*, 1996, **116/117**, 211.
56 R. I. Kaiser, T. N. Le, T. L. Nguyen, A. M. Mebel, N. Balucani, Y. T. Lee, F. Stahl, P. v. R. Schleyer and H. F. Schaefer III, *Faraday Discuss.*, 2001, **119**, 51.
57 E. Goos, H. Hippler, K. Hoyermann and B. Jürges, *Faraday Discuss.*, 2001, **119**, 243.
58 J. P. Hessler, S. Seifert, R. E. Winans and T. H. Fletcher, *Faraday Discuss.*, 2001, **119**, 395.
59 P. Lindstedt, L. Maurice and M. Meyer, *Faraday Discuss.*, 2001, **119**, 409.
60 M. Frenklach, D. W. Clary, W. C. Gardiner, Jr. and S. E. Stein, *Proc. Combust. Inst.*, 1984, **20**, 887.
61 J. A. Miller and C. F. Melius, *Combust. Flame*, 1992, **91**, 21.
62 A. D'Anna, A. Violi and A. D'Allessio, *Combust. Flame*, 2000, **121**, 418.
63 A. D'Anna and A. Violi, *Proc. Combust. Inst.*, 1998, **27**, 425.
64 P. Lindstedt, *Proc. Combust. Inst.*, 1998, **27**, 269.
65 C. F. Melius, J. A. Miller and E. M. Evleth, *Proc. Combust. Inst.*, 1992, **24**, 621.
66 U. Alkemade and K. H. Homann, *Z. Phys. Chem. Neue Folge*, 1989, **161**, 19.
67 S. E. Stein, J. A. Walker, M. M. Suryan and A. Fahr, *Proc. Combust. Inst.*, 1990, **23**, 85.
68 N. W. Moriarty, X. Krokidis, W. A. Lester, Jr. and M. Frenklach, *Second Joint Meeting of the U.S. Sections of the Combustion Institute*, Oakland, CA, 2001.
69 L. V. Moskaleva and M. C. Lin, *J. Comput. Chem.*, 2000, **21**, 415.
70 C. F. Melius, M. E. Colvin, N. M. Marinov, W. J. Pitz and S. M. Senkan, *Proc. Combust. Inst.*, 1996, **26**, 685.
71 C. J. Pope and J. A. Miller, *Proc. Combust. Inst.*, 2000, **28**, 1519.
72 A. M. Mebel, S. H. Lin, X. M. Yang and Y. T. Lee, *J. Phys. Chem. A*, 1997, **101**, 6781.
73 S. Scherer, Th. Just and P. Frank, *Proc. Combust. Inst.*, 2000, **28**, 1511.
74 A. Fahr and A. Nayak, *Int. J. Chem. Kinet.*, 2000, **32**, 118.
75 C. L. Morter, S. K. Farhat, J. D. Adamson, G. P. Glass and R. F. Curl, *J. Phys. Chem.*, 1994, **98**, 7029.
76 D. B. Atkinson and J. W. Hudgens, *J. Phys. Chem. A*, 1999, **103**, 4242.

List of Posters

Quantitative measurements of NO in flames by cavity ring-down spectroscopy **L. Pillier,*** **P. Desgroux,* A. Rida** and **P. Meunier**, *Université des Sciences et Technologies de Lille, France*

Studies of carbon atom reactions with H_2S, C_2H_2, C_2H_4 and C_6H_6: Overall rate constants and relative H atom production **A. Bergeat*** and **J.-C. Loison**, *Université Bordeaux I, France*

Rotationally resolved quenching and relaxation of CH ($A^2\Delta$, $v = 0$, N) in the presence of CO **P. Meden, M. Kind** and **F. Stuhl,*** *Ruhr-Universität Bochum, Germany*

Rate constants of allyl radical reactions with C_2H_2, CH_4, H_2 and C_3H_5-pyrolysis at temperatures between 1000 K and 1400 K **S. J. Isemer, K. Luther*** and **J. Troe**, *Universität Göttingen, Germany*

Principal component analysis based reduction of the GRI 3.0 methane combustion mechanism **I. Gy. Zsély*** and **T. Turányi**, *Eötvös University (ELTE), Budapest, Hungary*

The inverse Abel transform and negative intensity of small-angle X-ray scattering studies of soot inception and growth **J. P. Hessler,* S. Seifert, R. E. Winans** and **T. H. Fletcher**, *Argonne National Laboratory, USA*

The thermal deNOx process: NH_2 + NO **J. P. Cole*** and **G. G. Balint-Kurti**, *University of Bristol, UK*

The kinetics of the reaction between O_2 and three carbons (porous coal chars) at 650–900 °C **A. N. Hayhurst,* W. L. Hum, S. G. Taylor** and **A. Toole**, *University of Cambridge, UK*

Reaction and relaxation of vibrationally excited H_2O in collision with H atoms **P. W. Barnes,* I. R. Sims, I. W. M. Smith,* G. Lendvay** and **G. C. Schatz**, *University of Birmingham, UK*

Decomposition and isomerization of n-pentyl radicals: A master equation model for a 2-well reaction system **L. D. Jones,* L. C. Jitariu, I. H. Hillier, S. H. Robertson** and **M. J. Pilling**, *University of Manchester, UK*

Product branching ratios of the $C(^3P) + C_2H_3(^2A')$ and $CH(^2\Pi) + C_2H_2(^1\Sigma_g^+)$ reactions and photodissociation of $H_2CC=CH(^2B_1)$ at 193 and 242 nm: An *ab initio*/RRKM study **T. L. Nguyen, A. M. Mebel,* S. H. Lin** and **R. I. Kaiser**, *Institute of Atomic and Molecular Sciences, Taipei, Taiwan*

Electrochemistry in a flame **S. P. McCormack*** and **D. J. Caruana***, *University College London, UK*

Kinetics of sulfur containing species of importance in combustion **M. Blitz,* K. J. Hughes** and **M. J. Pilling**, *University of Leeds, UK*

H atom yields from the reactions of CH with H_2 and C_2 hydrocarbons **M. A. Blitz, D. G. Johnson, K. W. McKee, M. J. Pilling, H.-B. Qian** and **P. W. Seakins,*** *University of Leeds, UK*

Ab initio study of ammonium perchlorate initiation processes: The OH and ClO_x ($x = 1–3$) reactions **R. S. Zhu, Z. F. Xu** and **M. C. Lin,*** *Emory University, Atlanta, USA*

The dynamics of the H + DCl → HD + Cl gas phase reaction: Experimental and theoretical studies **A. Läuter,* F. J. Aoiz, T. Bohm, A. Hanf, V. J. Herrero, S. Dhanya, I. Tanarro, H.-R. Volpp** and **J. Wolfrum**, *Universität Heidelberg, Germany*

Simultaneous visualization of H_2CO-LIF, OH-LIF and temperature distributions in turbulent flames **A. Hoffmann,* Th. Kunzelmann, C. Schulz** and **J. Wolfrum**, *Universität Heidelberg, Germany*

Measurement of rate coefficients of Al + SF_6 and Al trimethylaluminium reactions **J. K. Parker,* N. L. Garland** and **H. H. Nelson**, *Naval Research Laboratory, Washington, DC, USA*

List of Participants

Mr R. F. Alvani, *University of Leeds, UK*
Mr R. Backreedy, *University of Leeds, UK*
Prof. G. G. Balint-Kurti, *University of Bristol, UK*
Mr P. W. Barnes, *University of Birmingham, UK*
Dr A. Bergeat, *Université de Bordeaux 1, France*
Dr M. Blitz, *University of Leeds, UK*
Mr T. Bohm, *University of Heidelberg, Germany*
Dr D. J. Caruana, *University College London, UK*
Prof. P. Casavecchia, *Università di Perugia, Italy*
Prof. S. Cheskis, *Tel Aviv University, Israel*
Mr J. P. Cole, *University of Bristol, UK*
Dr J. D. DeSain, *Sandia National Laboratories, USA*
Dr P. Desgroux, *Université des Sciences et Technologies de Lille, France*
Miss N. Ditzian, *Tel Aviv University, Israel*
Prof. G. Dixon-Lewis, *University of Leeds, UK*
Mr T. J. Frankcombe, *University of Queensland, Australia*
Prof. D. Golden, *Stanford University, USA*
Ms E. Goos, *Universität Karlsruhe, Germany*
Prof. P. Gray, *University of Cambridge, UK*
Prof. D. A. Greenhalgh, *Cranfield University, UK*
Prof. J. Griffiths, *University of Leeds, UK*
Miss C. L. Hall, *Royal Society of Chemistry, UK*
Mr A. Hanf, *University of Heidelberg, Germany*
Prof. A. N. Hayhurst, *University of Cambridge, UK*
Dr J. Hessler, *Argonne National Laboratory, USA*
Prof. H. Hippler, *Universität Karlsruhe, Germany*
Mr A. Hoffmann, *University of Heidelberg, Germany*
Prof. K. Hoyermann, *University of Göttingen, Germany*
Dr K. Hughes, *University of Leeds, UK*
Dr L. C. Jitariu, *University of Manchester, UK*
Mr G. P. Johnson, *University of Leeds, UK*
Dr J. M. Jones, *University of Leeds, UK*
Mr L. D. Jones, *University of Manchester, UK*
Dr R. I. Kaiser, *University of York, UK*
Prof. S. J. Klippenstein, *Sandia National Laboratories, USA*
Prof. K. Kohse-Höinghaus, *University of Bielefeld, Germany*
Miss A. Läuter, *University of Heidelberg, Germany*
Prof. M. C. Lin, *Emory University, USA*
Prof. P. Lindstedt, *Imperial College of Science, Technology and Medicine, UK*
Dr J.-C. Loison, *Université de Bordeaux 1, France*
Prof. K. Luther, *University of Göttingen, Germany*
Mr S. P. McCormack, *University College London, UK*
Mr K. W. McKee, *University of Leeds, UK*
Dr A. M. Mebel, *Academia Sinica, Taiwan*
Mr M. Meyer, *Imperial College of Science, Technology and Medicine, UK*
Dr J. A. Miller, *Sandia National Laboratories, USA*
Mrs P. A. Mohamed, *Royal Society of Chemistry, UK*
Dr C. Morley, *Shell Global Solutions*
Miss L. C. Palmer, *Royal Society of Chemistry, UK*
Dr J. K. Parker, *US Naval Research Laboratory, USA*
Mr A. Pekalski, *Delft University of Technology, Netherlands*
Miss L. Pillier, *Université des Sciences et Technologies de Lille, France*
Prof. M. J. Pilling, *University of Leeds, UK*
Prof. J. M. C. Plane, *University of East Anglia, UK*
Dr K. R. Ranson, *University of Leeds, UK*
Mr H. Schürmann, *Universität Stuttgart, Germany*

Dr G. H. E. Scott, *Royal Society of Chemistry, UK*
Dr P. W. Seakins, *University of Leeds, UK*
Mr J. P. Senosiain, *Stanford University, USA*
Dr D. B. Smith, *University of Leeds, UK*
Prof. I. W. M. Smith, *University of Birmingham, UK*
Prof. S. C. Smith, *University of Queensland, Australia*
Mr R. Sommarivo, *University of Leeds, UK*
Miss J. C. Stanton, *University of Leeds, UK*
Prof. F. J. Stuhl, *Ruhr-Universität Bochum, Germany*
Dr C. A. Taatjes, *Sandia National Laboratories, USA*
Prof. F. Temps, *University of Kiel, Germany*
Dr D. Thompson, *University of Sheffield, UK*
Dr A. S. Tomlin, *University of Leeds, UK*
Prof. J. Troe, *University of Göttingen, Germany*
Prof. P. J. Van Tiggelen, *Université Catholique de Louvain, Belgium*
Prof. J. Warnatz, *University of Heidelberg, Germany*
Dr B. C. Whitaker, *University of Leeds, UK*
Dr C. Whitehead, *University of Manchester, UK*
Prof. A. Williams, *University of Leeds, UK*
Prof. J. Wolfrum, *University of Heidelberg, Germany*
Mr R. M. Woolley, *University of Leeds, UK*
Ms S. Zhang, *University of Leeds, UK*
Mr I. G. Zsély, *Eötvös L. University, Hungary*

Index of Contributors*

Backreedy, R., **385**
Balint-Kurti, G. G., 255, 261, 266
Balucari, N., **27**, **51**
Bergeat, A., **27**, **67**, 125, 127, 137
Brockhinke, A., **275**
Calvo, T., **67**
Capozza, G., **27**
Caralp, F., **67**
Carlier, M., **305**
Cartechini, L., **27**
Casavecchia, P., **27**, 122, 123, 124, 125, 126, 128, 130, 131, 134
Chatterjee, D., **371**
Cheskis, S., 269, **321**, 353, 363, 364, 365, 366, 367, 370, 445
Cole, J. P., 255, 266
DeSain, J. D., **101**, 360
Desgroux, P., 272, **305**, 362, 363, 368
Deutschmann, O., **371**
Ditzian, N., **321**
Dorthe, G., **67**
Dupont, V. A., **337**
El Bakali, A., **305**
Fang, D. C., **207**
Fikri, M., **223**
Fillion, J.-H., **67**
Fletcher, T. H., **395**
Frankcombe, T. J., **159**, 452
Golden, D., 124, 140, 143, **173**, 256, 257, 261, 262, 263, 267, 271, 272, 445, 447, 451, 457
Goos, F., **243**
Gottwald, U., **433**
Greenhalgh, D. A., 354, 356, 361
Griffiths, J., 142, **287**, 355, 356, 357, 361
Hahn, D. K., **79**, **101**
Han, J. H., **173**
Harding, L. B., **207**
Hayhurst, A. N., 446, 449, 450, 457, 458, 459
Hein, K. R. G., **433**
Hessler, J., 261, 262, **395**, 446, 447, 449, 450, 452, 453, 454
Hippler, H., 121, **243**, 272, 354
Hoyemann, K., **243**, 272
Hughes, K. J., **337**, 368, 369, 370, 447, 448
Jones, J. M., **385**
Jürges, B., **243**
Kaiser, R. I., **51**, 123, 125, 127, 128, 131, 133, 134, 260, 273, 369
Klippenstein, S. J., **79**, **101**, 121, 128, 134, 137, **207**, 256, 260, 267, 268, 456
Kohse-Höinghaus, K., 124, 138, 257, 261, 272, **275**, 353, 354, 362, 364, 366, 452, 456
Le, T. N., **51**
Lee, Y. T., **51**
Lin, M. C., 121, 141, **191**, 261, 263, 265, 266, 268, 366, 446, 450, 458
Lindstedt, R. P., 361, 364, **409**, 445, 447, 451, 454, 456, 457, 458

Loison, J.-C., **67**, 136, 138, 139
Lozovsky, V. A., **321**
MacNamara, J. P., **287**
Maurice, L., **409**
Mebel, A. M., **51**, 134, **191**, 261, 262
Mercier, X., **305**
Meyer, M., **409**
Meyer, S., **223**
Miller, J. A., **79**, **101**, 139, 140, 141, 143, **207**, 258, 263, 264, 266, 270, 365, 367, 457, **461**
Mohamed, C., **287**
Monkhouse, P. B., **433**
Morley, C., 355, 356, 448, 459
Musgrave, C. B., **173**
Nguyen, T. L., **51**
Pan, J., **287**
Parker, J. K., 269
Pauwels, J.-F., **305**
Pillier, L., **305**
Pilling, M. J., 134, 139, 257, 260, 262, 264, 265, 358, 367, 446, 451, 453
Plane, J. M. C., 130, 138, 266, 267, 272, 368
Pourkashanian, M., **337**, **385**
Rahinov, I., **321**
Roggenbuck, J., **223**
Schaefer III, H. F., **51**
Schleyer, P. v. R., **51**
Schürmann, H., **433**, 458, 459
Seakins, P. W., 135, 137, 143, 369
Seifert, S., **395**
Senosiain, J. P., **173**, 451
Sheppard, G. W., **287**
Smith, D. B., 264, 268
Smith, I. W. M., 124, 133, 262
Smith, S. C., 134, 141, **159**, 258, 259, 260, 262, 269, 451
Stahl, F., **51**
Stuhl, F. J., 353
Taatjes, C. A., **101**, 142, 143, 268, 356, 358, 361
Temps, F., **223**, 268, 269, 271
Tomlin, A. S., **337**, 367, 447
Troe, J., 134, 140, 141, **145**, 255, 256, 257, 258, 259, 260, 261, 264, 265, 271, 370, 445, 455, 456
Unterberger, S., **433**
Ushakov, V. G., **145**
Van Tiggelen, P. J., 266
Volpi, G. G., **27**
Warnatz, J., **371**, 445, 446, 447, 448
Whitaker, B. C., 131, **287**, 353, 354, 361, 362, 448
Whitehead, C., 448
Williams, A., **385**, 448, 449, 450, 451, 452, 454
Winans, R. E., **395**
Wolfrum, J., **1**, 121, 142, 354, 366, 367, 369, 370, 449, 453, 459
Xia, W. S., **191**
Zsély, I. G., 363

* The page numbers in **bold** type indicate papers submitted for discussions.

General Discussions of the Faraday Society/Faraday Discussions of the Chemical Society

Date	Subject	Volume
1907	Osmotic Pressure	Trans. 3
1907	Hydrates in Solution	3
1910	The Constitution of Water	6
1911	High Temperature Work	7
1912	Magnetic Properties of Alloys	8
1913	Colloids and their Viscosity	9
1913	The Corrosion of Iron and Steel	9
1913	The Passivity of Metals	9
1914	Optical Rotary Power	10
1914	The Hardening of Metals	10
1915	The Transformation of Pure Iron	11
1916	Methods and Appliances for the Attainment of High Temperatures in a Laboratory	12
1916	Refractory Materials	12
1917	Training and Work of the Chemical Engineer	13
1917	Osmotic Pressure	13
1917	Pyrometers and Pyrometry	13
1918	The Setting of Cements and Plasters	14
1918	Electric Furnaces	14
1918	Co-ordination of Scientific Publication	14
1918	The Occlusion of Gases by Metals	14
1919	The Present Position of the Theory of Ionization	15
1919	The Examination of Materials by X-Rays	15
1920	The Microscope: Its Design, Construction and Applications	16
1920	Basic Slags: Their Production and Utilization in Agriculture	16
1920	Physics and Chemistry of Colloids	16
1920	Electrodeposition and Electroplating	16
1921	Capillarity	17
1921	The Failure of Metals under Internal and Prolonged Stress	17
1921	Physico-Chemical Problems Relating to the Soil	17
1921	Catalysis with special reference to Newer Theories of Chemical Action	17
1922	Some Properties of Powders with special reference to Grading by Elutriation	18
1922	The Generation and Utilization of Cold	18
1923	Alloys Resistant to Corrosion	19
1923	The Physical Chemistry of the Photographic Process	19
1923	The Electronic Theory of Valency	19
1923	Electrode Reactions and Equilibria	19
1923	Atmospheric Corrosion. First Report	19
1924	Investigation on Oppau Ammonium Sulphate-Nitrate	20
1924	Fluxes and Slags in Metal Melting and Working	20
1924	Physical and Physico-Chemical Problems relating to Textile Fibres	20
1924	The Physical Chemistry of Igneous Rock Formation	20
1924	Base Exchange in Soils	20
1925	The Physical Chemistry of Steel-Making Processes	21
1925	Photochemical Reactions of Liquids and Gases	21
1926	Explosive Reactions in Gaseous Media	22
1926	Physical Phenomena at Interfaces, with special reference to Molecular Orientation	22
1927	Atmospheric Corrosion, Second Report	23
1927	The Theory of Strong Electrolytes	23
1927	Cohesion and Related Problems	24
1928	Homogeneous Catalysis	24
1929	Crystal Structure and Chemical Constitution	25
1929	Atmospheric Corrosion of Metals, Third Report	25
1929	Molecular Spectra and Molecular Structure	26
1930	Colloid Science Applied to Biology	26
1931	Photochemical Processes	27
1932	The Adsorption of Gases by Solids	28
1932	The Colloid Aspect of Textile Materials	29

Date	Subject	Volume
1933	Liquid Crystals and Anisotropic Melts	29
1933	Free Radicals	30
1934	Dipole Moments	30
1934	Colloidal Electrolytes	31
1935	The Structure of Metallic Coatings, Films and Surfaces	31
1935	The Phenomena of Polymerization and Condensation	32
1936	Disperse Systems in Gases: Dust, Smoke and Fog	32
1936	Structure and Molecular Forces in (a) Pure Liquids, and (b) Solutions	33
1937	The Properties and Function of Membranes, Natural and Artificial	33
1937	Reaction Kinetics	34
1938	Chemical Reactions Involving Solids	34
1938	Luminescence	35
1939	Hydrocarbon Chemistry	35
1939	The Electrical Double Layer (owing to the outbreak of the war the meeting was abandoned, but the papers were printed in the *Transactions*)	35
1940	The Hydrogen Bond	36
1941	The Oil-Water Interface	37
1941	The Mechanism and Chemical Kinetics of Organic Reactions in Liquid Systems	37
1942	The Structure and Reactions of Rubber	38
1943	Modes of Drug Action	39
1944	Molecular Weight and Molecular Weight Distribution in High Polymers (Joint Meeting with the Plastics Group, Society of Chemical Industry)	40
1945	The Application of Infra-red Spectra to Chemical Problems	41
1945	Oxidation	42
1946	Dielectrics	42 A
1946	Swelling and Shrinking	42 B
1947	Electrode Processes	Disc. 1
1947	The Labile Molecule	2
1947	Surface Chemistry (Jointly with the Sociéité de Chimie Physique at Bordeaux Published by Butterworths Scientific Publications Ltd	
1947	Colloidal Electrolytes and Solutions	Trans. 43
1948	The Interaction of Water and Porous Materials	Disc. 3
1948	The Physical Chemistry of Process Metallurgy	4
1949	Crystal Growth	5*
1949	Lipo-proteins	6
1949	Chromatographic Analysis	7
1950	Heterogeneous Catalysis	8
1950	Physico-chemical Properties and Behaviour of Nuclear Acids	Trans. 46
1950	Spectroscopy and Molecular Structure and Optical Methods of Investigating Cell Structure	Disc. 9
1950	Electrical Double Layer	Trans. 47
1951	Hydrocarbons	Disc. 10
1951	The Size and Shape Factor in Colloidal Systems	11
1952	Radiation Chemistry	12
1952	The Physical Chemistry of Proteins	13
1952	The Reactivity of Free Radicals	14
1953	The Equilibrium Properties of Solutions on Non-electrolytes	15
1953	The Physical Chemistry of Dyeing and Tanning	16
1954	The Study of Fast Reactions	17
1954	Coagulation and Flocculation	18
1955	Microwave and Radio-frequency Spectroscopy	19
1955	Physical Chemistry of Enzymes	20
1956	Membrane Phenomena	21
1956	Physical Chemistry of Processes at High Pressures	22
1957	Molecular Mechanism of Rate Processes in Solids	23
1957	Interactions in Ionic Solutions	24
1958	Configurations and Interactions of Macromolecules and Liquid Crystals	25
1958	Ions of the Transition Elements	26
1959	Energy Transfer with special reference to Biological Systems	27
1959	Crystal Imperfections and the Chemical Reactivity of Solids	28
1960	Oxidation-Reduction Reactions in Ionizing Solvents	29
1960	The Physical Chemistry of Aerosols	30
1961	Radiation Effects in Inorganic Solids	31
1961	The Structure and Properties of Ionic Melts	32
1962	Inelastic Collisions of Atoms and Simple Molecules	33
1962	High Resolution Nuclear Magnetic Resonance	34
1963	The Structure of Electronically Excited Species in the Gas Phase	35
1963	Fundamental Processes in Radiation Chemistry	36
1964	Chemical Reactions in the Atmosphere	37
1964	Dislocations in Solids	38
1965	The Kinetics of Proton Transfer Processes	39

Date	Subject	Volume
1965	Intermolecular Forces	40
1966	The Role of the Absorbed State in Heterogeneous Catalysis	41
1966	Colloid Stability in Aqueous and Non-aqueous Media	42
1967	The Structure and Properties of Liquids	43
1967	Molecular Dynamics of the Chemical Reactions of Gases	44
1968	Electrode Reactions of Organic Compounds	45
1968	Homogeneous Catalysis with Special Reference to Hydrogenation and Oxidation	46
1969	Bonding in Metallo-organic Compounds	47
1969	Motions in Molecular Crystals	48
1970	Polymer Solutions	49
1970	The Vitreous State	50
1971	Electrical Conduction in Organic Solids	51
1971	Surface Chemistry of Oxides	52
1972	Reactions of Small Molecules in Excited States	53
1972	The Photoelectron Spectroscopy of Molecules	54
1973	Molecular Beam Scattering	55
1973	Intermediates in Electrochemical Reactions	56
1974	Gels and Gelling Processes	57
1974	Photo-effects in Adsorbed Species	58
1975	Physical Adsorption in Condensed Phases	59
1975	Electron Spectroscopy of Solids and Surfaces	60
1976	Precipitation	61
1977	Potential Energy Surfaces	62
1977	Radiation Effects in Liquids and Solids	63
1977	Ion–Ion and Ion–Solvent Interactions	64
1978	Colloid Stability	65
1978	Structures and Motion in Molecular Liquids	66
1979	Kinetics of State Selected Species	67
1979	Organization of Macromolecules in the Condensed Phase	68
1980	Phase Transitions in Molecular Solids	69
1980	Photoelectrochemistry	70
1981	High Resolution Spectroscopy	71
1981	Selectivity in Heterogeneous Catalysis	72
1982	Van der Waals Molecules	73
1982	Electron and Proton Transfer	74
1983	Intramolecular Kinetics	75
1983	Concentrated Colloidal Dispersions	76
1984	Interfacial Kinetics in Solution	77
1984	Radicals in Condensed Phases	78
1985	Polymer Liquid Crystals	79
1985	Physical Interactions and Energy Exchange at the Gas–Solid Interface	80
1986	Lipid Vesicles and Membranes	81
1986	Dynamics of Molecular Photofragmentation	82
1987	Brownian Motion	83
1987	Dynamics of Elementary Gas-phase Reactions	84
1988	Solvation	85
1988	Spectroscopy at Low Temperatures	86
1989	Catalysis by Well Characterised Materials	87
1989	Charge Transfer in Polymeric Systems	88
1990	Structure of Surfaces and Interfaces as studied using Synchrotron Radiation	89
1990	Colloidal Dispersions	90
1991	Structure and Dynamics of Reactive Transition States	91
1991	The Chemistry and Physics of Small Metallic Particles	92
1992	Structure and Activity of Enzymes	93
1992	The Liquid/Solid Interface at High Resolution	94
1993	Crystal Growth	95
1993	Dynamics at the Gas/Solid Interface	96
1994	Structure and Dynamics of Van der Waals Complexes	97
1994	Polymers at Surfaces and Interfaces	98
1994	Vibrational Optical Activity: From Fundamentals to Biological Applications	99
1995	Atmospheric Chemistry: Measurements, Mechanisms and Models	100
1995	Gels	101
1995	Unimolecular Reaction Dynamics	102
1996	Hydration Processes in Biological and Macromolecular Systems	103
1996	Complex Fluids at Interfaces	104
1996	Catalysis and Surface Science at High Resolution	105
1997	Solid State Chemistry: New Opportunities from Computer Simulations	106
1997	Interactions of Acoustic Waves with Thin Films and Interfaces	107*
1997	Dynamics of Electronically Excited States in Gaseous, Clusters and Condensed Media	108*
1998	Chemistry and Physics of Molecules and Grains in Space	109*
1998	Chemical Reaction Theory	110*

Date	Subject	Volume
1998	Molecular Interactions of Biomembranes	111*
1999	Physical Chemistry in the Mesoscopic Regime	112*
1999	Stereochemistry and Control in Molecular Reaction Dynamics	113*
1999	The Surface Science of Metal Oxides	114*
2000	Molecular Photoionisation	115*
2000	Bioelectrochemistry	116*
2000	Excited States at Surfaces	117*

* *Available for purchase, for current information on prices* etc. *please contact the Sales and Promotion Department, The Royal Society of Chemistry, Thomas Graham House, Science Park, Milton Road, Cambridge, UK CB4 0WF.*